Raptor

Research
and Management
Techniques

Raptor

Research
and Management
Techniques

SWAROVSKI
OPTIK

Edited by
DAVID M. BIRD
and
KEITH L. BILDSTEIN

Assistant Editors
DAVID R. BARBER
and
ANDREA ZIMMERMAN

ISBN 978-0-88839-639-6
Copyright © 2007 Raptor Research Foundation

Cataloging in Publication Data

Raptor research and management techniques / edited by David M. Bird ... [et al.].

First ed. published Washington, D.C. : Institute for Wildlife Research, National Wildlife Federation, 1987 under title: Raptor management techniques manual.

Includes bibliographies and index.
ISBN 978-0-88839-639-6

1. Birds of prey. 2. Birds of prey—Conservation.   3. Wildlife management.
I. Bird, David M. (David Michael), 1949-   II. Title: Raptor management techniques manual.

QL696.F3R366 2007   639.9'789   C2007-904971-0

Printed in China — SINOBLE

Copy editing: Theresa Laviolette
Production: Ingrid Luters
Cover design: Ingrid Luters

We acknowledge the financial support of the Government of Canada through the Book Publishing Industry Development Program (BPIDP) for our publishing activities.

Published simultaneously in Canada and the United States by

HANCOCK HOUSE PUBLISHERS LTD.
19313 Zero Avenue, Surrey, B.C. Canada V3S 9R9
(604) 538-1114 Fax (604) 538-2262

HANCOCK HOUSE PUBLISHERS
1431 Harrison Avenue, Blaine, WA U.S.A 98230-5005
(604) 538-1114 Fax (604) 538-2262

Website: **www.hancockhouse.com**
Email: **sales@hancockhouse.com**

Contents

Foreword

This is the RAPTOR RESEARCH FOUNDATION

Welcome to the 2nd edition of the *Raptor Management Techniques Manual*, now renamed *Raptor Research and Management Techniques*. I can think of no more appropriate undertaking for The Raptor Research Foundation, Inc. (RRF) than to update this terrific reference, first published by the National Wildlife Federation in 1987. RRF's purpose is to stimulate the dissemination of information concerning raptorial birds among interested parties worldwide and to promote a better public understanding and appreciation of the value of birds of prey. Thus, no other endeavor could be more central to RRF's *raison d'etre*, or more expressive of the manner in which knowledge within our profession is carefully passed from one generation to the next. The foundation of our professional capability is and always will be technique: the methods we apply to craft our investigations and management programs, to understand and conserve birds of prey.

Editors David Bird and Keith Bildstein, experts in their own right, have assembled a distinguished team of authors. The techniques they have synthesized are the product of hundreds of lifetimes of hard-won experience, thousands upon thousands of hours of trial and error, and tedious experimentation. As the ballet master, George Balanchine put it, *"Behind every good idea lies horrible, exhausting work. You knock your brains out and nothing comes. ... But after you've worked hard enough, the work gradually starts taking shape."* (Volkov 1985, page 199 *in* Balanchine's Tchaikovsky: Interviews with George Balanchine. Simon and Shuster, New York, New York, USA.). Take advantage of the experience gathered in these pages. For established practitioners, this is a tremendous resource to brush up on technique and review recent developments. For those at the start of their careers, this is a toolbox with which to build a life's body of work, implements shaped for your use by those who have walked the road you are standing on.

I am proud and delighted that RRF has taken on responsibility for this manual. I thank the National Wildlife Federation for the legacy they have passed to us, congratulate David and Keith on a job well done, and bow in appreciation to the authors who have made this 2nd edition a reality.

LEONARD YOUNG, President
The Raptor Research Foundation, Inc.

Preface

In 1987, the Raptor Information Center of the National Wildlife Federation published the *Raptor Management Techniques Manual*. The work, which was edited by Beth Giron Pendleton, Brian Millsap, Keith Cline, and David Bird, was a 420-page manual consisting of 19 chapters divided into three sections: Field Research Techniques, Management Techniques, and Laboratory Research Techniques. Each chapter was authored by one or more experts in the field, and each was reviewed by two independent referees. Priced at $25 U.S., the book sold out quickly. Although the *Raptor Management Techniques Manual* was published in binder format with the expectation that individual chapters would be updated and replaced as warranted, this never occurred. The Raptor Information Center was disbanded in the 1990s.

In 2000, the Raptor Research Foundation (RRF) approached the National Wildlife Federation and was given permission to pursue the publication of a thoroughly updated version of the manual. RRF then asked the two of us to solicit authors for individual chapters, edit the new work, and oversee its publication. The book before you, *Raptor Research and Management Techniques*, is the result of these efforts.

When we as editors took on this task, our aims and objectives were to produce a comprehensive work that reflected the state of the art in raptor research and management techniques, and to increase the geographic scope of the book beyond North America. We also wanted to produce a high-quality, attractive, and reasonably priced book that would be used globally by raptor researchers and conservationists and natural-resource managers. Unlike its predecessor, *Raptor Research and Management Techniques* is a bound work that is loosely modeled after the highly acclaimed *Bird Census Techniques*, second edition (2000) by Colin Bibby, Neil Burgess, David Hill, and Simon Mustoe. *Raptor Research and Management Techniques* is not intended to be an all-inclusive manual or detailed "how-to" book, but rather a review of the field with up-to-date information on various techniques that is designed to provide readers with a general overview of the field. That said, each chapter has numerous references that will direct readers to additional sources for details and cautions regarding various field and laboratory techniques and management tools.

The first four chapters, one each on the raptor literature, raptor systematics, raptor identification, and study design, data analysis, and the presentation of results, provide a general overview of the field of raptor research. The next ten chapters provide insights into field-study techniques, including surveying and monitoring, behavioral studies, diet analysis, habitat sampling, accessing nests and assessing nest success, capture and marking techniques, and spatial tracking. Four additional chapters provide information on the energetics, physiology, pathology, and toxicology of raptors; five more cover reducing management and researcher disturbance, mitigation, captive breeding, the augmentation of wild populations, and rehabilitation. The work concludes with chapters on public education and legal considerations. Although the book focuses on questions of importance to management and conservation, the scientific approach laid out at the beginning of the work, and the field and laboratory study techniques described thereafter, provide researchers with important tools for better understanding the basic biology of the birds as well.

We use the recommended English names of birds (Gill and Wright 2006, Birds of the World: recommended English names. Princeton University Press, Princeton, NJ, USA) throughout, together with their binomials (at first mention) in each chapter. The appendix provides an alphabetical list of the recommended English names of all diurnal raptors and other birds mentioned in the text, together with their binomials.

We view the publication of *Raptor Research and* *Management Techniques* as a way to enhance standardization in the field, and in so doing, increase our ability to compare our findings with those of others. We also view the book as a way to share both past successes and failures, and to speed improvement in our research and management techniques. Overall, we hope that like its predecessor, *Raptor Research and Management Techniques* will stand the test of time and help those who study and manage birds of prey protect them better.

DAVID M. BIRD and KEITH L. BILDSTEIN

We dedicate this volume in memory of our mentors,
Richard "Butch" Olendorff and Frances "Fran" Hamerstrom.

Acknowledgments

Raptor Research and Management Techniques is the culmination of more than four years of hard work by a large number of individuals and organizations. Without their expertise the book would not have been possible. Almost all participants worked as volunteers on this project. We thank them all for helping us bring this volume—which represents the discipline's communal knowledge on the subject—to fruition.

We particularly are indebted to four organizations, an optics company, and a publishing house, all of which directly and indirectly helped us achieve our goals. The National Wildlife Federation (NWF) was the driving force behind the predecessor of this work. NWF's Raptor Information Center conceived and published the widely acclaimed *Raptor Management Techniques Manual*, which forms the foundation upon which *Raptor Research and Management Techniques* was written and edited. NWF's willingness to allow us to use this seminal work as the basis for our own work was critical to initiating the project. The Raptor Research Foundation (RRF) provided the editors with the authority to use its good name when soliciting authors for the project, and acts as the work's principal sponsor. The Department of Natural Resource Sciences of McGill University and Hawk Mountain Sanctuary provided the editors and the assistant editors of this work the time and logistic support needed to prepare it for the publishing house. Swarovski Optics provided the editors with a generous grant to help defer travel costs incurred in soliciting authors, reporting to RRF, and meeting together when necessary to complete the project. Hancock House Publishers Ltd. was willing to work with the editors in producing a world-class presentation of the finished product. We thank all of these organizations and companies for their stalwart support and patience.

In addition to those mentioned above, a number of individuals worked overtime to ensure a well-written document. We thank all of the authors and co-authors of the work who took time from their professional and personal lives to, in some cases, meet deadlines, and in all cases, provide us with the essential text of the work. Adrian Aebischer, Nigel Barton, Rob Bennetts, Pete Bloom, Patricia Bright, Dale Clayton, Chris Farmer, Michael Fry, Lynda Gibbons, Laurie Goodrich, Carole Griffiths, Nigel Harcourt-Brown, Mike Hart, Elwood Hill, Grainger Hunt, Ron Jackman, Erkki Korpimaki, Brian Latta, Timothy Meehan, Mark Pokras, Alexandre Roulin, Karen Steenhof, William Stout, Russell Thorstrom, Michael Wallace, and Robert Zink served as technical referees for one or more of the chapters. Assistant editors, David Barber and Andrea Zimmerman, served admirably, both as copy editors and as content editors on the project. Kristen Naimoli and Michele Pilzer read and commented upon most of the chapters. Mike Wallace and Greg Septon are thanked for providing photographic material for the cover. Lindsay Zemba helped proof the page galleys. David Hancock, Theresa Laviolette, and Ingrid Luters at Hancock House shepherded our manuscript through to publication. Finally, we thank our family, friends, and loved ones for putting up with our distractions during the editorial process.

That said, we apologize in advance, first for any errors that have crept into the work, and second, to anyone who helped on the project, but whose efforts we have overlooked above.

DAVID M. BIRD
Avian Science and
Conservation Centre
McGill University
Ste. Anne-de-Bellevue, Quebec

KEITH L. BILDSTEIN
Acopian Center for
Conservation Learning
Hawk Mountain Sanctuary
Orwigsburg, Pennsylvania

The Raptor Literature

1

LLOYD F. KIFF
The Peregrine Fund,
5668 W. Flying Hawk Lane, Boise, ID 83709 U.S.A.

ROB G. BIJLSMA
Doldersummerweg 1, 7983 LD Wapse, The Netherlands

LUCIA LIU SEVERINGHAUS
Research Center for Biodiversity,
Academia Sinica, Taipei, Taiwan 115

JEVGENI SHERGALIN
Falconry Heritage Trust,
P.O. Box 19, Carmarthen, Dyfed SA335YL, U.K.

INTRODUCTION

We are currently experiencing a dramatic change in scholarly disciplines, as we shift from traditional print publications to electronic forms of communication. During this transition, many venerable journals are producing parallel electronic versions and others are completely discontinuing their print versions. Many libraries are discarding large quantities of paper copies of infrequently consulted publications and turning instead to electronic text, data storage, and information transfer. Simply put, the world of information storage and transfer is a moving target.

That said, this chapter provides a brief overview of important recent global raptor literature, lists the major technical journals with useful raptor content, and highlights the most important databases containing raptor literature. Emphasis is on identifying entry points into relevant raptor literature, rather than providing a thorough historical review. We focus on regions most familiar to us, and have touched lightly on the raptor literature of some parts of the world.

Raptor researchers suffer from two chronic problems: too little information and too much information. Traditionally, most researchers, regardless of their discipline, have suffered from a lack of access to the whole spectrum of global literature. Few libraries offer comprehensive coverage of all types of raptor literature, and even now, the major online abstracting services, although extremely valuable, do not yet provide access to the full text of most articles. Language differences also have posed perennial barriers to communication, and few, if any, abstracting services adequately cover the literature in all of the world's major languages.

Now, with a flood of information on its way onto the worldwide web, we run the risk of descending from the Information Age into a state of information chaos. As a result, raptor literature is becoming increasingly vast and amorphous. In his chapter on this topic in the first edition of this manual, LeFranc (1987) stated that approximately 370 and 1,030 raptor-related publications were listed in the 1970 and 1980 issues of *Wildlife Review*, respectively. By now, we suspect that at least three times as many useful raptor-related articles are being published annually. It is impossible for any but the world's largest research libraries to keep pace with this torrent of information, yet staying abreast of current studies is a prerequisite for effective research on raptors or any other topic.

Although the Internet makes it possible to gain access to an enormous amount of information, users may find it difficult to deal with the overwhelming mass of detail that has accumulated on the web, much of it trivial, redundant, or unrelated. For example, a recent (July 2006) search on the term "falcon" using Google yielded a total of 53,200,000 hits! Refining the search to "Peregrine Falcon" resulted in 2,860,000 matches. Narrowing the search to "Peregrine Falcon eggshells" still yielded 33,900 matches.

Clearly, plowing through this sea of information is unworkable for the average raptor researcher. Efficiency of information transfer depends upon its organization; thus, focused databases and indexes are worth their weight in gold. Such schemes are now appearing on many fronts, but someone must pay for the work. Thus, the most comprehensive literature-abstracting services require subscription fees, which can be prohibitively high for most individual users and smaller institutions. Some even require a fee to publish on the web. The Entomological Society of America, for example, recently began charging authors, who are willing to pay for the privilege, immediate web access for their papers. If this results in the cancellation of many print subscriptions, the price of this service will increase (Walker 2006).

Ultimately, this approach may lead to the demise of paper publications and traditional subscriptions. At the very least, a market-driven transition to open access (albeit for a fee) to all articles, at least those in major scientific journals, is likely. But as Worlock (2006) warns, *"Outside the consortia, and in the less-developed world, a genuine poverty of access is emerging as never before, with the scholarly rich and poor divided sharply on access and on the ability to stay abreast of the fast-moving research base."* We hope that solutions are found to such inequities. And indeed, some of the developing systems described in this chapter hold that promise.

OVERVIEW OF THE RAPTOR LITERATURE

Types of Literature

In general, scholarly literature falls into two broad categories, "primary literature," which presents original findings and ideas and is intended for a scientific audience, and "secondary literature," which consists of general works such as compilations, reviews, or other syntheses of information, derived from primary sources. The former includes books, journals, symposia volumes, dissertations, theses, and abstracts, as well as unpublished reports, which often are referred to as the "gray literature." Secondary literature publications are intended for both scientific and lay audiences, and include reference works on families and species, handbooks, encyclopedias, review articles, bibliographies, and most popular magazine articles. Appendix 1 lists journals that regularly publish papers about raptors.

For researchers, the secondary literature serves as an invaluable gateway to the primary literature. However, because of the inevitable errors in transcription, omissions, and misguided nuances of interpretation that find their way into handbooks and review volumes, researchers and reviewers always should consult the original sources of data cited in their papers whenever possible.

The Raptor Literature by Topic

General treatments. There are now scores of books on raptors on the market intended for a general audience, but probably the best introduction to the natural history and conservation of birds of prey is the one edited by Newton and Olsen (1990), which manages to be popular and authoritative. The two-volume set on hawks, eagles and falcons of the world by Brown and Amadon (1968) is a classic and, although a bit dated, remains an essential part of any raptor library. The *Handbook of Birds of the World* volumes treating diurnal birds of prey and owls (del Hoyo et al. 1994, 1999) provide good overviews of each raptor family, concise species accounts, and nice illustrations of all species.

It would be difficult to overestimate the importance of the body of work published by the World Working Group on Birds of Prey and Owls (WWGBPO). Now 30 years old, this group was originally part of the International Council for Bird Preservation (ICBP), which supported world conferences on raptors in 1975 and 1982 (Chancellor 1975, Newton and Chancellor 1985). Subsequently, the WWGBPO became independent of ICBP, and under the guidance of Bernd Meyburg, has met in different countries at intervals of a few years, with the proceedings of each meeting usually being edited by Meyburg and Robin Chancellor and published in ever-thickening volumes (Meyburg and Chancellor 1989, 1994; Chancellor et al. 1998, Chancellor and Meyburg 2000, 2004; Yosef et al. 2002). The WWGBPO, which now claims over 3,000 members worldwide, also published four volumes of *Bird of Prey Bulletin*, presenting the results of regional conferences, and a volume specifically devoted to eagles (Meyburg and Chancellor 1996). In aggregate, these publications provide the best available overview of global raptor conservation and research directions over the past three decades, and the meetings themselves have created a world community of raptor researchers.

Families and groups of raptors. Presently, the best general overviews of the diurnal birds of prey are the books by Ferguson-Lees and Christie (2001, 2005), which contain an enormous amount of useful information and attractive color plates, illustrating multiple plumages of each species. The odd disconnect between the in-text citations and the bibliographies of these books is disconcerting, but these two volumes remain useful as quick references. A similar volume on owls by König et al. (1999) is the best overview of those families, including detailed information on owl systematics based on recent molecular genetics studies.

Among the many general treatments of owls, those by Mikkola (1983), Voous (1988), and Duncan (2003) are particularly outstanding. World conferences on owls similar to those of the WWGBPO for all raptors have led to the publication of several information-rich proceedings volumes (Nero et al. 1987, Duncan et al. 1997, Newton et al. 2002).

There are a number of works on both Old and New World vultures, but the most prominent, by far, is the lavish book on African vultures by Mundy et al. (1992), which combines much original information and superb production values. The First International Symposium on Vultures (both Old and New World) resulted in a still useful book (Wilbur and Jackson 1983). A new book on the vultures of Georgia and the Caucasus (Gavashelishvili 2005) deserves mention, as does the one on European vultures by Baumgart (2001).

Genera. Aside from numerous popular books, there are relatively few published works on particular genera of raptors. Two excellent examples, however, are the overview by Simmons (2000) of the behavior and ecology of harriers (*Circus*), which also has important taxonomic implications, and Cade's comprehensive treatment of the genus *Falco* (Cade 1982), which is both attractive and informative. Among monographs of its type, Wattel's (1973) work on the systematics of the genus *Accipiter* was unusually thorough and still relevant.

Single species. There are many excellent books on single raptor species, some reporting on the results of studies extending for many years, and such monographs represent one of the strongest components of scientific raptor literature. Among several important (and ongoing) series of species monographs, two stand out from the rest, including those published by T. and A.D.

Poyser (now under the imprint of A&C Black), presently with nine monographs on diurnal raptors and two on owl species, and the *Neue Brehm-Bücherei* series, which was started in 1948 by Ziemsen Verlag in what was then East Germany and continued since 1992 by Westarp Wissenschaften after the reunion of both Germanies. This scholarly series includes monographs on at least 17 diurnal raptor and eight owl species, some of which are rather outdated, while others have been updated or entirely rewritten. Among titles in the former series, those by Newton (1986) on the Eurasian Sparrowhawk (*Accipiter nisus*) and Ratcliffe (1993) on the Peregrine Falcon (*Falco peregrinus*) are especially important. Arlequin Press also has produced a smaller, but important, series of monographs on several British raptor species (e.g., Carter 2001).

The complete list of important raptor species monographs is obviously too long to enumerate here, but a few worth special mention (and to illustrate their variety) include those on the California Condor (*Gymnogyps californianus*) (Koford 1953), Osprey (*Pandion haliaetus*) (Poole 1989), African Fish Eagle (*Haliaeetus vocifer*) (Brown 1980), Bald Eagle (*H. leucocephalus*) (Hunt et al. 1992), Bearded Vulture (*Gypaetus barbatus*) (Terrasse 2001), Spanish Imperial Eagle (*Aquila adalberti*) (Ferrer 2001), Verreaux's Eagle (*A. verreauxii*) (Gargett 1990), Eleonora's Falcon (*F. eleonorae*) (Walter 1979), Peregrine Falcon (Hickey 1969, Monneret 2000, Rockenbauch 1998, 2002), Black Shaheen Falcon (*F. p. peregrinator*) (Döttlinger 2002), Gyrfalcon (*F. rusticolus*) (Ford 1999, Potapov and Sale 2005), Barn Owl (*Tyto alba*) (Taylor 1994), and Eastern Screech-Owl (*Megascops asio*) (Gehlbach 1994).

Systematics. Until recently, the principal world authority on diurnal raptor taxonomy was the late Dean Amadon of the American Museum of Natural History. Somewhat by default, the 1968 Brown and Amadon volumes served as the best single source for diurnal raptor taxonomy until the publication of the revised edition of the falconiforms volume of Peters' *Checklist of Birds of the World*. The treatment there (Stresemann and Amadon 1979) was based on an early 1960s manuscript by Erwin Stresemann with subsequent modifications by Amadon. Later, Amadon and Bull (1988) suggested additional changes in diurnal raptor taxonomy and provided a global list of *Otus* species in the same volume. At the outset of the molecular age in systematics, Sibley and Monroe (1990) published a new world avian taxonomy, based largely on their findings using DNA hybridization techniques, and they recommended major changes in the phylogenetic arrangement of avian families. In a companion volume (Sibley and Ahlquist 1990), there is an extremely useful history of the classification of all avian groups, including raptors, based on traditional morphological characters.

The nomenclature, sequence, and limits of species in the respective volumes of *Handbook of Birds of the World* (del Hoyo et al. 1994, 1999) for falconiforms and strigiforms have been generally followed since their publication, but there are conspicuous departures in the books by Ferguson-Lees and Christie (2001, 2005), some of which, but not all, reflect advances in taxonomic knowledge. Even if the new family sequence suggested by Sibley has not enjoyed universal acceptance, he correctly predicted that molecular studies would soon rule the day in systematics. Among recent printed volumes, the world bird list *du jour* is the one edited by Dickinson (2003) with input from a respected committee of regional specialists. Their treatment is a transitional mixture based on traditional integration of morphological and behavioral characters and some newer findings from molecular genetics, primarily from studies of mitochondrial and nuclear DNA. For diurnal raptors, this list already has been rendered partially obsolete by major changes in generic- and species-level raptor taxonomy suggested by studies from several molecular labs (e.g., those of Helbig et al. [2005] and Lerner and Mindell [2005]). This is an extremely fast-moving field and perhaps the best strategy to keep up with new findings until a new authoritative list appears is to consult web-based databases, e.g., the Global Raptor Information Network (www.globalraptors.org).

For North American species, successive editions of the AOU Check-list (AOU 1998) have long been the undisputed authority for nomenclature and range descriptions since the first one was published in 1886. Periodic supplements to the latest edition of the checklist are posted on the AOU's website (http://www.aou.org), and an equivalent list of South American bird species is in preparation by an international committee headed by Van Remsen (www.aou. org/checklist/south.php3).

Bibliographies. Olendorff and Olendorff (1968-70) prepared one of the first comprehensive bibliographies on birds of prey in the modern era. It contained 7,492 citations, but covered only English-language titles. The senior author later collaborated with Dean Amadon and Saul Frank to produce an annotated bibliography of raptor books in English and western European languages (Olendorff et al. 1995) that includes informative annotations. The National Wildlife Federation published useful but now dated bibliographies on owls of the world (Clark et al. 1978), Bald Eagle (Lincer et al. 1979), Golden Eagle (*A. chrysaetos*) (LeFranc and Clark 1983), and Peregrine Falcon (Porter et al. 1987). The bibliography of German literature on raptors and owls for 1945-95 produced by Mammen et al. (1997) contained 6,940 entries, and updates and corrections are published at http://www.greifvogelmonitoring.de. By now, the best bibliographic resources on raptors are various online databases described in the last section, and the era of massive printed bibliographies is probably over.

Disease and medicine. Over the past two decades, the topic of raptor biomedicine has virtually become a sub-discipline of veterinary medicine, thanks to heightened interest in birds of prey by rehabilitators, conservationists, and falconers. Two of the leaders in this field have been Patrick Redig at The Raptor Center at the University of Minnesota (see Redig 1993) and John Cooper, a British pathologist with various appointments in Europe, South America and Africa, who has authored or edited several important volumes on raptor medicine (e.g., Cooper 2002, 2003). Other recent volumes on this topic worth mention are the work by Lumeij (2000), which contains an extensive bibliography, and the color atlas by Wernery et al. (2004). In addition, several researchers working in facilities on the Arabian Peninsula continue to publish many important studies in this field, particularly in the journal *Falco*.

Migration. The migration of raptors is one of the most interesting and observable aspects of their biology, and an increasing amount of attention has been paid to this topic in recent decades, especially with the emergence of numerous raptor observatories along the major migratory pathways in the world. Among the most important publications on raptor migration are reviews of the behavior and ecology of migrating raptors by Kerlinger (1989) and Bildstein (2006), and the broad global overviews by Zalles and Bildstein (2000) and Bildstein and Zalles (2005). In Israel, the study by Spaar (1996) and the comprehensive summary of 30 years of field research on migrating raptors by Shirihai et al. (2000) are especially useful.

Bernd and Chris Meyburg and their colleagues pioneered the use of satellite telemetry to study raptor migration in several Old World eagle species (e.g., Meyburg and Meyburg 1999, Meyburg et al. 2005), and other outstanding ongoing programs in Europe are mentioned in the section on the Palearctic Region below. In North America, the satellite telemetry studies by Mark

Martell and his colleagues on the Osprey (Martell et al. 2001) and those on Golden Eagles and Peregrine Falcons by Bill Seegar (e.g., Seegar et al. 1996) are especially noteworthy.

Some conservation topics. For endangered and threatened raptors globally, the most important summaries are those produced under the direction of BirdLife International biologists, Nigel Collar and Allison Stattersfield (Collar and Stuart 1985, Collar et al. 1992, 1994, 2001; Stattersfield and Cooper 2000). These works have set a high standard for their accuracy, thoroughness, and recommended conservation actions. More up-to-date information on threatened raptors can be found on the BirdLife International Globally Threatened Bird Species Database website (www.birdlife.org/data zone) and on the Global Raptor Information Network website.

The chronic problem of birds striking powerlines, or being electrocuted by them, was addressed by a still-current manual published by the Raptor Research Foundation (Avian Power Line Interaction Committee 1996) and an excellent symposium volume on this topic produced in Spain (Ferrer and Janss 1999). The problem of bird hazards to aircraft is one of relevance to raptor species, particularly along migration routes, and the proceedings of an international seminar on this topic in the Middle East were reported by Leshem et al. (1999). An earlier work by Leshem and Bahat (1994) provides a fascinating account of some solutions to this problem in Israel.

The reviews by Risebrough (1986) and Cooke et al. (1982) give excellent summaries of the effects of organochlorine contaminants, especially the eggshell-thinning metabolite, DDE, on raptors and other birds and should be required reading for anyone uninformed about the extent of these threats to bird populations. A particularly outstanding case history of how these contaminants have affected a raptor is that of Helander et al. (2002) on the White-tailed Eagle (*H. albicilla*) in Sweden, and the numerous papers by Ian Newton and his colleagues on Eurasian Sparrowhawks and other British raptors are also essential reading on this topic.

As summarized by Cade (2000), captive breeding and reintroduction projects have been an essential tool in the recovery of many formerly endangered raptor populations. Several techniques manuals on managing captive and released falcons have been published by The Peregrine Fund, including Sherrod et al. (1982), Cade et al. (1988), and Weaver and Cade (1991), as well as a similar manual on enhancing wild raptor populations, including owls, by Marti (2002).

Techniques. Standardization of field methods through the publication of manuals upgraded the quality and scope of raptor studies, enabled reliable between-study comparisons, and boosted atlas work and long-term censuses. Some of the most useful texts on techniques are those of Berthold et al. (1974), Ralph and Scott (1981), Hustings et al. (1985), Koskimies and Väisänen (1991), Gilbert et al. (1998), Bibby et al. (2000), and Südbeck et al. (2005). Manuals specifically targeted at raptors further contributed to standardization and quantification of field methods, including März (1987) and Bijlsma (1997). An earlier version of this manual (Giron Pendleton et al. 1987) published by the National Wildlife Federation quickly sold out, but lived on through numerous photocopies of its chapters by biologists and graduate students working on raptors.

The Raptor Literature by Region

Afrotropical. For raptor researchers, the most important journals for the whole African continent are *Bulletin of the African Bird Club*, *Gabar*, and *Ostrich*. The former journal, which is published in the United Kingdom, is the best source for new distributional and natural history information on African raptors. *Gabar* (known for a few years as *Journal of African Raptor Biology*) contains many papers of high quality, and *Ostrich* is one of the leading scholarly ornithological journals in the world. Studies of more global interest often are published in prominent European and American journals, including *Alauda*, *The Auk*, *Bulletin of the British Ornithologists' Club*, *Ibis*, *Journal of Avian Biology*, and *Journal of Ornithology*. Important regional journals include *The Babbler* (Botswana), *Journal of East Africa Natural History*, *Kenya Birds*, and *Scopus* (East Africa), *Mirafra* and *Promerops* (South Africa), *Malimbus* (West Africa), *Zambia Bird Report* (Zambia), and *Honeyguide* (Zimbabwe). The popular magazine, *Africa – Birds & Birding*, often contains raptor articles with original information and superb photographs. *Vulture News*, which is published in South African by the Vulture Study Group, has a global scope, but the majority of articles and news snippets are on African species.

The raptor volume of the monumental *Birds of Africa* series (Brown et al. 1982) still represents an excellent source on raptors of the entire continent, and the earlier atlas edited by Snow (1978) continues to be useful, albeit a bit dated by now. The two-volume atlas of southern African birds (Harrison et al. 1997) includes extensive species accounts by leading authorities on each raptor species, and it is one of the best examples of this genre anywhere.

Per capita, South Africa has one of the most productive communities of raptor researchers in the world, and it enjoys a wealth of excellent books on diurnal birds of prey and owls. One edition of the standard reference, Austin Roberts's *Birds of South Africa*, originally published in 1940, has been in print for the past 65 years. The latest (7th) edition (Hockey et al. 2005) is by far the most elaborate to date, having more detailed species accounts by specialists on each species, which provide thorough, up-to-date overviews on southern African raptors.

Alan Kemp produced an attractive overview of the owls of southern Africa (Kemp 1987) and with his wife, Meg, a concise guide to the diurnal birds of prey of the entire continent and the adjacent islands (Kemp and Kemp 1998). Another useful field guide dealing specifically with southern African raptors is Allan (1996). The general treatment of African raptors (including owls) by Brown (1970) still makes interesting reading, and the classic books by Peter Steyn on southern African raptor species (Steyn 1974, 1982, 1984) contain a wealth of information, coupled with many pleasing photographs. One of the most thorough analyses of the status of the raptors of a particular district in Africa was reported by Tarboton and Allan (1984), and the former author also produced nice volumes on southern African owls (Tarboton and Erasmus 1998) and on African diurnal prey in general (Tarboton 1990). An overview of ring recoveries of 38 diurnal and 3 nocturnal raptor species, based on 50 years of banding by SAFRING, is available for southern Africa (Oatley et al. 1998).

Elsewhere in Africa, the recent field guide by Borrow and Demey (2001) organized much new information on raptors and other birds of this sparsely studied region, and among several recent books on East African birds, those on Ugandan birds by Carswell et al. (2005) and the field guide to East African birds by Stevenson and Fanshawe (2002) are among the most useful. Beginning in the 1970s, Jean-Marc Thiollay has intensively studied diurnal raptors in West Africa, mainly in the palm savanna and gallery forests of the Lamto Reserve in central Ivory Coast (Thiollay 1976), but also including much of the Sahelian zone from Mali through Burkina Faso and Niger eastwards to Chad and Cameroon (Thiollay 1977). Thiollay's baseline information led to his recent startling discovery of recent severe population declines in nearly all raptor species in this large region (Thiollay 2001, 2006).

Since 1991, The Peregrine Fund has conducted research on Madagascar raptors, with particular emphasis on the endangered Madagascar Fish Eagle (*Haliaeetus vociferoides*) and on training local researchers. To date, 16 Master's degrees and three Ph.Ds have been earned by participants in this project, and 59 peer-reviewed papers have been produced. Particularly notable dissertations include those of Berkelman (1997), René de Roland (2000), and Tingay (2005).

Several recent workshops have been held in South Africa to create conservation plans for southern African vultures (Boshoff et al. 1998) and other raptors (Anderson and Kruger 2004), and a similar meeting on vulture conservation in East Africa occurred in 2004 (Virani and Muchai 2004).

Australasia. This region enjoys a rich selection of technical journals of interest to raptor researchers including the internationally important publications, *Emu* (published by Birds -Australia) and *Notornis* (published by the Ornithological Society of New Zealand). The Australasian Raptor Association produces a journal, *Boobook*, and a newsletter, *Circus*, which are devoted solely to raptor topics. Important regional journals include *Australian Field Ornithology* and *Corella*, both with articles drawn from all parts of the country, and *South Australian Birds* (South Australia), *Sunbird* (Queensland), and *VORG Notes* (Victoria) focusing on the avifauna of particular states.

Olsen's (1995) comprehensive book on Australian raptors is one of the finest examples of a country treatment of raptors, or for that matter, any group of birds, and is the logical starting point for anyone interested in Australian diurnal birds of prey. Several editions of the modest field guides to Australian birds of prey by Condon (1970) and another with nice color plates by Morris (1976) were published prior to the more recent and extensive guide by Debus (1998). The latter book provides a concise and useful introduction to Australian diurnal birds of prey, and the text (and some plates) was distilled from the exhaustive species accounts in the *Handbook of Australian, New Zealand, and Antarctic Birds* (Marchant and Higgins 1993). The more recent handbook volume on Western Australian birds (Johnstone and Storr 1998) also contains highly detailed information on the birds of prey in that state. Australia has enjoyed the riches of several decades of comprehensive avian atlas projects, resulting in large volumes (Blakers et al. 1984, Barrett et al. 2003) containing valuable information on the distribution and seasonal movements of raptors.

Non-technical books on Australian raptors include those by Cupper and Cupper (1981) on hawks and Hollands (1991, 2003) on owls (and other nocturnal birds) and eagles, hawks and falcons, respectively. The latter now is in its second edition and includes excellent photographs and concise species accounts, as well as entertaining anecdotal accounts of the author's pursuit of Australian raptors.

The Australasian Raptor Association has held conferences leading to two proceedings volumes (Olsen 1989, Czechura and Debus 1997) and, with BirdLife Australia, supported an important study on the relative abundance and seasonal movements of Australian Falconiformes from 1986-90 (Baker-Gabb and Steele 1999).

Comprehensive information on the relatively few raptors in New Zealand is found in the Marchant and Higgins (1993) handbook. Brief general accounts on the raptors of New Guinea, including several poorly studied and intriguing endemic species, are found in Coates (1985) and Beehler et al. (1986), and those of nearby "Wallacea" (Sulawesi, Moluccas, and Lesser Sundas) in Coates and Bishop (1997), but no substantive work dealing specifically with the raptors of these areas has been produced yet.

Indomalaysia. Important journals with raptor content for this region include *BirdingASIA* (formerly *Bulletin of the Oriental Bird Club*) and *Forktail*, both published by the Oriental Bird Club, based in the United Kingdom. Regional journals include *Journal of the Bombay Natural History Society*, *Journal of Indian Bird Records*, and *Pavo* (India), *Kukila* (Indonesia), and *Malayan Nature Journal* (Malaysia).

The monumental 10-volume handbook series produced by Salim Ali and S. Dillon Ripley during the 1970s and 1980s still serves as the best starting point for information on birds of the Indian subcontinent. The text of the second edition is presented succinctly in a compact edition (Ali and Ripley 1987). The recent books on birds of the Indian subcontinent by Grimmett et al. (1999) and southern Asia by Rasmussen and Anderton (2005) provide more up-to-date information on the status and natural history of Indomalayan raptors. Useful raptor information also can be found in several other recent avifaunal treatments, including those for the Malaysian Peninsula (Wells 1999), Philippines (Kennedy et al. 2000), and Sabah (Sheldon et al. 2001).

Although there is less published information on the raptors of Southeast Asia than for other tropical regions of the world, the recent creation of the Asian Raptor Research and Conservation Network (ARRCN) in 1998 through the efforts of Toru Yamazaki and his colleagues in Japan recently has created a thriving community of raptor researchers in this part of the world. ARRCN has held four raptor symposia in different countries, with more planned, and the proceedings and abstracts from these meetings (e.g., Ichinose et al. 2000) contain much valuable information. The ARRCN also published three issues of a journal, *Asian Raptors*, reporting original studies.

The finding that the pharmaceutical drug, diclofenac (a painkiller administered to aging livestock), is responsible for the drastic decline of three formerly abundant *Gyps* vultures in India, Pakistan, and Nepal (Oaks et al. 2004) led to a surge in research on these species and a whole new subset of raptor literature in the region. A recent paper by Cuthbert et al. (2006) appears to indicate similar problems for other vulture species, so this is a topic that may well see an increasing amount of research interest in the future.

Collar et al. (1999) compiled a Red Data Book on threatened birds of the Philippines, including raptors. The plight of one of these, the endangered Philippine Eagle (*Pithecophaga jeffreyi*), has generated much international interest and led to extensive literature on that species; a good recent overview can be found in Bueser et al. (2003). There also are numerous papers on various aspects of the status and biology of another globally endangered species, the Javan Hawk-Eagle (*Spizaetus bartelsi*) (van Balen et al. 1999, 2001).

Middle East and Northern Africa. The journal *Sandgrouse* covers the Middle East and parts of contiguous Central Asia, including important updates on the distribution and natural history of the region's raptors. The *Bulletin of the African Bird Club* fulfills a similar role for northern Africa, including the Arabian Peninsula. There have been regular updates on the status of the birds of Oman (Eriksen et al. 2003). Other important regional journals are *Podoces* (Iran), *Torgos* (Israel), *Oman Bird News* (Oman), *Yelkovan* (Turkey), and *Emirates Bird Report* (United Arab Emirates).

The most important contributions to the knowledge of raptors in the Middle East have come from Israel. The massive book on the birds of Israel by Shirihai (1996) is unusually thorough and contains excellent species accounts of raptors. The contributions of the International Birding & Research Center in Eilat, Israel to the knowledge of raptor migration through the Middle East also have been important. For North Africa, recent books on the birds of Algeria (Isenmann and

Maoli 2000), Morocco (Thévenot et al. 2003), and Tunisia (Isenmann et al. 2005) contain a wealth of information on raptors of the region, much of it previously unreported.

Nearctic. The most important sources of primary literature on North American raptors are the Ornithological Societies of North America (OSNA) journals, including *The Auk*, *The Condor*, *Journal of Field Ornithology*, *Journal of Raptor Research*, and *Wilson Journal of Ornithology* (formerly *The Wilson Bulletin*). Many state and provincial bird society journals (e.g., *Blue Jay*, *Chat*, *Florida Field Naturalist*, *Kingbird*, *Loon*, *Ontario Birds*, *Oriole*, and *Passenger Pigeon*) traditionally have been important outlets for natural history and distributional notes. Regional "naturalist" journals include *Canadian Field-Naturalist*, *Northwestern Naturalist*, and *Southwestern Naturalist*. North American raptor studies with broader significance also are frequently published in generalized biological journals, especially *Conservation Biology*, *Ecology*, *Journal of Wildlife Management*, and *Wildlife Society Bulletin*, and in ornithological journals published in other countries, including *Ibis*, *Journal of Avian Biology*, and *Journal of Ecology*. The Hawk Migration Association of North America's *Hawk Migration Studies* and the periodic reports of specific raptor observatories (e.g., Hawk Mountain Sanctuary, HawkWatch International, and the Golden Gate Raptor Observatory) provide excellent coverage of raptor migration trends across the continent.

For several decades after their publication, the two Arthur Cleveland Bent volumes on the life histories of North American birds of prey (Bent 1937, 1938) provided the best overview of North American raptor biology, despite their anecdotal tone. They were superseded by the still valuable two volumes on raptors in the *Handbook of North American Birds*, edited by Palmer (1988) on the diurnal birds of prey. The species accounts in the Bird of North America series, now available online and being updated regularly, are the best overviews on the biology of individual North American raptor species, and are effective gateways into the pertinent primary literature. Over the past century, there have been a number of useful pamphlets and small books on the raptors of particular states and provinces, but two that rise above the rest are the ones by Glinski (1998) on Arizona raptors and the recent volume on California raptors by Peeters and Peeters (2005). Excellent field guides on North American diurnal raptors have been produced by Clark and Wheeler (2001) and Wheeler (2003a, 2003b). There is much valuable raptor

information in state and provincial bird books, and, by now, there are excellent atlases of breeding bird distribution for most states and provinces, and even some counties. Among the many compilations on North American birds of prey intended for a general audience, those by Johnsgard (1990, 2002) on hawks, eagles, and falcons, and on owls, respectively, are the best.

Since the "endangered species" concept seized the public imagination in the late 1960s and early 1970s, it has been interesting to observe the trend toward an inverse relationship between the size of species populations and the amount of research conducted on them. Thus, the Spotted Owl (*Strix occidentalis*) is by now the best-studied strigid in North America, and there are more publications on Peregrine Falcons and Bald Eagles than on any other diurnal raptor species on the continent.

Biologists with Canadian and United States federal, provincial, and state government agencies have produced a myriad of valuable reports on raptors over the past three or four decades. These include long-term management studies of individual raptor species (e.g., the monographs on Peregrine Falcon ecology and management by Hayes and Buchanan [2002] and Craig and Enderson [2004], on Spotted Owls by Gutiérrez and Carey [1985] and Verner et al. [1992]), and a detailed conservation assessment of three other owl species by Hayward and Verner (1994). From 1975–1994, biologists on the staff of the Snake River Birds of Prey National Conservation Area produced a valuable, albeit somewhat overlooked, series of annual reports (e.g., Steenhof 1994) and related publications on the birds of prey of Idaho reporting the results of many original and long-term studies. Recovery plans for endangered species often contain valuable information, especially bibliographies, and the periodic Species Status Reports by the Committee on the Status of Endangered Wildlife in Canada (COSEWIC) are especially thorough.

Non-profit organizations also have made many significant contributions to the North American raptor literature. As mentioned above, the National Wildlife Federation published an earlier version of this manual (Giron Pendleton et al. 1987) as well as useful species bibliographies, and also held five regional workshops from 1987–1989 (e.g., Pendleton 1989), which contained much valuable information on the status and conservation of North American raptors. Another non-governmental organization (NGO), The Peregrine Fund, published a landmark volume on the Peregrine Falcon, based on papers given at a symposium on that species in

1985 (Cade et al. 1987). Several individuals have contributed to the North American raptor literature in an enduring way, including John and Frank Craighead, whose book *Hawks, Owls, and Wildlife* (Craighead and Craighead 1956) stimulated an interest in raptors by many young people who later became professional raptor biologists, and Frances Hamerstrom, whose writings (e.g., Hamerstrom 1986) and many personal contacts with young biologists with her husband, Fred, also were strong influences on multiple generations of American raptor enthusiasts (Corneli 2002).

In addition to publishing *Journal of Raptor Research* and *Raptor Research Reports*, the Raptor Research Foundation has produced several important symposium proceedings on North American raptors, including those on the Bald Eagle and Osprey (Bird et al. 1983), American Kestrel (*F. sparverius*) (Bird and Bowman 1987), raptors in urban habitats (Bird et al. 1996), and the Burrowing Owl (*Athene cunicularia*) (Lincer and Steenhof 1997). North American ornithological societies also have published important symposia proceedings on raptors of special conservation interest. Two by the Cooper Ornithological Society were edited by Block et al. (1994) on the Northern Goshawk (*A. gentilis*) and Forsman et al. (1996) on the Spotted Owl. A similar volume on California Spotted Owl population dynamics was published as an *Ornithological Monograph* by the American Ornithologists' Union (Franklin et al. 2004). An influential contribution to the topic of reversed sexual size dimorphism in raptors (Snyder and Wiley 1976) was published earlier in the same series.

Neotropical. The most important ornithological journals covering the Neotropical Region include *Bulletin of the British Ornithologists Club, Cotinga, Journal of Raptor Research*, and *Ornitologia Neotropical*. Excellent journals with raptor content, focusing mostly on the birds of particular countries include *El Hornero* and *Nuestras Aves* (Argentina), *Atualidades Ornitológicas, Boletim CEO* and *Revista Brasileira de Ornitologia* (Brazil), *Boletín Chileno de Ornitologica* (Chile), *Boletín SAO* and *Ornitologia Colombiana* (Colombia), *Zeledonia* (Costa Rica), *Acta Zoologica Mexicana* (Mexico), and *Journal of Caribbean Ornithology* (West Indies).

Much important information on Neotropical raptors can be found in the species accounts of country and regional avifaunal treatments. The books on Brazilian birds by Sick (1993), Argentine birds by Di Giacomo and Krapovickas (2005), Chilean birds by Housse

(1945), and the Suriname avifauna by Haverschmidt and Mees (1994) contain particularly detailed information on birds of prey.

The most recent comprehensive summaries of the conservation status of raptors in Mexico and South America are those by Bierregaard (1995 and 1998, respectively). There are few country-specific books on Neotropical raptors, except for those for Mexico by Urbina Torres (1996) and the more ambitious work by Márquez et al. (2005) for Colombia. Although it is now dated, the volume on Neotropical Falconiformes in the monumental *Catalogue of Birds of the Americas* (Hellmayr and Conover 1949) is still a rich source for the history of species-level taxonomy and earlier bibliographic sources.

The Peregrine Fund conducted the most ambitious single research project on Neotropical raptors to the present time at Tikal National Park, El Petén, Guatemala from 1988-96. During this period, detailed studies were made of 19 species of falconiforms and two species of owls, resulting in 36 peer-reviewed papers and the completion of seven Master's degrees. The theses by Gerhardt (1991) and Thorstrom (1993) made particularly important additions to our knowledge of two widely distributed, but poorly studied, Neotropical raptor genera. A bibliography of publications of The Peregrine Fund and its associates is posted on its website (www.peregrinefund.org), and PDF versions of all Maya Project summaries, including over 100 unpublished reports, are available upon request at library@peregrinefund.org.

The Peregrine Fund also created the listserver-based Neotropical Raptor Network, which has organized two conferences on Neotropical raptors, one in Panama in 2002 and another at Iguazú Falls, Argentina in June 2006. The abstracts from those meetings (available as PDFs from The Peregrine Fund) contain much exciting new information, especially on poorly studied species. Raptor aficionados held a similar symposium on Argentine raptor species in October 2004, and the abstracts can be obtained from Sergio Seipke (seipke@yahoo.com.ar).

The most active centers of raptor research in South America have been in Argentina, Brazil, Chile, and Ecuador, and numerous interesting dissertations and theses on raptors have been produced in these countries. Details on these studies, some of which remain unpublished, can often be found in bibliographies of the ornithological publications of these countries, including Oniki and Willis (2002), Friele et al. (2004), and Silva-

Aranguiz (2006). There also has been much research interest in the endemic Galapagos Hawk (*B. galapagoensis*), beginning with the Ph.D. dissertation of de Vries (1973), and followed later by a steady stream of North American researchers (e.g., Faaborg 1986).

Western Palearctic. The journals of the leading European ornithological societies contain a wealth of peer-reviewed papers on raptors and owls, including *Ardea, Bird Study, Ibis, Journal of Avian Biology* (formerly *Ornis Scandinavica*), *Journal of Ornithology* (formerly *Journal für Ornithologie*) and *Ornis Fennica*. An increasing number of scientific papers on raptors and owls is now being published in high-impact ecological, rather than ornithological, journals, such as *Behavioural Ecology and Sociobiology, Biological Conservation, Journal of Animal Ecology, Journal of Applied Ecology, Oecologia* and *Oikos*.

All European countries have one or more ornithological journals published in their native language, usually with English summaries (or English throughout), covering a specific nation, but sometimes casting a wider geographic net. Some examples include *Acrocephalus, Acta Ornithologica, Alauda, Ardeola, British Birds, Dansk Ornithologisk Forening Tidsskrift, Egretta, Fauna Norvegica, Limosa, Nos Oiseaux, Ornis Svecica, Ornithologische Anzeiger, Ornithologische Beobachter, Ornithologische Mitteilungen, Vogelwarte,* and *Vogelwelt*. In addition, high-quality papers on distribution, trends, reproduction, food and behavior of raptors and owls can be found in the many hundreds of regional and local journals. This major outlet of information is underused, partly because of poor accessibility and the linguistic diversity involved. However, many of these publications are covered by major abstracting services, including *Zoological Records, OWL,* and *Ornithologische Schriftenschau*. To give some idea of the magnitude of these sources, Hölzinger (1991) collated 851 ornithological periodicals for Central Europe alone! The abovementioned journals contained tens of thousands of papers on raptors and owls during the past few decades. In addition, several specialized raptor journals began in the 1980s, including *Biuletyn* (Polish, first published early 1980s), *Buteo* (Czech-Slovak, 1986), *Jahresbericht zum Monitoring Greifvögel und Eulen* (German, 1989), *De Takkeling* (Dutch, 1993), *Eulen-Rundblick* (German, 1993), *Rapaces de France* (annual supplement of *L'Oiseau*) and *Scottish Raptor Monitoring Report* (Scottish, 2003; preceded by the annual Raptor Round Up).

The quality of raptor work in Europe has steadily increased over the past century. Identification skills have improved with the publication of specialized raptor field guides, including those of Géroudet (1978), Porter et al. (1981), Clark (1999), Forsman (1999), and Génsbøl (2005) (the latest edition of a standard work first published in 1984 and now translated into several languages). The last attempts at condensing this information were made in 1971 and 1980, when volumes 4 (raptors) and 9 (owls) of the monumental *Handbuch der Vögel Mitteleuropas* (Glutz von Blotzheim et al. 1971, Glutz von Blotzheim and Bauer 1980) were published. Volumes 2 (raptors) and 4 (owls) of the *Birds of the Western Palearctic* (Cramp 1980, 1985), published in 1980 and 1985, already showed a less thorough coverage of the available literature, and despite heroic attempts to update this handbook series with *BWP Updates* (e.g., Sergio et al. 2001, Arroyo et al. 2004) in the 1990s (collated with all nine volumes of the Handbook and the Concise Edition on *BWPi DVD-ROM* in 2004), the exponential growth of raptor literature appears to have surpassed the feasibility of complete coverage in printed volumes.

Satellite-tracking became popular by the early 1990s, with, for example, 116 individuals of 14 raptor species being fitted with platform transmitter terminals (PTTs) within the framework of the Argos Program in 1992-2004 (overview in Meyburg and Meyburg 2006), and other important satellite-tracking studies include those that tracked the migrations of European Honey Buzzards (*Pernis apivorus*) (Hake et al. 2003), Western Marsh Harriers (*Circus aeruginosus*), Ospreys (www.roydennis.org), and Montagu's Harriers (*C. pygargus*) (www.grauwekiekendief.nl). This will improve and calibrate the information gathered in long-running ringing schemes, the latter summarized for raptors ringed in Sweden, Norway and Britain by Fransson and Pettersson (2001), Wernham et al. (2002), and Bakken et al. (2003), respectively.

Long-term trends and annual, age- and sex-specific variations in timing of migrating raptors are being monitored during broad front migration in The Netherlands (LWVT 2002) and at migratory bottlenecks in southern Sweden (Kjellén and Roos 2000), Randecker Maar in the Pre-Alps in southern Germany (Gatter 2000), Col d'Organbidexka in the French Pyrenees (http://www.organbidexka.org), the central Mediterranean (Agostini 2002), and the Strait of Gibraltar in southern Spain (Bernis 1980, with more recent information at www.seo.org).

The most pivotal and heavily cited overview of raptor ecology published in the second half of the 20th century is *Population Ecology of Raptors* by Ian Newton (1979). Having first-hand experience with several species, particularly the Eurasian Sparrowhawk, and gifted with a flowing style of writing, Newton synthesized the available information into an ecological framework, putting separate findings into perspective and pointing out new avenues of exploration. This book has been the source of inspiration for many, much like the works of Heinroth and Heinroth (1926), Uttendörfer (1939), and the Tinbergen brothers (Schuyl et al. 1936, Tinbergen 1946) were to researchers earlier.

Since the 1970s, distribution, abundance, trends and reproduction of raptors have been systematically sampled by tens of thousands of volunteers in every country in western and northern Europe. Nation-based overviews are available for Norway (Hagen 1952), Britain (Brown 1976), Denmark (Jørgensen 1989), Austria (Gamauf 1991), The Netherlands (Bijlsma 1993), Germany (Kostrzewa and Speer 2001), Serbia (Puzovic 2000), and France (Thiollay and Bretagnolle 2004). All these works show the great strides made by European "raptorphiles" during the past decades and an increasing concern about environmental problems.

In 1974, under the innocent title, *Birds of Prey in Europe*, M. Bijleveld published his overview of raptor destruction in Europe since the 18th century, with an emphasis on direct persecution (Bijleveld 1974). Since then, the threats to raptors have multiplied and diversified. The impact of persecution on raptor populations was a determinant of population size well into the early 20th century (and still is locally, e.g., in Malta, as documented by Fenech 1992), but it has been replaced by even greater threats like persistent chemicals and habitat destruction. Concurrently, raptor conservation has become a major topic. This also is apparent from the many proceedings of various raptor meetings, including those of the WWGBPO and the former International Council for Bird Preservation and the German series *Populationsökologie von Greifvögel- und Eulenarten* (Stubbe and Stubbe 1987, 1991, 1996, 2000, 2006), and such recent volumes as *Sea Eagle 2000* on the White-tailed Eagle (Helander et al. 2003), and *Birds of Prey in a Changing Environment* (Thompson et al. 2003).

At the same time, reintroduction programs were instigated in many countries to help threatened species regain a foothold, including the Red Kite (*Milvus milvus*), White-tailed Eagle, Bearded Vulture, Griffon Vulture (*Gyps fulvus*), Peregrine Falcon, and Eurasian Eagle-Owl (*Bubo bubo*) (Cade 2000). Large-scale conservation programs, including education, habitat and

nest protection, and research, are now operational for many species. For example, thousands of nests of Montagu's Harrier have been protected from destruction during harvest in farmland; nowadays their main breeding habitat is in The Netherlands (www.grauwekiekendief.nl), Germany (www.nabu.de), France (Leroux 2004), Hungary, Czech Republic and Poland (see Mischler 2002 for an overview of the state of the art).

Former USSR. About 2,000 articles on raptors have been published in the territory of the former USSR. Among the important regional ornithological journals are *Selevenia* (Kazakhstan), *Caucasian Ornithological Bulletin, Ornithologiya, Russian Journal of Ornithology*, and *Strepet* (Russia), and *Berkut* and *Branta* (Ukraine). Many articles on raptors, mainly large falcons, in the ex-USSR are published in English in the journal, *Falco*, the newsletter of the Middle East Falcon Research Group. All 27 numbers published so far are available at www.falcons.co.uk/default.asp?id=131.

Raptor Conservation is a bilingual (English-Russian), semi-annual newsletter on the raptors of eastern Europe and northern Asia. Five numbers are available at http://ecoclub.nsu.ru/raptors. Although its publication was discontinued, the bilingual (English-Russian) newsletter, *Raptor-Link*, published by Eugene Potapov from 1993-1996, contained useful information on birds of prey of the former USSR.

Among the most important monographs worth noting are the first volume in the series, *Birds of the Soviet Union*, first published in Russian in 1951 and later translated into English in Jerusalem (Dement'ev and Gladkov 1966), *Birds of Prey and Owls of Baraba and Northern Kulunda* (Danilov 1976), *Birds of Prey of the Forest* (Galushin 1980), *Eagles* (Bragin 1987), and *Eagles of Lake Baikal* (Ryabtsev 2000).

There have been a number of conferences on birds of prey in this region, including several on the ecology and conservation of birds of prey of northern Eurasia. Four of these conferences have been held by now, and the proceedings of three have been published (Galushin 1983, Flint 1983, Galushin and Khokhlov 1998, 1999; Belik 2003a, 2003b), and the latter included a special section on the Northern Goshawk. Each volume contains more than 100 abstracts on birds of prey. The table of contents in English for some of the conferences is available at http://my.tele2.ee/birds.

One volume in the series *Bird Migrations of Eastern Europe and Northern Asia* is devoted to Falconiformes and Gruiformes (Il'ichev 1982). This book is in Russian, but the footnotes for all tables, maps and diagrams are in English, and there is a 10-page bibliography. Other useful raptor books, also in Russian, are *Birds of Prey and Owls in Nature Reserves of the Russian Federation* (RSFSR) (Galushin and Krever, 1985) and *Methods of Study and Conservation of Birds of Prey (Methodological Recommendations)* (Priklonskiy et al. 1989). They also are available at www.raptors.ru/ library/books/methods_1989/Index.htm.

The 14th issue of the *Proceedings of Teberda State Nature Reserve* is a thematic collection of papers entitled *Birds of Prey and Owls of Northern Caucasia* (Polivanova and Khokhlov 1995). The articles are in Russian without English summaries. The proceedings of the workshop, *Rare Birds of Prey of the Northern Forest Zone of the European Part of Russia: Prospects on the Study and Means of Conservation*, held in Cherepovets from 11-14 September 2000, includes 18 abstracts (Galushin 2001). These articles also are in Russian without English summaries.

Other useful monographs include *Birds of Prey and Owls of Perm' Prikamie (Kama River Area)* (Shepel' 1992), which contains species accounts on all raptors of the Uralian area written in Russian. In 1999, the first issue in the series, *Threatened Bird Species of Russia and CIS*, was published by the Russian Bird Conservation Union (http://www.rbcu.ru) and includes about 30 articles on the Asian Imperial Eagle (*A. heliaca*) (Belik 1999).

The first volume in the series, *Life of our Birds and Mammals*, was devoted to owls, as a book entitled *Life of Owls* (Pukinskiy 1977). Later, the same author published the scientific-popular book on the Blakiston's Fish Owl (*B. blakistoni*), which was later translated into German and published in the former German Democratic Republic (Pukinskiy 1975). Full texts of a collection of papers on the Eurasian Eagle-Owl (Voronetskiy 1994) also are available at http://raptorsr.ru/library/index.html. The most complete species accounts on all owl species in the former USSR can be found in two volumes of the series, *Birds of Russia and Adjacent Countries*, (Pukinskiy 1992, Zubakin et al. 2005). The most up-to-date information on the numbers and distribution of owls was published in the collection of 69 papers, *Owls of Northern Eurasia* (Volkov et al. 2005). Although the papers are in Russian, each article contains an English summary. Interesting material on the diurnal birds of prey of Uzbekistan was published in the first volume of the book, *Birds of Uzbekistan* (Mitropolskiy et al. 1987). It too is in Russian and is available at http://ecoclub.nsu.ru/raptors/publicat/raptors/ Uzbek_bitds_1987.pdf.

About 60 major articles and short communications on raptor migration from the territory of the former USSR were translated by Jevgeni Shergalin into English and are available from the library of the Acopian Center for Conservation Learning, Hawk Mountain Sanctuary, Pennsylvania, or can be requested from the translator himself (zoolit@hotmail.co.uk). In addition, all of the major articles on the Cinereous Vulture (*Aegypius monachus*) were translated into English and are available at http://aegypiusrus.itgo.com and at http://aegypius.itgo.com. For Russian raptor articles in general, the best websites for downloading many raptor articles are http://www.raptors.ru and http://ecoclub.nsu.ru/raptors/RC. The former contains an electronic library with full-text versions of 298 articles.

Eastern Palearctic. Important journals with raptor content for this region are *BirdingASIA* and *Forktail*. The Chinese journal, *Acta Zoologica Sinica*, as well as *Ornithological Science*, which is published in Japan, are global in scope, although the majority of articles are on Asian topics. Important regional journals include *Hong Kong Bird Report* (Hong Kong), *Aquila chrysaetos*, *Bulletin of the Japanese Bird Banding Association*, *Japanese Journal of Ornithology*, *Journal of the Yamashina Institute of Ornithology*, *Strix* (Japan), and *Korean Journal of Ornithology* (South Korea).

The recent creation of the ARRCN has stimulated much more interest in raptors in this portion of the Palearctic, as a result of its well-attended symposia, active listserve community, and distribution of meeting abstracts.

Much of the raptor literature in eastern Asia concerns endangered species. A multinational symposium was held on the endangered Steller's Sea Eagle (*H. pelagicus*) and the White-tailed Eagle in Japan in 1999, leading to the publication of a useful proceedings volume (Ueta and McGrady 2000). As a result of the illegal falcon trade, there is much interest in Saker Falcons (*F. cherrug*) in Mongolia and nearby countries, and by now there are many papers on this topic (e.g., Gombobaatar et al. 2004). The Middle East Falcon Research Group organized the second international conference on the Saker Falcon and Houbara Bustard (*Chlamydotis undulata macqueenii*) at Ulaanbaatar, Mongolia in July 2000, and a proceedings volume containing 33 papers in Mongolian, Russian, and English was published (Anon. 2001). The full text of the proceedings is at www.falcons.co.uk/ mefrg/conference.htm.

There is a very active raptor community in Taiwan, although *Raptor Research of Taiwan* is the only journal focusing on raptors there. It is published biannually in Chinese by the Raptor Research Group of Taiwan (RRGT) and is not yet included in international abstracting services. *Zoological Studies*, an English quarterly journal published by Research Center for Biodiversity, Academia Sinica, occasionally contains scientific papers on raptor research, as does *Notes and Newsletter of Wildlifers* (NOW) published by National Pingtung University of Science and Technology.

The most important source of raptor sighting records in Taiwan is the membership of the Wild Bird Federation Taiwan (WBFT), which compiles and publishes all the raptor sighting records in Taiwan in its monthly bulletin, *Chinese Feathers*. Recent research results tend to be published in conference proceedings, such as *The Symposium on Ecology of Raptors in Taiwan, Proceedings of the Conference on Birds, Proceedings of the Taiwan and China Bi-coastal Bird Conference*, and *Proceedings of the International Symposium on Wildlife Conservation*. Some raptor papers occasionally can be found in *Bird Conservation Research Reports, Taiwanese Wild Birds, The Mikado Pheasant*, and *Wild Birds*, all serial or special publications of WBFT. The Changhua Wild Bird Society had a special interest in the spring migration of the Gray-faced Buzzard (*Butastur indicus*) and for several years reported their annual observations in *Bird Conservation Research Reports*. Diverse semi-popular magazines such as *Wildlife, Taipei Zoo Quarterly*, and *Taiwan Veterinary Journal* also contain a few reports on the raptors of Taiwan. Raptor researchers in Taiwan usually submit their study results to international journals, such as *Journal of Raptor Research, Ibis*, or *Wilson Journal of Ornithology*.

Various government agencies have been supporting raptor studies, and results of these studies are published in the *CAPD Forestry Series, Quarterly Journal of Chinese Forestry, Ecological Research Report, Council of Agriculture, Endemic Species Research, Natural Conservation Quarterly*, or the conservation research reports of different national parks.

Members of the RRGT have published three field guides of diurnal raptors in Taiwan (Hsiao 1996, 2001; Lin 2006), four books on Black Kites (*Milvus migrans*) (Shen 1993, 1998, 1999, 2004), three booklets (Chen et al. 2003, Chen 2004, Wang 2006), and organized several raptor workshops from 1998 to 2005. Other books include *Raptors of Taiwan* by Chung-Wei Yen (1982) and an illustrated handbook of owls by Chin-wen Tsai (2003).

There is an increasing amount of literature on raptors on the mainland of China that we have not had the opportunity to examine. *Raptors of China* (Weishu 1995) is worth special mention.

ABSTRACTING AND INDEXING SERVICES

There are now hundreds of on-line databases and indexing systems containing records of interest to raptor researchers. The sites vary in their coverage, ease of use, and the amount of information that is freely accessible. The list of websites provided here is necessarily arbitrary, but we have found them to be useful in our own work. The numerical data are from August 2006. Naturally, such estimates are constantly revised upward, and weaker systems fall by the wayside, but the data are included here for comparative purposes.

Some of the best electronic databases offer free access. Others are available only by subscription, sometimes for high fees. Some provide opportunities for obtaining the full texts of virtually any major paper for researchers and institutions with liberal financial resources. Regrettably, most have a strong Euro-American bias, and few do an adequate job of covering literature in many other important languages, including Arabic, Chinese, Hebrew, Japanese, and Russian.

Valiela and Martinetto (2005) discussed the relative merits and weakness of several of the major online schemes, and they rightly emphasized that none of the databases, taken singly or in combination, are as yet sufficiently comprehensive to provide truly adequate coverage of most research topics. Perhaps such a situation will eventually emerge, but in the meantime, electronic databases and web sources are still primarily useful as powerful tools that supplement traditional literature search methods, especially poring through the Literature Cited sections of papers by earlier authors.

Free Access Databases

Bookfinder.com (www.bookfinder.com). A commercial search engine listing over 100 million books for sale, incorporating the catalogues and databases of virtually every bookseller of any importance in Europe, North America, Australia, and South Africa. It is useful for locating books on raptors and as a source for bibliographic details.

Global Raptor Information Network (GRIN)

(www.globalraptors.org). Features a species-level database on the status and distribution of diurnal raptors, handbook-style species accounts, home-pages of raptor researchers, and a bibliographic database containing 36,000 citations with keywords on diurnal birds of prey. Free PDF copies of most of the listed articles are available upon request from library@peregrinefund.org. The site is maintained by The Peregrine Fund in Boise, Idaho.

Google Print (or Google Book Search) (http://print.google.com). Still in the "beta" stage, this is a project of breath-taking audacity with a stated goal of scanning the contents of as many books as possible and making the full texts searchable on line. At the start, Google plans to scan all or a large portion of the book collections of the University of California, University of Michigan, Harvard University, Stanford University, New York Public Library, and Oxford University. Non-copyrighted books will be completely viewable, but only "snippets" from books potentially or actually still in copyright will be freely accessible. Since affirmative action must be taken by parties who do not wish their copyrighted materials scanned, this has created tension between the publishing community and Google, and it seems likely that the future of this ambitious project may eventually be determined by the courts. If it survives in the form anticipated, or even only partly so, it will be an invaluable resource to researchers.

Google Scholar (http://scholar.google.com). A huge database assembled from peer-reviewed papers, theses, books, preprints, abstracts, technical reports, and popular articles, many of which are linked to full-text versions, or options for ordering the publications. During its initial stages of development, there has been much criticism over Google's unwillingness to disclose details on the parameters of the database, the sometimes puzzling gaps in coverage, and the inclusion of many non-scholarly references. Furthermore, the sheer vastness of the system often makes it difficult to focus narrowly on technical articles on a desired topic. For example, a search on "DDT and peregrine" yields 1,220 hits, but the results include many popular articles and press releases. Like other similar schemes, there are numerous typographical errors, apparent artifacts of the scanning process. Nevertheless, as this database continues to grow and becomes more refined, it will probably become the starting point for most scientific literature searches.

Ornithological Worldwide Literature (OWL) (http://egizoosrv.zoo.ox.ac.uk/OWL). An electronic database of citations with brief annotations from the worldwide ornithological literature, containing about 80,000 citations back to 1983. Formerly known as "Recent Ornithological Literature," OWL is a joint effort between the American Ornithologists' Union, Birds Australia, and the British Ornithologists' Union and is hosted by the Edward Grey Institute of Field Ornithology at Oxford University, U.K.

Ornithologische Schriftenschau (http://www.dda-web.de/index.php). This German-language service reviews the ornithological content of 340 national and international periodicals, especially those published in European countries, usually providing brief abstracts for most papers. A print version also can be obtained by subscription.

Raptor Information System (http://ris.wr.usgs.gov). A catalog of over 33,000 citations with keywords on birds of prey, including owls, with particular emphasis on raptor management, human impacts on raptors, the mitigation of impacts, and basic raptor biology. This database is particularly valuable for its coverage of the "gray literature," including in-house government reports, dissertations, and unpublished manuscripts. Maintained by the Resources Division of the U.S. Geological Survey Snake River Field Station in Boise, Idaho. The librarian may be reached at fresc_library@usgs.gov.

Searchable Ornithological Research Archives (SORA) (http://elibrary.unm.edu/sora). An open-access electronic journal archive initiated and maintained by Blair Wolf of the Cooper Ornithological Society. The contents of the site now include full-text versions of *The Auk* (1884-1999), *The Condor* (1899-2000), *Journal of Field Ornithology* (1930-2000), *North American Bird Bander* (1976-2000), *Ornithological Monographs* (1964-2005), *Ornitologia Neotropical* (1990-2002), *Pacific Coast Avifauna* (1900-1974), *Studies in Avian Biology* (1978-1999), *Western Birds* (1970-2004), and *Wilson Bulletin* (1889-1999).

Fee-based Databases and Indexes

BioOne® (http://www.bioone.org). A collaboration among scientific societies, libraries, academe, and the commercial sector which provides access to linked full text versions of interrelated journals focused on the biological, ecological and environmental sciences. Participating journals include those published by the Ornithological Societies of North America, The Wildlife Society and numerous other titles of interest to raptor researchers.

Blackwell Synergy (www.blackwell-synergy.com). An online journal service including citations, abstracts and, in some cases, fully linked texts of about 900,000 articles from nearly 900 scholarly journals, including several leading ornithological titles (*Ibis, Journal of Avian Biology, Journal of Field Ornithology, Journal of Ornithology*). Although primarily a subscription service, free access is provided to many abstracts and full-text versions of articles, especially from older issues. Together with Google, Blackwell Synergy may soon be able to launch a typical Google search but filter the result set to the scholarly research content from participating publishers.

Current Contents/Life Sciences (http://scientific. thomson.com/products/ccc). Current Contents provides online access to complete bibliographic information from articles, editorials, meeting abstracts, and commentaries in current issues of 1,370 life sciences journals and books. The site is marketed by Thomson ISI, a company offering a wide array of other information products, including leading bibliographic software programs (ProCite®, EndNote®, Reference Manager®). Archived files are available back to 1990.

IngentaConnect (www.ingentaconnect.com). Access to an online database of over 20 million citations from over 30,000 academic publications and online access to full-text versions of many articles through online purchase of individual articles or through subscriptions to publications. A well-designed system, but contains fewer journals of interest to raptor biologists than Blackwell Synergy.

JSTOR (www.jstor.org). A non-profit scheme designed to maintain an archive of scanned images of back issues of major journals, including many of interest to raptor biologists, namely, *American Midland Naturalist, American Naturalist, Avian Diseases, Bio-Science, Biotropica, Condor, Conservation Biology, Evolution*, and the journals published by the British Ecological Society and the Ecological Society of America. More journal titles will be added to the archives, including *The Auk, Journal of Field Ornithology, Journal of Wildlife Management, Wildlife Monographs*, and *Wildlife Society Bulletin*. Current issues are not covered, so there is typically a gap of two to five years between the most recent issues and their availability on JSTOR.

OCLC (www.oclc.org). The world's largest library cataloging service, now used by 55,000 libraries in 110 countries and territories. The "WorldCat" database is maintained by more than 9,000 member institutions, and it contains over 67 million records of every form of human expression, ranging from stone tablets to electronic books, CDs, and DVDs. It remains the single best source for bibliographic information on books in virtually every language.

Scirus (www.scirus.com). This is purportedly the most comprehensive science-specific search engine on the Internet, covering over 250 million science-specific web pages, including many non-journal sources, from over 214 million websites. Unlike Google Scholar, it purportedly filters out non-scientific sites and finds peer-reviewed PDF and PostScript files overlooked by most other search engines. Sciurus is maintained by the giant publishing house, Elsevier, and it includes Bio-Med Central, an independent online publishing house that publishes several journal titles with occasional raptor content.

UMI Dissertations Services (www.umi.com/prod ucts_umi/dissertations). This is the best source for dissertations and theses, with over two million entries covering over 1,000 North American graduate schools and European universities. The citations for Ph.D. dissertations from 1980 onward contain abstracts, as do Master's theses from 1988 forward. Full texts are offered digitally through ProQuest Digital Dissertations or in traditional paper versions through Dissertation Express.

Wildlife & Ecology Studies Worldwide (www.nisc.com). Provides a large index to the literature on wild vertebrates, including 400,000 bibliographic records, many with abstracts, from 1935 to the present. Includes Wildlife Review Abstracts, Swiss Wildlife Information Service, U.S. Fish & Wildlife Reference Service's Wildlife Database (containing many unpublished reports and surveys), BIODOC (Neotropical literature), the World Conservation Union publications database, and the Afro-Tropical Bird Information Retrieval database. A product of National Information Services Corporation (NISC), which provides access through their Web search service, Biblioline.

Zoological Record (http://scientific.thomson. com/products/zr). This is the world's oldest continuing database of bibliographic records on animal biology. It has been published continuously since 1864 and now contains 1.7 million records in electronic format. Zoological Record screens 5,000 serials and many other literature sources to add 72,000 records to the database annually. The present online version covers the literature back to 1978, but will soon provide the original bibliographic and taxonomic indexing data from the print volumes back to 1864. The *Aves* section is most pertinent to raptor research. Formerly maintained by a non-profit consortium, BIOSIS, Zoological Record and related products are now owned by Thomson ISI.

LITERATURE CITED

AGOSTINI, N. 2002. La migrazione dei rapaci in Italia. *Manuelo pratico di Ornitologia* 3:157–182.

ALI, S. AND S.D. RIPLEY. 1987. Compact handbook of the birds of India and Pakistan: together with those of Bangladesh, Nepal, Bhutan, and Sri Lanka, 2nd Ed. Oxford University Press, New York, NY U.S.A.

ALLAN, D. 1996. A photographic guide to birds of prey of southern, central and East Africa. New Holland Ltd., London, United Kingdom.

AMADON, D. AND J. BULL. 1988. Hawks and owls of the world: a distributional and taxonomic list, with the genus *Otus* by J.T. Marshall and B.F. King. *Proc. West. Found. Vertebr. Zool.* 3:294–357.

ANDERSON, M.D. AND R. KRUGER. 2004. Raptor conservation in the Northern Cape Province, 3rd Ed. Northern Cape Department of Tourism, Environment and Conservation & Eskom, Upington, Kalahari, South Africa.

AOU COMMITTEE ON CLASSIFICATION AND NOMENCLATURE. 1998. Check-list of North American birds: the species of birds of North America from the Arctic through Panama, including the West Indies and Hawaiian Islands, 7th Ed. American Ornithologists' Union, Washington, DC U.S.A.

ARROYO, B., J.T. GARCIA AND V. BRETAGNOLLE. 2004. *Circus pygargus* Montagu's Harrier. *BWP Update* 6:39–53.

AVIAN POWER LINE INTERACTION COMMITTEE. 1996. Suggested practices for raptor protection on power lines: the state of the art in 1996. Raptor Research Foundation, Washington, DC U.S.A.

BAKER-GABB, D. AND W.K. STEELE. 1999. The relative abundance, distribution and seasonal movements of Australian Falconiformes, 1986–90. *Birds Aust. Rep. Ser.* 6:1–107.

BAKKEN, V., O. RUNDE AND E. TJØRVE. 2003. Norwegian bird ringing atlas, Vol. 1: divers – auks. Ringmerkningssentralen, Stavanger Museum, Stavanger, Norway.

BARRETT, G., A. SILCOCKS, S. BARRY, R. CUNNINGHAM AND R. POULTER. 2003. The new atlas of Australian birds. Royal Australasian Ornithologists Union, Victoria, Australia.

BAUMGART, W. 2001. Europas geier. AULA-Verlag, Wiebelsheim, Germany.

BEEHLER, B.M., T.J. PRATT AND D.A. ZIMMERMAN. 1986. Birds of New Guinea. Princeton University Press, Princeton, NJ U.S.A.

BELIK, V.P. [ED.]. 1999. [The Imperial Eagle: distribution, population status and conservation perspectives within Russia.] Russian Bird Conservation Union, Moscow, Russia. http://ecoclub.nsu.ru/raptors/publicat/aquila_hel.shtm (last accessed 21 December 2006).

———, V.P. [ED.]. 2003a. [Materials of the 4th Conference on Raptors of Northern Eurasia, Penza, 1–3 February 2003.] Rostov State Pedagogical University, Rostov, Russia. http://raptors.ru/library/index.html (last accessed 21 December 2006).

——— [ED.]. 2003b. [Goshawk in ecosystems.] Rostov State Pedagogical University, Rostov, Russia.

BENT, A.C. 1937. Life histories of North American birds of prey. Order Falconiformes (Pt. 1). *U. S. Nat. Mus. Bull. 167.* U.S. Government Printing Office, Washington, DC U.S.A.

———. 1938. Life histories of North American birds of prey (Pt. 2). *U.S. Nat. Mus. Bull. 170.* U.S. Government Printing Office, Washington, DC U.S.A.

BERKELMAN, J. 1997. Habitat requirements and foraging ecology of the Madagascar Fish-eagle. Ph.D. dissertation, Virginia Polytechnic Institute and State University, Blacksburg, VA U.S.A.

BERNIS, F. 1980. La migración de las aves en el Estrecho de Gibraltar, Vol. 1: aves planeadoras. Universidad Complutense de Madrid, Madrid, Spain.

BERTHOLD, P., E. BEZZEL AND G. THIELCKE. 1974. Praktische Vogelkunde. Empfehlungen für die Arbeit von Avifaunisten und Feldornithologen. Kilda-Verlag, Greven, Germany.

BIBBY, C.J., N.D. BURGESS, D.A. HILL AND S.H. MUSTOE. 2000. Bird census techniques, 2nd Ed. Academic Press, San Diego, CA U.S.A.

BIERREGAARD, R.O., Jr. 1995. The status of raptor conservation and our knowledge of the resident diurnal birds of prey of Mexico. *Trans. N. Am. Wildl. Nat. Resour. Conf.* 60:203–213.

———. 1998. Conservation status of birds of prey in the South American tropics. *J. Raptor Res.* 32:19–27.

BIJLEVELD, M. 1974. Birds of prey in Europe. Macmillan Press Ltd., London, United Kingdom.

BIJLSMA, R. 1993. Ecological atlas of Netherlands raptors. Schuyt & Co., Haarlem, The Netherlands.

———. 1997. Manual for field research in raptors. KNNV Uitgeverij, Utrecht, The Netherlands.

BILDSTEIN, K.L. 2006. Migrating raptors of the world: their ecology and conservation. Cornell University Press, Ithaca, NY U.S.A.

——— AND J.I. ZALLES. 2005. Old World vs. New World long-distance migration in accipiters, buteos, and falcons. Pages 116–154 in R. Greenberg and P. P. Marra [EDS.], Birds of two worlds: the ecology and evolution of migration. John Hopkins University Press, Baltimore, MD U.S.A.

BIRD, D.M. AND R. BOWMAN [EDS.]. 1987. The ancestral kestrel. *Raptor Res. Rep.* 6:1–178.

———, N.R. SEYMOUR AND J.M. GERRARD [EDS.]. 1983. Biology and management of Bald Eagles and Ospreys: proceedings of 1st international symposium on Bald Eagles and Ospreys, Montreal, 28–29 October, 1981. Macdonald Raptor Research Centre of McGill University and Raptor Research Foundation, Inc., Ste. Anne de Bellevue, Quebec, Canada.

———, D.E. VARLAND AND J.J.NEGRO [EDS.]. 1996. Raptors in human landscapes: adaptations to built and cultivated environments. Academic Press, San Diego, CA U.S.A.

BLAKERS, M., S.J.J.F. DAVIES AND P.N. REILLY. 1984. The atlas of Australian birds. Royal Australasian Ornithologists Union/Melbourne University Press, Melbourne, Australia.

BLOCK, W.M., M.L. MORRISON, AND M.H. REISER [EDS.]. 1994. The Northern Goshawk: ecology and management. *Stud. Avian Biol.* 16:1–136.

BORROW, N. AND R. DEMEY. 2001. A guide to the birds of western Africa. Princeton University Press, Princeton, NJ U.S.A.

BOSHOFF, A.F., M.D. ANDERSON AND W.D. BORELLO [EDS.]. 1998. Vultures in the 21st century: proceedings of a workshop on vulture research and conservation in southern Africa. Vulture Study Group, Johannesburg, South Africa.

BRAGIN, E. 1987. [Eagles.] Kainar Press, Alma-Ata, Kazakhstan.

BROWN, L. 1970. African birds of prey. Houghton Mifflin, Boston, MA U.S.A.

———. 1976. British birds of prey. Collins, London, United Kingdom.

———. 1980. The African Fish Eagle. Purnell & Sons, Cape Town, South Africa.

——— AND D. AMADON. 1968. Eagles, hawks and falcons of the world, Vols. 1–2. McGraw-Hill, New York, NY U.S.A.

BROWN, L.H., E.K. URBAN, AND K. NEWMAN. 1982. The birds Africa, Vol. 1. Ostriches to falcons. Academic Press, London, United Kingdom.

BUESER, G.L.L., K.G. BUESER, D.S. AFAN, D.I. SALVADOR, J.W. GRIER, R.S. KENNEDY AND H.C. MIRANDA. 2003. Distribution and nesting density of the Philippine Eagle *Pithecophaga jeffreyi* on Mindanao Island, Philippines: what do we know after 100 years? *Ibis* 145:130–135.

CADE, T.J. 1982. Falcons of the world. Comstock/Cornell University Press, Ithaca, NY U.S.A.

———. 2000. Progress in translocation of diurnal raptors. Pages 343–372 *in* R. D. Chancellor and B.-U. Meyburg [EDS.], Raptors at risk. World Working Group on Birds of Prey and Owls/Pica Press, Berlin, Germany.

———, J.H. ENDERSON, C.G. THELANDER AND C.H. WHITE [EDS.]. 1988. Peregrine Falcon populations: their management and recovery. The Peregrine Fund, Inc., Boise, ID U.S.A.

CARSWELL, M., D. POMEROY, J. REYNOLDS, AND H. TUSHABE. 2005. The bird atlas of Uganda. British Ornithologists' Club and British Ornithologists' Union, Oxford, United Kingdom.

CARTER, I. 2001. The Red Kite. Arlequin Press, Chelmsford, United Kingdom.

CHANCELLOR, R.D. [ED.]. 1975. Proceedings of the world conference on birds of prey, 1975. International Council for Bird Preservation, Cambridge, United Kingdom.

——— AND B.-U. MEYBURG [EDS.]. 2000. Raptors at risk: proceedings of the fifth world conference on birds of prey and owls, 1998. World Working Group on Birds of Prey and Owls, Berlin, and Hancock House, Blaine, WA U.S.A.

——— AND B.-U. MEYBURG [EDS.]. 2004. Raptors worldwide: proceedings of the VI world conference on birds of prey and owls, Budapest, Hungary, 18–23 May 2003. World Working Group on Birds of Prey and Owls, Berlin, Germany and MME/BirdLife Hungary, Budapest, Hungary.

———, B.-U. MEYBURG AND J.J. FERRARO [EDS.]. 1998. Holarctic birds of prey: proceedings of an international conference, 1995. World Working Group on Birds of Prey and Owls, Berlin, Germany and ADENA, Barcelona, Spain.

CHEN, S.C. 2004. Handbook on surveying migratory raptors in Taiwan. Raptor Research Group Taiwan, Taipei, Taiwan.

———, Y.Y. CHANG AND M.H. TSAO. 2003. Identification guide to the raptors of Taiwan. Raptor Research Group Taiwan, Taipei, Taiwan.

CLARK, R.J., D.G. SMITH AND L.H. KELSO. 1978. Working bibliography of owls of the world. National Wildlife Federation Scientific and Technical Series no. 1.

CLARK, W.S. 1999. A field guide to the raptors of Europe, the Middle East, and North Africa. Oxford University Press, Oxford, United Kingdom.

——— AND B.K. WHEELER. 2001. A field guide to hawks of North America, 2nd Ed. Houghton Mifflin, Boston, MA U.S.A.

COATES, B.J. 1985. The birds of Papua New Guinea, including the Bismarck Archipelago and Bougainville, Vol. 1. Non-passerines. Dove Publications, Alderley, Queensland, Australia.

——— AND K.D. BISHOP. 1997. A guide to the birds of Wallacea, Sulawesi, the Moluccas, and Lesser Sunda Islands, Indonesia. Dove Publications, Alderley, Queensland, Australia.

COLLAR, N.J. AND S.N. STUART. 1985. Threatened birds of Africa and related islands: the ICBP/IUCN Red Data Book. International Council for Bird Preservation, Cambridge, United Kingdom.

———, L.P. GONZAGA, N. KRABBE, A. MADROÑO NIETO, L.G. NARANJO, T.A. PARKER III AND D.C. WEGE. 1992. Threatened birds of the Americas: the ICBP/IUCN Red Data Book, 3rd Ed., pt. 2. Smithsonian Institution Press, Washington, DC U.S.A, and International Council for Bird Preservation, Cambridge, United Kingdom.

———, M.J. CROSBY AND A.J. STATTERSFIELD. 1994. Birds to watch 2: the world list of threatened birds. BirdLife Conservation Series 4. BirdLife International, Cambridge, United Kingdom.

———, A.D. MALLARI AND B.R. TABARANZA, Jr. 1999. Threatened birds of the Philippines. The Haribon Foundation/BirdLife International Red Data Book. Bookmark, Inc., Makati City, Philippines.

———, A.V. ANDREEV, S. CHAN, M.J. CROSBY, S. SUBRAMANYA AND J.A. TOBIAS. 2001. Threatened birds of Asia: the BirdLife International Red Data Book. Parts A & B. BirdLife International, Cambridge, United Kingdom.

CONDON, H. T. 1970. Field guide to the hawks of Australia, 4th Ed. Bird Observers Club, Melbourne, Australia.

COOKE, A.S., A.A. BELL AND M.B. HAAS. 1982. Predatory birds, pesticides, and pollution. National Environment Research Council, Institute of Terrestrial Ecology, Monks Wood Experimental Station, Huntingdon, Cambridgeshire, United Kingdom.

COOPER, J.E. 2002. Birds of prey: health and disease, 3rd Ed. Blackwell Science, Oxford, United Kingdom.

———. 2003. Captive birds in health and disease. Hancock House, Surrey, British Columbia, Canada.

CORNELI, H.M. 2002. Mice in the freezer, owls on the porch: the lives of naturalists Frederick and Frances Hamerstrom. University of Wisconsin Press, Madison, WI U.S.A.

CRAIG, G.R. AND J.H. ENDERSON. 2004. Peregrine Falcon biology and management in Colorado 1973–2001. *Colo. Div. Wildl. Tech. Publ. 43.*

CRAIGHEAD, J.J. AND F.C. CRAIGHEAD, Jr. 1956. Hawks, owls and wildlife. Stackpole Books, Harrisburg, PA U.S.A.

CRAMP, S. [ED.]. 1980. Handbook of the birds of Europe, the Middle East and North Africa, Vol. 2. Hawks to bustards. Oxford University Press, Oxford, United Kingdom.

——— [ED.]. 1985. Handbook of the birds of Europe, the Middle East and North Africa, Vol. 4. Terns to woodpeckers. Oxford University Press, Oxford, United Kingdom.

CUPPER, J. AND L. CUPPER. 1981. Hawks in focus: a study of Australia's birds of prey. Jaclin Enterprises, Mildura, Australia.

CUTHBERT, R., R.E. GREEN, S. RANADE, S. SARAVANAN, D.J. PAIN, V. PRAKASH AND A.A. CUNNINGHAM. 2006. Rapid population declines of Egyptian Vulture (*Neophron percnopterus*) and Red-headed Vulture (*Sarcogyps calvus*) in India. *Anim. Conserv.* 9:349–354.

CZECHURA, G. AND S. DEBUS. 1997. Australian raptor studies II: proceedings of the second Australasian Raptor Association conference, Currumbin, Queensland, 8–9 April 1996. *Birds Aust. Monogr.* 3:1–125.

DANILOV, O.N. 1976. Birds of prey and owls of Baraba and northern Kulunda. Nauka, Novosibirsk, Russia.

DE VRIES, T. 1973. The Galapagos Hawk: an eco-geographical study with special reference to its systematic position. Ph.D. dissertation, Vrije Universiteit te Amsterdam, Amsterdam, Netherlands.

DEBUS, S. 1998. The birds of prey of Australia: a field guide. Oxford University Press, Oxford, United Kingdom.

DEL HOYO, J., A. ELLIOTT AND J. SARGATAL [EDS.]. 1994. Handbook of birds of the world, Vol. 2. New World vultures to guineafowl. Lynx Edicions, Barcelona, Spain.

———, A. ELLIOTT AND J. SARGATAL [EDS.]. 1999. Handbook of birds of the world, Vol. 5. Barn-owls to hummingbirds. Lynx Edicions, Barcelona, Spain.

DEMENT'EV, G.P. AND N.A. GLADKOV [EDS.]. 1966. Birds of the Soviet Union, Vol. 1. Israel Program for Scientific Translations, Jerusalem, Israel.

DICKINSON, E. D. [ED.]. 2003. The Howard and Moore complete checklist of birds of the world. Princeton University Press, Princeton, NJ U.S.A.

DI GIACOMO, A.G. AND S. KRAPOVICKAS [EDS.]. 2005. Inventario de la biodiversidad de la Reserva Ecológica El Bagual, Formosa, Argentina. Temas de Naturaliza y Conservacion 4, Aves Argentina/AOP, Buenos Aires, Argentina.

DÖTTLINGER, H. 2002. The Black Shaheen Falcon (*Falco peregrinus peregrinator* Sundevall 1837): its morphology, geographic variation and the history and ecology of the Sri Lanka (Ceylon) population. Ph.D. dissertation, University of Kent, Canterbury, United Kingdom.

DUNCAN, J.R. 2003. Owls of the world: their lives, behavior, and survival. Firefly Books, Buffalo, NY U.S.A.

———, D.H. JOHNSON AND T.H. NICHOLS [EDS.]. 1997. Biology and conservation of owls in the Northern Hemisphere: 2nd international symposium. USDA Forest Service, General Technical Report NC-190, North Central Forest Experiment Station, St. Paul, MN U.S.A.

ERIKSEN, J., D.E. SARGEANT AND R. VICTOR. 2003. Oman bird list: the official list of the birds of the Sultanate of Oman. Centre for Environmental Studies and Research, Sultan Qaboos University, Sultanate of Oman.

FAABORG, J. 1986. Reproductive success and survivorship of the Galapagos Hawk, *Buteo galapagoensis*: potential costs and benefits of cooperative polyandry. *Ibis* 128:337–347.

FENECH, N. 1992. Fatal flight. The Maltese obsession with killing birds. Quiller Press, London, United Kingdom.

FERGUSON-LEES, J. AND D.A. CHRISTIE. 2001. Raptors of the world. Houghton Mifflin, Boston, MA U.S.A.

——— AND D.A. CHRISTIE. 2005. Raptors of the world. Princeton University Press, Princeton, NJ U.S.A.

FERRER, M. 2001. The Spanish Imperial Eagle. Lynx Edicions, Barcelona, Spain.

——— AND G.F.E. JANSS [EDS.]. 1999. Birds and power lines: collision, electrocution, and breeding. Quercus, Madrid, Spain.

FLINT, V.E. [ED.]. 1983. [Conservation of birds of prey.] Nauka, Moscow, Russia.

FORD, E.B. [ED.]. 1999. Gyrfalcon. John Murray Publishing, London, United Kingdom.

FORSMAN, D. 1999. The raptors of Europe and the Middle East: a handbook of field identification. T. & A.D. Poyser, London, United Kingdom.

FORSMAN, E.D., S. DeSTEFANO, M. G. RAPHAEL AND R.J. GUTIÉRREZ [EDS.]. 1996. Demography of the Northern Spotted Owl. Studies in Avian Biology no. 17. Cooper Ornithological Society, Lawrence, KS U.S.A.

FRANKLIN, A.B., J. GUTIÉRREZ, J.D. NICHOLS, M.E. SEAMANS, G.C. WHITE, G.S. ZIMMERMAN, J.E. HINES, T.E. MUNTON, W.S. LAHAYE, J.A. BLAKESLEY, G.N. STEGER, B.R. NOON, D.W.H. SHAW, J.J. KEANE, T.L. McDONALD AND S. BRITTING. 2004. Population dynamics of the California Spotted Owl (*Strix occidentalis*): a meta-analysis. *Ornithol. Monogr.* 54:1–54.

FRANSSON, T. AND J. PETTERSSON. 2001. Swedish bird ringing atlas, Vol. 1: divers – raptors. Naturhistoriska riksmuseet and Sveriges Ornitologiska Förening, Stockholm, Sweden.

FREILE, J.F., J.M. CARRIÓN, F. PRIETO-ALBUJA AND F. ORTIZ-CRESPO. 2004. Listado bibliográfico sobre las aves del Ecuador: 1834–2001. *Boletines Bibliográficos sobre la Biodiversidad del Ecuador* 3:1–511.

GALUSHIN, V.M. 1980. [Birds of prey of the forest.] Lesnaya Promyshlennost, Moscow, Russia.

———. 1983. [Ecology of birds of prey.] Nauka, Moscow, Russia.

——— [ED.]. 2001. [Rare birds of prey of the northern forest zone of the European part of Russia: prospects on the study and means of conservation.] Darwin State Nature Reserve, Cherepovets, Russia.

——— AND A.N. KHOKHLOV [EDS.]. 1998. [The 3rd conference on birds of prey of eastern Europe and northern Asia, Kislovodsk, 15–18 September 1998. Pt. 1.] Russian Bird Conservation Union, Stavropol, Russia. http://ornithology.chat.ru/ (last accessed 21 December 2006).

——— AND A.N. KHOKHLOV [EDS.]. 1999. [The 3rd conference on birds of prey of eastern Europe and northern Asia, Kislovodsk, 15–18 September 1998. Pt. 2.] Russian Bird Conservation Union, Stavropol, Russia. http://ornithology.chat.ru/ (last accessed 21 December 2006).

——— AND V.G. KREVER [EDS.]. 1985. [Birds of prey and owls in nature reserves of the Russian Federation.] Central Scientific-Research Laboratory on Hunting, Moscow, Russia.

GÁLVEZ, R.A., L. GAVASHELISHVILI AND Z. JAVAKHISHVILI. 2005. Raptors and owls of Georgia. Georgian Center for the Conservation of Wildlife and Buneba Print, Tiblisi, Georgia.

GAMAUF, A. 1991. Greifvögel in Österreich. Bestand – Bedrohung – Gesetz. Monographien Bd. 29. Bundesministerium für Umwelt, Jugend und Familie, Vienna, Austria.

GARGETT, V. 1990. The Black Eagle: a study. Acorn Books, Randburg, South Africa.

GATTER, W. 2000. Vogelzug und Vogelbestände in Mitteleuropa. 30 Jahre Beobachtung des Tagzugs am Randecker Maar. AULA-Verlag, Wiebelsheim, Germany.

GAVASHELISHVILI, L. 2005. Vultures of Georgia and the Caucasus. Georgian Center for the Conservation of Wildlife, Tbilisi, Georgia.

GEHLBACH, F.R. 1994. The Eastern Screech Owl: life history, ecology, and behavior in the suburbs and countryside. Texas A & M University Press, College Station, TX U.S.A.

GÉNSBØL, B. 1984. Rovfuglene i Europa, Nordafrika og Mellemøsten. G.E.C. Gads Forlag, Copenhagen, Denmark.

GERHARDT, R. 1991. Mottled Owls (*Ciccaba virgata*): response to calls, breeding biology, home range, and food habits. M.Sc.

thesis, Boise State University, Boise, ID U.S.A.

GÉROUDET, P. 1978. Les rapaces diurnes et nocturnes d'Europe, 4th Ed. Édition Delachaux et Niestlé, Neuchâtel, Switzerland.

GILBERT, G., D.W. GIBBONS AND J. EVANS. 1998. Bird monitoring methods: a manual of techniques for key UK species. Royal Society for the Protection of Birds, Sandy, United Kingdom.

GIRON PENDLETON, B. A., B.A. MILLSAP, K.W. CLINE AND D.M. BIRD [EDS.]. 1987. Raptor management techniques manual. National Wildlife Federation, Washington, DC U.S.A.

GLINSKI, R.L. [ED.]. 1998. The raptors of Arizona. University of Arizona Press, Tucson AZ U.S.A.

GLUTZ VON BLOTZHEIM, U.N. BAUER AND K.M. BAUER. 1980. Handbuch der Vögel Mitteleuropas. Band 9. Columbiformes-Piciformes. Akademische Verlagsgesellschaft, Wiesbaden, Germany.

———, U.N. BAUER, K.M. BAUER,AND E. BEZZEL. 1971. Handbuch der Vögel Mitteleuropas. Band 4. Falconiformes. Akademische Verlagsgesellschaft, Frankfurt, Germany.

GOMBOBAATAR, S., D. SUMIYA, O. SHAGARSUREN, E. POTAPOV AND N.C. FOX. 2004. Saker Falcon (Falco cherrug milvipes Jerdon) mortality in central Mongolia and population threats. Mongolian J. Biol. Sci. 2:13–22.

GRIMMETT, R., C. INSKIPP AND T. INSKIPP. 1999. A guide to the birds of India, Pakistan, Nepal, Bangladesh, Bhutan, Sri Lanka, and the Maldives. Princeton University Press, Princeton, NJ U.S.A.

GUTIÉRREZ, R.J. AND A.B. CAREY. 1985. Ecology and management of the Spotted Owl in the Pacific Northwest. USDA Forest Service, General Technical Report PNW-185, Pacific Northwest Forest and Range Experiment Station, Portland, OR U.S.A.

HAGEN, Y. 1952. Rovfuglene og viltpleien. Gyldendal Norsk Forlag, Oslo, Norway.

HAKE, M., N. KJELLEN AND T. ALERSTAM. 2003. Age-dependent migration strategy in Honey Buzzards Pernis apivorus tracked by satellite. Oikos 103:385–396.

HAMERSTROM, F. 1986. Harrier: hawk of the marshes. Smithsonian Institution Press, Washington, DC U.S.A.

HARRISON, J.A., D.G. ALLAN, L.G. UNDERHILL, M. HERREMANS, A. J. TREE, V. PARKER AND C.J. BROWN [EDS.]. 1997. The atlas of southern African birds, Vols. 1–2. BirdLife South Africa, Johannesburg, South Africa.

HAVERSCHMIDT, F. AND G.F. MEES. 1994. The birds of Suriname. Vaco Press, Uitgeversmaatschappij, Paramribo, Suriname.

HAYES, G.E. AND J.B. BUCHANAN. 2002. Washington State status report for the Peregrine Falcon. Washington Department of Fish and Wildlife, Wildlife Program, Olympia, WA U.S.A.

HAYWARD, G.D. AND J. VERNER. 1994. Flammulated, Boreal, and Great Gray Owls in the United States: a technical conservation assessment. USDA Forest Service, General Technical Report RM-253, Rocky Mountain Forest Range and Experiment Station, Fort Collins, CO U.S.A.

HEBERT, E. AND E. REESE. 1995. Avian collision and electrocution: an annotated bibliography. California Energy Commission, Sacramento, CA U.S.A.

HEIDENREICH, M. 1997. Birds of prey: medicine and management. Blackwell Scientific Publications, Oxford, United Kingdom.

HEINROTH, O. AND M. HEINROTH. 1926. Die Vögel Mitteleuropas, Band 2. Bermühler Verlag, Berlin-Lichterfelde, Germany.

HELANDER, B., A. OLSSON, A. BIGNERT, L. ASPLUND AND K. LITZÉN. 2002. The role of DDE, PCB, coplanar PCB and eggshell parameters for the reproduction in the White-tailed Sea Eagle (Haliaeetus albicilla) in Sweden. Ambio 31:386–403.

———, M. MARQUISS AND W. BOWERMAN [EDS.]. 2003. SEA EAGLE 2000. Swedish Society for Nature Conservation/SNF and Åtta.45 Tryckeri AB, Stockholm, Sweden.

HELBIG, A., A. KOCUM, I. SEIBOLD AND M.J. BRAUN. 2005. A multigene phylogeny of aquiline eagles (Aves: Accipitriformes) reveals extensive paraphyly at the genus level. Mol. Phylogenet. Evol. 35:147–164.

HELLMAYR, C.E. AND B. CONOVER. 1949. Catalogue of birds of the Americas and the adjacent islands. Pt. 1, no. 4. Cathartidae, Accipitridae, Pandionidae, Falconidae. Publ. Field Mus. Nat. Hist., Zool. Ser. 13:1–358.

HICKEY, J.J. [ED.]. 1969. Peregrine Falcon populations: their biology and decline. University of Wisconsin Press, Madison, WI U.S.A.

HOCKEY, P.A.R., W.R.J. DEAN AND N. SLABBERT. 2005. Roberts birds of southern Africa, 7th Ed. John Voelcker Bird Book Fund, Cape Town, South Africa.

HOLLANDS, D. 1991. Birds of the night: owls, frogmouths, and nightjars of Australia. A.H. & A.W. Reed, Balgowlah, Australia.

———. 2003. Eagles, hawks and falcons of Australia, 2nd Ed. Bloomings Books, Melbourne, Australia.

HÖLZINGER, J. 1991. Die Vögel Baden-Württembergs. Band 7. Bibliographie. Eugen Ulmer Verlag, Stuttgart, Germany.

HOUSSE, É. 1945. Las aves de Chile en su clasificación moderna, su vida y costumbres. Ediciones de la Universidad de Chile, Santiago, Chile.

HSIAO, C.L. 1996. Diurnal raptors of Taiwan. Taiwan Fonghuanggu Bird Park, Luku, Taiwan.

———. 2001. Field guide to raptor watching in Taiwan. Morning Star Publishing, Inc., Taichung, Taiwan.

HUNT, W.G., D.E. DRISCOLL, E.W. BIANCHI AND R.E. JACKMAN. 1992. Ecology of Bald Eagles in Arizona. Pts. A-E. BioSystems Analysis, Santa Cruz, CA U.S.A.

HUSTINGS, M.F.H., R.G.M. KWAK, P.F.M. OPDAM AND M.J.S.M. REIJNEN [EDS.]. 1985. [Bird census techniques. Backgrounds, guidelines and reporting.] Pudoc, Wageningen/Nederlandse Vereniging tot Bescherming van Vogels, Zeist, Netherlands.

ICHINOSE, H., T. INOUE AND T. YAMAZAKI [EDS.]. 2000. Asian raptor research & conservation: proceedings of the first symposium on raptors of Asia, Lake Biwa Museum, Shiga, Japan, December 12–13, 1998. Committee for the Symposium on Raptors of South-East Asia/EINS, Shiga, Japan.

IL'ICHEV, V.D. [ED.]. 1982. [Migrations of birds of eastern Europe and northern Asia. Falconiformes-Gruiformes.] Nauka, Moscow, Russia.

ISENMANN, P. AND A. MAOLI. 2000. Birds of Algeria. Société d'Études Ornithologiques de France, Paris, France.

———, T. GAULTIER, A.E. HILI, H. AZAFRAF, H. DLENSI AND M. SMART. 2005. Birds of Tunisia. Société d'Études Ornithologiques de France, Paris, France.

JOHNSGARD, P.A. 1990. Hawks, eagles and falcons of North America: biology and natural history. Smithsonian Institution Press, Washington, DC U.S.A.

———. 2002. North American owls: biology and natural history, 2nd Ed. Smithsonian Institution Press, Washington, DC U.S.A.

JOHNSTONE, R.E. AND G.M. STORR. 1998. Handbook of Western Australian birds, Vol. 1. Non-passerines (Emu to Dollarbird). Western Australian Museum, Perth, Australia.

JØRGENSEN, H.E. 1989. Danmarks Rovfugle – en statusoversigt.

Hans Erik Jørgensen, Fredrikshus, Denmark.

KEMP, A. 1987. The owls of southern Africa. Struik Winchester, Cape Town, South Africa.

——— AND M. KEMP. 1998. SASOL birds of prey of Africa and its islands. New Holland Publishers, London, United Kingdom.

KENNEDY, R.S., P.C. GONZALES, E.D. DICKINSON, H. MIRANDA AND T.H. FISHER. 2000. A guide to the birds of the Philippines. Oxford University Press, New York, NY U.S.A.

KERLINGER, P. 1989. Flight strategies of migrating hawks. University of Chicago Press, Chicago, IL U.S.A.

KJELLÉN. N. AND G. ROOS. 2000. Population trends in Swedish raptors demonstrated by migration counts at Falsterbo, Sweden 1942–97. Bird Study 47:195–211.

KOFORD, C.B. 1953. The California Condor. Nat. Audubon Soc. Res. Rep. 4:1–154.

KÖNIG, C., F. WEICK AND J.H. BECKING. 1999. Owls: a guide to owls of the world. Yale University Press, New Haven, CT U.S.A.

KOSKIMIES, P. AND R.A. VÄISÄNEN. 1991. Monitoring bird populations: a manual of methods applied in Finland. Zoological Museum, Helsinki, Finland.

KOSTRZEWA, A. AND G. SPEER [EDS.]. 2001. Greifvögel in Deutschland. Bestand, Situation, Schutz, 2nd Ed. AULA-Verlag, Wiebelsheim, Germany.

LEFRANC, M.N., Jr. 1987. Introduction to the raptor literature. Pages 1–11 in B. A. Giron Pendleton, B. A. Millsap, K. W. Cline, and D. M. Bird [EDS.], Raptor management techniques manual. National Wildlife Federation, Washington, DC U.S.A.

——— AND W.S. CLARK. 1983. Working bibliography of the Golden Eagle and the genus AQUILA. National Wildlife Federation Scientific & Technical Series No. 7, Washington, DC U.S.A.

LERNER, H.R.L. AND D.P. MINDELL. 2005. Phylogeny of eagles, Old World vultures, and other Accipitridae based on nuclear and mitochondrial DNA. Mol. Phylogenet. Evol. 37:327–346.

LEROUX, A. 2004. Le Busard cendré. Édition Belin, Paris, France.

LESHEM, Y. AND O. BAHAT. 1994. [Flying with the birds.] Miâsrad habòtaòhon, Tel-Aviv, Israel.

———, Y. MANDELIK AND J. SHAMOUN-BARANES. 1999. Migrating birds know no boundaries: proceedings of the international seminar on birds and flight safety in the Middle East, Israel, 25–29 April 1999. International Center for the Study of Bird Migration, Latrun, Tel Aviv, Israel.

LIN, W.H. 2006. A field guide to the raptors of Taiwan. Yuan-Liou Publishing Co., Taipei, Taiwan.

LINCER, J.L. AND K. STEENHOF [EDS.]. 1997. The Burrowing Owl: its biology and management, including the proceedings of the first international Burrowing Owl symposium. Raptor Res. Rep. 9:1–177.

———, W.S. CLARK AND M.N. LEFRANC, Jr. 1979. Working bibliography of the Bald Eagle. National Wildlife Federation Scientific and Technical Series no. 2.

LUMEIJ, J.T. 2000. Raptor biomedicine III, including a bibliography of birds of prey. Zoological Education Network, Lake Worth, FL U.S.A.

LWVT (LANDELIJKE WERKGROEP VOGELTREKTELLEN). 2002. [Bird migration over The Netherlands 1976–1993.] Schuyt & Co., Haarlem, The Netherlands.

MAMMEN, U., K. GEDEON, D. LÄMMEL AND M. STUBBE. 1997. Bibliographie deutschsprachiger Literatur über Greifvögel und Eulen von 1945 bis 1995. Jahresbericht zum Monitoring Greifvögel und Eulen Europas, 2. Ergebnisband:1–189.

MARCHANT, S. AND P.J. HIGGINS [EDS.]. 1993. Handbook of Australian, New Zealand and Antarctic birds, Vol. 2. Raptors to lapwings. Oxford University Press, Melbourne, Australia.

MÁRQUEZ, C., F. GAST, V.H. VANEGAS AND M. BECHARD. 2005. Aves rapaces diurnas de Colombia. Instituto de Investigación de Recursos Biológicos Alexander von Humboldt, Bogotá DC, Colombia.

MARTELL, M., C.J. HENNY, P.E. NYE, AND M.J. SOLENSKY. 2001. Fall migration routes, timing, and wintering sites of North American Ospreys as determined by satellite telemetry. Condor 103:715–724.

MÄRZ, R. 1987. Gewöll- und Rupfungskunde, 3rd Ed. Akademie-Verlag, Berlin, Germany.

MEYBURG, B.-U. AND R.D. CHANCELLOR [EDS.]. 1989. Raptors in the modern world: proceedings of the 3rd world conference on birds of prey and owls, 1987. World Working Group on Birds of Prey and Owls, Berlin, Germany.

——— AND R.D. CHANCELLOR [EDS.]. 1994. Raptor conservation today: proceedings of the 4th world conference on birds of prey and owls, 1992. World Working Group on Birds of Prey and Owls, Berlin, Germany and Pica Press, London, United Kingdom.

——— AND R.D. CHANCELLOR [EDS.]. 1996. Eagle studies. World Working Group on Birds of Prey and Owls, Berlin, Germany.

——— AND C. MEYBURG. 1999. The study of raptor migration in the Old World using satellite telemetry. Pages 2292–3006 in N. J. Adams and R. H. Slotow [EDS.], Proceedings of the 22nd international ornithological congress. BirdLife South Africa, Johannesburg, South Africa.

——— AND C. MEYBURG. 2006. Fortschritte der Satelliten-Telemetrie: Technische Neuerungen beim Monitoring von Greifvögeln und einige Ergebnisbeispiele. Populationsökologie Greifvogel- u. Eulenarten 5:75–94

———, T. BELKA, S. DANKO, J. WÓJKCIAK, G. HEISE, T. BLOHM AND H. MATTHESE. 2005. Age at first breeding, philopatry, longevity, and causes of mortality in the Lesser Spotted Eagle Aquila pomarina. Limicola 19:153–179.

MIKKOLA, H. 1983. Owls of Europe. T. & A.D. Poyser, Calton, United Kingdom.

MISCHLER, T. [ED.]. 2002. Sonderheft Wiesenweihe. Ornithol. Anz. 41:81–216.

MITROPOL'SKIY, O.V., E.R. FOTTELER AND G.P.TRETYAKOV. 1987. [Birds of Uzbekistan.] Fan, Tashkent, Uzbekistan. http://eco-club.nsu.ru/raptors/publicat/raptors/Uzbek_bitds_1987.pdf (last accessed 21 December 2006).

MONADJEM, A., M.D. ANDERSON, S.E. PIPER AND A.F. BOSHOFF [EDS.]. 2004. The vultures of southern Africa – quo vadis? Proceedings of a workshop on vulture research and conservation in southern Africa. Bird of Prey Working Group, Endangered Wildlife Trust, Johannesburg, South Africa.

MONNERET, R.-J. 2000. Le faucon pèlerin: description, mœurs, observation, protection, mythologie Delachaux et Niestlé, Paris, France.

MORRIS, F.T. 1976. Birds of prey of Australia: a field guide. Lansdowne Editions, Melbourne, Australia.

MUNDY. P., D. BUTCHART, J. LEDGER AND S. PIPER. 1992. The vultures of Africa. Acorn Books, Randburg, South Africa.

NERO, R.W., R.J. CLARK, R.J. KNAPTON AND R.H. HAMRE [EDS.]. 1987. Biology and conservation of northern forest owls: symposium proceedings, 3–7 February 1987, Winnipeg, Manitoba.

USDA Forest Service, General Technical Report RM-142, Rocky Mountain Forest Range and Experiment Station, Fort Collins, CO U.S.A.

NEWTON, I. 1979. Population ecology of raptors. Buteo Books, Vermillion, SD U.S.A.

———. 1986. The sparrowhawk. T. & A.D. Poyser, Calton, United Kingdom.

——— AND R.D. CHANCELLOR [EDS.]. 1985. Conservation studies on raptors: proceedings of the 2nd ICBP world conference on birds of prey, 1982. International Council for Bird Preservation Technical Publication no. 5. ICBP, Cambridge, United Kingdom.

——— AND P. OLSEN [EDS.]. 1990. Birds of prey. Facts on File Publications, New York, NY U.S.A.

———, R. KAVANAGH, J. OLSEN AND I. TAYLOR [EDS.]. 2002. Ecology and conservation of owls. CSIRO Publishing, Collingwood, Australia.

OAKS, J.L., M. GILBERT, M.Z. VIRANI, R.T. WATSON, C.U. METEYER, B. RIDEOUT, H.L. SHIVAPRASED, S. AHMED, M.J.I. CHAUDRY, M. ARSHAD, S. MAHMOOD, A. ALI AND A.A. KHAN. 2004. Diclofenac residues as the cause of vulture population decline in Pakistan. *Nature* 427:630–633.

OATLEY, T.B., H.D. OSCHADLEUS, R.A. NAVARRO AND L.G. UNDERHILL. 1998. Review of ring recoveries of birds of prey in southern Africa: 1948–1998. Endangered Wildlife Trust, Johannesburg, South Africa.

OLENDORFF, R.R. AND S.E. OLENDORFF. 1968–70. An extensive bibliography on falconry, eagles, hawks, falcons and other diurnal birds of prey. Parts 1–4. Published by the authors, Fort Collins, CO U.S.A.

———, D. AMADON AND S. FRANK. 1995. Books on hawks and owls: an annotated bibliography. *Proc. West. Found. Vert. Zool.* 6:1–89.

OLSEN, P. [ED.]. 1989. Australian raptor studies. Australasian Raptor Association, Victoria, Australia.

——— [ED.]. 1995. Australian birds of prey. New South Wales University Press, Sydney, Australia.

ONIKI, Y. AND E.P. WILLIS. 2002. Bibliography of Brazilian birds: 1500–2002. Instituto de Estudos de Natureza, Publication no. 33:1–531.

PALMER, R.S. [ED.]. 1988. Handbook of North American birds. Vol. 4. Parts 1 and 2. Yale University Press, New Haven, CT U.S.A.

PEETERS, H. AND P. PEETERS. 2005. Raptors of California. University of California Press, Berkeley, CA U.S.A.

PENDLETON, B.G. [ED.]. 1989. Proceedings of the western raptor management symposium and workshop, 26–28 October 1987, Boise, ID. National Wildlife Federation Scientific and Technical Series no. 12.

POLIVANOVA, N.N. AND A.N. KHOKHLOV [EDS.]. 1995. [Birds of prey and owls of northern Caucasia]. Teberda State Nature Reserve, Stavropol, Russia.

POOLE, A.F. 1989. Ospreys: a natural and unnatural history. Cambridge University Press, Cambridge, United Kingdom.

PORTER, R.D., M.A. JENKINS AND A.L. GASKI. 1987. Working bibliography of the Peregrine Falcon. National Wildlife Federation Scientific and Technical Series no. 9.

PORTER, R.F., I. WILLIS, S. CHRISTENSEN AND B.P. NIELSEN. 1981. Flight identification of European raptors. T. & A.D. Poyser, Calton, United Kingdom.

POTAPOV, E. [ED.]. 2001. Proceedings of the II international conference on the Saker Falcon and Houbara Bustard, Ulaanbaatar, Mongolia, 1–4 July 2000. Middle East Falcon Research Group, Abu Dhabi, United Arab Emirates.

——— AND R. SALE. 2005. The Gyrfalcon. A&C Black, London, United Kingdom.

PRIKLONSKIY, S.G., V.M. GALUSHIN AND V.G. KREVER [EDS.]. 1989. [Methods of study and conservation of birds of prey (methodological recommendations).] Central Scientific-Research Laboratory on Hunting, Moscow, Russia. http://www.raptors.ru/library/books/methods_1989/Index.htm (last accessed 21 December 2006).

PUKINSKIY, Y.B. 1975. [Through the Taiga Bikin River (in search of Blakiston's Fish Owl).] Mysl', Moscow, Russia.

———. 1977. [Life of owls.] Leningrad University Press, Leningrad, Russia.

——— [ED.]. 1992. [Birds of Russia and adjacent countries. Strigiformes.] KMK Press, Moscow, Russia.

PUZOVIC, S. 2000. Atlas of the birds of prey of Serbia: their breeding distribution and abundance 1977–1996. Institute for Protection of Nature of Serbia, Beograd, Serbia.

RALPH, C.J. AND J.M. SCOTT [EDS.]. 1981. Estimating numbers of terrestrial birds. *Stud. Avian Biol.* 6:1–630.

RASMUSSEN, P.C. AND J.C. ANDERTON. 2005. Birds of South Asia: the Ripley guide, Vol. 1. Smithsonian Institution Press, Washington, DC U.S.A. and Lynx Edicions, Barcelona, Spain.

RATCLIFFE, D. 1993. The Peregrine Falcon, 2nd Ed. T. & A.D. Poyser, London, United Kingdom.

REDIG, P.T. [ED.]. 1993. Raptor biomedicine. University of Minnesota Press, Minneapolis, MN U.S.A.

RENÉ DE ROLAND, L.-A. 2000. Contribution à l'étude biologique, ecologique et ethologique de trois espèces d'*Accipiter* dans la Presqu'ile de Masoala. Ph.D. dissertation, Universite de Antananarivo, Antananarivo, Malagasy Republic.

RISEBROUGH, R.W. 1986. Pesticides and bird populations. Pages 397–427 *in* D. M. Power [ED.], Current ornithology, Vol. 3. Plenum Press, New York, NY U.S.A.

ROCKENBAUCH, D. 1998. Der Wanderfalke im Deutschland und umliegenden Gebieten. Band 1. Verbreitung, Bestand, Gefährdung, und Schutz. Verlag Christine Hölzinger, Ludwigsburg, Germany.

———. 2002. Der Wanderfalke im Deutschland und umliegenden Gebieten. Band 2. Jahresablauf und Brutbiologie, Beringungsergebnisse, Jagdverhalten und Ernährung, Verschiedenes. Verlag Christine Hölzinger, Ludwigsburg, Germany.

RYABTSEV, V. 2000. [Eagles of Lake Baikal.] Taltsy Press, Irkutsk, Russia. http://gatchina3000.narod.ru/literatura/_other/baikal/4_08_5.htm (last accessed 21 December 2006).

SCHUYL, G., L. TINBERGEN AND N. TINBERGEN. 1936. Ethologische Beobachtungen an Baumfalken (*Falco s. subbuteo*) L.). *J. Ornithol.* 84:387–433.

SEEGAR, W.S., P.N. CUTCHIS AND J.S. WALL. 1996. Fifteen years of satellite tracking development and application to wildlife research and conservation. *John Hopkins APL Tech. Dig.* 17:305–315.

SERGIO, F., R.G. BIJLSMA, G. BOGLIANA AND I. WYLLIE. 2001. *Falco subbuteo* Hobby. *BWP Update* 3:133–156.

SHELDON, F.H., R.G. MOYLE AND J. KENNARD. 2001. Ornithology of Sabah: history, gazetteer, annotated checklist, and bibliography. *Ornithol. Monogr.* 52:1–285.

SHEN, C.C. 1993. The story of the Black Kite. Morning Star Publishing, Inc., Taichung, Taiwan.

———. 1998. Black Kite wants to go home. Morning Star Publishing Inc., Taichung, Taiwan.

———. 1999. Contemplating the Black Kite. Keelung Wild Bird Society, Taiwan.

———. 2004. Searching for the lost Black Kite. Morning Star Publishing Inc., Taichung, Taiwan.

SHEPEL', A.I. 1992. [Birds of prey and owls of Perm' Prikamie (Kama River area).] Irkutsk University Press, Irkutsk, Russia.

SHERROD S.K., W.R. HEINRICH, W.A. BURNHAM, J.H. BARCLAY, AND T.J. CADE. 1982. Hacking: a method for releasing Peregrine Falcons and other birds of prey, 2nd Ed. The Peregrine Fund, Cornell University, Ithaca, NY U.S.A.

SHIRIHAI, H. 1996. The birds of Israel. Academic Press, San Diego, CA U.S.A.

———, R. YOSEF, D. ALON, G.M. KIRWAN AND R. SPAAR. 2000. Raptor migration in Israel and the Middle East: a summary of 30 years of field research. International Birding & Research Center in Eilat/Israel Ornithological Center, SPNI, Eilat, Israel.

SIBLEY, C. AND J.E. AHLQUIST. 1990. Phylogeny and classification of birds. Yale University Press, New Haven, CT U.S.A.

——— AND B.L. MONROE, Jr. 1990. Distribution and taxonomy of birds of the world. Yale University Press, New Haven, CT U.S.A.

SICK, H. 1993. Birds in Brazil. Princeton University Press, Princeton, NJ U.S.A.

SILVA-ARANGUIZ, E. 2006. Recopilacion de la literature ornitologica Chilena desde 1847 hasta 2006. www.bio.puc.cl/auco/artic01/ornito01.htm (last accessed 21 December 2006).

SIMMONS, R.E.L. 2000. Harriers of the world: their behaviour and ecology. Oxford University Press, Oxford, United Kingdom.

SNOW, D.W. [ED.]. 1978. The atlas of speciation of African non-passerine birds. Trustees of the British Museum (Natural History), London, United Kingdom.

SNYDER, N.F.R. AND J.W. WILEY. 1976. Sexual size dimorphism in hawks and owls of North America. Ornithol. Monogr. 20:1–96.

SPAAR, R. 1996. Flight behaviour of migrating raptors in southern Israel. Schweizerische Vogelwarte, Sempach, Switzerland.

STATTERSFIELD, A.J. AND D.R. COOPER. 2000. Threatened birds of the world: the official source for birds on the IUCN Red List. BirdLife International and Lynx Edicions, Barcelona, Spain.

STEENHOF, K. [ED.]. 1994. Snake River Birds of Prey National Conservation Area 1994 annual report. USDI, Bureau of Land Management, Boise District, Idaho/National Biological Survey, Raptor Research and Technical Assistance Center, Boise, ID U.S.A.

STEVENSON, T. AND J. FANSHAWE. 2002. Field guide to the birds of East Africa: Kenya, Tanzania, Uganda, Rwanda, Burundi. T. & A.D. Poyser, London, United Kingdom.

STEYN, P. 1974. Eagle days – a study of African eagles at the nest. Macdonald & Jane's, London, United Kingdom.

———. 1982. Birds of prey of southern Africa. David Philip, Cape Town, South Africa.

———. 1984. A delight of owls – African owls observed. Tanager Books, Dover, DE U.S.A.

STRESEMANN, E. AND D. AMADON. 1979. Order Falconiformes. Pages 271–425 in E. Mayr and G. W. Cottrell [EDS.], Check-list of birds of the world, Vol. 1, 2nd Ed. Museum of Comparative Zoology, Harvard University, Cambridge, MA U.S.A.

STUBBE, M. AND A. STUBBE [EDS.]. 1987, 1991, 1996, 2000, 2006. Populationsökologie von Greifvogel- und Eulenarten, 1–5. Martin-Luther-Universität Halle-Wittinberg, Halle/Saale, Germany.

SÜDBECK, P., H. ANDRETZKE, S. FISCHER, K. GEDEON, T. SCHIKORE, K. SCHRÖDER AND C. SUDFELDT [EDS.]. 2005. Methodenstandards zur Erfassung der Brutvögel Deutschlands. Dachverband Deutscher Avifaunisten, Radolfzell, Germany.

TARBOTON, W. 1990. African birds of prey. Cornell University Press, Ithaca, NY U.S.A.

——— AND D. ALLAN. 1984. The status and conservation of birds of prey in the Transvaal. Transvaal Museum, Pretoria, South Africa.

——— AND R. ERASMUS. 1998. SASOL owls and owling in southern Africa. Struik Winchester, Cape Town, South Africa.

TAYLOR, I. 1994. Barn Owls: predator–prey relationships and conservation. Cambridge University Press, Cambridge, United Kingdom.

TERRASSE, J.F. 2001. Le Gypaète Barbu. Delachaux et Niestlé, Lausanne, Paris, France.

THÉVENOT, M., R. VERNON AND P. BERGIER. 2003. The birds of Morocco. BOU Checklist Series no. 20. British Ornithologists' Union/British Ornithologists' Club, Tring, Herts., United Kingdom.

THIOLLAY, J.-M. 1976. Les rapaces d'une zone de contact savane-forêt en Côte-d'Ivoire: modalités et succès de la reproduction. Alauda 44:175–300.

THIOLLAY, J.-M. 1977. Distribution saisonnière des rapaces diurnes en Afrique occidentale. Oiseau Rev. Fr. Ornithol. 47:253–294.

———. 2001. Long-term changes of raptor populations in northern Cameroon. J. Raptor Res. 35:173–186.

———. 2006. The decline of raptors in West Africa: long-term assessment and the role of protected areas. Ibis 148:240–254.

——— AND V. BRETAGNOLLE [EDS.]. 2004. Rapaces nicheurs de France. Distribution, effectifs et conservation. Delachaux et Niestlé, Paris, France.

THOMPSON, D.B.A., S.M. REDPATH, A.H. FIELDING, M. MARQUISS AND C.A. GALBRAITH [EDS.]. 2003. Birds of prey in a changing environment. The Natural History of Scotland Series. Government Printing Office, Edinburgh, Scotland.

THORSTROM, R. 1993. Breeding biology of two species of forest-falcons (Micrastur) in northeastern Guatemala. M.Sc. thesis, Boise State University, Boise, ID U.S.A.

TINBERGEN, L. 1946. [The sparrow-hawk (Accipiter nisus L.) as a predator of passerine birds.] Ardea 34:1–213.

TINGAY, R. 2005. Historical distribution, contemporary status and cooperative breeding in the Madagascar Fish-eagle: implications for conservation. Ph.D. dissertation, University of Nottingham, Nottingham, United Kingdom.

TSAI, C. W. 2003. A guide to owls. Owl Publishing House, Co., Taipei, Taiwan.

UETA, M. AND M.J. MCGRADY [EDS.]. 2000. First symposium on Steller's and White-tailed Sea Eagles in East Asia: proceedings of the international workshop and symposium, 9–15 February 1999, Tokyo and Hokkaido, Japan. Wild Bird Society of Japan, Tokyo, Japan.

URBINA TORRES, F. 1996. Aves rapaces de Mexico. Centro Investigaciones Biologicas UAEM, Cuernavaca, Morelos, Mexico.

UTTENDÖRFER, O. 1939. Die Ernährung der deutschen Raubvögel und Eulen und ihre Bedeutung in der heimsichen Natur. Neumann-Neudamm, Melsungen, Germany.

VALIELA, I. AND P. MARTINETTO. 2005. The relative ineffectiveness of bibliographic search engines. *Bioscience* 55:688–692.

VAN BALEN, S., V. NIJMAN AND R. SÖZER. 1999. Distribution and conservation of the Javan Hawk-eagle *Spizaetus bartelsi*. *Bird Conserv. Int.* 9:333–349.

———, V. NIJMAN, AND R. SÖZER. 2001. Conservation of the endemic Javan Hawk-Eagle *Spizaetus bartelsi* Stresemann, 1924 (Aves: Falconiformes): density, age-structure and population numbers. *Contrib. Zool.* 70:161–173.

VERNER, J., K.S. MCKELVEY, B.R. NOON, R.J. GUTIÉRREZ, G.I. GOULD, Jr., AND T.W. BECK. 1992. The California Spotted Owl: a technical assessment of its current status. USDA Forest Service, General Technical Report PSW-133, Pacific Southwest Research Station, Albany CA U.S.A.

VIRANI, M.Z. AND M. MUCHAI. 2004. Vulture conservation in the Masai Mara National Reserve, Kenya: proceedings and recommendations of a seminar and workshop held at the Masai Mara National Reserve, 23 June 2004. *Ornithol. Res. Rep.* 57:1–19.

VOLKOV, S., V.V. MOROZOV AND A.V. SHARIKOV [EDS.]. 2005. [Owls of northern Eurasia.] Russian Bird Conservation Union, Working Group on Falconiformes and Strigiformes of Northern Eurasia, Institute of Problems of Ecology and Evolution of Russian Academy of Sciences, Moscow, Russia.

VOOUS, K.H. 1988. Owls of the Northern Hemisphere. MIT Press, Cambridge, MA U.S.A.

VORONETSKIY, V. [ED.]. 1994. [Eagle Owl in Russia, Belarus and Ukraine.] Moscow State University Press, Moscow, Russia.

WALKER, T.J. 2006. Authors willing to pay for instant web access. www.nature.com/nature/debates/e-access/Articles/walker.html (last accessed 21 December 2006).

WALTER, H. 1979. Eleonora's Falcon: adaptations to prey and habitat in a social raptor. University of Chicago Press, Chicago, IL U.S.A.

WANG, C.C. 2006. Handbook on raptor watching in Kenting National Park. Kenting National Park Administration, Kenting, Taiwan.

WATTEL, J. 1973. Geographic differentiation in the genus *Accipiter*. *Pub. Nuttall Ornithol. Club 13.*

WEAVER, J.D. AND T.J. CADE. 1991. Falcon propagation: a manual for captive breeding, Revised Ed. The Peregrine Fund, Inc., Boise, ID U.S.A.

WEISHU, X.C. 1995. Raptors of China. China Forestry Publishing House, Beijing, China.

WELLS, D.R. 1999. The birds of the Thai-Malay Peninsula, covering Burma and Thailand south of the eleventh parallel, Peninsula Malaysia and Singapore, Vol. 1. Non-passerines. Academic Press, San Diego, CA U.S.A.

WERNERY, R., U. WERNERY, J. KINNE AND J. SAMOUR. 2004. Colour atlas of falcon medicine. Schlütersche, Hanover, Germany.

WERNHAM, C., M. TOMS, J. MARCHANT, J. CLARK, G. SIRIWARDENA AND S. BAILLIE [EDS.]. 2002. The migration atlas: movements of the birds of Britain and Ireland. T. & A.D. Poyser, London, United Kingdom.

WHEELER, B.K. 2003a. Raptors of eastern North America. Princeton University Press, Princeton, NJ U.S.A.

———. 2003b. Raptors of western North America. Princeton University Press, Princeton, NJ U.S.A.

——— AND W.S. CLARK. 1996. A photographic guide to North American raptors. Academic Press, London, United Kingdom.

WILBUR, S.R. AND J.A. JACKSON [EDS.]. 1983. Vulture biology and management. University of California Press, Berkeley, CA U.S.A.

YEN, C.W. 1982. Raptors of Taiwan. Center for Environmental Sciences, Tunghai University, Taichung, Taiwan.

YOSEF, R., M.L. MILLER AND D. PEPLER [EDS.]. 2002. Raptors in the new millennium: proceedings of the joint meeting of the Raptor Research Foundation and the World Working Group on Birds of Prey and Owls. International Birding & Research Centre in Eilat, Eilat, Israel.

ZALLES, J.I. AND K.L. BILDSTEIN [EDS.]. 2000. Raptor watch: a global directory of raptor migration sites. *Birdlife Conservation Series no. 9.* BirdLife International, Cambridge, United Kingdom; and Hawk Mountain Sanctuary, Kempton, PA U.S.A.

ZUBAKIN, V.A. ET AL. [EDS.]. 2005. [Birds of Russia and adjacent regions: Strigiformes, Apodiformes, Coraciiformes, Upupiformes, Piciformes.] KMK Press, Moscow, Russia.

Appendix 1. Journals of interest to raptor researchers.

[a] E = electronic, P = print

Title	Country	Publisher	Geographical Emphasis	Topical Emphasis	Issues Annually	Article Language	Medium[a]
Acrocephalus	Slovenia	BirdLife Slovenia	SE Europe & E Mediterranean	Ornithology	6	Slovenia	P
Acta Ornithoecologica	Germany	Schriftleitung Acta Ornithoecologica	Germany	Ornithology	2	German	P
Acta Ornithologica	Poland	Museum and Institute of Zoology (Warsaw)	Global	Ornithology	2	English	P
Acta Zoologica Mexicana (nueva serie)	Mexico	Instituto de Ecología A. C.	Mexico	Zoology	3	Spanish, English, French, Portuguese	P
Acta Zoologica Sinica	People's Republic of China	Science Press, Beijing	Global	Zoology	6	Chinese or English	P
Africa — Birds & Birding	Republic of South Africa	Africa Geographic	Southern Africa	Ornithology/Birding	6	English	P
Afring News	Republic of South Africa	Avian Demography Unit, Capetown	Africa	Banding	2	English	E/P
Airo	Portugal	Sociedade Portuguesa para o Estudo das Aves	Iberian Peninsula & Canary Islands	Ornithology	1	Portuguese or English	P
Alabama Birdlife	USA	Alabama Ornithological Society	Alabama	Ornithology	Occasional	English	P
Alauda	France	Sociéte d'Études Ornithologiques de France	Global	Ornithology	4	French	P
Alula	Finland	Alula	Global	Ornithology/Birding	4	English or Finnish	P
American Midland Naturalist	USA	University of Notre Dame	North America	General natural history	4	English	P
Anales del Instituto de Biología, Serie Zoología	Mexico	Instituto de Biología, UNAM	Mexico	Zoology	2	Spanish	E/P
Animal Behaviour	United Kingdom	Elsevier	Global	Behavior	12	English	P
Anser	Sweden	Skånes Ornitologiska Förening	Sweden	Ornithology	4	Swedish	P
Anuari d'Ornitologia de Catalunya	Spain	Institut Català d'Ornitologia	Spain (Catalonia)	Ornithology	1	Catalan	P
Anuari Ornitològic de les Balears	Mallorca	Grup Balear d'Ornitologia I Defensa de la Naturalesa	Balearic Islands	Ornithology	1	Spanish	P
Anuário Ornitológico	Portugal	Sociedade Portuguesa para o Estudo das Aves	Portugal	Ornithology	1	Portuguese	P

Title	Country	Publisher	Geographical Emphasis	Topical Emphasis	Issues Annually	Article Language	Medium[a]
Anuario Ornitologico de Navarra	Spain	GOROSTI, Sociedad de Ciencias Naturales de Navarra	Northern Spain	Ornithology	1	Spanish	P
Anzeiger des Vereins Thüringer Ornithologen	Germany	Vereins Thüringer Ornithologen	Global	Ornithology	Occasional	German	P
Apus	Germany	Beiträge zur Avifauna Sachsen-Anhalts	Sachsen-Anhalts, Germany	Ornithology	6	German	P
Aquila	Hungary	Instituti Ornithologici Hungarici	Hungary	Ornithology	1	Hungarian or English	P
Aquila chrysaetos	Japan	Society for Research on the Golden Eagle	Japan	Raptors	1	Japanese	P
Ardea	Netherlands	Netherlands Ornithologists' Union	Global	Ornithology	2	English	E/P
Ardeola	Spain	Sociedad España de Ornitología	Global	Ornithology	2	Spanish or English	P
Asian Raptors	Malaysia	Asian Raptor Research and Conservation Network	Oriental Region	Raptors	Occasional	English	P
Atualidades Ornitológicas	Brazil	Atualidades Ornitológicas	Brazil	Ornithology	6	Portuguese	E/P
Auk, The	USA	American Ornithologists's Union	Global	Ornithology	4	English	E/P
Australian Field Ornithology	Australia	Bird Observers Club of Australia	Australia	Ornithology	4	English	P
Aves	Belgium	Société d'Études Ornithologiques	Global	Ornithology	4	French	P
Aves Ichnusae	Italy	Gruppo Ornitologico Sardo	Sardinia	Ornithology	Occasional	Italian	P
Avian Diseases	USA	American Association of Avian Pathologists	Global	Avian medicine	4	English	P
Avian Ecology and Behaviour	Russia	Biological Station "Rybachy" of the Zoological Institute, Russian Academy of Sciences	Global	Ornithology	2	English	P
Avian Ecology and Conservation	Canada	Society of Canadian Ornithologists/Bird Study Canada	Global	Ornithology	Occasional	English	E
Avian Pathology	United Kingdom	World Veterinary Poultry Association	Global	Avian medicine	6	English	E/P
Aviculture Magazine	United Kingdom	Avicultural Society	Global	Aviculture	4	English	P
Avifaunistik in Bayern	Germany	Ornithologische Gesellschaft in Bayern	Germany	Ornithology	2	German	P

Title	Country	Publisher	Geographical Emphasis	Topical Emphasis	Issues Annually	Article Language	Medium[a]
Avocetta	Italy	CISO Centro Italiano Studi Ornitologici	Global	Ornithology	2	Italian or English	P
Babbler, The	Botswana	BirdLife Botswana	Botswana	Ornithology	2	English	P
Behavioral Ecology	USA	International Society for Behavioral Ecology	Global	General ecology	6	English	E/P
Behavioral Ecology and Sociobiology	Germany	Springer Berlin/Heidelberg	Global	General ecology	6	English	E/P
Berkut	Ukraine	Ukrainian Journal of Ornithology	Western Europe to Russian Far East	Ornithology	2	Ukrainian,Russian, English, German	E/P
Berliner Ornithologischer Bericht	Germany	Berliner Ornithologische Arbeitsgemeinschaft	Germany	Ornithology	2	German	P
Bièvre, Le	France	Le Centre Ornithologique Rhône Alpes	France	Ornithology	Occasional	France	P
Bird Behavior	USA	Cognizant Communication Corporation	Global	Ornithology	2	English	P
Bird Conservation International	United Kingdom	BirdLife International/ Cambridge University Press	Global	Conservation	4	English	E/P
Bird Observer	USA	Bird Observer of Eastern Massachusetts	Massachusetts	Ornithology	6	English	P
Bird Populations	USA	Institute for Bird Populations	Global	Avian populations	Occasional	English	P
Bird Study	United Kingdom	British Trust for Ornithology	Global	Ornithology	3	English	E/P
Bird Trends	Canada	Canadian Wildlife Service	North America	Avian populations	Occasional	English	E/P
Birding	USA	American Birding Association	North America	Distribution/ identification	4	English	P
BirdingASIA	United Kingdom	Oriental Bird Club	Asia	Ornithology	2	English	P
Birds of North America	USA	Cornell Laboratory of Ornithology and American Ornithologists' Union	North America & Hawaii	Ornithology	Continuous	English	E/P
Bliki	Iceland	Icelandic Institute of Natural History	Iceland	Ornithology	Occasional	Danish or English	P
Blue Jay	Canada	Nature Saskatchewan	Saskatchewan	Ornithology	4	English	P
Bluebird, The	USA	Audubon Society of Missouri	Missouri	Ornithology	4	English	P

Title	Country	Publisher	Geographical Emphasis	Topical Emphasis	Issues Annually	Article Language	Medium[a]
Boletim CEO	Brazil	Centro de Estudos Ornitológicos São Paulo - SP	Brazil	Ornithology	2	Portuguese	P
Boletín Chileno de Ornitologia	Chile	Unión de Ornitólogos de Chile	Neotropical Region (mostly Chile)	Ornithology	Annual	Spanish	P
Boletín SAO	Colombia	Sociedad Antioqueña de Ornitología	Colombia	Ornithology	2	Spanish	P
Boletin Zeledonia	Costa Rica	Asociación Ornitológica de Costa Rica	Costa Rica	Ornithology	2	Spanish	E
Boobook	Australia	Australasian Raptor Association	Australia	Raptors	2	English	E/P
British Birds	United Kingdom	British Birds 2000 Ltd.	British Isles	Ornithology	12	English	P
British Columbia Birds	Canada	British Columbia Field Ornithologists	British Columbia	Ornithology	1	English	P
Bulletin of the African Bird Club	United Kingdom	African Bird Club	Africa	Ornithology	2	English	P
Bulletin of the British Ornithologists' Club	United Kingdom	British Ornithologists' Club	Global	Ornithology	4	English	P
Bulletin of the Japanese Bird Banding Society	Japan	Japanese Bird Banding Society	Japan	Banding	2	Japanese	P
Bulletin of the Oklahoma Ornithological Society	USA	Oklahoma Ornithological Society	Oklahoma	Ornithology	4	English	P
Bulletin of the Texas Ornithological Society	USA	Texas Ornithological Society	Texas	Ornithology	2	English	P
Buteo	Czech Republic/Slovakia	Czech Society for Ornithology	Czech Republic/Slovakia	Raptors	1	Czech, Slovak, or English	P
Butlleti del Grup Català d'Anellament	Spain	Grup Català d'Anellament (Catalan Ringing Group)	Spain	Banding	2	Catalan, Spanish, English	P
Caldasia	Colombia	Instituto de Ciencias, Facultad de Ciencias, Universidad Nacional de Colombia	Neotropical Region (mostly Colombia)	Ornithology	2	Spanish or English	E/P
Canadian Field-Naturalist	Canada	Ottawa Field-Naturalists' Club	North America	Ornithology	4	English	P
Cassinia	USA	Delaware Valley Ornithological Club	New Jersey, Pennsylvania, & Delaware	Ornithology	Occasional	English	P
Charadrius	Germany	Zeitschrift für Vogelkunde, Vogelschutz und Naturschutz im Rheinland und in Westfalen	Germany	Ornithology	4	German	P
Chat, The	USA	Carolina Bird Club	North & South Carolina	Ornithology	4	English	E/P

Title	Country	Publisher	Geographical Emphasis	Topical Emphasis	Issues Annually	Article Language	Medium[a]
Ciconia	France	Ligue pour la Protection des Oiseaux, Delegation Alsace	France (Alsace)	Vertebrate natural history	3	French	P
Cimbebasia	Namibia	National Museum of Namibia	Namibia	General natural history	Occasional	English	P
Cinclus	Germany	Bund für Vogelschutz und Vogelkunde e V.	Germany	Ornithology	2	German	P
Colorado Birds	USA	Colorado Field Ornithologists	Colorado	Ornithology	4	English	P
Condor, The	USA	Cooper Ornithological Society	Global	Ornithology	4	English	P
Connecticut Warbler	USA	Connecticut Ornithological Society	Connecticut	Ornithology	4	English	P
Conservation Biology	USA	Society for Conservation Biology	Global	Conservation biology	6	English	E/P
Corax	Germany	Ornithologischen Arbeitsgemeinschaft	Germany	Ornithology	2	German	P
Corella	Australia	Australian Bird Study Association, Inc.	Australia	Ornithology	4	English	P
Cotinga	United Kingdom	Neotropical Bird Club	Neotropical Region	Ornithology	2	English, Spanish, Portuguese	P
Dansk Ornithologisk Forenings Tidsskrift	Denmark	Dansk Ornithologisk Forening	Global (mostly Denmark)	Ornithology	4	Danish or English	P
Dutch Birding	Netherlands	Dutch Magazine Association	Western Palearctic	Ornithology	6	English or Dutch	P
Ecological Applications	USA	Ecological Society of America	Global	General ecology	6	English	E/P
Ecological Monographs	USA	Ecological Society of America	Global	General ecology	4	English	P
Ecology	USA	Ecological Society of America	Global	General ecology	12	English	P
Egretta	Austria	Vogelkundliche Nachrichten aus Oesterreich	Central Europe (mostly Austria)	Ornithology	2	German	P
Emirates Bird Report	United Kingdom	Ornithological Society of the Middle East	United Arab Emirates	Ornithology	1	English	P
Emu – Austral Ornithology	Australia	Birds Australia (Royal Australasian Ornithologists Union)	Global	Ornithology	4	English	E/P
Falco	United Kingdom	Middle East Falcon Research Group	Global	Raptors	2	English	P

Title	Country	Publisher	Geographical Emphasis	Topical Emphasis	Issues Annually	Article Language	Medium[a]
Falco	Slovenia	Association IXOBRYCHUS	Slovenia	Ornithology	Occasional	Slovene, English, Italian/Croat	P
Field Notes of Rhode Island Birds	USA	Audubon Society of Rhode Island	Rhode Island	Ornithology	6	English	P
Florida Field Naturalist	USA	Florida Ornithological Society	Florida	Ornithology	4	English	P
Folia Zoologica	Czech Republic	Academy of Sciences of the Czech Republic	Global	Vertebrate zoology	4	English	E/P
Forktail	United Kingdom	Oriental Bird Club	Oriental Region	Ornithology	1	English	P
Foundation for the Conservation of the Bearded Vulture Annual Report	Netherlands	Foundation for the Conservation of the Bearded Vulture	Europe	Bearded Vulture	1	English	P
Gabar	Republic of South Africa	Endangered Wildlife Trust	Africa	Raptors	2	English	P
Garcilla, La	Spain	Sociedad España de Ornitología	Spain	Ornithology	4	Spanish	P
Gibraltar Bird Report	Gibraltar	Gibraltar Ornithological and Natural History Society	Gibraltar	Ornithology	1	English	P
Great Basin Birds	USA	Great Basin Bird Observatory	Great Basin	Ornithology	1	English	P
Hamburger Avifaunistische Beiträge	Germany	Arbeitskreis an der Staatlichen Vogelschutzwarte Hamburg	Germany	Ornithology	2	German	P
Hawk Migration Studies	USA	Hawk Migration Association of North America	North America	Raptors	2	English	E/P
Héron, Le	France	Groupe Ornithologique et Naturaliste du Nord/Pas de Calais	France	General natural history	3	French	P
Hirundo	Estonia	Estonian Ornithological Society	Estonia	Ornithology	2	Estonian or English	P
Honeyguide, The	Zimbabwe	BirdLife Zimbabwe	Zimbabwe	Ornithology	2	English	P
Hong Kong Bird Report	Hong Kong	Hong Kong Bird Watching Society	Hong Kong	Ornithology	1	English	P
Hornero, El	Argentina	Asociación Ornitologica del Plata	Neotropical Region	Ornithology	2	Spanish or English	P
Huitzil - Journal of Mexican Ornithology	Mexico	Huitzil/CIPAMEX - BirdLife Mexico	Mexico	Ornithology	Continuous	Spanish	E
Iberis	Gibraltar	Gibraltar Ornithological and Natural History Society	Gibraltar	General natural history	1	English	P

Title	Country	Publisher	Geographical Emphasis	Topical Emphasis	Issues Annually	Article Language	Medium[a]
Ibis, The	United Kingdom	British Ornithologists' Union	Global	Ornithology	4	English	E/P
Iheringia, Seríe Zoologia	Brazil	Fundação Zoobotânica do Rio Grande do Sul	Brazil	Zoology	4	Portuguese or English	E/P
Indiana Audubon Quarterly	USA	Indiana Audubon Society	Indiana	Ornithology	4	English	P
International Hawkwatcher	USA	Donald Heintzelman	Global	Raptors	Occasional	English	P
Iowa Bird Life	USA	Iowa Ornithologists' Union	Iowa	Ornithology	4	English	P
Irish Birds	Ireland	BirdWatch Ireland	Ireland	Ornithology	1	English	P
Japanese Journal of Ornithology	Japan	Ornithological Society of Japan	Japan	Ornithology	4	Japanese	P
Journal of Animal Ecology	United Kingdom	British Ecological Society	Global	General ecology	6	English	E/P
Journal of Applied Ecology	United Kingdom	British Ecological Society	Global	General ecology	6	English	E/P
Journal of Avian Biology	Sweden	Scandinavian Ornithologists' Union	Global	Ornithology	6	English	E/P
Journal of Caribbean Ornithology	USA	Society for the Conservation and Study of Caribbean Birds	Caribbean region	Ornithology	1	English	P
Journal of East African Natural History	Kenya	National Museums of Kenya & Nature Kenya	East Africa	Natural history	2	English	P
Journal of Ecology	United Kingdom	British Ecological Society	Global	General ecology	6	English	E/P
Journal of Field Ornithology	USA	Association of Field Ornithologists	Global	Ornithology	4	English	P
Journal of Indian Bird Records and Conservation	India	Harini Nature Conservation Foundation	Indian subcontinent	Ornithology	1	English	E
Journal of Ornithology	Germany	Deutsch Ornithologen-Gesellschaft	Global	Ornithology	4	German or English	E/P
Journal of Raptor Research	USA	Raptor Research Foundation	Global	Raptors	4	English	P
Journal of the Bombay Natural History Society	India	Bombay Natural History Society	India	General natural history	3	English	P
Journal of Wildlife Management	USA	The Wildlife Society	Global	General wildlife biology	4	English	E/P

Title	Country	Publisher	Geographical Emphasis	Topical Emphasis	Issues Annually	Article Language	Medium[a]
Journal of the Yamashina Institute for Ornithology	Japan	Yamashina Institute for Ornithology	Japan	Ornithology	2	English or Japanese	P
Kansas Ornithological Society Bulletin	USA	Kansas Ornithological Society	Kansas	Ornithology	4	English	P
Kentucky Warbler, The	USA	Kentucky Ornithological Society	Kentucky	Ornithology	4	English	P
Kingbird, The	USA	Federation of New York State Bird Clubs	New York	Ornithology	4	English	P
Korean Journal of Ornithology	Korea	Ornithological Society of Korea	Korea	Ornithology	2	English	P
Kukila	Indonesia	Indonesian Ornithological Society	Indonesia	Ornithology	Occasional	English	P
Larus	Croatia	Hrvatska Akademija Znanosti I Umjetnosti	Global	Ornithology	1	Croatian or English	P
Limosa	Netherlands	Nederlandse Ornithologische Unie	Netherlands	Ornithology	4	Dutch	P
Loon, The	USA	Minnesota Ornithologists' Union	Minnesota	Ornithology	4	English	P
Malimbus	France	West African Ornithological Society	West Africa	Ornithology	2	English/French	P
Maryland Birdlife	USA	Maryland Ornithological Society	Maryland	Ornithology	Occasional	English	P
Meadowlark, The	USA	Illinois Ornithological Society	Illinois	Ornithology	4	English	P
Michigan Birds and Natural History	USA	Michigan Audubon Society	Michigan	Ornithology	4	English	P
Migrant, The	USA	Tennessee Ornithological Society	Tennessee	Ornithology	4	English	P
Mirafra	Republic of South Africa	Free State Bird Club	Central South Africa	Ornithology	4	English	P
Mississippi Kite, The	USA	Mississippi Ornithological Society	Mississippi	Ornithology	4	English	P
Populationsökologie von Greifvogel und Eulenarten	Germany	Monitoring Greifvögel Eulen Europas	Europe	Ornithology	Occasional	German	P
Museum Heineanum Ornithologische Jahresberichte	Germany	Museum Heineanum	Germany	Ornithology	1	German	P
N.B. Naturalist	Canada	New Brunswick Federation of Naturalists	New Brunswick	Ornithology	4	English or French	P

Title	Country	Publisher	Geographical Emphasis	Topical Emphasis	Issues Annually	Article Language	Medium[a]
Natura Croatica	Croatia	Croatian Natural History Museum	Global (mainly Croatia)	General natural history	4	English	P
Nature Alberta	Canada	Federation of Alberta Naturalists	Alberta	General natural history	4	English	P
Nebraska Bird Review, The	USA	Nebraska Ornithologists' Union	Nebraska	Ornithology	4	English	P
New Hampshire Bird Records	USA	Audubon Society of New Hampshire	New Hampshire	Ornithology	4	English	P
New Jersey Birds	USA	New Jersey Audubon Society	New Jersey	Ornithology	4	English	P
NMOS Bulletin	USA	New Mexico Audubon Society	New Mexico	Ornithology	4	English	P
North American Bird Bander	USA	Eastern, Inland, and Western Bird Banding Associations	North America	Banding	4	English	P
North American Birds	USA	American Birding Association	North America	Ornithology	4	English	P
Northeastern Naturalist	USA	Eagle Hill Foundation	Northeastern USA	General natural history	4	English	P
Northwestern Naturalist	USA	Society for Northwestern Vertebrate Biology	Pacific Northwest	Vertebrate natural history	3	English	P
Nos Oiseaux	Switzerland	Societe Romande pour l'Etude de la Protection des Oiseaux	Switzerland	Ornithology	4	French	P
Notatki Ornitologiczne	Poland	Kwartalnik Sekcji Ornitologicznej	Global	Ornithology	4	Polish or English	P
Notornis	New Zealand	Ornithological Society of New Zealand	New Zealand	Ornithology	4	English	E/P
Nova Scotia Birds	Canada	Nova Scotia Bird Society	Nova Scotia	Ornithology	4	English	P
Nuestras Aves	Argentina	Asociación Ornitologica del Plata	Argentina	Ornithology	2	Spanish	P
Ohio Cardinal	USA	Ohio Ornithological Society	Ohio	Ornithology	4	English	P
Oikos	Norway	Nordic Ecological Society	Global	General ecology	12	English	P
Oman Bird News	Oman	Oman Bird Records Committee	Oman	Ornithology	Occasional	English	P
Ontario Birds	Canada	Ontario Field Ornithologists	Ontario	Ornithology	3	English	P

Title	Country	Publisher	Geographical Emphasis	Topical Emphasis	Issues Annually	Article Language	Medium[a]
Oregon Birds	USA	Oregon Field Ornithologists	Oregon	Ornithology	4	English	P
Oriole, The	USA	Georgia Ornithological Society	Georgia	Ornithology	4	English	P
Ornis Fennica	Finland	Finnish Ornithological Society	Global	Ornithology	4	English	P
Ornis Hungarica	Hungary	BirdLife Hungary	Hungary & East-central Europe	Ornithology	2	Hungarian or English	P
Ornis Norvegica	Norway	Norsk Ornitologisk Forening	Norway	Ornithology	2	English	P
Ornis Svecica	Sweden	Sveriges Ornitologiska Förening	Sweden	Ornithology	4	Swedish	P
Ornithological Monographs	USA	American Ornithologists' Union	Global	Ornithology	Occasional	English	P
Ornithological Science	Japan	Ornithological Society of Japan	Global	Ornithology	2	English	P
Ornithologische Beobachter, Der	Switzerland	Schweizer Gesellschaft für Vogelkunde	Switzerland	Ornithology	4	German	P
Ornithologische Gesellschaft Basel Jahresbericht	Switzerland	Ornithologische Gesellschaft Basel	Switzerland	Ornithology	1	German	P
Ornithologische Mitteilungen	Germany	Ornithologische Mitteilungen	Global (mostly Europe)	Ornithology	12	German	P
Ornithologischer Anzieger	Germany	Ornithological Society in Bavaria	Germany	Ornithology	2-3	German or English	P
Ornithologischer Jahresbericht Helgoland	Germany	Ornithologischer Arbeitsgemeinschaft Helgoland	Helgoland (Germany)	Ornithology	1	German	P
Ornithos	France	Ligue pour la Protection des Oiseaux	France	Ornithology	6	French	P
Ornitologia Colombiana	Colombia	Asociación Colombiana de Ornitologia Colombia	Colombia	Ornithology	Occasional	Spanish or English	E
Ornitologia Neotropical	Canada	Neotropical Ornithological Society	Neotropical Region	Ornithology	4	English/Spanish/Portuguese	P
Oryx	United Kingdom	Fauna & Flora International	Global	Conservation biology	4	English	P
Osprey	Canada	Newfoundland and Labrador Natural History Society	Newfoundland and Labrador	General natural history	4	English	P
Ostrich: Journal of African Ornithology	Republic of South Africa	National Inquiry Services Centre/BirdLife South Africa	Africa	Ornithology	4	English	P

Title	Country	Publisher	Geographical Emphasis	Topical Emphasis	Issues Annually	Article Language	Medium[a]
Passenger Pigeon, The	USA	Wisconsin Society for Ornithology	Wisconsin	Ornithology	4	English	P
Pavo	India	Society of Animal Morphologists & Physiologists	India	Ornithology	4	English	P
Pennsylvania Birds	USA	Pennsylvania Society for Ornithology	Pennsylvania	Ornithology	4	English	P
QuébecOiseaux	Canada	l'Association québecoise des groups d'ornithologues	Quebec	Ornithology	4	French	P
Raptors Conservation	Russia	Siberian Environmental Center & Center for Field Studies	Eastern Europe and northern Asia	Raptors	2	Russian or English	E/P
Raven, The	USA	Virginia Ornithological Society	Virginia	Ornithology	4	English	P
Redstart, The	USA	Brooks Bird Club	West Virginia	Ornithology	4	English	P
Revista Brasileira de Ornitologia	Brazil	Sociedade Brasileira de Ornitologia	Neotropical Region (mostly Brazil)	Ornithology	2	Portuguese, Spanish, English	P
Revista Catalana d'Ornitologia	Spain	Institut Català d'Ornitologia	Spain (Catalonia)	Ornithology	Continuous	Catalan	E
Revista de Anillamiento	Spain	Sociedad España de Ornitología	Spain	Banding	1	Spanish	P
Ring, The	Poland	Polish Zoological Society	Global	Banding	4	English	P
Ringing & Migration	United Kingdom	British Trust for Ornithology	United Kingdom	Banding & migration	2	English	P
Ringmerkaren	Norway	Norsk Ornitologisk Forening	Norway	Banding	1	Norwegian	P
Rivista Italiana di Ornitologia	Italy	Società Italiana di Scienze Naturali	Global	Ornithology	2	Italian or English	P
Sandgrouse	United Kingdom	Ornithological Society of the Middle East	Middle East	Ornithology	2	English	P
Scopus	Kenya	Bird Committee of the East African Natural History Society	East Africa	Ornithology	1-2	English	P
Scottish Bird Report	United Kingdom	Scottish Ornithologists' Club	Scotland	Ornithology	1	English	P
Scottish Birds	United Kingdom	Scottish Ornithologists' Club	Scotland	Ornithology	1	English	P
Scottish Raptor Monitoring Report	United Kingdom	Scottish Ornithologists' Club	Scotland	Raptors	1	English	P

Title	Country	Publisher	Geographical Emphasis	Topical Emphasis	Issues Annually	Article Language	Medium[a]
South Australian Ornithologist	Australia	Ornithological Association of South Australia	South Australia	Ornithology	2	English	P
South Dakota Bird Notes	USA	South Dakota Ornithologists' Union	South Dakota	Ornithology	4	English	P
Southeastern Naturalist	USA	Eagle Hill Foundation	Southeastern USA	General natural history	4	English	P
Southwestern Naturalist	USA	Southwestern Association of Naturalists	Southwestern USA, Mexico, & Central America	General natural history	4	English	P
Strix	Japan	Wild Bird Society of Japan	Japan	Ornithology	1	Japanese	P
Studies in Avian Biology	USA	Cooper Ornithological Society	Global	Ornithology	Occasional	English	P
Subbuteo: The Belarusian Ornithological Bulletin	Belarus	West Belarusian Society for Bird Preservation	Belarus	Ornithology	1	Russian	P
Sunbird, The	Australia	Birds Queensland	Queensland	Ornithology	Occasional	English	P
Sylvia	Czech Republic	Czech Society for Ornithology	Czech Republic/Slovakia	Ornithology	1	Czech, Slovak, or English	P
Systematic Biology	United Kingdom	Society of Systematic Biologists	Global	Systematics	6	English	E/P
Takkeling, De	Netherlands	Dutch Raptor Working Group	Netherlands	Raptors	3	Dutch	P
Túzok	Hungary	BirdLife Hungary	Hungary	Ornithology/Birding	4	Hungarian	P
Utah Birds	USA	Utah Ornithological Society	Utah	Ornithology	4	English	P
Virginia Birds	USA	Virginia Ornithological Society	Virginia	Ornithology	4	English	P
Vogelkundliche Berichte aus Niedersachsen	Germany	Niedersaechsische Ornithologische Vereingung	Lower Saxony (Germany)	Ornithology	2	German	P
Vogelwarte, Die	Germany	Vogelwarte Helgoland & Vogelwarte Radolfzell	Global	Ornithology	4	German or English	P
Vogelwelt, Die: Beiträge zur Vogelkunde	Germany	AULA-Verlag GmbH	Global	Ornithology	4	German	P
Vulture News	Republic of South Africa	Endangered Wildlife Trust	Global	Vultures	2	English	P
Washington Birds	USA	Washington Ornithological Society	Washington	Ornithology	2	English	P

Title	Country	Publisher	Geographical Emphasis	Topical Emphasis	Issues Annually	Article Language	Medium[a]
Western Birds	USA	Western Field Ornithologists	Western U.S. and Mexico	Ornithology	4	English	P
Western North American Naturalist	USA	Brigham Young University	Western North America	General natural history	4	English	P
Wildlife Monographs	USA	The Wildlife Society	Global	General wildlife biology	Occasional	English	E/P
Wildlife Research	Australia	CSIRO Publishing	Global	General wildlife biology	8	English	P
Wildlife Society Bulletin	USA	The Wildlife Society	Global	General wildlife biology	4	English	E/P
Wilson Journal of Ornithology, The	USA	Wilson Ornithological Society	Global	Ornithology	4	English	P
Yelkovan	Turkey	Ornithological Society of Turkey	Turkey	Ornithology	Continuous	English	E

Raptor Identification, Ageing, and Sexing

2

WILLIAM S. CLARK
2301 S. Whitehouse Circle, Harlingen, TX 78550 USA

To successfully conduct raptor research or, indeed, any ornithological research, researchers must be able to identify their subjects accurately to species and, in many studies, determine their age and sex. This is true for (1) field studies, including observations and counts, (2) capture for ringing (i.e., banding), color marking, and radio tracking and telemetry, and (3) the examination and measurement of museum specimens.

Identification is more difficult for Falconiformes than for Strigiformes, as there are more species (more than 300 Falconiformes versus about 200 Strigiformes) and more variation in plumages within species. Most of this chapter applies to diurnal raptors. The following paragraphs discuss plumages, including field marks and unusual plumages; field guides, with cautions about their use; the use of molt as a tool in ageing; the use of behavior in field identification; and a section on identification in the hand. Important references are listed at the end of the chapter.

INITIAL POINTERS

Accurate Identification is Critical

The validity of the results and conclusions of any raptor research depends upon the accurate identification of the subjects involved. Therefore, workers must acquire or sharpen their skills in species identification, including sexing and ageing, to produce the best research. Fortunately, for some researchers, good bird field guides, often including field and photo guides specifically for raptors, as well as an ever-growing list of published articles on identification, ageing, and sexing of raptors, are readily available.

Why Are Raptors Difficult to Identify?

Diurnal raptors are difficult to identify because most species have a variety of plumages, including different plumages for immatures, sexes, and color morphs; and many exhibit considerable individual variation. Many of these plumages are similar to those of other species. Another cause of difficulty is that many bird field guides don't show the range of variation in plumages, don't include the latest information regarding important field marks, and don't portray the shapes of flying and perched raptors accurately. This is true even in an era with access to many wonderful photographs.

Optical and Photographic Equipment

We now have better binoculars, telescopes, and cameras to aid our raptor research. Although some researchers are not able to afford top-of-the-line equipment, lower priced equipment is often very good and adequate for most research needs. Thus, researchers are able to get much better views of their subjects and use more subtle field marks to identify them and determine age and sex. Also, we can take high-quality photographs of raptors, especially in those cases where the identification could not be made in the field. Raptor researchers are urged to always bring their cameras in the field with them so that they can photograph raptors of questionable identification for later verification. More than once, I have changed the identification of a raptor after viewing photographs of them later.

FIELD IDENTIFICATION

The topics below should be read and studied to better understand how to identify raptors correctly under field conditions.

Age Terminology

Use of proper age terminology helps us to understand molt, plumages, and ageing. The best age terminology is one that corresponds one-to-one with annual changes in plumage; those that change with the calendar year are confusing because the age of the bird changes on 1 January, but the bird does not change in appearance.

Nestlings begin with two sets of down, descriptions of which are not within the scope of this chapter. While still in the nest, young raptors acquire their first, or juvenal plumage (note that the spelling is *juvenal* plumage in North America and *juvenile* plumage elsewhere). In most species, juvenal plumage is worn for 7 months to almost a year. In temperate regions, raptors typically fledge in summer and begin the molt from their juvenal plumage into their second plumage the next spring. In tropical areas, this molt usually begins 8 or 9 months after fledging, which can be at any month of the year depending on the timing of breeding and is usually determined by the timing of regional wet and dry seasons.

Depending on the size of the raptor, annual molt takes between 3 and 10 months to complete. In some species, usually the smaller ones, the resulting plumage is the adult or Definitive Basic plumage (Humphrey and Parks 1959). Most falcons have only one immature or juvenal plumage. Many accipitrid raptors, especially larger buzzards, vultures, and eagles, however, have more than one immature plumage. In North America, the latter immature plumages are called Basic I, then Basic II, etc. until Definitive Basic (Adult) plumage is achieved (Humphrey and Parks 1959). Howell et al. (2003) refers to the latter immature plumages as Basic II, Basic III, etc. In other geographic areas, most call them Second plumage, Third plumage, etc. The second and subsequent plumages are acquired by annual molts.

Note that many field guides and authors use the term immature for juvenile. The term subadult has been used to refer to at least three different age categories. The use of subadult should be avoided.

Field Marks

Field marks are the characters of a bird, in our case, a raptor, that can be used by the observer to identify it to species and, often, its age and sex. Field marks include plumage characters such as the white head and tail of adult Bald Eagles (*Haliaeetus leucocephalus*), the wide white line running through the underwings of immature Steppe Eagles (*Aquila nipalensis*), or the bold black-and-white plumage of adult male Pied Harriers (*Circus melanoleucus*). For most raptors, more than one field mark is needed for identity, and the more field marks correctly seen, the more certain the identity. Other field marks are the shape and length of wings and tail on soaring raptors (Fig. 1), head projection beyond the wings on flying raptors, and, for many species, the positions of the wing tip relative to the tail tip on perched raptors (Fig. 2). Wing attitude of soaring and gliding raptors also can be field marks, along with behavior patterns, such as kiting and hovering or the wing flex of vultures. Some field marks, such as pale wing panels, are useful only on flying individuals, whereas the color of the shoulders can apply only to perched individuals.

Field Guides

Field guides, especially raptor field guides and photo guides, are one of the best sources for field marks used to identify raptors in the field. Unfortunately, many of the general bird guides are inadequate for raptors, although they are useful for most other species of birds. Bird guides often contain errors in age and sex characteristics, fail to show the range of variation in plumages, and incorrectly depict the shape of flying and perched individuals. That said, some of the newer bird field guides, including Hollom et al. (1988), Fjeldså and Krabbe (1990), Jonsson (1993), Zimmerman et al. (1996), Mullarney et al. (1999), Sibley (2000), and Rasmussen and Alderton (2005), depict wing and tail shapes correctly, and their perched raptors look like their real-life subjects. Additional general bird guides, including Barlow and Wacher (1997), Grimmett et al. (1999), and Stevenson and Fanshawe (2002), adequately describe plumage and field marks, but don't depict wing, tail, and body shapes correctly.

Figure 1. Adult Verreaux's Eagle (*Aquila verreauxii*). Wing shape is an important field mark, as shown on this African eagle. *(W.S. Clark, Kenya)*

Figure 2. Adult Peregrine Falcon (*Falco peregrinus*). In the Americas, the Peregrine Falcon is the only falcon that, when perched, shows wingtips reaching the tail tip. *(W.S. Clark, Saskatchewan)*

On the other hand, many raptor guides vary from very good to excellent. The very first field guide to show accurate wing and tail shapes in flight was *Flight Identification of European Raptors* (Porter et al. 1981). The authors, including the artist, are to be commended for this classic work. Although this guide is somewhat out of date, does not include perched raptors, and has only black-and-white drawings and photos, it is highly recommended. Following the lead of Porter et al. (1981), other raptor field and photo guides include Wheeler and Clark (1995), Morioka et al. (1995), DeBus (1998), Forsman (1999), Clark (1999a), Clark and Wheeler (2001), Coates (2001), Wheeler (2003a,b), and Ligouri (2005). Two other photo guides with good photos, but little information are Allen (1996) and Kemp and Kemp (1998).

The two most recent global handbooks for raptors, del Hoyo et al. (1994) and Ferguson-Lees and Christie (2001), have some information on raptor identification, but their illustrations were produced primarily from museum specimens, often with simplistic "cookie cutter" wing and body shapes that don't resemble their real-life counterparts. The new world raptor field guide by the latter authors (Ferguson-Lees and Christie 2005) uses most of the same museum-specimen plates. Several continental handbooks, including Cramp and Simmons (1980) for Europe, Palmer (1988) for North America, and Marchant and Higgins (1993) for Australia, contain much information and useful illustrations on raptor identification.

Other important sources of information for field identification include the many articles on the subject that have appeared in the peer-reviewed literature. There are too many of these to list all of them, but examples include Watson (1987), Brown (1989), Clark and Wheeler (1989, 1995), Clark et al. (1990), Shirihai and Doherty (1990), Clark and Schmitt (1993, 1998), Clark and Shirihai (1995), Debus (1996), Forsman (1996a,b), Alström (1997), Forsman and Shirihai (1997), Corso and Clark (1998), Clark (1999b), Corso (2000), and Rasmussen et al. (2001).

Methods of Flight

Raptors use one of four methods for flying. Recognizing which method they are using is important in identification. Raptors soar to gain altitude in rising air, usually in a thermal or a deflection updraft. When soaring, their wings are spread to the maximum with outer primaries often recognized as fingers and often with wrists pushed forward somewhat. Their tails also are usually spread. The shape of soaring raptors is constant and is an excellent field mark. Further, when they are in a thermal or deflection updraft, they are usually visible for some time, aiding in their identification. Birds in soaring flight should be, and in many cases are, depicted in field guides. Gliding is used by raptors to travel overland after they have gained height. In gliding flight, a raptor's wrists are pushed forward more and their wingtips are pulled back from the soaring position, such that they are somewhat more pointed and don't show fingers. The amount to which the wings are pulled back varies with the angle of glide, from slightly in a shallow glide, which is most often used by migrating raptors, to almost completely folded to the body in hunting raptors that are stooping on potential prey. Hovering and kiting are additional methods of flight that are used by some raptors for hunting. In both flight patterns the bird is fixed over ground or water while looking for prey. Hovering, which is more properly called wind hovering, occurs when a raptor faces into the wind and flaps its wings to remain in the same place. Kiting is when the raptor does not flap but holds its wings steady to remain in the same position. Not all raptors hover and kite, and flight behavior itself can be used to help identify a raptor to species. Flapping or powered flight is used to move from one place to another, often when thermals and deflection updrafts are not available. The wing-beat rate can be used to indicate size, with larger raptors beating their wings more slowly than smaller raptors. Some species can be identified by the shape of the wingtips at the apex and nadir of the wing strokes. The Northern Goshawk (*Accipiter gentilis*), for example, shows very pointed wingtips in powered flight.

Variation in Appearance Due to Light Conditions

Raptors and other birds usually appear differently under the varying light conditions that occur throughout the day and the year and during different weather conditions (e.g., sunny, overcast, and rainy periods). Far too little has been written about this variation and its effect on field marks. For example, the sun gives a reddish cast to pale areas on birds early and late in the day. In mid-day, especially when it is sunny and the ground is reflective, snow or pale desert, the reflected light allows better definition on underwings and undertails of flying raptors. When the surface is dark grass or forest, much less light is reflected and the underwing and undertail appear

much darker. Wet birds and those flying against overcast, whitish skies have an overall darker appearance.

Size

Although many believe that it is possible to do so, humans are not capable of judging the size of singly flying raptors accurately. Thus, size of a raptor flying by itself is not a field mark. However, the relative sizes of two or more raptors or a raptor and another bird, such as a Common Raven (*Corvus corax*), flying together can be used successfully, as we can judge relative sizes.

Distance

Raptors flying at a distance are hard to identify because their field marks, especially colors, are difficult to discern, and because their plumage appears more black or white.

Jizz

Jizz is use of subconscious clues to identify raptors and other birds, usually at a distance when field marks are not visible. The term is thought to come from the phrase "General impression, size and shape," and most likely was coined during World War II to describe the technique used to distinguish aircraft flying to England from the continent. Dunne et al. (1988) describe the method in more detail for North American raptors. Accuracy in identification by JIZZ depends on the experience and skill level of the observer and, in most cases, is much less than that derived from the use of standard field marks.

Flight-Feather Molt

Accipitrid raptors molt their primary feathers beginning at P1 (innermost), with the molt proceeding outward sequentially to the outermost primary, P10 (Edelstam 1984). They molt their secondary feathers beginning at three molt centers, S1 (outermost), S5, and S12 in the smallest hawks to S22 on large vultures (Miller 1941). Molt proceeds sequentially inward from S1 and S5 and outward from the innermost center.

Primary molt of falconid raptors begins at P4 and proceeds both outward and inward sequentially to P10 and P1, respectively. Their secondary molt begins at S5 and proceeds sequentially inward and outward to inner secondary and S1, respectively (Edelstam 1984).

Rectrix (i.e., tail-feather) molt in raptors begins almost always with T1 (the central or deck feathers). There is a great deal of variation in the order of replacement, although T2 and T6 are usually replaced next. While T5 is the last to be replaced in some species, T4 is the last replaced in others. Asymmetry occurs more often in tail-feather than in wing-feather molt.

The wing and tail molt of all falconid raptors, and that of smaller accipitrid raptors, usually is complete (i.e., occurs within a single year), and subsequent molts in subsequent years typically begin at the same molt centers. However, larger accipitrid raptors don't complete wing molt annually, and many don't complete the tail molt annually, either. (See below for the use of incomplete molt in ageing immatures in these species.)

Body Feather and Covert Molt

Molt of body feathers begins slowly for juveniles of many species not long after fledging. Pyle (2005) describes this process as "pre-formative molt." Molt begins actively 7 to 10 months after fledging, starting at the head and proceeding down the neck and through the body caudally. Wing- and tail-covert molt begins after body-feather molt is well underway. Body feather and covert molt is complete for all but the larger species. Even so, a few feathers may not be replaced every year, particularly among the uppertail or upperwing coverts.

Molt and Its Use in Ageing

Molt of flight feathers can be an aid in determining the age of immatures of species that take more than one year of molt to attain adult plumage. This is true for most of the larger accipitrid raptors. Although not all primaries are replaced annually in these raptors, P1 is always replaced annually. Thus, molt can occur simultaneously in two to three locations in these feathers (Clark 2004a). Juveniles show no molt (Fig. 3a). Second-plumage raptors (Basic II) have new inner primaries and old, retained juvenile outer primaries (Fig. 3b). Third-plumage raptors show two "waves" of molt, with new inner primaries and new outer primaries, with retained juvenile P10 in most large eagles (Fig. 3c). In some raptors, this is adult plumage, but in others, especially large eagles, body and tail feathers are still immature at this time. Fourth plumage in most eagles begins to resemble adult plumage, but with noticeable immature characters. The primary molt is like that of adults, with three waves of primary molt (Fig. 3d). Secondary

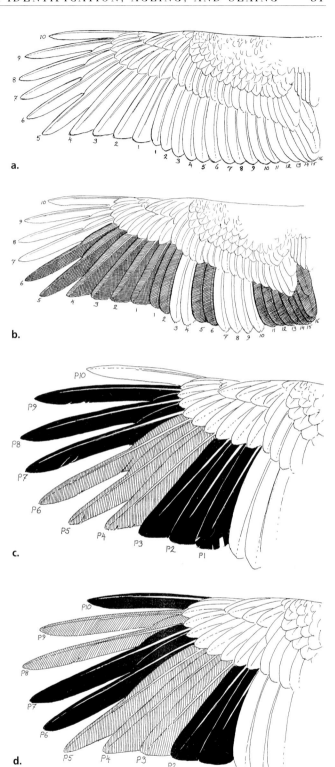

Figure 3. a. Juvenile remiges. Juveniles show no obvious molt; all feathers are the same age and show the same amount of wear. Secondaries have rather pointed tips. **b.** Second plumage (Basic II). Inner primaries have been replaced in sequence P1 outwards, and outer ones are retained juvenile. New secondaries of eagles are shorter than those of juveniles (but see Fig. 4). **c.** Third plumage (Basic III). First wave of molt has progressed to P9 and second wave to P3. **d.** Adult. Adults show from two to four waves of primary molt.

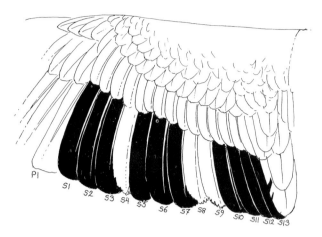

Figure 4. New secondaries of some raptors (e.g., Buteos) are longer than juvenile ones.

molt is useful for ageing immatures of larger accipitrid raptors, with juveniles showing no molt. The secondaries of juveniles are distinctive with pointed, narrower tips than those of subsequent plumages; the new secondaries are shorter than the juvenile ones in some species (e.g., eagles [Fig. 3a]), but are longer in others (e.g., Buteos [Fig.4]). In a few species the new and juvenile secondaries are the same length. In the second plumage, the secondaries are a mix of new feathers and old retained juveniles feathers (Fig. 4). Third-plumage eagles usually show all non-juvenile feathers, but occasionally S4, S8, or S9, or some combination of these, are not replaced. In some species, especially eagles, the tail feather pattern varies with each age and can be used for ageing. Best results are obtained by using field marks in the molt of all three feather types: primary, secondary, and tail.

Pyle (1997) describes molt in owls.

Unusual Plumages

Besides individual variation, raptors can show abnormal plumages, especially partial albinism, dilute plumage (also called leucism or shizochronism), and melanism, including erythrism. There are few, if any, records of complete albinism in raptors. Partial albinos however, occur regularly, especially in some species such as the Red-tailed Hawk (*Buteo jamaicensis*) (Wheeler and Clark 1995, Clark and Wheeler 2001). Individuals so affected show a variable number of all-white feathers, from few to almost all; and most show pigment in the eyes, cere, and beak, but often lack it in the talons. A similar pale condition, dilute plumage, occurs when most to all of their feathers have a reduced

amount of dark pigmentation (melanin); usually dark brown feathers then appear much paler, even tawny, or "café-au-lait." At the opposite extreme, a few individuals exhibit an excess of pigmentation, resulting in a darker bird overall. In some species, especially buzzards, where it occurs regularly, individuals with this condition are referred to as "dark morphs." In species lacking regular dark morphs, the condition is usually referred to as melanism, to indicate that it is an aberrant plumage that is not usually passed on genetically (e.g., see Clark 1998 for Ospreys [*Pandion haliaetus*]). Erythrism, in which there is an excess of the reddish pigment, erythrin, also occurs in raptors but much less often than melanism.

Range Maps

Range maps in field guides and handbooks should be used with caution, as they have major limitations. They don't show density information, nor do they show habitat preferences. Also, they offer no information on whether a species is detected easily. Lastly, ranges of birds are dynamic and change over time, especially with changing land use. Range maps in regional or smaller area guides and handbooks can be more useful, at least compared with continent-wide guides and handbooks.

In-hand Identification

In many field activities it is necessary to have the raptors in hand (e.g., banding or ringing, attaching radio or satellite transmitters for telemetry, taking measurement data, collecting blood or feather samples for analysis), or a combination of these, and other reasons. Hand-held raptors should be easier to identify, age, and sex than those spotted in the field (Fig. 5). One can see plumage details well and take measurements, which is especially helpful in determining sex. Most bird and raptor field and photo guides are generally sufficient for in-hand identification. Even so, several specialty, in-hand guides for raptors have been published, including Baker (1993) for Europe and Clark and Yosef (1998) for the Middle East. In North America, the Bird Banding Lab maintains a downloadable manual for ageing and sexing raptors by Hull and Bloom (2001) at www.pwrc. usgs.gov/bbl/resource.htm. Several raptor banding stations, including the Golden Gate Raptor Observatory in California (www.GGRO.org), have their own raptor in-hand identification, ageing, and sexing manuals (Culliney and Hull 2005).

Figure 5. Adult Common Buzzard (*Buteo buteo buteo*). Raptors in hand can more easily aged and sexed by measurements, molt, and plumage. *(W.S. Clark, Israel)*

There also are many articles on ageing individual species of raptors in-hand, including the Bald Eagle (Clark 2001), Egyptian Vulture (*Neophron percnopterus*) (Clark and Schmitt 1998), White-tailed Hawk (*B. albicaudatus*)(Clark and Wheeler 1989), Roughleg (*B. lagopus*) (Clark and Bloom 2005), Steppe Eagle (Clark 1996), Asian Imperial Eagle (*A. heliaca*) (Clark 2004b), Golden Eagle (*A. chrysaetos*) (Bloom and Clark 2001), as well as others on sexing individual species of raptors, including Bald Eagle (Bortolotti 1984a), and Golden Eagle (Bortolotti 1984b).

Ageing and Sexing Owls

Plumage differences can be used to age and sex some species of owls (Fig. 6). Pyle (1997) uses plumage differences and measurements to sex several North American owls and molt of flight feathers to age them. David Brinker has developed a discriminate function analysis for sexing Northern Saw-whet Owls (*Aegolius acadicus*) that is available on-line at www.projectowlnet.org.

Sound Recordings

Field studies on raptors sometimes involve using recorded vocalizations to bring individual birds closer to the observer or to verify their identification. Professional quality recordings of raptor vocalizations can be purchased from the Macaulay Library at the Cornell Lab of Ornithology (www.birds.cornell.edu/macaulay library) or the Borror Laboratory of Bioacoustics at The Ohio State University (blb.biosci.ohio-state.edu). Both labs have a searchable database of recordings on their

Figure 6. Juvenile Snowy Owls (*Bubo scandiaca*). Bird field guides are sufficient for identification of owls. Sexes of Snowy Owls differ in plumage, with juvenile females (left) being more heavily marked than males (right). *(W.S. Clark, British Columbia)*

websites. A search on Google™ of "Bird sound recordings" also can be useful.

ON-LINE REFERENCES

The Internet is an excellent source of raptor ID references and raptor images. Particularly useful sites include:

(1) SORA (Searchable Ornithological Research Archive), an open access electronic journal archive and the product of collaboration between ornithological organizations and the University of New Mexico libraries and IT department, at http://elibrary.unm.edu/sora. The

archive provides access to back issues of *The Auk* (1884–1999), *The Condor* (1899–2000), *Journal of Field Ornithology* (1930–1999), *The Wilson Bulletin* (1889–1999), *Pacific Coast Avifauna* (1900–1974) and *Studies in Avian Biology* (1978–1999).

(2) The Global Raptor Information Network (GRIN) provides information on diurnal raptors (hawks, eagles, and falcons) and facilitates communication between raptor researchers and organizations interested in the conservation of these species. This site also includes information on identifying species of raptors (www.globalraptors.org/grin/indexAlt.asp).

(3) The Raptor Information System is a key-worded catalog of over 40,000 references about the biology, management, and identification of birds of prey (http://ris.wr.usgs.gov).

(4) Ornithological Worldwide Literature (OWL) is a compilation of citations and abstracts from the worldwide scientific literature about owls that includes information on identification. The site includes considerable coverage of the "gray literature," much of which is not abstracted in commercial databases (http://egizoosrv.zoo.ox.ac.uk/owl).

(5) Hawk wing photos. The University of Puget Sound provides photos of the spread underwings of specimens at www.ups.edu/biology/museum/wings_Accipitridae.html.

The journal *North American Bird Bander* also will be available online soon.

SUMMARY

Accurate identification of raptors is key to successful raptor research. Recent advances and improvements in optics, together with increased knowledge of raptor field marks for species identification, as well as for ageing and sexing within species, continue to make this increasingly possible for most species of birds of prey.

LITERATURE CITED

ALLEN, D. 1996. A photographic guide to birds of prey of southern, central, and East Africa. New Holland, Cape Town, South Africa.

ALSTRÖM, P. 1997. Field identification of Asian *Gyps* vultures. *Orient. Bird Club Bull.* 25:32–49.

BAKER, K. 1993. Identification guide to European non-passerines: BTO Guide 24. British Trust for Ornithology, Thetford, United Kingdom.

BARLOW, C. AND T. WACHER. 1997. A field guide to birds of The Gambia and Senegal. Pica Press, Robertsbridge, United Kingdom.

BLOOM, P. AND W. S. CLARK. 2001. Molt and sequence of plumages of Golden Eagles, and a technique for in-hand ageing. *N. Am. Bird Band.* 26:97–116.

BORTOLOTTI, G.R. 1984a. Sexual size dimorphism and age-related size variation in Bald Eagles. *J. Wildl. Manage.* 48:72–81.

———. 1984b. Age and sex size variation in Golden Eagles. *J. Field Ornithol.* 55:54–66.

BROWN, C.J. 1989. Plumages and measurements of Bearded Vultures in Southern Africa. *Ostrich* 60:165–171.

CLARK, W.S. 1996. Ageing Steppe Eagles. *Birding World* 9:269–274.

———. 1998. First North American record of a melanistic Osprey. *Wilson Bull.* 110:289–290.

———. 1999a. A field guide to the raptors of Europe, the Middle East, and North Africa. Oxford University Press, Oxford, United Kingdom.

———. 1999b. Plumage differences and taxonomic status of three similar *Circaetus* snake-eagles. *Bull. Br. Ornithol. Club* 119:56–59.

———. 2001. Aging Bald Eagles. *Birding* 33:18–28.

———. 2004a. Wave molt of the primaries of accipitrid raptors, and its use in ageing immatures. Pages 795–804 *in* R.D. Chancellor and B.-U. Meyburg [EDS.], Raptors Worldwide. World Working Group on Birds of Prey and Owls, Berlin, Germany.

———. 2004b. Immature plumages of the Eastern Imperial Eagles *Aquila heliaca*. Pages 569–574 *in* R.D. Chancellor and B.-U. Meyburg [EDS.], Raptors Worldwide. World Working Group on Birds of Prey and Owls, Berlin, Germany.

CLARK, W.S. AND P.H. BLOOM. 2005. Basic II and Basic III plumages of Rough-legged Hawks. *J. Field Ornithol.* 76:83–89.

——— AND N. J. SCHMITT. 1993. Field identification of the Rufous-bellied Eagle. *Forktail* 8:7–9.

——— AND N.J. SCHMITT. 1998. Ageing Egyptian Vultures. *Alula* 4:122–127.

——— AND H. SHIRIHAI. 1995. Identification of Barbary Falcon. *Birding World* 8:336–343.

——— AND B.K. WHEELER. 1989. Field identification of the White-tailed Hawk. *Birding* 21:190–195.

——— AND B.K. WHEELER. 1995. The field identification of Common and Great Black Hawks. *Birding* 27:33–37.

——— AND B.K. WHEELER. 2001. A field guide to hawks of North America, 2nd Ed. Houghton Mifflin Co., Boston, MA U.S.A.

——— AND R. YOSEF. 1998. In-hand identification guide to Palearctic raptors. International Birdwatching Center - Eilat, Eilat, Israel.

———, R. FRUMKIN, AND H. SHIRIHAI. 1990. Field identification of the Sooty Falcon. *Br. Birds* 83:47–54.

COATES, B.J. 2001. Birds of New Guinea and the Bismarck Archipelago. Dove Publications, Alderley, Australia.

CORSO, A. 2000. Identification of European Lanner. *Birding World* 13:200–213.

——— AND W.S. CLARK. 1998. Identification of Amur Falcon. *Birding World* 11:261–268.

CRAMP, S. AND K.E.L. SIMMONS. 1980. The Birds of the Western Palearctic, Vol. 2. Oxford University Press, Oxford, United Kingdom.

CULLINEY, S. AND B. HULL. 2005. GGRO Banders' Raptor Identification Guide, 2nd Ed. Golden Gate Raptor Observatory, San Francisco, CA U.S.A.

DEBUS, S.J.S. 1996. Problems in identifying juvenile Square-tailed Kite. *Aust. Bird Watcher* 16:260–264.

———. 1998. The birds of prey of Australia: a field guide to Australian raptors. Oxford University Press, South Melbourne, Australia.

DEL HOYO, J., A. ELLIOTT, AND J. SARGATAL [EDS.]. 1994. Handbook of the Birds of the World, Vol. 2. Lynx Edicions, Barcelona, Spain.

DUNNE, P., D.A. SIBLEY, AND C. SUTTON. 1988. Hawks in flight. Houghton Mifflin Company, Boston, MA U.S.A.

EDELSTAM, C. 1984. Patterns of moult in large birds of prey. *Ann. Zool. Fenn.* 21:271–276.

FERGUSON-LEES, J. AND D.A. CHRISTIE. 2001. Raptors of the world. Houghton Mifflin Co., Boston, MA U.S.A.

——— AND D.A. CHRISTIE. 2005. Raptors of the world. Princeton University Press, Princeton, NJ U.S.A.

FJELDSÅ, J. AND N. KRABBE. 1990. Birds of the high Andes. Zoological Museum, University of Copenhagen and Apollo Books, Copenhagen and Svendborg, Denmark.

FORSMAN, D. 1996a. Identification of Spotted Eagle. *Alula* 2:16–21.

———. 1996b. Identification of Lesser Spotted Eagle. *Alula* 2:64–67.

———. 1999. The raptors of Europe and the Middle East. T. & A.D. Poyser Ltd., London, United Kingdom.

——— AND H. SHIRIHAI. 1997. Identification, ageing and sexing of Honey-buzzards. *Dutch Birding* 19:1–7.

GRIMMETT, R., C. INSKIPP, AND T. INSKIPP. 1999. A guide to the birds of India, Pakistan, Nepal, Bangladesh, Bhutan, Sri Lanka, and the Maldives. Princeton University Press, Princeton, NJ U.S.A.

HOLLOM, P.A.D., R.F. PORTER, S. CHRISTENSEN, AND I. WILLIS. 1988. Birds of the Middle East and North Africa: a companion guide. T. & A.D. Poyser Ltd., Calton, United Kingdom.

HOWELL, S.N.G., C. CORBIN, P. PYLE, AND D.I. ROGERS. 2003. The first basic problem: a review of molt and plumage homologies. *Condor* 105:635–653.

HULL, B. AND P.H. BLOOM. 2001. The North American banders' manual for raptor banding techniques. North American Banding Council, Patuxent, MD U.S.A.

HUMPHREY, P.H. AND K.C. PARKES. 1959. An approach to the study of molts and plumages. *Auk* 76:1–31.

JONSSON, L. 1993. Birds of Europe. Princeton University Press, Princeton, NJ U.S.A.

KEMP, A. AND M. KEMP. 1998. Birds of prey of Africa and its islands. New Holland, London, United Kingdom.

LIGUORI, J. 2005. Hawks from every angle: how to identify raptors in flight. Princeton University Press, Princeton, NJ U.S.A.

MARCHANT, S. AND P.J. HIGGINS. 1993. Handbook of Australian, New Zealand, and Antarctic birds. Oxford University Press, Melbourne, Australia.

MILLER, A.H. 1941. The significance of molt centers among the secondary remiges in the Falconiformes. *Condor* 43:113–115.

MORIOKA, T., N. YAMAGATA, T. KANOUCHI, AND T. KAWATA. 1995. The birds of prey in Japan. [In Japanese with English summaries.] Bun-ichi Sogu Shuppan Company, Tokyo, Japan.

MULLARNEY, K., L. SVENSSON, D. ZETTERSTROM, AND P. GRANT. 1999. Birds of Europe. Harper Collins, London, United Kingdom.

PALMER, R. [ED.]. 1988. Handbook of North American birds, Vols. 4 and 5. Yale University Press, New Haven, CT U.S.A.

PORTER, R.F., I. WILLIS, S. CHRISTENSEN, AND B.P. NIELSEN. 1981. Flight identification of European raptors, 3rd Ed. T. & A.D. Poyser Ltd., London, United Kingdom.

PYLE, P. 1997. Flight-feather molt patterns and age in North American owls. *Monogr. Avian Biol.* 2:1–32.

———. 2005. First-cycle molts in North American Falconiformes. *J. Raptor Res.* 39:378–385.

RASMUSSEN, P.C. AND J.C. ALDERTON. 2005. Birds of South Asia, Vol. 1: field guide. Lynx Edicions, Barcelona, Spain.

———, W.S. CLARK, S.J. PARRY, AND N.J. SCHMITT. 2001. Field identification of "Long-billed Vultures" (Indian and Slender-billed Vultures). *Bull. Orient. Bird Club* 34:24–29.

SHIRIHAI, H. AND P. DOHERTY. 1990. Steppe Buzzard plumages. *Birding World* 3:10–14.

SIBLEY, D. 2000. The Sibley guide to birds. Alfred A. Knopf, New York, NY U.S.A.

STEVENSON, T. AND J. FANSHAWE. 2002. A field guide to the birds of East Africa: Kenya, Tanzania, Uganda, Rwanda, Burundi. T. & A.D. Poyser Ltd., London, United Kingdom.

WATSON, R.T. 1987. Flight identification of the Bateleur age classes: a conservation incentive. *Bokmakierie* 39:37–39.

WHEELER, B.K. 2003a. Raptors of eastern North America. Princeton University Press, Princeton, NJ U.S.A.

———. 2003b. Raptors of western North America. Princeton University Press, Princeton, NJ U.S.A.

——— AND W.S. CLARK. 1995. A photographic guide to North American raptors. Academic Press, London, United Kingdom.

ZIMMERMAN, D.A., D.A. TURNER, AND D.J. PEARSON. 1996. Birds of Kenya and northern Tanzania. Christopher Helm, London, United Kingdom.

Systematics

3

MICHAEL WINK

Universität Heidelberg, Institut für Pharmazie
und Molekulare Biotechnologie,
INF 364, 69120 Heidelberg, Germany

INTRODUCTION

Systematics is the branch of biology that deals with the classification of living organisms, describing their diversity and interrelationships. It can be divided into three parts:

■ **Taxonomy** is the description and naming of new taxa (a taxon is any specifically defined group of organisms). Taxonomic groups are used to categorize similar taxa for identification, such as field guides. Taxa do not necessarily mirror evolutionary relationships. Taxonomists have not agreed universally on a single species concept (Mayr 1969, Sibley and Ahlquist 1990). The oldest is the *Typological or morphological species concept*. This concept combines a group of organisms into a single species if they conform sufficiently to certain fixed properties or differ anatomically from other populations of organisms. For many years, the *Biological species concept* was favored in ornithology. According to this species concept, a species consists of "groups of actually or potentially interbreeding natural populations which are reproductively isolated from other such groups." In the *Phylogenetic* or *Evolutionary species concept*, a species comprises a group of organisms that shares an ancestor and can be separated from others by distinctive characters. This species concept describes lineages that maintain their integrity with respect to other lineages through both time and space. At some point in the progress of such groups, members may diverge from one another: when such a divergence becomes sufficiently clear, the two populations are regarded as separate species. See Otte and Endler (1989) for additional details regarding various species concepts.

■ **Classification** is the organization of information about diversity that arranges it into a convenient hierarchical system of classification, such as the Linnaean system.

■ **Phylogenetics** is the field of biology concerned with identifying and understanding the evolutionary relationships among the many different kinds of life on earth. It is the basis for evolutionary systematics. Phylogeny is the determination of the ancestral relationships of organisms, and the group's evolutionary history.

Classification of plants and animals is a basic discipline of biology, and the *Systema Naturae* of Linné in 1753 was a landmark in this field. Traditionally, systematists and taxonomists have used morphological and anatomical characters to define species and subspecies. More recently, they also have used behavior, vocalizations, and biochemistry. The new era of molecular biology has provided a broad set of genetic tools that complement existing methods.

It is likely that biologists will soon establish an improved taxonomy and classification of most orders of the living world that is based on phylogenetic relationships and not solely on similarity. Many morphological characters can be formed by convergent evolution, and anatomical similarity alone can result in misleading classifications. Genetic characters, which are more numerous overall than the former, can help to clarify systematics. Today a dream of Charles Darwin may become a reality. In 1857 Darwin wrote to his friend Thomas H. Huxley: *"In regard to classification, & all the endless disputes about the 'natural system' which no two authors define in the same way, I believe it ought, in accordance with my heterodox notions, to be simply genealogical. The time will come I believe, though I shall not live to see it, when we shall have fairly true genealogical trees of each kingdom of nature..."*

In this chapter, I introduce the methods used in taxonomy, classification, phylogeny, and systematics and then discuss in detail the newer DNA methods.

PRINCIPAL METHODS

Comparative Nonmolecular Characters

An array of details can be recorded about an organism, and each detail can be used as a character for comparison with the same homologous character (i.e., a character inherited from a common ancestor) in other organisms. These characters can be tabulated and analyzed by cladistics, a method that groups organisms on the basis of common ancestry into clades that represent monophyletic groups (cf. Wiley 1981, Wiley et al. 1991). The intrinsic characters below have been used in systematic studies. The list is not exhaustive, nor will it ever be as innovations continue to extend the range of characters that can be documented and improvements are made in assessing the usefulness of characters in classification.

Measurable characters. A comprehensive set of measurements that can be taken on live, freshly killed or dried museum specimens of raptors has been described by Biggs et al. (1978). Those measurements found to be practical in extensive field and museum work are described and illustrated in Figs. 1 and 2.

Workers should practice such taking such measurements in general before doing so on the organisms they

are studying, and should repeat all measurements to determine the extent of both intra- and inter-recorder variability. When comparing measurements from fresh and dry specimens, shrinkage in the latter should be accounted for. Special attention should be given to recording body mass, body temperature and neural (brain) mass. Brain mass values can be important when behavioral and sociological data are compared. Scaling

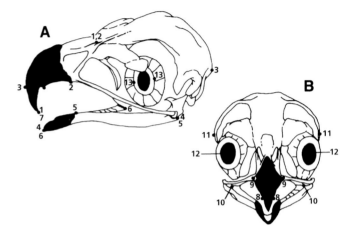

Figure 1. Views of a raptor skull. A — dorsolateral, and B — anterior, with measuring points taken by calipers and indicated by numbers and detailed below.

1. Bill chord — from the suture at the bill-skull junction to the tip of the mandible.

2. Bill depth — from center of the suture at the bill-skull junction to the junction of the cutting edge of the upper mandible and cere (or skin) along the gape.

3. Skull length — from the center back of the skull to the front edge of the upper mandible with the calipers held parallel to the plane of the top of the head.

4. Jaw length — from the posterior point of the ramus of the lower jaw to the tip of the lower mandible.

5. Jaw-bill length — from the posterior point of the ramus of the lower jaw to the junction of the cutting edge of the dorsal surface of the lower mandible and the skin forming the edge of the gape.

6. Gape length — from the back of the fold of the gape, with the mouth almost closed, to the tip of the lower mandible.

7. Tooth depth (for those species with tomial teeth) — from the tip of the upper mandible to the tip of the longest tomial tooth.

8. Tooth width (for those species with tomial teeth) — between the tips of the tomial teeth.

9. Bill width — between the junctions of the cutting edges of the upper mandible and the cere (or skin) on each side of the gape.

10. Gape width — between the back points or fold of the gape when the mouth is closed.

11. Skull width — between the widest points of the skull behind the eyes, with calipers vertical to the plane of the top of the head.

12. Eye spacing — as the width between the centers of the eyes, with the calipers as close to the surface of the eyes and the recorder's eyes as far away as possible.

13. Eye diameter — between the outer edges of the colored (iris) area of the eye, corresponding to the inner edge of the ring of sclerical ossicles. *Figure originally from Kemp (1987).*

Figure 2. Diagrammatic layout of a raptor with measuring points taken by various methods and indicated by numbers and detailed below.

14. Wing length — Taken with a wing rule, from the front of the folded wrist to the tip of the longest primary, with the feather flattened and checking that it is not affected by molt.

15. Secondary length — Taken with wing rule, from the front of the folded wrist to the tip of the outermost secondary, with the feather flattened and checking that it is not affected by molt.

16. Alula length — Taken with a wing rule, from the proximal side of the protuberance on the carpometcarpus to which the alula is attached, to the tip of the longest alula feather, with the feather flattened and checking that it is not affected by molt.

17. — **Ulna length index** — Taken with a wing rule, from the front of the folded wrist to the inner surface of the elbow joint (inner surface of the distal humerus head).

18. — **Humerus length index** — Taken with a wing rule, from the outer edge of the elbow joint (posterior surface of proximal ulna head) to the anterior edge of the distal end of the coracoid (forming a point at the anterior edge of the shoulder).

19. — **Femur length index** — Taken with calipers, from the top of the exterior proximal crest of the femur to the anterior center of the tibiotarsal-tarsometatarsal joint.

20. Tibiotarsal length — Taken with calipers, from the anterior center of the tibiotarsal-tarsometatarsal joint.

21. Tarsometatarsal length — Taken with calipers, from the posterior center of the tibiotarsal-tarsometatarsal joint to the dorsal base of the center toe (point is located by flexion of the toe).

22. Foot volume — Recorded by displacement of water when immersing the foot and tarsus up to the tibiotarsal-tarsometatarsal joint.

23. Toe lengths — Taken with calipers, along the dorsal surface of each straightened toe, from the junction with the tarsometatarsus (found by flexion of the toe) to the claw-skin junction.

24. Claw chords — Taken with calipers, from the dorsal surface of the claw at the junction with the skin to the tip of the claw.

25. Tail lengths — Taken with a wing rule, from the feather-skin junction of the central pair of rectrices to their tips (center tail length), and to the tip of one of the outer pair of rectrices (outer tail length), with the feather flattened and checking that it is not affected by molt.

Figure originally from Kemp (1987)

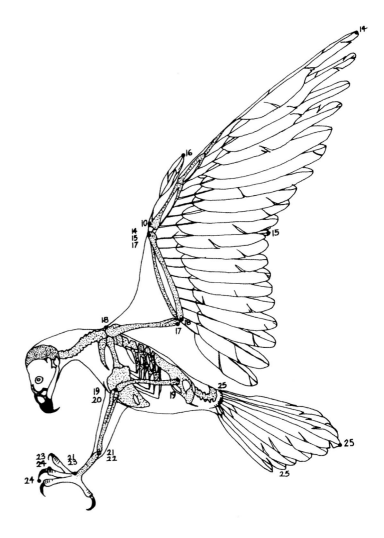

relationships based on these parameters have great potential in predicting a wide range of attributes for homoeotherms, particularly those concerning life-history indexes, growth, and energetics (Calder 1983, 1984).

Other important parameters include mass and wing area. Mass should be recorded as accurately as possible (preferably with electronic balances), and the accuracy of the weighing instrument should be noted. Adjustments to the mass, to account for differences in body condition or the presence of food in the crop or stomach, can be attempted. Wing area can be drawn directly onto scaled, gridded paper or photographed together with an appropriate scale. To ensure comparable measurements, the wing should be extended with the leading edge forming a straight line perpendicular to the body. Tracing should include secondaries and tertials.

Anatomical characters. Anatomical characters incorporate any external or internal structural attributes of an organism. External characters include plumage color, structure, and form; soft-body-part colors and extent; bill and foot form; and the sizes and proportions of these characters. Internal characters most often refer to skeletal or myological attributes as well as details of organ systems. A comprehensive introduction to raptor morphology is provided by Jollie (1976, 1977a,b). Obviously, characters that differ with age and sex must be described separately for each class. This is a regular requirement for raptors, in which juvenile and sub-adult plumages are common and sexes differ at least in size if not in other characters. Age and sex differences posed many taxonomic problems to earlier workers (e.g., Finch-Davies 1919). Molecular methods now are available to sex individuals not only of "difficult" species, but also individuals at early stages of development (see below).

In addition to anatomical characters in full grown specimens, the ontogeny of these characters at various stages (embryological and post-hatching) often provides important systematic insight. Embryology rarely has been explored in raptor studies (Desai and Malhotra 1980, Bird et al. 1984), even though it may shed light on evolution of such important characters as fused phalanges (Olson 1982). Post-hatching development is regularly recorded and used in suggesting relationships (e.g., of *Circaetus* and *Terathopius* eagles [Brown and Amadon 1968] that have been confirmed by DNA sequence data [Wink and Sauer-Gürth 2000]).

The morphology of chromosomes (karyology) is an established technique whose application to birds has been reviewed (Shields 1982). Only the basic chromosome number and size has been determined for some raptors (e.g., Belterman and de Boer 1984, de Boer and Sinoo 1984, Schmutz et al. 1993). Studies of centromere position and arm proportions, or more advanced studies of chromatin banding within chromosome arms, are rare or lacking (Harris and Walters 1982, Shields 1983, Bed'hom et al. 2003).

Behavioral characters. Detailed ethograms have not been recorded for any species of raptor, although Walter (1983) has suggested a method for tackling this problem. The use of communication characters (primarily visual and vocal in raptors and in most other birds) has been advocated for assessing differences among species. These characters can be documented and analyzed using tape recorders, cameras and video equipment. Basic display patterns, such as pendulum flights by eagles, have been used in systematic studies (e.g., Brown and Amadon 1968), but detailed recording and analysis of these characters have yet to be achieved for most species. Maintenance patterns, such as scratching and stretching postures, too, deserve further attention. Locomotion, feeding and hunting patterns, with special attention given to any ritualization of these behaviors, also warrant study.

Molecular Characters

Biochemical characters. Biochemical characters involve either estimation or direct documentation of protein or nucleic-acid structural diversity. Such characters involve collection and storage of suitable tissues, followed by some form of laboratory analysis. Simpler techniques are most relevant to comparisons among closely related species or populations within species where larger sample sizes are available. More complex techniques may be applied to an array of species, from which only a few samples of each are available.

Historically, the first molecules to be analyzed were proteins. Amino-acid sequences were used to infer overall phylogenetic relationships. Microcomplement fixation and allozyme analysis were employed at the species and subspecies level (Prager et al. 1974, Brush 1979, Avise 1994). Except for allozyme analysis, the resolution of the protein methods was low and the analysis time-consuming. Because DNA methods are faster and more informative, protein methods, including allozyme analysis, largely have been replaced by them (see below).

Sample Preparation and Storage

The deposition of voucher specimens is strongly advocated to ensure the results of any study against subsequent changes in the systematic or taxonomic status of the organism involved. Ideally, such specimens should store as diverse a set of characters of the organism as possible. Specimens can include an entire carcass (providing information on plumage, skeletal, myological, and general anatomy), an entire clutch of eggs (preferably containing embryos), any slides of chromosome karyotypes, preserved tissues, DNA preparations, images (film or video) of nests, behavior (such as displays), and developmental stages, and tape recordings of vocalizations. Specimens also may include tissues preserved for later biochemical studies. All specimens should be deposited with an institution whose charter includes maintenance of material for perpetuity, such as an established museum.

The fresher the tissue is for analysis, the better. Freshly killed birds should be rapidly dissected and the appropriate tissues used immediately or stored appropriately to prevent degradation of proteins and DNA. Sometimes, tissues are quick-frozen in liquid nitrogen and stored at temperatures at -80°C, or in liquid nitrogen. Some techniques require less demanding storage conditions. One should first decide on the analytical procedures to be employed and then determine the optimum collection and storage system for that technique.

For DNA studies and, in particular, intra-species studies, the more samples (5 to 10 is a minimum) per species, the better. For the detection of gene flow between populations, a higher number of samples is required (sometimes up to several hundred). A comprehensive and complete sampling (e.g., covering all members of a genus, family or order, or within-species samples from all major populations) is one of the main keys to a successful study.

In birds, blood tissue provides high-quality DNA and is better than feathers or feces. Museum skins sometimes can be a source for DNA, but DNA therein often is highly degraded. Blood (50 to 200 µl are sufficient) and tissues should be stored in 70% ethanol or, if possible, an ethylenediaminetetraacetic acid (EDTA) buffer containing 10% EDTA, 1% sodium lauryl sulfate (SDS), 0.5% NaF, 0.5% thymol in 100 mM Tris, pH 7.4 (Arctander 1988). Using the EDTA buffer, blood tissue can be stored for long periods at ambient temperatures without refrigeration. Because polymerase chain reaction (PCR) methods can pick up tiny amounts of contaminations, separate needles and pipette tips must be used for each bird.

DNA METHODS

The principles of the major DNA methods (Table 1) useful for studying the systematics of raptors are described briefly below.

DNA as a Logbook of Life

The field of DNA analysis is developing rapidly, and new techniques are being devised constantly (Hoelzel 1992, Avise 1994, Hillis et al. 1996, Mindell 1997, Karp et al. 1998, Hall 2001, Storch et al. 2001, Frankham et al. 2002, Beebe and Rowe 2004). For many groups of organisms, genetic data already are available that help identify an individual to a species (sometimes called *DNA-barcoding*). Molecular methods can answer questions in population genetics, such as immigration and dispersal rates and gene flow, and assess the connectivity between the breeding and wintering grounds of species.

Because DNA methods and DNA markers are important tools in systematics and taxonomy, the following paragraphs review some of the important background information on DNA (Griffiths et al. 1999, Klug and Cummings 1999, Alberts et al. 2002).

The genetic information of all organisms is encoded in DNA. DNA is built from four nucleotides: adenine (A), guanine (G), thymine (T) and cytosine (C). The genetic message is fixed in the specific sequences of A, T, G and C. DNA consists of two complementary strands organized as a double helix. In the nucleus of eukaryotic cells (i.e., cells with compartmentalized internal structure), linear DNA is arranged in separate

Table 1. Important methods of molecular biology that are useful in evolutionary and phylogeographic studies.

Method[a]	DNA Type	Adequate for Studying
Sequencing	mtDNA[a], ncDNA[a]	Phylogeny, taxonomy, phylogeography
STR-Analysis[a]	Microsatellites	Population genetics, tracing of individuals, paternity, and pedigree
SNP-Analysis[a]	Point mutations in all genes	Population genetics, tracing of individuals, paternity, and pedigree
AFLP[a]	Nuclear genome	Population genetics, gene mapping
ISSR[a]	Nuclear genome	Phylogeny, population genetics, hybridizations, gene mapping
DNA fingerprinting	Satellite DNA (VNTR, STR)	Paternity, tracing of individuals
Sexing	Sex chromosome	Molecular sexing

[a]STR = short tandem repeats; SNP = single nucleotide polymorphisms; VNTR = variable number tandem repeats; mtDNA = mitochondrial DNA, ncDNA = nuclear DNA; AFLP = amplified fragment length polymorphisms; ISSR = inter-simple sequence repeats.

chromosomes. At one time or another, all cells of the body, except gametes, have a double set of chromosomes and are termed diploid.

Eukaryotic cells carry DNA in the nuclear genome (ncDNA), but also in their mitochondria (mtDNA). Algae and plants have a third genome in form of chloroplast DNA (cpDNA). Mitochondrial DNA, which is a circular molecule similar to the DNA found in bacteria, was derived from endosymbiotic bacteria that were taken up by the ancestral eukaryotic cell some 1.4 billion years ago. The mitochondrial genome of animals consists of 16,000 to 19,000 base pairs (bp) and contains 13 genes that code for enzymes involved in the respiratory chain, 22 genes for transfer RNA (tRNA) and two genes for ribosomal RNA (rRNA) (Table 2). Mitochondria have a short stretch of non-coding DNA, the D-loop or control region (or origin of replication), that is four to six times more variable than protein-coding genes such as cytochrome b. A typical animal cell has between 100 and 1000 mitochondria, each of which harbors 5 to 10 copies of mtDNA. This makes mtDNA an especially frequent molecule in cells, although it accounts for only 1% of all cellular DNA. MtDNA, therefore, provides an important source of genetic material for DNA studies. MtDNA is inherited maternally and can be regarded as clonal in nature (Avise 1994, Hillis et al. 1996, Mindell 1997, Karp et al. 1998, Hall 2001, Storch et al. 2001).

The nuclear genome of birds and other vertebrates typically consists of more than one billion base pairs. Only 25% of the genome represents genes and gene-related sequences used by the organism. About 2% of the DNA actually encodes proteins. Seventy-five percent of the genome consists of extragenic DNA, with highly repetitive sections, such as long interspersed nuclear elements (LINE), short interspersed nuclear elements (SINE), and mini- and microsatellite DNA. In animal genomes, there are abundant sequences that consist of almost identical elements of 15 to 100 bases that are tandemly repeated 5 to 50 times (i.e., so-called mini-satellite DNA or VNTR — *variable number tandem repeats*). Mini-satellites show many point mutations and vary in the lengths of their repetitive elements in each DNA locus. Another abundant repetitive element consists of tandem repeats of two (sometimes up to five) nucleotides, such as (GC)n or (CA)n, that are repeated 10 to 50 times (so-called STR; *short tandem repeats*, or microsatellites).

The genome of vertebrates has more than 20,000 distinct loci of STR sequences that usually consist of polymorphic alleles. The repetitive elements are highly variable in length, a phenomenon that is caused by uneven crossing-over during meiotic recombination and slippage of DNA polymerase during replication. VNTR- and STR-loci are hot spots of evolution and are inherited co-dominantly. The chance that two individu-

Table 2. Composition of mitochondrial DNA.

DNA	Number of Elements	Substitution Rate
16S rRNA	1	Low
12S rRNA	1	Low
tRNA	22	Low
Cytochrome b	1	Medium
Cytochrome oxidase (CO), subunits I-III	3	Medium
NADH dehydrogenase (ND), subunits I-VII	7	Medium
ATP synthase, subunits a, b	2	Medium
D-loop (ori)	1	High

als have identical sets of VNTR and STR profiles is less than one in a million. Therefore, these genetic elements are ideal markers when a high degree of genetic resolution is required (Avise 1994, Hillis et al. 1996, Mindell 1997, Karp et al. 1998, Hall 2001, Storch et al. 2001).

DNA has several repair and copy-reading enzymes that help conserve its molecular structure. Even so, mutations do occur. Point mutations and chromosomal recombination are abundant, and the genomes of any two individuals of the same species are not identical. About one million single nucleotide differences are common among individuals of the same species. Mutations can be regarded as evolutionary landmarks that are transmitted to later generations, and as such can be used to trace the origin of any organism.

The total number of mutations to the original sequence increases with time. This is the base for the molecular-clock concept (Zuckerkandl and Pauling 1965), which is helpful in many areas of evolutionary research. Molecular clocks are not as precise as physical clocks and are better thought of as defining relative time windows. Molecular clocks based on amino-acid exchanges are nearly linear with time, whereas nucleotide clocks are linear only initially. Over time the latter level out because of multiple substitutions that occur during longer divergence times. Many protein-coding mitochondrial genes have an estimated divergence rate of 2% substitutions per million years (Wilson et al. 1987, Tarr and Fleischer 1993), and may be reliable up to about five million years, after which divergences will be underestimated because of the plateau effect caused by multiple substitutions. Older events can be evaluated using non-synonymous substitutions or amino acid changes that are linear over a much longer time period. Mutation rates differ between coding and non-coding DNA regions. Mutations in synonymous codon positions do not influence the fitness of the organism. Consequently, they are not a target of selection processes (i.e., they are neutral mutations and show higher apparent mutation rates). Mutation rates also differ between nuclear and mitochondrial genes; protein coding mtDNA evolves 10 to 20 times faster than protein coding nuclear genes.

The genome can be regarded as the "logbook of life" in which previous evolutionary events are fixed in terms of mutations. In vertebrates, an estimated 100,000 to 10 million nucleotide differences exist among individuals belonging to the same species. Differences between closely related species are in the range of 10 to 100 million nucleotide substitutions. Presently, it is not possible to detect all genetic polymorphisms. Instead DNA markers are analyzed as representatives for the whole genome. Depending on the biological question to be answered, different methods have been developed that are appropriate to study a given problem (Table 1).

Before the availability of rapid DNA sequencing, *DNA-DNA hybridization* was a widely used tool in molecular systematics. Sibley and Ahlquist (1990) used this method to formulate phylogenetic relationships in birds (Sibley and Monroe 1990). Many workers have since criticized DNA-DNA hybridization on methodological grounds (see Ericson et al. 2006). Today DNA-DNA hybridization can best be regarded as a historical method that has been replaced by a wide set of more versatile techniques based on PCR and DNA sequencing.

DNA Sequencing

The analysis of nucleotide sequences of marker genes is a powerful method for reconstructing the phylogeny of organisms (Hillis et al. 1996, Mindell 1997, Karp et al. 1998, Griffiths et al. 1999, Hall 2001, Storch et al. 2001, Frankham et al. 2002, Beebe and Rowe 2004). These methods are now being employed by researchers studying all parts of the living kingdom with one aim being to assemble the tree of life. Raptors are only a small group in this effort, but some progress already has been made (see below). Phylogenetic information is fundamental for taxonomy and systematics as they allow establishment of a natural, genealogical classification. The general procedure to produce and analyze DNA sequences in a phylogenetic and phylogeographic context is outlined in Fig. 3. Many researchers send their sequences to public databases such as GENBANK (National Institutes of Health) or EMBL (European Bioinformatics Institute). For a detailed description of methods and concepts see Hillis et al. (1996), Mindell (1997), Karp et al. (1998), Hall (2001), Storch et al. (2001), Frankham et al. (2002) and Beebe and Rowe (2004).

Studies of the phylogeny of raptors typically focus on the nucleotide sequences of conserved marker genes, such as protein coding mtDNA (Table 2) and of coding ncDNA (such as RAG1) (Griffiths et al. 2004) or non-coding DNA (such as intron regions of protein coding genes, including LDH, and ODC-6). Species that evolved several million years ago exhibit enough divergence among geographically separated lineages to permit useful analyses. In these cases, sequencing of mitochondrial genes, such as cytochrome b, ND or CO (Table 2) often helps to identify the mitochondrial line-

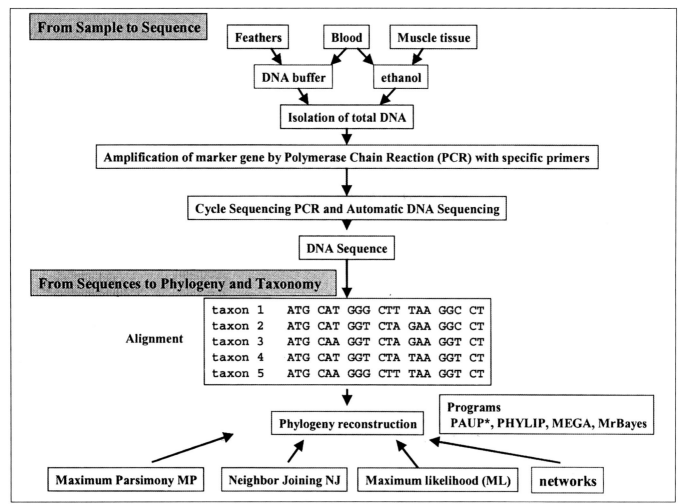

Figure 3. From sample to DNA sequence and phylogeny.

ages of groups (so-called haplotype). Because the mitochondrial D-loop is more variable, this stretch of DNA may provide an even higher resolution, but, because of its variability, it can sometimes be difficult to amplify and sequence the D-loop region by PCR.

Haplotypes of individual raptors caught during migration or on the wintering grounds sometimes can be used to determine the bird's geographic origin. In a best-case scenario, the breeding populations have distinctive haplotypes with little gene flow between lineages (Fig. 4a). If the haplotype of the migrating bird matches that of a breeding population, its origin can be inferred with some certainty (Fig. 4c). In a worst-case scenario (Fig. 4b), populations have several haplotypes and share them with neighboring populations, suggesting considerable gene flow among them. In this case, an intelligent guess is possible only if the haplotype of a migrant has to be allocated.

In relatively new species, intraspecies-sequence variation is quite small, and it can be difficult to establish an informative genetic population map using this variation. When this happens, DNA methods with a higher resolution are required.

Application of DNA sequences to study taxonomy and systematics of raptors. Diurnal raptors have been grouped into five families; Accipitridae, Pandionidae, Sagittariidae, Falconidae, and Catharthidae, and placed in the order Falconiformes (del Hoyo et al. 1994) or the infraorders Falconides and Ciconiides (Sibley and Monroe 1990). Whether Falconiformes is a monophyletic group remains an open question that many researchers are currently attempting to resolve. However, morphological and molecular data (based on several nuclear and mitochondrial genes) provide evidence that at least Falconidae do not share direct ancestry with Accipitridae, Sagittariidae, Pandionidae, and Cathartidae (Wink 1995, Wink et al. 1998b, Fain and Houde 2004, and Ericson et al. 2006).

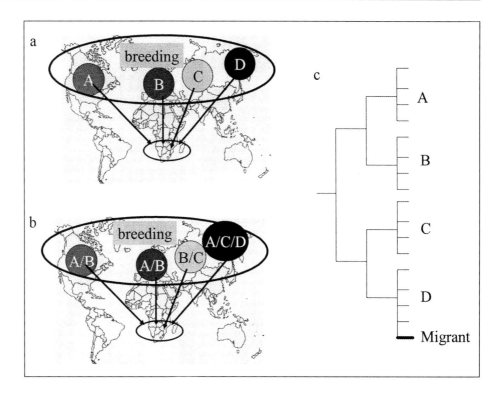

Figure 4. Distribution of haplotypes in geographically distinctive populations of a species that breeds in the northern hemisphere but winters in southern Africa.

Best-case scenario (a): all populations have unique haplotypes (C); therefore, a bird found in the wintering quarters (here with the haplotype D) unambiguously can be attributed to its breeding population D (C).

Realistic scenario (b): individual populations have more than one haplotype and share them with neighboring populations. A bird with the haplotype A can thus have originated from North America, Europe or East Asia.

(c) Cladogram of haplotypes found in scenario A.

DNA sequences have become an important tool for taxonomy and evolutionary studies including owls and diurnal raptors. Recent work in this area includes that of Wink (1995), Seibold and Helbig (1995a,b, 1996), Wink et al. (1996), Griffiths (1997), Matsuda et al. (1998), Wink and Heidrich (1999), Haring et al. (1999, 2001), Wink and Sauer-Gürth (2000, 2004), Groombridge et al. (2002), Riesing et al. (2003), Hendrickson et al. (2003), Godoy et al. (2004), Griffiths et al. (2004), Kruckenhauser et al. (2004), Pearlstine (2004), Roques et al. (2004), Roulin and Wink (2004), Gamauf et al. (2005), Helbig et al. (2005), Lerner and Mindell (2005), and Nittinger et al. (2005).

PCR Methods (Fingerprinting)

The analysis of genetic differences within a species demands methods that have a high degree of resolution. Sequences of mtDNA are sometimes uninformative at an intraspecific level. Because mtDNA is inherited maternally, hybridization and introgression can mask an unambiguous allocation of individuals to species, lineages and populations. To overcome these problems, molecular markers of ncDNA that are inherited by both sexes and that have a higher degree of resolution are more appropriate. These methods involve the amplification of polymorphic DNA markers by PCR and their separation by high-resolution gel electrophoresis (often on agarose, better on polyacrylamide gels) or by capillary electrophoresis (using a DNA sequencer).

DNA fingerprinting with VNTR or oligonucleotide probes has been employed to trace individuals for paternity and pedigree studies (Hoelzel 1992, Swatschek et al. 1993, 1994; Karp et al. 1998); it also can be applied to estimate adult mortality rates (Wink et al. 1999). Classical fingerprinting often has been replaced by microsatellite analysis (see below), that is more reliable and can be better automated.

Microsatellite (STR) Analysis

Each raptor has two alleles for each locus: one derived from the father, the other from the mother (Fig. 5). These alleles can be identical (homozygote) or not (heterozygote). As mentioned above, a vertebrate genome may contain more than 20,000 microsatellite loci that are characterized by 10- to 20- fold repeats of short-sequence elements, such as CA, TA, GACA, etc. The alleles of these loci show a high degree of length polymorphism. For each polymorphic STR locus, several alleles exist that differ in the number of tandem repeats; thus, they can be distinguished by size.

Because the sequences that flank microsatellite loci vary between species, special efforts are needed to identify sequences that can be used to amplify the STR loci. Several protocols have been published on generating species-specific STR sequences. A typical STR analysis is schematically illustrated in Fig. 5. A single locus provides information for two alleles; usually more than 8 to 10 polymorphic loci are needed to identify an individual unambiguously. For pedigree and population studies, 10 or more polymorphic STR loci are required. To reduce the number of PCR and sequencing runs, it is useful to establish a multiplex PCR system that allows the parallel analysis of several loci in one run.

Allele frequencies can be determined to characterize populations. If unique alleles can be identified within a population, they can help to assign an unknown individual to such a population. If unique alleles are not available, allele distributions are tabulated and allele frequencies calculated for any locus and population.

The presence and absence of alleles can be recorded in a 1/0 matrix and evaluated by cluster analysis (e.g., unweighted pair-group method with arithmetic means [UPGMA], neighbor-joining) and other programs (such as STRUCTURE or GENELAND). The result is a phenogram as shown in Fig. 4c. Individuals with similar patterns are clustered together in a clade. Examples for the use of STR-markers are found in Gautschi et al. (2000, 2003a,b), Nesje and Roed (2000a), Nesje et al. (2000), Nichols et al. (2001), Martinez-Cruz et al. (2002, 2004),

Mira et al. (2002), Hille et al. (2003), Kretzmann et al. (2003), Sonsthagen et al. (2004), Topinka and May (2004), Busch et al. (2005), and Wink et al. (2006).

Single Nucleotide Polymorphisms (SNP)

Information from single-nucleotide polymorphisms (SNPs) can be used to build a genetic map of populations, so long as at least 30 loci are determined for each individual. SNPs are analyzed in a similar way as STR data (i.e., via a 0/1 matrix) and have a similar resolution power (Lopez-Herraez et al. 2005). Because SNP marker systems have yet to be established for individual species, they are not available for raptors at this time. However, because SNP analysis can be automated via DNA chips and mass spectrometry, this method is likely to become an important tool in the future.

Amplified Fragment Length Polymorphism (AFLP) and Inter-Simple Sequence Repeats (ISSR)

If information on PCR primers of microsatellites is not available, genomic-fingerprint methods including amplified fragment length polymorphism (AFLP) and inter-simple sequence repeats (ISSR) provide an alternative.

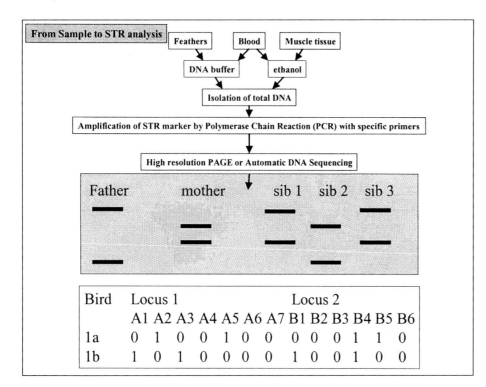

Figure 5. Illustration of the inheritance of STR markers. Microsatellite PCR products are analyzed by polyacrylamide gel electrophoresis (PAGE). The lower box illustrates a 1/0 matrix that can be constructed from STR data. It can be analyzed by phenetic methods that produce phenograms (similar to the cladogram shown in Fig. 4[c]).

Bird	Locus 1							Locus 2					
	A1	A2	A3	A4	A5	A6	A7	B1	B2	B3	B4	B5	B6
1a	0	1	0	0	1	0	0	0	0	0	1	1	0
1b	1	0	1	0	0	0	0	1	0	0	1	0	0

AFLP combines restriction-length analysis with PCR, making it a convenient and powerful tool. In the first step, DNA is digested by two restriction enzymes, MseI and PstI, which produce sticky ends. These sticky ends are ligated with oligonucleotide adaptors that recognize the restriction site and which also carry a PCR recognition sequence. Using specific PCR primers for the MseI and PstI adaptors, PCR fragments can be generated that relate to restriction fragments. These can then be separated by high-resolution polyacrylamide gel electrophoresis (PAGE) or capillary electrophoresis (Fig. 6). The result is a complex fingerprint that can be detailed in a 0/1 matrix and analyzed by cluster methods. AFLP loci are inherited co-dominantly. Examples for the application of AFLP analysis can be found in de Knijff et al. (2001) and Irwin et al. (2005).

ISSR produces fingerprints similar to AFLP. The procedure involves fewer experimental steps than AFLP and, therefore, is easier to carry out. ISSR uses a single PCR primer, whose sequence is identical to common microsatellite motives, such as (CA)10. Because such loci are widely present in genomes and occur in both orientations, a single primer is enough to amplify between 10 and 80 loci (i.e., DNA stretches between adjacent microsatellite loci) simultaneously. Because the PCR products differ in size they need to be analyzed by high resolution PAGE or capillary electrophoresis

(Fig. 7). The ISSR loci are inherited co-dominantly and, since some of them are polymorphic, they provide information on the genomic makeup of an individual. In practice, several ISSR primers are used, so that several hundred loci are available for analysis. The advantage of ISSR is that the primers work universally in most animals and plants. There is no need to define PCR primers for an individual species, such as in microsatellite analysis. The results are plotted in a 1/0 matrix and evaluated by cluster analysis (e.g., UPGMA) that places individuals together based on the similarity of their ISSR band patterns.

ISSR can reveal population specific DNA bands, which can be useful to trace back individual birds to individual populations (Wink et al. 2002). Beause ISSR loci are inherited by both sexes, this method also allows the analysis of hybrids and of sex (Wink et al. 1998a, 2000). ISSR markers also can be used to infer phylogenies of closely related taxa, such as genera (Wink et al. 2002, Treutlein et al. 2003a,b).

Molecular Sexing

Another useful molecular method for raptor systematic work is molecular sexing. This technique allows the sexing of birds, which can be difficult in monomorphic species outside the breeding season and in nestlings. In

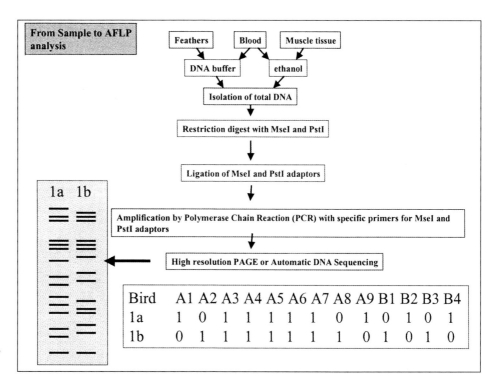

Figure 6. Illustration of the AFLP method.

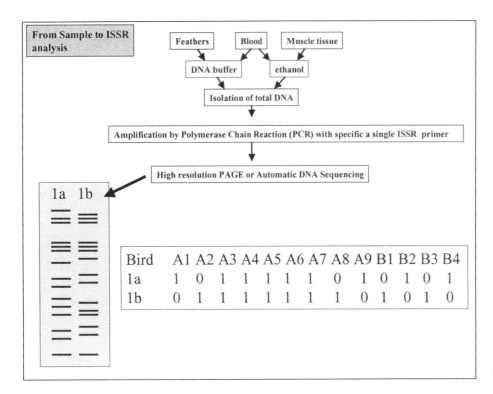

Figure 7. Illustration of the ISSR method.

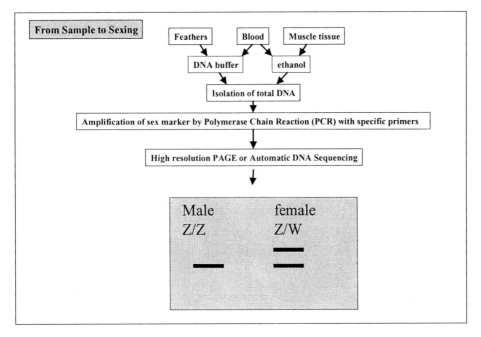

Figure 8. From sample to molecular sexing.

birds, sex chromosomes are in the opposite order as in mammals: females have heterogametic ZW chromosomes, whereas males are homogametic WW. PCR methods have been developed that target introns of the CHD gene present on the sex chromosomes. Because the alleles differ in size, two PCR products can be obtained in females as opposed to one in males (Fig. 8). Using high resolution PAGE, molecular sexing has been successful with all species of birds that have been examined to date (Kahn et al. 1998, Morrison and Maltbie 1999, Höfle et al. 2000, Nesje and Roed 2000b, Becker and Wink 2003, Ristow and Wink 2004, Ristow et al. 2004).

Consequences for Taxonomy and Systematics

According to the rules of cladistics, only monophyletic groups that are derived from a common ancestor should form taxonomic units such as genera, tribes, or families. Because convergent and adaptive characters have been used in traditional taxonomy, not all current taxonomic units are monophyletic. Molecular phylogenies that are less prone to convergence offer the opportunity to detect para- and polyphyletic groups. Examples of the latter found within raptors are vultures and eagles. It had been recognized earlier that New World vultures differ from Old World vultures. Within the Old World vultures, two major and unrelated clades could be determined unambiguously by DNA methods (Wink 1995). Thus, vultures are a polyphyletic assemblage that are adapted to a special lifestyle and have convergently developed certain characters. Eagles of the genus *Aquila* are paraphyletic in that eagles of the genus *Hieraaetus* and *Lophaetus* are not included in the genus despite ancestral relationships to it (Wink et al. 1996, Wink 2000, Wink and Sauer-Gürth 2000, 2004; Helbig et al. 2005). As a consequence, molecular phylogenies will help redefine monophyletic taxa; but this may lead to a change in genus names.

Another point of possible concern is that cryptic species (superficially identical sibling species) have been overlooked because of similar anatomy. Cryptic species appear to be more common in nocturnal than diurnal raptors (Olsen et al. 2002). The use of DNA barcoding will probably result in identifying new species of raptors and owls in the future.

Molecular data also can be used to determine the systematic status of species or subspecies. If subspecies show a high degree of both morphological and genetic differentiation, it is reasonable to treat them as "good" species (Helbig et al. 2002). Recent examples of taxa researchers have suggested should be split into two or more species include: Asian Imperial Eagle (*A. heliaca*) and Spanish Imperial Eagle (*A. adalberti*), and Bonelli's Eagle (*Hieraaetus fasciatus*) and African Hawk-Eagle (*H. spilogaster*) (Wink et al. 1996, Cardia et al. 2000, Wink 2000, Wink and Sauer-Gürth 2000, 2004; Helbig et al. 2005). These changes already have been widely accepted.

SUMMARY

The systematics of raptors has been studied in many ways. Molecular methods in use in evolutionary research also can be applied to taxonomy, phylogenetics, phylogeography, and the population genetics of raptors. These new methods are powerful tools that can supplement information obtained from morphology, geography, behavior, vocalizations, breeding biology, banding, telemetry, and isotope analysis.

LITERATURE CITED

ALBERTS, B., A. JOHNSON, J. LEWIS, M. RAFF, K. ROBERTS AND P. WALTER. 2002. Molecular biology of the cell, 4th Ed. Garland Science, New York, NY U.S.A.

ARCTANDER, P. 1988. Comparative studies of avian DNA by restriction fragment length polymorphism analysis: convenient procedures based on blood samples from live birds. *J. Ornithol.* 129:205–216.

AVISE, J. C. 1994. Molecular markers, natural history and evolution. Chapman and Hall, London, United Kingdom.

BECKER, P. AND M. WINK. 2003. Influences of sex, sex composition of brood and hatching order on mass growth in Common Terns *Sterna hirundo. Behav. Ecol. Sociobiol.* 54:136–146.

BED'HOM, B., P. COULLIN, W. GUILLIER-GENCIK, S. MOULIN, A. BERNHEIM AND V. VOLOBOUEV. 2003. Characterization of the atypical karyotype of the Black-winged Kite *Elanus caeruleus* (Falconiformes: Accipitridae) by means of classical and molecular cytogenetic techniques. *Chromosome Res.* 11:335–343.

BEEBE, T. AND G. ROWE. 2004. Introduction to molecular ecology. Oxford University Press, Oxford, United Kingdom.

BELTERMAN, R.H.R. AND L.E.M. DE BOER. 1984. A karyological study of 55 species of birds, including karyotypes of 39 species new to cytology. *Genetica* 65:39–82.

BIGGS, H.C., R. BIGGS AND A.C. KEMP. 1978. Measurement of raptors. Pages 77–82 in A. C. Kemp [Ed.], Proceedings from the Symposium on African Predatory Birds, Northern Transvaal Ornithological Society, Pretoria, South Africa.

BIRD, D.M., J. GAUTIER AND V. MONTPETIT. 1984. Embryonic growth of American Kestrels. *Auk* 101:392–396.

BROWN, L.H. AND D. AMADON. 1968. Eagles, hawks and falcons of the world, Vols. I & II. Hamlyn House, Feltham, Middlesex,

United Kingdom.

BRUSH, A.H. 1979. Comparison of egg-white proteins: effect of electrophoretic conditions. *Biochem. Syst. Ecol.* 7:155–165.

BUSCH, J.D., T.E. KATZNER, E. BRAGIN AND P. KEIM. 2005. Tetranucleotide microsatellites for *Aquila* and *Haliaeetus* eagles. *Mol. Ecol. Notes* 5:39–41.

CALDER, W.A., III. 1983. Ecological scaling: mammals and birds. *Annu. Rev. Ecol. Syst.* 14:213–230.

———. 1984. Size, function, and life history. Harvard University Press, Cambridge, MA U.S.A.

CARDIA, P., B. FRÁGUAS, M. PAIS, T. GUILLEMAUD, L. PALMA, M. L. CANCELA, N. FERRAND AND M. WINK. 2000. Preliminary genetic analysis of some western Palaearctic populations of Bonelli's Eagle, *Hieraaetus fasciatus*. Pages 845–851 *in* R.D. Chancellor and B.-U. Meyburg [EDS.], Raptors at risk. World Working Group on Birds of Prey and Owls, Berlin, Germany, and Hancock House Publishers, Surrey, British Columbia, Canada, and Blaine, WA U.S.A.

DE BOER, L.E.M. AND R.P. SINOO. 1984. A karyological study of Accipitridae (Aves: Falconiformes), with karyotypic descriptions of 16 species new to cytology. *Genetica* 65:89–107.

DE KNIJFF, P, F. DENKERS, N.D. VAN SWELM AND M. KUIPER. 2001. Genetic affinities within the Herring Gull *Larus argentatus* assemblage revealed by AFLP genotyping. *J. Mol. Evol.* 52:85–93.

DEL HOYO, J., A. ELLIOTT AND J. SARGATAL [EDS.]. 1994. Handbook of the birds of the world, Vol. 2. Lynx Edicions, Barcelona, Spain.

DESAI, J.H. AND A.K. MALHOTRA. 1980. Embryonic development of Pariah Kite *Milvus migrans govinda. J. Yamishina Inst. Ornithol.* 12:82–88.

ERICSON, P.G.P., C.L. ANDERSON, T. BRITTON, A. ELZANOWSKI, U.S. JOHANSON, M. KÄLLERSJÖ, J.I. OHLSON, T.J. PARSONS, D. ZUCCON AND G. MAYR. 2006. Diversification of Neoaves: integration of molecular sequence data and fossils. *Biol. Lett.* doi:10.1098/rsbl.2006.0523. www.systbot.uu.se/staff/c_anderson/pdf/neoaves.pdf (last accessed 5 January 2007).

FAIN, M.G. and P. HOUDE. 2004. Parallel radiations in the primary clades of birds. *Evolution* 58:2558–2573.

FINCH-DAVIES, C.G. 1919. Some notes on *Hieraaetus ayresi* Gurney Sen. (*Lophotriorchis lucani* Sharpe *et actorum*). *Ibis* 11:167–179.

FRANKHAM, R, J.D. BALLOU AND D.A. BRISCOE. 2002. Introduction to conservation genetics. Cambridge University Press, Cambridge, United Kingdom.

GAMAUF, A., J.-O. GJERSHAUG, N. ROV, K. KVALOY AND E. HARING. 2005. Species or subspecies? The dilemma of taxonomic ranking of some South-East Asian hawk-eagles (genus *Spizaetus*). *Bird Conserv. Int.* 15:99–117.

GAUTSCHI, B., I. TENZER, J.P. MULLER, AND B. SCHMID. 2000. Isolation and characterization of microsatellite loci in the Bearded Vulture (*Gypaetus barbatus*) and cross-amplification in three Old World vulture species. *Mol. Ecol.* 9:2193–2195.

———, G. JACOB, J.J. NEGRO, J.A. GODOY, J.P. MULLER AND B. SCHMID. 2003a. Analysis of relatedness and determination of the source of founders in the captive Bearded Vulture, *Gypaetus barbatus*, population. *Conserv. Genet.* 4:479–490.

———, J.P. MULLER, B. SCHMID AND J.A. SHYKOFF. 2003b. Effective number of breeders and maintenance of genetic diversity in the captive Bearded Vulture population. *Heredity* 91:9–16.

GODOY, J.A., J.J. NEGRO, F. HIRALDO AND J.A. DONAZAR. 2004. Phylogeography, genetic structure and diversity in the endangered Bearded Vulture (*Gypaetus barbatus*, L.) as revealed by mitochondrial DNA. *Mol. Ecol.* 13:371–390.

GRIFFITHS, A.J.F., W.M. GELBART, J.H. MILLER AND R.C. LEWONTIN. 1999. Modern genetic analysis. W.H. Freeman and Company, New York, NY U.S.A.

GRIFFITHS, C.S. 1997. Correlation of functional domains and rates of nucleotide substitution in cytochrome b. *Mol. Phylog. Evol.* 7:352–365.

———, G.F. BARROWCLOUGH, J.G. GROTH AND L. MERTZ. 2004. Phylogeny of the Falconidae (Aves): a comparison of the efficacy of morphological, mitochondrial, and nuclear data. *Mol. Phylogen. Evol.* 32:101–109.

GROOMBRIDGE, J J., C.G. JONES, M.K. BAYES, A.J. VAN ZYL, J. CARRILLO, R.A. NICHOLS AND M.W. BRUFORD. 2002. A molecular phylogeny of African kestrels with reference to divergence across the Indian Ocean. *Mol. Phylogen. Evol.* 25:267–277.

HALL, B.G. 2001. Phylogenetic trees made easy. Sinauer Associates, Sunderland, MA U.S.A.

HARING, E., M.J. RIESING, W. PINSKER AND A. GAMAUF. 1999. Evolution of a pseudo-control region in the mitochondrial genome of Palearctic buzzards (genus *Buteo*). *J. Zool. Syst. Evol. Res.* 37:185–194.

———, L. KRUCKENHAUSER, A. GAMAUF, M.J. RIESING AND W. PINSKER. 2001. The complete sequence of the mitochondrial genome of *Buteo buteo* (Aves, Accipitridae) indicates an early split in the phylogeny of raptors. *Mol. Biol. Evol.* 18:1892–1904.

HARRIS, T. AND C. WALTERS. 1982. Chromosomal sexing of the Black-shouldered Kite (*Elanus caeruleus*) (Aves: Accipitridae). *Genetica* 60:19–20.

HELBIG, A.J., A.G. KNOX, D.T. PARKIN, G. SANGSTER AND M. COLLINSON. 2002. Guidelines for assigning species rank. *Ibis* 144:518–525.

———, A. KOCUM, I. SEIBOLD AND M.J. BRAUN. 2005. A multi-gene phylogeny of aquiline eagles (Aves: Accipitriformes) reveals extensive paraphyly at the genus level. *Mol. Phylogen. Evol.* 35:147–164.

HENDRICKSON, S.L., R. BLEIWEISS, J.C. MATHEUS, L.S. DE MATHEUS, N.L. JÁCOME AND E. PAVEZ. 2003. Low genetic variability in the geographically widespread Andean Condor. *Condor* 105:1–12.

HILLE, S.M., M. NESJE AND G. SEGELBACHER. 2003. Genetic structure of kestrel populations and colonization of the Cape Verde archipelago. *Mol. Ecol.* 12:2145–2151.

HILLIS, D.M., C. MORITZ AND B.K. MABLE. 1996. Molecular systematics. Sinauer Associates, Sunderland, MA U.S.A.

HOELZEL, A.R. 1992. Molecular genetic analysis of populations. Oxford University Press, Oxford, United Kingdom.

HÖFLE, U., J.M. BLANCO, H. SAUER-GÜRTH AND M. WINK. 2000. Molecular sex determination in Spanish Imperial Eagle (*Aquila adalberti*) nestlings and sex related variation in morphometric, haematological and biochemical parameters. Pages 289–293 *in* J. T. Lumeij, J. D. Temple, P. T. Redig, M. Lierz, and J. E. Cooper [EDS.], Raptor biomedicine III. Zoological Education Network, Lake Worth, FL U.S.A.

HULL, J.M. AND D.J. GIRMAN. 2005. Effects of Holocene climate change on the historical demography of migrating Sharp-shinned Hawks (*Accipiter striatus velox*) in North America.

Mol. Ecol. 14:159–170.

IRWIN, D.E., S. BENSCH, J.H. IRWIN AND T.D. PRICE. 2005. Speciation by distance in a ring species. *Science* 307:414–416.

JOLLIE, M. 1976. A contribution to the morphology and phylogeny of the Falconiformes, part 1. *Evol. Theory* 1:285–298.

———. 1977a. A contribution to the morphology and phylogeny of the Falconiformes, parts 2–3. *Evol. Theory* 2:209–300.

———. 1977b. A contribution to the morphology and phylogeny of the Falconiformes, part 4. *Evol. Theory* 3:1–141.

KAHN, N.W., J.S. JOHN AND T.W. QUINN. 1998. Chromosome-specific intron-size differences in the avian CHD gene provide an efficient method for sex identification in birds. *Auk* 115:1074–1078.

KARP, A., P.G. ISAAC AND D.S. INGRAM [EDS.]. 1998. Molecular tools for screening biodiversity. Chapman and Hall, London, United Kingdom.

KLUG ,W. S. AND M. R. CUMMINGS. 1999. Essentials of genetics, 3rd Ed. Prentice Hall, Upper Saddle River, NJ U.S.A.

KRETZMANN, M.B., N. CAPOTE, B. GAUTSCHI, J.A. GODOY, J.A. DONAZAR AND J.J. NEGRO. 2003. Genetically distinct island populations of the Egyptian Vulture (*Neophron percnopterus*). *Conserv. Genet.* 4:697–706.

KRUCKENHAUSER, L., E. HARING, W. PINSKER, M.J. RIESING, H. WINKLER, M. WINK AND A. GAMAUF. 2004. Genetic versus morphological differentiation of old world buzzards (genus *Buteo*; Accipitridae). *Zool. Scr.* 33:197–211.

LEARNER, H.R. AND D.P. MINDELL. 2005. Phylogeny of eagles, Old World Vultures, and other Accipitridae based on nuclear and mitochondrial DNA. *Mol. Phylogenet. Evol.* 37:327–346.

LOPEZ-HERRAEZ, D., H. SCHÄFER, J. MOSNER, R. FRIES AND M. WINK. 2005. Comparison of the exclusion power between microsatellite and single nucleotide polymorphism (SNP) markers in individual identification and parental analysis in a Galloway cattle population. *Z. Naturforsch.* 60C:637–643.

MARTINEZ-CRUZ, B., V.A. DAVID, J.A. GODOY, J.J. NEGRO, S.J. O'BRIEN AND W.E. JOHNSON. 2002. Eighteen polymorphic microsatellite markers for the highly endangered Spanish Imperial Eagle (*Aquila adalberti*) and related species. *Mol. Ecol. Notes* 2:323–326.

———, J.A. GODOY, AND J.J. NEGRO. 2004. Population genetics after fragmentation: the case of the endangered Spanish Imperial Eagle (*Aquila adalberti*). *Mol. Ecol.* 13:2243–2255.

MASUDA, R., M. NORO, N. KUROSE, C. NISHIDA-UMEHARA, N. TAKECHI, T. YAMAZAKI, M. KOSUGE AND M. YOSHIDA. 1998. Genetic characteristics of endangered Japanese Golden Eagles (*Aquila chrysaetos japonica*) based on mitochondrial DNA D-loop sequences and karyotypes. *Zoo Biol.* 17:111–121.

MAYR, E. 1969. Principles of systematic zoology. McGraw-Hill Book Co., New York, NY U.S.A.

MINDELL, D.P. 1997. Avian molecular evolution and systematics. Academic Press, San Diego, CA U.S.A.

MIRA, S., C. BILLOT, T. GUILLEMAUD, L. PALMA AND M.L. CANCELA. 2002. Isolation and characterization of polymorphic microsatellite markers in Eurasian Vulture *Gyps fulvus*. *Mol. Ecol. Notes* 2:557–558.

MORRISON, J.L. AND M. MALTBIE. 1999. Methods for gender determination of Crested Caracaras. *J. Raptor Res.* 33:128–133.

NESJE, M. AND K.H. ROED. 2000a. Microsatellite DNA markers from the Gyrfalcon (*Falco rusticolus*) and their use in other raptor species. *Mol. Ecol.* 9:1438–1440.

——— AND K.H. ROED. 2000b. Sex identification in falcons using microsatellite DNA markers. *Hereditas* 132:261–263.

———, K.H. ROED, J.T. LIFJELD, P. LINDBERG AND O. F. STEEN. 2000. Genetic relationships in the Peregrine Falcon (*Falco peregrinus*) analysed by microsatellite DNA markers. *Mol. Ecol.* 9:53–60.

NICHOLS, R.A., M.W. BRUFORD AND J.J. GROOMBRIDGE. 2001. Sustaining genetic variation in a small population: evidence from the Mauritius Kestrel. *Mol. Ecol.* 10:593–602.

NITTINGER, F., E. HARING, W. PINSKER, M. WINK AND A. GAMAUF. 2005. Out of Africa: phylogenetic relationships between *Falco biarmicus* and other hierofalcons (Aves Falconidae). *J. Zool. Syst. Evol. Res.* 43:321–331.

OLSEN, J., M. WINK, H. SAUER-GÜRTH AND S. TROST. 2002. A new Ninox owl from Sumba, Indonesia. *Emu* 102:223–232.

OLSON, S.L. 1982. The distribution of fused phalanges of the inner toe in the Accipitridae. *Bull. Brit. Ornithol. Club* 102:8–12.

OTTE, D. AND J.A. ENDLER [EDS.]. 1989. Speciation and its consequences. Sinauer Associates, Sunderland, MA U.S.A.

PEARLSTINE, E.V. 2004. Variation in mitochondrial DNA of four species of migratory raptors. *J. Raptor Res.* 38:250–255.

PRAGER, E.M., A.M. BRUSH, R.A. NOLAN, M. NAKANISHI AND A.C. WILSON. 1974. Slow evolution of transferrin and albumin in birds according to micro-complement fixation analysis. *J. Mol. Evol.* 3:243–262.

RIESING, M.J., L. KRUCKENHAUSER, A. GAMAUF AND E. HARING. 2003. Molecular phylogeny of the genus *Buteo* (Aves: Accipitridae) based on mitochondrial marker sequences. *Mol. Phylogen. Evol.* 27: 328–342.

RISTOW, D. AND M. WINK. 2004. Seasonal variation in sex ratio of fledging Eleonora's Falcon, *Falco eleonorae*. *J. Raptor Res.* 38:320–325.

———, L. WITTE AND M. WINK. 2004. Sex determination of nestlings in Eleonora's Falcon (*Falco eleonorae*): plumage characteristics and molecular sexing. Pages 459–466 *in* R. D. Chancellor and B.-U. Meyburg [EDS.], Raptors worldwide. World Working Group on Birds of Prey and Owls, Berlin, Germany.

ROQUES, S., J.A. GODOY, J.J. NEGRO AND F. HIRALDO. 2004. Organization and variation of the mitochondrial control region in two vulture species, *Gypaetus barbatus* and *Neophron percnopterus*. *J. Heredity* 95:332–337.

ROULIN, A. AND M. WINK. 2004. Predator–prey polymorphism: relationships and the evolution of colour a comparative analysis in diurnal raptors. *Biol. J. Linn. Soc.* 81:565–578.

SCHMUTZ, S.M., J.S. MOKER AND T.D. THUE. 1993. Chromosomes of five North American buteonine hawks. *J. Raptor Res.* 27:196–202.

SEIBOLD, I. AND A. HELBIG. 1995a. Evolutionary history of New and Old World vultures inferred from nucleotide sequences of the mitochondrial cytochrome b gene. *Phil. Transact. Roy. Soc. London Series B* 350:163–178.

——— AND A.J. HELBIG. 1995b. Systematic position of the Osprey *Pandion haliaetus* according to mitochondrial DNA sequences. *Vogelwelt* 116:209–217.

——— AND A. HELBIG. 1996. Phylogenetic relationships of the sea eagles (genus *Haliaeetus*): reconstructions based on morphology, allozymes and mitochondrial DNA sequences. *J. Zool. Syst. Evol. Res.* 34:103–112.

———, A.J. HELBIG, B.-U. MEYBURG, J. NEGRO AND M. WINK.

1996. Genetic differentiation and molecular phylogeny of European Aquila eagles according to cytochrome b nucleotide sequences. Pages 1–15 *in* B.-U. Meyburg and R. D. Chancellor [EDS.], Eagle studies. World Working Group on Birds of Prey and Owls, Berlin, Germany.

SHIELDS, G.F. 1982. Comparative avian cytogenetics: a review. *Condor* 84:45–58.

SIBLEY, C.G. AND J.E. AHLQUIST. 1990. Phylogeny and classification of birds. Yale University Press, New Haven, CT U.S.A.

——— AND B.L. MONROE. 1990. Distribution and taxonomy of birds of the world. Yale University Press, New Haven, CT U.S.A.

SONSTHAGEN, S.A., S.L. TALBOT AND C.M. WHITE. 2004. Gene flow and genetic characterization of Northern Goshawks breeding in Utah. *Condor* 106:826–836.

STORCH V, U. WELSCH AND M. WINK. 2001. Evolutionsbiologie. Springer, Heidelberg, Germany.

SWATSCHEK, I., D. RISTOW, W. SCHARLAU, C. WINK AND M. WINK. 1993. Populationsgenetik und vaterschaftsanalyse beim Eleonorenfalken (*Falco eleonorae*). *J. Ornithol.* 134:137–143.

———, F. FELDMANN, D. RISTOW, W. SCHARLAU, C. WINK AND M. WINK. 1994. DNA-fingerprinting of Eleonora's Falcon. Pages 677–682 *in* B.-U. Meyburg and R. D. Chancellor [EDS.], Raptor conservation today. World Working Group on Birds of Prey and Owls, Berlin, Germany.

TARR, C.L. AND R.C. FLEISCHER. 1993. Mitochondrial DNA variation and evolutionary relationships in the Amakihi complex. *Auk* 110:825–831.

TOPINKA, J.R. AND B. MAY. 2004. Development of polymorphic microsatellite loci in the Northern Goshawk (*Accipiter gentilis*) and cross-amplification in other raptor species. *Conserv. Genet.* 5:861–864.

TREUTLEIN J., G.F. SMITH, B.-E. VAN WYK AND M. WINK. 2003a. Evidence for the polyphyly of *Haworthia* (Asphodelaceae, subfamily Alooideae; Asparagales) inferred from nucleotide sequences of rbcL, matK, ITS1 and genomic fingerprinting with ISSR-PCR. *Plant Biol.* 5:513–521.

———, G.F. SMITH, B.-E. VAN WYK AND M. WINK. 2003b. Phylogenetic relationships in the *Asphodelaceae* (subfamily *Alooideae*) inferred from chloroplast DNA sequences (*rbcL*, *matK*) and from genomic fingerprinting (ISSR). *Taxon* 52:193–207.

VALI, U. 2002. Mitochondrial pseudo-control region in old world eagles (genus *Aquila*). *Mol. Ecol.* 11:2189–2194.

WALTER, H. 1983. The raptor actigram: a general alphanumeric notation for raptor field data. *J. Raptor Res.* 17:1–9.

WILEY, E.O. 1981. Phylogenetics: the theory and practice of phylogenetic systematics. Wiley Interscience, New York, NY U.S.A.

———, D. SIEGEL-CAUSEY, D.R. BROOKS AND V.A. FUNK. 1991. The compleat cladist: a primer of phylogeny procedures. *Univ. Kans. Nat. Hist. Mus. Spec. Publ.* 19.

WILSON, A.C., H. OCHMAN AND E.M. PRAGER. 1987. Molecular time scale for evolution. *Trends Genet.* 3:241–247.

WINK, M. 1995. Phylogeny of Old and New World vultures (Aves: Accipitridae and Cathartidae) inferred from nucleotide sequences of the mitochondrial cytochrome b gene. *Z. Naturforsch* 50C:868–882.

———. 2000. Advances in DNA studies of diurnal and nocturnal raptors. Pages 831–844 *in* R. D. Chancellor and B.-U. Meyburg [EDS.], Raptors at risk. World Working Group on Birds of Prey and Owls, Berlin, Germany, and Hancock House Publishers, Surrey, British Columbia, Canada, and Blaine, WA U.S.A.

——— AND P. HEIDRICH. 1999. Molecular evolution and systematics of owls (Strigiformes). Pages 39–57 *in* C. König, F. Weick, and J. H. Becking [EDS.], Owls of the world. Pica Press, Kent, United Kingdom.

——— AND H. SAUER-GÜRTH. 2000. Advances in the molecular systematics of African raptors. Pages 135–147 *in* R. D. Chancellor and B.-U. Meyburg [EDS.], Raptors at risk. World Working Group on Birds of Prey and Owls, Berlin, Germany, and Hancock House Publishers, Surrey, British Columbia, Canada, and Blaine, WA U.S.A.

——— AND H. SAUER-GÜRTH. 2004. Phylogenetic relationships in diurnal raptors based on nucleotide sequences of mitochondrial and nuclear marker genes. Pages 483–498 *in* R. D. Chancellor and B.-U. Meyburg [EDS.], Raptors worldwide. World Working Group on Birds of Prey and Owls, Berlin, Germany.

———, P. HEIDRICH AND C. FENTZLOFF. 1996. A mtDNA phylogeny of sea eagles (genus *Haliaeetus*) based on nucleotide sequences of the cytochrome b gene. *Biochem. Syst. Ecol.* 24:783–791.

———, H. SAUER-GÜRTH, F. MARTINEZ, G. DOVAL, G. BLANCO AND O. HATZOFE. 1998a. Use of GACA-PCR for molecular sexing of Old World vultures (Aves: Accipitridae). *Mol. Ecol.* 7:779–782.

———, I. SEIBOLD, F. LOTFIKHAH AND W. BEDNAREK. 1998b. Molecular systematics of holarctic raptors (Order Falconiformes). Pages 29–48 *in* R. D. Chancellor, B.-U. Meyburg, and J. J. Ferrero [EDS.], Holarctic birds of prey. Adenex and World Working Group on Birds of Prey and Owls, Berlin, Germany.

———, H. STAUDTER, Y. BRAGIN, R. PFEFFER AND R. KENWARD. 1999. The use of DNA fingerprinting to determine annual survival rates in Saker Falcons (*Falco cherrug*). *J. Ornithol.* 140:481–489.

———, D. GUICKING AND U. FRITZ. 2000. Molecular evidence for hybrid origin of *Mauremys iversoni* Pritchard and McCord, 1991 and *Mauremys pritchardi* McCord, 1997 (Reptilia: Testudines: Bataguridae). *Zool Abh Staatl Mus Tierkunde Dresden* 51:41–49.

———, H. SAUER-GÜRTH AND E. GWINNER. 2002. A molecular phylogeny of Stonechats and related turdids inferred from mitochondrial DNA sequences and genomic fingerprinting by ISSR-PCR. *Brit. Birds* 95:349–355.

———, M. PREUSCH AND J. GERLACH. 2006. Genetische Charakterisierung südwestdeutscher Wanderfalken (*Falco peregrinus*). *DFO Jahrbuch* 37–47.

XIANG, W., S. YI, Y. XIAO-DONG, T. MIN-QIAN, W. LI, Y. YE-FE AND L. QING-WEI. 2004. Comparative study of mitochondrial tRNA gene sequence and secondary structure among fifteen predatory birds. *Acta Genetica Sinica* 31:411–419.

ZUCKERKANDL, E. AND L. PAULING. 1965. Molecules as documents of evolutionary history. *J. Theor. Biol.* 8:357–66.

Study Design, Data Management, Analysis, and Presentation

4

JAMES C. BEDNARZ

Department of Biological Sciences, Arkansas State University
Jonesboro, AR 72467 U.S.A.

INTRODUCTION

Below, I cover a variety of mandatory, often relatively mundane, and generally not particularly exciting tasks that are required to accomplish meaningful research on raptors. However, attention to research design, data management, implementation of reasonable analytical approaches, and the publication or presentation of research results probably represent the most fundamentally important aspects of the effort to advance science in any area of interest.

The stark dichotomy between the tasks of data collection related to raptors, which often involve working in remote wilderness conditions while engaging in activities characterized as extreme outdoor adventure (e.g., rappelling down cliffs, handling eagles), and sitting in an office managing data sets makes the latter seem banal at times. Our ultimate purpose in conducting research on raptors, however, is to enhance our understanding of these unique animals and their spectacular adaptations, and to ensure their conservation. To accomplish this we must perform the latter as well as the former. To do this effectively, and to allow researchers more time to participate in the more exciting aspects of raptor research, the tasks of data management and analysis and write-up should be executed with maximum efficiency.

Here, I provide guidance in this regard and suggest more in-depth treatments of various aspects of the broad areas of research design, data management, analysis, and presentation of results to aid raptor biologists wishing to increase their efficiency. The chapter is designed to be especially useful for individuals relatively new to ornithological or ecological research (e.g., graduate students), but also may represent a worthwhile read to more experienced researchers who want to evaluate continually and improve their research efficiencies. Specifically, this chapter represents a brief outline of "how to conduct raptor science."

Why Study Raptors?

The first recommendation that I will offer is that you should consider working with another model rather than a raptor! I say this because science is basically the pursuit of new knowledge, and raptors, by their very nature, are inherently difficult to study (i.e., to obtain knowledge about and from). Raptors can be hard to find, hard to observe, in part because they occur in extremely low densities overall (sometimes <1 pair/100 km^2), and in part because many are found in nearly inaccessible situations (e.g., on the tops of the tallest trees or on huge vertical cliffs). By committing to study raptors, one of the greatest challenges is obtaining a large enough sample size from which to say anything meaningful. For many basic questions related to biology or ecology, it would be more productive to study a small, abundant bird or mammal. And, one should at least consider this before investing further in studying raptors. By electing to spend time, or a lifetime, studying birds of prey, raptor researchers "choose" to become inherently challenged scientists.

On the other hand, I also would argue that raptors can and do often make ideal models to study a number of interesting biological and ecological questions. Their appearance, their action-filled and risky life of predation, and the many mysteries of their lives represent intrinsic values that give raptors appeal as research subjects. Moreover, birds of prey commonly are used as national and cultural symbols and mascots, and are of great interest to the public, especially related to bird watching and falconry. Many people simply want to know more about these "cool" birds and are willing to buy books and watch videos about raptors. This public demand for knowledge about raptors provides raptor researchers with a large audience. Certainly the fascination the public has for raptors has led to support for laws (e.g., the Bald and Golden Eagle Protection Act in the U.S.A.) to conserve them. These laws, in turn, require knowledge of raptors to guide the implementation of conservation and management programs.

I would further argue that raptors might provide one of the best models for studying certain questions in ecology. Jaksic (1985), for example, has made a strong case that assemblages of raptors may represent some of the best model systems available to study the influence of competition among species and several additional aspects of community ecology. The lack of predators of many raptors eliminates the potentially major confounding influence of predation when addressing community ecology questions. Also, the fact that many raptors rely on well-studied vertebrates for their food resources allows biologists to more thoroughly document and understand trophic relationships than may be accomplished on many small birds, mammals, and predatory insects that consume smaller insects that are difficult to identify; prey groups for which there is poor consensus on how to assess density effectively. Extensive work on small birds (e.g., Bibby et al. 2000) and mammals (Lancia et al. 2005) has resulted in the development of a variety of methods that may produce reliable estimates of population densities or abundance. As the availability of food resources is key to understanding the ecology of any organism (e.g., Lack 1968), raptors, potentially, may provide a more informative model animal than many alternative smaller organisms.

Another area in which raptors provide a useful research model is in the investigation of brood reduction and sibling interactions. Because raptors represent a group of birds in which nestlings are equipped with

weapons (talons) that can kill nest mates, they have the capacity for intense intra-brood aggression that could lead to siblicide, and provide one of the few animal models that may be studied to understand such interactions (Mock and Parker 1997). Other topics in which raptors may provide one of the more effective research models, include predation ecology, migration strategies, reversed sexual size dimorphism in higher vertebrates, and the evolution of various forms of cooperative breeding, especially cooperative polyandry (Kimball et al. 2003). My point is that whereas raptors, in general, make a poor model to conduct research into basic biology, they also may provide one of the better research models to investigate some key and contemporary behavioral ecology questions of substantial interest to science. Moreover, sharing the top of the food chain with humans renders raptors invaluable for research on the biomagnification and impact of various pollutants in our environment (see Chapter 18).

THE TWO KEYS TO SUCCESS IN RAPTOR RESEARCH

If you do decide to conduct raptor research, you should do it in a way that maximizes your potential success. I submit that there are two primary and fundamental elements to successful research in raptor science, and that these elements also apply to success in any ecological study. These are (1) an innovative research idea, approach, or both and (2) sample size, sample size, sample size.

To advance the discipline, new and novel ideas are required to guide the collection of data and to move our paradigms (i.e., basic scientific theories and methods) and general knowledge of raptors forward. Science by its very nature and emphasis on replication forces us all to conduct mostly normal science (*sensu* Kuhn 1962). There always will be a need to replicate studies on firmly held paradigms or conduct an investigation on a local question that has been investigated thoroughly elsewhere. Some examples of normal science would include: (1) comparing the diets of two or more species of raptor co-existing in a specific area, (2) determining the "habitat requirements" of a species of conservation concern, and (3) examining most aspects of raptor productivity or nesting biology. In such cases, raptor scientists are using long-held paradigms and are filling in small gaps in our knowledge base. I maintain that conducting normal science is productive and necessary, and as such is a worthwhile endeavor for all raptor biologists. As raptors are generally poorly known in many parts of the world, much of our work requires careful description of their natural history, and this, clearly, is normal science. I would classify most undergraduate and Master's theses as normal science.

By successfully conducting normal science, a biologist does achieve some level of success. However, truly major advances in the discipline require novel ideas that challenge long-established paradigms and that stimulate cutting-edge and exciting investigations by the scientific community. This is revolutionary science (Kuhn 1962). To be most successful in science and to advance our disciplines most dramatically, we all should try to participate in revolutionary science as well as normal science. Recent examples of thinking "outside the box" that could be classified as revolutionary science include Brandes and Ombalski's (2004) use of laminar fluid-flow models to understand and predict migration pathways used by Golden Eagles (*Aquila chrysaetos*) and Ellis and Lish's (2006) ideas on specific adaptations of the patterns and pigment deposition in eagle rectrices. In my own thinking, I am beginning to question our heavy reliance on the importance of habitat features (which are usually taken to mean vegetation features) to the population viability of several raptors. An alternative paradigm may involve the fact that raptors are highly site-faithful and that their population viability and success may be more tied to gaining experience and improving hunting skills (see Dekker and Taylor 2005) on one given territory regardless of specific habitat features there, and that our long-held paradigm that habitat (i.e., vegetation, topography, etc.) is vitally important to conservation of raptors may not be true in all cases (also see Ahlering and Faaborg 2006). Another interesting idea that warrants investigation is the long-term influence of sibling interactions and competition in raptor nests—does the alpha chick enjoy the lifelong benefits of being a "winner?" Is the runt in each brood destined to be a "loser," evolutionarily and otherwise? Importantly, I would encourage all biologists alike to strive to conduct revolutionary science in raptors by testing long-held assumptions, challenging conventional wisdom, employing innovative tests of hypotheses, developing new paradigms, and thinking outside the box.

Scientific Method

Designing a well-reasoned study is vital to the effective

completion of any fieldwork and subsequent data analysis. Although there are many ways of acquiring knowledge (Kerlinger 1973), the most commonly accepted approach in the ecological sciences is use of the hypothetico-deductive method. This approach was developed through the 1900s and was popularized through the works of Popper (1959, 1968) and others. In brief, the hypothetico-deductive approach involves identifying a research problem or question, developing alternative explanatory research hypotheses, deriving logical and testable predictions from the hypotheses, and then implementing the experimental test. Tests of hypotheses may be either observational or manipulative, although manipulative experiments generally are more powerful (see Diamond 1986). There are several excellent review papers (Romesburg 1981, James and McCulloch 1985, Eberhardt and Thomas 1991, Sinclair 1991, Ford 2000, Garton et al. 2005, and others) that describe the scientific method as it applies to raptor research and related disciplines, and I will not repeat that information here. I recommend that all raptor researchers read Romesburg (1981) and Garton et al. (2005) at a minimum. Raptor biologists have been slow to adopt the hypothetico-deductive method into the practice of raptor science (see Guthery et al. 2004); perhaps, in part, because there is still much basic natural history to describe in raptors worldwide. However, I advocate that the use of the hypothetico-deductive approach is long overdue in raptor biology, and that all raptor scientists should be conducting problem-based research by testing research hypotheses. Descriptive natural history data should and can be easily collected simultaneously by taking detailed observation notes while testing both basic and applied hypotheses related to raptor biology and conservation.

In implementing this scientific method, I find that there often is confusion, especially among students, between research hypotheses and statistical hypotheses (Guthery et al. 2001). Good explanations of differences between these terms are provided by James and McCulloch (1985) and Ratti and Garton (1996). Specifically, a research hypothesis is an explanatory answer or conceptual model that answers the research question. A statistical hypothesis is a derived testable prediction that guides the collection of specific data. For example, we may hypothesize that human disturbance is the cause of low reproductive success in a population of Ferruginous Hawks (*Buteo regalis*). A derived testable prediction that may be used to evaluate this hypothesis might be that experimentally applied pedestrian intrusions at a random sample of nests would result in significantly lower reproductive success (i.e., fledglings per nest) than at a comparable control sample of nests. The null statistical hypothesis in this case would be the expectation of no statistical difference in the mean number of fledglings produced between experimentally disturbed and control nests. In both their understanding and presentation of research results, biologists should clearly distinguish between research hypotheses and testable predictions or statistical hypotheses.

In designing studies, all raptor researchers need to be aware and careful of the potential of pseudoreplication and the inappropriate use of inferential statistics. (I recommend a thorough reading and understanding of Hurlbert 1984.) Pseudoreplication is the use of traditional null-hypothesis statistics to test for treatment effects from experiments in which either the treatments are not replicated (although samples are often replicated) or the replicates are not statistically independent. For example, a researcher may be interested in assessing the effects of petroleum exploration on nesting raptors by comparing the reproductive success in an area of development to that in a similar area where there is no development. Use of statistics to compare the reproductive success of large samples of nests in these two areas would be an obvious example of pseudoreplication. In this case, samples are replicated (i.e., nests), but the treatment is not (i.e., one area of development compared to one undisturbed reference area). The use of statistics may be appropriate to compare the impact of this development relative to the single reference area, but it would be inappropriate for the authors of this study to extrapolate their results to other areas of petroleum exploration.

I cannot overemphasize that a well-reasoned and carefully developed study design guides the collection and subsequent management of data. Specifically, the test predictions should clearly identify the key data that should be collected. Also, at this early stage of the research process, investigators should consider the appropriate statistical tests to be used. Are parametric or nonparametric methods more appropriate (see Potvin and Roff 1993, Smith 1995, Johnson 1995)? Often with a clean experimental design, researchers can develop an analysis of variance (ANOVA) model designating specific experimental effects and covariates (ANCOVA) that can be considered in the model before the research is implemented.

For example, below I present a potential ANOVA model associated with a proposed experimental design for a hypothetical study on the effects of pedestrian and ATV traffic on nesting Ferruginous Hawks. For this study, the dependent variable would be the distance from an experimental disturbance at which the hawk flees the nest. The potential effects terms and error term included would be as follows:

$$F = \mu + Ai + Bj(i) + Ck(ij) + Dl(ijk) + E(ijkl)m,$$

where F = flee distance from disturbance, A = disturbance type (i = pedestrian or ATV disturbance or no disturbance [control]), B = breeding stage (j = incubation or brood-rearing period), C = nest substrate type (k = cliff nest or tree nest), D = vegetation type (l = open-grassland or shrubland), E = error term, and m = replicates.

Also, at this stage investigators should consider the applicability of using information-theoretic methods in which a set of alternative models are evaluated based on Akaike Information Criteria (AIC; Anderson et al. 2000, Burnham and Anderson 2002). I recommend that in addition to peer evaluation during the research project development stage that a statistician be consulted before field data are collected.

The Magic Window

Once you have a good idea or have developed a new approach to test an old idea, then the other key ingredient for success in ecological or raptor research is to collect an adequate sample size of data. This is especially challenging with raptors because of their inherent attributes: they are wary of human observers, exist at low densities, are wide-ranging, and often occur in inaccessible locations. Moreover, with almost every raptor research project that I have been involved with there is a limited "magic window" during which data can be collected most effectively. That window may be limited to a few weeks during the breeding season or to just a few hours when conditions are right to capture the critical individual(s) for which data are needed. For example, for most temperate stick-nesting raptors there is a critical window for nest finding spanning the period when the hawks begin building their nests and when the trees leaf out. Depending on the species and circumstances, that window may be 3 weeks or less, after which it becomes very difficult to find occupied nests. Thus, the researcher's sample size depends on their effectiveness in that nest-finding window. I also have found, especially during migration, that there are often prime periods to capture and mark raptors. In other words, if the research depends on marked individuals, the researcher must identify those conditions and then take maximum advantage of capturing and marking birds during this window of opportunity. Thus, raptor researchers must be aware of their magic window of data collection and do all they can to take advantage of that period. This requires successful researchers to be extremely efficient and organized and to be willing to work those extra hours (e.g., when raptors are catchable) and those extra days and weekends (when your prime window of collection of key data [e.g., finding of nests] is relatively limited). To my knowledge, raptors never take a holiday, and data-collection opportunities are often lost if the raptor researcher takes a break during the magic window. Simply put, a strong work ethic is required by successful raptor biologists to take advantage of these fleeting periods of opportunity to augment sample sizes.

Finally, in my experience, most raptor science manuscripts are rejected from scientific journals because of "inadequate sample size." The only fix to this problem is to work hard and maximize the efficiency of the available resources during the magic window of data collection. Time devoted to the management of data and maximizing data collection efficiency also will aid researchers in taking the best advantage of the data collection window.

Organization

The existence of magic windows for data collection means that organizing time in the field is of paramount importance. Researchers should devise general season-long and more specific week-to-week plans of required tasks or activities. In my own research projects, we have commonly used over-sized calendars, chalkboards, or dry boards to plan the specific research tasks that need to be accomplished within the next 7 to 10 days. Such planning takes into account priorities such as checking all occupied nests at 3-day intervals, regular replacement of tapes and batteries at time-lapse video cameras located at nests, maintaining standard intervals of monitoring radio-instrumented hawks, and other required tasks demanded by the research study design. Whether the project is large (>10 investigators and technicians) or small (a single graduate student), I recommend that key project investigators should take time to review study needs and priorities at least weekly, and develop a task plan for the coming 7 to 10 days. The plan should emphasize priority and time-critical tasks, allocate time for lower-priority tasks as available, and involve input of all members of the study team. Also, the plan should allow for contingencies (e.g., nest or transmitter failures, inclement weather that cancels field work) and be adjusted when those events arise. Importantly, this plan of tasks should be written down.

DATA

I believe that one lost art in this day and age is the practice of writing accurate and complete field notes. I have been in the field checking on nests or research sites of interest with several graduate students, who never once jotted down what we saw in a field notebook. Do they remember the data (e.g., that we saw two chicks about three weeks old in a nest and the adult flew in with a frog) and record this vital information later? Can that researcher remember what they saw and record that information accurately later in the evening? Or, do memories fade and become confused as additional information is observed and the researcher tries to retain more facts before "downloading" the information into their field notebook. Every raptor researcher should maintain a complete and accurate field notebook with all facts recorded as soon as possible after observation. Exceptions would be data recorded on pre-prepared data sheets (see below).

Field researchers should obtain a suitable field notebook before the first scheduled field day. In my lab, we use low-cost, "Rite in the Rain" all-weather No. 350 field books (J.L. Darling Corp., Tacoma, WA U.S.A), which have bright yellow covers and waterproof paper. I generally employ a system of taking notes similar to that which was originally developed by Joseph Grinnell and is known as the "Grinnellian system" (Herman 1986). In brief, each page should have the current date on the top line; location information should be given on the second line (use multiple lines if necessary), followed by field observations (Fig. 1). Importantly, field notes should be entered immediately after observations are made as needed during the course of a field day. Observations should be recorded on sequential pages in chronological order, each page with the date indicated on the top line (undated field-note pages can often lead to confusion as to which date applies to which page). Include every detail and record the maximum amount of quantitative information possible (e.g., numbers, estimated distances, directions, duration of events or behavioral patterns in seconds or minutes as appropriate). I encourage frequent use of sketches of locations,

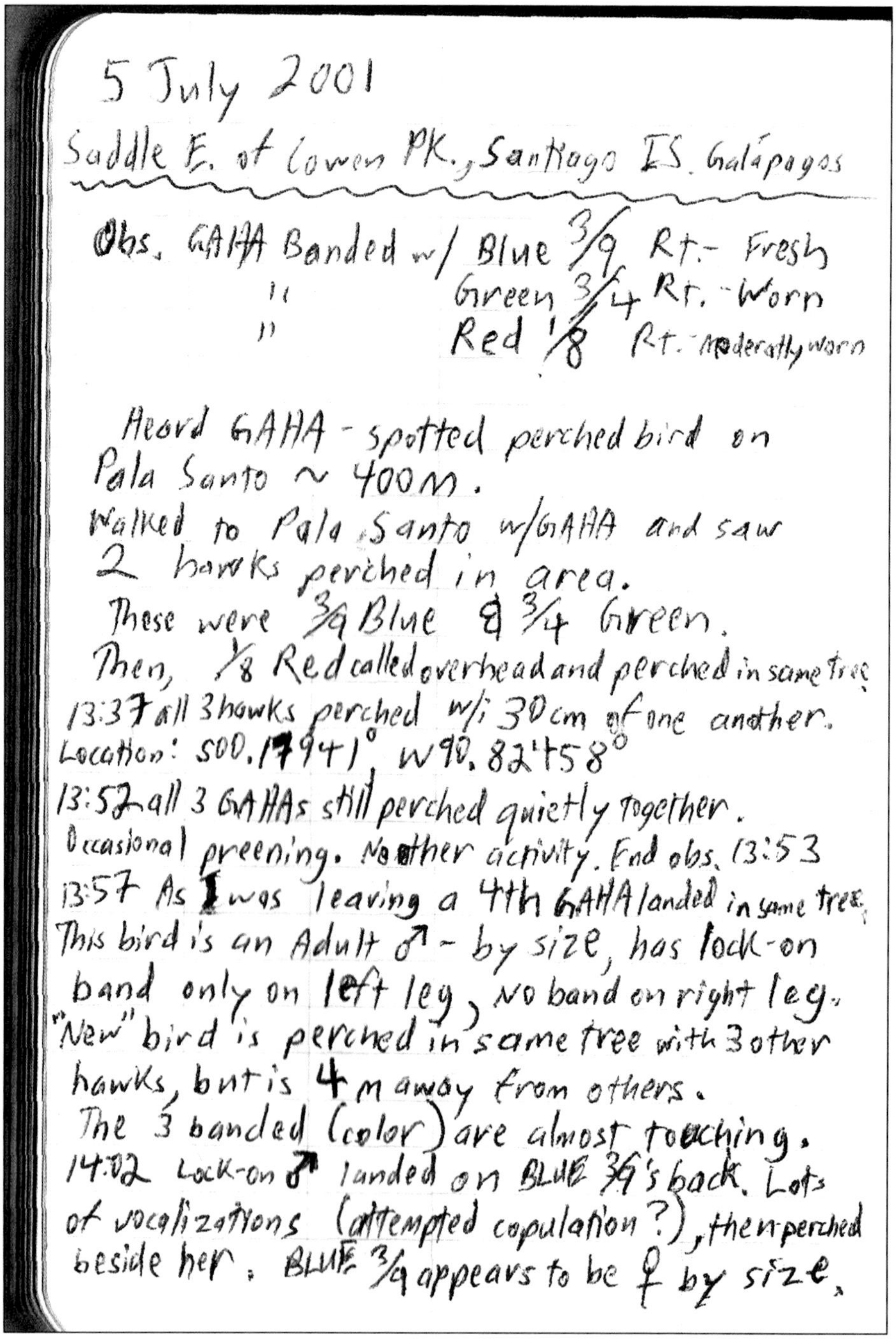

5 July 2001
Saddle E. of Cowen PK., Santiago IS. Galápagos

Obs. GAHA Banded w/ Blue 3/9 Rt. - Fresh
    "   Green 3/4 Rt. - Worn
    "   Red 1/8 Rt. Moderately worn

Heard GAHA - spotted perched bird on
Pala Santo ~ 400M.
Walked to Pala Santo w/GAHA and saw
2 hawks perched in area.
These were 3/9 Blue & 3/4 Green.
Then, 1/8 Red called overhead and perched in same tree.
13:37 all 3 hawks perched w/i 30cm of one another.
Location: S00.1794 1° W90.82'158°
13:52 all 3 GAHAs still perched quietly together.
Occasional preening. No other activity. End obs. 13:53
13:57 As I was leaving a 4th GAHA landed in same tree.
This bird is an Adult ♂ - by size, has lock-on
band only on left leg, No band on right leg.
"New" bird is perched in same tree with 3 other
hawks, but is 4m away from others.
The 3 banded (color) are almost touching.
14:02 Lock-on ♂ landed on BLUE 3/9's back. Lots
of vocalizations (attempted copulation?), then perched
beside her. BLUE 3/9 appears to be ♀ by size.

Figure 1. One page of field notes recorded by J. Bednarz during an investigation of Galápagos Hawks (*Buteo galapagoensis*) on Santiago Island on 5 July 2001. The assigned task that resulted in these notes was to obtain band reads (Acraft Co., Edmonton, Alberta, Canada) from the Peregrino Galápagos Hawk group that traditionally inhabits a territory on Cowen Peak.

cliff sites, maps, unique characteristics of birds, etc. Your field notebook also is a good place to record the names and contact information for people you encounter in the field (e.g., a landowner of the property that contains a raptor nest that you are monitoring), appointments you make while in the field, list of work tasks to be done, research supplies needed, and other vital information you need to accomplish the research. On the back pages of your field book, you might staple or attach critical research information that you need

while conducting field research, such as a list of nest locations, band combinations of known study birds, frequencies of radio-tagged hawks, or available color band combinations for newly captured birds.

At the end of the field season, I carefully read through all my field notes again, number all pages starting at page one, and make an index of general topics (Tables 1 and 2). A thorough reading of the field notes at the end of a study season gives you a good sense of the successes and setbacks encountered, allows you to

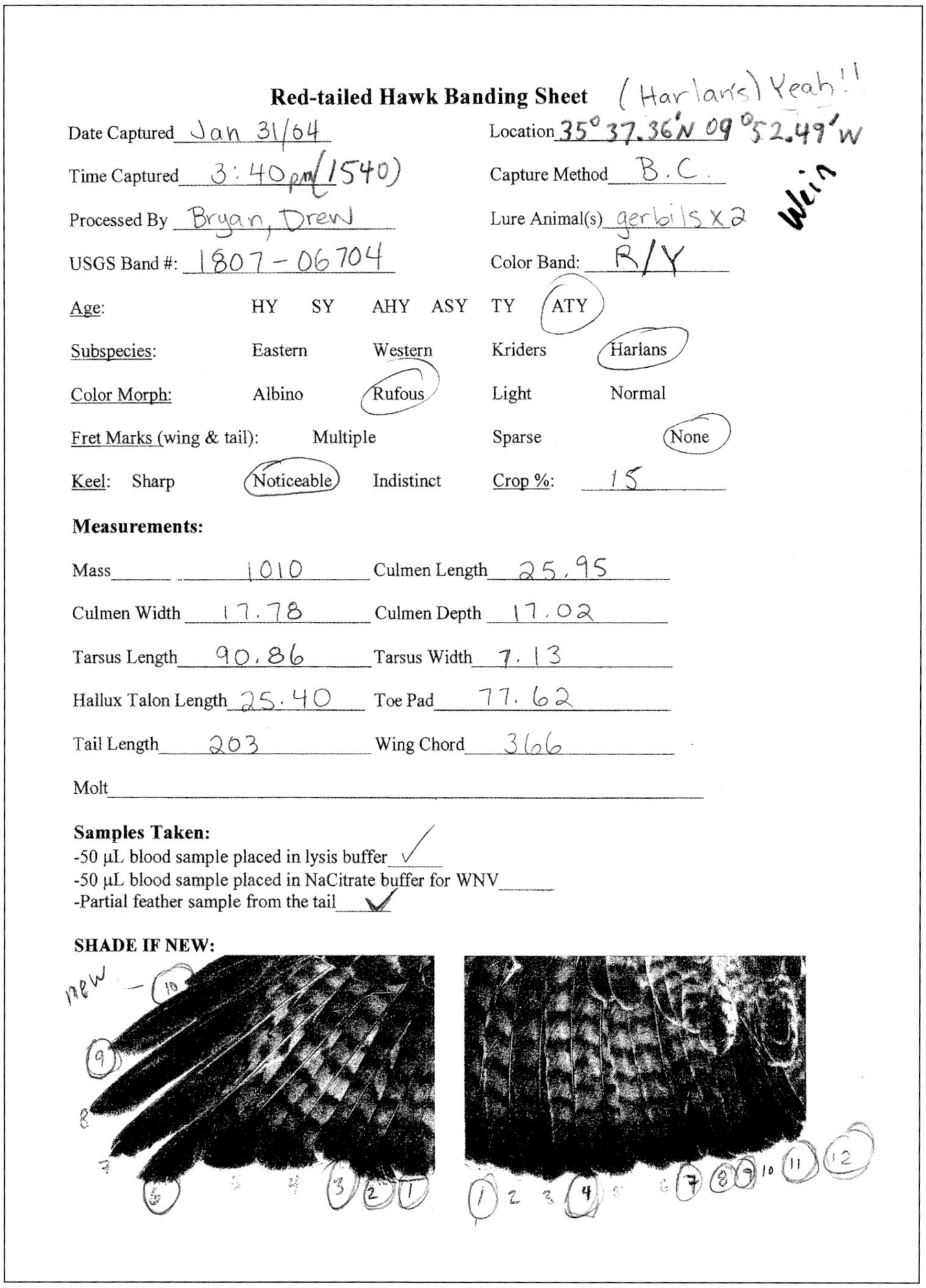

Red-tailed Hawk Banding Sheet (Harlans) Yeah!!

Date Captured Jan 31/04 — Location 35° 37.36'N 09° 52.49'W

Time Captured 3:40 pm (1540) — Capture Method B.C.

Processed By Bryan, Drew — Lure Animal(s) gerbils x 2

USGS Band #: 1807-06704 — Color Band: R/Y

Age: HY SY AHY ASY TY (ATY)

Subspecies: Eastern Western Kriders (Harlans)

Color Morph: Albino (Rufous) Light Normal

Fret Marks (wing & tail): Multiple Sparse (None)

Keel: Sharp (Noticeable) Indistinct Crop %: 15

Measurements:

Mass 1010 — Culmen Length 25.95

Culmen Width 17.78 — Culmen Depth 17.02

Tarsus Length 90.86 — Tarsus Width 7.13

Hallux Talon Length 25.40 — Toe Pad 77.62

Tail Length 203 — Wing Chord 366

Molt

Samples Taken:

-50 μL blood sample placed in lysis buffer ✓

-50 μL blood sample placed in NaCitrate buffer for WNV

-Partial feather sample from the tail ✓

SHADE IF NEW:

Figure 2. A completed, pre-printed data sheet for capturing and marking Red-tailed Hawks (*Buteo jamaicensis*) in the winter period.

identify discrepancies that still may be rectified, brings to mind important information that you may have forgotten, and helps you consolidate the information needed for preparation of end-of-season reports or planning of the next field season. The preparation of an index of field notes is a huge time-saver. This is particularly so when additional data are needed to conduct a new analysis, when an emerging issue needs to be addressed, or when one needs to contact an individual (e.g., a landowner) encountered in the field.

In addition to complete field notes, use of pre-printed data sheets is an effective way to ensure that you collect all data desired. Pre-printed data sheets also provide an excellent means to organize data for later processing. Pre-printed sheets or cards may be used for recording data when processing trapped hawks (Fig. 2), visiting nests, taking telemetry fixes, and recording vegetation types or habitat sampling. I strongly recommend use of pre-printed data sheets whenever possible. If pre-printed data sheets are well prepared, all data related to a

Table 1. Abbreviated field notebook index for a research project at the Los Medaños area of New Mexico, U.S.A. in 1987.

Topic	Field Notebook Pages	Topic	Field Notebook Pages	Topic	Field Notebook Pages
Band		Radiotelemtry		Tethered prey blind watch	21, 23, 25, 26, 39, 40,
Destroyed	148	Hawk No. 322	3, 8, 9, 11, 13, 17		42, 300, 301, 302, 303
Recovered	91, 127, 318	Hawk No. 755	3, 7, 18, 19, 27, 33, 35,		
			38, 46, 53, 58, 65, 75,	Transmitter	
Barn Owl nests	73, 74, 84, 93, 99, 107,		77, 78, 86, 90, etc.	Mounted	23, 41, 52, 60, 82, 181,
	114, 152, 153, 195, 211,		*(Pages related to*		188, 189, 255, 281, 284,
	234, 280		*telemetry listed for 15*		306
			additional hawks as	Recovered	4, 58, 66, 67, 218, 224,
Burrowing Owl nests	24, 96, 106, 128, 141,		*above.)*		296
	142, 172, 178, 198, 200,				
	201, 205, 231, etc.	Raptor aggressive		Vegetation transect	288, 294, 295, 296
		interaction	18, 19, 39, 97, 249, 250		
Dead raptor	8, 53, 55, 66, 127, 233,			People and contacts	
	236, 289, 318, 321	Raptor captures		Dee Armstrong	170
		American Kestrel	305	Jack Barnitz	160
Emlen census	116, 122, 133, 150, 159,	Barn Owls	93, 211, 247	Larry Blum	37, 44
	170	Great Horned Owls	21, 23, 38, 88, 190, 255,	Marc Bluhm	68
			266, 268, 277, 304, 307	John Brininstool	286
Great Horned Owl nests	see Table 2	Harris's Hawks	1, 2, 23, 41, 52, 57,	Joneen Cockman	285
			60,82, 181, 188, 189,	Tim Fischer	19, 122, 170, 285, 311
Harris's Hawks			255, 280, 306	Tay Gerstel	134, 184, 187, 197, 202,
Banded bird obs.	4, 11,15, 22, 23, 24, 27,	Red-tailed Hawks	21, 25		203, 208, 209
	33, 42, 53, 56, 59,	Screech Owl	120	Stuart Jones	115
	72,87, 92, 95, 104, etc.	Swainson's Hawk	187	Jess Juen	1, 9, 36, 40, 112,
Copulation	45, 49, 59, 86, 99			Bob Kehrman	37, 48
Nest blind watch	106, 108, 109, 110, 115,	Raptor		Bill Iko	81
	116, 119, 120, 123, 126,	Census	6, 7, 10, 13, 16, 20, 22,	David Ligon	192, 193, 197
	127, 129, etc.		27, 29, 33, 34, 42, 45,	Danna Stretch	124, 126, 127, 128
Nest building	47		49, 53, 54, 57, etc.	Steve West	124, 125
Observations	11, 16, 24, 25, 33, 43,	Hunting	49, 274, 302, 315, 319,	Don York	113
	52, 56, 62, 74, 79, 89,		325		
	104, 147, 148, etc.	Nest	169		
		Trapping	1, 4, 20, 22, 25, 31, 38,		
Harris's Hawk nests	see Table 2		39, 41, 43, 51, 55, 56,		
			59, 62, 63, 76, etc.		
Injured raptor	18, 161, 194, 305				
		Raptor with prey	19, 27, 29, 30, 31, 35,		
Laparotomy	192, 195, 196		39, 40, 45, 46, 48, 59,		
			63, 65, 72, 86, etc.		
Nest					
Platform	48, 50, 51, 72, 78, 109,	Raven nest	92, 116, 127, 128, 129,		
	112, 167, 171, 172, 173,		131, 135, 136, 137, 138,		
	193, 213		139, 140, 141, etc.		
Predation	123, 130				
		Screech Owl nest	74, 93, 114, 118, 120,		
Other bird observations	56, 60, 85, 89, 90, 91,		125, 152		
	137, 201, 267, 281, 283,				
	293, 294, 302, etc.	Swainson's Hawk			
		Mist netting	184, 187, 197		
Rabbit census	12, 14, 28, 34, 46, 69,	Rehabilitated hawk	285		
	97, 108, 124, 133, 168,				
	174, 185, 212, etc.	Swainson's Hawk nests	see table 2		

Table 2. Abbreviated field note book index of 57 Harris's Hawk (*Parabuteo unicinctus*) nests monitored during the raptor research project at the Los Medaños Area of New Mexico in 1987. We constructed similar nest indexes of all notes related to 17 Great Horned Owl (*Bubo virginianus*) and 30 Swainson's Hawk (*Buteo swainsoni*) nests (not shown).

Topic	Field Notebook Pages	Topic	Field Notebook Pages	Topic	Field Notebook Pages
Harris's Hawk nest no.		14	73, 103, 109, 110 133, 154		169, 171, 182, 183
1	72, 92, 107, 158, 170	15	102, 105, 107, 134, 150	37	79, 104, 122, 126, 129, 151
2	102, 134, 161, 183	16	115, 120, 127, 132, 148, 151, 157, 159,	39	132, 151, 152, 182
4	98, 99, 109, 110, 115, 120, 162, 170,		160, 166	40	70, 100, 164, 183, 192
	179	17	81, 103, 126, 129, 130, 133, 153, 160,	41	70, 102, 104, 108, 139, 150
5	67		179	42	103, 108, 132, 165, 182
6	61, 71, 102, 107, 109, 112, 116, 131,	20	80, 103, 120, 123, 129, 134	43	80, 107, 149, 150, 165 182
	151	26	75, 98, 113, 161	51	101, 111, 124, 149
7	78, 99, 122, 124	28	84, 103, 126, 130, 133, 148, 164	54	88, 104, 129, 131, 151, 161
8	96, 105, 109, 131, 148	29	70, 101, 102, 104, 109, 112, 117, 121,	56	100, 109, 151, 153, 158, 160, 163, 169,
10	75, 98, 112, 119, 196		126, 130, 134, 152, 156, 159		171, 178, 183, 184, 186
11	73, 103, 111	30	100, 117, 119, 161, 186	57	103, 132, 165
12	80, 100, 149, 151, 165	35	71, 100, 117, 150, 152, 156, 159, 163,		

specific data-collection activity should be entered on one or multiple sheets. In wet environments, making copies of blank data forms on "rite-in-the rain" paper is strongly advised. An advantage of using pre-printed data sheets is that individual data sheets can be manually sorted (e.g., by species, by date, by nest) in various ways to facilitate efficient data entry. If data need to be sorted in different ways or stored in multiple locations, copies can be easily made to facilitate this type of data management.

Periodically, legible copies should be made of all completed data sheets and field notes (I recommend at approximately 2-week intervals) to avoid the catastrophe of data loss. Moreover, these data should be stored in a safe location away from the field location (e.g., at a university or agency office). Loss of a week or two of data is a serious setback; loss of a season of data is disastrous.

DATA ENTRY

It is always advisable to enter data into a computer file as soon as possible. On a number of research projects in which housing and computers are available, the data should be entered during the evenings of fieldwork, whenever possible. One advantage of this approach is that a duplicate data set based on the original notes or data sheets is now immediately available, which minimizes the potential of those data being lost. In many cases, this optimal approach is not available because data are collected in a remote field location, investiga-

tors are living out of tents, computers are not available, or field workers are simply fully occupied by the demands of field work each evening.

Most data may be entered in a computer spreadsheet (e.g., Microsoft EXCEL), which allows for versatile management and transfer into most other programs including most statistical packages. The general format for data entry should be variables labeled on the top of columns and each observation or sample should be entered across the row (Table 3). I encourage the use of the maximum possible "identifier" variables that precede the data columns. Identifier variables basically identify what the observation is (e.g., subject individual, site name, date, year, and all attributes of that observation [gender, age, experimental vs. control, etc.]). In the example data set (Table 3) the variables — Year, No. of Males, Territory Name, Min. Observ., Start Date, and End Date — could be classified as identifier variables. The identifier variables are useful in subsequent manipulation of data and in implementing analyses. Year is one identifier variable that relatively new researchers tend to overlook, but is really a must as Year is typically a key analytical or confounding variable in most field research. In the example provided (Table 3), one of the key questions of interest was "does frequency of prey delivery differ by the number of males (i.e., No. of Males) in the group?" In an ANOVA type analysis this would be a main-effect variable. However, the data also could be examined for the effect of year, observation time, and the influence of territory site by employing a time-series, mixed-model analysis.

Table 3. Example of a summary data set on the prey delivered at Galápagos Hawk (*Buteo galapagoensis*) nests in 1999 and 2000 on Santiago Island, Galápagos, Ecuador.

Year	No. of Males	Territory Name	Min Observ.	Start Date	End Date	Centi-pede	Lizard	Rat	Dove	Mice	Snake	Sea-bird	Finch	Goat	Small Unid.	Total Prey
1999	2	Cave	3620	5-May	10-May	7	2	2	0	1	0	0	0	0	0	12
2000	2	Cave	3125	16-May	21-May	0	0	0	0	0	0	0	0	0	0	0
1999	1	Coast	3620	12-May	17-May	15	0	1	0	0	2	0	0	0	0	18
1999	2	Cowan 2	3630	21-May	26-May	8	6	0	0	0	11	0	0	0	0	25
2000	2	Cowan 2	3014	25-May	30-May	1	0	0	3	0	0	0	0	0	0	4
2000	1	Espino	3030	8-Jun	13-Jun	3	1	0	3	0	0	0	1	0	0	8
1999	3	Guayabillo	3645	29-May	2-Jun	12	0	1	0	1	0	0	0	0	0	14
2000	1	Gully	3261	21-Jun	26-Jun	4	4	0	0	0	0	0	0	0	2	10
2000	2	Lagoon	3851	1-Jul	6-Jul	8	2	0	0	0	0	0	0	0	0	10
2000	2	Lava	3090	12-Jul	17-Jul	0	1	0	0	0	0	0	0	0	0	1
1999	3	Malgueno	3770	4-Jun	9-Jun	2	0	0	0	0	0	0	0	0	0	2
2000	3	Mordor	3400	23-Jul	28-Jul	2	0	0	0	0	0	0	0	2	0	4
1999	2	Peak	3809	23-Jun	28-Jun	3	0	1	0	0	0	0	0	2	0	6
1999	3	Peregrino	3705	30-Jun	4-Jul	3	0	0	0	0	0	1	0	0	0	4
2000	3	Peregrino	3025	30-Jul	3-Aug	0	0	0	1	0	0	0	0	0	0	1
2000	2	Red Mtn	2162	5-Aug	10-Aug	3	0	0	0	0	0	0	0	0	0	3
2000	2	Shangri La	3155	11-Aug	16-Aug	4	1	0	0	0	0	0	0	0	0	5
1999	2	Valley	3635	11-Jul	16-Jul	1	1	1	0	0	0	1	0	0	0	4
2000	2	Valley	3071	17-Aug	22-Aug	0	0	0	0	0	0	0	0	0	0	0

Spatial data are readily displayed and analyzed with the relatively recent availability of Geographic Information System (GIS) software. The most frequently used software related to biological analyses is ArcView or ArcGIS (ESRI, Redlands, CA U.S.A). The low cost of Global Positioning System (GPS) receivers (<$150 U.S.) readily allows researchers to collect relatively accurate data on spatial coordinates. All field researchers should have and use a GPS receiver to collect location information. These data generally can be input into an EXCEL spreadsheet in a manner similar to that described above. Two columns with UTM or degree location coordinates should be included for each observation in these spreadsheet files. Files can then be converted to "dbf" files and uploaded into ArcView or similar software packages for spatial displays on maps or aerial images.

DATA ANALYSIS

Most students and practicing biologists have been trained in significance or null-hypothesis-testing statistical techniques. These analytical techniques have been aggressively criticized in journals recently (e.g., Johnson 1999, Anderson et al. 2000) because these approaches have emphasized the testing of trivial "straw-man" or "silly-null" hypotheses and the analyses often are uninformative. A variety of alternative approaches have been offered including emphasis on reporting estimates of effects (e.g., providing means and confidence intervals), use of Bayesian inference approaches (Johnson 1999), and the information-theoretic (I-T) approach (Anderson et al. 2000). The Bayesian approach has not been well accepted as a tool to evaluate data patterns in the ecological disciplines, in part because the mathematics involved are relatively complex and the lack of available "canned" programs to calculate Bayesian probabilities. I do not cover this approach further here, but refer interested readers to Ellison (1996) for an introduction with an ecological orientation. The I-T approach (Anderson et al. 2000, Burnham and Anderson 2002) recently has gained some popularity and, in my view, has both advantages and disadvantages over traditional null-hypothesis testing. In brief, the I-T approach involves examining alternative models that affect a selected response variable (traditionally considered a dependent variable) based on several potential explanatory variables (independent variables). Then, formal likelihood measures (e.g., often Akaike Information Criteria [AIC]) may be used to evaluate the fit of the data to various alternative models. Currently, some referees and editors advocate almost exclusive use of the I-T approach. However, assessments by Guthery et al. (2001, 2005) review I-T approaches and point out several limitations with these analyses, as well as the fact that such approaches can be misused (also see Anderson and Burnham 2002) in much the same manner as null-hypothesis testing.

Individual researchers need to consider alternative approaches as possible analytical tools (e.g., null-hypothesis testing, effects estimation, I-T modeling). I agree with Guthery et al. (2001, 2005) that I-T approaches tend to be more exploratory in nature and that this technique in most cases is probably not the best analytical approach in which to test patterns in data for a well-developed field or lab experiment in which potential causal and response variables are well-defined. In the latter case, I advocate the use of traditional null-testing statistics, especially ANOVA, a technique that is both robust and in which the results can be understood readily. However, Anderson et al. (2000) seem to imply that the I-T approach can be used as a rigorous "test" of alternative models. At least in most uses that I have seen, I question this assessment because explanatory variables are often selected arbitrarily or as a matter of convenience, they typically include relatively easy-to-measure available variables, and relationships with the response variable may go in either direction (positive or negative) producing an acceptable model (this is not an *a priori* test of a clearly stated research hypothesis). Therefore, most uses of I-T approaches seem to be best suited to exploring relationships rather than testing a specific research hypothesis. Moreover, if two or more alternative models fit the data well (similar AIC values), there is no acceptable way to discriminate which model is best, except by subjective argument. That said, the I-T approach does have value in identifying possible meaningful relationships between response variables (e.g., survivorship) and a suite of possible explanatory variables (e.g., age, year, and selected cover/vegetation or behavioral variables). Although often misused (Anderson and Burnham 2002), the I-T approach also could be used to evaluate the relative merit of competing explanatory research hypotheses if vacuous models are eliminated *a priori* from the analysis (Guthery et al. 2005).

Data as entered in spreadsheets (described above) may be easily imported or "cut-and-pasted" into statistical software packages such as SAS, Minitab, or Systat. If a study is well designed, most data may be analyzed using parametric or non-parametric analyses (see Potvin and Roff 1993, Johnson 1995, Smith 1995). The I-T analysis can often be accomplished based on output values from SAS and other canned statistical programs (see Anderson et al. 2000, Burnham and Anderson 2002).

It is not my purpose here to review the standard null-testing statistical procedures. A brief review of the common statistical analytical techniques used in wildlife studies is provided by Bart and Notz (2005) and more extensive treatments can be found in statistical textbooks, such as Sokal and Rohlf (1995).

SUCCESSFUL PRESENTATION AND PUBLICATION OF RESULTS

Once data are collected and analyses are mostly complete, the final and most important step in raptor science is the presentation of the results. There are three primary means for presenting data: (1) preparing a manuscript for publication, (2) giving an oral presentation at a scientific meeting, and (3) giving a poster presentation at a scientific meeting. Of these, the most challenging to accomplish is the publication of the results in manuscript form, which undergoes rigorous peer evaluation. There are additional detailed resources addressing various aspects of how to present your research in final form and how to write a scientific paper (e.g., Day 1998 and see below). My purpose here is to hit some key points that may be especially useful in the successful publication of manuscripts about raptors, and regarding specific points not well covered in other resources.

Although I focus primarily on manuscript presentation, many of the basic guidelines for presentation of research results, such as keep it simple and eloquent, and make it clear, apply equally to oral and poster presentations.

Develop an Outline

The first step that I firmly recommend is to prepare an outline for your manuscript. This could be done in classic outline format (i.e., designating topics with numbers and letters indicating levels of importance) or simply writing down major headings (e.g., Introduction, Methods) and developing a list of items in logical order that you wish to address under each of these major headings. The outline allows you to "brainstorm" about how to approach the write-up of your research and see a logical sequence in the proposed topics to address. The outline should provide a framework to help you organize your thoughts and materials, as well as to highlight deficiencies or areas where you will need to do more literature research or analysis before you begin writing. The outline serves as an adaptable guideline that will enable you to better see adjustments that will make your manuscript more logical, complete, and effective. Therefore, expect to cut-and-paste and move topics around, and to add and eliminate topics until you are satisfied with the proposed framework of the manuscript. As you develop the outline, make sure to follow the same sequence of topics (e.g., provide information on observations of birds first, reproductive success next, and relationships with vegetative structure last) in each major section of the paper (i.e., Introduction, Methods, etc.).

General Guidelines for Manuscript Preparation

At this stage of preparing a scientific manuscript, I recommend selecting an appropriate "target" journal. The selection of a target journal is a topic that requires careful deliberation, regarding the stature of the journal, time to decision of acceptance or rejection and to potential publication, probable quality of the referees and review process, interest of the readers of the journal, dissemination potential related to the topic of your manuscript, and other factors. For relatively new scientists, I encourage you to discuss the selection of a journal with senior investigators involved in the project or with academic advisors active in raptor science. Once a target journal is selected, I strongly recommend that authors review the manuscript preparation guidelines for that journal and adhere carefully to those guidelines

as they prepare their manuscript. Most scientific journals or their sponsoring societies maintain a web site where manuscript preparation guidelines can be found. I also recommend obtaining copies of recently published articles or issues of that journal and carefully reviewing them for format and style. Typically, journals also publish their manuscript guidelines periodically in the journal issues. *The Journal of Raptor Research*, for example, publishes information for contributors annually in the December issue (e.g., *J. Raptor Res.* 39:480–483).

Numerous books and other resources have been published to provide a how-to guidance on preparing a scientific manuscript for publication (e.g., Day 1998, Gustavii 2003). For a simple and straightforward writing style guide, I recommend the 4th Edition of Strunk and White (1999). This brief book provides excellent advice regarding effective writing (scientific or otherwise). This style manual advises use of simple, eloquent and active voice in writing, which is also most effective in scientific prose. Text written in passive voice is usually wordy, unclear, and somewhat awkward. Always strive for both brevity and completeness, which often equals clarity.

Introduction. In some respects this is the most important section of the manuscript, and in many ways the most difficult to write. I have seen many otherwise excellent manuscripts rejected simply because the author set the stage poorly with their Introduction. Pay careful attention to the development of this section, and do not hesitate to re-write this section again, and again, if necessary. Specifically, you need to develop the context for the research. Why is this study important to advancing our understanding of raptor biology or to ensuring their conservation? The Introduction should answer this question clearly and provide the appropriate background citations to support your case that this research is a meaningful contribution. Cite only the "best," most current, and most-relevant references. Avoid being too scholarly and exhausting: do not provide excessive citations in the Introduction and elsewhere in the manuscript. The Introduction should be relatively easy to compose if your research idea and study design were developed with scientific rigor prior to the initiation of data collection.

Methods. The Study Area and Methods is usually the easiest section to compose. Although somewhat falling out of favor these days, in part because of the relatively high costs of producing figures and the difficulty that authors have in developing a suitable map, I feel that study-area figures are extremely informative. An appropriate "picture" is always worth a thousand words, probably more when it comes to setting the stage for describing a research study. Thus, use of an informative study-area figure can provide maximum content (it can and should be used to illustrate key spatial relationships such as distributional patterns of cover types, locations of nests, or other key features relative to the study) and is probably among the most "cost-effective" approaches to provide supportive documentation for a field study.

The Methods section is a straightforward description of the techniques used by the author(s) of the manuscript. The key point here is to provide enough detail that readers would be able to duplicate your study or experiment. The Methods section can be shortened by citing other papers that clearly describe the techniques used in the current study. By relying on citations, you would only need to describe clearly any modifications employed beyond what was described in the original reference of that technique.

Results. This is a critical section of any manuscript, but, generally, is easy to write. Following the original outline of topics to cover in the Results (see Develop an Outline), I recommend first preparing working tables and figures for possible inclusion in the results. The working figures and tables provide an outline for the text of the Results. Text should not repeat data presented in tables, but should briefly describe primary patterns in the data that are evident in tables and figures. All tables and figures should be cited in the text. If a figure or table is not cited, then it is not needed and should be omitted. If data are few in a working table, these may be more concisely presented in the text. Key means, medians, estimates of variation, and statistical results should be provided parenthetically in the text of the Results. During the course of writing the Results, each working figure and table should be evaluated and revisions should be made to improve clarity for the final version of the manuscript.

Tables should rarely provide raw data, but rather summaries of statistical data (e.g., means, confidence intervals, sample sizes). Tables need to be clear, straightforward, simple, and easy to interpret. Avoid excessive clutter and footnotes in tables. Also, eliminate redundancy and minimize the use of acronyms or cryptic variable codes in tables and text. Sometime in the 1970s, somebody "decided" that the use of cryptic acronyms to label individual vegetation structural or other variables was a concise approach for presenting such results. Unfortunately, this confusing and ineffec-

tive presentation approach carries on today. As such, papers often include the analysis of scores of vegetation variables, and the jumble of confusing acronyms (e.g., PDFCC = Percent Deciduous Forest Canopy Cover) is almost impossible to follow unless the reader makes a cheat-sheet of the codes to refer to as the paper is read. All but a few dedicated readers are willing to make this effort to sort through the confusion of cryptic acronyms. I strongly recommend that authors avoid the use of cryptic codes for data variables and use an abbreviated, but descriptive variable name. Consider again PDFCC, for example. If the author analyzed 40 vegetation variables with similarly awkward codes; the text and tables would be extremely difficult to comprehend. For this variable, a clearer label might be "Tree Canopy Cover." Minimize the use of acronyms and cryptic codes throughout manuscripts.

I strongly recommend figures over tables, as I feel visual representations can leave very effective and lasting impressions of the results in the readers' minds. Tufte (1983) offers some guidance on the visual display of quantitative data. Some examples of particularly effective figures include the following in the *J. Raptor Res.* (39:356, Fig. 1; 39:369, Fig. 2; 39:397, Fig. 1; 39:448. Fig. 1; 39:464, Fig. 1; 39:470, Fig. 3; 40:14, Fig. 9; 40:18, Fig.14; 40:68, Fig. 2). Always give consideration as to whether the data in any of your working tables can be presented more effectively in the form of a figure.

Discussion. The Discussion should address the same sequence of general topics that was set forth in the Introduction and other sections of the manuscript. Probably the first items to address in your Discussion are the research questions and hypotheses introduced in the Introduction of the paper. Assess how your data support or refute the hypotheses that you set out to test. Discuss and acknowledge any inconsistencies in the results. Then review any potential biases in your methodology and comment on the seriousness of these biases. Do these weaknesses potentially affect any of your interpretations? Compare your results with those of current and relevant literature objectively. And again, generally avoid being too scholarly. There is no need to compare your results to every paper remotely addressing the same question(s). Simply stick to the most relevant papers.

If you have unexpected or surprising results, it is fair to suggest reasonable and logical explanations. Support these hypotheses with whatever *post hoc* data you have available and consistent patterns reported in the literature. Do not, however, go off the deep end. If you have no supporting evidence or reasonable logic to support your speculation, do not go there. Excessive speculation will get your manuscript rejected almost every time. The Discussion section does provide you with the opportunity to develop new hypotheses, but your data should be consistent with these new ideas. Be prudent with speculation; restrict it to the development of one or two alternative hypotheses at the most. Finally, it is often worthwhile to highlight interesting patterns that may have emerged from the data as a spin-off from collecting data on your primary research questions.

Authorship

One issue of importance to scientists in all disciplines is the question of how to assign authorship to reflect the contributions of individuals to science. Although the topic largely has been ignored in the past (e.g., Tarnow 1999), ethical guidelines on who qualifies to be an author and the order of authors are now available from a number of sources (e.g., Day 1998, Macrina 2005). Ideally, general principles and philosophies of assigning authorship should be discussed by potential collaborators (e.g., a graduate student and graduate advisor) before the research begins. However, actual assignment of authorship can only be done fairly after the research, including the preparation of the manuscript, is near completion.

It is a clearly accepted ethical norm that only individuals that have substantially contributed to the development and execution of the research should be included as authors. Substantial contributions are typically described as those that have an effect on the direction, scope, or depth of the research (Macrina 2005), or involve significant contribution to the concept, design, and interpretation of the study (Tarnow 1999). Honorary authorship, especially by individuals who merely facilitated funding, inclusion of project or program directors that were not directly involved in the research, or listing technicians that simply collected data violate acceptable ethical standards.

Moreover, all authors should accept responsibility for the content and the integrity of the science reported in a published paper. Thus, to the extent reasonable, all listed authors must understand and defend the basic aspects of the work, and take responsibility for errors, flawed interpretations, and the consequences resulting from publication (including any bad science). For example, consider a group of authors that published

data, based on a pseudoreplicated and confounded study design further masked by a vague presentation of results, that suggested a rare species of raptor benefited by logging operations. Then resource managers, on the basis of this publication, undertook a good-faith effort to initiate aggressive timber harvest operations "to benefit" this rare raptor, which subsequently resulted in near total demise of the species. Although this hypothetical scenario would expose multiple shortcomings in the publication processing of this specific paper (e.g., superficial reviews by referees, poor decisions and oversight by editors, lack of critical assessment of the research by resource managers), the major responsibility for the near loss of this species would reside with the authors who published the paper. Certainly, honorary authors or individuals with poor understanding of the research should not be included in the authorship line because they cannot defend or critically evaluate the science, potentially leading to the publication of unreliable results and conservation disasters similar to the scenario that I described above.

Dickson et al. (1978), Schmidt (1987), and others offer guidelines for assigning authorship in papers published in basic and applied ecology. Their guidelines have been elaborated by J. F. Piatt (unpubl. manuscript), and I briefly summarize his suggestions here. There is general agreement that scientific investigations can be broken down into five basic areas:

1. **conception** — including original study idea, development of proposals, and acquisition of funding;
2. **design** — development of study design, intricacies of data-collection protocols, and related logistic matters;
3. **execution and data collection** — the actual work of data collection and the administrative and logistic efforts needed to support the field or laboratory activities;
4. **data analysis** — including all aspects of data manipulation such as data entry, verification, and analysis;
5. **writing** — including synthesis and interpretation, which most often represents the most intense intellectual development of any paper (generally the first draft of any manuscript is written by the lead author). As a guideline, for an individual to be considered as a potential co-author they should make significant contributions in at least two of the five key areas of

research described above. One area that all co-authors should be involved with is manuscript preparation. At a minimum, all co-authors should carefully review and provide critical input on the validity and interpretation on an early draft of the manuscript. Later, before submission, all co-authors should be comfortable with and accept responsibility for the science and results presented in the manuscript. This "approval step" may be accomplished informally and given verbally, or more formally with all co-authors providing written consent via a letter or e-mail message.

Order of authors should be based on the importance of significant and practical contributions to the overall research. Piatt (unpubl. manuscript) offers a suggested approach to assess the relative contribution of each individual by having all co-authors estimate their contribution to each of the five key areas of producing a research manuscript. All potential co-authors should review their estimated contributions and revise these estimates until they reach a consensus. This "quantitative" assessment could indicate differences between co-authors and individuals that should be listed in the acknowledgments if there is a distinct break between scores of major and minor contributors. The contributions of closely ranked individuals ideally should be discussed and further resolved by agreement of all co-authors. In some fields, including the medical sciences and molecular biology it is customary for the leader of the research group or laboratory — assuming they are actively involved in that specific research — to be last author on papers, a position considered to be second in importance and prestige following that of the first author. Generally, in the wildlife field this is not the convention and the order of authorship reflects the overall contribution of each co-author, with the first author providing the greatest contribution and so on. However, this philosophy is changing with the convergence of disciplines, and many laboratory raptor biologists ascribe added significance to the last individual listed on an author byline.

CONCLUSION

In some respects, the task of presentation, particularly the publication of results, is the most challenging aspect of conducting raptor research. This often is the most humbling aspect, especially when one receives frank criticism from one's peers, as well as the most gratifying aspect of raptor science. Keep in mind that your study is not complete until it has been published. In a way, the research project really never even took place unless the results are published in a scientific journal. If the data never see the light of publication, you in essence wasted much of your time conducting the research, wasted the funder's money, and most likely unnecessarily disturbed the birds (by using them as research subjects, which may have involved observations, trapping, banding, "disturbing" nesting activities, etc.). On the other hand, perhaps one of the most gratifying aspects of science is the recognition of the importance of your work as reflected by publication in a peer-reviewed scientific journal and the knowledge that you have made a lasting contribution to the knowledge base and probably to the conservation of raptors.

ACKNOWLEDGMENTS

I thank Keith Bildstein for suggesting this as a chapter in *Raptor Research and Management Techniques*. I developed the ideas expressed in this contribution through my experiences working with many graduate and undergraduate students and interns at Hawk Mountain Sanctuary, Boise State University, and Arkansas State University, and I thank all of these individuals for their direct and indirect contributions. Keith Bildstein, T.J. Benson, Rich Grippo, Nick Anich, and David Bird provided a variety of suggestions that I have incorporated into this chapter in one way or another.

LITERATURE CITED

AHLERING, M.A. AND J. FAABORG. 2006. Avian habitat management meets conspecific attraction: if you build it, will they come? *Auk* 123:301–312.

ANDERSON, D.R. AND K.P. BURNHAM. 2002. Avoiding pitfalls when using information-theoretic methods. *J. Wildl. Manage.* 66:912–918.

———, K.P. BURNHAM AND W.L. THOMPSON. 2000. Null hypothesis testing: problems, prevalence, and an alternative. *J. Wildl. Manage.* 64:912–923.

BART, J.R. AND W.I. NOTZ. 2005. Analysis of data in wildlife biology. Pages 72–105 *in* C. E. Braun [ED.], Techniques for wildlife investigations and management. The Wildlife Society, Bethesda, MD U.S.A.

BIBBY, C.J., N.D. BURGESS, D. JILL AND S. MUSTOE. 2000. Bird census techniques, 2nd Ed. Hancock House Publishers, Blaine, WA U.S.A.

BRANDES, D. AND D.W. OMBALSKI. 2004. Modeling raptor migration pathways using a fluid-flow analogy. *J. Raptor Res.* 38:195–207.

BURNHAM, K.P. AND D.R. ANDERSON. 2002. Model selection and multimodel inference. Springer, New York, NY U.S.A.

DAY, R.A. 1998. How to write and publish a scientific paper, 5th Ed. Oryx Press. Phoenix, AZ U.S.A.

DEKKER, D. AND R. TAYLOR. 2005. A change in foraging success and cooperative hunting by a breeding pair of Peregrine Falcons and their fledglings. *J. Raptor Res.* 39:394–403.

DIAMOND, J. 1986. Overview: laboratory experiments, field experiments, and natural experiments. Pages 3–22 *in* J. Diamond [ED.], Community ecology. Harper & Row Publishers, Inc., New York, NY U.S.A.

DICKSON, J.G., R.N. CONNER AND K.T. ADAIR. 1978. Guidelines for authorship of scientific articles. *Wildl. Soc. Bull.* 6:260–261.

EBERHARDT, L.L. AND J.M. THOMAS. 1991. Designing environmental field studies. *Ecol. Monogr.* 61:53–73.

ELLIS, D. H. AND J. W. LISH. 2006. Thinking about feathers: adaptations of Golden Eagle rectrices. *J. Raptor Res.* 40:1–28.

ELLISON, A.M. 1996. An introduction to Bayesian inference for ecological research and environmental decision-making. *Ecol. Appl.* 6:1036–1046.

FORD, E.D. 2000. Scientific method for ecological research. Cambridge University Press, Cambridge, United Kingdom.

GARTON, E.O., J.J. RATTI AND J.H. GIUDICE. 2005. Research and experimental design. Pages 43–71 *in* C. E. Braun [ED.], Techniques for wildlife investigations and management. The Wildlife Society, Bethesda, MD U.S.A.

GUSTAVII, B. 2003. How to write and illustrate a scientific paper. Cambridge University Press, Cambridge, United Kingdom.

GUTHERY, F.S., L.A. BRENNAN, M.J. PETERSON AND J.J. LUSK. 2005. Information theory in wildlife science: critique and viewpoint. *J. Wildl. Manage.* 69:457–465.

———, J.J. LUSK AND M.J. PETERSON. 2001. The fall of the null hypothesis: liabilities and opportunities. *J. Wildl. Manage.* 65:379–384.

———, J.J. LUSK AND M.J. PETERSON. 2004. In my opinion: hypotheses in wildlife science. *Wildl. Soc. Bull.* 32:1325–1332.

HERMAN, S.G. 1986. The naturalist's field journal: a manual of instruction based on a system established by Joseph Grinnell. Buteo Books, Vermillion, SD U.S.A.

HURLBERT, S.T. 1984. Pseudoreplication and the design of ecological field experiments. *Ecol. Monogr.* 54:187–211.

JAKSIC, F.M. 1985. Toward raptor community ecology: behavior bases of assemblage structure. *Raptor Res.* 19:107–112.

JAMES, F.C. AND C.E. MCCULLOCH. 1985. Data analysis and the design of experiments in ornithology. Pages 1–63 *in* R. F. Johnston [ED.], Current ornithology, Vol. 2. Plenum Press, New York, NY U.S.A.

JOHNSON, D.H. 1995. Statistical sirens: the allure of nonparametrics. *Ecology* 76:1997–1998.

———. 1999. The insignificance of statistical significance testing. *J. Wildl. Manage.* 63:763–772.

KERLINGER, F.N. 1973. Foundations of behavioral research, 2nd Ed. Holt, Rinehart and Winston, Inc., New York, NY U.S.A.

KIMBALL, R.T., P.G. PARKER AND J.C. BEDNARZ. 2003. The occurrence and evolution of cooperative breeding among the diurnal raptors (Accipitridae and Falconidae). *Auk* 120:717–729.

KUHN, T. 1962. The structure of scientific revolutions. University of Chicago Press, Chicago, IL U.S.A.

LACK, D. 1968. Ecological adaptations for breeding in birds. Methuen, London, United Kingdom.

LANCIA, R.A., W.L. KENDALL, K.H. POLLOCK AND J.D. NICHOLS. 2005. Estimating the number of animals in wildlife populations. Pages 106–153 *in* C. E. Braun [ED.], Techniques for wildlife investigations and management. The Wildlife Society, Bethesda, MD U.S.A.

MACRINA, F.L. 2005. Scientific integrity. American Society of Microbiology Press, Washington, DC U.S.A.

MOCK, D.W. AND G.A. PARKER. 1997. The evolution of sibling rivalry. Oxford University Press, Oxford, United Kingdom.

PIATT, J.F. Guidelines for assigning authorship on scientific publications. Unpubl. Manuscript. U.S. Geological Survey, Alaska Science Center, Anchorage, AK U.S.A.

POPPER, K.R. 1959. The logic of scientific discovery. Hutchinson and Co., London, United Kingdom.

———. 1968. Conjectures and refutations: the growth of scientific knowledge, 2nd Ed. Harper & Row, New York, NY U.S.A.

POTVIN, C. AND D. A. ROFF. 1993. Distribution-free and robust statistical methods: viable alternatives to parametric statistics? *Ecology* 74:1617–1628.

RATTI, J.T. AND E.O. GARTON. 1996. Research and experimental design. Pages 1–23 *in* T. A. Bookhout [ED.], Research and management techniques for wildlife and habitats. The Wildlife Society, Bethesda, MD U.S.A.

ROMESBURG, H.C. 1981. Wildlife science: gaining reliable knowledge. *J. Wildl. Manage.* 45:293–313.

SCHMIDT, R.H. 1987. A worksheet for authorship of scientific articles. *Bull. Ecol. Soc. Amer.* 68:8–10.

SMITH, S.M. 1995. Distribution-free and robust statistical methods: viable alternatives to parametric statistics? *Ecology* 76:1997–1998.

SINCLAIR, A.R.E. 1991. Science and the practice of wildlife management. *J. Wildl. Manage.* 55:767–773.

SOKAL, R.R. AND F.J. ROHLF. 1995. Biometry, 3rd Ed. W. H. Freeman and Company, New York, NY U.S.A.

STRUNK, W. AND E.B. WHITE. 1999. Elements of style. MacMillan Publ. Co., New York, NY U.S.A.

TARNOW, E. 1999. The authorship list in science: junior physicists' perceptions of who appears and why. *Sci. Engineering Ethics* 5.1:73–88.

TUFTE, E.R. 1983. The visual display of quantitative information. Graphic Press, Cheshire, CT U.S.A.

Survey Techniques

5

DAVID E. ANDERSEN
U.S. Geological Survey,
Minnesota Cooperative Fish and Wildlife Research Unit,
200 Hodson Hall, 1980 Folwell Avenue, St. Paul, MN 55108 U.S.A.

INTRODUCTION

Compared with most other groups of birds and many other vertebrates, raptors often are widely dispersed, and many of their populations exist at relatively low densities across the landscapes in which they occur. Although many species are relatively easy to detect either by sight or by sound, conducting surveys for raptors can be difficult and require a substantial commitment of resources. In spite of these difficulties, investigators have expended considerable effort counting birds of prey (Fuller and Mosher 1987), and have used information derived from their surveys to estimate population size or trend, locate nests and monitor reproduction, assess the population status or distribution of species, monitor raptor populations of particular conservation concern, investigate behavioral ecology, and evaluate methods used to detect and count raptors.

In their review of raptor survey techniques, Fuller and Mosher (1987:37) defined "survey" after Ralph (1981) as *"(1) the process of finding individuals in relation to geographic areas (e.g., continental, regional, local) or habitat features (e.g., physiography, vegetation); and (2) an enumeration ... of abundance of individuals in an area from which inferences about the population can be made."* Inherent in this definition is that observers must be able to ascertain the presence of individual raptors, either directly through visual or aural observation or indirectly by finding recently refurbished nests or prey remains. In addition, surveys inherently have a spatial aspect, in that they are conducted over a discrete area.

How raptor surveys are planned and conducted depends on survey objectives. For example, surveys intended to locate nests (i.e., things that do not move within the same season) may need to be designed differently from surveys intended to estimate the population size of wintering raptors, which may or may not exhibit site fidelity to local areas. Thus, survey objectives should be clearly defined, and surveys should be designed to meet those objectives.

Finally, surveys need to provide reliable results. Poorly designed or implemented surveys that result in imprecise or biased estimates of population size, for example, have limited use. Extending results from such surveys to other areas or comparing such results with those from surveys conducted elsewhere or to address different objectives also may have limited value.

The primary objective of this chapter, then, is to provide an overview of sampling procedures and general survey methods used to count raptors. Specific topics include (1) survey objectives, (2) survey design considerations, and (3) the application of wildlife survey methods to raptors. Because considerations for counting migrating raptors are presented elsewhere in this book (see Chapter 6), this chapter focuses on surveys for raptors during non-migratory periods. Many of the survey considerations discussed herein also apply to surveys for raptor nests. Sources of information for this overview include the summary of raptor-survey techniques by Fuller and Mosher (1987), literature published since 1987 compiled by searching electronic databases (Wildlife and Ecology Studies Worldwide, Raptor Information System, and Web of Science), and my general familiarity with raptor literature. The chapter is not a complete list or summary of all of the published literature, but rather an overview of raptor-survey methods and results.

SURVEY OBJECTIVES

Objectives of raptor surveys need to be clear and explicit before surveys are conducted. Survey designers also need to consider how their data will be used. Fuller and Mosher (1987) identified two objectives for raptor surveys: determining raptor distribution and determining abundance (both absolute density and relative abundance). In addition, surveys often are used to locate raptors to study population dynamics and other aspects of raptor ecology (e.g., raptor-habitat relations and reproduction), and to provide information for management and conservation.

Surveys to assess raptor distribution occur at a variety of spatial scales, from local study areas to large geographic regions, and entail locating raptors across a defined area, often stratified by habitat type, topography, or other environmental characteristics. Because raptors often occur at low densities, surveys to assess distribution at larger spatial scales usually involve sampling representative subdivisions of larger areas.

Surveys to determine raptor abundance fall into two categories, those designed to determine population size or density and those designed to compare relative abundance, either spatially or temporally. Population size is the number of individuals in a population, where population can be defined biologically (e.g., as a group of interacting individuals of the same species) or spatially (e.g., a group of individuals using a particular area during a defined time period). Density is the number of individuals or groups of individuals, including pairs, per unit area. A complete enumeration of raptors within a defined area, or census, often is not practical because raptors can be difficult to detect (i.e., the probability of detecting a raptor is less than one) and because they are often widely spaced. Finally, delineation of the boundaries and location of the area searched to determine density can influence interpretation of survey results (Smallwood 1998).

In practice, raptor density is generally estimated based on surveys designed to sample a representative portion of the raptor population or area of interest. Sampling biological populations is discussed in detail in Williams et al. (2002) and Schreuder et al. (2004). Sampling considerations pertinent to raptor population surveys are discussed below. One primary concern is that sample surveys be designed so that results can be used appropriately to meet survey objectives (e.g., to estimate raptor density in a particular study area or to make comparisons between or among populations or study areas).

In addition to assessing raptor distributions and abundances, data derived from surveys also are used in population modeling and monitoring, to investigate raptor ecology, including assessing raptor-habitat relations, and to provide information from which to base conservation strategies and activities. For example, Bustamante and Seoane (2004) used raptor surveys to compare distribution predicted from statistical models with existing distribution maps for four species of raptors in southern Spain. Meyer (1994) evaluated whether counts at communal roosts of Swallow-tailed Kites (*Elanoides forficatus*) could be used to monitor kite populations and develop conservation strategies, and Currie et al. (2004) based conservation strategies for Seychelles Scops Owls (*Otus insularis*) on surveys designed to assess distribution and abundance. In these and other cases, surveys must be designed to meet explicit study objectives, and study design should facilitate obtaining reliable survey results. When designing surveys for raptors, it is generally wise to discuss design and statistical considerations with a biostatistician prior to data collection.

SURVEY DESIGN CONSIDERATIONS

Factors Affecting Detection

Many factors potentially affect detection of raptors during surveys. These include attributes of the birds themselves (e.g., species, age, sex, behavior, group size, etc.), environmental conditions when surveys are conducted (e.g., weather, degree of illumination, and, in aural counts, factors affecting sound transmission, etc.), temporal variables that affect behavior or distribution (e.g., time of day or time of year), habitat characteristics (e.g., forested versus open landscapes, distribution of perches, etc.), and attributes of observers (e.g., experience, visual or aural acuity, etc.). Results from raptor surveys that do not consider these factors may be difficult to interpret or may not be compared appropriately to results from surveys conducted under different conditions, at different places, or at different times.

Probability of detecting individual raptors varies and can be influenced by raptor size and color (visual-based surveys), type and intensity of vocalization (aural-based surveys), behavior, and factors related to sex and age. Under the same conditions, larger raptors whose coloration contrasts with their background are easier to see than smaller raptors that blend in with their surroundings. Similarly, raptors that call loudly and frequently are easier to hear than raptors that call quietly or less frequently. Behavior also influences detection probability. For example, moving raptors are generally more likely to be detected visually than perched raptors. Sex and age can influence raptor behavior, and consequently, detection probability. In species where males and females exhibit a division of tasks during breeding, the member of the pair primarily responsible for hunting, which is usually the male, may be more likely to be detected away from a nest. In contrast, birds associated with nests may be more likely to respond vocally to conspecific calls broadcast near the nest. Similarly, fledglings may be more detectible than adults while calling when begging for food.

Environmental conditions during surveys can influence raptor detectability directly and indirectly. Weather conditions can influence detection probability directly through effects on visibility (e.g., fog, snow, or rain) and on sound transmission (aurally based surveys). Indirect effects can occur when weather influences raptor behavior, such as when it triggers roosting behavior or is conducive to soaring. As with most wildlife surveys, surveys for raptors are generally conducted during specified environmental conditions (e.g., Andersen et al. 1985), which minimize variation among surveys attributable to environmental conditions.

Detectability of raptors often changes both through the day and through the year. Many raptors exhibit activity patterns where they are more active, away from cover, or more vocal during some times during the day than others. For example, in the temperate zone, soaring raptors, including buteoine hawks and vultures, may not leave roosts until mid-morning, when thermals form. Similarly, raptors may be detected with higher probability at some times of the year, or even within the same season, than at other times. Red-shouldered Hawks (*Buteo lineatus*), for example, respond to conspecific call broadcasts more readily during the breeding season than at other times of the year, and within the breeding season are more likely to respond during courtship than during incubation (McLeod and Andersen 1998).

Habitat characteristics can substantially affect detection of raptors on surveys. Forest-dwelling raptors, even species that are relatively large-bodied and strikingly colored, are notoriously difficult to detect on visually based surveys. Distribution of perches can influence the distribution of raptors that hunt from perches (Janes 1984), thereby affecting raptor detection probability. Perch distribution can affect surveys conducted along roads paralleled by power or communication lines that afford perching by raptors. Similarly, habitat characteristics can influence raptor detection in aurally based surveys by influencing sound transmission of both call broadcasts and raptor vocalizations. The presence of foliage, for example, substantially influences sound transmission in deciduous forested habitat.

The attributes of observers also can influence raptor-detection probability. The number of observers, their experience level, and their visual and aural acuity all can influence the ability to detect raptors. Only a few raptor surveys have considered attributes of observers (e.g., McLeod and Andersen 1998, Ayers and Anderson 1999), but observer effects have been well documented in surveys for other birds (e.g., Ralph et al. 1993).

Many additional factors can affect the detection of raptors on surveys. Investigators need to be aware of factors that may influence detectability, control for these factors when possible, and recognize potential influence of various factors on survey results and interpretation, especially when comparing results among surveys (Andersen et al. 1985).

Sampling and Sample Size

In raptor surveys there are generally two populations, one biological and one statistical, that must be considered. As indicated above, a biological population is a collection of raptors, and the objective of raptor surveys is to better understand this biological population. In contrast, a statistical population is a collection of sampling units, each of which can be evaluated to ascertain presence, abundance, or some other aspect related to raptors. Sampling is a method of measuring attributes of a portion of a statistical population and using observed attributes of the evaluated portion to make inference about an entire statistical population. How a sample of the statistical population is obtained determines whether inference can be made to the entire population, or whether survey results apply only to that portion of the population in the sample.

The nature of sample units used in raptor surveys depends on many of the factors identified above, and especially on the scale of the survey. For example, if the objective of a particular survey is to estimate abundance or distribution of raptors across a broad area, sample units might be sections of coastline (e.g., Jacobson and Hodges 1999) or large plots (e.g., Hargis and Wood-bridge 2006). If surveys are designed to assess abundance at smaller spatial scales such as a well defined study area or landscape, sample units might be routes (e.g., Andersen et al. 1985) or fixed points (e.g., Henne-man et al., in press). In either case, sampling entails examining a portion of all sample units contained in the statistical population, and extending the information derived from this sample to the entire population. How this sample is derived is of particular importance, and needs to be considered prior to conducting surveys and when survey results are presented.

Mendenhall et al. (1971) and Cochran (1977) de-scribe sampling methods in detail, Schreuder et al. (2004) provides an overview of sampling methods relat-ed to natural resources, and Ralph and Scott (1981) describe sampling methods specific to birds. Regardless of survey objectives and spatial scale, investigators should consult with a statistician prior to conducting surveys for raptors. Simple random sampling occurs when all sample units have a finite and equal chance of being included in the sample. At small spatial scales, raptor surveys might be designed with simple random sampling if raptor distribution and habitat across the survey area are homogeneous. Raptors have been sur-veyed using simple random sampling where raptor den-sities are relatively high (e.g., Henneman et al., in press), but other forms of sampling for raptors are used more frequently. Stratified random sampling results when sample units can be grouped based on factors related to raptor distribution; such as habitat composi-tion of the study area, political boundaries that may reflect different management scenarios, etc. The pri-mary advantages of stratified random sampling are an increase in precision of estimates and increased effi-ciency in sampling. Sampling effort within strata can be allocated based on stratum size, raptor density, cost of conducting surveys, to minimize variance of final esti-mates, or combinations of these considerations. A pro-posed bioregional monitoring strategy for Northern Goshawks (*Accipiter gentilis*) is based on stratified ran-dom sampling considering goshawk density and acces-sibility of sample units (Hargis and Woodbridge 2006). Kochert and Steenhof (2004) also used this approach to estimate the number of nesting pairs of Prairie Falcons (*Falco mexicanus*) in southern Idaho.

Systematic sampling involves randomly selecting a starting point to sample, and then sampling additional points based on a regular spatial pattern. Advantages of systematic sampling are that (1) the entire range of vari-ability usually is represented in the sample, (2) logistic efficiency sometimes can be increased, and (3) preci-sion of estimates can be increased compared to simple random sampling (Cochran 1977). Disadvantages are that (1) estimates of precision can be difficult to obtain, and (2) if a spatial pattern in sampling units parallels the spatial pattern in sampling, the resulting estimates can be biased. Systematic sampling within sample units has been proposed as part of a bioregional monitoring effort for Northern Goshawks (Hargis and Woodbridge 2006).

More elaborate sampling strategies include cluster sampling (Mendenhall et al. 1971, Cochran 1977), where a cluster is a group of smaller sampling units that are close together. Double sampling (Mendenhall et al. 1971, Cochran 1977, Bart and Earnst 2002) involves sampling at two spatial scales and using information from one scale to improve estimates at another scale. Haines and Pollock (1998) used this approach to esti-mate abundance of Bald Eagle (*Haliaeetus leuco-cephalus*) nests. These and other sampling strategies have potential statistical and logistical advantages, but should be undertaken only after consulting a statistician.

Finally, sampling related to monitoring presents additional considerations. Considerations for initial selection of samples are similar to those discussed above. Sample units examined in subsequent surveys can be the same, a different random sample, or a com-bination of the same and different sample units (Schreuder et al. 2004). If the same sample units are examined repeatedly through time (e.g., annually), then power to detect changes through time is higher (i.e., variance is lower) than if a new sample is drawn at each sampling occasion. This is because sequential observa-tions at individual sample units are positively correlat-ed (Schreuder et al. 2004). However, as the time inter-val between the original and a subsequent sampling occasion increases, one has less and less confidence that observed changes in the sample reflects changes in the target population. Without independent evidence that the sample units being monitored continue to represent the target population through time, surveys may reflect only changes in raptors (or attributes thereof) that occur in those sample units, and not the larger population. Again, clear objectives and consultations with a statistician versed in monitoring prior to initiating a monitoring protocol are essential.

Kinds of Surveys

Raptor surveys can be grouped into several categories based on distribution of the study population and sample units. At smaller spatial scales (e.g., study areas of 10s to 1000s of hectares) surveys often are designed to count all raptors present (e.g., Craighead and Craighead 1956), with adjustments sometimes made for detection probability of less than one (e.g., Anthony et al. 1999). Alternatively, mark-resighting methods may be used to estimate population size on study areas (e.g., Manly et al. 1999). Studies employing such techniques are generally designed to investigate raptor population ecology, and incorporate surveys to identify and describe study populations. Under such design considerations, the sample unit is essentially the study area, and this approach is often used in studies of raptor nesting ecology (e.g., Borges et al. 2004).

At larger spatial scales (1,000s of hectares to regional or continental scales), transects (e.g., survey routes along roads or trails; Andersen et al. 1985, Vinuela 1997) or plots (e.g., Phillips et al. 1984, Hargis and Woodbridge 2006) are the sample units, and detections of raptors in these units are used to make inference about a larger raptor population. Raptor surveys based on transects (e.g., Kenward et al. 2000) are relatively common in the published literature, with fewer examples of surveys based on plots (e.g., Grier 1977, Schmutz 1984, Lehman et al. 1998). However, survey results often can be applied only to the area actually surveyed, and cannot be extrapolated to a larger area if selection of survey locations was not random (see above).

At larger spatial scales, counts or detections at points have been used to document raptor presence (e.g., Kennedy and Stahlecker 1993), community diversity (Manosa and Pedrocchi 1997), and to estimate occupancy (e.g., Mosher et al. 1990, McLeod and Andersen 1998). Advances in statistical methods (e.g., Geissler and Fuller 1987, MacKenzie et al. 2002) have made it possible to use repeated sampling of the same points as a population monitoring tool. These techniques are just beginning to be applied to raptors (e.g., Olson et al. 2005, Seamans 2005, Hargis and Woodbridge 2006, Henneman et al., in press,), but are generally applicable to surveys where raptor detection probability is imperfect (MacKenzie et al. 2002, 2003, 2004; Royle and Nichols 2003). Points, or routes consisting of a series of points, are the sample units, and if sample units are selected to be representative of a larger population, survey results can be extended to a variety of spatial scales.

RAPTOR SURVEYS

Raptor surveys can be conducted from the ground (e.g., McLeod and Andersen 1998) or water (e.g., Garrett et al. 1993), and air (e.g., White et al. 1995) or, in limited cases, through remote sensing (e.g., radar; Harmata et al. 2000).

Surveys from the Ground or Water

Raptor surveys from the ground or water generally involve traversing a specified route along roads or trails (e.g., Andersen et al. 1985, Vinuela 1997) or along a shoreline (e.g., Castellanos et al. 1997), while searching a specified area, such as known colonial breeding sites (e.g., Martinez et al. 1997), or visiting pre-identified points (e.g., McLeod and Andersen 1998) and assessing the presence of raptors through direct observation or indirect evidence, such as the presence of nests. Fuller and Mosher (1987) summarized considerations for designing and conducting ground-based raptor surveys and the general applications, advantages, and limitations for different categories of surveys. I provide a brief description of types of ground- and water-based surveys, and summarize survey results and considerations published since Fuller and Mosher's review.

Surveys for raptors often have been conducted along roads where raptors are observed and counted from vehicles (e.g., Andersen et al. 1985). Surveys along roads have been used to describe raptor distribution (e.g., Yosef et al. 1999, Bak et al. 2001), diversity (e.g., Ross et al. 2003), relative abundance in relation to land-use practices (e.g., Sorley and Andersen 1994, Yahner and Rohrbaugh 1998, Williams et al. 2000), and habitat use at broad spatial scales (e.g., Garner and Bednarz 2000, Olson and Arsenault 2000). Studies of raptor behavior (e.g., Manosa et al. 1998, Rejt 2001), food habits (e.g., Dekker 1995, Kaltenecker et al. 1998) or population dynamics (e.g., Kerlinger and Lein 1988, Hiraldo et al. 1995, Bridgeford and Bridgeford 2003) also have been based on surveys along roads. Surveys from roads also have been used to locate nests in natural (e.g., Travaini et al. 1994, Woodbridge et al. 1995,

Goldstein 2000) or urban landscapes (e.g., Stout et al. 1998), or to assess conservation status (e.g., Herremans and Herremans-Tonnoeyr 2000, Thiollay and Rahman 2002, Prakash et al. 2003, Sanchez-Zapata et al. 2003) or raptor responses to epizootics (e.g., Seery and Matiatos 2000). Surveys for raptors along roads have been used to describe raptor abundance at specific times of the year (e.g., Andersen et al. 1985, Goldstein and Hibbitts 2004), at different spatial scales (e.g., Sorley and Andersen 1994, Belka et al. 1996, Ferguson 2004), and to assess changes in raptor abundance through time (e.g., Hubbard et al. 1988, Herremans and Herremans-Tonnoeyr 2001, Thiollay 2001).

Ground-based surveys of plots or study areas searched for the presence of raptors have been used to monitor colonial-nesting raptors (e.g., Martinez et al. 1997), and to find raptors to assess breeding ecology (e.g., Gerhardt et al. 1994), habitat use (e.g., Thome et al. 1999), and communal roosting (e.g., Kaltenecker 2001). Surveys for raptor nests often are conducted on foot (e.g., Joy et al. 1994), but may incorporate a suite of survey techniques used to find raptors (e.g., Andersen 1995), including surveys from horseback or all-terrain cycle (e.g., Andersen 1995) and call broadcast and aerial surveys (e.g., McLeod et al. 2000). A combination of ground-based survey techniques often are used to find raptors in a study area (e.g., Craighead and Craighead 1956) and searches from foot often are used to find raptors or their nests in historical nesting areas or habitat patches thought likely to harbor them (e.g., Clough 2001).

Surveys from watercraft have been used to estimate raptor population size (e.g., Anthony et al. 1999) and relative abundance (e.g., Frere et al. 1999), and to find vocalizing owls (e.g., Erdman et al. 1997) and raptors that nest near shorelines or forage on aquatic prey to monitor reproduction (e.g., Gerrard et al. 1990) or to study behavior (e.g., Flemming et al. 1992, Garrett at al. 1993), and to assess responses of migrating raptors to local prey abundance (e.g., Restani et al. 2000). Surveys from watercraft also have been used to assess raptor breeding population change following perturbation (e.g., Murphy et al. 1997) and to document population recovery (e.g., Castellanos et al. 1997, Wilson et al. 2000).

Raptor Surveys from the Air

Aircraft (primarily airplanes and helicopters, but also ultralight aircraft [e.g., Leshem 1989]) have been used to conduct surveys for raptors. Safety and design con-

siderations for aerial surveys are summarized and outlined in Fuller and Mosher (1987). Aerial surveys, which have been used most frequently to find and identify raptor nests (e.g., Sharp et al. 2001) and nesting aggregations (Simmons 2002), are generally most useful for species with prominent nests, such as eagles (McIntyre 2002) and cliff-nesting falcons (Gaucher et al. 1995). In North America, aerial surveys have been used extensively to find nests and monitor reproduction of Bald Eagles (Jacobson and Hodges 1999) and Ospreys (*Pandion haliaetus*) (Ewins and Miller 1994). Similar aerial surveys have been used to find nests of large raptors in Australia (Mooney 1988, Sharp et al. 2001), Africa (Tarboton and Benson 1988, Hustler and Howells 1988), and Asia (Utekhina 1994). Aerial surveys to find prominent nests of smaller, often cliff-nesting raptors have been conducted in North America (Wilson et al. 2000), Africa (Simmons 2002), Central America (Thorstrom et al. 2002), and the Middle East (Gaucher et al. 1995).

In open habitats, nests in isolated trees, on cliff faces, and on other prominent locations are readily detected from the air (Ayers and Anderson 1999, Wilson et al. 2000). Surveys from aircraft also have been used to find nests in less-open habitats as well (Cook and Anderson 1990), and to supplement ground-based nest searches (Dickinson and Arnold 1996, McLeod et al. 2000) for tree-nesting raptors, but detectability of nests on these surveys generally has not been estimated (but see Anthony et al. 1999, Ayers and Anderson 1999, Bowman and Schempf 1999).

To a lesser extent, aerial surveys also have been used to find and count raptors outside of the breeding season (e.g., Kaltenecker and Bechard 1994, Lish 1997). Even so, because individual raptors can be difficult to detect from the air and because they often are widely dispersed, aerial surveys have not been used extensively.

Raptor Counts at Fixed Locations

The most widely reported survey technique that involves counting raptors from fixed locations is counting raptors as they pass sites where they concentrate during migration (Kjellén and Roos 2000; Chapter 6). Beyond counts at raptor migration sites, raptor surveys based on counts at specified points have been used to assess population status or trend, often in conjunction with surveys designed to monitor status of bird communities over broad geographic areas (Arrowood et al. 2001, Ross et al. 2003). As with most raptor surveys, those based on counts at points have occurred primarily during the breeding season (e.g., Steenhof et al. 1999, Kochert and Steenhof 2004), although counts at points have been used to assess winter raptor distribution and abundance in the Netherlands (Sierdsema et al. 1995).

Surveys of raptors over broad geographic areas based on counts at points have been conducted in several regions where very little information exists regarding raptor populations. Such counts have been used to assess status, abundance, and distribution of raptors across large areas in Asia (Thiollay 1989a, Thiollay 1998), Asia Minor (Vaassen 2000), South America (Thiollay 1989b, Manosa and Pedrocchi 1997), and Africa (Thiollay 2001). In North America, trends in abundance of some raptors are discernable at broad geographic scales based on surveys at points along routes established to monitor breeding birds (i.e., the Breeding Bird Survey [Sauer et al. 2004]).

At smaller spatial scales, Debus (1997) incorporated counts at points into a survey of raptors in an Australian park, and Sykes et al. (1999) used counts at points to describe distribution and abundance of Swallow-tailed Kites in Florida. Lehman et al. (1998) and Steenhof et al. (1999) incorporated counts at points into surveys for raptors at the Snake River Birds of Prey National Conservation Area in Idaho. Herremans and Herremans-Tonnoeyr (2000) incorporated point counts into a study of raptor distribution in two landscapes in Botswana.

Surveys also have been based on counting vocalizing raptors (Lane et al. 2001) and on broadcasting calls (McLeod et al. 2000) at points to solicit responses from nocturnal (Takats et al. 2001, Crozier et al. 2003) and diurnal raptors (Kennedy and Stahlecker 1993). Surveys for owls often are based on listening at pre-determined points for vocalizations (e.g., Lane et al. 2001, Takats et al. 2001), or broadcasting conspecific calls to elicit responses and, presumably, to increase the probability of detection (Whelton 1989). Broadcasting or mimicking (Forsman et al. 1996) owl vocalizations has been used to locate nests to assess population dynamics (LaHaye et al. 1997), distribution (Mazur et al. 1997) and range expansion (Wright and Hayward 1998), and diet (Seamans and Gutiérrez 1999). Broadcasts also have been incorporated into surveys designed to estimate owl population trends (Shyry et al. 2001, Takats et al. 2001).

Surveys based on detecting vocalizing diurnal raptors have been used most frequently in forested habitats (Fig. 1). Kimmel and Yahner (1990) and Kennedy and Stahlecker (1993) described survey methodology for Northern Goshawks using call broadcasts, and there

Figure 1. Broadcasting conspecific calls has been used extensively to survey forest-dwelling raptors. Recently developed statistical methods allow for the use of resulting detection data in population monitoring. *(Photo by David E. Andersen.)*

have been numerous subsequent applications (Watson et al. 1999) and extensions (McClaren et al. 2003, Roberson et al. 2005, Hargis and Woodbridge 2006) of this technique.

Broadcasting conspecific or competitor calls has been used extensively to survey forest-dwelling raptors in North America (Rosenfield et al. 1988, Johnson and Chambers 1990, Mosher et al. 1990, Kennedy et al. 1995, Mosher and Fuller 1996, Bosakowski and Smith 1998, McLeod and Andersen 1998, Watson et al. 1999, Dykstra et al. 2001, Gosse and Montevecchi 2001) and, to a lesser extent, in Europe (Cerasoli and Penteriani 1992, Sanchez-Zapata and Calvo 1999, Salvati et al. 2000) and Australia (Debus 1997, Fulton 2002). Listening for spontaneous vocalizations of diurnal raptors near nests (Stewart et al. 1996, Penteriani 1999, Dewey et al. 2003) without broadcasting calls also has been used to detect birds of prey.

Surveying Raptors Remotely

There are few published examples of using remotely sensed data to estimate raptor abundance or distribution. Harmata et al. (2000) used radar to assess timing and passage rate of birds, including raptors, during fall migration in Montana, U.S.A. Several others including Kjellén et al. (2001) and Gudmundsson et al. (2002) used radar images to study migrating birds, including raptors. However, species identification can be difficult during migration and migrants do not cross large water bodies where radar may be most effective in detecting birds (Gauthreaux and Belser 2003). Boonstra et al. (1995) had some success detecting raptors using far-infrared thermal imaging, and Leshem (1989) reported on using radar in conjunction with ground observations and motorized gliders to assess raptor migration in Israel. Overall, available remote-sensing technology so far has not been used extensively as a survey tool for raptors.

SUMMARY

Fuller and Mosher (1987) summarized existing information and provided a background regarding objectives of raptor surveys and factors that influence survey design. Since then, several papers have reported results of raptor surveys, and survey methods have advanced considerably. Surveys incorporating call broadcasts have been applied extensively since 1987, and recent statistical advances provide a framework for survey analyses based on occupancy (MacKenzie et al. 2002, 2003, 2004; Royle and Nichols 2003) and factors related to occupancy. Although these methods only recently have been applied to raptor surveys (Olson et al. 2005, Seamans 2005, Hargis and Woodbridge 2006, Henneman et al., in press), their use is likely to increase in the future.

Many of the considerations regarding raptor surveys summarized by Fuller and Mosher (1987) are still primary issues that need to be considered when designing and conducting surveys. First, survey objectives need to be clearly set prior to survey implementation. Survey objectives include determining distribution and abundance, finding raptors to study population dynamics and other aspects of raptor ecology, and providing information from which to base management and conservation decisions. Second, survey techniques must address factors that affect raptor detection, including attributes of raptors themselves (e.g., behavior that makes raptors more or less detectable), environmental conditions, temporal patterns of raptor behavior or dis-

tribution, habitat characteristics, and attributes of observers. Third, survey design must address sampling considerations including what constitutes a sample unit, what is the appropriate sample size, and at what temporal and spatial scale surveys need to be conducted. Only surveys that appropriately address these factors are likely to provide reliable results that relate directly to survey objectives.

Surveys for raptors are a part of almost all raptor research and monitoring efforts, as finding and locating their nests or other evidence of their presence is a necessary component of most field studies. By clearly identifying survey objectives and incorporating survey techniques that appropriately address survey objectives, results of raptor surveys are more likely to provide reliable results that can be extended beyond single efforts and compared spatially and temporally.

Because survey objectives can vary considerably, and because logistical considerations affect conduct of surveys, surveys need to be designed differently for different purposes and in some instances, for different locations. Furthermore, in many instances, raptors exhibit characteristics that make them difficult to survey. Foremost among these is that raptors often occur at low biological densities, making it difficult to apply sampling strategies that result in precise estimates of abundance. Since Fuller and Mosher (1987) presented their review, survey methods for raptors have been developed considerably, in part because of a growing need for such work. Indeed, it is more important than ever that raptor surveys be well designed and implemented, so that resulting information can be used confidently, both to understand raptor ecology and to guide effective raptor management and conservation.

ACKNOWLEDGMENTS

B. Martinez compiled literature from a variety of sources as part of the basis for this manuscript, and C.W. Boal and K. Steenhof reviewed previous drafts of this manuscript and offered suggestions for improvement.

LITERATURE CITED

ANDERSEN, D.E. 1995. Productivity, food habits, and behavior of Swainson's Hawks breeding in southeast Colorado. *J. Raptor Res.* 29:158–165.

———, O.J. RONGSTAD AND W.R. MYTTON. 1985. Line transect analysis of raptor abundance along roads. *Wildl. Soc. Bull.*

13:533–539.

ANTHONY, R.G., M.G. GARRETT AND F.B. ISAACS. 1999. Double-survey estimates of Bald Eagle populations in Oregon. *J. Wildl. Manage.* 63:794–802.

ARROWOOD, P.C., C.A. FINLEY AND B.C. THOMPSON. 2001. Analyses of Burrowing Owl populations in New Mexico. *J. Raptor Res.* 35:362–370.

AYERS, L.W. AND S.H. ANDERSON. 1999. An aerial sightability model for estimating Ferruginous Hawk population size. *J. Wildl. Manage.* 63:85–97.

BAK, J.M., K.G. BOYKIN, B.C. THOMPSON AND D.L. DANIEL. 2001. Distribution of wintering Ferruginous Hawks (*Buteo regalis*) in relation to black-tailed prairie dog (*Cynomys ludovicianus*) colonies in southern New Mexico and northern Chihuahua. *J. Raptor Res.* 35:124–129.

BART, J. AND S. EARNST. 2002. Double sampling to estimate density and population trends in birds. *Auk* 119:36–45.

BELKA, T., O. SREIBR AND V. MRLÍK. 1996. Roadside counts of birds of prey in southeastern Europe and Asia Minor. *Buteo* 8:131–136.

BOONSTRA, R., J.M. EADIE, C.J. KREBS AND S. BOUTIN. 1995. Limitations of far infrared thermal imaging in locating birds. *J. Field Ornithol.* 66:192–198.

BORGES, S.H., L.M. HENRIQUES AND A. CARVALHAES. 2004. Density and habitat use by owls in two Amazonian forest types. *J. Field Ornithol.* 75:176–182.

BOSAKOWSKI, T. AND D.G. SMITH. 1998. Response of a forest raptor community to broadcasts of heterospecific and conspecific calls during the breeding season. *Can. Field-Nat.* 112:198–203.

BOWMAN, T.D. AND P.F. SCHEMPF. 1999. Detection of Bald Eagles during aerial surveys in Prince William Sound, Alaska. *J. Raptor Res.* 33:299–304.

BRIDGEFORD, P. AND M. BRIDGEFORD. 2003. Ten years of monitoring breeding Lappet-faced Vultures *Torgos tracheliotos* in the Namib-Naukluft Park, Namibia. *Vulture News* 48:3–11.

BUSTAMANTE, J. AND J. SEOANE. 2004. Predicting the distribution of four species of raptors (Aves: Accipitridae) in southern Spain: statistical models work better than existing maps. *J. Biogeography* 31:295–306.

CASTELLANOS, A., F. JARAMILLO, F. SALINAS, A. ORTEGA-RUBIO AND C. ARGUELLES. 1997. Peregrine Falcon recovery along the west central coast of the Baja California peninsula, Mexico. *J. Raptor Res.* 31:1–6.

CERASOLI, M. AND V. PENTERIANI. 1992. Effectiveness of censusing woodland birds of prey by playback. *Avocetta* 16:35–39.

CLOUGH, L. 2001. Nesting habitat selection and productivity of Northern Goshawks in west-central Montana. *Intermountain J. Sci.* 7:129.

COCHRAN, W.G. 1977. Sampling techniques, 3rd Ed. John Wiley & Sons, Inc., New York, NY U.S.A.

COOK, J.G. AND S.H. ANDERSON. 1990. Use of helicopters for surveys of nesting Red-shouldered Hawks. *Prairie Nat.* 22:49–53.

CRAIGHEAD, J.J. AND F.C. CRAIGHEAD, JR. 1956. Hawks, owls, and wildlife. Stackpole Co., Harrisburg, PA U.S.A.

CROZIER, M.L., M.E. SEAMANS AND R.J. GUTIÉRREZ. 2003. Forest owls detected in the central Sierra Nevada. *West. Birds* 34:149–156.

CURRIE, D., R. FANCHETTE, J. MILLETT, C. HOAREAU AND N.J. SHAH. 2004. The distribution and population of the Seychelles (barelegged) Scops Owl *Otus insularis* on Mahe: consequences for conservation. *Ibis* 146:27–37.

DEBUS, S.J.S. 1997. A survey of the raptors of Jervis Bay National Park. *Aust. Birds* 30:29–44.

DEKKER, D. 1995. Prey capture by Peregrine Falcons wintering on southern Vancouver Island, British Columbia. *J. Raptor Res.* 29:26–29.

DEWEY, S.R., P.L. KENNEDY AND R.M. STEPHENS. 2003. Are dawn vocalization surveys effective for monitoring goshawk nest-area occupancy? *J. Wildl. Manage.* 67:390–397.

DICKINSON, V.M. AND K.A. ARNOLD. 1996. Breeding biology of the Crested Caracara in south Texas. *Wilson Bull.* 108:516–523.

DYKSTRA, C.R., F.B. DANIEL, J.L. HAYS AND M.M. SIMON. 2001. Correlation of Red-shouldered Hawk abundance and macrohabitat characteristics in southern Ohio. *Condor* 103:652–656.

ERDMAN, T.C., T.O. MEYER, J.H. SMITH AND D.M. ERDMAN. 1997. Autumn populations and movements of migrant Northern Saw-whet Owls (*Aegolius acadicus*) at Little Suamico, Wisconsin. Pages 167–172 *in* J. R. Duncan, D. H. Johnson, and T. H. Nicholls [EDS.], Proceedings of the Second International Symposium on Biology and Conservation of Owls of the Northern Hemisphere. USDA Forest Service, North Central Research Station, St. Paul, MN USA.

EWINS, P.J. AND M.J.R. MILLER. 1994. How accurate are aerial surveys for determining productivity of Ospreys? *J. Raptor Res.* 33:295–298.

FERGUSON, H.L. 2004. Relative abundance and diversity of winter raptors in Spokane County, eastern Washington. *J. Raptor Res.* 38:181–186.

FLEMMING, S.P., P.C. SMITH, N.R. SEYMOUR AND R.P. BANCROFT. 1992. Ospreys use local enhancement and flock foraging to locate prey. *Auk* 109:649–654.

FORSMAN, E.D., S.G. SOVERN, D.E. SEAMAN, K.J. MAURICE, M. TAYLOR AND J.J. ZISA. 1996. Demography of the Northern Spotted Owl on the Olympic Peninsula and east slope of the Cascade Range, Washington. *Stud. Avian Biol.* 17:21–30.

FRERE, E., A. TRAVAINI, A. PARERA AND A. SCHIAVINI. 1999. Striated Caracara (*Phalcoboenus australis*) population at Staten and Ano Nuevo Islands. *J. Raptor Res.* 33:268–269.

FULLER, M.R. AND J.A. MOSHER. 1987. Raptor survey techniques. Pages 37–65 *in* B. A. Giron Pendleton, B. A. Millsap, K. W. Cline, and D. M. Bird [EDS.], Raptor management techniques manual. National Wildlife Federation, Washington, DC U.S.A.

FULTON, G.R. 2002. Avifauna of Mount Tomah Botanic Gardens and upper Stockyard Gully in the Blue Mountains, New South Wales. *Corella* 26:1–12.

GARNER, H.D. AND J.C. BEDNARZ. 2000. Habitat use by Red-tailed Hawks wintering in the Delta Region of Arkansas. *J. Raptor Res.* 34:26–32.

GARRETT, M.G., J.W. WATSON AND R.G. ANTHONY. 1993. Bald Eagle home range and habitat use in the Columbia River estuary. *J. Wildl. Manage.* 57:19–27.

GAUCHER, P., J.-M. THIOLLAY AND X. EICHAKER. 1995. The Sooty Falcon *Falco concolor* on the Red Sea coast of Saudi Arabia - distribution, numbers and conservation. *Ibis* 137:29–34.

GAUTHREAUX, S.A., JR. AND C.G. BELSER. 2003. Radar ornithology and biological conservation. *Auk* 120:266–277.

GEISSLER, P.H. AND M.R. FULLER. 1987. Estimation of the proportion of an area occupied by an animal species. Pages 533–538 *in* Proceedings of the section on survey research methods of the American Statistical Association. American Statistical Association, Alexandria, VA U.S.A.

GERHARDT, R.P., N.B. GONZALEZ, D.M. GERHARDT AND C.J. FLAT-

TENS. 1994. Breeding biology and home range of two *Ciccaba* owls. *Wilson Bull.* 106:629–639.

GERRARD, J.M., G.R. BORTOLOTTI, E.H. DZUS, P.N. GERRARD AND D.W.A. WHITFIELD. 1990. Boat census of Bald Eagles during the breeding season. *Wilson Bull.* 102:720–726.

GOLDSTEIN, M. I. 2000. Nest-site characteristics of Crested Caracaras in La Pampa, Argentina. *J. Raptor Res.* 34:330–333.

——— AND T. J. HIBBITTS. 2004. Summer roadside raptor surveys in the western Pampas of Argentina. *J. Raptor Res.* 38:152–157.

GOSSE, J.W. AND W.A. MONTEVECCHI. 2001. Relative abundances of forest birds of prey in western Newfoundland. *Can. Field-Nat.* 115:57–63.

GRIER, J.W. 1977. Quadrat sampling of a nesting population of Bald Eagles. *J. Wildl. Manage.* 41:438–443.

GUDMUNDSSON G.A., T. ALERSTAM, M. GREEN AND A. HEDENSTROEM. 2002. Radar observations of arctic bird migration at the Northwest Passage, Canada. *Arctic* 55:21–43.

HAINES, D.E. AND K.H. POLLOCK. 1998. Estimating the number of active and successful Bald Eagle nests; an application of the dual frame method. *Environ. Ecol. Stat.* 5:245–256.

HARGIS, C.D. AND B. WOODBRIDGE. 2006. A design for monitoring Northern Goshawks at the bioregional scale. *Stud. Avian Biol.* 31:274–287.

HARMATA, A.R., K.M. PODRUZNY, J.R. ZELENAK AND M.L. MORRISON. 2000. Passage rates and timing of bird migration in Montana. *Am. Midl. Nat.* 143:30–40.

HENNEMAN, C., M.A. MCLEOD AND D.E. ANDERSEN. Presence/absence surveys to assess Red-shouldered Hawk occupancy in central Minnesota. *J. Wildl. Manage.* In press.

HERREMANS, M. AND D. HERREMANS-TONNOEYR. 2000. Land use and the conservation status of raptors in Botswana. *Biol. Conserv.* 94:31–41.

——— AND D. HERREMANS-TONNOEYR. 2001. Roadside abundance of raptors in the western Cape Province, South Africa: a three-decade comparison. *Ostrich* 72:96–100.

HIRALDO, F., J.A. DONAZAR, O. CEBALLOS, A. TRAVAINI, J. BUSTAMANTE AND M. FUNES. 1995. Breeding biology of a Grey Eagle-buzzard population in Patagonia. *Wilson Bull.* 107:675–685.

HUBBARD, J.P., J.W. SHIPMAN AND S.O. WILLIAMS, JR. 1988. An analysis of vehicular counts of roadside raptors in New Mexico, 1974–1985. Pages 204–209 *in* R. L. Glinski, B.A. Giron Pendleton, M.B. Moss, M.N. LeFranc, JR., B.A. Millsap, S.W. Hoffman, C.E. Ruibal, D.L. Karhe and D.L. Ownens [EDS.], Proceedings of the southwest raptor management symposium and workshop. National Wildlife Federation Scientific and Technical Series No. 11. National Wildlife Federation, Washington, DC U.S.A.

HUSTLER, K. AND W.W. HOWELLS. 1988. The effect of primary production on breeding success and habitat selection in the African Hawk-eagle. *Ostrich* 58:135–138.

JACOBSON, M.J. AND J.I. HODGES. 1999. Population trend of adult Bald Eagles in southeast Alaska, 1967–97. *J. Raptor Res.* 33:295–298.

JANES, S. W. 1984. Influences of territory composition and interspecific competition on Red-tailed Hawk reproductive success. *Ecology* 65:862–870.

JOHNSON, G. AND R.E. CHAMBERS. 1990. Response to conspecific, roadside playback recordings: an index of Red-shouldered Hawk breeding density. *N. Y. State Mus. Bull.* 471:71–76.

JOY, S.M., R.T. REYNOLDS, R.L. KNIGHT AND R.W. HOFFMAN. 1994.

Feeding ecology of Sharp-shinned Hawks nesting in deciduous and coniferous forests in Colorado. *Condor* 96:455–467.

KALTENECKER, G.S. 2001. Continued monitoring of Boise's wintering Bald Eagles, and monitoring of the Dead Dog Creek Bald Eagle roost site, winters 1997/1998 and 1998/1999. *Idaho Tech. Bull.* (01).

——— AND M.J. BECHARD. 1994. Accuracy of aerial surveys for wintering Bald Eagles. *J. Raptor Res.* 28:59.

———, K. STEENHOF, M.J. BECHARD AND J.C. MUNGER. 1998. Winter foraging ecology of Bald Eagles on a regulated river in southwest Idaho. *J. Raptor Res.* 32:215–220.

KENNEDY, P.L. AND D.W. STAHLECKER. 1993. Responsiveness of nesting Northern Goshawks to taped broadcasts of three conspecific calls. *J. Raptor Res.* 27:74–75.

———, D.E. CROWE AND T.F. DEAN. 1995. Breeding biology of the Zone-tailed Hawk at the limit of its distribution. *J. Raptor Res.* 29:110–116.

KENWARD, R.E., S.S. WALLS, K.H. HODDER, M. PAHKALA, S.N. FREEMAN AND V.R. SIMPSON. 2000. The prevalence of non-breeders in raptor populations: evidence from rings, radio-tags and transect surveys. *Oikos* 91:271–279.

KERLINGER, P. AND M.R. LEIN. 1988. Population ecology of Snowy Owls during winter on the Great Plains of North America. *Condor* 90:866–874.

KIMMEL, J.T. AND R.H. YAHNER. 1990. Response of Northern Goshawks to taped conspecific and Great Horned Owl calls. *J. Raptor Res.* 24:107–112.

KJELLÉN, N. AND G. ROOS. 2000. Population trends in Swedish raptors demonstrated by migration counts at Falsterbo, Sweden 1942–97. *Bird Study* 47:195–211.

———, M. HAKE AND T. ALERSTAM. 2001. Timing and speed of migration in male, female and juvenile Ospreys *Pandion haliaetus* between Sweden and Africa as revealed by field observations, radar and satellite tracking. *J. Avian Biol.* 32:57–67.

KOCHERT, M.N. AND K. STEENHOF. 2004. Abundance and productivity of Prairie Falcons and Golden Eagles in the Snake River Birds of Prey National Conservation Area: 2003 annual report (Final Draft). U.S. Geological Survey, Forest and Rangeland Ecosystem Science Center, Snake River Field Station, Boise, ID U.S.A.

LAHAYE, W.S., R.J. GUTIÉRREZ AND D.R. CALL. 1997. Nest-site selection and reproductive success of California Spotted Owls. *Wilson Bull.* 109:42–51.

LANE, W.H., D.E. ANDERSEN AND T.H. NICHOLLS. 2001. Distribution, abundance, and habitat use of singing male Boreal Owls in northeast Minnesota. *J. Raptor Res.* 35:130–140.

LEHMAN, R.N., L.B. CARPENTER, K. STEENHOF AND M.N. KOCHERT. 1998. Assessing relative abundance and reproductive success of shrubsteppe raptors. *J. Field Ornithol.* 69:244–256.

LESHEM, Y. 1989. Following raptor migration from the ground, motorized glider and radar at a junction of three continents. Pages 43–52 *in* B.-U. Meyburg and R.D. Chancellor [EDS.], Raptors in the modern world: proceedings of the III world conference on birds of prey and owls. World Working Group on Birds of Prey and Owls, Berlin, Germany.

LISH, J.W. 1997. Diet, population size, and high-use areas of Bald Eagles wintering at Grand Lake, Oklahoma. *Okla. Ornithol. Soc. Bull.* 30:1–6.

MACKENZIE, D.I., J.D. NICHOLS, G.B. LACHMAN, S. DROEGE, J.A. ROYLE AND C.A. LANGTIMM. 2002. Estimating site occupancy rates when detection probabilities are less than one. *Ecology*

83:2248–2255.

———, J.D. NICHOLS, J.E. HINES, M.G. KNUTSON AND A.B. FRANKLIN. 2003. Estimating site occupancy, colonization, and local extinction when a species is detected imperfectly. *Ecology* 84:2200–2207.

———, L.L. BAILEY AND J.D. NICHOLS. 2004. Investigating species co-occurrence patterns when species are detected imperfectly. *J. Anim. Ecol.* 73:546–555.

MANLY, B.F.J., L.L. MCDONALD AND T.L. MCDONALD. 1999. The robustness of mark-recapture methods: a case study of the Northern Spotted Owl. *J. Agric. Biol. Environ. Stat.* 4:78–101.

MANOSA, S. AND V. PEDROCCHI. 1997. A raptor survey in the Brazilian Atlantic rainforest. *J. Raptor Res.* 31:203–207.

———, J. REAL AND J. CODINA. 1998. Selection of settlement areas by juvenile Bonelli's Eagle in Catalonia. *J. Raptor Res.* 32:208–214.

MARTINEZ, F., R. F. RODRIGUEZ, AND G. BLANCO. 1997. Effects of monitoring frequency on estimates of abundance, age distribution, and productivity of colonial Griffon Vultures. *J. Field Ornithol.* 68:392–399.

MAZUR, K.M., P.C. JAMES, M.J. FITZSIMMONS, G. LANGEN AND R.H.M. ESPIE. 1997. Habitat associations of the Barred Owl in the boreal forest of Saskatchewan, Canada. *J. Raptor Res.* 31:253–259.

MCCLAREN, E.L., P.L. KENNEDY AND P.L. CHAPMAN. 2003. Efficacy of male goshawk food-delivery calls in broadcast surveys on Vancouver Island. *J. Raptor Res.* 37:198–208.

MCINTYRE, C.L. 2002. Patterns in nesting area occupancy and reproductive success of Golden Eagles (*Aquila chrysaetos*) in Denali National Park and Preserve, Alaska, 1988–99. *J. Raptor Res.* 36:50–54.

MCLEOD, M.A. AND D.E. ANDERSEN. 1998. Red-shouldered Hawk broadcast surveys: factors affecting detection of responses and population trends. *J. Wildl. Manage.* 62:1385–1397.

———, B.A. BELLEMAN, D.E. ANDERSEN AND G. OEHLERT. 2000. Red-shouldered Hawk nest site selection in north-central Minnesota. *Wilson Bull.* 112:203–213.

MENDENHALL, W., L. OTT AND R.L. SCHAEFFER. 1971. Elementary survey sampling. Duxbury Press, Belmont, CA U.S.A.

MEYER, K.D. 1994. Communal roosts of American Swallow-tailed Kites: implications for monitoring and conservation. *J. Raptor Res.* 28:62.

MOONEY, N. 1988. Efficiency of fixed-winged aircraft for surveying eagle nests. *Australas. Raptor Assoc. News* 9:28–29.

MOSHER, J.A. AND M.R. FULLER. 1996. Surveying woodland hawks with broadcasts of Great Horned Owl vocalizations. *Wildl. Soc. Bull.* 24:531–536.

———, M.R. FULLER AND M. KOPENY. 1990. Surveying woodland hawks by broadcast of conspecific vocalizations. *J. Field Ornithol.* 61:453–461.

MURPHY, S.M., R.H. DAY, J.A. WIENS AND K.R. PARKER. 1997. Effects of the Exxon Valdez oil spill on birds: comparisons of pre- and post-spill surveys in Prince William Sound, Alaska. *Condor* 99:299–313.

OLSON, C.V. AND D.P. ARSENAULT. 2000. Differential winter distribution of Rough-legged Hawks (*Buteo lagopus*) by sex in western North America. *J. Raptor Res.* 34:157–166.

OLSON, G.S., R.G. ANTHONY, E.D. FORSMAN, S.H. ACKERS, P.J. LOSCHL, J.A. REID, K.M. DUGGER, E.M. GLENN AND W.J. RIPPLE. 2005. Modeling of site occupancy dynamics for Northern Spotted Owls, with emphasis on the effects of Barred Owls. *J.*

Wildl. Manage. 69:918–932.

PENTERIANI, V. 1999. Dawn and morning goshawk courtship vocalizations as a method for detecting nest sites. *J. Wildl. Manage.*63:511–516.

PHILLIPS, R.L., T.P. MCENEANEY AND A.E. BESKE. 1984. Population densities of breeding Golden Eagles in Wyoming. *Wildl. Soc. Bull.* 12:269–273.

PRAKASH, V., D.J. PAIN, A.A. CUNNINGHAM, P.F. DONALD, N. PRAKASH, A. VERMA, R. GARGI, S. SIVAKUMAR AND A.R. RAHMANI. 2003. Catastrophic collapse of Indian White-backed *Gyps bengalensis* and Long-billed *Gyps indicus* vulture populations. *Biol. Conserv.* 109:381–390.

RALPH, C.J. 1981. Terminology used in estimating numbers of terrestrial birds. Pages 502–578 *in* C.J. Ralph and J.M. Scott [EDS.], Estimating numbers of terrestrial birds. *Stud. Avian Biol.* 6.

——— AND J.M. SCOTT [EDS.]. 1981. Estimating numbers of terrestrial birds. *Stud. Avian Biol.* 6.

———, G.R. GEUPEL, P. PYLE, T.E. MARTIN AND D.F. DESANTE. 1993. Handbook of field methods for monitoring landbirds. USDA Forest Service General Technical Report PSW-GTR-144, Pacific Southwest Research Station, Albany, CA U.S.A.

REJT, L. 2001. Feeding activity and seasonal changes in prey composition of urban Peregrine Falcons *Falco peregrinus*. *Acta Ornithologica* 36:165–169.

RESTANI, M., A.R. HARMATA AND E.M. MADDEN. 2000. Numerical and functional responses of migrant Bald Eagles exploiting a seasonally concentrated food source. *Condor* 102:561–568.

ROBERSON, A.M., D.E. ANDERSEN AND P.L. KENNEDY. 2005. Do breeding phase and detection distance influence the effective area surveyed for Northern Goshawks? *J. Wildl. Manage.* 69:1240–1250.

ROSENFIELD, R.N., J. BIELEFELDT AND R.K. ANDERSON. 1988. Effectiveness of broadcast calls for detecting breeding Cooper's Hawks. *Wildl. Soc. Bull.* 16:210–212.

ROSS, B.D., D.S. KLUTE, G.S. KELLER, R.H. YAHNER AND J. KARISH. 2003. Inventory of birds at six national parks in Pennsylvania. *J. PA Acad. Sci.* 77:20–40.

ROYLE, J.A. AND J.D. NICHOLS. 2003. Estimating abundance from repeated presence-absence data or point counts. *Ecology* 84:777–790.

SALVATI, L., A. MANGANARO AND S. FATTORINI. 2000. Responsiveness of nesting Eurasian Kestrels *Falco tinnunculus* to call playbacks. *J. Raptor Res.* 34:319–321.

SANCHEZ-ZAPATA, J.A. AND J.F. CALVO. 1999. Raptor distribution in relation to landscape composition in semi-arid Mediterranean habitats. *J. Appl. Ecol.* 36:254–262.

———, M. CARRETE, A. GRAVILOV, S. SKLYARENKO, O. CEBALLOS, J.A. DONAZAR AND F. HIRALDO. 2003. Land use changes and raptor conservation in steppe habitats of eastern Kazakhstan. *Biol. Conserv.* 111:71–77.

SAUER, J.R., J.E. HINES AND J. FALLON. 2004. The North American Breeding Bird Survey, results and analysis 1966–2003. Version 2004.1. U.S. Geological Survey, Patuxent Wildlife Research Center, Laurel, MD U.S.A.

SCHMUTZ, J.K. 1984. Ferruginous and Swainson's Hawk abundance and distribution in relation to land use in southeastern Alberta. *J. Wildl. Manage.* 48:1180–1187.

SCHREUDER, H.T., R. ERNST AND H. RAMIREZ-MADONADO. 2004. Statistical techniques for sampling and monitoring natural resources. USDA Forest Service General Technical Report

RMRS-GTR-126, Rocky Mountain Research Station. Fort Collins, CO U.S.A.

SEAMANS, M.E. 2005. Population biology of the California Spotted Owl in the central Sierra Nevada. Ph.D. dissertation, University of Minnesota, St. Paul, MN U.S.A.

——— AND R.J. GUTIÉRREZ. 1999. Diet composition and reproductive success of Mexican Spotted Owls. *J. Raptor Res.* 33:143–148.

SEERY, D.B. AND D.J. MATIATOS. 2000. Response of wintering buteos to plague epizootics in prairie dogs. *West. N. Am. Nat.* 60:420–425.

SHARP, A., M. NORTON AND A. MARKS. 2001. Breeding activity, nest site selection and nest spacing of Wedge-tailed Eagles (*Aquila audax*) in western New South Wales. *Emu* 101:323–328.

SHYRY, D.T., T.I. WELLICOME, J.K. SCHMUTZ, G.L. ERICKSON, D.L. SCOBIE, R.F. RUSSELL AND R.G. MARTIN. 2001. Burrowing Owl population-trend surveys in southern Alberta: 1991–2000. *J. Raptor Res.* 35:310–315.

SIERDSEMA, H., W. HAGEMEIJER, F. HUSTINGS AND T. VERSTRAEL. 1995. Point transect counts of wintering birds in The Netherlands 1978–1992. *Ring* 17:46–60.

SIMMONS, R.E. 2002. A helicopter survey of Cape Vultures *Gyps coprotheres*, Black Eagles *Aquila verreauxii* and other cliff-nesting birds of the Waterberg Plateau, Namibia, 2001. *Lanioturdus* 34:23–29.

SMALLWOOD, S.K. 1998. On the evidence for listing Northern Goshawks (*Accipiter gentilis*) under the Endangered Species Act: a reply to Kennedy. *J. Raptor Res.* 32:323–329.

SORLEY, C.S. AND D.E. ANDERSEN. 1994. Raptor abundance in south-central Kenya in relation to land-use patterns. *Afr. J. Ecol.* 32:30–38.

STEENHOF, K., M.N. KOCHERT, L.B. CARPENTER AND R.N. LEHMAN. 1999. Long-term Prairie Falcon population changes in relation to prey abundance, weather, land uses, and habitat conditions. *Condor* 101:28–41.

STEWART, A.C., R.W. CAMPBELL AND S. DICKIN. 1996. Use of dawn vocalizations for detecting breeding Cooper's Hawks in an urban environment. *Wildl. Soc. Bull.* 24:291–293.

STOUT, W.E., R.K. ANDERSON AND J.M. PAPP. 1998. Urban, suburban and rural Red-tailed Hawk nesting habitat and populations in southeast Wisconsin. *J. Raptor Res.* 32:221–228.

SYKES, P.W., C.B. KEPLER, K.L. LITZENBERGER, H.R. SANSING, E.T.R. LEWIS AND J.S. HATFIELD. 1999. Density and habitat of breeding Swallow-tailed Kites in the lower Suwannee ecosystem, Florida. *J. Field Ornithol.* 70:321–336.

TAKATS, D.L., C.M. FRANCIS, G. HOLROYD, J.R. DUNCAN, K.M. MAZUR, R.J. CANNINGS, W.HARRIS AND D. HOLT. 2001. Guidelines for nocturnal owl monitoring in North America. Beaverhill Bird Observatory and Bird Studies Canada, Edmonton, Alberta.

TARBOTON, W.R. AND P.C. BENSON. 1988. Aerial counting of Cape Vultures. *S. Afr. J. Wildl. Res.* 18:93–96.

THIOLLAY, J.-M. 1989a. Censusing of diurnal raptors in a primary rain forest: comparative methods and species detectability. *J. Raptor Res.* 23:72–84.

———. 1989b. Area requirements for the conservation of rain forest raptors and game birds in French Guiana. *Conserv. Biol.* 3:128–137.

———. 1998. Current status and conservation of Falconiformes in tropical Asia. *J. Raptor Res.* 32:40–55.

———. 2001. Long-term changes of raptor populations in northern Cameroon. *J. Raptor Res.* 35:173–186.

——— AND Z. RAHMAN. 2002. The raptor community of central Sulawesi: habitat selection and conservation status. *Biol. Conserv.* 107:111–122.

THOME, D.M., C.J. ZABEL AND L.V. DILLER. 1999. Forest stand characteristics and reproduction of Northern Spotted Owls in managed north-coastal California forests. *J. Wildl. Manage.* 63:44–59.

THORSTROM, R., R. WATSON, A. BAKER, S. AYERS AND D. L. ANDERSON. 2002. Preliminary ground and aerial surveys for Orange-breasted Falcons in Central America. *J. Raptor Res.* 36:39–44.

TRAVAINI, A., J.A. DONAZAR, O. CEBALLOS, M. FUNES, A. RODRIGUEZ, J. BUSTAMANTE, M. DELIBES AND F. HIRALDO. 1994. Nest-site characteristics of four raptor species in the Argentinian Patagonia. *Wilson Bull.* 106:753–757.

UTEKHINA, I.G. 1994. Productivity at Steller's Sea Eagle and Osprey nests on the Magadan State Nature Reserve, Magadan, Russia. *J. Raptor Res.* 28:66.

VAASSEN, E.W.A.M. 2000. Habitat choice, activity pattern, and hunting method of wintering raptors in the Goksu Delta, southern Turkey. *De Takkeling* 8:142–162.

VINUELA, J. 1997. Road transects as a large-scale census method for raptors: the case of the Red Kite *Milvus milvus* in Spain. *Bird Study* 44:155–165.

WATSON, J.W., D.W. HAYS AND D.J. PIERCE. 1999. Efficacy of Northern Goshawk broadcast surveys in Washington state. *J. Wildl. Manage.* 63:98–106.

WHELTON, B.D. 1989. Distribution of the Boreal Owl in eastern Washington and Oregon. *Condor* 91:712–716.

WHITE, C.M., R.J. RITCHIE AND B.A. COOPER. 1995. Density and productivity of Bald Eagles in Prince William Sound, Alaska, after the *Exxon Valdez* oil spill. Pages 762–779 *in* P.G. Wells, J.N. Butler and J.S. Hughes, [EDS.], *Exxon Valdez* oil spill: fate and effects in Alaskan waters, ASTM STP 1219. American Society for Testing and Materials, Philadelphia, PA U.S.A.

WILLIAMS, B.K., J.D. NICHOLS AND M.J. CONROY. 2002. Analysis and management of animal populations. Academic Press, San Diego, CA U.S.A.

WILLIAMS, C.K., R.D. APPLEGATE, R.S. LUTZ AND D.H. RUSCH. 2000. A comparison of raptor densities and habitat use in Kansas cropland and rangeland ecosystems. *J. Raptor Res.* 34:203–209.

WILSON, U.W., A. MCMILLAN AND F.C. DOBLER. 2000. Nesting, population trend and breeding success of Peregrine Falcons on the Washington outer coast, 1980–98. *J. Raptor Res.* 34:67–74.

WOODBRIDGE, B., K.K. FINLEY AND P.H. BLOOM. 1995. Reproductive performance, age structure, and natal dispersal of Swainson's Hawks in the Butte Valley, California. *J. Raptor Res.* 29:187–192.

WRIGHT, A.L. AND G.D. HAYWARD. 1998. Barred Owl range expansion into the central Idaho wilderness. *J. Raptor Res.* 32:77–81.

YAHNER, R.H. AND R.W.J. ROHRBAUGH. 1998. A comparison of raptor use of reclaimed surface mines and agricultural habitats in Pennsylvania. *J. Raptor Res.* 32:178–180.

YOSEF, R., J. BOULOS AND O. TUBBESHAT. 1999. The Lesser Kestrel (*Falco naumanni*) at Dana Nature Reserve, Jordan. *J. Raptor Res.* 33:341–342.

Migration Counts and Monitoring

6

Keith L. Bildstein
Acopian Center for Conservation Learning,
Hawk Mountain Sanctuary
410 Summer Valley Road, Orwigsburg, PA 17961 U.S.A.

Jeff P. Smith
HawkWatch International
1800 S. West Temple, Suite 226,
Salt Lake City, UT 84115-1851 U.S.A.

Reuven Yosef
International Birding and Research Center in Eilat
PO Box 744, Eilat 88000 Israel

Long-distance raptor migration has fascinated humanity for thousands of years. Palearctic accounts of the phenomenon date from the Old Testament (Job 39:26–29). Western Hemisphere accounts date from within 30 years of European settlement (Baughman 1947). Today, premiere raptor-migration watchsites, such as those in Eilat, Israel (International Birdwatching Center Eilat 1987), and at Hawk Mountain Sanctuary, U.S.A. (Allen et al. 1995, Bildstein and Compton 2000), attract tens of thousands of visitors annually (Fig. 1). In North America, the Hawk Migration Association of North America — an organization of more than 400 members — is devoted entirely to the study and conservation of migrating raptors.

Because of long-standing interest in raptor migration, specialists in the field know much about the flight mechanics and geography of the phenomenon (Kerlinger 1989, Zalles and Bildstein 2000, Bildstein 2006).

Figure 1. Hawk Mountain Sanctuary (top) and Eilat, Israel (below). The view at Hawk Mountain Sanctuary is to the east along the Kittatinny Ridge from the North Lookout. Populations of 16 species of North American breeders have been monitored at the site since 1934. The view at Eilat is to the south, toward the Gulf of Aqaba from near Mt. Yoash. Populations of 38 species of European and Asian breeders have been monitored at the site since 1977.
(Hawk Mountain photo by M. Linkevich; Eilat photo by K. Bildstein)

And indeed, in many ways, the movements of the world's more than 183 species of migratory raptors are better documented than those of any other avian taxon (Zalles and Bildstein 2000, Bildstein 2006). Studies of migrating raptors have made major contributions, both to avian ecology (Newton 1979) and to conservation biology (e.g., Newton and Chancellor 1985, Senner et al. 1986, Meyburg and Chancellor 1994, Chancellor et al. 1998, Yosef et al. 2002, Thompson et al. 2003, Yosef and Fornasari 2004). The status of raptor-migration science is especially solid with regard to spatial and temporal patterns of migration, particularly along major migratory corridors in North America, the Western Palearctic, and portions of the Middle East (Shirihai et al. 2000, Zalles and Bildstein 2000, Hoffman et al. 2002, Bildstein 2006). On the other hand, much remains to be learned of raptor migration elsewhere, as well as about the principal causes and consequences of raptor migration.

Kerlinger (1989) and Bildstein (2006) provide thorough reviews of many aspects of raptor-migration science, including the principal methods of study used to date (Appendix 1). Zalles and Bildstein (2000), Bildstein and Zalles (2005), and Bildstein (2006) detail the patterns and processes of the global geography of the flight. Bildstein (1998a) reviews the status of raptor-migration science through the mid-1990s.

In this chapter we detail the rationale and methods involved in sampling the visible migration of raptors at established raptor-migration watchsites (including the means by which watchsites are identified), guidelines for data recording, information on the ways in which migration-count data can be stored for later analysis, and how resulting status and trends data can be communicated to the scientific community. We then discuss migration counts within the perspective of long-term monitoring, presenting and exploring the use of such counts as indexes of regional population trends. We offer an operational definition of environmental monitoring, and conclude by outlining a procedure for designing long-term monitoring efforts at watchsites.

RAPTOR MIGRATION WATCHSITES

Raptors are secretive, wide-ranging, highly mobile avian predators whose populations can be both logistically difficult and financially prohibitive to survey and monitor (Fuller and Mosher 1981, 1987). One potentially cost-effective method for monitoring regional populations of raptors is sampling their numbers during migration at one or more migration watchsites along traditional migration corridors (Bildstein 1998b, Zalles and Bildstein 2000).

Counts of migrating raptors at established watchsites have been used to study raptor migration ecology since the late nineteenth century (Kerlinger 1989, Bildstein 2006). Recently, counts of visible raptor migration at watchsites (hereafter referred to as migration counts) have helped determine the conservation status of migratory populations of raptors (Carson 1962, Hickey 1969, Bednarz et al. 1990, Bildstein 1998b, Hoffman and Smith 2003, Yosef and Fornasari 2004). In addition to their value in monitoring regional populations of raptors, migration counts have helped identify principal migration routes, assess the phenology of raptor migration, and determine raptor flight dynamics and other aspects of raptor behavior (Smith 1980, 1985a,b; Kerlinger 1989, Zalles and Bildstein 2000, Bildstein and Zalles 2001, Hoffman et al. 2002).

Indeed, because they are cost-effective and relatively easy to implement, migration counts are one of the most commonly used methods in raptor migration science (Kerlinger 1989, Bildstein 1998b). Conducted over time, migration counts have been used to determine daily and seasonal timing of migration, species diversity, and the volume of migration as a function of weather (Haugh 1972, Kerlinger 1989). In addition, direct visual observations associated with migration counts have yielded valuable information on the behavior of migrating raptors, including the relative use of flight patterns (e.g., soaring versus flapping flight), flocking behavior, interspecies interactions, roosting behavior, and weather effects (Kerlinger 1989, Allen et al. 1996, Yates et al. 2001).

Although the use of migration counts to indicate raptor population trends is not without its limitations, and although statistical methods regarding their analyses continue to be modified (Hussell 1985, Fuller and Titus 1990, Titus et al. 1990, Hoffman and Smith 2003), preliminary evaluations of the usefulness of such counts for determining population trends are encouraging (Bednarz and Kerlinger 1989, Dunn and Hussell 1995). Properly collected and analyzed, such data can provide valuable information regarding population fluctuations in these species (Bednarz et al. 1990, Bildstein 1998b, Hoffman and Smith 2003, Yosef and Fornasari 2004).

MIGRATION-COUNT TECHNIQUES

Identifying Objectives

Both long- and short-term studies of raptor populations benefit greatly from careful planning and attention to study design (Fuller and Mosher 1987, Titus et al 1989, Fish 2001). The first step in designing a count effort is to define its objectives. Is the goal of data collection to monitor the passage of all species of raptors in the region, or only certain species? Is the focus of the effort on autumn migration, spring migration, or both? Goldsmith (1991), Spellerberg (1991), and Fish (2001) provide valuable suggestions with regard to identifying objectives of monitoring programs.

Choosing a Site

Once objectives have been established it is necessary to identify a watchsite: the place from which migrating raptors are seen and counted. Watchsites include sites from which migratory raptors can be counted as they migrate past, as well as sites from which they can be observed entering or departing nighttime roosts. Most watchsites are along principal migration corridors, routes that raptors regularly use during their long-distance movements. Identifying these routes is the first step in determining where to locate a watchsite. Although many raptors migrate across broad fronts (cf. Bednarz and Kerlinger 1989), many concentrate during migration along "leading lines" and "diversion lines." As originally described by Geyr von Schweppenburg (1963), leading lines are narrow and relatively long geographical and topographical features that intersect with the principal axis of migration in a region, and whose properties attract migrants to them and induce them to change their direction of travel to follow the leading line. In addition to mountain ridges and their associated deflection updrafts, leading lines include rivers and associated riparian areas that often attract and concentrate large numbers of potential prey items for migrating raptors. Diversion lines, in contrast, are geographical and topographical features along which migrants concentrate not because they are attracted to them, but because they are trying to avoid what lies beyond them (i.e., large bodies of water).

Reviewing the literature on the subject of leading lines and raptor-migration corridors helps one determine the migration routes and timing of raptor migration for a given region. Kerlinger (1989) provides the most up-to-date coverage of raptor flight dynamics available. Anecdotal information and preliminary counts at several potential watchsites can play an important role in identifying points of concentration. Local or regional guides to bird fauna often imply or suggest where currently unconfirmed concentrations of migrants may be passing. Talking to local inhabitants and others who know the area surrounding possible sites also can help you determine when and where large numbers of flying raptors can be seen. Once likely sites have been identified, field reconnaissance will be needed to determine precisely where and when migrating raptors can be seen.

Once a concentration point has been found, it is necessary to establish the best vantage point for counting birds. Ideally, watchsites should have as wide a field of view of the surrounding landscape as possible. Field of view and local relief (height relative to that of the surrounding landscape) determine the amount of sky that can be seen from a counting station. Dunne et al. (1984) recommend a minimum 180° field of view. Potential visibility, however, is not the only concern in determining location. Other factors include site accessibility and safety. Good accessibility, for example, is critical to ensure the logistic feasibility for intensive and prolonged monitoring, and is particularly important if the watchsite is to be used for conservation education as well as monitoring.

In some instances, more than one count point per watchsite may be appropriate. The objectives of a particular study — for instance, determining migration volume as a function of distance from the coast — may necessitate the use of simultaneous counts at several sites, as may flight lines that shift predictably in response to local weather.

Spotting Migrants

Much of what follows has been taken from Dunne et al. (1984) and Brett (1991). Both are useful and informative references on the subject of conducting raptor migration counts. Another useful source is the Hawk Migration Association of North America's (HMANA) *A Beginner's Guide to Hawkwatching* (1982).

Raptors are best spotted by methodically scanning the sky in the direction from which migrants are expected. Observers should scan along the horizon, or below the horizon if the watchsite is at a high-elevation point, beginning perpendicular to the direction of flight and moving upstream until facing directly into the flight line. Then, they should move their binoculars up slightly less than one field of view and repeat the procedure two or three times on the same side of the flight line, and then repeat the same systematic procedure on the opposite side of the flight line (e.g., the other side of the ridgeline). Observers should systematically scan with both binoculars and unaided eyes to ensure effective coverage of both close and distant migrants. Scanning should cover at least 180°, laterally and vertically. Between scans, counters should look to their sides as well as directly overhead to look for birds they may have missed during their scans. Multiple observers, if available, can effectively rotate responsibilities between binocular and unaided-eye scanning, and simultaneously cover different sections of the overall flight line. Observers also should watch for aggregations of resident raptors because they often indicate flight conditions that may be conducive to migration.

Migrating raptors often fly overhead or laterally at some distance from the counter. Establishing a focal point that approximates that distance will help improve detectability. Human eyes typically focus at a distance of 6–7 m if there is nothing in particular upon which to focus. Clouds, distant landmarks, and passing airplanes all provide the observer's eyes with a frame of reference for distant focusing. Particular care must be taken to ensure effective focus and scanning against clear blue skies.

Because correct identification of migrants is important, appropriate optical equipment (i.e., binoculars and telescopes) is essential for migration counts. Binoculars with 8x to 10x magnification are considered best, although 7x binoculars also can be used. Because large areas of sky must be searched, binoculars with wide-angle lens and wide fields of view are recommended. Telescopes with 15x to 20x magnification are considered sufficient magnification. And in fact, heat-wave distortion and tripod vibrations often compromise the use of telescopes with higher magnification. Unless telescopes are routinely available during all observations, they should be used sparingly and only to confirm identifications of distant birds, not for spotting migrants, as variability in their use can impart significant detectability bias. In general, observers should avoid spending too much time staring into a telescope trying to positively identify every distant migrant, as they may inadvertently miss counting far closer, more easily identified birds.

It is important to consider observer fatigue when determining the materials needed to count migrants. Factors such as binocular weight (heavy binoculars

induce arm fatigue more rapidly than lighter ones) and direct sunlight (glare causes eye strain) are important considerations. Counters should dress appropriately, and a comfortable place should be provided for them to sit from time to time. A storage site for field equipment near the watchsite also is useful.

Identifying Migrants

Many raptors are difficult to identify at the species level, especially when they are flying at great speeds and altitudes. When apparent, plumage color and pattern, overall size, general configuration, and characteristic field marks are good ways to identify a raptor. For many species, differences in plumage can be used to determine age and gender. Determining the size of flying migrants is tricky, especially when the distance to the bird is difficult to gauge. Identifying a bird to species usually involves using a combination of cues, including flight pattern, wing-to-tail ratio, head-to-body ratio, wing shape in relation to wind speed, flight profile, etc. Silhouette recognition and the overall *gestalt*, or "GISS" (general impression, size and shape), of a bird can help place individuals in groups that will aid in their identification (e.g., accipiters, buteos, falcons, vultures, eagles, etc., all of which have recognizable *gestalts*).

Field guides that describe migrating raptors in terms of their characteristic field marks are especially useful in this regard. North American field guides include *The Mountain and the Migration* (Brett 1991), *Hawk watch: A Guide for Beginners* (Dunne et al. 1984), *Hawks in Flight: The Flight Identification of North American Raptors* (Dunne et al. 1988), *A Field Guide to Hawks of North America, second ed.* (Clark and Wheeler 2001), *A Photographic Guide to North American Raptors* (Wheeler and Clark 1995), *Hawks from Every Angle* (Liguori 2005), and *Raptors of Eastern North America* and *Raptors of Western North America* (Wheeler 2003a,b). Palearctic guides include *Flight Identification of European Raptors* (Porter et al. 1976), *Collins Guide to the Birds of Prey of Britain and Europe* (Génsbøl 1984), *The Raptors of Europe and The Middle East: A Handbook of Field Identification* (Forsman 1999), and *A Field Guide to the Raptors of Europe, The Middle East, and North Africa* (Clark 1999). Although all of these guides were written for northern temperate-zone audiences, many of the species described are likely to be seen at tropical and southern hemisphere watchsites as well. For those in need of a global guide, *Raptors of the World: A Field Guide* (Ferguson-Lees and

Christie 2005), is quite useful. See Chapter 2 for additional information on ageing, sexing, and identifying raptors.

Partial migration is the most common form of raptor migration (Kerlinger 1989), and individuals migrating past a watchsite often have resident counterparts in the area. Although there is no simple way to differentiate between residents and migrants, consistency of flight direction and altitude often indicate a migrating bird. In addition, resident birds often exhibit distinct behavioral patterns, such as territorial defense or displays, and extended periods of perching and hunting behavior. For some species, the migratory status of individuals in the region is unclear. Watchsites that keep records of the movements of such species can provide important life-history information about these birds.

Counting Migrants

In most cases, counting migrants is relatively straightforward; however, four specific complications warrant mention. First, there are times when the number of migrants is so large that counting and recording every individual becomes difficult. At such times, counters will need to estimate the number of passing migrants. Counting birds in large flocks by mentally dividing the flock into groups of 5, 10 or, if necessary, 20 or 50 migrants is a useful technique at such times, however the accuracy of estimates declines rapidly as the number of birds in a group increases. Another technique is to focus your efforts on an estimated 10 or 20 percent of the flock, and to carefully count all of the birds within that subset. Total numbers can then be estimated by extrapolation (Bibby et al. 1992). Another technique is to use a series of digital photographs to count migrants. This last approach, however, is labor-intensive and requires careful timing to avoid duplicative counts (Smith 1980, 1985a).

Flocking species (e.g., Turkey Vultures [*Cathartes aura*], European Honey Buzzards [*Pernis apivorus*], Black Kites [*Milvus migrans*], Levant Sparrowhawks [*Accipiter brevipes*], Common Buzzards [*Buteo buteo*], Broad-winged Hawks [*B. platypterus*], and Swainson's Hawks [*B. swainsoni*]), present additional complications associated with counting large numbers of migrating birds. These species often form swirling aggregations, or "kettles," of hundreds to thousands of birds while exploiting the same thermal or mountain updraft. Under these circumstances, birds are best counted as they begin "streaming" in long skeins along the princi-

pal axis of migration, rather than while they are "kettling" (Dunne et al. 1984). Practice counting and estimation exercises available on *Wildlife Counts* (www.wildlifecounts.com) and other population-estimation software are useful training tools for counters assigned to flocking species. When two or more species are likely to pass in large numbers, simultaneously assigning one or more counters to each species also is helpful.

One critical tool for counting large numbers of migrants is a hand-held, mechanical tally device that can be operated while looking through binoculars. With practice, an individual can operate two tally devices in each hand, and keep track of four species simultaneously, if necessary. Multiple-unit tally counters also can be useful in these situations. Unfortunately, there is little more to recommend regarding how to count extremely large numbers of raptors at migration watchsites because so little has been written about the subject. Watchsites with large numbers of migrants are encouraged to develop and test their own means of counting birds accurately and communicate their results to other workers.

The third complicating factor applies at count sites at water-crossing bottlenecks, such as at the tips of peninsulas. Due to the reluctance of many raptors to cross large bodies of water (Kerlinger 1989), individual migrants may approach and retreat from the peninsula several times before actually making the crossing. Compared with monitoring sites where the migratory flow is consistently unidirectional, these cases either require customized counting strategies that minimize double-counting (e.g., simultaneously tracking both southbound and northbound movements and estimating net southbound flow by subtracting northbound from total southbound counts on a daily basis [C. Lott, pers. comm.]) or explicit recognition that the resulting "counts" represent an activity index rather than an actual estimate of the numbers of individuals passing through (Fish 1995).

The fourth complicating factor concerns situations where raptors migrate across broad coastal plains or otherwise open landscapes in which topographic leading lines do not concentrate their movements along a consistent pathway and, therefore, flight lines shift regularly depending upon wind conditions or variations in thermal development. In such cases, a monitoring set-up involving multiple observation sites that effectively sample across the typical expanse of flight lines may be necessary to provide robust and consistent indexes of migration activity. Two primary examples where multi-site "picket-line," transect monitoring strategies have been employed successfully are Veracruz, Mexico (Ruelas-Inzunza et al. 2000) and northern Israel (Leshem and Yom-Tov 1996).

Recording the Count and Additional Data

Basic information recorded in migration counts includes the numbers of individuals seen and their identity to species, or at least to genus if the birds are too far off or are moving too quickly to allow identification to species. Workers also should record flight behavior, the date and times of observation effort (including both the time spent observing and the number of observers), and local weather at the time of observation. Flight-behavior information should include predominant direction of flight and the estimated altitude of migrants (i.e., below eye level, at eye level, and above eye level; birds seen easily without optical equipment, at limit of optical equipment, as small specks, etc.). Sites with considerable vertical relief both below and above eye level sometimes estimate line-of-sight distance to the flight using the same basic categories to estimate distance (i.e., birds seen easily without optical equipment, etc.) that are used to estimate flight altitude.

Weather data should include visibility (estimates of clarity of view plus notes about occurrence of visibility-reducing haze, dust, smoke or fog, if relevant), percent cloud cover, presence and type of precipitation when relevant, wind direction and speed, ambient temperature, relative humidity, and barometric pressure. Consistency across years in the type of weather data that are collected is important. When possible, regional weather parameters should be obtained from the local weather service. All count and additional data should be recorded hourly, with additional weather data collected as needed if conditions change rapidly within an hour. It also is helpful to record notes in a daily journal about the passage of cold fronts, major precipitation events, and reasons for missing observation days or portions of days due to inclement weather or other factors, when such are not readily evident from data recorded during actual observations.

Additional data relevant to migration behavior (e.g., flocking, flight style, altitude of the flight [Kerlinger and Gauthreaux 1985], agonistic behavior [Klem et al. 1985], feeding behavior [Shelley and Benz 1985], etc.) should be recorded whenever possible (Dunne et al. 1984). If feasible, and whenever the objectives of a

study require it, the gender and ages of migrants should be recorded as well (Bednarz and Kerlinger 1989). Many migration watchsites provide conditions favorable to the migration of other large soaring birds, such as pelicans, storks, and anhingas. Counts of these, as well as other taxa, also should be made if possible (cf. Willimont et al. 1988). Recording the passage of unusual migrants constitutes additional valuable information. Considerations should be made for collecting additional data in ways that do not compromise the validity of the overall count (e.g., by having a person other than the counter or counters record pertinent notes).

Daily record forms on which all relevant data for each day are recorded can form the basis of a permanent archive of migration count data. The use of standardized forms also is helpful in long-term studies of raptor migration, or for monitoring the status of regional populations (Bednarz and Kerlinger 1989, Titus et al. 1989). HMANA provides an excellent daily report form for recording counts and observations of migrating raptors (Fig. 2). The HMANA form, on which relevant data are recorded hourly, was specifically designed to facilitate the transfer of accumulated data to computerized databases.

Because missing data will affect the interpretation of results, recording all the data called for in the standardized form is especially important. Illegible field notes affect the interpretation of results as well (Fuller and Titus 1990). In the field, some observers prefer using a field notebook or a field version of the standardized form. This allows them to record data quickly without bothering to keep a neat form that will be used as a permanent record. If this is done, it is essential that data be transferred to the permanent record on the same day they were collected, and while the counter's memory of events is still detailed and accurate. As with other types of long-term studies (see, for example, Ralph et al. 1993), proofreading and correcting forms at the end of each count day can help reduce errors in recording, and increase the reliability of the observations.

HanDBase (www.ddhsoftware.com) and other mobile relational databases designed for Palm and Pocket PC devices also can be used to eliminate the need for pen-and-paper data recording in the field and paper-to-electronic database transcription in the office. One potential downside is that data can be lost if the electronic equipment fails during data collection, particularly during periods of extreme weather.

HAWK MIGRATION ASSOCIATION OF NORTH AMERICA

HMANA DAILY REPORT FORM

LOCATION _____

OBSERVER(S) _____ MO____ DAY____ YR____

ADDRESS _____

TIME (STD)	5-6	6-7	7-8	8-9	9-10	10-11	11-12	12-1	1-2	2-3	3-4	4-5	5-6	6-7			
Wind Speed																	
Wind Dir. (From)																	
Temp. (Deg. C)																	
Humidity																	
Bar. Pressure																	
Cloud Cover																	
Visibility																	
Precipitation																	
Flight Direction																	
Height of Flight																	
No. of Observers																Total	
Dur. of Obs. (min)																	
Black Vulture																BV	
Turkey Vulture																TV	
Osprey																OS	
Swallow-tailed Kite																SK	
White-tailed Kite																WK	
Mississippi Kite																MK	
Hook-billed Kite																HK	
Bald Eagle																BE	
Northern Harrier																NH	
Sharp-shinned																SS	
Cooper's Hawk																CH	
Northern Goshawk																NG	
Red-shouldered																RS	
Broad-winged																BW	
Short-tailed Hawk																ST	
Swainson's Hawk																SW	
Red-tailed Hawk																RT	
Ferruginous Hawk																FH	
White-tailed Hawk																WT	
Zone-tailed Hawk																ZT	
Harris' Hawk																HH	
Rough-legged																RL	
Golden Eagle																GE	
American Kestrel																AK	
Merlin																ML	
Peregrine Falcon																PG	
Gyrfalcon																GY	
Prairie Falcon																PR	
Crested Caracara																CC	
Unid. Vulture																UV	
Unid. Accipiter																UA	
Unid. Buteo																UB	
Unid. Eagle																UE	
Unid. Falcon																UF	
Unid. Raptor																UU	
Other (From Back)																OO	
TOTAL																TH	

(Comments column on right side)

Figure 2. Daily Report Form from the Hawk Migration Association of North America. Forms of this type have been in use since the 1970s across much of North America. Note that all data are recorded hourly. An Excel version of this form is available at www.hmana.org.

Sources of Variability in Count Data

Variability refers to day-to-day and season-to-season fluctuations in count totals. There are many reasons for such variability (for reviews see Hussell 1985, Bednarz and Kerlinger 1989, Fuller and Titus 1990). For our purposes, we divide potential sources of variability into two categories: those intrinsic to the migration itself (e.g., weather during migration, fluctuations in the size of source populations, etc.), and those intrinsic to the count methods used (e.g., observer bias and observation effort). Observer bias refers to the rates of detection of migrating raptors on the part of an individual, also called observer efficiency or detectability, and to the individual's propensity for making errors while collecting data. Observation effort refers to the amount of time actually spent counting, either in terms of days during the season, or hours during a specific day, and to the number of counters present.

Observer fatigue and attentiveness affect efficiency (Sattler and Bart 1984). As is true for other types of raptor population studies (Fuller and Mosher 1987), rates of detection can be determined by the degree to which an observer is familiar with a species' flight behavior. At a watchsite in Veracruz, Mexico, for example, second-year counters record lower percentages of unidentified raptors than do first-year counters (E. Ruelas, pers. comm.). Differences in methods of data collection among individual observers are a source of considerable bias that also affects the data collected in migration counts (Bednarz and Kerlinger 1989). Finally, one of the more intractable forms of observer bias occurs when few (i.e., one or two) observers are used each season and are then changed when the season changes. When this happens, variance due to different observers cannot be partitioned from variance due to year.

Another factor that can bias a count is the rate of detectability of a particular species. Some species, such as American Kestrels (*Falco sparverius*), are more difficult to detect than other migrants because of their smaller size (Sattler and Bart 1984).

All things being equal, the number of birds counted versus those that actually pass a given point in space is proportional to the time spent counting. Therefore, daily count totals depend upon the number of hours spent counting, and the total count for a season will depend on the number of days in which counts were conducted. Although several well-known watchsites conduct counts every day of a migration season (see Titus et al. 1990), this is not the only way to schedule counting effort within a season (Titus et al. 1989). (See "Sam-

pling Considerations" below for additional information on temporal aspects of migration count efforts.)

Continual and consistent training, clear explanations of objectives, proper guidance, and standardized data-collection and recording protocols can serve to reduce observer bias (Fuller and Mosher 1987, Bednarz and Kerlinger 1989). Consistency in day-to-day and season-to-season counting schedules can reduce variability due to differences in observation effort, as well as make the data comparable over long periods (Bednarz and Kerlinger 1989).

Sampling Considerations

Migration counts are samples of particular raptor populations (Titus et al. 1989, Dunn and Hussell 1995). Unlike a census, which aims to count all individuals in a specified area (Ralph 1981), samples represent only a portion of the total population. The portion that is recorded depends in part on the logistic circumstances of the particular study and in part on the sampling scheme used to collect the data.

Two considerations determine sampling frameworks: those that are spatial and those that are temporal. Spatial considerations entail determining the places from which samples are to be taken; in the case of migration counts, choosing the exact site where counts will occur. In some cases, watchsite workers have little control over this because there will be a limited number of locations, perhaps only one, adequate for conducting counts. Although it often is difficult to quantify accurately, shifting count sites even relatively short distances (e.g., 100 m) can significantly affect the portion of the observable flight that is recorded. Thus, interannual consistency in both count site location and observation effort is important to ensure comparability across years. Temporal considerations entail determining when samples will be taken in a particular location. In the case of migration counts, temporal considerations are those associated with differences in the degree of counting coverage over the course of a migration season: the number of days or hours during which counts take place (Pendleton 1989). The simplest type of temporal sampling scheme is complete coverage, which entails conducting full-day counts each and every day of the migration season, weather permitting. Systematic or "even sampling" refers to a periodic spacing of count days (i.e., counting once every certain number of days) throughout the season (Titus et al. 1989). Stratified sampling refers to dividing the migration season into time frames of approximately the same length (e.g., blocks of 15 days) (Titus et al. 1989); counts are then conducted systematically in each stratum. Bednarz and Kerlinger (1989), Titus et al. (1989, 1990), and Lewis and Gould (2000) discuss the benefits and costs of various statistical analyses of data collected by means of these different sampling schemes.

Careful sampling design is needed to obtain useful estimates of population abundance. Bednarz and Kerlinger (1989) recommend that complete coverage be attempted if logistic conditions, such as availability of funds and personnel, permit it. The larger the number of samples, in this case the more days in which counts were conducted, the more reliable are results from statistical analyses, such as determining population trends (Bednarz and Kerlinger 1989, Pendleton 1989, Lewis and Gould 2000). Regardless of the sampling scheme used in a particular count effort, sampling schemes should be consistent among years in order to ensure that data can be compared reliably.

Summarizing Count Data

The audience and the objectives of the summary will determine the way in which data are eventually presented. The easiest way to summarize each year's data is to list the total count for the season of each species. Graphic summaries that demonstrate changes in a species' daily total count over the course of the season can be obtained by plotting dates on the "x" axis and count results on the "y" axis. By using time intervals (weeks, months, etc.) in the "x" axis, histograms can be used for the same purpose. Changes in migration volume during the day can be summarized in much the same way, using hours instead of dates on the "x" axis. When examining seasonal or diurnal (diel) patterns of variation in flight magnitude within a given year, and in cases where variability in daily observation effort is significant within the period of interest, a more accurate picture may be derived by standardizing daily counts based on daily effort (e.g., counts per hour of observation). Similarly, when analyzing seasonal or diel variation across years in cases where interannual variation in observation effort is significant, a more accurate picture may be derived by standardizing daily, or time-interval, counts as the proportion of that year's total flight. See Allen et al. (1995) for details.

It is helpful to include a measure of observation effort, such as total number of hours or days over which counts took place, and the average number of hours per day in summaries of migration count data. Unusual circumstances that may have affected the count during a particular season also should be cited, such as uncommon weather events.

Several international and regional publications include migration count summaries. In the Western Hemisphere, *HMANA Hawk Migration Studies*, the journal of the Hawk Migration Association of North America, publishes regional count totals twice a year. The *Journal of Raptor Research* also includes papers that summarize raptor migration count data and information on raptor migration in general. Local and regional ornithological journals also are potential publication venues for such data.

Archiving Migration Count Data

Establishing a formal system of managing and storing data generated by migration counts facilitates access to data by watchsite workers, as well as data transfer among watchsites and off-site researchers. Systematic summaries and consistent filing guidelines make the information contained in the data easier to find and to report.

Chronologically archiving permanent record forms makes it easy to find count data from a particular day, set of days, or from an entire season. Seasonal summaries can be placed in these files as well. Each season's file should be arranged chronologically by year. Clearly labeling each file to include the months and year in which counts took place provides an effective way of keeping the files in order. Duplicate archives of all permanent records (both paper and electronic) also should be maintained as a form of record security. Calamities, such as floods, fires, and storms can easily ruin years of work and resources. The duplicate archive should be kept in a different geographic location (i.e., another city or town). Along with recording and archiving the basic count, observation effort, and weather data, it is also very helpful to maintain "metadata" that clearly describe site protocols, including all variables recorded, the observation techniques employed, the qualifications of all observers involved, how observer duties were assigned and conducted, and the nature of any preseason observer training.

MONITORING TECHNIQUES

Monitoring — *to watch, observe, check, especially for a special purpose* (1986, Webster's Ninth New Collegiate Dictionary)

Monitoring ecological or biological events consists of collecting data systematically in order to detect changes in the parameters being measured. There is considerable variation in the terminology used to refer to studies of this nature (Spellerberg 1991) (Appendix 2). We use the term "monitoring" to refer to any study in which data are collected consistently and in the same manner over a certain period of time, regardless of the intent of the study. Thus, a monitoring program results in an accumulated time-series database, to which different statistical analyses, descriptive or analytic, can be applied for many purposes.

Typically, migration counts at watchsites are used to monitor one of two things: regional population trends of migratory raptors or the status of raptor migration. Monitoring raptor population trends entails detecting changes in the abundances of migratory raptors. Monitoring raptor migration also entails determining the reasons for changes in raptor migration, including assessing the potential impacts of habitat and climate change.

In order to discuss the use of migration counts for monitoring population trends, it is useful to place such counts in the context of bird-population studies in general. Studies of bird populations can be grouped into two categories: those concerned with population size, and those concerned with demographic parameters (i.e., natality, mortality, and age-class or size-class distribution) (Spellerberg 1991, Butcher et al. 1993). Studies of population size rely on three main measures: absolute abundance, relative abundance, and density (Jones 1986a).

Density refers to the number of individuals per unit area. Relative abundance measures the number of individuals of a particular species as a percentage of the total number of individuals in a given community; both are associated with particular spatial units (Jones 1986a). Absolute abundance refers to the total number of individuals in a given population and is seldom measured by biologists due to the excessive amount of resources and time required. Instead, biologists usually employ indexes of total population size that are not ascribed to a particular geographic area (e.g., number of raptors counted as a function of the number of days in which counts took place) (Jones 1986a). Because it often is difficult to determine the origins of migrating

raptors (Fuller and Mosher 1981, but see for example Meehan et al. 2001 and Hoffman et al. 2002), migration counts are used to estimate only absolute abundance, not density or relative abundance.

Using recorded fluctuations in numbers counted to track changes in the abundances of migratory raptors is the aim of population-trend monitoring. With regard to migration counts of raptors, a trend can be defined as a "statistically significant change in counts over (a certain) period," that implies a change in the numbers (i.e., abundance) of raptors being monitored (Titus et al. 1990). Trends, however, are only one of the types of time-series data of interest to ecologists. Cycles, regular periodic fluctuations, and "noise," or stochastic fluctuations, also need to be considered (Usher 1991).

Population-trend monitoring is sometimes used to refer, specifically, to a process aimed at determining a change in abundance of a certain magnitude (e.g., a 50% change during 25 years) (cf. Finch and Stangel 1993). Used in this sense, the distinguishing characteristic of monitoring is that it sets limits, or thresholds, beyond which change is deemed worthy of conservation attention.

Several recent publications deal with general aspects of monitoring bird populations. These include *Status and Management of Neotropical Migratory Birds* (Finch and Stangel 1993), *Handbook of Field Methods for Monitoring Landbirds* (Ralph et al. 1993), and *Bird Census Techniques* (Bibby et al. 1992), all of which provide detailed descriptions of methods used in bird population studies. Sauer and Droege (1990) offer an extensive treatment of the statistical analysis of surveys, including migration count data. Ralph and Scott (1981) provide an excellent reference on the subject of monitoring bird populations in general.

Establishing a Monitoring Program

The most critical aspect of any monitoring plan is its design. Appropriate design increases the effectiveness and reduces the costs of a monitoring program by providing a flexible, systematic, and logical approach to the program (Jones 1986b). There are many approaches to designing monitoring programs (cf. Spellerberg 1991, Usher 1991 or Ralph et al. 1993). One of the simplest focuses on asking three basic questions before fieldwork begins: why, what, and how (Roberts 1991). "Why" refers to the objectives of a study, "what" refers to the data that need to be collected, and "how" refers to the methods used to collect and analyze the data.

■ *Why?* In its simplest form, the objectives of a monitoring program are the questions that are being asked of the data (Roberts 1991). The answers that are being sought will determine what data need to be collected and what methods will be used to collect them. Since the cost of collecting all possible data is high (Hellawell 1991), it is often practical to collect only those data necessary to answer the questions being posed.

■ *What?* The basic data collected at migration watchsites are numbers and types of migrants. Ancillary data include meteorological conditions and factors related to observation effort (see above). However, the particular species that will be counted at a watchsite need to be chosen before monitoring begins. Different species have different detectability rates (Sattler and Bart 1984), mainly due to differences in size (smaller birds being less likely to be detected) and flight dynamics (birds flying closer to the ground being less likely to be detected). It also is important to recognize that limits of logistic feasibility may preclude effective full-season monitoring of some species at some sites. For example, in western North America, heavy snow cover limits the seasonal duration of autumn monitoring at high-elevation, ridgetop monitoring sites, precluding effective full-season monitoring of late-season migrants such as Roughlegs (*B. lagopus*) and Bald Eagles (*Haliaeetus leucocephalus*).

At first, only a very general declaration of objectives is needed. General objectives can then be modified or refined according to the particulars of a monitoring program, such as logistic and resource limitations, the restrictions arising from study design, etc. And indeed, there are cases where a certain amount of data collection without a clear idea of how they are to be used can be helpful in determining what questions should and can be posed (Roberts 1991).

There also are many cases where data have been recorded for one purpose at one time and proved to be useful in answering another question at a later date, a phenomenon Spellerberg (1991) termed "retrospective" monitoring. Therefore, even if the objective of a given monitoring program does not contemplate other uses for the data at the moment it is being carried out, the study design should be such that future data may be compared with those that are presently being collected. When standardizing data collection procedures for a given monitoring program, possible future uses for the data should be considered.

■ *How?* Study design entails considering the methodologies that are to be employed at the watchsite.

The statistical validity of migration count data depends largely on the degree to which data collection is standardized and on the sampling scheme used. Standardizing data collection requires a good understanding of the sources of migration-count variability and can mean more intensive training and frequent supervision to ensure that data-recording guidelines are being followed properly and consistently. Even if complete sampling is desired, lack of personnel might negate this possibility, thus making a systematic sampling schedule necessary. Another detail worth emphasizing and considering carefully when planning a raptor migration count is that if the primary objective is to provide robust data for assessing population trends, standardized annual effort across multiple decades is essential.

In some cases, the unique flight dynamics associated with specific monitoring sites may require site-specific sampling methods. For example, the complexity of multi-directional movements at peninsula watchsites often necessitates special counting procedures that produce activity indexes rather than counts representing estimates of actual numbers of individuals. In such cases, it is necessary to recognize that the data collected will not be directly comparable to those collected at sites where uni-directional flow is the rule. Although this precludes direct integration of such datasets into multi-site regional assessments, qualitative comparisons are still possible. In other cases, decisions about adjusting methods to better fit site-specific characteristics may involve tradeoffs. If watchsite coordinators and sponsors consider it more important to maximize statistical power for detecting trends at that site, then adjusting count methods to increase the accuracy and precision of site-specific annual indexes may be the best approach. Alternatively, if the primary motivation for conducting a given count is to serve as one node in a regional monitoring network, then maximizing methodological consistency across sites may be more important, even if it results in reduced site-specific precision.

Interpretation of migration-count data can entail a good deal of statistical analysis. For these analyses to be valid, data must conform to the assumptions inherent in particular statistical methods. Consulting a professional statistician may be necessary to determine the appropriateness of sampling schedules, as well as to determine if resulting data conform to the assumptions of the statistical tests that will be used to analyze them (Lewis and Gould 2000). In addition, to maximize the accuracy and precision of migration-count data for detecting true population trends, especially in cases where sampling effort varies within or among seasons, it may be necessary to employ complex multivariate statistical models to derive robust annual indexes of migration activity to form the basis for analysis of long-term trends (e.g., see Hussell 1985, Hussell and Brown 1992).

Fish (2001) provides a valuable review of questions to be asked and considerations to be addressed with regard to establishing a raptor-migration monitoring effort.

Exploratory Monitoring

Exploratory monitoring serves several purposes. It can help determine exactly what questions can be answered at a particular watchsite. It also can help determine where in a watchsite it is best to conduct counts from; it can establish the duration of the migration season, as well as peaks of passage for certain species; it is an excellent way to train counters, as well as to establish standard data collection methods that are appropriate for the site, etc. Data gathered during this exploratory phase can be used in trial statistical analyses, as well as to consolidate data-management procedures, and to determine the best way to summarize data at the end of a season. It also provides an opportunity to identify logistic problems and resource limitations that are likely to affect long-term monitoring efforts.

One aspect of exploratory monitoring deserves particular attention: the determination of count location. In some instances, a single counting point is self-evident (e.g., a mountain-top watchsite or one in a narrow mountain pass). In others, possible counting points will be spread out over several kilometers (e.g., at coastal-plain or broad, intermountain-valley watchsites). Transects of preliminary counting points can be established, either at uniform intervals, along lines likely to offer good views of migrating raptors, or stratified according to meteorological or topographic parameters.

It may be necessary to monitor flights for several years from different locations to determine the best place from which to conduct long-term monitoring. Bednarz and Kerlinger (1989), for example, suggest that 5 years may be needed to determine adequately the timing of migratory movements at a particular watchsite.

ACKNOWLEDGMENTS

We thank Laurie Goodrich, Kyle McCarty, Kristen Naimoli, Michele Pilzer, David Barber, David Bird, and Andrea Zimmerman for comments and suggestions on earlier drafts of the manuscript. This is Hawk Mountain Sanctuary contribution to conservation science number 117.

LITERATURE CITED

ALLEN, P., L.J. GOODRICH AND K.L. BILDSTEIN. 1995. Hawk Mountain's million-bird database. *Birding* 27:24–32.

————, L.J. GOODRICH AND K.L. BILDSTEIN. 1996. Within- and among-year effects of cold fronts on migrating raptors at Hawk Mountain, Pennsylvania, 1934–1991. *Auk* 113:329–338.

BAUGHMAN, J.L. 1947. A very early notice of hawk migration. *Auk* 64:304.

BEDNARZ, J.C. AND P. KERLINGER. 1989. Monitoring hawk populations by counting migrants. Pages 328–342 *in* B. Pendleton [ED.], Proceedings of the Northeast Raptor Management Symposium and Workshop. National Wildlife Federation, Washington, DC U.S.A.

————, D. KLEM, L.J. GOODRICH AND S.E. SENNER. 1990. Migration counts at Hawk Mountain, Pennsylvania, as indicators of population trends, 1934–1986. *Auk* 107:96–109.

BIBBY, C.J., N.D. BURGESS AND D.A. HILL. 1992. Bird census techniques. Academic Press, London, United Kingdom.

BILDSTEIN, K.L. 1998a. Linking raptor-migration science to mainstream ecology and conservation: an ambitious agenda for the 21st century. Pages 583–610 *in* R.D. Chancellor, B.-U. Meyburg, and J.J. Ferrero [EDS.], Holarctic birds of prey. World Working Group Birds of Prey and Owls, Berlin, Germany, and ADENEX, Merida, Spain.

————. 1998b. Long-term counts of migrating raptors: a role for volunteers in wildlife research. *J. Wildl. Manage.* 62:435–445.

————. 2006. Migrating raptors of the world: their ecology and conservation. Cornell University Press, Ithaca, NY U.S.A.

———— AND R.A. COMPTON. 2000. Mountaintop science: the history of conservation ornithology at Hawk Mountain Sanctuary. Pages 153–181 *in* W.E. Davis, JR. and J.A. Jackson [EDS.], Contributions to the history of North American ornithology, Vol. II. Memoirs of the Nuttall Ornithological Club, Boston, MA U.S.A.

———— AND J.I. ZALLES. 2001. Raptor migration along the Mesoamerican land corridor. Pages 119–141 *in* K.L. Bildstein and D. Klem, JR. [EDS.], Hawkwatching in the Americas, Hawk Migration Association of North America, North Wales, PA U.S.A.

———— AND J.I. ZALLES. 2005. Old World versus New World long-distance migration in accipiters, buteos, and falcons: the interplay of migration ability and global biogeography. Pages 154–167 *in* R. Greenberg and P. Marra [EDS.], Birds of two worlds: the ecology and evolution of migratory birds. Johns Hopkins University Press, Baltimore, MD U.S.A.

BRETT, J. 1991. The mountain and the migration. Cornell University Press, Ithaca, NY U.S.A.

BUTCHER, G.S., B. PETERJOHN AND C.J. RALPH. 1993. Overview of national bird population monitoring programs and databases. Pages 192–203 *in* D.M. Finch and P.W. Stangel [EDS.], Status and management of Neotropical migratory birds. USDA Forest Service General Technical Report RM-229, Rocky Mountain Forest and Range Experiment Station, Fort Collins, CO U.S.A.

CARSON, R. 1962. Silent spring. Houghton Mifflin Company, Boston, MA U.S.A.

CHANCELLOR, R.D., B.-U. MEYBURG AND J.J. FERRERO [EDS.]. 1998. Holarctic birds of prey. World Working Group on Birds of Prey and Owls, Berlin, Germany, and ADENEX, Merida, Spain.

CLARK, W.S. 1999. A field guide to the raptors of Europe, the Middle East, and North Africa. Oxford University Press, Oxford, United Kingdom.

—— AND B.K. WHEELER. 2001. A field guide to hawks of North America, 2nd Ed. Houghton Mifflin, Boston, MA U.S.A.

DUNN, E.H. AND D.J.T. HUSSELL. 1995. Using migration counts to monitor landbird populations: review and evaluation of status. *Curr. Ornithol.* 12:43–88.

DUNNE, P., D. KELLER AND R. KOCHENBERGER. 1984. Hawk watch: a guide for beginners. Cape May Bird Observatory, Cape May Point, NJ U.S.A.

——, D.A. SIBLEY, AND C.C. SUTTON. 1988. Hawks in flight: the flight identification of North American migrant raptors. Houghton Mifflin, Boston, MA U.S.A.

FERGUSON-LEES, J. AND D.A. CHRISTIE. 2005. Raptors of the world: a field guide. Princeton University Press, Princeton, NJ U.S.A.

FINCH, D.M. AND P.W. STANGEL. 1993. Status and management of Neotropical migratory birds. USDA Forest Service General Technical Report RM-229, Rocky Mountain Forest and Range Experiment Station, Fort Collins, CO U.S.A.

FISH, A.M. 1995. How to measure a hawk migration—evolution of the quadrant system at the Golden Gate (Abstract). *J. Raptor Res.* 29:56.

——. 2001. More than one way to count a hawk: toward site-specific documentation of raptor-migration count field methods. Pages 161–168 *in* K.L. Bildstein and D. Klem, JR. [EDS.], Hawkwatching in the Americas. Hawk Migration Association of North America, North Wales, PA U.S.A.

FORSMAN, D. 1999. The raptors of Europe and the Middle East: a handbook of field identification. T. & A.D. Poyser. London, United Kingdom.

FULLER, M.R. AND J. MOSHER. 1981. Methods of detecting and counting raptors: a review. Estimating numbers of terrestrial birds. *Stud. Avian Biol.* 6:235–246.

—— AND J. MOSHER. 1987. Raptor survey techniques. Pages 37–65 *in* B. A. Giron Pendleton, B.A. Millsap, K.W. Cline and D.M. Bird [EDS.], Raptor management techniques manual. National Wildlife Federation, Washington, DC U.S.A.

—— AND K. TITUS. 1990. Sources of migrant hawk counts for monitoring raptor populations. Pages 41–46 *in* J.R. Sauer and S. Droege [EDS.], Survey designs and statistical methods for the estimation of avian population trends. Biological Report 90(1), USDI Fish and Wildlife Service, Washington, DC U.S.A.

——, W.S. SEEGAR AND L.S. SCHUECK. 1998. Routes and travel rates of migrating Peregrine Falcons *Falco peregrinus* and Swainson's Hawks *Buteo swainsoni* in the Western Hemisphere. *J. Avian Biol.* 29:433–440.

GAUTHREAUX, S.A., JR. AND C.G. BELSER. 2001. How to use Doppler weather surveillance radar to study hawk migration. Pages 149–160 *in* K.L. Bildstein and D. Klem, JR. [EDS.], Hawkwatching in the Americas. Hawk Migration Association of North America, North Wales, PA U.S.A.

GÉNSBØL, B. 1984. Collins guide to the birds of prey of Britain and Europe, North Africa and the Middle East. Collins, London, United Kingdom.

GEYR VON SCHWEPPENBURG, H.F. 1963. Zut Terminologie und Theorie der Leitlinie. *J. Ornithol.* 104:191–204.

GOLDSMITH, F.B. 1991. Monitoring for conservation and ecology. Chapman and Hall, London, United Kingdom.

HAUGH, J.R. 1972. A study of hawk migration in eastern North America. *Search* 2:1–60.

HAWK MIGRATION ASSOCIATION OF NORTH AMERICA. 1982. A beginner's guide to hawkwatching. Hawk Migration Association of North America, Medford, MA U.S.A.

HELLAWELL, J.M. 1991. Development of a rationale for monitoring. Pages 1–14 *in* F. B. Goldsmith [ED.], Monitoring for conservation and ecology. Chapman and Hall, London, United Kingdom.

HICKEY, J.J. [ED.]. 1969. Peregrine Falcon populations: their biology and decline. University of Wisconsin Press, Madison, WI U.S.A.

HOFFMAN, S.W. AND J.P. SMITH. 2003. Population trends of migratory raptors in western North America, 1977–2001. *Condor* 105:397–419.

——, J.P. SMITH AND T.D. MEEHAN. 2002. Breeding grounds, winter ranges, and migratory routes of raptors in the mountain West. *J. Raptor Res.* 36:97–110.

HUSSELL, D.J.T. 1985. Analysis of hawk migration counts for monitoring population levels. Pages 243–254 *in* M. Harwood [ED.], Proceedings of Hawk Migration Conference IV. Hawk Migration Association of North America, Lynchburg, VA USA.

—— AND L. BROWN. 1992. Population changes in diurnally-migrating raptors at Duluth, Minnesota (1974–1989) and Grimsby, Ontario (1975–1990). Ontario Ministry of Natural Resources, Maple, Ontario, Canada.

INTERNATIONAL BIRDWATCHING CENTER EILAT. 1987. Eilat as an intercontinental highway for migrating birds. International Birdwatching Center - Eilat, Eilat, Israel.

JONES, K.B. 1986a. Data types. Pages 11–28 *in* A. Y. R. Cooperrider, R.J. Boyd, and H.R. Stuart [EDS.], Inventory and monitoring of wildlife habitat. USDI Bureau of Land Management Service Center, Denver, CO U.S.A.

——. 1986b. The inventory and monitoring process. Pages 1–10 *in* A. Y. R. Cooperrider, R.J. Boyd, and H.R. Stuart [EDS.], Inventory and monitoring of wildlife habitat. USDI Bureau of Land Management Service Center, Denver, CO U.S.A.

KENWARD, R.E. 2001. A manual for wildlife radio tagging. Academic Press, San Diego, CA U.S.A.

KERLINGER, P. 1989. Flight strategies of migrating hawks. University of Chicago Press, Chicago, IL U.S.A.

—— AND S.A. GAUTHREAUX. 1985. Flight behavior of raptors during spring migration in Texas studied with radar and visual observations. *J. Field Ornithol.* 56:394–402.

KLEM, D., JR., B.S. HILLEGASS AND D.A. PETERS. 1985. Raptors killing raptors. *Wilson Bull.* 97:230–231.

LESHEM, Y. AND Y. YOM-TOV. 1996. The magnitude and timing of migration by soaring raptors, pelicans, and storks over Israel. *Ibis* 138:188–203.

LEWIS, S.A. AND W.R. GOULD. 2000. Survey effort effects on power to detect trends in raptor migration counts. *Wildl. Soc. Bull.* 28:317–329.

LIGUORI, J. 2005. Hawks from every angle. Princeton University Press, Princeton, NJ U.S.A.

LOTT, C.A., T.D. MEEHAN AND J.A. HEATH. 2003. Estimating the latitudinal origins of migratory birds using hydrogen and sulfur stable isotopes in feathers: influence of marine prey base. *Oecologia* 134:505–510.

MARTELL, M.S., C.J. HENNY, P.E. NYE AND M.J. SOLENSKY. 2001. Fall migration routes, timing, and wintering sites of North American Ospreys as determined by satellite telemetry. *Condor* 103:715–724.

MEEHAN, T.D., C.A. LOTT, Z.D. SHARP, R.B. SMITH, R.N. ROSENFIELD, A.C. STEWART AND R.K. MURPHY. 2001. Using hydrogen isotope geochemistry to estimate the natal latitudes of immature Cooper's Hawks migrating through the Florida Keys. *Condor* 103:11–20.

MEYBURG, B.-U., AND R.D. CHANCELLOR [EDS.]. 1994. Raptor conservation today. World Working Group on Birds of Prey and Owls, Berlin, Germany.

———— AND C. MEYBURG. 1999. The study of raptor migration in the old world using satellite telemetry. Pages 1–20 *in* N.J. Adams and R.H. Slotow [EDS.], Proceedings of the 22nd International Ornithological Congress, Durban, South Africa, 16–22 August 1998. BirdLife South Africa, Johannesburg, South Africa.

NEWTON, I. 1979. Population ecology of raptors. Buteo Books, Vermillion, SD U.S.A.

———— AND R.D. CHANCELLOR. 1985. Conservation studies on raptors. ICBP Technical Publication 5, International Council for Bird Preservation, Cambridge, United Kingdom.

PENDLETON, G.W. 1989. Statistical considerations in designing raptor surveys. Pages 275–280 *in* B. G. Pendleton [ED.], Proceedings Northeast raptor management symposium and workshop. National Wildlife Federation, Washington, DC U.S.A.

PORTER, R.F., I. WILLIS, S. CHRISTENSEN AND B.P. NIELSEN. 1976. Flight identification of European raptors, 2nd Ed. Buteo Books, Vermillion, SD U.S.A.

RALPH, C.J. 1981. Terminology used in estimating numbers of birds. *Stud. Avian Biol.* 6:577–578.

———— AND J.M. SCOTT [EDS.]. 1981. Estimating numbers of terrestrial birds. *Stud. Avian Biol.* 6:1–630.

————, G.R. GEUPEL, P. PYLE, T.E. MARTIN AND D.F. DESANTE. 1993. Handbook of field methods for monitoring landbirds. USDA Forest Service General Technical Report PSW-GTR-144, Pacific Southwest Research Station, Albany, CA U.S.A.

ROBERTS, K.A. 1991. Field monitoring: confessions of an addict. Pages 179–211 *in* F.B. Goldsmith [ED.], Monitoring for conservation and ecology. Chapman and Hall, London, United Kingdom.

RUELAS-INZUNZA, E., S.W. HOFFMAN, L.J. GOODRICH AND R. TINGAY. 2000. Conservation strategies for the world's largest known raptor migration flyway. Pages 591–596 *in* R.D. Chancellor and B.-U. Meyburg [EDS.], Raptors at risk. World Working Group for Birds of Prey and Owls, Berlin, Germany, and Hancock House Publishers, Surrey, British Columbia, Canada, and Blaine, WA U.S.A.

SATTLER, G. AND J. BART. 1984. Reliability of counts of migrating raptors: an experimental analysis. *J. Field Ornithol.* 55:415–423.

SAUER, J.R. AND S. DROEGE. 1990. Survey designs and statistical methods for the estimation of avian population trends. Biolog-

ical Report 90(1). USDI Fish and Wildlife Service, Washington, DC U.S.A.

SENNER, S.E., C.M. WHITE AND J.R. PARRISH. 1986. Raptor conservation in the next 50 years. (Raptor Research Report 5). Raptor Research Foundation, Hastings, MN and Hawk Mountain Sanctuary, Kempton, PA U.S.A.

SHELLEY, E. AND S. BENZ. 1985. Observations of the aerial hunting, food carrying, and crop size of migrant raptors. Pages 299–301 *in* I. Newton and R.D. Chancellor [EDS.], Conservation studies on raptors. ICBP Technical Publication 5. International Council for Bird Preservation, Cambridge, United Kingdom.

SHIRIHAI, H., E. YOSEF, D. ALON, G.M. KIRWAN AND R. SPAAR. 2000. Raptor migration in Israel and the Middle East. International Birding and Research Center - Eilat, Eilat, Israel, and SPNI, Tel Aviv, Israel.

SMITH, N.G. 1980. Hawk and vulture migration in the Neotropics. Pages 51–65 *in* A. Keast and E. S. Morton [EDS.], Migrant birds in the Neotropics: ecology, behavior and conservation. Smithsonian Institution Press, Washington, DC U.S.A.

————. 1985a. Dynamics of the transisthmian migration of raptors between Central and South America. Pages 271–290 *in* I. Newton and R. D. Chancellor [EDS.], Conservation studies on raptors. ICBP Technical Publication 5. International Council for Bird Preservation, Cambridge, United Kingdom.

————. 1985b. Thermals, cloud streets, trade winds, and tropical storms: how migrating raptors make the most of atmospheric energy in Central America. Pages 51–65 *in* M. Harwood [ED.], Proceedings of Hawk Migration Conference IV. Hawk Migration Association of North America, Lynchburg, VA U.S.A.

SPAAR, R. 1995. Flight behavior of Steppe Buzzards (*Buteo buteo vulpinus*) during spring migration in southern Israel: a tracking-radar study. *Isr. J. Zool.* 41:489–500.

SPELLERBERG, I.F. 1991. Monitoring ecological change. Cambridge University Press, Cambridge, United Kingdom.

THOMPSON, D.B.A., S.M. REDPATH, A.H. FIELDING, M. MARQUISS AND C.A. GALBRAITH. 2003. Birds of prey in a changing environment. The Stationary Office, Edinburgh, United Kingdom.

TITUS, K., M.R. FULLER AND D. JACOBS. 1990. Detecting trends in hawk migration count data. Pages 105–113 *in* J.R. Sauer and S. Droege [EDS.], Survey designs and statistical methods for the estimation of avian population trends. Biological Report 90(1). USDI Fish and Wildlife Service, Washington, DC U.S.A.

————, M.R. FULLER AND J.L. RUOS. 1989. Considerations for monitoring raptor population trends. Pages 19–32 *in* B.-U. Meyburg and R.D. Chancellor [EDS.], Raptors in the modern world. World Working Group for Birds of Prey and Owls, Berlin, Germany.

USHER, M.B. 1991. Scientific requirements of a monitoring programme. Pages 15–32 *in* F.B. Goldsmith [ED.], Monitoring for conservation and ecology. Chapman and Hall, London, United Kingdom.

WHEELER, B.K. 2003a. Raptors of eastern North America. Princeton University Press, Princeton, NJ U.S.A.

————. 2003b. Raptors of western North America. Princeton University Press, Princeton, NJ U.S.A.

———— AND W.S. CLARK. 1995. A photographic guide to North American raptors. Academic Press, NY U.S.A.

WILLIMONT, L.A., S.E. SENNER AND L.J. GOODRICH. 1988. Fall migration of Ruby-throated Hummingbirds in the northeastern United States. *Wilson Bull.* 100:482–48.

YATES, R.E., B.R. MCCLELLAND, P.T. MCCLELLAND, C.H. KEY AND R.E. BENNETTS. 2001. The influence of weather on Golden Eagle migration in northwestern Montana. *J. Raptor Res.* 35:81–90.

YOSEF, R. AND L. FORNASARI. 2004. Simultaneous decline in Steppe Eagle (*Aquila nipalensis*) populations and Levant Sparrowhawk (*Accipiter brevipes*) reproductive success: coincidence or a Chernobyl legacy? *Ostrich* 75:20–24.

———, M.L. MILLER AND D. PEPLER. 2002. Raptors in the new millennium. International Birding and Research Center - Eilat, Eilat, Israel.

ZALLES, J.I. AND K.L. BILDSTEIN [EDS.]. 2000. Raptor watch: a global directory of raptor migration sites. Birdlife International, Cambridge, United Kingdom, and Hawk Mountain Sanctuary, Kempton, PA U.S.A.

Appendix 1. Techniques for studying raptor migration (with representative pertinent references).
(After Kerlinger 1989 and Bildstein 2006)

1. Raptor migration counts. Common and widespread; inexpensive and relatively easy to conduct. Documents occurrence, timing, and volume of migration at a site; can be used to document habitat use. Biased towards low-flying migrants; data are affected by a variety of factors including observer fatigue, number of observers, weather, etc. (Bednarz et al. 1990, Shirihai et al. 2000, Hoffman and Smith 2003).

2. Trapping and banding. Common, but labor intensive. Determines origins and destinations of migrants, and migratory pathways; can be used for measuring anatomy and physiology, for monitoring migrant health, and for determining causes of mortality. Low band-return and recovery rates result in small sample sizes; potential age- and gender-class biases. Enables application of other cutting-edge techniques, including satellite tracking and stable-isotope analysis of feathers (see below) to determine migrant origins and document migration routes (Hoffman et al. 2002).

3. Marking. Uncommon and inexpensive, but labor intensive depending upon capture effort. Documents habitat use and movements of individuals. Low resighting rates; removal of markers by birds can affect results (see Chapter 13).

4. Conventional tracking. Uncommon to rare; expensive and labor intensive. Determines habitat use, time of stay, and behavior at stopovers along entire portions of the migratory journey. Following migrants usually presents difficulties (Kenward 2001).

5. Satellite tracking. Increasingly common, but extremely expensive. Documents long-distance movement of individuals, sometimes across multiple years. As of mid-2004, transmitter size restricts use of the technique to large (> 500-g) raptors (Fuller et al. 1998, Meyburg and Meyburg 1999, Martell et al. 2001).

6. Motorgliders and aircraft. Rare, expensive, and labor intensive. Documents flight behavior and determines geographic distribution of migrants. Affects flying behavior of migrants; biased towards high-flying migrants (Kerlinger 1989).

7. Visual observations of behavior. Uncommon to rare, although inexpensive and adaptable. Used to document flight behavior. Biased towards low-flying migrants.

8. Photography and cinematography. Rare and, historically, expensive and labor intensive. Documents flight behavior and is used to verify counts made by ground observers. Care must be taken when comparing images (Smith 1980, 1985a).

9. Radar. Uncommon, and relatively expensive and labor intensive. Documents flight behavior and geographic distribution. Mobility somewhat limited; results sometimes biased to high-flying migrants. Currently simultaneous visual observations are needed to verify identity of migrants (Spaar 1995, Leshem and Yom-Tov 1996, Gauthreaux and Belser 2001).

10. Stable-isotope analysis of feathers. Rapidly advancing new field of inquiry; large samples required, but easily obtained through migration trapping; expensive; relatively few laboratories established for processing, but number growing. Used to identify approximate natal origins of juvenile birds sampled on migration or on wintering grounds (Meehan et al. 2001, Lott et al. 2003, C. Lott and J. Smith, pers. comm.; see Chapter 14 this volume).

Appendix 2. Monitoring and surveillance defined.

Monitoring. Intermittent (regular or irregular) surveillance carried out to ascertain the extent of compliance with a predetermined standard or the degree of deviation from an excepted norm (Hellawell 1991).

Monitoring. A systematic collection of data on a particular parameter used to determine changes in its status within a certain time frame (Roberts 1991).

Surveillance. An extended program of surveys, undertaken to provide a time series, to ascertain the variability or range of states or values that might be encountered over time, or both (Hellawell 1991).

Behavioral Studies

7

GIORGIA GAIBANI AND DAVIDE CSERMELY
Dipartimento di Biologia Evolutiva e Funzionale,
Sezione Museo di Storia Naturale
Università di Parma, Via Farini 90, 43100 Parma, Italy

ETHOLOGY: THE SCIENTIFIC APPROACH TO ANIMAL BEHAVIOR

A flying raptor is fascinating. Whether you are observing an eagle's display flight, an aerial transfer of food between two courting harriers, or a successful hunt by a Verreaux's Eagle (*Aquila verreauxii*), each is an impressive experience. Beyond the aesthetics of these displays lie many things that are of interest to biologists. Most of these deal with the ecological aspects of raptor biology, others with behavioral aspects. The latter, in particular, remain unstudied among raptor biologists.

This chapter provides raptor biologists with an introduction to behavior-study techniques, including methods and equipment used for descriptive and experimental analysis of behavior, both in the field and in the lab.

Some Introductory Concepts

Ethology, or Comparative Psychology, is a relatively young but growing discipline. Its name, literally, means "the study of behavior." Ethology usually is considered the legacy of Konrad Lorenz, Niko Tinbergen, and a few other animal behaviorists. The mark that these early investigators made on this discipline was that behavior is a product of natural selection, just like any phenotypic character. Natural selection acted, in the past, in shaping the behavior that is now observed in the present. Therefore, in ethological studies, it is important to consider the behavior in relation to its adaptive function in each species. Consequently, behavior usually can be better understood in free-ranging animals than in captive individuals.

The study of behavior's proximal causes was the start of a vigorous debate between North American animal behaviorists, who concentrated on the possibility of behavior modification (i.e., the ability to learn) rather than natural selection, and European ethologists, who speculated about causation and experimentally tested the adaptive function of specific behavior. After several decades the debate was settled, as both schools realized that all behavior, being the result of evolution, is comprised of both innate and experiential components.

Formulating the Hypothesis

Raptor biologists should keep in mind that the behavior of the birds they study is a central aspect of their biology and that behavior has the same degree of importance as other biological and ecological patterns. The seemingly limited repertoire of active behavior patterns displayed, interrupted by long periods of inactivity, can lower the appeal of behavioral studies. However, as any patient observer will soon realize, birds of prey display many types of behavior, and the study of ethology is critical to understanding the biology of raptors.

Before setting up an ethological study, the researcher must determine the experiment's starting point as well as its goals. Although the behavior itself is what an individual does, an ethological study should not restrict itself to the simple description of which behaviors are displayed, but also should ask questions regarding what, who, why, where and when the patterns in question occur (Lehner 1996).

What is simply an accurate description of behavior, which is made up of a sequence of behaviors. Taken together, these behaviors form the behavioral repertoire of individuals performed in particular contexts, roughly corresponding to what ethologists call an ethogram. Today this is considered to be a list of behaviors displayed by a certain species, or a behavior repertoire.

Who refers to the identification of the individual performing the behavior. This is important, not simply to avoid repetitive recordings, but because behavior can differ between and within sex, age class, and species. It also is important to know the identity of individuals near the bird performing the act, as well as the entity to whom that act is directed.

How refers to the description of the motor patterns used by the individual to perform a goal-oriented behavior, such as how to fly from one perch to another.

Why refers to either motivation or adaptation. Motivation refers to the individual performing the behavior, whereas adaptation has an evolutionary or ecological implication. Although seemingly separate concepts, they often are connected.

Where deals with the spatial aspect of the behavior. It refers to the geographic location where the behavior is performed, the location within the ecosystem, or the relative position of the individual performing the behavior in relation to other individuals.

When refers to the temporal component of the behavior. It includes the frequency of occurrence with respect to day, year, and lifetime, as well as timing of such a pattern within a behavioral sequence.

The "what" question usually is the starting point of any ethological investigation, but generally all of the questions above should be addressed. Whereas the "where" and "when" questions tend to follow the "what" question logically enough, the "how" and, above all, "why" questions often are the most difficult to answer. What follows provides an overview of the steps needed to collect useful information for behavioral studies.

DATA COLLECTION

In order to test hypotheses and achieve the study's goals it is essential that you obtain data that can be analyzed

statistically and compared with those of other researchers. Therefore, it is necessary to assess and to plan precisely what you will study and how you will do so before starting data collection.

Level of Investigation

The choice of what to study ranges from the analysis of various types of behavior exhibited by one species to the analysis of a specific behavior in several species. In both cases, the study species should have some basic characteristics (Lehner 1996) including:

Suitability. The species must perform behavior patterns in a repeated, observable way. Suitability is increased if birds are individually recognizable or are marked to make them so.

Availability. Individuals must be accessible and observations should be carried out without affecting behavior. If the study involves captive individuals, the appropriate permits must be obtained for trapping or holding them in captivity.

Adaptability. If the study requires captivity, the species must be able to adapt to this context without altering the behavior in question.

Background information. The researcher must do a thorough literature search on the species to know how to best approach individuals and how to plan the study.

Based on the study's goals, the researcher should decide whether to conduct the study in the wild, in captivity, or both. In studies involving captive individuals, the researcher can manipulate the environment and control many variables. However, there is the risk of alterations in behavior due to the unnatural environment. In contrast, the researcher can observe a natural behavior in its entirety when studying the subject in the wild, but has to adapt to the animal's rhythm or activity cycle, and might not to be able to control the often numerous natural variables. Both studies in captivity and in the wild have disadvantages and advantages, and it is ideal to study a behavior or species in both circumstances.

One also should decide whether to simply describe the behavior (a **descriptive study**) or to collect data to test one or more hypotheses (an **analytic study**). If the latter, it is necessary to decide whether to collect data by simple behavioral observations (a **measurative study**) or by means of environmental or animal manipulation (a **manipulative study**), or both. There are many intermediate situations between these two extreme situations. In fact, Lehner (1996) points out that, *"we can categorize ethological research along a continuum from* *descriptive field studies to manipulative laboratory experiments."*

Before beginning the study, the researcher should consider carefully which methods to use. Below, we list a series of recording and sampling rules to help in the choice concerning how to carry out a study. These rules can be used in ethological research, in both descriptive and in measurative and manipulative analytical studies.

Describing Behavior

In every ethological study, the description of a behavior must be clear and precise in order to obtain data that are comparable with those collected by other researchers. Therefore, before beginning the study, it is important to choose *a priori* the behavioral categories to observe and record, and to define them clearly and precisely. A preliminary study and the drawing of an ethogram can be a great help. A good example of an actigram — or standardized form of an ethogram — for raptors can be found in Walter (1983). Martin and Bateson (1993) suggested that it is also important to consider that two types of behavior patterns can be identified:

An "**event**" is a relatively brief behavior pattern, such as a discrete body movement or vocalization, which can be approximated as a point in time. Often, the most relevant feature of an event is its frequency of occurrence.

A "**state**" is a relatively long behavior pattern, such as a prolonged activity, body posture or proximity measure. Often, the most relevant feature of a state is its duration.

Choice of behavioral categories. Each behavior is represented by a continuum of several movements and postures, making it difficult to obtain a definitive measurement. Consequently, before starting to collect data, it is often advisable to split any behavior into categories in order to make collection easier and more precise. For example, to describe and measure the hunting behavior of a bird of prey, it is better to divide this activity into its various components: prey search, pursuit, capture, manipulation, and ingestion. Although the type and number of categories are related to the type of behavior, the study's goals, and the level of investigation, Martin and Bateson (1993) suggested features that should characterize these categories:

Number. The number of categories should provide a sufficiently detailed description of the behavior in relation to the research goals.

Definition. Each category should be defined in a clear, precise, and comprehensive way, describing what is meant to be included in that category (ostensive definition) and describing the method used to measure it (operational definition).

Independence. The categories should be independent, so that each behavior pattern can be ascribed to only one category.

Homogeneity. All behavior patterns assigned to the same category should exhibit the same properties.

Cresswell (1996) defined behavior patterns in his study on Eurasian Sparrowhawks (*Accipiter nisus*), Peregrine Falcons (*Falco peregrinus*) and Merlins (*F. columbarius*) very precisely:

"(1) Hunting. Purposeful flight in an area of potential prey in a manner that led to, or could potentially lead to, an attack. For Sparrowhawks, this was rapid low contour-hugging, an approach flight that used cover or direct dashes at prey. For Peregrines, this included any flight through, or with, groups of prey, except when the potential prey was mobbing. For Merlins, it included only periods of flight in which attacks were recorded.

(2) Hunting/moving. Any flight in an area of potential prey that could not be classified as hunting. Merlins, for example, would use the same very low and rapid hunting flight to move between perches between long periods of activity as well as during definite hunting periods with many attacks.

(3) Perching. Either on the ground or on an object. Perching did not include any time spent feeding or caching prey.

(4) Feeding. Plucking or eating prey."

Types of behavioral description. There are basically two types of behavioral descriptions (Martin and Bateson 1993, Lehner 1996): empirical and functional descriptions.

An **empirical description** (i.e., a description based on structure) describes the behavior according to how it is subdivided, annotating a series of postures and body movements. An example is "the Golden Eagle (*A. chrysaetos*) flies, maintaining its wings open and still." This type of description is particularly useful during preliminary studies and when drawing up ethograms. However, it can be redundant and of little use in the other contexts.

A **functional description** (i.e., a description based on the function) describes the behavior according to the functional outcome that follows a series of postures and movements. An example of description based on function is "the Golden Eagle is gliding." Caution should be exercised, however, as a functional description can induce the observer to subjectively interpret observed behavior. Using the previous example, an observer could write, "the Golden Eagle is in a hunting glide" or simply "the Golden Eagle is hunting," attributing purpose to the Golden Eagle's behavior. Interpreting the aim of the behavior during data recording sometimes results in incorrect or incomplete information and can create confusion during data processing. During the description of a behavior, it is very important that the researcher does not attribute adjectives or definitions that can implicitly or explicitly give an indication of the behavior's causes or aims.

Although the distinction between these two behavioral descriptions is important, it is not always clear, and it is sometimes appropriate to use both description types within the same study.

Sampling Rules

Sampling is at the core of any ethological study. The sampling rules used depend on several variables specific to each study. These include the experimental design, the type of behavioral unit (events, states, or both) to be recorded, the observational accuracy available, and, above all, the research question.

Ad libitum sampling. This method is useful to record events and states. In fact, this sampling rule allows the researcher to record all behavior patterns exhibited during the sampling period by all individuals visible during that period. In other words, the researcher records all that is observable, without limitation to the number of subjects or behavior patterns seen. The recording of all that is observable has two problems. The first is that the observer will be inclined to record the more frequent and more striking behaviors (i.e., those attracting attention more than others), whereas they may overlook rare behavior. The second problem is that this method is very exacting. This sampling rule is of little use for collecting quantitative data, but is particularly useful during preliminary studies, or when compiling an ethogram.

Focal-animal sampling. This method involves recording the occurrence and the duration of all types of behavior patterns exhibited by a single focal individual. In this case, the limiting factor is that only one individual is observed, whereas there is no restriction on the number of behaviors recorded. Sometimes the

researcher may choose to record the behaviors of a focal-pair or a focal-group, but at such times recording can become more difficult and the researcher runs the risk of not recording important information. Focal individuals can be chosen randomly or on the basis of certain characteristics. The focal-animal sampling method is useful for recording both events and states. Tolonen and Korpimäki (1994) studied parental effort in several pairs of Common Kestrel (*F. tinnunculus*) using this method. Behavioral categories tied to male activity (sitting, directional flight, active flight-hunting and soaring) were recorded by using continuous observation of each focal male for 6–8 hours during courtship and incubation, or for 4–6 hours during the nestling period.

All-animal sampling. With all-animal sampling, the observer records the occurrence of a certain behavior or a category of behaviors exhibited by a group of individuals. Thus, the limiting factor is the number of behavioral event or states to observe, whereas there is no restriction on the number of individuals. This method can be used to record both events and states. Sergio (2003) assessed the effect of weather on the foraging performance of Black Kites (*Milvus migrans*) by observing the entire colony and recording each hunting attempt, and relative outcome, during each observation session.

Recording Rules

Within a study, a one-sampling rule (focal-animal sampling, all-animal sampling, or *ad libitum* sampling) is usually combined with one of the following recording rules (continuous-recording sampling or time sampling).

Continuous-recording sampling. This involves several methods consisting of recording various parameters of a behavior or behavioral categories during a specific sampling period: time of beginning and ending, sequence and duration. The data obtained are numerous and precise, and the effort required by the observer is quite high.

All-occurrences sampling. This method is also called "event-sampling" or "complete record sampling." It records the frequency and the rate of all occurrences of a certain behavior or behavioral category. Usually, it is used to record events, and is useful to assess synchrony or the rate of an easily observable behavior pattern that does not occur frequently. Given the practical difficulty of recording all occurrences of specific behaviors or states, the all-occurrences sampling method is often associated with focal-animal sampling. An example is seen in Mougeot (2000), where territorial intrusions and copulation patterns in 26 pairs of Red Kites (*M. milvus*) were investigated. During each observation period, which lasted on average 1.6 hours, Mougeot observed one focal pair, continuously identifying and recording the occurrences of various behavior patterns (interaction with conspecifics, copulation, male prey deliveries, time spent by male and female within the breeding territory).

Sequence sampling. Sequence sampling is mainly used to study behavior patterns, displayed by an individual, pair, or group in sequence (e.g., courtship displays, hunt displays). During sequence sampling, the observer records all behavior exhibited, noting time and frequency of individual events and states. Usually, the start and end of each sequence-sampling period is determined by the start and end of the sequence. Sampling duration must be chosen in relation to the type and occurrence frequency of the behavioral sequences. This method can be used to record both events and states. Edut and Eilam (2004) studied the protean behavior of the social (Guenther's) vole (*Microtus socialis*) and of the common spiny mouse (*Acomys cahirinus*) under Barn Owl (*Tyto alba*) attack. Within each 3-hour test period, the continuous recording of both owl and rodent

behavioral sequences started on the first owl attack and ended with the capture of a rodent.

Sociometric matrix. This is a method for tabulating data useful for measuring the synchrony and sequence of behaviors of individuals in a group. Csermely and Agostini (1993) investigated the social and agonistic interactions of an acquainted group of rehabilitated Barn Owls. They identified and recorded seven behaviors displayed by the active bird (initiator) in the interaction and eight behaviors displayed by the passive bird (recipient). Each interaction was characterized by the dyads of behavior performed by both initiator and recipient birds. Interactions could then be summed in a matrix of 56 (7 x 8) cells (Table 1). The matrix usually is read from left to right, with the frequency of the initiator's behavior listed in rows and that of the follower in columns.

It is important to keep in mind that the sociometric matrix is a method used only to organize data and is not the same as a contingency table, despite their similar appearances.

Time sampling methods. These methods consist of recording behavioral events periodically, instead of continuously, during the sampling period. The sampling period is usually subdivided into several intervals during which behaviors are recorded. These methods col-

Table 1. An example of a sociometric matrix using data from Csermely and Agostini (1993). The behaviors considered for the initiating bird are listed in the column on the left and those for the recipient bird in the heading across the top. In this example, each cell indicates the frequency of the behavior transition recorded for each interaction. For instance, the interaction in which the initiating bird displaced the recipient bird, which reacted with physical contact (PC), was recorded eight times. A total of 202 interactions was recorded in this session. (TH, threatening; AP, approach; AL, allopreening; PC, physical contact; DI, displacement; BB, beak-beak contact; AG, aggression; RE, retreat; NR, no reaction.)

BEHAVIOR	TH	AL	PC	DI	BB	AG	RE	NR	TOTAL
Threatening	0	0	0	0	1	0	1	4	6
Approach	2	3	7	0	18	3	0	0	33
Allopreening	0	7	0	0	5	0	0	23	35
Physical contact	0	0	0	1	4	1	5	26	37
Displacement	0	3	8	0	2	1	3	11	28
Beak-beak contact	0	2	3	0	0	0	19	31	55
Aggression	0	0	0	0	0	0	6	2	8
TOTAL	**2**	**15**	**18**	**1**	**30**	**5**	**34**	**97**	**202**

lect less information than continuous-recording sampling methods, but they are less demanding of the observer, are particularly useful if the observer is not an expert, and also allow the observation of several subjects or behavior at the same time.

One-zero sampling. This method is also called "fixed-interval time-span sampling" or the "Hansen system." The observer scores whether a certain behavior occurred (1) or not (0) during very short sampling intervals of 10 to 60 seconds each, in which the observation period is split. This method can be used to record both events and states, but is usually used to record states and, above all, to study behaviors that begin and end quickly. The length of the sample intervals and the time between sampling intervals must be chosen carefully with respect to the type of behavior or behavioral categories studied. Usually, the shorter the sample interval, the more accurate is the documentation of the behavior in question. Because the simultaneous recording of many behavioral categories is difficult, the length of the sampling interval will be a compromise between length of the observation and number of behavior patterns recorded. The greater limitation of this method is that it does not measure actual frequency and duration. It is worth noting that some authors (e.g., Altmann 1974) believe that this method should not be used because it is not always reliable.

Instantaneous and scan sampling. Instantaneous sampling also is called "point sampling," "fixed-interval time point sampling," "on-the-dot sampling," or "time sampling." The observer records the behavior displayed by one individual at a fixed point sampled within the sampling period. This method is useful for recording states, but not events because both events and time-points are instantaneous and it is unlikely that they will occur at the same time.

Scan sampling is a form of instantaneous sampling where the observer records the behavior displayed by several individuals at fixed-point samples. This method is important to estimate the percentage of time that an individual spends in particular activities. In a laboratory experiment, Palokangas et al. (1994) tested whether female Common Kestrels preferred brightly ornamented males. Each female had to choose between two males caged in front of her. During the 15-minute tests, the researcher recorded which male the female was looking at every minute.

Regardless of the method, it is important that the duration of each sampling is always the same to allow the comparison with other data collected by the researcher with data in other ethological studies. The length of the sampling period depends on the type of behavior studied and on its frequency of occurrence. If, during the sampling period, the subject is out of sight of the observer, it is necessary to estimate this duration and to consider it during data processing. When this occurs, we suggest consulting the guidelines proposed by Lehner (1996).

Finally, it is necessary to emphasize that it is a good idea to carry out preliminary observations before choosing sampling and recording methods in order to have an accurate overview of which behaviors to study. Moreover, during the first phase of data collection, the recording observer's efficiency tends to improve (observer drift; Martin and Bateson 1993). Consequently, it is advisable to familiarize yourself with the collection method before beginning the experiment. This will help mitigate the possibility of changing data reliability over the course of the study.

Once the data have been collected, it may be very difficult to explain why a behavior is displayed. Animal behavior is affected by many factors including habitat, season, hormones, genetics, and phylogeny. Consequently, when planning an ethological study, the researcher has to take these factors into account.

EXAMPLES OF DATA COLLECTION IN THE FIELD AND IN CAPTIVITY

There are many different ways to investigate the behavior in birds of prey. Every species has its own set of adaptations and can respond differently to the same environmental stimulus. This section reviews the tools that can be used to study raptor behavior both in the wild and in captivity. This section is not exhaustive, and should be considered introductory for those who wish to set up an ethological study.

Mate Choice

Mate choice is one of the most investigated behaviors in ethology. In most studies, the attention focuses on female choice, but, nevertheless, it is important to know which factors affect male choice, principally in monogamous species, where both partners are involved in parental care, and which often is the case in birds of prey.

Mate choice can be influenced by several of the partner's characteristics: age, phenotype (e.g., body

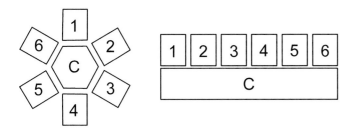

Figure 1. An experiment investigating mate choice in which a bird in cage C "chooses" among birds in cages 1–6. The choice can be assessed in each situation by measuring which bird looked toward or "visited" longer or more frequently than the others.

size, body symmetry, plumage color and brightness), parasite load, hunting efficiency, territory characteristics, etc. However, because most of these factors are interrelated (e.g., territory quality and hunting efficiency, plumage brightness, general health, and parasite load, etc.), it is necessary to consider these relationships when designing the study so as to distinguish the influence exerted by each factor and to draw the correct conclusions.

In works conducted in the wild, it may be necessary to tag the individuals with proper identification (e.g., using color rings, wing-tags, dyed feathers, or radio-transmitters) and to gather morphometric data, physiological (e.g., blood) samples, or both. Capturing birds of prey and tagging them are discussed in detail in Chapters 12 and 13.

Standardized observations (focal-animal sampling or all-animal sampling) allow the researcher to verify if certain behaviors affect mate choice. Village (1985) recorded the individual arrival dates of a Common Kestrel population arriving in spring. All birds were tagged and aged (first-year or older) in previous years in order to recognize them and to distinguish migrants from overwintering kestrels. The date when the last-arriving partner was seen for the first time was considered the date of pairing. Pairing dates revealed assortative mating based on age and arrival time in each territory.

Tagging of individuals also helps record the possible occurrence of extra-pair copulations or polygamy (polygyny or polyandry), which are important factors to be considered in mate-choice studies.

Although courtship in birds of prey often involves acrobatic display flights, and whereas mate quality is evaluated on the basis of several factors, including male hunting skill, mate preference can be examined in cap-

tivity as well. The experimental structure for doing so usually consists of several cages, one containing the "choosing bird" and, in front of it, several others containing the birds to be evaluated and chosen. The latter cages are either in a row or in a radial position (Fig. 1). The choice can be made by simply visiting each cage or, in the second case, by turning the body to watch the preferred mate. Such a test is ultimately a replication of lek behavior displayed by many animals (Höglund and Alatalo 1995).

Individuals are evaluated based on body characters or behavior. In each case, it is important to limit the number of variables by which individuals differ. Palokangas et al. (1994) tested the Common Kestrel female's preference for brightly feathered males. The test was carried out in a room divided by a wall; each half contained one male, unable to see the other. The female was placed in a small box in the middle of the test room and was able to see the males through a one-way window from the box. Each female had 15 minutes in which to evaluate both males. During this time the researchers recorded which male the female was looking toward every minute (instantaneous sampling).

Parental Care

Because most of them are monogamous and raise altricial nestlings, birds of prey can be interesting subjects when it comes to parental care and parental effort. To quantify parental care several variables should be measured: parental and offspring survival over time, time spent by parents incubating the eggs and brooding the young, food provisioning rate (measured as the number of prey items delivered to the nest per time unit), and defense behavior. Tolonen and Korpimäki (1995) studied the nest defense behavior of Common Kestrels towards a stuffed pine marten (*Martes martes*) placed under a cover on a nest box roof. After removal of the cover, defense behavior was recorded for 5 minutes, with the activity of the male and female recorded separately. The intensity of the behavior, classified into six categories, was evaluated, and data recording started when at least one member of the pair overtly reacted to the predator.

Social Behavior

Some raptors are social or at least gregarious. Some, including Eleonora's Falcon (*F. eleonorae*), exhibit social feeding strategies, others, including Lesser

Kestrel (*F. naumanni*), nest in colonies, and still others, including Red Kite, roost communally. Any study focusing on social behavior requires individual recognition, either by plumage characteristics or by markers (color rings, wing-tags, dyed feathers, or transmitters for radio-tracking). Hiraldo et al. (1993) tagged 46 Red Kites wintering in communal roosts with radio- and wing-tags and defined four categories of individuals on the basis of age (young or adult) and status (wintering or resident), and then conducted all-animal sampling to determine the time of departure from the roost, the flight direction, whether kites flew alone or in groups, and if there was a group leader. The researchers considered feeding duration from the previous day as the basis for their foraging success: high success for more than 30 minutes of feeding and low success for less than 5 minutes of feeding. The data recorded did not confirm the hypothesis that roost sites act as food information centers (i.e., sites where kites get information from mates about food locations [cf. Ward and Zahavi 1973]).

In captivity, such as during physical rehabilitation, birds of prey often can be kept in groups without showing apparent behavioral alterations due to unnatural density. In this context, a detailed analysis of their behavioral repertoire and the behavioral transitions occurring when birds interact is a useful tool to anticipate negative effects of forced cohabitation.

Interactions between individuals can be evaluated by recording the behavioral transitions of birds kept in the same cage. Each bird must be identifiable, for instance by color rings or wing-tags, and the observer should first create a list of behavioral categories that are displayed when the birds interact. Once this is done, the observation sessions can start. These should be carried out for a sufficiently long period (e.g., one to two hours), and distributed temporally in such a way as to cover the entire activity period of the birds over a few days. Csermely and Agostini (1993) investigated a group of rehabilitated Barn Owls by initially recording the social-agonistic interactions within the already acquainted group and then by observing possible modifications due to the introduction of a strange conspecific. The authors recorded the identity of the interacting birds and the behavior patterns of both the initiator and the follower.

The data were then transposed into a sociometric matrix to analyze both the interacting birds and the behavioral transitions. The first matrix allowed the researchers to rank the birds by aggression frequency (Table 2), leading them to compile a "social" hierarchy, while the second matrix allowed them to ascertain the probability that a certain action causes a certain type of response (Table 1). This later allowed the researchers to describe the probability that a certain pattern displayed by the initiating bird would cause a certain reaction (Fig. 2).

Table 2. A hypothetical example of a sociometric matrix used to rank individuals by interaction frequency, usually agonistic behavior. The initiating bird's identity is listed in the column on the left and those of recipient birds in the heading across the top. In this example, bird C initiated seven interactions with bird A. Bird B is the most frequent initiator (total frequency = 21), bird A is the most frequent receiver (total frequency = 19), bird F was not involved in any interactions. The data can be used to establish a social hierarchy among the six individuals.

INITIATIOR	RECIPIENT						TOTAL
	A	B	C	D	E	F	
A	——	5	2	0	0	0	7
B	9	——	12	0	0	0	21
C	7	3	——	1	2	0	13
D	3	0	4	——	3	0	10
E	0	0	0	1	——	0	1
F	0	0	0	0	0	——	0
TOTAL	19	8	18	2	5	0	52

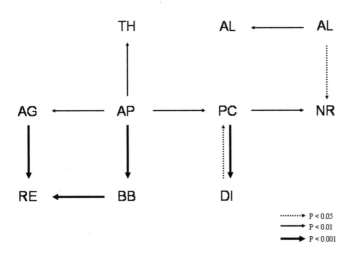

Figure 2. An example of results of a study in which behavioral sequences between individuals were assessed. The probability that any behavior displayed by the initiating bird causes reaction in the follower can be ascertained with statistical analysis and summarized graphically with a diagram similar to a flow chart. The arrows (including thickness and solid versus dashed) show the direction and probability of each sequence occurrence (AG, aggression; AL, allopreening; AP, approach; BB, bill–bill interaction; DI, displacement; NR, no reaction; PC, physical contact; RE, retreat; TH, threatening [from Csermely and Agostini 1993]).

Territorial Behavior

Territorial behavior is usually studied by observing a focal bird and recording its behavior against an intruder (agonistic display or physical aggression). Obviously, it is necessary to recognize the individuals. When radio-transmitters are used, it is important to note that radio-tracking alone is not sufficient, as it can only provide information on movements and home-range size. A detailed analysis of territorial behavior requires direct observation. Newton and Marquiss (1991) trapped and removed female and male Eurasian Sparrowhawks from their territories to verify whether their possible replacement could be attributed to movement of resident non-breeders or of neighboring individuals. In order for the researchers to answer this question, birds were banded and monitored in every territory in the surrounding area. This way, the authors verified that non-breeders of both sexes were likely present in the population and that spacing behavior was involved in limiting breeding density.

Assessing agonistic interactions in captivity can be easily performed by direct visual observation, or video recording from a blind. Preliminary observations are necessary to identify the repertoire of behaviors displayed during interactions. Observation sessions, carried out at different times of the day, provide the frequency and identity of interacting birds as well as the patterns performed by attacking and receiving birds. It also is advisable to distinguish observations carried out in the presence or absence of food, as food, being a resource to be defended, could likely be an important source of aggression.

As discussed above, the data are transposed into a sociometric matrix to assess the interacting bird dyads that are more frequent than expected, and the significant behavioral transitions. In the first instance one can obtain a social hierarchy by calculating the dominance index of each bird (Crook and Butterfield 1970) to establish a more or less linear hierarchy. In the second instance one can produce a diagram similar to a flow-chart that describes the probability that a pattern displayed by the initiating bird causes a reactive pattern in the receiving bird (Fig. 2).

Csermely and Brocchieri (1990) studied the interactions among captive Common Buzzards (*Buteo buteo*) after rehabilitation, identifying 12 social-agonistic behavior patterns and three types of vocalization. The most frequent behavior patterns recorded for the initiating bird were related to agonistic interactions, such as threatening, leg-strike, run-toward, piracy, whereas the attacked or retreating bird reacted principally performing retreat or run-toward. When food was present in the pen, piracy and run-toward were used more frequently far from the food source, whereas griffon-posture was observed most often over it.

Predatory Behavior

The study of hunting behavior in the wild often is exacting, as it is difficult to follow hunting birds. Consequently, many studies assess hunting behavior indirectly from prey deliveries to the nest or from prey remains in the nest or beneath perches. Nevertheless, only direct observations can provide information about foraging behavior, such as the rate of successful hunts, usually calculated as the proportion of successes over capture attempts.

Jenkins (2000) studied the relationship between hunting success and nesting habitat in 16 pairs of African Peregrine Falcons (*F. p. minor*). After splitting hunting behavior into several categories, he observed both partners of each pair using focal-animal sampling. He recorded hunting attempts, hunting mode (perch hunt or strikes made from air) and types of prey captured. Observation periods were classified according to season (breeding or non-breeding) and time of day. Jenkins concluded that the height of nest cliffs affected foraging success.

Cresswell et al. (2003) tested whether free-ranging Eurasian Sparrowhawks preferentially attacked vigilant or non-vigilant (i.e., feeding) prey models presented in pairs, using two types of models: a stuffed 3-week old Red Junglefowl (domestic chicken, *Gallus gallus*) and a resin-cast model of an adult European Greenfinch (*Carduelis chloris*). Half of the models of each type were mounted in a head-up position to mimic a scanning wild bird, and half were mounted in a head-down position to mimic a feeding bird. Models were placed on flexible wires planted in the ground in low vegetation. Each pair of models was connected to a camera trap. The resulting photos recorded which model was hit, and from which direction the attack occurred.

Predation also can be studied in captivity. Doing so allows the researcher to control variables and to observe behavior much more closely than in the field. Captive studies allow the researcher to investigate hunting and capture techniques, prey recognition, different responses to stimuli coming from different prey types, and maturation and refinement of the behavior sequence in the case of captive-bred birds.

To study predation behavior in captivity, it is best to have individually penned birds so as to avoid competitive interactions. The pen should be large enough, relative to the body size of the study species, to allow as natural an attack as possible. It also is advisable to equip the pen with a limited number of perches (one perch located at one end of the pen works best) so that the predation attempt begins from a fixed starting point. The prey item is placed in a small pen or enclosure opposite the perch, either on the ground or on a tabletop. The pen should be designed to prevent the prey from escaping in case the bird of prey does not attack immediately. The front should be transparent so that the predator can follow the prey's movements, but at the same time marked in some manner (painted stripes, etc.) to ensure that the bird is aware of its presence and height.

A blind, possibly equipped with a one-way window, should be placed as close as possible to the aviary, preferably immediately behind the prey enclosure, so that the researcher can record the test by direct observation or video recording. Recorded behavior patterns can range from exploratory flight, preening or movement on the perch (both interpreted as conflict patterns), to true predation-behavior sequence. Observations can

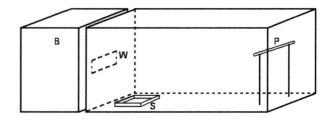

Figure 3. A diagrammatic representation of a pen for predation tests. A surface containing the prey (S) is opposite the single perch (P) upon which the bird stands. Next to the prey is a blind (B) equipped with a one-way window (W), behind which the observer records the behavior of both predator and prey.

involve the description of the attack glide, the type of landing (directly on the prey or next to it), the capture "tool" (beak, talons or both), which part of the body is grasped, and the prey's attempts to escape. Detailed descriptions of pens and recorded behavioral events are found in Csermely et al. (1989, 1991; Fig. 3).

Behavioral descriptions should be paired with temporal measurements, such as the latency from the start of the test, the frequency of each behavior and, sometimes, as in the case of preening, cumulative duration. After a successful attempt, the time elapsed until ingestion should be measured (considered latency to ingestion), and how and where in the pen it is performed are other important behaviors to be recorded. These behaviors can be recorded using check-sheets and stopwatches during both direct observation and videotape playing. An event recorder also is useful, as it automatically tracks frequency, duration, and latency of behavior.

The ethical issue of using live prey in such studies should be kept in mind and appropriate permissions sought. Captive birds of prey are easily trained to feed and to prey on dead items (Csermely 1993, Csermely and Gaibani 1998, Shifferman and Eilam 2004) and their predatory sequence is similar to that displayed when preying on live prey. Although captive studies usually are conducted to increase our knowledge of predation behavior, they also are useful in assessing the predatory abilities of rehabilitated birds. In the latter case, the use of live prey is usually necessary even from the viewpoint of the animal's well-being (cf. Csermely 2000).

BEHAVIOR AND CONSERVATION

Several authors (e.g., Caro 1998, Gosling and Sutherland 2000) have pointed out the need for ethology in conservation biology in order to ensure successful management strategies. In particular, Gosling and Sutherland (2000) state that, *"studies of behavior and conservation have a great deal to offer each other. This crossplay can happen at a number of levels. For example, the high priority given to conservation helps provide a justification for theoretically based studies of behavior and this may become increasingly important to justify research spending. Studies of behavior also can provide essential new insight into intractable conservation problems. Perhaps most important, it can also be argued that an evolutionary understanding of the behavior of individuals in populations allows us to predict responses under changed conditions with greater confidence than in the case of higher-level processes."*

In order to effectively protect a bird of prey within its habitat, it is necessary to understand its nest-site and prey preferences including the behaviors associated with these preferences. To establish a protected area of adequate size, we also must know the movement behavior of the species in question, as well as factors involved in both the intra- and inter-sexual and inter-species competition for resources. Ethological studies often require large amounts of time and money, but a project failing due to bad planning is economically more disadvantageous.

Unfortunately, and in spite of several common goals, ethology and conservation management still interact in only a limited way. Too often, ethological studies do not find application within conservation, and conservation projects often are planned without sufficient thought regarding a raptor's behavior. When used with conservation in mind, behavioral analyses can help increase success in raptor management, both in the wild and in captivity. Below we list examples of why this is important.

Raptor Rehabilitation

In rehabilitation centers, veterinary care is obviously of primary concern. Even so, such care often is not sufficient to guarantee successful rehabilitation. Conditions in captivity can be extremely stressful for raptors and can slow or even prevent their rehabilitation. It is therefore important to consider the behavioral aspects of each species in order to estimate the minimal dimensions of the aviaries, the maximum density of animals inside the aviaries, which species can cohabit, how the food must be supplied, etc.

Captive Breeding and Release to the Wild

For birds of prey at high risk of extinction, wild populations can be bolstered by offspring from the successful breeding of captive populations. The study of breeding behavior both in the wild and in captivity is very important to ensure adequate environmental conditions and to adjust rearing techniques to successfully breed captive pairs. At the same time, ethological studies also can assess whether captive-bred young behave normally and are likely to be capable of survival and reproduction in the wild. When nestlings are reared by hand, imprinting, or an imprinting-like social bond, can pose serious problems, particularly in Falconiformes (cf. Jones 1981). One way to reduce or avoid this problem is by feeding orphaned nestlings using a puppet that resembles the head of an adult. Thus, nestlings do not become wrongly imprinted on humans, and so avoid any complications in future breeding behavior. The common use of hand puppets is a useful consequence of ethological studies (Gosling and Sutherland 2000). Alternatively, hand-raising young in groups allows them to imprint on one another, thereby reducing its irreversibility (D. M. Bird, unpubl. data).

Furthermore, restocking or reintroduction projects do not end with the release of individuals. On the contrary, these projects should include long-term post-release monitoring to assess their success (cf. Csermely 2000). Behavioral studies of released birds allow researchers to determine which problems are related to the new conditions and where to look for solutions.

Specifically, applied ethology can be used to teach or to condition raptors to avoid potential threats they may encounter in the wild. One example of this is the experiment that Wallace (1997) conducted on California Condors (*Gymnogyps californianus*) at the Los Angeles Zoo. Young condors reared in captivity were conditioned using electrified wires on mock power poles not to perch on power poles once released.

CONCLUSION

In summary, behavioral studies have much to offer raptor management and conservation. Although often overlooked, this important topic promises to play a relevant role in protecting birds of prey, both in captivity and in the wild.

LITERATURE CITED

ALTMANN, J. 1974. Observational study of behavior: sampling methods. *Behaviour* 49:227–267.

CARO, T. [ED.]. 1998. Behavioural ecology and conservation biology. Oxford University Press, New York, NY U.S.A.

CRESSWELL, W. 1996. Surprise as a winter strategy in Sparrowhawks *Accipiter nisus*, Peregrines *Falco peregrinus* and Merlins *F. columbarius*. *Ibis* 138:684–692.

———, J. LIND, U. KABY, J.L. QUINN AND S. JAKOBSSON. 2003. Does an opportunistic predator preferentially attack nonvigilant prey? *Anim. Behav.* 66:643–648.

CROOK, J.H. AND P.A. BUTTERFIELD. 1970. Gender role in the social system of *Quelea*. Pages 211–248 in J.H. Crook [ED.], Social behaviour in birds and mammals. Academic Press, London, United Kingdom.

CSERMELY, D. 1993. Duration of rehabilitation period and familiarity with the prey affect the predatory behaviour of captive wild kestrels, *Falco tinnunculus*. *Boll. Zool.* 60:211–214.

——— 2000. Rehabilitation of birds of prey and their survival after release. Pages 303–312 in J.T. Lumeij, J.D. Remple, P.T. Redig, M. Lierz and J.E. Cooper [EDS.], Raptor biomedicine III, including bibliography of diseases of birds of prey. Zoological Education Network, Inc., Lake Worth, FL U.S.A.

——— AND N. AGOSTINI. 1993. A note on the social behaviour of rehabilitating wild Barn Owls (*Tyto alba*). *Ornis Hungarica* 3:13–22.

——— AND L. BROCCHIERI. 1990. Type of housing affects social and feeding behaviour of captive buzzards (*Buteo buteo*). *Appl. Anim. Behav. Sci.* 28:301.

——— AND G. GAIBANI. 1998. Is the foot squeezing pressure by two raptor species a tool used to subdue their prey? *Condor* 100:757–763.

———, D. MAINARDI AND N. AGOSTINI. 1989. The predatory behaviour of captive wild kestrels, *Falco tinnunculus*. *Boll. Zool.* 56:317–320.

———, D. MAINARDI AND N. AGOSTINI. 1991. Predatory behaviour in captive wild buzzards (*Buteo buteo*). *Birds Prey Bull.* 4:133–142.

EDUT, S. AND D. EILAM. 2004. Protean behavior under Barn Owl attack: voles alternate between freezing and fleeing and spiny mice flee in alternating patterns. *Behav. Brain Res.* 155:207–216.

GOSLING, L.M. AND W.J. SUTHERLAND [EDS.]. 2000. Behaviour and conservation. Cambridge University Press, Cambridge, United Kingdom.

HIRALDO, F., B. HEREDIA AND J.C. ALONSO. 1993. Communal roosting of wintering Red Kites *Milvus milvus* (Aves, Accipitridae): social feeding strategies for the exploitation of food resources. *Ethology* 93:117–124.

HÖGLUND, J. AND R.V. ALATALO. 1995. Leks. Princeton University Press, Princeton, NJ U.S.A.

JENKINS, A.R. 2000. Hunting mode and success of African peregrines *Falco peregrinus minor*: does nesting habitat quality affect foraging efficiency? *Ibis* 142:235–246.

JONES, C.G. 1981. Abnormal and maladaptive behaviour in captive raptors. Pages 53–59 in J.E. Cooper and A.G. Greenwood [EDS.], Recent advances in the study of raptor diseases. Chiron Publications, Keighley, United Kingdom.

LEHNER, P.N. 1996. Handbook of ethological methods, 2nd Ed. Cambridge University Press, Cambridge, United Kingdom.

MARTIN, P. AND P. BATESON. 1993. Measuring behaviour: an introductory guide, 2nd Ed. Cambridge University Press, Cambridge, United Kingdom.

MOUGEOT, F. 2000. Territorial intrusions and copulation patterns in Red Kites, *Milvus milvus*, in relation to breeding density. *Anim. Behav.* 59:633–642.

NEWTON, I. AND M. MARQUISS. 1991. Removal experiments and the limitation of breeding density in Sparrowhawks. *J. Anim. Ecol.* 60:535–544.

PALOKANGAS, P., E. KORPIMÄKI, H. HAKKARAINEN, E. HUHTA, P. TOLONEN AND R.V. ALATALO. 1994. Female kestrels gain reproductive success by choosing brightly ornamented males. *Anim. Behav.* 47:443–448.

SERGIO, F. 2003. From individual behaviour to population pattern: weather-dependent foraging and breeding performance in Black Kites. *Anim. Behav.* 66:1109–1117.

SHIFFERMAN, E. AND D. EILAM. 2004. Movement and direction of movement of a simulated prey affect the success rate in Barn Owl *Tyto alba* attack. *J. Avian Biol.* 35:111–116.

TOLONEN, P. AND E. KORPIMÄKI. 1994. Determinants of parental effort: a behavioural study in the Eurasian Kestrel, *Falco tinnunculus*. *Behav. Ecol. Sociobiol.* 35:355–362.

——— AND E. KORPIMÄKI. 1995. Parental effort of kestrels (*Falco tinnunculus*) in nest defense: effects of laying time, brood size, and varying survival prospects of offspring. *Behav. Ecol.* 6:435–441.

VILLAGE, A. 1985. Spring arrival times and assortative mating of kestrels in South Scotland. *Anim. Ecol.* 54:857–868.

WALLACE, M. 1997. Carcasses, people, and power lines. *Zoo View* 31:12–17.

WALTER, H. 1983. The raptor actigram: a general alphanumeric notation for raptor field data. *Raptor Res.* 17:1–8.

WARD, P. AND A. ZAHAVI. 1973. The importance of certain assemblage of birds as "information-centres" for food-finding. *Ibis* 115:517–534.

Food Habits

8

CARL D. MARTI

Raptor Research Center,
Boise State University, Boise, Idaho 83725 U.S.A.

MARC BECHARD

Department of Biology,
Boise State University, Boise, Idaho 83725 U.S.A.

FABIAN M. JAKSIC

Center for Advanced Studies in Ecology and Biodiversity,
Catholic University of Chile,
P. O. Box 114-D, Santiago CP 6513677, Chile

INTRODUCTION

Wildlife managers first became interested in raptor food habits in their attempts to assess the impact of raptors on game animals and livestock (Fisher 1893, Errington 1930), but ecologists soon found other reasons to understand raptor diets. What a raptor eats, and how, when, and where it obtains its food not only are significant in understanding the ecological relationships of the raptor itself, but also for understanding community ecology. Besides helping researchers understand raptor niches and how they relate to community structure, studying raptor diets can provide valuable information on prey distribution, abundance, behavior, and vulnerability (Johnson 1981, Johnsgard 1990, 2002; del Hoyo et al. 1994, 1999). The debate on whether raptors can limit the densities of their prey continues today; Valkama et al. (2005) provided a comprehensive review of the literature on this topic with an emphasis on Europe that also includes an overview of North America.

In this chapter we present methods of analyzing and interpreting raptor diets and discuss related precautions, advantages and disadvantages, and biases. We present analytical techniques for the collection of prey in raptor diets including pellet analysis, stomach-content analysis, examination of uneaten prey in nests, direct and photographic observation of prey delivered to nests, and confinement of nestling raptors in order to prolong data-collection intervals. Procedures for identifying prey and interpreting and characterizing raptor diets through dietary diversity, rarefaction, prey-weight, dietary-overlap, and stable-isotope techniques are demonstrated as well as guidelines for assessing adequate sample sizes. Methods for evaluating the composition, density, and vulnerability of prey populations are closely related to studies of raptor food habits, but are beyond the scope of this chapter. See Fitzner et al. (1977), Otis et al. (1978), Burnham et al. (1980), Schemnitz (1980), Call (1981), Johnson (1981), Hutto (1990), and Valkama et al. (2005) for an entry into this literature. Also valuable to the subject of this chapter are bibliographies containing references to the foods of raptors. Olendorff and Olendorff (1968), Earhart and Johnson (1970), Clark et al. (1978), Sherrod (1978), Pardinas and Cirignolli (2002), and Valkama et al. (2005) provide a wide range of such information.

ANALYTICAL PROCEDURES

Below we discuss the advantages and disadvantages of

each technique as a guide to its selection for a particular question. Regardless of the method selected, sampling is a very important consideration in food-habits studies; inadequate samples can produce misleading conclusions (Errington 1932). Information should be collected from more than one bird, nest, and, depending on the study objectives, more than one season or year (Korpimäki et al. 1994). Non-representative food-habits data may be obtained if the sample size is too small, if a prey species is locally or temporally abundant (e.g., during a population irruption), or if an individual or pair specializes on certain prey (i.e., behaves idiosyncratically).

Despite its importance, determining the adequate sample size prior to beginning a study may be difficult. Valid descriptions of diets that have a high variability in prey require more and larger samples than descriptions of diets with homogeneity of prey. Studies of seasonal changes in diet and inter- or intra-population dietary variation also require more samples. Investigators must ask whether it is important to document even those prey species eaten in very small proportions of the diet or whether it is more important to know which species are the mainstays of the raptor's diet, either numerically or by biomass. The answers to these questions will depend upon study objectives. See Morrison (1988) and Gotelli and Colwell (2001) and below for discussions on quantifying, evaluating, and justifying the size and nature of data sets, and Eckblad (1991) for general help on determining how many samples must be taken in biological studies, and our simulations included below relating sample size with dietary diversity and richness. Other statistical considerations are vital as well, and are similar to those in most biological situations (Sokal and Rohlf 1995).

Regurgitated Pellet Analysis

Most raptors, the Osprey (*Pandion haliaetus*) being a notable exception, produce pellets consisting of the less digestible remains of their prey including bones, teeth, scales, hair, feathers, keratin, and chitin. These materials are compacted by the stomach and regurgitated, usually daily. Identification of remains in pellets can provide both qualitative and quantitative information about the diet of a raptor. Although this method has been used for more than a century (Fisher 1893), Errington's (1930, 1932) extensive studies on raptor feeding did much to promote its use. Some early critics dismissed the technique of pellet analysis entirely (Brooks 1929), but it is now widely accepted as valid for most species.

In general, pellet analysis is most reliable for owls (Errington 1932, Glading et al. 1943), and is generally less reliable for falconiforms because many of the latter species dismember prey prior to swallowing and may not ingest all portions (Craighead and Craighead 1956, Cade 1982). Falconiforms also digest bone to a greater extent than do owls (Duke et al. 1975, Cummings et al. 1976). Owls tend to swallow prey whole or in large portions, with less rejection of identifiable remains (Errington 1932, Duke et al. 1975). Errington (1932) believed that only young owls digested bones significantly, but Raczynski and Ruprecht (1974) and Lowe (1980) reported considerable bone loss attributed to digestion in adults; neither study, however, provided enough details on the analytical procedures to allow evaluation of accuracy. Others have reported that not all food fed to captive owls was represented in pellets (Errington 1932, Glading et al. 1943, Southern 1969). Nevertheless, Mikkola (1983) found very close correlation between food eaten and remains in pellets, and Duke et al. (1975) and Cummings et al. (1976) indicated that very little, if any, bone digestion occurs in owls.

Insectivorous raptors present a different problem. Even though the entire prey is usually swallowed, chitinous portions may be broken into small fragments that are difficult to identify. Chitin digestion, however, appears to be slight at least in American Kestrels (*Falco sparverius*) and Eastern Screech Owls (*Megascops asio*) (Akaki and Duke 1999).

Pellets containing remains of prey too large for a single meal (e.g., rabbits or hares eaten by eagles, large buteos, or owls in the genus *Bubo*) pose a problem of quantification. Did the raptor feed once on a large prey item and leave a portion, with the result that remains in a pellet represent only a part of the prey? Or, did the raptor return later and consume the rest, so that all or most of the identifiable remains are in several pellets? Evidence shows that some raptors do return to large kills for several meals (Bowles 1916, Brown and Amadon 1968), but the number of larger prey species eaten may be greatly underestimated when pellet analysis alone is used to determine food habits. Large prey items brought to nestlings have a greater chance of being consumed totally. The remains may be distributed in pellets of several siblings and, in some cases, those of the adults as well (Bond 1936, Collopy 1983a).

The most profitable strategy for collecting pellets is to search nest sites and roosts. Larger samples can be obtained, species of raptor verified, and seasonal or yearly trends in prey consumption both determined from serial collections at the same site. Accumulation of data by this method is not uniformly successful with all raptors. Some species remain at one roost for long periods (e.g., Barn Owl [*Tyto alba*] and Long-eared Owl [*Asio otus*]), facilitating the collection of a large number of pellets (Marks and Marti 1984). However, many other species regurgitate their pellets over wide areas (e.g., Northern Harrier [*Circus cyaneus*] and Short-eared Owl [*A. flammeus*]), making collection of an adequate sample difficult (Errington 1932, Craighead and Craighead 1956, Southern 1969, Ziesemer 1981). It is important for statistical testing to collect pellets at as many nests, roosts, or both as possible to reduce problems associated with the lack of independent sampling.

Pellets of some species are distinctive in size and shape, but many are not. Guides to pellet identification for owl species are available (Wilson 1938, Burton 1984), but no method is foolproof for separating pellets of different species by appearance alone. To ensure that pellets are identified to species, only fresh pellets should be collected at nests, roosts, and perches known to be occupied by the raptor under study. The same nest sites often are used by different species at different times, so all old material should be removed and discarded prior to collecting new pellets for study.

Food-habits data are most valuable when the approximate date of deposition is known; hence, the knowledge of how long pellets persist in the wild is important. Moisture, invertebrates, and fungi rapidly break down pellets in exposed situations (Philips and Dindall 1979); most pellets in open environments decompose in less than 1 year (Wilson 1938, Fairley 1967, Marti 1974). In protected places, such as cavities, caves, or buildings, they may last much longer. Experiments to determine the rate of pellet decay in the local area of study might be necessary if there is doubt about how long pellets persist.

The method selected for pellet dissection depends upon the number of pellets to be analyzed and the objectives of the analysis. If the quantity is small or if the objective is to obtain immediate practical management information (e.g., to determine the principal food of a raptor or its impact upon a certain species of prey), pellets may be dissected individually by hand. Hair and feathers are teased away from bones, teeth, and other identifiable remains. Forceps and a dissecting needle are helpful aids for this. If quantities of pellets are large, or if better resolution of diet is required, hard remains should be separated from hair and feathers more carefully. This can be done by soaking and washing pellets with water. A more effective technique is to dissolve hair and feathers with sodium hydroxide (Schueler 1972). A modification of this procedure works well: dissolve 100 ml of NaOH crystals in 1 l of water, and then combine a sample of pellets with two to three times as much of this solution by volume. Two to four hours of soaking with occasional gentle stirring will sufficiently dissolve hair so that washing the solution through a screen (1/4 in mesh [6.35 mm]) will completely free the bones. Washing should be done over a pan to catch any fragments that pass through the screen, and the residue can then be washed, decanted, and added to the sample. Even very small, delicate bones are unharmed by this process, and the likelihood of finding smaller fragments is much greater than with dry dissection of pellets. Pellets must not be left in the NaOH solution more than 4 hours because teeth may become dislodged, reducing the chance of specific identification of mammalian prey remains. Chitinous materials also are unaffected by NaOH and are easily recovered, but any hair or feathers will be dissolved. Thus, this technique should not be used if the intent is to identify prey by the use of hair or feathers.

Skulls and dentaries are the most useful remains for identifying and counting mammalian prey, and a hand lens or low-power dissecting scope will be necessary in many cases to examine these prey remains. Limb bones and pelvic girdles also are helpful, especially for counting larger prey. Keys may aid in identifying small mammals (Stains 1959, Glass 1973, DeBlase and Martin 1974). Reference collections and investigator experience, though, usually are better than keys because skulls in pellets often are broken and may be missing diagnostic parts needed by keys. Thus, side-by-side comparison with skulls from reference collections is preferred. Mammalian hair from pellets also may be used to identify prey from raptors that digest bone or do not swallow it. Hair has little value however, for quantifying the prey consumed. Adorjan and Kolenosky (1969) and Moore et al. (1974) developed keys for identifying mammalian hair, and Korschgen (1980) gives instructions for preparing reference slides for hair. Feathers in pellets create similar but even greater problems than hairs. Feathers recovered from pellets typically require cleaning before they can be identified. Sabo and Laybourne (1994) provide techniques for feather preparation and also clues useful in identifying individual feathers.

Small mammals usually are enumerated in pellet samples by counting skulls and considering dentaries and leg bones as a backup, especially if decapitation of prey is suspected. For larger mammal species, fragments should be assembled from a sample (skulls, dentaries, pelvic bones, and heads of limb bones) and then pieced together to estimate how many individuals were consumed (see Mollhagen et al. [1972] for more details). This procedure assumes that all parts of the prey were eaten and that all pellets containing the remains were recovered. Thus, counts based on this method most likely will be conservative. If possible, an additional technique should be used as a check.

Identifying bird prey is possible from feathers, beaks, and feet but often is difficult to accomplish without a large reference collection. Skulls, sterna, and synsacra are most useful for counting birds in pellets. Experts with access to extensive reference collections may be able to identify bone fragments and individual feathers to genus or species.

Bones of amphibians and bones and scales of fish and reptiles should be retained for identification. Collections of fish opercula at and around nests have been used to identify the prey of Osprey (Newsome 1977, Prevost 1977, Van Daele and Van Daele 1982). Comparison with reference material and consultation with experts on these taxa are recommended for identification.

Insects and other invertebrate prey also pose problems. The exoskeleton of arthropods is the only portion not digested by raptors, but often it is highly fragmented, making keys of little value as identification aids. Again, a good reference collection and consultation with experts are the best approaches to identifying those remains.

Pellet analysis offers advantages over other techniques—a large sample often may be acquired with relatively little expense, time, or disturbance of the raptors, and both seasonal and yearly trends in diet can be obtained, often from the same birds. Disadvantages are that pellets of some raptors, particularly falconiforms, do not always contain remains of a significant portion of prey eaten. For this reason, less confidence is possible from analysis of most falconiform pellets and from pellets of large owls preying on large prey. Available evidence indicates that pellet analysis is an excellent technique for medium-sized and smaller owls, e.g., Boreal Owl (*Aegolius funereus*) (Korpimäki 1988) and Eurasian Pygmy Owl (*Glaucidium passerinum*) (Kellomäki 1977), but slightly less reliable for insectivorous owls, e.g., Burrowing Owl (*Athene cunicularia*) and Flammulated Owl (*Otus flammeolus*), because their prey remains may be very small and pellets consisting

of insect parts decompose rapidly (Marti 1974). Pellet analysis also appears to be a good method to study diet variation of Common Kestrels (*Falco tinnunculus*) using small rodents as their main foods, but also including many insects as alternative prey (Korpimäki 1985, Itämies and Korpimäki 1987). Although some investigations of falconiform diets have used pellet analysis exclusively (see references in Sherrod 1978), we recommend that a second method be used to check the accuracy of data from pellet analysis. On the other hand, Ritchie (1982) recommended using pellet analysis to complement studies based primarily upon prey remains in nests.

Contents of the Digestive System

Most early studies of raptor food habits were based upon examination of prey remains in raptor stomachs (Fisher 1893, McAtee 1935). This technique has no place in modern research or management practice except where a source of dead raptors, such as road kills, is available. Killing enough raptors to obtain a sample size sufficient to characterize diet is highly undesirable because the populations of most raptors are relatively small. The quantity of data obtained from an individual raptor using this technique is minimal compared with all the other available methods. The procedure for stomach analysis is simply to open the stomachs and crops of dead raptors and examine the contents. Identification and quantification of prey are similar to the processes described under pellet analysis. If analysis cannot be done immediately, stomachs can be frozen or preserved in 10% formalin until examined (Korschgen 1980).

If it is essential to examine stomach contents of live raptors, an emetic technique should be considered (Tomback 1975). Pulin and Lefebvre (1995) employed an antimony potassium tartrate (tartar emetic) on 137 bird species from 29 families. This technique apparently has not been tried on raptors and its safety is not known. Rosenberg and Cooper (1990) recommended flushing the digestive tract or forcing regurgitation with warm water instead of an emetic.

Another alternative for studying freshly eaten food without killing raptors is to massage food out of the crops of nestling or captured falconiforms (owls do not have crops) (Errington 1932). Workers with little experience in handling young raptors should avoid this practice because of the possibility of damaging the esophagus (Sherrod 1978).

Uneaten Prey Remains

Examination of nests for uneaten prey has proved useful by itself or in conjunction with other techniques (Craighead and Craighead 1956, Smith and Murphy 1973, Collopy 1983a). In one study, Bureau of Land Management (BLM) crews (USDI 1979) entered nests of several falconiform species every 4 to 6 days to collect all inedible prey remains and pellets. Fresh prey was marked by collecting the head, feet, and tail, and the remainder was left in the nest. Each collection was then examined for diagnostic remains to ascertain the species and number of prey represented. Collopy (1983a) collected similar materials from Golden Eagle (*Aquila chrysaetos*) nests. He found that these samples were not significantly different in species composition from what he saw in direct observation of the nests, but that they did seriously underestimate biomass of prey eaten compared with direct observation. Rutz (2003) radio-tracked male Northern Goshawks (*Accipiter gentilis*) in order to locate all kills the birds made. He showed that the remains of some prey species are harder to find by visual scanning and may result in biased dietary determination.

Several important considerations must be noted when collecting and interpreting prey remains in raptor nests. Larger, heavier bones may persist longer in the nest and cause overestimation of larger prey types. K. Steenhof (pers. comm.) suggests that collection intervals of 5 days or less help reduce this problem. Bones of smaller prey may be consumed at a higher rate (Mollhagen et al. 1972) or lost in the nest structure, causing underestimation of their contribution to a diet. Snyder and Wiley (1976) found similar circumstances at Red-shouldered Hawk (*Buteo lineatus*) nests. According to Bielefeldt et al. (1992), indirect collection of Cooper's Hawk (*A. cooperii*) prey remains near nests (92% birds) overestimated the proportion of avian items in comparison with direct observation of prey deliveries to nestlings (51–68% birds); most avian items brought to nestlings in their Wisconsin study, as elsewhere, were young birds. Thus, they suggest that other studies relying on indirect methods and using prey species' adult mass to calculate avian biomass probably have been biased toward birds among prey remains.

One potentially serious problem associated with collecting prey remains from nests is disturbance of the raptors. Caution must be taken to avoid keeping adults away from nests when weather conditions are detrimental to the young and to avoid any other excessive interference with normal behavior at the nest (Chapter 19). Another danger is that repeated visits may increase the likelihood of leading predators to the nests of some raptors.

Prey remains also may be recovered at plucking posts for some species, especially falcons, accipiters, and owls in the genus *Glaucidium*. Special care should be taken in interpreting such materials, particularly when using this method in conjunction with pellet analysis. Reynolds and Meslow (1984) collected pellets and other prey remains every 3 to 6 days at Cooper's Hawk nests and associated plucking sites, and Boal and Mannan (1994) used the same method in studying Northern Goshawks. They attempted to reconstruct and count each kind of prey by matching rectrices, remiges, and bills of birds, and fur, skull fragments, and feet of mammals from all material collected at each visit. Ziesemer (1981) discovered a bias in numbers of different prey types recovered by searching for plucking posts — birds were more readily found because of scattered feathers and prey larger than a single meal were often missed because of scavenging by mammals.

Some raptors store excess prey, which also can be a source of food habits information. Korpimäki (1987a) found that Boreal Owls stored prey mainly during the breeding season in the nest cavity, but Eurasian Pygmy Owls store prey mainly in the winter (Solheim 1984). Food storing also has been documented in the Northern Hawk-Owl (*Surnia ulula*) (Ritchie 1980), and Barn Owl (Marti et al. 2005), Eleonora's Falcon (*F. eleonora*) (Vaughan 1961), Merlin (*F. columbarius*) (Pitcher et al. 1979), and American Kestrel (Collopy 1977).

Direct Observation

Direct visual observations, while requiring a great deal of investigator time, offer some advantages over other techniques. This method is used most often at nests with the observer concealed in a nearby blind (Collopy 1983a, Sitter 1983, Younk and Bechard 1994, Rosenfield et al. 1995, Real 1996, Dykstra et al. 2003, Meyer et al. 2004). Others have used direct observation of foraging raptors, often from a vehicle and with the aid of a spotting scope (Wakeley 1978, Bunn et al. 1982, Beissinger 1983, Collopy 1983b). The most satisfactory approach is to observe continually all day or night. This approach will usually include a significant amount of time when no prey deliveries are made. If shorter periods of observation are used, they should be rotated randomly to include all hours when the species is active.

Several investigators preferred direct observation to other methods (Snyder and Wiley 1976, Collopy 1983a,

Sitter 1983) and it may be the best technique to use for species whose pellets do not provide accurate representation of their diet. Southern (1969) discovered by observing Tawny Owls (*Strix aluco*) that they were feeding earthworms to their young, a fact that had not been apparent from pellet analysis. Collopy (1983a) found that observation provided the best means of estimating biomass of prey consumed; both the number and size of prey can be accurately determined.

Direct observation from blinds can provide some of the most complete and accurate information on the diets of many raptors, as well as useful data on behavior. The chief drawback is the great amount of observer time required, often under uncomfortable conditions, to obtain an adequate sample. Blinds should be constructed in short periods over several days to reduce disturbances. The best time to build blinds is before a traditionally used site is occupied, keeping in mind that the birds may not select that site in a particular year. Some species and even some individuals are sensitive to disturbance and may not tolerate blinds placed near the nest, whereas others will accept blinds as close as 2 m (Geer and Perrins 1981). Size of prey involved is another consideration in distance from blind to nest; insectivorous species will necessitate close placement of blinds in order to identify prey, but the prey of eagles can be identified up to 40 m away (Collopy 1983a). R. Reynolds (pers. comm.) cautions that estimating the size of small vertebrate prey by observation is difficult. Sitter (1983) preferred to observe Prairie Falcons (*F. mexicanus*) from about 15 m distant and slightly above the nest. R. Glinski (pers. comm.) placed blinds slightly below the nest to reduce disturbance. Regardless of the distance between blind and nest, binoculars or spotting scopes are usually needed to identify prey.

Cavity-nesting species also can be observed directly, but some modification of the site may be necessary and this technique should be used only with great caution. Southern (1969) used nest boxes with a partially cut-away side so that prey delivered to the young could be seen. Smith et al. (1972) installed a one-way mirror in an American Kestrel nest cavity, and one of us did the same in a Barn Owl nest box with blind attached (Marti 1989).

Nocturnal species, obviously, are harder to observe. Night-vision scopes or goggles (image intensifiers) provide the most satisfactory answer to this problem but are expensive; DeLong (1982) used one with good results at nests of Long-eared Owls. A simpler and less costly approach is to illuminate the nest with artificial light. Southern (1969) found that a red light placed at Tawny Owl nests did not disturb the birds, and a six-volt, clear flashlight bulb produced no behavioral changes in Barn Owls when placed just outside nest cavities or even within a nest box (Marti 1989). At distances of 10 to 60 m, aided with 7 x 50 binoculars, adult prey deliveries to nestlings could be monitored but the prey could not be identified. Prey was easily identified however, when deliveries were observed through a one-way mirror in the back of a Barn Owl nest box illuminated as described above.

Non-breeding raptors are harder to observe for documenting prey captured because of their mobility and, in many species, secretive habits. Roth and Lima (2003), employing radio-tracking to follow Cooper's Hawks in winter, were able to observe 179 attacks — 35 of which were successful — and identify the prey captured.

Confining Nestlings

Additional food-habits information has been acquired for 4 to 10 weeks beyond normal fledging times by tethering young raptors on the ground near their nests so that prey brought by the adults could be studied more easily (Errington 1932); tethers were similar to falconry jesses. Losses of young raptors to predators while using this method (as high as 50%) prompted Petersen and Keir (1976) to tether young on platforms off the ground. Care should be taken to adjust the length of the tether so that the young cannot hang over the edge of the platform. Selleck and Glading (1943) placed cages over young raptors in their nests. This forced adults to leave prey outside so it could be identified and counted. These workers found that the cage-nest technique worked well for Barn Owls but not as well for Northern Harriers, because of behavioral differences in prey delivery between the two species. Sulkava (1964) used this technique with success on Northern Goshawks in Finland.

These methods may be useful in studies of raptor species for which food-habits data are otherwise difficult to obtain, but they should be used sparingly and with great care. Increased predation upon the young, abandonment by the adults, and interference with normal behavioral development are inherent dangers.

Photographic and Digital Image Recording

Several generations of systems, from film to digital, have been described for monitoring wildlife activity including the use of cameras automatically triggered by photocells (Dodge and Snyder 1960, Osterberg 1962, Cowardin and Ashe 1965, Browder et al. 1995, Danielson et al. 1996), cameras triggered by observers in blinds (Wille and Kam 1983), and automatic sampling using time-lapse cameras or video recorders (see references below).

Single-lens, reflex, 35-mm cameras, the first camera type employed for raptor food-habits monitoring, have many accessories helpful in remote or automatic operation (e.g., auto-winders, telephoto and close-up lenses, bulk-film backs, and radio-controlled shutter releases), or both. Users have reported that the 35-mm format provides good resolution for identifying prey, but the cost of equipment, film, and film processing is high. Another drawback of this technique, one shared with other similar techniques, is that many photographs are under- or over-exposed and others do not show prey clearly enough to allow identification.

Another monitoring option is to sample automatically by using a time-lapse camera set to take one or more frames at constant intervals throughout the sampling period. Time-lapse photography has been used to study raptor diets since the early 1970s when Temple (1972) described one of the first portable systems using a super-eight camera that could be installed at raptor nests and programmed to expose images at set time intervals, usually one frame every 1 to 5 minutes. Similar systems were used to study a variety of nesting raptor species (Enderson et al. 1972, Franklin 1988, Tømmeraas 1989, Hunt et al. 1992). However, super-eight cameras are no longer easily available and film is difficult to find and have processed.

A number of video-camera systems can be used for recording the diets of diurnal raptors (Kristan et al. 1996, Delaney et al. 1998, Booms and Fuller 2003a). Lewis et al. (2004a) designed a video-surveillance system to document the diet of Northern Goshawks consisting of a miniature video camera, time-lapse video recorder, and a portable 13-cm television, powered with a single, deep-cycle marine battery.

Recent advances in time-lapse video surveillance systems have made videography a far more useful technique for recording diets of raptors. If the species of interest is sensitive to disturbance, cameras can be placed so that recording equipment and power sources are well away from its nest and visits to replace batteries and tapes can be made daily or at intervals of two to three days. Time-lapse videography is versatile and accommodates options for capturing images from real-time (20 frames/second) to 960-hour time-lapse (0.25

frames/second) on a standard 8-hour VHS videotape. To maximize the number of frames of each prey delivery while maximizing the interval between visits to change videotapes, the systems can be programmed to record at various frames/second and at specified times of the day.

Solar-powered surveillance systems are useful if routine replacement of batteries is difficult. Booms and Fuller (2003a) used solar-powered, time-lapse Sentinel All-Weather Video Surveillance Systems (Sandpiper Technologies Inc., Manteca, California) to record prey deliveries to Gyrfalcon (*F. rusticolus*) nests in Greenland. Video cameras were mounted within 1 m of nests and all other equipment was installed at the bases of nest cliffs where a time-lapse VCR was used to record images from the camera. The recording unit was placed in a location that allowed easy and safe access to change tapes while not being detected by the adult birds. Cameras were installed during mid- to late incubation, and nests were not visited again until after young had fledged.

Solar-powered, radio-frequency linked, transmitting video camera systems also are available for use with species that are sensitive to repeated disturbances near their nesting areas. These systems transmit video signals from the nest site to a remote receiver and disturbance at nest sites is minimal because personnel do not need to visit nesting areas to change videotapes or batteries. Kristan et al. (1996) used such a system that performed reliably up to 8 km, to document prey delivered to Osprey nests in California. While the cost of the system was approximately $6,100 (U.S.), the savings in personnel time were substantial.

Video systems using miniature, infrared-sensitive video cameras equipped with infrared light-emitting diodes and time-lapse video recorders have proved to be effective in documenting the dietary habits of several species of owls. Proudfoot and Beasom (1997) used such a camera and light source to record prey deliveries to nests of Ferruginous Pygmy-Owls (*G. brasilianum*) and Delaney et al. (1998) used a similar system to study Mexican Spotted Owls (*S. occidentalis*). A useful range up to 3 m in total darkness was possible with the aid of six infrared light-emitting diodes. Video images were recorded using time-lapse VHS recorders connected to cameras via coaxial cables. Each tape provided 24-hour coverage when recording at approximately five frames/second. These camera systems were powered by either 12-volt, deep-cycle marine batteries or 12-volt, sealed-gel-cell batteries. The latter are rugged and reduce the potential for spillage during backpack transport. Oleyar et al. (2003) described an inexpensive camera system designed to study the diet of Flammulated Owls. This system used a miniature pinhole, infrared camera and a single infrared-emitting diode connected to an 8-mm camcorder to record prey deliveries on tape. The camera system was powered by three batteries: a 6-volt camcorder battery, a 1.5-volt battery for the infrared diode, and a 9-volt battery for the camera. Cameras were turned on each night and allowed to record until the batteries failed, which was generally at about two hours.

Images recorded on videotapes can be viewed using VCR equipment and a color TV monitor. Many VCRs allow frames to be replayed at different speeds and each frame can be frozen for inspection.

Comparing Collection Methodologies

It is obvious from the information presented above that different raptor species require different methods for collecting unbiased food-habits material. A number of investigators have used multiple methods on the same species and offer insights on which method is best, and when it may be appropriate to use more than one method of collection. Pavez et al. (1992), Real (1996), and Sequin et al. (1998) made direct observations at nests of Black-chested Buzzard-Eagles (*Geranoaetus melanoleucus*), Bonelli's Eagles (*Hieraaetus fasciatus*), and Golden Eagles, respectively, and compared prey counted by observation with prey identified in pellets and uneaten remains in the nest. For Black-chested Buzzard-Eagles, pellet contents under-represented birds whereas insects were over-represented by observation and under-represented by prey remains. In the case of Bonelli's Eagle, prey remains were collected under two regimens—fresh remains while nestlings were in the nest and old remains collected after breeding finished. Pellets also were collected; using old prey remains was the only method that differed significantly from observations and Real (1996) concluded that pellet analysis was the most efficient method for studying Bonelli's Eagle diet. Sequin et al. (1998) recommended that combining pellet contents and prey remains is the best procedure if direct observations cannot be made. Mersmann et al. (1992) compared three techniques for studying Bald Eagles (*Haliaeetus leucocephalus*). Direct observations resulted in biases toward easily identified species such as eels, but also permitted documenting consumption of small soft-bodied fish that were not well detected by other methods. Using captive eagles,

Mersmann et al. (1992) discovered that fish were under-represented in the pellets, but that most birds and mammals eaten were detected. Analysis of food remains of the captive eagles over-represented birds, medium-sized mammals, and large, bony fish; small mammals and small fish were under-represented.

Sharp et al. (2002) and Marchesi et al. (2002) compared diets obtained through pellet analysis and uneaten prey remains for Wedge-tailed Eagles (*Aquila audax*) and Eurasian Eagle-Owls (*Bubo bubo*), respectively. Sharp et al. (2002) concluded that combining data from the two methods may result in a biased diet determination and recommended that results for the two techniques be reported separately. On the other hand, Marchesi et al. (2002) recommended combining data from the two techniques, but indicating the relative contribution of each method in the pooled sample; they found that prey remains over-represented birds and large prey in general, under-represented mammals, and failed to detect fish. Pellets gave a more realistic picture of diet but failed to detect many birds identified in prey remains.

Studying Barn Owl diets, Taylor (1994) compared prey delivered to nests as recorded by continual photographic monitoring with contents of pellets produced during the same period; results of the two techniques agreed closely. Comparison of prey remain collections, pellet contents, and prey delivery videography showed that videography provided the most complete descriptions and least biased data on the diets of Northern Goshawks and Gyrfalcons (Booms and Fuller 2003b, Lewis et al. 2004b). Additionally, Lewis et al. (2004b) felt that videography equipment and its maintenance is cost-effective compared to human-resource costs associated with prolonged direct observations made from blinds.

INTERPRETATION OF RAPTOR DIETS

Quantification

Raptor diets can be quantified in a number of ways depending upon the needs and objectives of the analysis. One common method is to calculate the percentage of occurrence by number for each prey category in the total sample. In cases where it is not possible to count the number of each prey, diets may be quantified by giving the percentage of samples (e.g., pellets or nest contents) in which each kind of prey occurred. Diets also can be quantified by the relative contribution of the various prey types to the total weight (biomass) of prey consumed. Both frequency and biomass methods have value. For example, frequency data provide useful information on the relative impact a raptor has upon various prey species, whereas biomass determination may give a more accurate evaluation of the relative importance of prey species to the diet of a raptor (i.e., one rabbit provides the equivalent energy of many mice).

Frequency by number of prey (species or other taxon) is calculated by dividing the number of individuals in each identifiable category of prey by the total number of prey in the sample. When prey are identified by hair or feather analysis, obviously it is not possible to count the number of individuals in a sample. In these and other cases where it is not possible to count numbers of individual prey, frequency of occurrence may be used. This may be calculated, for example, by dividing the total number of pellets in a sample into the number of pellets in which each kind of prey was found; the disadvantage of analyzing dietary data using this approach is that these data cannot be used to calculate niche metrics, which are described below.

Biomass of prey in a diet sample usually is estimated by multiplying the number of individuals of each prey species by the mean weight of that prey. Biomass is then expressed as the proportion each prey species (or other taxa) contributed to the total weight consumed. Several sources provide tables of weights for this purpose (Smith and Murphy 1973, Marti 1974, Brough 1983, Steenhof 1983, Dunning 1984), but locally obtained prey specimens, when available, may provide more accurate weight information. In many cases prey should be assigned to different weight categories according to age and sex for more accurate estimates of dietary biomass. If raptors select other than average-sized prey of a particular species, biomass estimates derived in the above manner will be biased (Santibáñez and Jaksic 1999). Sometimes greater accuracy may be obtained by measuring or estimating weights of prey actually eaten, as determined through direct observation, examination of whole prey in nests, or photographic techniques. Prey weights also can be estimated from measurements of skeletal remains in uneaten prey remains (Diller and Johnson 1982, Woffinden and Murphy 1982) and pellets (Boonstra 1977, Goszczynski 1977, Morris 1979, Nilsson 1984). Fairley and Smal (1988) provide correction factors for more accurate estimation of the mass of prey eaten from measurements of

bones found in pellets. Norrdahl and Korpimäki (2002) warned that body mass of some small mammals can vary considerably among years, especially in species that undergo cyclic population fluctuations. If this is occurring, it must be accounted for in estimating biomass of prey consumed by raptors.

Wijnandts (1984) obtained weights of prey delivered to nestlings by placing nests containing nestling Long-eared Owls on platforms equipped with electronic balances. He reported that accuracy depended upon wind speed and stability of the supporting tree but was usually within ± 2 g. This technique would seem to be applicable to many raptor species.

Diversity

Diversity is an expression of community structure wherein groups of organisms (identified to species or higher taxa) are characterized by the number of categories in the group and the relative number of individuals in each category (Magurran 1988). Measures of diversity are employed to examine the structure of assemblages such as the prey species in a raptor's diet. Properly used, diversity indexes allow the summarization of large quantities of data as a single value. These indexes have been used as a quantitative measure of niche breadth (Pielou 1972, Hurtubia 1973) and, as such, to characterize and compare raptor diets (Jaksic et al. 1982, Marks and Marti 1984, Steenhof and Kochert 1985, Bellocq 2000). Korpimäki (1987b, 1992) related the variation in diet diversity to variation in breeding density and reproductive success.

Below we use the terms diversity and food-niche breadth synonymously. Diversity has two components, richness (the number of prey categories, species or other) and evenness (how uniformly represented the various kinds of prey are) (Margalef 1958, Pielou 1966). A raptor's diet has high diversity (i.e., represents a broader food niche) if many species are included in nearly equal numbers. Conversely, a collection consisting of few species or with species represented in very different abundances has low diversity (represents a more narrow food niche).

Several assumptions, some stringent, must be met in collecting data for calculating diversity indexes. See discussions of these in Pielou (1969), Brower and Zar (1984), and Hair (1980). Much has been published on the relative value of different diversity indexes, including opinions by some authors that these indexes have no value (Hurlbert 1971). Others, though, found them very useful (Hill 1973). A comprehensive coverage of the problems in measuring diversity is not appropriate here, but see Greene and Jaksic (1983), Kinako (1983) and Ghent (1991) for background, criticisms, and precautions in using these indexes.

Greene and Jaksic (1983) present information directly useful for the interpretation of diversity indexes. Not surprisingly, they found that high resolution of categories (identification of prey to species or genus) compared to low resolution (identification of prey to order or class) yields greater niche breadths, and that high resolution more consistently measures the extent to which raptors affect various prey populations. Low resolution of prey, though, may be useful in comparing functional niches; broader niches at this level, in comparisons among raptor species, may indicate a more versatile predator (e.g., a predator able to consume prey presenting many different kinds of problems in capture and handling).

Many measures of diversity have been devised and are in current use (Washington 1984). See Brower and Zar (1984), Hair (1980), and Ghent (1991) present and compare many of the commonly used indexes. Only a few of the most widely employed indexes are covered here (examples of the calculation of these and the following evenness indexes are in Appendix 1).

Simpson (1949) was the first to devise an index incorporating both richness and evenness:

$$D = \Sigma\, p_i^2,$$

where p_i is the relative proportion of each member of the assemblage being investigated. This index yields values from zero to one. When calculated with this formula, Simpson's index actually measures dominance (i.e., larger values indicate lower diversity in the assemblage) (Whittaker 1965). For example, a raptor diet heavily dominated by one or two kinds of prey will yield values close to one in the Simpson's index, whereas a diet containing a more even distribution of prey types (higher diversity) will yield a value closer to zero. In order to convert Simpson's index to a more interpretable measure of diversity (i.e., where larger values of the index reflect greater diversity), it is common to calculate $1/D$ (Levins 1968) or $1-D$ (Odum 1983). Ghent (1991) recommended using Simpson's index because it is the simplest diversity index that adequately performs its task.

Shannon's index (Shannon and Weaver 1949) is another measure of diversity widely used in ecology.

The formula is:

$$H' = -\Sigma\, p_i \log p_i,$$

where p_i represents the proportion of each species in the sample. The larger the value obtained for H' (or antilog H'), the greater the diversity of the sample. Any logarithmic base can be used as long as consistency is maintained throughout. However, indexes calculated with different logarithmic bases must be converted to the same base before comparisons between them are meaningful. Brower and Zar (1984) list appropriate conversion factors. The antilog of H' is more readily interpretable as a measure of diversity than H' because it is linearly related to the number of prey categories in the sample (Hill 1973, Alatalo and Alatalo 1977).

Even though both Simpson's and Shannon's indexes measure richness and evenness, DeJong (1975) found that Shannon's index places nearly twice as much weight on the richness component than does Simpson's. Conversely, Simpson's is influenced by evenness much more than Shannon's.

Colwell and Futuyma (1971) developed a standardized measure of food-niche breadth (FNB) that permits meaningful comparisons between diets of different species or the same species in different geographic areas:

$$FNB_{sta} = (B_{obs} - B_{min})\,/\,(B_{max} - B_{min}),$$

where B_{obs} is the reciprocal of Simpson's Index, B_{min} is the minimum niche breadth possible (equals one), and B_{max} is the maximum breadth possible ($= N$). See Jaksic and Braker (1983) and Marti (1988) for examples of its use in comparing food-niche breadth among geographical areas where differing numbers of prey were available to widespread raptors.

No easy way exists to determine what constitutes an adequate sample size for calculating dietary diversity. Larger samples are more likely to include rare prey, thus increasing the measure of diversity (although the lack of including rare prey has little effect on Shannon's index [Brower and Zar 1984]). Many factors, though, complicate the situation: density, number of species, and availability of prey. For example, a large diet sample that yields a narrow estimate of food-niche breadth might indicate that only a small number of prey species was available to the predator. Conversely, it might indicate that a larger assemblage of available prey species contained one or a few prey that were particularly abundant or vulnerable to the predator. Competition, either by exploitation or interference, also could affect how a predator exploits prey species and thus alter its dietary diversity. Extensive literature exists on the influence of competition upon food-niche breadth, but coverage of it is beyond the scope of this chapter.

One means of determining the sample size needed to accurately reflect the number of prey types in a raptor's diet is to plot the number of new prey species occurring per sample as a function of sample size; when an asymptote is reached, a sufficient sample size has been obtained (Heck et al. 1975, Gotelli and Colwell 2001). As sample size increases, more species will be recorded with the sampling curve rising rapidly at first and then more slowly as increasingly rare species are included. See Green and Young (1993) for formulas to estimate the sample size needed to detect rare species.

We provide several populations (Appendix 2) to illustrate the required sample size on estimating species richness and diet diversity of the populations from which samples are drawn. Two of these are simulated populations; the other is a sample of actual dietary data from a population of Barn Owls. From each population, we drew random samples with replacement ranging from 5 to 500 individuals in increments of five. Each sample size was repeated 100 times after which the mean number of prey types (richness) and mean sample diversity (reciprocal of Simpson's index) were calculated. The results in Fig. 1 illustrate that when species richness is very low (five, population A, Appendix 2), a sample of less than 20 individuals will include all potential species. When species richness doubles to 10 (population B, Appendix 2), a sample of about 50 individuals is required to include all potential species (Fig. 2). The simulated populations A and B have maximum evenness (i.e., all prey species are present in exactly the same numbers). In contrast, population C (Appendix 2) has 29 prey species but is dominated by two species and only six species are common; a sample size of less than 20 will include six species, but many prey species are rare and a sample size of 1,000 only includes about 50% of the potential prey types.

When trying to estimate diversity, the situation is reversed. The two populations with maximum evenness (A and B) require sample sizes of more than 100 to approach an asymptote. The yield of additional information when sample sizes are more than 100 is slight, and samples of even 500 individuals do not quite reach an estimation of the population's true diversity (Figs. 1

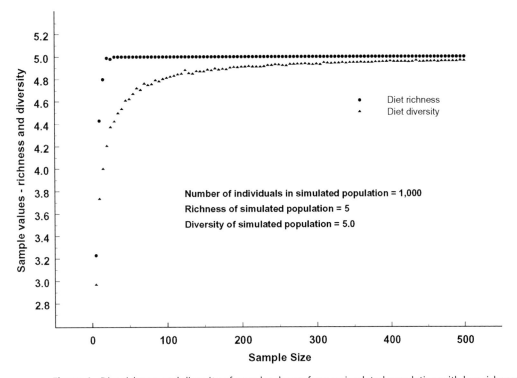

Figure 1. Diet richness and diversity of samples drawn from a simulated population with low richness and high evenness to illustrate the sample size needed to adequately characterize that of the sampled population.

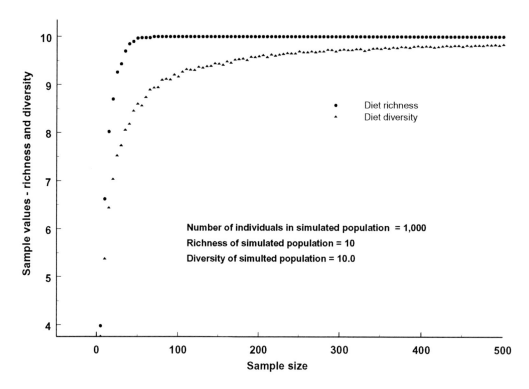

Figure 2. Diet richness and diversity of samples drawn from a simulated population with higher richness and high evenness to illustrate the sample size needed to adequately characterize that of the sampled population.

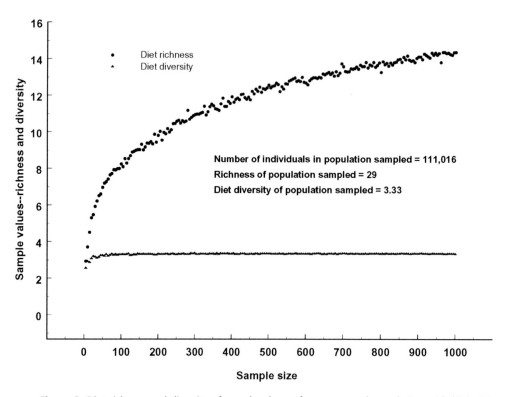

Figure 3. Diet richness and diversity of samples drawn from an actual population with high richness and low evenness to illustrate the sample size needed to adequately characterize that of the sampled population.

and 2). In contrast, population C needs samples of only 50 to 100 individuals to correctly estimate the diversity of the population and larger samples offer no additional information about diversity (Fig. 3).

Often, biologists are interested in identifying the common or dominant prey in a raptor's diet (i.e., those that make significant contributions of energy). Prey species taken rarely are of incidental interest. They show the widest range of the raptor's diet, but contribute little to the energy intake. In such situations, we suggest that samples of around 100 prey individuals are sufficient to give a reasonable approximation of a raptor's diet. This is not to say that samples that small are always ample. If the goal is to understand variation (e.g., geographic or temporal), many samples of 100 or more from different individual raptors or from different times (seasons, years) will be needed.

As noted, diversity indexes include both the richness and evenness of a sample, but it is often desirable to provide separate measures of the two components. Richness is simply expressed as the number of species (or other taxa) in a raptor's diet, and several approaches to measuring evenness or equitability have been developed (Pielou 1969, Hurlbert 1971, Hill 1973). Frequently used is Pielou's (1969):

$$J' = H' / H\,max',$$

where H' is the diversity value calculated from Shannon's index, and $H\,max'$ is the logarithm of the number of species (species richness) employing the same logarithmic base used in the calculation of H'. Because species richness (i.e., the number of prey actually eaten by the raptor) is often underestimated in a dietary sample, J' tends to overestimate evenness. Alatalo (1981) modified Hill's (1973) ratio to develop a more interpretable measure of evenness:

$$F = (N_2 - 1) / (N_1 - 1),$$

where N_1 is the antilog of Shannon's index (H') and N_2 is the reciprocal of Simpson's index ($1/D$). Alatalo (1981) cautioned that there is no single mathematical definition of evenness; each measure weights different properties of abundance distributions in different ways.

Another technique for comparing dietary prey frequencies with relative availability of prey is Ivlev's (1961) selectivity index:

$$S = (r - p) / (r + p),$$

where r is the proportion of prey taken by the predator and p is the proportion of the same prey available to the raptor. This index ranges from -1 to $+1$. Values near $+1$, 0, and -1 indicate a prey type taken above, at, and below its availability, respectively. This method has been applied to an experimental study of prey selection in a raptor (Marti and Hogue 1979). Ivlev's approach is useful however, only to compare prey species one at a time and does not allow simultaneous comparison of the entire spectrum of prey in a diet with its availability.

The chief drawback to the indexes described above and many other diversity indexes is that they assume that all resources are equally available. Measures that consider resource availability have been developed and should be considered for use if adequate data on prey availability can be obtained (Petraitis 1979, 1981; Feinsinger et al. 1981, Bechard 1982). One problem remains even with these measures: does the raptor perceive relative availability of prey in the same way the investigator does? This is similar to a problem in the measurement of dietary diversity by any method: do the prey categories (species or other) chosen by the investigator correspond to real differences among prey as perceived by raptors? Prey choice has been studied by identifying prey captured by raptors and comparing it with estimates of the availability of prey in the vicinity by live- or snap-trapping small mammals, censusing birds, or both (Kellomäki 1977, Koivunen et al. 1996a, 1996b).

Index of Relative Importance

The index of relative importance (IRI) is another composite measure combining three means of characterizing a diet sample: (1) the number of prey in a sample, (2) the volume or mass of each kind of prey in a sample, (3) and the frequency of occurrence for a kind of prey in a sample (i.e., the percentage of pellets in a sample of pellets that contain the prey in question). Introduced in the fishery literature (Pinkas 1971, Pinkas et al. 1971), it rarely has been used for terrestrial predators, but Hart et al. (2002) recently promoted its use for a wider taxonomic array including birds. IRI is calculated as:

$$IRI = (N + V)F,$$

where N = numerical percentage, V = volumetric percentage, and F = frequency of occurrence percentage. Martin et al. (1996) substituted mass for volume in their analysis of the diets of feral cats using this formula. Hart et al. (2002) applied the method to Barn Owls, the only application we know of for a raptor, but it may be a technique potentially valuable to raptor biologists.

Rarefaction

Rarefaction is a statistical method for estimating the number of species expected to be present in a random sample of individuals taken from any given collection and is a powerful standardization technique (Gotelli and Colwell 2001). Rarefaction is an appropriate tool for defining community structure and has been used in comparing species richness among communities in various ecosystems. Estimating community diversity by rarefaction provides an alternative that avoids some of the difficulties of calculating species richness by scaling down all collections to the same sample size (Hurlbert 1971, Heck et al. 1975).

Because a larger sample should contain more species, it may often be of interest to estimate how many species would be expected in smaller samples from the same population. From the number of individuals of each species in an original collection, a series can be calculated that reflects the numbers of species present in each smaller subset randomly drawn from the original collection. This method estimates not only species richness, but also the confidence limits for this parameter (Heck et al. 1975). Doing this allows you to compare statistically raptor diets with different species richness. The technique also allows for the generation of a rarefaction curve the shape of which is a graphic display of accumulation rates of relative abundance; therefore, the evenness of diets can be compared by examining the steepness of the curves and their intersection (James and Rathbun 1981). In general, the steeper the rarefaction curve is, the higher the evenness.

Studies of food-web structure, especially when attempting to determine the putative association between a factor such as productivity and a measure of food-web connectivity, depend heavily on using rarefaction procedures. For instance, Arim and Jaksic (2005) knew that the total number of prey identified affected the number of trophic links estimated per species and controlled for the effect of variation in sample size with a rarefaction procedure. Considering the types of prey present in raptor diets, several rarefaction procedures may be conducted (e.g., one for vertebrate and another for invertebrate prey), and the expected richness from both rarefactions can then be added. For more omnivorous raptors, even a third prey type might be used. A rarefaction calculator is available online: www2.biology.ualberta.ca/jbrzusto/rarefact.php (last accessed 11 January 2007).

Mean Prey Weight

Diets of predatory birds also can be quantified by estimating the mean mass for all prey in a diet sample. This grand mean is calculated by multiplying the total number of each kind of prey by the mean mass for that species, then summing these totals and dividing the sum by the total mass of prey individuals in the sample. Estimating the grand mean mass of prey is subject to several potential problems. Frequencies of prey masses in a sample of raptor food cannot be assumed to follow normal distributions because the masses of prey eaten often are skewed to one side of the mean. Also, mean mass of prey calculated in the manner described is sensitive to very large or very small prey, even if they occur in low frequencies. Problems caused by these conditions can by minimized by log-transformation of the mean masses of individual prey species prior to calculating the grand mean prey mass. The re-transformed mean (antilog) of the log-transformed masses is called the geometric mean (Sokal and Rohlf 1995).

Estimation of mean prey mass also is subject to the same problems and biases discussed in biomass quantification above. Despite this, this approach has been used successfully to characterize and compare the diets of many raptors (Storer 1966, Jaksic et al. 1981, Marks and Marti 1984, Steenhof and Kochert 1985).

Dietary Overlap

Another useful technique for making comparisons between two raptor diets is dietary overlap or similarity—the degree of joint use of prey species. Dietary overlap may be used in comparing diets of different species, comparing diets of the same species in different areas or times, and other similar comparisons. An objective measure of overlap is required to quantify such comparisons; many methods have been proposed (Levins 1968, Schoener 1968, Pianka 1973, Hurlbert 1978), but considerable disagreement still exists about which measure is superior (Ricklefs and Lau 1980, Slobodchikoff and Schulz 1980, Linton et al. 1981). The interpretation of overlap also lacks unanimity, especially in regard to its use as a measure of competition. Although niche overlap has been widely used as an indicator of competition (MacArthur and Levins 1967, Cody 1974, May 1975), such use has been criticized (Colwell and Futuyma 1971, Pianka 1974, Abrams 1980). High overlap in the diets of two or more raptors could be an indication of competition or the result of abundant food resources being exploited by both species without competition (Lack 1946, Pianka 1974). Low overlap, on the other hand, has been viewed as an indicator of divergence caused by prior competitive interactions (Lawlor 1980). Changes in dietary overlap may reveal more about competition than the degree of overlap (Schoener 1982, Steenhof and Kochert 1985). Korpimäki (1987) found that when diets of Long-eared Owls and Common Kestrels overlapped, it decreased the reproductive success of both when they were breeding close together. Schoener (1982) in his review of dietary overlap studies concluded that changes in overlap often occurred between seasons and from year to year; most cases showed less overlap in lean times. Pianka's (1973) index has been widely used in comparing raptor diets (Jaksic et al. 1981, Steenhof and Kochert 1985, Marti et al. 1993a,b) and is calculated as:

$$O = \Sigma\, p_{ij}\, p_{ik} \sqrt{\ } \sqrt{\ } (\, \Sigma\, p_{ij}^2,\ \Sigma\, p_{ik}^2),$$

where p_{ij} and p_{ik} are proportions of prey species (or other prey taxa) in the diets of raptors j and k, respectively. Values obtained range from zero (indicating no overlap) to one (indicating complete overlap). An illustration of the calculation of this overlap index is included in Appendix 1.

Several investigators have devised methods of weighting availability or abundance for more accurate calculation of the joint use of resources by two species (Colwell and Futuyma 1971, Hanski 1978, Hurlbert 1978). Although few raptor studies will have data adequate to make use of these methods, investigators exploring resource overlap should be aware that they exist.

Community Trophic Ecology

The techniques discussed above can be useful in understanding how trophic factors contribute to the structure of ecological communities (Jaksic et al. 1981, Jaksic and Delibes 1987, Jaksic 1988, Bosakowski and Smith 1992, Marti et al. 1993a,b; Korpimäki and Marti 1995, Aumann 2001). Similarly, they may be used to compare the ecological roles of two species (Marks and Marti 1984, Donazar et al. 1989, Marti and Kochert 1995,

Burton and Olsen 2000, Hamer et al. 2001). In addition, studies of food-web structure that attempt to disentangle the roles of predation and competition versus exogenous factors such as climate, still rely heavily on these apparently old-fashioned tools (Lima et al. 2002, Arim and Jaksic 2004).

Potential Use of Stable-Isotopes in Diet Analyses of Raptors

The analysis of trophic relationships in bird assemblages through conventional dietary assessments (e.g., stomach contents, prey remains, pellets, and feces) can be difficult, daunting, and biased because the determination of prey composition depends heavily on digestibility and on the nature of prey items (i.e., hard-versus soft-bodied). To resolve this bias, a complementary approach based on the use of stable isotopes has been gaining use. This approach relies on the ratios of stable isotopes of nitrogen (15N/14N, conventionally expressed as δ15N), and of carbon (13C/12C, or δ13C) in consumer proteins reflecting those of their prey in a predictable manner (DeNiro and Epstein 1978, 1981; Peterson and Fry 1987).

In the case of nitrogen, δ15N signature shows a stepwise enrichment at each successive level within a food chain (Hobson et al. 1994, Sydeman et al. 1997). As a result, predators occupying relatively high trophic positions have correspondingly elevated δ15N values. For carbon, δ13C values also may show a tendency to increase with trophic level, but to a lesser extent than that of δ15N (Hobson and Welch 1992). Nevertheless, the δ13C value can provide information about the source of carbon entering a food chain, for example, distinguishing between marine and freshwater systems (Mizutani et al. 1990) or discriminating between inshore versus benthic feeding and pelagic feeding in seabirds (Hobson et al. 1994).

In recent decades, the application of stable-isotopic analysis to studies of avian nutritional ecology and movement has increased tremendously. One of the important advances in this field has been the development of nondestructive sampling approaches that involve the isotopic analysis of bird feathers (Mizutani et al. 1990, Hobson and Clark 1992). Multiple stable-isotope analyses applied to investigations of entire seabird assemblages have yielded important insights into intra- and inter-specific trophic relationships, and have resolved trophic interactions on both spatial and temporal scales (Hobson et al. 1994). Dual-isotope multiple-source mixing models have been developed to quantify the proportions of various prey categories in the diet of carnivorous mammals (Ben-David et al. 1997), seabirds (Hobson 1995, Schmutz and Hobson 1998), and birds across a terrestrial-marine landscape (Harding and Stevens 2001), thus emphasizing the utility of stable isotopes in studies of diet and community trophic structure. To date, no such analyses have been attempted with raptors, but the information to be garnered could be important. For additional information on stable-isotope analyses, see Chapter 14, part C.

CONCLUSIONS

We cannot overemphasize that high-quality food-habits data are obtainable only with a correspondingly large investment of time, effort, and resources. Standardization (as much as is possible under field conditions) of data collection methods is highly desirable in order to make results comparable with other studies, and reporting of methods and results must include sufficient detail so that a study can be evaluated and compared with others. We emphasize that no matter how highly technical and sophisticated community analyses become, they will still depend on rather low-technology tools such as the ones discussed above. In other words, unless data are collected and analyzed in an unbiased manner, subsequent sophisticated analyses will not produce valid results.

ACKNOWLEDGMENTS

We thank J.A. Mosher and R.L. Glinski for their contributions on field observations of raptor feeding and B.A. Millsap for providing several references included in this chapter. K. Steenhof and R.T. Reynolds reviewed the first edition of this chapter, and an anonymous reviewer suggested valuable additions to this edition. Bret Harvey wrote the computer code used to generate the figures illustrating sample sizes needed to estimate prey richness and diet diversity in diet collections. We thank them for their contributions. FMJ acknowledges the support of grant FONDAP-FONDECYT 1501-0001 to the Center for Advanced Studies in Ecology and Biodiversity. CDM thanks the Raptor Research Center, Boise State University for providing logistical support during the writing of this chapter.

LITERATURE CITED

ABRAMS, P. 1980. Some comments on measuring niche overlap. *Ecology* 61:44–49.

ADORJAN, A.S. AND G.B. KOLENOSKY. 1969. A manual for the identification of hairs of selected Ontario mammals. *Ont. Dep. Lands For. Res. Rep. Wildl.* 90.

ALATALO, R.V. 1981. Problems in the measurement of evenness in ecology. *Oikos* 37:199–204.

———— AND R. ALATALO. 1977. Components of diversity: multivariate analysis with interaction. *Ecology* 58:900–906.

AKAKI, C. AND G.E. DUKE. 1999. Apparent chitin digestibilities in the Eastern Screech-Owl (*Otus asio*) and American Kestrels (*Falco sparverius*). *J. Exper. Zool.* 283:387–393.

ARIM, M. AND F.M. JAKSIC. 2005. Productivity and food web structure: association between productivity and link richness among top predators. *J. Anim. Ecol.* 74:31–40.

AUMANN, T. 2001. An intraspecific and interspecific comparison of raptor diets in the south-west of the Northern Territory, Australia. *Wildl. Res.* 28:379–393.

BECHARD, M.J. 1982. Effect of vegetative cover on foraging site selection by Swainson's Hawk. *Condor* 84:153–159.

BECK, T.W. AND R.A. SMITH. 1987. Nesting chronology of the Great Gray Owl at an artificial nest site in the Sierra Nevada. *J. Raptor Res.* 21:116–118.

BEISSINGER, S.R. 1983. Hunting behavior, prey selection, and energetics of Snail Kites in Guyana: consumer choice by a specialist. *Auk* 100:84–92.

BELLOCQ, M.I. 2000. A review of the trophic ecology of the Barn Owl in Argentina. *J. Raptor Res.* 34:108–119.

BEN-DAVID, M.R., W. FLYNN AND D.M. SCHELL. 1997. Annual and seasonal changes in diets of martens: evidence from stable isotope analysis. *Oecologia* 111:280–291.

BIELEFELDT. J., R.N. ROSENFIELD AND J.M. PAPP. 1992. Unfounded assumptions about diet of the Cooper's Hawk. *Condor* 94:427–436.

BOAL, C.W. AND R.W. MANNAN. 1994. Northern Goshawk diets in ponderosa pine forests on the Kaibab Plateau. *Stud. Avian Biol.* 16:97–102.

BOND, R.M. 1936. Eating habits of falcons with special reference to pellet analysis. *Condor* 38:72–76.

BOOMS, T.L. AND M.R. FULLER. 2003a. Time-lapse video system used to study nesting Gyrfalcons. *J. Field Ornithol.* 74:416–422.

———— AND M.R. FULLER. 2003b. Gyrfalcon diet in central west Greenland during the nesting period. *Condor* 105:528–537.

BOONSTRA, R. 1977. Predation on *Microtus townsendii* populations: impact and vulnerability. *Can. J. Zool.* 55:1631–1643.

BOSAKOWSKI, T. AND D.G. SMITH. 1992. Comparative diets of sympatric nesting raptors in the eastern deciduous forest biome. *Can. J. Zool.* 70:984–992.

BOWLES, J.H. 1916. Notes on the feeding habits of the Dusky Horned Owl. *Oologist* 33:151–152.

BROOKS, A. 1929. Pellets of hawks and owls are misleading. *Can. Field-Nat.* 43:160–161.

BROUGH, T. 1983. Average weights of birds. Minist. Agric., Fish. and Food, Surrey, United Kingdom.

BROWDER, R.G., R.C. BROWDER AND G.C. GARMAN. 1995. An inexpensive and automatic multiple-exposure photographic system.

J. Field Ornithol. 66:37–43.

BROWER, J.E. AND J.H. ZAR. 1984. Field and laboratory methods for general ecology. W.C. Brown, Dubuque, IA U.S.A.

BROWN, L.H. AND D. AMADON. 1968. Eagles, hawks and falcons of the world, Vols. I and II. Country Life Books, United Kingdom.

BUNN, D.S., A.B. WARBURTON AND R.D.S. WILSON. 1982. The Barn Owl. Buteo Books, Vermillion, SD U.S.A.

BURNHAM, K.P., D.R. ANDERSON AND J.L. LAAKE. 1980. Estimation of density from line transect sampling of biological populations. *Wildl. Monogr.* 72.

BURTON, A.M. AND P. OLSEN. 2000. Niche partitioning by two sympatric goshawks in the Australian wet tropics: ranging behaviour. *Emu* 100:216–226.

BURTON, J.A. [ED.]. 1984. Owls of the world. Tanager Books, Dover, NH U.S.A.

CADE, T.J. 1982. The falcons of the world. Cornell University Press, Ithaca, NY U.S.A.

CAIN, S.L. 1985. Nesting activity time budgets of Bald Eagles in southeast Alaska. M.S. thesis, University of Montana, Missoula, MT U.S.A.

CALL, M.W. 1981. Terrestrial wildlife inventories—some methods and concepts. USDI Bureau of Land Management Tech. Note 349. Denver, CO U.S.A.

CLARK, R.J., D.G. SMITH AND L.H. KELSO. 1978. Working bibliography of owls of the world. National Wildlife Federation Science Technical Series no. 1. National Wildlife Federation, Washington, DC U.S.A.

CODY, M.L. 1974. Competition and the structure of bird communities. Princeton University Press, Princeton, NJ U.S.A.

COLLOPY, M.W. 1977. Food caching by female American Kestrels in winter. *Condor* 79:63–68.

———. 1983a. A comparison of direct observations and collections of prey remains in determining the diet of Golden Eagles. *J. Wildl. Manage.* 47:360–368.

———. 1983b. Foraging behavior and success of Golden Eagles. *Auk* 100:747–749.

COLWELL, R.K. AND D.J. FUTUYMA. 1971. On the measurement of niche breadth and overlap. *Ecology* 52:567–576.

COWARDIN, L.M. AND J.E. ASHE. 1965. An automatic camera device for measuring waterfowl use. *J. Wildlife Manage.* 29:636–640.

CRAIGHEAD, J.J. AND F.C. CRAIGHEAD, JR. 1956. Hawks, owls and wildlife. Stackpole Co., Harrisburg, PA U.S.A.

CUMMINGS, J.H., G.E. DUKE AND A.A. JEGERS. 1976. Corrosion of bone by solutions simulating raptor gastric juice. *Raptor Res.* 10:55–57.

DANIELSON, W.R., R.M. DEGRAFF AND T.K. FULLER. 1996. An inexpensive compact automatic camera system for wildlife research. *J. Field Ornithol.* 67:414–421.

DEBLASE, A.F. AND R.E. MARTIN. 1974. A manual of mammalogy. W.C. Brown, Dubuque, IA U.S.A.

DEJONG, T.M. 1975. A comparison of three diversity indexes based on their components of richness and evenness. *Oikos* 26:222–227.

DEL HOYO, J., A. ELLIOTT AND J. SARGATAL [EDS.]. 1994. Handbook of the birds of the world, Vol. 2. New World vultures to guineafowl. Lynx Edicions, Barcelona, Spain.

———, A. ELLIOTT AND J. SARGATAL [EDS.]. 1992. Handbook of the bird of the world, Vol. 5. Barn-owls to hummingbirds. Lynx Edicions, Barcelona, Spain.

DELANEY, D.K., T.G. GRUBB AND D.K. GARCELON. 1998. An infrared video camera system for monitoring diurnal and nocturnal raptors. *J. Raptor Res.* 32:290–296.

DELONG, T.R. 1982. Effect of ambient conditions on nocturnal nest behavior in Long-eared Owls. M.S. thesis, Brigham Young University, Provo, UT U.S.A.

DENIRO, M.J. AND S. EPSTEIN. 1978. Influence of diet on the distribution of carbon isotopes in animals. *Geochim. Cosmochim. Acta* 42:495–506.

——— AND S. EPSTEIN. 1981. Influence of diet on the distribution of nitrogen in animals. *Geochim. Cosmochim. Acta* 45:341–351.

DILLER, L.V. AND D.R. JOHNSON. 1982. Ecology of reptiles in the Snake River Birds of Prey Area. Final Report submitted to USDI Bureau of Land Management, Boise, ID U.S.A.

DODGE, W.E. AND D.P. SNYDER. 1960. An automatic camera device for recording wildlife activity. *J. Wildl. Manage.* 24:340–342.

DONAZAR, J.A., F. HIRALDO, M. DELIBES AND R.R. ESTRELLA. 1989. Comparative food habits of the Eagle Owl *Bubo bubo* and the Great Horned Owl *Bubo virginianus* in six Palearctic and Nearctic biomes. *Ornis Scand.* 20:298–306.

DUKE, G.E., A.A. JEGERS, G. LOFF AND O.A. EVANSON. 1975. Gastric digestion in some raptors. *Comp. Biochem. Physiol.* 50A:649–656.

DUNNING, J.B. 1984. Body weights of 686 species of North American birds. *West. Bird-Banding Assoc. Monogr.* 1.

DYKSTRA, C.R., J.L. HAYS, M.M. SIMON AND F.B. DANIEL. 2003. Behavior and prey of nesting Red-shouldered Hawks in southwestern Ohio. *J. Raptor Res.* 37:177–187.

EARHART, C.M. AND N.K. JOHNSON. 1970. Size dimorphism and food habits of North American owls. *Condor* 72:251–264.

ECKBLAD, J.W. 1991. How many samples should be taken? *BioScience* 41:346–348.

ENDERSON, J.H., S.A. TEMPLE AND L.G. SWARTZ. 1972. Time-lapse photographic records of nesting Peregrine Falcons. *Living Bird* 11:113–128.

ERRINGTON, P.L. 1930. The pellet analysis method of raptor food habits study. *Condor* 32:292–296.

———. 1932. Technique of raptor food habits study. *Condor* 34:75–86.

FAIRLEY, J.S. 1967. Food of long-eared owls in north-east Ireland. *Br. Birds* 60:130–135.

———, C. M., AND C. M. SMAL. 1988. Correction factors in the analysis of the pellets of the Barn Owl *Tyto alba* in Ireland. *Proc. R. Ir. Acad. Sect. B Biol. Geol. Chem.* 88:119–133.

FEINSINGER, P., E.E. SPEARS AND R.W. POOLE. 1981. A simple measure of niche breadth. *Ecology* 62:27–32.

FISHER, A.K. 1893. The hawks and owls of the United States in their relation to agriculture. *U.S. Dep. Agric. Div. Ornithol. Mammal. Bull.* 3.

FITZNER, R.E., L.E. ROGERS AND D.W. URESK. 1977. Techniques useful for determining raptor prey-species abundance. *Raptor Res.* 11:67–71.

FRANKLIN, A.B. 1988. Breeding biology of the Great Grey Owl in southeastern Idaho and northwest Wyoming. *Condor* 90:689–696.

GEER, T.A. AND C.M. PERRINS. 1981. Notes on observing nesting accipiters. *Raptor Res.* 15:45–48.

GHENT, A.W. 1991. Insights into diversity and niche breadth analyses from exact small-sample tests of the equal abundance hypothesis. *Am. Midl. Nat.* 126:213–255.

GLADING, B., D.F. TILLOTSON AND D.M. SELLECK. 1943. Raptor pellets as indicators of food habits. *Calif. Fish Game* 29:92–121.

GLASS, B.P. 1973. A key to the skulls of North American mammals. Oklahoma State University, Stillwater, OK U.S.A.

GOSZCZYNSKI, J. 1977. Connections between predatory birds and mammals and their prey. *Acta Theriol.* 22, 30:399–430.

GOTELLI, N.J. AND R.K. COLWELL. 2001. Quantifying biodiversity: procedures and pitfalls in the measurement and comparison of species richness. *Ecol. Letters* 4:379–391.

GREEN, R.H. AND R.C. YOUNG. 1993. Sampling to detect rare species. *Ecol. Appl.* 3:351–356.

GREENE, H.W. AND F.M. JAKSIC. 1983. Food-niche relationships among sympatric predators: effects of level of prey identification. *Oikos* 40:151–154.

HAIR, J.D. 1980. Measurement of ecological diversity. Pages 265–275 *in* S.D. Schemnitz [ED.], Wildlife management techniques manual, 4th Ed. The Wildlife Society, Washington, DC U.S.A.

HAMER, T.E., D.L. HAYS, C.M. SENGER, M. CLYDE AND E.D. FORSMAN. 2001. Diets of Northern Barred Owls and Northern Spotted Owls in an area of sympatry. *J. Raptor Res.* 35:221–227.

HANSKI, I. 1978. Some comments on the measurement of niche metrics. *Ecology* 59:168–174.

HARDING, E.K. AND E. STEVENS. 2001. Using stable isotopes to assess seasonal patterns of avian predation across a terrestrial-marine landscape. *Oecologia* 129:436–444.

HART, R.K., M.C. CALVER AND C.R. DICKMAN. 2002. The index of relative importance: an alternative approach to reducing bias in descriptive studies of animal diets. *Wildl. Res.* 29:415–421.

HECK, K.L., JR., G. VAN BELLE AND D. SIMBERLOFF. 1975. Explicit calculation of the rarefaction diversity measurement and the determination of sufficient sample size. *Ecology* 56:1459–1461.

HILL, M.O. 1973. Diversity and evenness: a unifying notation and its consequences. *Ecology* 54:427–432.

HOBSON, K.A. 1995. Reconstructing avian diets using stable-carbon and nitrogen isotope analysis of egg components: patterns of isotopic fractionation and turnover. *Condor* 97:752–762.

——— AND R.W. CLARK. 1992. Assessing avian diets using stable isotopes II: factors influencing diet-tissue fractioning. *Condor* 94:189–197.

——— AND H. E. WELCH. 1992. Determination of trophic relationships within a high Arctic marine food web using $\partial 13C$ and $\partial 15N$ analysis. *Mar. Ecol. Prog. Ser.* 84:9–18.

———, J.F. PIATT, AND J. PITOCCHELLI. 1994. Using stable isotopes to determine seabird trophic relationships. *J. Anim. Ecol.* 63:786–798.

HUNT, W.G., J.M. JENKINS, R.E. JACKMAN, C.G. THELANDER AND A.T. CERSTELL. 1992. Foraging ecology of Bald Eagles on a regulated river. *J. Raptor Res.* 26:243–256.

HURLBERT, S.H. 1971. The nonconcept of species diversity: a critique and alternative parameters. *Ecology* 52:577–586.

———. 1978. The measurement of niche overlap and some relatives. *Ecology* 59:67–77.

HURTUBIA, J. 1973. Trophic diversity measurement in sympatric predatory species. *Ecology* 54:885–890.

HUTTO, R.L. 1990. Measuring the availability of food resources. *Stud. Avian Biol.* 13:20–28.

ITÄMIES, J. AND E. KORPIMÄKI. 1987. Insect food of the Kestrel, *Falco tinnunculus*, during breeding in western Finland. *Aquilo Ser. Zool.* 25:21–31.

IVLEV, V.S. 1961. Experimental ecology of the feeding of fishes. Yale University Press, New Haven, CT U.S.A.

JAKSIC, F.M. 1988. Trophic structure of some Nearctic, Neotropical and Palearctic owl assemblages: potential roles of diet opportunism, interspecific interference and resource depression. *J. Raptor Res.* 22:44–52.

——— AND H.E. BRAKER. 1983. Food-niche relationships and guild structure of diurnal birds of prey: competition versus opportunism. *Can. J. Zool.* 61:2230–2241.

——— AND M. DELIBES. 1987. A comparative analysis of food-niche relationships and trophic guild structure in two assemblages of vertebrate predators differing in species richness: causes, correlations, and consequences. *Oecologia* 71:461–472.

———, H.W. GREENE AND J.L. YANEZ. 1981. The guild structure of a community of predatory vertebrates in central Chile. *Oecologia* 49:21–28.

———, J.E. JIMÉNEZ AND P. FEINSINGER. 1990. Dynamics of guild structure among avian predators: competition or opportunism? *Acta XX Congressus Internationalis Ornithologici* 20:1480–1488.

———, R.L. SEIB AND C.M. HERRERA. 1982. Predation by the Barn Owl (*Tyto alba*) in mediterranean habitats of Chile, Spain and California: a comparative approach. *Am. Midl. Nat.* 107:151–162.

JAMES, F.C. AND S. RATHBUN. 1981. Rarefaction, relative abundance, and diversity of avian communities. *Auk* 98:785–800.

JOHNSGARD, P.A. 1990. Hawks, eagles, and falcons of North America. Smithsonian Institution Press, Washington, DC U.S.A.

———. 2002. North American owls. Smithsonian Institution Press, Washington, DC U.S.A.

JOHNSON, D.R. 1981. The study of raptor populations. University of Idaho Press, Moscow, ID U.S.A.

KELLOMAKI, E. 1977. Food of the Pygmy Owl *Glaucidium passerinum* in the breeding season. *Ornis Fenn.* 54:1–29.

KINAKO, P.D.S. 1983. Mathematical elegance and ecological naivety of diversity indexes. *Afr. J. Ecol.* 21:93–99.

KOIVUNEN, V., E. KORPIMÄKI AND H. HAKKARAINEN. 1996a. Differential avian predation on sex and size classes of small mammals: doomed surplus or dominant individuals? *Ann. Zool. Fennici* 33:293–301.

———, E. KORPIMÄKI, H. HAKKARAINEN AND K. NORRDAHL. 1996b. Prey choice of Tengmalm's Owls (*Aegolius funereus funereus*): preference for substandard individuals? *Can. J. Zool.* 74:816–823.

KORPIMÄKI, E. 1985. Diet of the Kestrel *Falco tinnunculus* in the breeding season. *Ornis Fenn.* 62:130–137.

———. 1987a. Prey caching of breeding Tengmalm's Owls *Aegolius funereus* as a buffer against temporary food shortage. *Ibis* 129:499–510.

———. 1987b. Dietary shifts, niche relationships and reproductive output of coexisting Kestrels and Long-eared Owls. *Oecologia* 74:277–285.

———. 1988. Diet of breeding Tengmalm's Owls, *Aegolius funereus*: long-term changes and year-to-year variation under cyclic food conditions. *Ornis Fenn.* 65:21–30.

———. 1992. Diet composition, prey choice, and breeding success of Long-eared Owls: effects of multiannual fluctuations in food abundance. *Can J. Zool.* 70:2372–2381.

——— AND C.D. MARTI. 1995. Geographical trends in trophic

characteristics of mammal-eating and bird-eating raptors in Europe and North America. *Auk* 112:1004–1023.

———, P. TOLONEN AND J. VALKAMA. 1994. Functional responses and load-size effect in central place forager: data from the kestrel and some general comments. *Oikos* 69:504–510.

KORSCHGEN, L.J. 1980. Procedures for food-habits analyses. Pages 13–127 *in* S.D. Schemnitz [ED.]. Wildlife management techniques manual, 4th Ed. The Wildlife Society, Washington, DC U.S.A.

KRISTAN, D.M., R.T. GOLIGHTLY AND S.M. TOMKIEWICZ. 1996. A solar-powered transmitting video camera for monitoring raptor nests. *Wildlife Soc. Bull.* 24:284–290.

LACK, D. 1946. Competition for food by birds of prey. *J. Anim. Ecol.* 15:123–129.

LAWLOR, L.R. 1980. Overlap, similarity, and competition coefficients. *Ecology* 61:245–251.

LEVINS, R. 1968. Evolution in changing environments. Princeton University Press, Princeton, NJ U.S.A.

LEWIS, S.B., P. DESIMONE, M.R. FULLER AND K. TITUS. 2004a. A video surveillance system for monitoring raptor nests in a temperate rainforest environment. *Northwest Sci.* 78:70–74.

———, M.R. FULLER AND K. TITUS. 2004b. A comparison of three methods for assessing raptor diet during the breeding season. *Wildlife Soc. Bull.* 32:373–385.

LIMA, M., N.C. STENSETH AND F.M. JAKSIC. 2002. Food web structure and climate effects on the dynamics of small mammals and owls in semiarid Chile. *Ecol. Let.* 5:273–284.

LINTON, L.R., R.W. DAVIES AND F.J. WRONA. 1981. Resource utilization indexes: an assessment. *J. Anim. Ecol.* 50:283–292.

LOWE, V.P.W. 1980. Variation in digestion of prey by the Tawny Owl (*Strix aluco*). *J. Zool., London.* 192:283–293.

MACARTHUR, R. AND R. LEVINS. 1967. The limiting similarity, convergence, and divergence of coexisting species. *Am. Nat.* 101:377–385.

MAGURRAN, A.E. 1988. Ecological diversity and its measurement. Princeton University Press, Princeton, NJ U.S.A.

MARCHESI, L., P. PEDRINI AND F. SERGIO. 2002. Biases associated with diet study methods in the Eurasian Eagle-Owl. *J. Raptor Res.* 36:11–16.

MARGALEF, D.R. 1958. Information theory in ecology. *Gen. Syst.* 3:36–71.

MARKS, J.S. AND C.D. MARTI. 1984. Feeding ecology of sympatric Barn Owls and Long-eared Owls in Idaho. *Ornis Scand.* 15:135–143.

MARTI, C.D. 1974. Feeding ecology of four sympatric owls. *Condor* 76:45–61.

———. 1988. A long-term study of food-niche dynamics in the Common Barn-Owl: comparisons within and between populations. *Can. J. Zool.* 66:1803–1812.

———. 1989. Food sharing by sibling Common Barn-Owls. *Wilson Bull.* 101:132–134.

——— AND J.G. HOGUE. 1979. Selection of prey by size in Screech Owls. *Auk* 96:319–327.

——— AND M.N. KOCHERT. 1995. Are Red-tailed Hawks and Great Horned Owls diurnal-nocturnal dietary counterparts? *Wilson Bull.* 107:615–628.

———, E. KORPIMÄKI AND F.M. JAKSIC. 1993a. Trophic structure of raptor communities: a three-continent comparison and synthesis. Pages 47–137 *in* D. M. Power [ED.], Current ornithology, Vol. 10. Plenum Press, NY U.S.A.

———, A.F. POOLE AND L.R. BEVIER. 2005. Barn Owl (*Tyto alba*). *In* The birds of North America Online (A. Poole, ed.). Ithaca: Cornell Laboratory of Ornithology; The Birds of North American Online database: <http://bna.birds.cornell.edu/BNA/account/Barn_Owl/> (1 June 2007).

———, K. STEENHOF AND M.N. KOCHERT. 1993b. Community trophic structure: the roles of diet, body size, and activity time in vertebrate predators. *Oikos* 67:6–18.

MARTIN, G.R., L.E. TWIGG AND D.J. ROBINSON. 1996. Comparison of the diet of feral cats from rural and pastoral Western Australia. *Wildl. Res.* 23:475–484.

MAY, R.M. 1975. Some notes on estimating the competition matrix, α. *Ecology* 56:737–741.

MCATEE, W.L. 1935. Food habits of common hawks. *U. S. Dep. Agric. Circ.* 370.

MERSMANN, T.J., D.A. BUEHLER, J.D. FRASER AND J.K.D. SEEGAR. 1992. Assessing bias in studies of Bald Eagle food habits. *J. Wildl. Manage.* 56:73–78.

MEYER, K.D., S.M. MCGEHEE AND M.W. COLLOPY. 2004. Food deliveries at Swallow-tailed Kite nests in southern Florida. *Condor* 106:171–176.

MIKKOLA, H. 1983. Owls of Europe. Buteo Books, Vermillion, SD U.S.A.

MIZUTANI, H., M. FUKUDA, Y. KANABYA AND E. WADA. 1990. Carbon isotope ratio reveals feeding behavior of cormorants. *Auk* 107:400–403.

MOLLHAGEN, T.R., R.W. WILEY AND R.L. PACKARD. 1972. Prey remains in Golden Eagle nests: Texas and New Mexico. *J. Wildl. Manage.* 36:784–792.

MOORE, T.D., L.E. SPENCE AND C.E. DUGNOLLE. 1974. Identification of the dorsal guard hairs of some mammals of Wyoming. Wyoming Game and Fish Department, Cheyenne, WY U.S.A.

MORRIS, P. 1979. Rats in the diet of the Barn Owl (*Tyto alba*). *J. Zool., Lond.* 189:540–545.

MORRISON, M.L. 1988. On sample sizes and reliable information. *Condor* 90:275–278.

NEWSOME, G.E. 1977. Use of opercular bones to identify and estimate lengths of prey consumed by piscivores. *Can. J. Zool.* 55:733–736.

NILSSON, I.N. 1984. Prey weight, food overlap, and reproductive output of potentially competing Long-eared and Tawny owls. *Ornis Scand.* 15:176–182.

NORRDAHL, K. AND E. KORPIMÄKI. 2002. Changes in individual quality during a 3-year population cycle of voles. *Oecologia* 130:239–249.

ODUM, E.P. 1983. Basic ecology. W.B. Saunders, Philadelphia, PA U.S.A.

OLENDORFF, R.R. AND S.E. OLENDORFF. 1968. An extensive bibliography of falconry, eagles, hawks, falcons, and other diurnal birds of prey. Published by the authors, Ft. Collins, CO U.S.A.

OLEYAR, M.D., C.D. MARTI AND M. MIKA. 2003. Vertebrate prey in the diet of Flammulated Owls in northern Utah. *J. Raptor Res.* 37:244–246.

OSTERBERG, D.M. 1962. Activity of small mammals as recorded by a photographic device. *J. Mammal.* 43:219–229.

OTIS, D.L., K.P. BURNHAM, G.C. WHITE AND D.R. ANDERSON. 1978. Statistical inference from capture data on closed animal populations. *Wildl. Monogr.* 62.

PARDINAS, U.F.J. AND S. CIRIGNOLI. 2002. Bibliografia comentada sobre los analisis de egagropilas de aves rapaces en Argentina.

Ornitol. Neotrop. 13:31–59.

PAVEZ, E.F., C.A. GONZÁLEZ AND J.E. JIMÉNEZ. 1992. Diet shifts of Black-chested Eagles (*Geranoaetus melanoleucus*) from native prey to European rabbits. *J. Raptor Res.* 26:27–32.

PETERSON, B.J. AND B. FRY. 1987. Stable isotopes in ecosystem studies. *Annu. Rev. Ecol. Syst.* 18:293–320.

PETERSEN, L.R. AND J.R. KEIR. 1976. Tether platforms-an improved technique for raptor food habits study. *Raptor Res.* 10:21–28.

PETRAITIS, P.S. 1979. Likelihood measures of niche breadth and overlap. *Ecology* 60:703–710.

———. 1981. Algebraic and graphical relationships among niche breadth measures. *Ecology* 62:545–548.

PHILIPS, J.R. AND D.L. DINDAL. 1979. Decomposition of raptor pellets. *Raptor Res.* 13:102–111.

PIANKA, E.R. 1973. The structure of lizard communities. *Annu. Rev. Ecol. Syst.* 4:53–74.

———. 1974. Niche overlap and diffuse competition. *Proc. Natl. Acad. Sci. U.S.A.* 71:2141–2145.

PIELOU, E.C. 1966. Species diversity and pattern diversity in the study of ecological succession. *J. Theor. Biol.* 10:370–383.

———. 1969. An introduction to mathematical ecology. Wiley-Interscience, New York, NY U.S.A.

———. 1972. Niche width and niche overlap: a method for measuring them. *Ecology* 53:687–692.

PINKAS, L. 1971. Food habits study. *Fish. Bull.* 152:5–10.

———, M.S. OLIPHANT, AND I.L.K. INVERSON. 1971. Food habits of albacore, bluefin tuna, and bonito in California waters. *Fish. Bull.* 152:11–105.

PITCHER, E., P. WIDENER AND S.J. MARTIN. 1979. Winter food caching in Richardson's Merlin *Falco columbarius*. *Raptor Res.* 13:39–40.

PREVOST, Y.A. 1977. Feeding ecology of Ospreys in Antigonish county, Nova Scotia. M.S. thesis, McGill University, Montreal, Quebec, Canada.

PROUDFOOT, G.A. AND S.L. BEASOM. 1997. Food habits of nesting Ferruginous Pygmy-Owls in southern Texas. *Wilson Bull.* 109:741–748.

PULIN, B. AND G. LEFEBVRE. 1995. Additional information on the use of tartar emetic in determining the diet of tropical birds. *Condor* 97:897–902.

RACZYNKI, J. AND A.L. RUPRECHT. 1974. The effect of digestion on the osteological composition of owl pellets. *Acta Ornithol.* 14:25–38.

REAL, J. 1996. Biases in diet study methods in the Bonelli's Eagle. *J. Wildl. Manage.* 60:632–638.

REYNOLDS, R.T. AND E.C. MESLOW. 1984. Partitioning of food and niche characteristics of coexisting *Accipiter* during breeding. *Auk* 101:761–779.

RICKELFS, R.E. AND M. LAU. 1980. Bias and dispersion of overlap indexes: results of some Monte Carlo simulations. *Ecology* 61:1019–1024.

RITCHIE, R.J. 1980. Food caching behavior of nesting wild Hawk Owls. *Raptor Res.* 14:59–60.

———. 1982. Porcupine quill and beetles in Peregrine castings, Yukon River, Alaska. *Raptor Res.* 16:59–60.

ROSENBERG, K.V. AND R.J. COOPER. 1990. Approaches to avian diet analysis. *Stud. Avian Biol.* 13:80–90.

ROSENFIELD, R.N., J.W. SCHNEIDER, J.M. PAPP AND W.S. SEEGAR. 1995. Prey of Peregrine Falcons breeding in West Greenland. *Condor* 97:763–770.

ROTH, T.C., II. AND S. L. LIMA. 2003. Hunting behavior and diet of Cooper's Hawks: an urban view of the small-bird-in-winter paradigm. *Condor* 105:474–483.

RUTZ, C. 2003. Assessing the breeding season diet of Goshawks *Accipiter gentilis*: biases of plucking analysis quantified by means of continuous radio-monitoring. *J. Zool. London.* 259:209–217.

SABO, B.A. AND R.C. LAYBOURNE. 1994. Preparation of avian material recovered from pellets and as prey remains. *J. Raptor Res.* 28:192–193.

SANTIBÁÑEZ, D. AND F.M. JAKSIC. 1999. Prey size matters at the upper tail of the distribution: a case study in northcentral Chile. *J. Raptor Res.* 33:170–172.

SCHEMNITZ, S. D. [ED]. 1980. Wildlife management techniques manual. The Wildlife Society, Washington, DC U.S.A.

SCHMUTZ, J.A. AND K.A. HOBSON. 1998. Geographic, temporal, and age-specific variation in diets of Glaucous Gulls of western Alaska. *Condor* 100:119–130.

SCHOENER, T.W. 1968. The *Anolis* lizards of Bimini: resource partitioning in a complex fauna. *Ecology* 49:704–726.

———. 1982. The controversy over interspecific competition. *Am. Sci.* 70:586–595.

SCHUELER, F. W. 1972. A new method of preparing owl pellets: boiling in NaOH. *Bird-Banding* 43:142.

SELLECK, D.M. AND B. GLADING. 1943. Food habits of nesting Barn Owls and Marsh Hawks at Dune Lakes, California; as determined by the "cage nest" method. *Calif. Fish Game* 29:122–131.

SEQUIN, J.F., P. BAYLE, J.C. THIBAULT, J. TORRE AND J.D. VIGNE. 1998. A comparison of methods to evaluate the diet of Golden Eagles in Corsica. *J. Raptor Res.* 32:314–318.

SHANNON, C.E. AND W. WEAVER. 1949. The mathematical theory of communication. University of Illinois Press, Urbana, IL U.S.A.

SHARP, A., L. GIBSON, M. NORTON, A. MARKS, B. RYAN AND L. SEMERARO. 2002. An evaluation of the use of regurgitated pellets and skeletal material to quantify the diet of Wedge-tailed Eagles, *Aquila audax*. *Emu* 102:181–185.

SHERROD, S.K. 1978. Diets of North American falconiformes. *Raptor Res.* 12:49–121.

SIMPSON, E.H. 1949. Measurement of diversity. *Nature* 163:688.

SITTER, G. 1983. Feeding activity and behavior of Prairie Falcons in the Snake River Birds of Prey Natural Area in southwestern Idaho. M.S. thesis, University of Idaho, Moscow, ID U.S.A.

SLOBODCHIKOFF, C.N. AND W.C. SCHULZ. 1980. Measures of niche overlap. *Ecology* 61:1051-1055.

SMITH, D.G. AND I.R. MURPHY. 1973. Breeding ecology of raptors in the eastern Great Basin of Utah. *Brigham Young Univ. Sci. Bull. Biol. Ser.* 18:1–76.

———, C.R. WILSON, AND H.H. FROST. 1972. The biology of the American Kestrel in central Utah. *Southwest. Nat.* 17:73–83.

SNYDER, N.F.R. AND I.W. WILEY. 1976. Sexual size dimorphism in hawks and owls of North America. *Ornithol. Monogr.* 20.

SOKAL, R.R. AND F.I. ROHLF. 1995. Biometry. W.H. Freeman, New York, NY U.S.A.

SOLHEIM, R. 1984. Caching behaviour, prey choice and surplus killing by Pygmy Owls *Glaucidium passerinum* during winter, a functional response of a generalist predator. *Ann. Zool. Fennici* 21:301–308.

SOUTHERN, H.N. 1969. Prey taken by Tawny Owls during the breeding season. *Ibis* 111:293–299.

STAINS, H.I. 1959. Use of the calcaneum in studies of taxonomy and food habits. *J. Mammal.* 40:392–401.

STEENHOF, K. 1983. Prey weights for computing percent biomass in raptor diets. *Raptor Res.* 17:15–27.

———. AND M.N. KOCHERT. 1985. Dietary shifts of sympatric buteos during a prey decline. *Oecologia* 66:6–16.

STORER, R.W. 1966. Sexual dimorphism and food habits in three North American accipiters. *Auk* 83:423–436.

SULKAVA, S. 1964. Zur Nahrungsbiologie des Habichts, *Accipiter gentilis* (L.). *Aquilo Ser. Zool.* 3:1–103.

SYDEMAN, W.J., K.A. HOBSON, P. PYLE AND E.B. MCLAREN. 1997. Trophic relationships among seabirds in central California: combined stable isotope and conventional dietary approach. *Condor* 99:327–336.

TAYLOR, I. 1994. Barn Owls. Cambridge University Press, Cambridge, United Kingdom.

TEMPLE, S.A. 1972. A portable time-lapse camera for recording wildlife activity. *J. Wildl. Manage.* 36:944–947.

TOMBACK, D.F. 1975. An emetic technique to investigate food preferences. *Auk* 92:581–583.

TØMMERAAS, P.J. 1989. A time-lapse nest study of a pair of Gyrfalcons *Falco rusticolus* from their arrival at the nesting ledge to the completion of egg laying. *Fauna Nor. Ser. C* 12:52–63.

UNITED STATES DEPARTMENT OF THE INTERIOR. 1979. Snake River birds of prey special research report to the Secretary of the Interior. USDI Bureau of Land Management, Boise, ID U.S.A.

VALKAMA, J., E. KORPIMÄKI, B. ARROYO, P. BEJA, V. BRETAGNOLLE, E. BRO, R. KENWARD, S. MAÑOSA, S.M. REDPATH, S. THIRGOOD AND J. VIÑUELA. 2005. Birds of prey as limiting factors of game-bird populations in Europe: a review. *Biol. Rev.* 80:171–203.

VAN DAELE, L.J. AND H.A. VAN DAELE. 1982. Factors affecting the productivity of Ospreys nesting in west-central Idaho. *Condor* 84:292–299.

VAUGHAN, R. 1961. *Falco eleonora. Ibis* 103a:114–128.

WAKELEY, I.S. 1978. Hunting methods and factors affecting their use by Ferruginous Hawks. *Condor* 80:327–333.

WASHINGTON, H.G. 1984. Diversity, biotic, and similarity indexes: a review with special relevance to aquatic ecosystems. *Water Res.* 18:653–694.

WHITTAKER, R.H. 1965. Dominance and diversity in land plant communities. *Science* 147:250–260.

WIJNANDTS, H. 1984. Ecological energetics of the Long-eared Owl (*Asio otus*). *Ardea* 72:1–92.

WILLE, K. AND K. KAM. 1983. Food of the White-tailed Eagle *Haliaeetus albicilla* in Greenland. *Holarct. Ecol.* 6:81–88.

WILSON, K.A. 1938. Owl studies at Ann Arbor, Michigan. *Auk* 55:187–197.

WOFFINDEN, N.D. AND I.R. MURPHY. 1982. Age and weight estimation of leporid prey remains from raptor nests. *Raptor Res.* 16:77–79.

YOUNK, J.V. AND M.J. BECHARD. 1994. Breeding ecology of the Northern Goshawk in high-elevation aspen forests of northern Nevada. *Stud. Avian Biol.* 16:119–121.

ZIESEMER, F. 1981. Methods of assessing Goshawk predation. Pages 44–151 *in* R.E. Kenward and I.M. Lindsay [EDS.], Understanding the Goshawk. Int. Assoc. Falconry Conserv. Birds Prey, Fleury en Biére, France.

Appendix 1. Sample calculations of diversity and evenness indexes for a hypothetical raptor diet and calculation of overlap between two hypothetical raptor diets.

Diet A (raptor j)

	Prey species	Prey abundance	Relative abundance
	(n_i)	(p_i)	$\log_e p_i$
A	105	0.40	-0.91
B	98	0.37	-0.99
C	32	0.12	-2.09
D	25	0.10	-2.34
E	1	0.004	-5.52
Totals	261	1.00	—

Diet B (raptor k)

	Prey species	Prey abundance	Relative abundance
	(n_i)	(p_i)	$\log_e p_i$
A	52	0.18	-1.73
B	40	0.14	-1.99
C	115	0.39	-0.94
D	87	0.30	-2.34
E	0	0.0	-1.22
Totals	294	1.00	—

Calculation of diet diversity

Diet diversity (food-niche breadth) for the data in diet A according to the reciprocal of Simpson's Index:

$$D = 1/\Sigma p_i^2$$
$$= 1/((0.402)^2 + (0.375)^2 + (0.123)^2 + (0.096)^2 + (0.004)^2)$$
$$= 1/(0.162 + 0.141 + 0.015 + 0.009 + 0.00002)$$
$$= 1/0.33$$
$$= 3.03$$

Diet diversity (food-niche breadth) for the data in diet A according to Shannon's Index:

$$H' = - \Sigma p_i \log p_i$$
$$= - [(0.402 \log 0.402) + (0.375 \log 0.375) + (0.123 \log 0.123) +$$
$$(0.096 \log 0.096) + (0.004 \log 0.004)]$$
$$= - [(0.402 (-0.911)) + (0.375 (- 0.994)) + (0.123 (-2.095)) +$$
$$(0.096 (-2.343)) + (0.004 -5.521))]$$
$$= - [-0.366 - .0373 - 0.258 - 0.225 - 0.022]$$
$$= 1.24$$

Calculation of diet evenness

Diet evenness for the data in diet A according to Pielou's Index:

$$J' H' / H max'$$
$$= 1.24 / H max'$$
$$= 1.24 / 1.609$$
$$= 0.77$$

Diet evenness for the data in diet A according to Alatalo's modification of Hill's Index:

$$F = (N_2 - 1) / (N_1 - 1)$$
$$= (1/D - 1) / (\text{antilog } H' - 1)$$
$$= (1/0.327 - 1) / (3.45 - 1)$$
$$= 2.06 / 2.45$$
$$= 0.84$$

Calculation of diet overlap

Diet overlap between diets A and B according to Piankas' Index:
$$O = \Sigma p_{ij} p_{ik} / \sqrt{(\Sigma pij^2, \Sigma pik^2)}$$
$$= ((0.40 \times 0.18) + (0.37 \times 0.14) + (0.12 \times 0.39) + (0.1 \times 0.3) + (0.004 \times 0)) /$$
$$\sqrt{((0.16 + 0.14 + 0.01 + 0.01 + 0.00002) \times (0.03 + 0.02 + 0.15 + 0.09))}$$
$$= (0.07 + 0.05 + 0.5 + 0.03) / \sqrt{(0.33 \times 0.29)}$$
$$= 0.2 / 0.09$$
$$= 0.2 / 0.31$$
$$= 0.64$$
$$0.64 \times 100 = 64\% \text{ overlap in diet}$$

Appendix 2. Diet samples to illustrate calculation of diversity, evenness, and diet overlap.

Population A (simulated)

Species	Number of individuals
1	200
2	200
3	200
4	200
5	200

Species richness = 5
Diet diversity = 5.0 (1/D)
Number of individuals in population = 1,000

Population B (simulated)

Species	Number of individuals
1	100
2	100
3	100
4	100
5	100
6	100
7	100
8	100
9	100
10	100

Species richness = 10
Diet diversity = 10.0 (1/D)
Number of individuals in population = 1,000

Population C (actual diet information from Utah Barn Owls [*Tyto alba*])

Species	Number of individuals
Sorex vagrans	4,223
Eptesicus fuscus	7
Myotis spp.	8
Sylvilagus nuttalli	3
Thomomys talpoides	649
Perognathus parvus	2
Reithrodontomys megalotis	6,517
Peromyscus maniculatus	6,853
Microtus montanus	41,527
Microtus pennsylvanicus	42,718
Ondatra zibethicus	40
Rattus norvegicus	308
Mus musculus	6,193
Mustela frenata	1
Rallus limicola	4
Porzana carolina	76
Charadrius vociferus	1
Recurvirostra americana	1
Gallinago gallinago	16
Columba livia	23
Tyto alba	1
Cistothorus palustris	36
Sturnus vulgaris	382
Sturnella neglecta	11
unidentified icterid	198
Passer domesticus	146
unidentified medium passerine	455
unidentified small passerine	603
unidentified insect	14

Species richness = 29
Diet diversity = 3.33 (1/D)
Number of individuals in population = 111,016

Habitat Sampling

9

Luis Tapia

Department of Animal Biology, Faculty of Biology,
University of Santiago de Compostela, Santiago de Compostela,
Campus sur, s/n 15782, Galicia, Spain

Patricia L. Kennedy

Eastern Oregon Agricultural Research Center and
Department of Fisheries & Wildlife,
Oregon State University, P. O. Box E, Union, OR 97883 U.S.A.

R. William Mannan

Biological Sciences East, School of Natural Resources,
University of Arizona, Tucson, AZ 85721 U.S.A.

INTRODUCTION

Understanding why animals are not distributed randomly across the landscape has been a main objective of ecology for some time (Cody 1985, Wiens 1989a). Students of the relationships between organisms and their habitats usually assume that individuals select where they choose to live (Cody 1985), and that it is possible to find correlations between the distribution, abundance, and demography of organisms and environmental variables (Buckland and Elston 1993, Morrison et al. 1998, Rushton et al. 2004, Guisan and Thuiller 2005). The search for correlations of this kind is common in studies of "habitat selection" (Anderson and Gutzwiller 1994, Litvaitis et al. 1994, Garshelis 2000, Jones 2001), although less attention has been given to the patterns of behavior that underlie choosing a place to live.

Because managing populations largely depends on managing or maintaining habitat (Anderson and Gutzwiller 1994), habitat use often is a basic element in conservation and management plans (Anderson et al. 1994, Edwards et al. 1996, Norris 2004). The assumption underlying these plans is that species will reproduce or survive better in habitats they prefer. Although it is an integral part of wildlife management, the process of evaluating what constitutes appropriate habitat for a given species or population can be difficult to achieve and often is beset with problems. Many of the problems involved have been recognized, and published discussions of them have prompted a host of evolving sampling designs and methods (Anderson and Gutzwiller 1994, Litvaitis et al. 1994, Garshelis 2000, Jones 2001, Hirzel et al. 2002, Guisan and Thuiller 2005, MacKenzie et al. 2006). In this chapter, we review the scope and objectives of habitat studies in raptors, and the methods for quantifying raptor habitats. We emphasize that studying the habitats of raptors essentially is no different than studying the habitats of any other group of organisms; thus, literature on habitat studies from almost any species is useful when designing a study on raptors.

TERMINOLOGY

Habitat terminology is not well defined. For example, the semantic and empirical distinctions between the terms "habitat use" (i.e., where individuals are) and "habitat selection" (i.e., where they choose to be) often are unclear (Garshelis 2000, Jones 2001). Any discussion of habitat sampling must be based on clearly defined terms. We recommend the following, based on Hall et al. (1997), Morrison et al. (1998) and Kennedy (2003) for studies of raptor habitat.

Habitat: the resources and conditions present in an area that produce occupancy by raptors. This is a synonym for the "niche" of the raptor according to the Grinnellian concept of the niche.

Habitat use: the way in which a raptor uses a collection of physical and biological components (i.e., resources) within a defined area and time.

Habitat abundance: the amount of habitat within a defined area and time.

Habitat availability: the amount of habitat that is exploitable by a raptor within a defined area and time.

Habitat selection: an hierarchical process involving a presumed series of innate or learned responses, or both, made by raptors regarding what habitat to use at different scales of the environment.

Habitat preference: the consequence of a raptor's habitat selection process, resulting in disproportionate use of some areas over others.

Habitat quality: the relative ability of habitats to provide conditions appropriate for raptor survival and reproduction.

Landscape: a mosaic of environmental patches across which raptors move, settle, reproduce and die. In principle, the landscape containing a raptor population can be mapped as a mosaic of suitable and unsuitable patches. Each map must be at a scale appropriate to the raptor under study.

OBJECTIVES

When designing an ecological study, the first step is to develop a clear list of objectives (Starfield 1997). Objectives should provide information about the intent of the study and the level of acceptable uncertainty. Moreover, appropriate objectives, combined with a good introduction, should describe clearly how the study would enhance understanding in ecology or

implementation of management actions. The following questions should be among those considered when developing objectives: What question is being asked and how does it advance understanding of ecological processes, or the requirements of the species under investigation? What is the focus of the study? Is it a population, a species (all populations) or a community that is being studied? What temporal and spatial scales are being considered?

Most habitat studies are searches for patterns, and not experimental tests about hypothesized underlying ecological processes. Because of this, objectives usually are expressed in the form of a question, or statistical hypothesis. If an explanation about a process involved in habitat selection is being tested via the hypothetical-deductive method (e.g., field experiments), then a statement indicating that the study involves the test of research hypotheses, as defined by Romesburg (1981), would be an appropriate objective.

CONSIDERATIONS FOR STUDY DESIGN

Excellent overviews of the basic principles of study design can be found in Ford (2000), Quinn and Keough (2002), and Williams et al. (2002). Important elements of design that should be considered at the outset of any habitat study involving a search for patterns include the proposed scope of inference of the study, and random and adequate sampling procedures. Below we describe several conceptual and practical elements that are central to studies of habitat and potentially influence study design.

Temporal and Spatial Scales

Factors that explain ecological processes usually are scale-dependent (Wiens et al. 1987, Mitchell et al. 2001, Sergio et al. 2003). Populations, for example, usually are influenced by how habitat is distributed across the landscape in both space and time (Wiens 1989b, Levin 1992, Corsi et al. 2000, Martínez et al. 2003). Study designs must be consistent with the abilities of the subject species to perceive and move among existing habitat patches, and investigators should consider the various scales at which habitat features may have influence (Litvaitis et al. 1994, Pribil and Picman 1997, Morrison et al. 1998, Rotenberry and Knick 1999, Sánchez-Zapata and Calvo 1999, Mitchell et al. 2001). There are at least three levels of spatial scale used by raptors during the breeding season: the nest area, the post-fledging family area (PFA), and the foraging area (Fig. 1). The nest area (or nest site), which typically is defined as the area immediately around the nest, often contains alternative nests and may be reused in consecutive years. The PFA surrounds the nest area and is defined as the area used by the family group from the time the young fledge until they no longer are dependent on the adults for food. The foraging area is the area used by the provisioning adults and typically encompasses the remainder of the home range during the breeding season. Below we use the Northern Goshawk (*Accipiter gentilis*) to illustrate the relative sizes of these areas and how interpretation of Northern Goshawk habitat can vary depending on the scale used to define nesting habitat.

In North America, nest areas of Northern Goshawks typically are less than 20 ha (DeStefano et al. 2006, Squires and Kennedy 2006,). Mean PFA size ranges from 60 to 170 ha depending on local environmental conditions (Kennedy et al. 1994, McClaren et al. 2005), and home ranges during the breeding season vary between 570 and 5,300 ha, depending on sex, habitat characteristics, and choice of home-range estimator (Squires and Kennedy 2006).

McGrath et al. (2003) evaluated goshawk nesting habitat empirically at various spatial scales to develop models that could be used to assess the effects of forest management on suitability of nesting habitat. Their work compared nesting habitat on four study areas in the inland Pacific Northwest during 1992-1995 and used four stand structures that represent different stages of stand development following disturbance. Eight habitat scales ranging from 1 to 170 ha (PFA scale; they did not analyze foraging habitat) surrounding 82 nests and 95 random sites were analyzed. A few key points are relevant to this chapter: (1) the ability to discriminate goshawk nest sites from available habitat decreased as landscape scale increased; (2) at the 1-ha scale, the stem exclusion stage of stand development (onset of self-thinning, no regeneration and the beginning of crown class differentiation into dominant and subordinate species) was preferred, whereas understory re-initiation (colonization of the forest floor by advanced regeneration and continued overstory competition) and old-growth (irregular senescence of overstory trees and recruitment of understory trees into the overstory) phases were used in proportion to their availability; (3) at larger scales, the middle stages of stand development consisting of stem exclusion and understo-

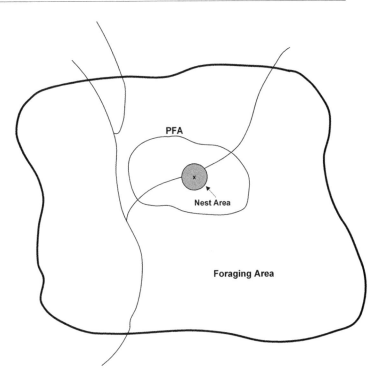

Figure 1. Conceptual diagram of three levels of spatial organization at a raptor nest in a drainage, including nest area, post-fledging area (PFA), and foraging area (from Squires and Kennedy 2006). (See text for definitions of area types.)

ry re-initiation (both with canopy closure > 50%) were preferred, suggesting that the types of habitats used increased as scale increased.

The influence of habitat features at different spatial scales is likely to be species-specific, and can change with body size, mobility and life history requirements. Thus, the commonly used terms "macrohabitat" and "microhabitat" are relative; a macrohabitat feature for a relatively wide-ranging, mobile species may be characterized on a much larger geographic scale than a macrohabitat feature for a less mobile species. However, even raptors of limited mobility can move rapidly over large areas. The accuracy with which a raptor can be placed at a particular point and time is an important consideration for habitat assessment at the microscale. For example, determining how a vegetation type is used depends on the accuracy of the bird's location (Withey et al. 2001) and how accurately sites are sampled relative to the size and distribution of patches of vegetation.

Raptors also exhibit temporal variation in habitat preferences. Studies of habitat and subsequent descriptions of a species habitat in management plans should account for these temporal changes. A relatively long-time scale would be an examination of the effects of plant succession and disturbances (measured in years) on the habitat of a raptor. In contrast, a short time-scale is exemplified by studies that measure vegetation in conjunction with momentary behavioral events such as an attempt to catch prey.

Some species use particular habitats during specific periods of the year, and only an assessment of habitat use during a complete annual cycle would describe the species habitat preferences. Northern Goshawks, for example, occupy a much broader range of habitats during the winter than during the nesting season. Because of this, breeding season habitat evaluations do not fully describe Northern Goshawk habitat patterns. A hotly debated management question regarding this species is whether or not it is a specialist that depends upon mature forest habitat. In their recent review of Northern Goshawk ecology, Squires and Kennedy (2006) addressed this question and concluded that the answer depends upon the season and residency patterns of the birds. This species is a partial migrant, meaning that some individuals occupy nest territories year-round whereas others undergo seasonal movements to wintering areas (Berthold 1993). The evidence suggests that Northern Goshawks prefer mature forests for nesting, and that some individuals have winter home ranges that include their nest areas (Boal et al. 2003, 2005) and, as such, may have year-round preferences for mature forest. That said, the limited data on patterns of winter habitat use by migratory birds suggest that Northern Goshawks use non-forested as well as forested habitat once they leave their nesting territory.

Selecting and Measuring Environmental Features in Habitat Studies

Habitat characteristics. Biologists often measure many characteristics of the environment that are associated with the presence or absence or the abundance of specific organisms, and infer that these characteristics, or features to which they are related, are "life requirements" and important elements of habitat for these organisms. Vegetation, for example, can provide shelter for small mammals, which, in turn, provide food for raptors (Preston 1990, Madders 2000, Ontiveros et al. 2005). Because of this, it is important to determine which environmental features to measure before beginning a habitat study (Anderson and Gutzwiller 1994). Because there are many potential features to measure, and because it takes time to collect and analyze data, the number of features to be sampled must be limited. Choice of features to measure should be based on a

thorough review of the literature on what is known about the species of interest (or close relatives), consultation with experts, and, in some instances, preliminary sampling (Anderson and Gutzwiller 1994). Features chosen should meet the objectives of the study, and be significant in terms of biological and conservation interest (Morrison et al. 1998, Morrison 2001, MacKenzie et al. 2006).

Habitat features that can be measured as part of a study of raptor habitat are numerous and vary with the kind of environment in which the study is conducted. Physiographic features such as slope, elevation, vegetation cover, distribution of water, human development, or soil type are relevant in many environments (Sutherland and Green 2004). In forest environments, species, size, density, and form of trees and vertical structure are common measures. Detailed measurements of raptor habitat in more open environments are less commonly described in the literature, but variables frequently include descriptions of ground vegetation, visibility, and the number of perches.

Techniques for measuring features of raptor habitats often are the same as those used by foresters, range managers, and other professional land managers. Employing widely used, standard techniques has two advantages: baseline information, collected with these techniques, already exists for many areas, and features of raptor habitat are expressed in terms familiar to land managers (Mosher et al. 1987). A disadvantage of land-management measurements is that many are arbitrary categorical variables used in one country or region (e.g., the U.S. Forest Service tree-density classes). As such, their use in raptor habitat studies limits comparisons across broad geographic areas and can reduce the likelihood of identifying habitat variables that are important range-wide (Penteriani 2002). In addition, management variables tend to be microscale variables, and additional methods generally are needed to obtain macroscale information (Oldemeyer and Regelin 1980, Bullock 1998, Morrison et al. 1998).

One important component of habitat structure is spatial heterogeneity or patchiness. This variable integrates not only absolute values of vegetation or physiography, but also their distribution in space. Habitat heterogeneity can be viewed at both coarse-grained (e.g. between cover types) or a fined-grained (within cover types) scales, and can be expressed in both vertical and horizontal dimensions. The choice of scale and the method for assessing heterogeneity or patchiness always should be organism-specific and should not be based on the perceptions of the investigator (Morrison et al. 1998). Many methods for measuring heterogeneity have been developed for use at a variety of scales (Anderson and Gutzwiller 1994).

Use versus availability. Studies of habitat preference and selection often necessitate designs and sampling schemes that assess available habitat, habitat not used, or the extent and manner of use by a species. For instance, biologists usually infer that certain features are available to a bird if the features in question occur within the individual's home range. In fact, investigators generally do not know if a bird is precluded from using certain features of the habitat due to phenomena not related to habitat (Cody 1981, 1985), including nest predation, interspecific competition, intra-specific attraction, and human disturbance (Newton 1998, Sutherland and Green 2004). The extent to which such phenomena affect both the habitat choices made by individuals and inferences drawn from correlative studies is unknown. Inferences about relationships among structural features of habitat (e.g., vegetation cover) and use of other features (e.g., prey) sometimes can be made by combining behavioral and habitat analyses (Bechard 1982, Bustamante et al. 1997, Selas 1997b, Thirgood et al. 2002, Amar and Redpath 2005). Inferences can be strengthened by experimentally manipulating features (e.g., prey, nest sites) hypothesized to be relevant to selection and monitoring the bird's response to this manipulation (Marcström et a!. 1988). Finally, both the spatial and temporal scales of the study can influence the perception of habitat availability (Orians and Wittenberger 1991, Levin 1992, Anderson and Gutzwiller 1994, Sutherland and Green 2004).

Raptor populations and distributions. To assess the habitat needs of a species, researchers commonly study habitat use and infer selection or preference from it. Presumably, species should reproduce or survive better in habitats they prefer. This approach assumes that such preferences relate to fitness and, hence, to population growth (Garshelis 2000). Traditionally, measures of presence-absence and abundance have been considered appropriate surrogate measures for fitness in the study of habitat requirements of terrestrial vertebrates, including raptors (Litvaitis et al. 1994, MacKenzie 2005, but see Van Horne 1983). Because of recent analytical developments in modeling (Guisan and Thuiller 2005, MacKenzie et al. 2006), presence-absence data can be used in a variety of contexts, including identifying habitats that are of value to species of conservation concern (MacKenzie 2005). Because presence-absence data are easier to collect, they often are preferred over abundance data (Pearce and Ferrier 2001, MacKenzie 2005). The use of abundance data in habitat studies has advantages, however (Gibbons et al. 1994). A positive relationship between distribution and abundance has been demonstrated for numerous taxa, and this relationship has been used to evaluate the status of species of conservation concern (Kennedy 1997, DeStefano 2005). In areas where habitats for breeding are relatively scarce, the relationship between abundance and distribution appear to be less well defined (Venier and Fahrig 1996).

Assessing habitat quality. High-quality habitats for a given species presumably have the resources required to sustain relatively high rates of survival and reproduction. Directly measuring the required resources present in an area (e.g., number of prey items or nest sites) is one way to assess habitat quality, but it requires that resources needed by the species in question be known, and that the resources measured are available for use (see above). Another approach for assessing habitat quality is based on indicators of population health. As noted in the previous section, information on presence-absence and abundance of raptors is common in investigations of their habitats. However, the presence of a raptor in an area, although potentially indicating that the area constitutes habitat for that species, indicates little about its quality. In contrast, measures of abundance of a given species in an area often are indicative of the relative quality of the area as habitat, but may be misleading in some situations (Van Horne 1983), as measures of abundance alone cannot distinguish between habitat sources and sinks (cf. Pulliam 1988).

Perhaps the best indicators for assessing habitat quality for a given species are estimates of productivity and survival, or combinations of both (e.g., rate of population change, λ). Unfortunately, these measures are difficult to obtain in short-term studies. Estimating survival is especially problematic (e.g., Diffendorfer 1998) and usually requires monitoring marked animals (e.g., banded or radio-tagged) over extended periods. Yet the value of long-term banding programs is high, especially in assessing habitat quality, and they can be done at both small- (less expensive) and large- (more expensive) spatial scales.

One example of a long-term study of marked animals over a broad spatial scale involves the massive research effort focused on the Northern Spotted Owl (*Strix occidentalis caurina*) in the northwestern United States. Anthony et al. (2006) analyzed demographic data collected from 1985 to 2003 on Northern Spotted

Owls from 14 study areas, covering about 12% of the entire range of the subspecies, in Washington, Oregon, and California. The meta-analysis presented was based on 32,054 captures and re-sightings of 11,342 banded individuals, and was designed to assess population trends, and related issues of habitat quality, throughout the subspecies range. Obviously, this kind of research effort is rare, in part because of the high financial costs associated with such work. Such studies usually focus on high-profile threatened or endangered species.

Long-term banding programs also have been successful on relatively small spatial scales. For example, marking and monitoring Merlins (*Falco columbarius*) for 10 years in a single city, Saskatoon, Saskatchewan, Canada, allowed researchers to assess population viability (James et al. 1994) and lifetime reproductive performance in relation to nest-site quality (Espie et al. 2004). Similarly, marking and re-sighting Cooper's Hawks (*A. cooperii*) at 40 to 80 breeding sites for up to 10 years in another city, Tucson, Arizona, U.S.A., permitted examinations of questions about ecological traps (Boal 1997), source-sink dynamics (R. W. Mannan et al., pers. comm.), and natal-habitat imprinting (Mannan et al. 2007, unpubl. data).

APPLICATION

Raptor populations sometimes are limited at the microscale level (Bevers and Flater 1999) by the availability of breeding or roosting habitat (Newton 1979), and much research has been conducted on habitat use and preference of nest and roost sites (Thompson et al. 1990, Reynolds et al. 1992, Mañosa 1993, Cerasoli and Penteriani 1996, Gil-Sánchez et al. 1996, Iverson et al. 1996, Selas 1997a, Mariné and Dalmau 2000, Martínez and Calvo 2000, Finn et al. 2002, Penteriani 2002, Poirazidis et al. 2004, Squires and Kennedy 2006). Investigations of habitat associated with nesting activities that occur at larger spatial scales such as the PFAs (Daw and DeStefano 2001, McGrath et al. 2003), foraging areas (Bosakowski and Speiser 1994, Sergio et al. 2003, Boal et al. 2005, Tapia et al. 2007), and areas used during natal dispersal (Ferrer and Harte 1997, Mañosa et al. 1998, Balbontín 2005) are less common.

Below we illustrate how habitat has been measured in studies focusing on raptors and use examples from the finest scale, called activity points, working progressively through larger spatial scales, called activity sites, and activity areas.

Activity points. Nest structure and substrate during the breeding period is an example of an activity point that often is measured in studies of woodland raptors (Cerasoli and Penteriani 1996, Siders and Kennedy 1996, Selas 1997a, Reich et al. 2004). There are many measurements possible at nest trees, perches, roost sites, and foraging sites (Table 1). Locating these sites (i.e., activity points) in forested environments can be difficult and often requires intensive field observations (Rutz 2003, Leyhe and Ritchison 2004) or radiotelemetry that results in visual observations (e.g., Mannan et al. 2004). In non-forested habitats, finding sites where raptors hunt and perch is easier (Leyhe and Ritchison 2004), but researchers should not assume 100% detection, even in these open habitats (MacKenzie et al. 2006, P. Kennedy et al., unpubl. data).

Most, but not all, raptor studies in open habitats have focused on macroscale habitat measurements. A few have provided detailed microhabitat measurements surrounding activity points (Salamolard 1997, Martínez et al. 1999, Arroyo et al. 2002). For example, ledges or small caves on cliffs are important to many raptors during the breeding season (Cade et al. 1988, Donázar et al. 1989, Donázar et al. 1993, Ratcliffe 1993, Thiollay 1994, Carrete et al. 2000, Rico-Alcázar 2001, McIntyre 2002), and many characteristics have been measured and described at cliff sites (Table 1).

Habitat measurements also can be made at activity points outside of the breeding season. For example, measurements related to land-cover types (permanent pasture, crops, plowed, woodland) or different kind of perches (trees, poles, on the ground) (Plumpton and Andersen 1997, Canavelli et al. 2003) can be useful in explaining patterns of non-breeding habitat use.

Activity sites. Plots of various sizes surrounding activity points are examples of activity sites (Hubert 1993, McLeod et al. 2000). Physiographic features often measured in activity sites include forest structure and composition, elevation, slope, aspect, soil type, and distance to water and forest openings (Table 2). Some raptors perch on fence posts and roost and nest in buildings (Bird et al. 1996, Leyhe and Ritchison 2004), and distance to human dwellings also can be an important measure. Distance measures often extend beyond the bounds of measured plots, but are collected along with data at this scale.

Uncertainty about the relative importance of various habitat features can lead biologists to take many measurements at activity sites (Mosher et al. 1987). However, as mentioned above, a broad, "shotgun" approach to

Table 1. Variables that are regularly measured at raptor activity points in forests, open country, and at cliffs.

Variable	Comments and References
NESTS IN FORESTED HABITATS	
Stick nest or cavity	
Nest dimensions	Length, width, and depth (m) of the body of the nest and the nest cup (Lokemoen and Duebbert 1976, Schmutz et al. 1980).
Nest access distance	Measured as nest circumference minus sum of the diameters of support branches divided by the number of support branches (Bednarz and Dinsmore 1982, Morris et al. 1982).
Surface area of nest	Measured on the top of nest in cm^2 (Morris et al. 1982).
Nest volume	Measured both for the nest and the nest bowl in cm^3 (Morris et al. 1982).
Nest minus trunk difference	Distance between the nest and the main trunk estimated in m (Bednarz and Dinsmore 1982).
Number of supporting branches	(Bednarz and Dinsmore 1982).
Size of support branches	By size categories (Bednarz and Dinsmore 1982).
Cavity measurements	For variables related to cavity measures (cavity diameter, cavity depth, opening dimensions, opening exposure, number of tree cavities, etc.) see Korpimaki (1984) Mariné and Dalmau (2000), and Rolstad et al. (2000).
Visibility about point	
Nest concealment	Historically measured at the nest with a spherical densiometer, standard photograph, or categorical estimates (Moore and Henny 1983). Recent developments in image analysis may be useful (Ortega et al. 2002, Luscier et al. 2006).
Nest canopy coverage	Measure of the canopy coverage above the nest with a spherical densiometer (Moore and Henny 1983, Siders and Kennedy 1996). Recent developments in image analysis may be useful (Ortega et al. 2002, Luscier et al. 2006).
Vegetation openness above and around nest	Green and Morrison (1983).
Nest tree	
Point dbh	Diameter at breast height (dbh), measured in cm using a dbh tape or Biltmore (Morris et al. 1982, Hubert 1993).
Height of tree	Usually measured in m using a clinometer (Haga type altimeter) (Reynolds et al. 1982, Rosenfield et al. 1998).
Tree species	To describe usage or to determine preference (Rottemborn 2000).
Age of tree	Estimated using a site index table or increment borer (Tjernberg 1983, Selas 1996, Siders and Kennedy 1996, Selas 1997a).
Height of nest, perch, roost	Measured directly with a meter tape when in the tree banding young, or with a clinometer to the nearest tenth meter (Titus and Mosher 1981, Cerasoli and Penteriani 1996).
Percent nest height or relative nest height.	Calculated as nest height/nest tree height x 100 (Titus and Mosher 1981, Morris et al. 1982, Cerasoli and Penteriani 1996, Rosenfield et al. 1998).
Percent canopy height	Calculated as nest height/mean canopy height x 100 (Devereux and Mosher 1984).
Slope	Measured as a percent (Selas 1996, Rosenfield et al. 1998).
Elevation	Elevation of nest site (m) taken from altimeter, topographic maps or GIS database (Garner 1999).
Altitude category	Nest stand plots and control plots: assigned to the lower, middle or upper altitude zone (Selas 1996).
Nest-tree health	Estimated percent dead or diseased, or alive or dead (Moore and Henny 1983, Devereux and Mosher 1984).
Nest distance to	Landscape features that might influence nest preferences (Speiser and Bosakowski 1987, Iverson et al. 1996, Penteriani 2002). Typically measured in m or km.
Perch or roost point	
Perch type	Pole, tree, fencepost, windmill, etc (Preston 1980, Holmes et al. 1993).
Number of perches	Count of the number and types of perches in a given area (Janes 1985, Holmes et al. 1993).
Microclimate	Temperature, light, wind speed, etc. (Barrows 1981, Keister et al. 1985).

Table 1. *(continued)*

Variable	Comments and References
Perch or roost protection	Ranked variable for protection from the weather (Hayward and Garton 1984).
Distance to trunk	Distance from perch or roost point to trunk along limb (Hayward and Garton 1984).
Tree species, perch type	Description of substrate (Marion and Ryder 1975, Steenhof et al. 1980, Hayward and Garton 1984, Leyhe and Ritchison 2004).
Point dimensions	Similar to variables measured for nest trees (Steenhof et al. 1980).
Point distance to	Landscape features that might influence perch or roost preferences (Thompson et al. 1990, Rottenborn 2000). Typically measured in m or km.

NEST IN OPEN HABITATS

Variable	Comments and References
Plant species at site	(Bullock 1998, Sutherland 2000).
Nest site visibility	Measuring the distance from which the nest contents are no longer visible along equally spaced transects from the nest (Simmons and Smith 1985, Amat and Masero 2004).

NEST ON CLIFFS

Variable	Comments and References
Nest (or scrape) location	Describe as: on ledge, crevice, stick nests, in pothole or cave (Cade 1960, Ratcliffe 1993).
Nest location measurements	Length, width, height, depth of ledge or cavity (Squibb and Hunt 1983, Ratcliffe 1993).
Nest materials	Describe substrate: e.g., sand, gravel, dirt, vegetation, etc. (Cade 1960, Ratcliffe 1993).
Rock type	Describe (e.g., granite, shale, soil, etc.) (Cade 1960, Ratcliffe 1993, Gainzarain et al. 2002, Hirzel et al. 2004).
Overhang	Categorize and describe (e.g., overhang > 90° vertical, open <90°) (Squibb and Hunt, 1983), or use a tape measure and clinometer to measure size and angles.
Vegetation near nest	Describe type and proximity of plants to nest (scrape) (Ratcliffe 1993).
Vegetation, plant community at base of and at top of clif	List species or describe community (Cade 1960, Ratcliffe 1993, Martínez and Calvo 2000, Martínez et al. 2003).
Nest height on cliff (or percent nest height) and above water	Measure (meter tape, rope length, transit, photographic comparison with topographic maps) or estimate (Cade 1960, Burnham and Mattox 1984, Donázar et al. 1993).
Distance from top (brink) and base of cliff to nest	Measure or estimate (Cade 1960, Ratcliffe 1962, Ratcliffe 1993).
Exposure of nest	Direction that nest (scrape, opening) faces (Ratcliffe 1962, Ratcliffe 1993).
Altitude of cliff	Height of site above sea level, often taken from topographic maps (Ratcliffe 1962, Burnham and Mattox, 1984, Gainzarain et al. 2002).
Orientation of cliff	Direction (aspect) the cliff faces, measure with compass (or estimated from topographic map) as angle perpendicular to main cliff face (Ratcliffe 1962, Donázar et al. 1993).
Height and length of cliff	Can be measured using meter tape, rope, distance and angle height with clinometer, range finder, or from topo graphic maps or air photo) or estimated and placed in categories (Ratcliffe 1993, Ontiveros 1999, Martínez and Calvo 2000).
Cliff relief	Highest point on cliff minus lowest point (Donázar et al. 1993).
Slope of cliff	Measure (clinometer) or place in categories (e.g., >90°, 80–90°, etc.) (Ratcliffe 1962, Ratcliffe 1993).
Relation of cliff to surrounding topography	General description (Ratcliffe 1993, Martínez and Calvo, 2000).
Distance and direction across valley, height (slope) of opposite valley	Estimate in field or use topographic maps (Donázar et al. 1993) or GIS.
Distance to human activity	Estimate in field or use topographic maps (Donázar et al. 1993, Ratcliffe 1993, Ontiveros 1999, Martínez and Calvo 2000) or GIS.
Distance to the nearest-neighbor nest	Estimate, or use topographic maps (Gil-Sánchez et al. 1996, Martínez and Calvo 2000) or GIS.

Table 2. Vegetation structure and floristic variables measured at raptor activity sites in wooded habitat.

Variable	Comments and References
Plant species richness, diversity index	Record the plant species within a plot (Titus and Mosher 1981) accounting for detectability (MacKenzie et al. 2006).
Tree species importance values	Record tree species relative density and frequency to compute importance values (Morris and Lemon 1983).
Tree-stem density by size, class, and species	Measured directly by recording the dbh of all trees in the plot by species. Provides data from which different size classes may be constructed or importance values calculated (Titus and Mosher 1981, Morris and Lemon 1983, Selas 1996, Selas 1997a). Plotless sampling techniques will provide estimates of some of the same information (Reynolds et al. 1992, Siders and Kennedy 1996).
Shrub and understory density	Either estimated by an index or census the plot according to dbh or height criteria (Titus and Mosher 1981, Morris and Lemon 1983, Rosenfield et al. 1998). Numerous techniques are available (Oldemeyer and Regelin, 1980, Bullock 1998).
Distance between trees (m)	(Siders and Kennedy 1996, Penteriani and Faivre 1997).
Tree density	Number of trees per hectare (Rosenfield et al. 1998, Garner 1999), or by size class (Siders and Kennedy 1996).
Mean dbh	Mean diameter (cm) at breast height of trees in study plot (Mañosa 1993, Rosenfield et al. 1998).
Basal area (m^2/ha)	May be calculated from tree dbh per unit area (Morris and Lemon 1983, Mañosa 1993, Cade 1997, Rosenfield et al. 1998) or estimated using an angle gauge.
Tree height class	Tally of trees by height class (see revision of Penteriani 2002).
Tree structure class	Used to classify dead or dying trees (Devereux and Mosher 1984, Selas 1996).
Crown volume	Determines volume by height and shapes categories (Moore and Henny 1983).
Crown depth	Expressed as a percent of tree height (Reynolds et al. 1992).
Tree strata	Discrete number of layers of canopy and understory (Reynolds et al. 1992).
Canopy volume (m^3)	(Penteriani and Faivre 1997).
Canopy cover	Measure of area potentially covered by multiple trees due to crown overlap. Typically expressed as percent cover (Reynolds et al. 1982, 1992, Penteriani and Faivre 1997).
Closure of canopy, understory, ground cover	Estimated using a GRS densitometer (K. A. Stumpf, unpubl. data). Recent developments in image analysis may be useful (Ortega et al. 2002, Luscier et al. 2006).
Percent cover in perch stand	Similar to variable measured for nest stand (Leyhe and Ritchison 2004).
Tree density in perch stand	Similar to variable measured for nest stand (Leyhe and Ritchison 2004).
Vegetation height in perch stand	Height of total ground and shrub vegetation surrounding point in m (Leyhe and Ritchison, 2004).
Vegetation profile	A density board may be used to estimate the amount of vegetation at height intervals (Nudds 1977, Bullock 1998). Numerous variables and categories can be created or more quantitative approaches can be used (Blondel and Cuvillier 1977).
Inter-tree heterogeneity	Index of mean inter-tree distance and variability (Roth 1976).
Horizontal diversity and habitat heterogeneity	For examples and methods see Litvaitis et al. 1994.

data collection may not be the best design strategy. Variable selection should be based on the study objectives, and should be significant in terms of biological and conservation interest. Variables that often describe significant features of nest sites of woodland raptors include tree-stem density by size class, canopy closure and basal area (Selas 1996, Siders and Kennedy 1996, Daw and DeStefano 2001). These measures usually relate to stand age. Shrub and ground cover variables, which are considered less important (but see Boal et al. 2005), may characterize significant features around hunting perches (Farrel 1981, Leyhe and Ritchison 2004).

The choice of sample plot size, shape, and distribution is fundamental to field studies and the raptor-habitat literature illustrates many choices. Size of plots can vary from 0.04 ha (Armstrong and Euler 1982, Siders and Kennedy 1996, Rosenfield et al. 1998) to 0.75 ha (Tjernberg 1983, Poirazidis et al. 2004) to 64 ha (P. Kennedy et al., unpubl. data).

Activity areas. Activity areas are similar to activity sites, but encompass larger areas. For example, an activity area might be identified as a plot large enough to include a substantial portion of a home range (e.g., a 1-km radius). Habitat features measured in activity sites also can be measured in activity areas (McGrady et al. 2002, Bosch et al. 2005, Tapia et al. in press), although measures taken in activity areas tend to be coarser. For example, vegetation in activity areas might be described by listing dominant plants or proportional coverage of vegetative communities. Areas also could be described by proportional coverage of land-use or land-cover types (Mosher et al. 1987, Table 3).

Table 3. Physiographic, land-cover, and land-use type variables measured at raptor activity sites and areas.

Variable	Comments and References
Altitude	Measured using a surveying altimeter or topographic map (Donázar et al. 1993, Penteriani and Faivre 1997, Martínez et al. 2003), or obtained from analysis of the variable using a digital elevation model with digital cartography and GIS (Tapia el al. 2004, López-López et al. 2006).
Slope gradient in degrees, and slope exposure (%)	Measured with clinometer, abney rule, or level; (Titus and Mosher 1981, Reynolds et al. 1982, Penteriani 2002); or obtained from analysis of the variable using a digital-elevation model with digital cartography and GIS (Tapia et al. 2004, López-López et al. 2006).
Aspect	The direction toward which a point or site faces; the direction away from the slope; the direction of most open vegetation (Titus and Mosher 1981, Reynolds et al. 1982, Selas 1996).
Type of water	Categories (temporary versus permanent, stream, river, pond, lake, size categories, 1 ha, 1.1-5 ha, etc.) (Reynolds et al. 1982).
Distance to water or other landscape feature	Measurement with a tape or paced; can record as seasonal water or permanent (Morris and Lemon 1983); or obtained from analysis of the variable using a digital elevation model with digital cartography and GIS.
Soil-woods or land productivity index	See Newton et al. (1981) for examples relating raptor use to land cover and land productivity indexes.
Land cover or land use	Probably the most commonly obtained set of macro-variables in raptor habitat studies. Usually categorized by general habitat type at the activity site (e.g., pasture, cropland, woodlot, water, field/forest edge). Used in many studies (see Bullock 1998, Sutherland and Green 2004).
Amount of land cover	Measured in ha, km^2 or categories. May be delineated based on habitat use as determined by radio-telemetry (Selas and Rafoss 1999, Newton 1986), direct observation (Tapia et al. 2004), or indirectly by measuring plots delineated by home range boundaries or circles centered on a point (Moorman and Chapman 1996).
Relief index	An index of topographic variation based on the number of contour lines crossed by transects radiating from activity point (González et al. 1990, Donázar et al. 1993).
Baxter-Wolfe interspersion index	Determines the number of changes in habitat type occurring along transects (Litvaitis et al. 1994).
Area of cover type	Satellite imaging involves computer-driven interpretation of available satellite images. The resolution of these images is determined by pixel size (Andries et al. 1994).
Human disturbance	Number of and distance to human settlements, buildings, roads, etc. (Tapia et al. 2004, 2007, Balbontín 2005).
Cover-type and prey-base associations	Index of abundance of prey associated with cover types and raptor use (Bechard 1982, Thirgood et al. 2003, Ontiveros et al. 2005).

The rapid development of remote-sensing techniques and Geographic Information Systems (GIS) has facilitated handling and management of environmental data at increasingly larger spatial scales (Koeln et al. 1994, Bullock 1998, Corsi et al. 2000). Remote sensing is useful for collecting macrohabitat features of activity areas, such as slope, elevation and other physiographic features. However, it does not replace field observations because many vegetation variables (e.g., stand structure, range condition) cannot be obtained accurately from remotely sensed data. Even for measurements that can be measured accurately with remotely sensed data, ground truthing is required to quantify the level of accuracy for a particular landscape.

Prey abundance. Prey abundance and availability are known to limit raptor populations (Newton 1979, Newton 1998, Dewey and Kennedy 2001). As a result, raptor habitat requirements often are linked to the distribution of their prey. Because it is difficult to observe predatory behavior in most raptors, the influence of prey on habitat use by raptors often is inferred by comparing, typically at the scale of activity areas, measures of prey abundance and raptor use among categories of vegetation or land use (Graham and Redpath 1995, Marzluff et al. 1997, Selas 1997b, Bakaloudis et al. 1998, Ontiveros et al. 2005). Use of land or vegetation types by raptors often is positively associated with prey abundance (Selas and Steel 1998, Ontiveros et al. 2005), but such relationships can be confounded by density of vegetation. That predation is sometimes more intense in areas where vegetation is less dense, regardless of prey abundance (Bechard 1982, Thirgood et al. 2003, Ontiveros et al. 2005) illustrates the need to distinguish, whenever possible, prey abundance from prey availability (Mosher et al. 1987).

DATA ANALYSIS

Statistical methods by which habitat preference is inferred are highly variable and differ in their precision and applicability (Alldredge and Ratti 1986, 1992, Titus 1990, Manly et al. 1993, MacKenzie et al. 2006). Analytical techniques for examining the multivariate nature of wildlife-habitat relationships (Corsi et al. 2000) include Generalized Linear Models, Bayesian approaches, classification trees and multivariate statistical methods such as Multiple Regression, Canonical Correlation Analysis, Principal Component Analysis, and Discriminant Function Analysis (Donázar et al.

1989, Kostrezewa 1996, Morrison et al. 1998, González-Oreja 2003).

Habitat features may have linear or nonlinear effects and these effects can be additive or multiplicative on the abundance of a species. Analytical techniques that enable examination of complex associations may be desirable over methods that assume simple linear relationships (e.g. simple correlation). On the other hand, inadequate sample size and many predictor variables often are problems when multivariate methods are used (over-parameterized model) (Morrison et al. 1998). Interpretations based on complex models and inadequate samples can be misleading. Required sample sizes largely are related to existing variation in the system being studied and effect size, but a crude estimate of the minimum sample size needed for multivariate analyses is 20 observations, plus 3 to 5 additional observations for each variable in the analysis. Morrison et al. (1998) suggested that an additional 5 to 10 observations for each variable provide a more conservative target for an adequate sample size. Even so, large sample sizes do not compensate for poorly designed studies or biased sampling. Recently, many biologists have shifted away from using statistical significance (and arbitrary or *a priori* p-values) as the defining point for biological significance and instead develop multiple, competing *a priori* models which are then evaluated by model selection techniques such as Akaike Information Criterion (AIC) (Anderson et al. 2000, Jongman et al. 2001). Whatever approach is adopted, the requirements and limitations of the statistical techniques employed should be understood before embarking on such a study (Manly 1993, Morrison et al. 1998).

Modeling the distribution of raptors and other vertebrate species (i.e., generating atlases) has become more common in recent years (Bustamante 1997, Sánchez-Zapata and Calvo 1999, Sergio et al. 2003, Rushton et al. 2004). Although the value of such atlases is somewhat limited (Donald and Fuller 1998, Sutherland 2000), they can be useful in predicting the presence of a species, and often play a role in assessing habitat suitability (Osborne and Tigar 1992, Tobalske and Tobalske 1999, Jaber and Guisan 2001, Bustamante and Seoane 2004, Tapia et al. 2004, in press). To illustrate this, we use an example where both the historical and present distributions of Golden Eagles (*Aquila chrysaetos*) were modelled in the province of Ourense (7,278 km^2, southeast of Galicia in northwestern Spain).

Current distribution of eagles was estimated by searching the province for breeding pairs each spring from 1997 to 2002. The historical distribution was estimated by reviewing available published information, as well as historic field data provided by biologists and gamekeepers. Several environmental variables were selected to model habitat attributes, namely land use, degree of humanization, topographic irregularity, and habitat heterogeneity. These parameters were represented on a 10-km^2 grid. Values for these environmental variables were obtained from 1:50,000 digital cartography with the aid of GIS software. The distribution of Golden Eagles was modelled for three periods: current (1997-2002), historical (the 1960s and 1970s), and current and historical periods pooled. Stepwise logistic regression analysis was then performed for each period with presence-absence of Golden Eagle as the dependent variable. It was assumed that the distribution of the Golden Eagle in Ourense is known with full precision, with no false absences (Hirzel et al. 2002, Bustamante and Seoane 2004). At the spatial scale considered, the best predictors of habitat suitability for breeding were topographical variables indicative of rugged relief. Cartographic models derived from these analyses showed estimated probability of occurrence of eagles within each 10-km^2 grid square.

The model allows managers to: (1) simulate the effects of silviculture, mining or fires within each grid, thus enabling effective assessment of environmental impacts; (2) identify shrublands to manage for enhancing prey density and prey availability; (3) annually identify areas to monitor for the presence of potential hazards for Golden Eagles (e.g., wind farms, power lines, etc.); (4) regulate outdoor recreation potentially hazardous for eagles; and (5) catalog the cliffs and rocky outcrops potentially suitable for nesting. Information about the location of nesting areas must be updated annually to allow the generation of new models, predict range expansions or contractions or identify suitable sites for reintroductions, and provide a basis for design of protected areas.

CONCLUSIONS

We conclude with a quotation from the first version of this work that remains as true today as it was in 1987: *"As evidenced by the recent literature, raptor habitat research is becoming more rigorous. Questions require more accurate and precise answers, and statistical support is mandatory in many cases, whether the objectives are ecological interpretation or application to manage-ment. Manuscripts and reports to agencies will be subjected to increasingly critical review of methodology and statistical analyses. Because of sample size problems and regionally limited applicability, researchers should consider opportunities for collaborative studies, and adopt proven techniques and measurements. Comparability of data will increase their value and make it possible to apply more complex statistical analyses to large, shared data bases"* (Mosher et al. 1987:93).

ACKNOWLEDGMENTS

During the writing of this chapter Luis Tapia was supported by a post-doctoral grant from the Galician Government (Xunta de Galicia). We also thank David Bird who waited patiently for this manuscript.

LITERATURE CITED

ALLDREDGE, J.R. AND J.T. RATTI. 1986. Comparison of some statistical techniques for analysis of resource selection. *J. Wildl. Manage.* 50:157–165.

———— AND J.T. RATTI. 1992. Further comparison of some statistical techniques for analysis of resource selection. *J. Wildl. Manage.* 56:1–9.

AMAR, A. AND S.M. REDPATH. 2005. Habitat use by Hen Harriers *Circus cyaneus* in Orkney: implications of land-use change for this declining population. *Ibis* 147:37–47.

AMAT, J.A. AND J.A. MASERO. 2004. Predation risk on incubating adults constrains the choice of thermally favourable nest sites in a plover. *Anim. Behav.* 67:293–300.

ANDERSON, S.H. AND K.J. GUTZWILLER. 1994. Habitat evaluation methods. Pages 254–271 in T.A. Bookhout [ED.], Research and management techniques for wildlife and habitats. The Wildlife Society, Bethesda, MD U.S.A.

ANDERSON, D.R., K.P. BURNHAM AND W.L. THOMPSON. 2000. Null hypothesis testing: problems, prevalence, and an alternative. *J. Wildl. Manage.* 64:912–923.

ANDRIES, A.M., H. GULINCK AND M. HERREMANS. 1994. Spatial modeling of the Barn Owl *Tyto alba* habitat using landscape characteristics derived from SPOT data. *Ecography* 17:278–287.

ANTHONY, R.G., E.D. FORSMAN, A.B. FRANKLIN, D.R. ANDERSON, K.P. BURNHAM, G.C. WHITE, C.J. SCHWARZ, J.D. NICHOLS, J.E. HINES, G.S. OLSEN, S.H. ACKERS, L.S. ANDREWS, B.L. BISWELL, P.C. CARLSON, L.V. DILLER, K.M. DUGGER, K.E. FEHRING, T.L. FLEMING, R.P. GERHARDT, S.A. GREMEL, R.J. GUTIERREZ, P.J. HAPPE, D.L. HERTER, J.M. HIGLEY, R.B. HORN, L.L. IRWIN, P.J. LOSCHL, J.A. REID AND S.G. SOVERN. 2006. Status and trends in demography of Northern Spotted Owls, 1985–2003. *Wildl. Monogr.* 163:1–48.

ARMSTRONG, E. AND D. EULER. 1982. Habitat usage of two woodland *Buteo* species in central Ontario. *Can. Field-Nat.* 97:200–207.

ARROYO, B., J. GARCÍA AND V. BRETAGNOLLE. 2002. Conservation of the Montagu's Harrier (*Circus pygargus*) in agricultural areas. *Anim. Conserv.* 5:283–290.

BAKALOUDIS, D.E., C.G. VLACHOS AND G.J. HOLLOWAY. 1998. Habitat use by Short-toed Eagles *Circaetus gallicus* and their reptilian prey during the breeding season in Dadia Forest (northeastern Greece). *J. Appl. Ecol.* 35:821–828.

BALBONTÍN, J. 2005. Identifying suitable habitat for dispersal in Bonelli's Eagle: an important issue in halting its decline in Europe. *Biol. Conserv.* 126:74–83.

BARROWS, C.W. 1981. Roost selection by Spotted Owls: an adaptation to heat stress. *Condor* 83:302–309.

BECHARD, M.J. 1982. Effect of vegetative cover on foraging site selection by Swainson's Hawk. *Condor* 84:153–159.

BEDNARZ, J.C. AND J.J. DINSMORE. 1982. Nest-sites and habitat of Red-shouldered and Red-tailed hawks in Iowa. *Wilson Bull.* 94:31–45.

BERTHOLD, P. 1993. Bird migration: a general survey. Oxford University Press, New York, NY U.S.A.

BEVERS, M. AND C.H. FLATHER. 1999. The distribution and abundance of populations limited at multiple spatial scales. *J. Anim. Ecol.* 68:976–987.

BIRD, D.M., D.E. VARDLAND AND J.J. NEGRO [EDS.]. 1996. Raptors in human landscapes: adaptations to built and cultivated environments. Academic Press, San Diego, CA U.S.A.

BLONDEL, J. AND R. CUVILLIER. 1977. Une methode simple et rapide pour decride les habitats d'oiseaux: le stratiscope. *Oikos* 29:326–331.

BOAL, C.W. 1997. An urban environment as an ecological trap for Cooper's Hawks. Ph.D. dissertation, University of Arizona, Tucson, AZ U.S.A.

———, D.E. ANDERSEN AND P.L. KENNEDY. 2003. Home range and residency status of Northern Goshawks breeding in Minnesota. *Condor* 105:811–816.

———, D.E. ANDERSEN AND P.L. KENNEDY. 2005. Foraging and nesting habitat of Northern Goshawks in the Laurentian Mixed Forest Province, Minnesota. *J. Wildl. Manage.* 69:1516–1527.

BOSAKOWSKI, T. AND R. SPEISER. 1994. Macrohabitat selection by nesting Northern Goshawks: implications for managing eastern forests. *Stud. Avian Biol.* 16:46–49.

BOSCH, J., A. BORRAS AND J. FREIXAS. 2005. Nesting habitat selection of Booted Eagle *Hieraaetus pennatus* in Central Catalonia. *Ardeola* 52:225–233.

BUCKLAND, S.T. AND D.A. ELSTON. 1993. Empirical models for spatial distribution of wildlife. *J. Appl. Ecol.* 30:478–495.

BULLOCK, J. 1998. Plants. Pages 111–138 *in* W.J. Sutherland [ED.], Ecological census techniques. Cambridge University Press, Cambridge, United Kingdom.

BURNHAM, W.A. AND W.G. MATTOX. 1984. Biology of the Peregrine and Gyrfalcon in Greenland. *BioScience* 14:1–25.

BUSTAMANTE, J. 1997. Predictive models for Lesser Kestrel (*Falco naumanni*) distribution, abundance and extinction in southern Spain. *Biol. Conserv.* 80:153–160.

——— AND J. SEOANE. 2004. Predicting the distribution of four species of raptors (Aves: Accipitridae) in southern Spain: statistical models work better than existing maps. *J. Biogeogr.* 31:295–306.

———, J.A. DONÁZAR, F. HIRALDO, O. CEBALLOS AND A. TRAVAINI. 1997. Differential habitat selection by immature and adult Grey Eagle-buzzards *Geranoaetus melanoleucus*. *Ibis* 139:322–330.

CADE, B.S. 1997. Comparison of tree basal area and canopy cover in

habitat models: subalpine forest. *J. Wildl. Manage.* 61:326–335.

CADE, T.J. 1960. Ecology of the Peregrine and Gyrfalcon populations in Alaska. *Univ. Calif. Publ. Zool.* 63:151–290.

———, J.H. ENDERSON, C.G. THELANDER AND C.M. WHITE [EDS.]. 1988. Peregrine Falcon populations: their management and recovery. The Peregrine Fund, Boise, ID U.S.A.

CANAVELLI, S.B., M.J. BECHARD, B. WOODBRIDGE, M.N., KOCHERT, J.J. MACEDA AND M.E. ZACCAGNINI. 2003. Habitat use by Swainson's Hawks on their austral wintering grounds in Argentina. *J. Raptor Res.* 37:125–134.

CARRETE, M., J.A. SÁNCHEZ-ZAPATA AND J.F. CALVO. 2000. Breeding densities and habitat attributes of Golden Eagles in southeastern Spain. *J. Raptor Res.* 34:48–52.

CERASOLI, M. AND V. PENTERIANI. 1996. Nest-site and aerial meeting point selection by Common Buzzards (*Buteo buteo*) in Central Italy. *J. Raptor Res.* 30:130–135.

CODY, M.L. 1981. Habitat selection in birds: the roles of vegetation structure, competitors and productivity. *BioScience* 31:107–113.

———. 1985. Habitat selection in birds. Academic Press, Orlando, FL U.S.A.

CORSI, F., J. LEEUW AND A. SKIDMORE. 2000. Modeling species distribution with GIS. Pages 389–434 *in* L. Boitani and T.K. Fuller [EDS.], Research techniques in animal ecology: controversies and consequences. Columbia University Press, New York, NY U.S.A.

DAW, S.K. AND S. DESTEFANO. 2001. Forest characteristics of Northern Goshawk nest stands and post-fledging areas in Oregon. *J. Wildl. Manage.* 65:59–65.

DESTEFANO, S. 2005. A review of the status and distribution of Northern Goshawks in New England. *J. Raptor Res.* 39:342–350.

———, M.T. MCGRATH, S.K. DAW AND S.M. DESIMONE. 2006. Ecology and habitat of breeding Northern Goshawks in the inland Pacific Northwest: a summary of research in the 1990s. *Stud. Avian Biol.* 31:75–84.

DEVEREUX, J.C. AND J.A. MOSHER. 1984. Breeding ecology of Barred Owls in central Appalachians. *J. Raptor Res.* 18:49–58.

DEWEY, S.R. AND P.L. KENNEDY. 2001. Effects of supplemental food on parental care strategies and juvenile survival of Northern Goshawks. *Auk* 118:352–365.

DIFFENDORFER, J.E. 1998. Testing models of source-sink dynamics and balanced dispersal. *Oikos* 81:417–433.

DONALD, P. F. AND R. J. FULLER. 1998. Ornithological atlas data: a review of uses and limitations. *Bird Study* 45:129–145.

DONÁZAR, J.A., O. CEBALLOS AND C. FERNÁNDEZ. 1989. Factors influencing the distribution and abundance of seven cliff-nesting raptors: a multivariate study. Pages 545–551 *in* B.-U. Meyburg and R.D. Chancellor [EDS.], Raptors in the modern world. World Working Group for Birds of Prey and Owls, London, United Kingdom.

———, F. HIRALDO AND J. BUSTAMANTE. 1993. Factors influencing nest site selection, breeding density and breeding success in the Bearded Vulture (*Gypaetus barbatus*). *J. Appl. Ecol.* 30:504–514.

EDWARDS, T.C., E.T. DESHLER, D. FOSTER AND G.G. MOISEN. 1996. Adequacy of wildlife habitat relation models for estimating spatial distributions of terrestrial vertebrates. *Conserv. Biol.* 10:263–270.

ESPIE, R.H.M., P.C. JAMES, L.W. OLIPHANT, I.G. WARKENTIN AND D.J. LIESKE. 2004. Influence of nest-site and individual quality in breeding performance in Merlins *Falco columbarius*. *Ibis* 148:623–631.

FARREL, L. [ED.]. 1981. Heathland management. Nature Conservancy Council, Peterborough, United Kingdom.

FERRER, M. AND M. HARTE. 1997. Habitat selection by immature Spanish Imperial Eagles during the dispersal period. *J. Appl. Ecol.* 34:1359–1364.

FINN, S.P., D.E. VARLAND AND J.M. MARZLUFF. 2002. Does Northern Goshawk breeding occupancy vary with nest-stand characteristics on the Olympic Peninsula, Washington? *J. Raptor Res.* 36:265–279.

FORD, E.D. 2000. Scientific method for ecological research. Cambridge University Press, Cambridge, United Kingdom.

GAINZARAIN, J.A., R. ARAMBARRI AND A. F. RODRÍGUEZ. 2002. Population size and factors affecting the density of Peregrine Falcon (*Falco peregrinus*) in Spain. *Ardeola* 49:67–74.

GARNER, H.D. 1999. Distribution and habitat use of Sharp-shinned and Cooper's Hawks in Arkansas. *J. Raptor Res.* 33:329–332.

GARSHELIS, D.L. 2000. Delusions in habitat evaluation: measuring use, selection, and importance. Pages 111–164 *in* L. Boitani and T.K. Fuller [EDS.], Research techniques in animal ecology: controversies and consequences, Columbia University Press, New York, NY U.S.A.

GIBBONS, D., S. GATES, R.E. GREEN, R.J. FULLER AND R.M. FULLER. 1994. Buzzards *Buteo buteo* and Ravens *Corvus corax* in the uplands of Britain: limits to distribution and abundance. *Ibis* 137:S75–S84.

GIL-SÁNCHEZ, J.M., F. MOLINO GARRIDO AND S. VALENZUELA SERRANO. 1996. Selección de hábitat de nidificación por el Águila perdicera (*Hieraaetus fasciatus*) en Granada (SE de España). *Ardeola* 43:189–197.

GONZÁLEZ, L.M., J. BUSTAMANTE AND F. HIRALDO. 1990. Factors influencing the present distribution of the Spanish Imperial Eagle *Aquila adalberti*. *Biol. Conserv.* 51:311–319.

GONZÁLEZ-OREJA, J.A. 2003. Aplicación de análisis multivariantes al estudio de las relaciones entre las aves y sus hábitats: un ejemplo con paseriformes montanos no forestales. *Ardeola* 50:47–58.

GRAHAM, I.M. AND S.M. REDPATH. 1995. The diet and breeding density of Common Buzzards *Buteo buteo* in relation to indices of prey abundance. *Bird Study* 42:165–173.

GREEN, G.A. AND M.L. MORRISON. 1983. Nest site characteristics of sympatric Ferruginous and Swainson's Hawks. *Murrelet* 64:20–22.

GUISAN, A. AND W. THUILLER. 2005. Predicting species distribution: offering more than simple habitat models. *Ecol. Letters* 8:993–1009.

HALL, L.S., P.R. KRAUSMAN AND M.L. MORRISON. 1997. The habitat concept and a plea for standard terminology. *Wildl. Soc. Bull.* 25:173–182.

HAYWARD, G.D. AND E.O. GARTON. 1984. Roost habitat selection by three small forest owls. *Wilson Bull.* 96:690–692.

HIRZEL, A.H., J. HAUSER, D. CHESSEL AND N. PERRIN. 2002. Ecological-niche factor analysis: how to compute habitat-suitability maps without absence data? *Ecology* 83:2027–2036.

———, A.H., B. POSSE, P.A. OGGIER, Y. CRETTENAND, C. GLENZ AND R. ARLETTAZ. 2004. Ecological requirements of reintroduced species and the implications for release policy: the case of the Bearded Vulture. *J. Appl. Ecol.* 41:1103–1116.

HOLMES, T.L., R.L. KNIGHT AND G.R. CRAIG. 1993. Responses of wintering grassland raptors to human disturbance. *Wildl. Soc. Bull.* 21:461–468.

HUBERT, C. 1993. Nest-site habitat selected by Common Buzzard (*Buteo buteo*) in Southwestern France. *J. Raptor Res.* 27:102–105.

IVERSON, G.C., G.D. HAYWARD, K. TITUS, E. DEGAYNER, R.E. LOWELL, D.C. CROCKER-BEDLFORD, P.F. SCHEMPF AND J. LINDELL. 1996. Conservation assessment for the Northern Goshawk in southeast Alaska. USDA Forest Service General Technical Report, PNW-GTR-387, Pacific Northwest Research Station, Portland, OR U.S.A.

JABER, C. AND A. GUISAN. 2001. Modeling the distribution of bats in relation to landscape structure in a temperate mountain environment. *J. Appl. Ecol.* 38:1169–1181.

JAMES, P.C., I.G. WARKENTIN AND L.W. OLIPHANT. 1994. Population viability analysis of urban Merlins. [Abstract]. *J. Raptor Res.* 28:47.

JANES, S.W. 1985. Habitat selection in raptorial birds. Pages 159–188 *in* M.L. Cody [ED.], Habitat selection in birds. Academic Press, San Diego, CA U.S.A.

JONES, J. 2001. Habitat selection in avian ecology: a critical review. *Auk* 118:557–562.

JONGMAN, R.H.G., C.J.F. TER BRAAK AND O.F.R. VAN TONGEREN. 2001. Data analysis in community and landscape ecology. Cambridge University Press, Cambridge, United Kingdom.

KEISTER, G.P., R.G. ANTHONY AND H.R. HOLBO. 1985. A model of energy consumption in Bald Eagles: an evaluation of night communal roosting. *Wilson Bull.* 97:148–160.

KENNEDY, P.L. 1997. The Northern Goshawk (*Accipiter gentilis atricapillus*): is there evidence of a population decline? Special issue on responses of forest raptors to management: a holarctic perspective. *J. Raptor Res.* 31:95–106.

———. 2003. Northern Goshawk conservation assessment for Region 2, USDA Forest Service. www.fs.fed.us/r2/projects/scp/assessments/northerngoshawk.pdf (accessed 1 November 2006).

———, J.M. WARD, G.A. RINKER AND J.A. GESSAMAN. 1994. Postfledging areas in Northern Goshawk home ranges. *Stud. Avian Biol.* 16:75–82.

KOELN, G.T., L.M. COWARDIN AND L.S. LAURENCE. 1994. Geographic information systems. Pages 540–566 in T. A. Bookhout [ED.], Research and management techniques for wildlife and habitats. The Wildlife Society, Bethesda, MD U.S.A.

KORPIMAKI, E. 1984. Clutch size and breeding success of Tengmalm's Owl *Aegolius funereus* in natural cavities and nest boxes. *Ornis Fenn.* 61:80–83.

KOSTREZEWA, A. 1996. A comparative study of nest-site occupancy and breeding performance as indicators for nesting-habitat quality in three European raptor species. *Ethol. Ecol. Evol.* 8:1–18.

LEVIN, S.A. 1992. The problem of pattern and scale in ecology. *Ecology* 73:1943–1967.

LEYHE, J.E. AND G. RITCHISON. 2004. Perch sites and hunting behavior of Red-tailed Hawks (*Buteo jamaicensis*). *J. Raptor Res.* 38:19–25.

LITVAITIS, J.A., K. TITUS AND E.M. ANDERSON. 1994. Measuring vertebrate use of terrestrial habitats and foods. Pages 254–271 *in* T.A. Bookhout [ED.], Research and management techniques for wildlife and habitats. The Wildlife Society, Bethesda, MD U.S.A.

LOKEMOEN, J.T. AND H.F. DUEBBERT. 1976. Ferruginous Hawk nesting ecology and raptor populations in northern South Dakota. *Condor* 78:464–470.

LÓPEZ-LÓPEZ, P., C. GARCÍA-RIPOLLÉS, J.M. AGUILAR, F. GARCÍA-LÓPEZ AND J. VERDEJO. 2006. Modeling breeding habitat preferences of Bonelli's Eagle (*Hieraaetus fasciatus*) in relation to topography, disturbance, climate and land use at different spatial scales. *J. Ornithol.* 147:97–106.

LUSCIER, J. D., W. L. THOMPSON, J. M. WILSON, B. E. GORHAM, AND L. D. DRAGUT. 2006. Using digital photographs and object-based image analysis to estimate percent ground cover in vegetation plots. *Frontiers Ecol. Environ.* 4:408–413

MACKENZIE, D.I. 2005. What are the issues with presence-absence data for wildlife managers? *J. Wildl. Manage.* 69:849–860.

———, J. D. NICHOLS, J.A. ROYLE, K.H. POLLOCK, L.L. BAILEY AND J.E. HINES. 2006. Occupancy estimation and modeling. Academic Press, Boston, MA U.S.A.

MADDERS, M. 2000. Habitat selection and foraging success of Hen Harriers *Circus cyaneus* in west Scotland. *Bird Study* 47:32–40.

MANLY, B., L. MCDONALD AND D. THOMAS. 1993. Resource selection by animals. Chapman and Hall, London, United Kingdom.

MANNAN, R.W., W.A. ESTES AND W.J. MATTER. 2004. Movements and survival of fledgling Cooper's Hawks in an urban environment. *J. Raptor Res.* 38:26–34.

———, R.N. MANNAN, C.A. SCHMIDT, W.A. ESTES-ZUMPF AND C.W. BOAL. 2007. Influence of natal experience on nest site selection by urban-nesting Cooper's Hawks. *J. Wildl. Manage.* 71:64–68.

MAÑOSA, S. 1993. Selección de hábitat de nidificación en el Azor (*Accipiter gentilis*): recomendaciones para su gestión. *Alytes* 6:125–136.

———, J. REAL AND J. CODINA. 1998. Selection of settlement areas by juvenile Bonelli's Eagle in Catalonia. *J. Raptor Res.* 32:208–214.

MARCSTRÖM, V., R.E. KENWARD AND E. ENGRE. 1988. The impact of predation on boreal Tetraonids during the vole cycles: an experimental study. *J. Anim. Ecol.* 57:895–872.

MARINÉ, R. AND J. DALMAU. 2000. Uso del hábitat por el Mochuelo boreal (*Aegolius funereus*) en Andorra (Pirineo Oriental) durante el período reproductor. *Ardeola* 47:29–36.

MARION, W.R. AND R.A. RYDER. 1975. Perch-site preferences of four diurnal raptors in northeastern Colorado. *Condor* 77:350–352.

MARTÍNEZ, J.E. AND J.F. CALVO. 2000. Selección de hábitat de nidificación por el Búho real (*Bubo bubo*) en ambientes mediterráneos semiáridos. *Ardeola* 4:215–220.

MARTÍNEZ, J.A., G. LÓPEZ, F. FALCO, A. CAMPO AND A. DE LA VEGA. 1999. Hábitat de caza y nidificación del Aguilucho cenizo (*Circus pygargus*) en el Parque Natural de la Mata-Torrevieja (Alicante, SE de España): efectos de la estructura de la vegetación y de la densidad de presas. *Ardeola* 46:205–212.

———, D. SERRANO AND I. ZUBEROGOITIA. 2003. Predictive models of habitat preferences of Eurasian Eagle Owl *Bubo bubo*: a multiscale approach. *Ecography* 26:21–28.

MARZLUFF, J.M., B.A. KIMSEY, L.S. SCHUEK, M.E. MCFADZEN, M.S. VEKASY AND J.C. BEDNARZ. 1997. The influence of habitat, prey abundance, sex, and breeding success on ranging behavior of Prairie Falcons. *Condor* 99:567–584.

MCCLAREN, E.L., P.L. KENNEDY AND D.D. DOYLE. 2005. Northern Goshawk (*Accipiter gentilis laingi*) post-fledging areas on Vancouver Island, British Columbia. *J. Raptor Res.* 39:253–263.

MCGRADY, M.J., J.R. GRANT, I.P. BAINBRIDGE AND D.R.A. MCLEOD. 2002. A model of Golden Eagle (*Aquila chrysaetos*) ranging behaviour. *J. Raptor Res.* 36(Suppl.):62–69.

MCGRATH, M.T., S. DESTEFANO, R.A. RIGGS, L.L. IRWIN AND G.J. ROLOFF. 2003. Spatially explicit influences on Northern Goshawk nesting habitat in the interior Pacific Northwest. *Wildl. Monogr.* 154:1–63.

MCINTYRE, C.L. 2002. Patterns in nesting area occupancy and reproductive success of Golden Eagles (*Aquila chrysaetos*) in Denali National Park and Preserve, Alaska, 1988–99. *J. Raptor Res.* 36(Suppl.):50–54.

MCLEOD, M.A., B.A. BELLEMAN, D.E. ANDERSEN AND G.W. OEHLERT. 2000. Red-shouldered Hawk nest site selection in north-central Minnesota. *Wilson Bull.* 112:203–213.

MITCHELL, M.S., R.A. LANCIA AND J.A. GERWIN. 2001. Using landscape-level data to predict the distribution of birds on a managed forest: effects of scale. *Ecol. Appl.* 1:1692–1708.

MOORE, K.R. AND C.H. HENNY. 1983. Nest site characteristics of three coexisting *Accipiter* hawks in northeastern Oregon. *Raptor Res.* 17:65–76.

MOORMAN, C.E. AND B.R. CHAPMAN. 1996. Nest-site selection of Red-shouldered and Red-tailed Hawks in a managed forest. *Wilson Bull.* 108:357–368.

MORRIS, M.M.J., B.L. PENAK, R.E. LEMON AND D.M. BIRD. 1982. Characteristics of Red-shouldered Hawk, *Buteo lineatus*, nest sites in southwestern Quebec. *Can. Field. Nat.* 96:139–142.

——— AND R.E. LEMON. 1983. Characteristics of vegetation and topography near Red-shouldered Hawk nests in southwestern Quebec. *J. Wildl. Manage.* 47:138–145.

MORRISON, M.L. 2001. A proposed research emphasis to overcome the limits of wildlife-habitat relationship studies. *J. Wildl. Manage.* 65:613–623.

———, B. G. MARCOT AND R. W. MANNAN. 1998. Wildlife-habitat relationships: concepts and applications. University of Wisconsin Press, Madison, WI U.S.A.

MOSHER, J.A., K. TITUS AND M.R. FULLER. 1987. Habitat sampling, measurement and evaluation. Pages 81–97 *in* B.A. Giron Pendleton, B.A. Millsap, K.W. Cline, and D.M. Bird [EDS.], Raptor management techniques manual. National Wildlife Federation, Washington, DC U.S.A.

NEWTON, I. 1979. Population ecology of raptors. T. & A.D. Poyser, London, United Kingdom.

———. 1986. The Sparrowhawk. T. & A.D. Poyser, Calton, United Kingdom.

———. 1998. Population limitation in birds. Academic Press, London, United Kingdom.

———, M. MARQUISS, AND D. MOSS. 1981. Distribution and breeding of Red Kites in relation to land-use in Wales. *J. Appl. Ecol.* 19:681–706.

NORRIS, K. 2004. Managing threatened species: the ecological toolbox, evolutionary theory and declining-population paradigm. *J. Appl. Ecol.* 41:413–426.

NUDDS, T.D. 1977. Quantifying the vegetative structure of wildlife cover. *Wildl. Soc. Bull.* 5:113–117.

OLDEMEYER, J.L. AND W.L. REGELIN. 1980. Comparison of 9 methods of estimating density of shrubs and saplings in Alaska. *J. Wildl. Manage.* 38:280–282.

ONTIVEROS, D. 1999. Selection of nest cliffs by Bonelli's Eagle (*Hieraaetus fasciatus*) in southern Spain. *J. Raptor Res.* 33:110–116.

————, J.M. Pleguezuelos and J. Caro. 2005. Prey density, prey detectability and food habits: the case of Bonelli's Eagle and the conservation measures. *Biol. Conserv.* 123:19–25.

Orians, G.H. and J.F. Wittenberger. 1991. Spatial and temporal scales in habitat selection. *Am. Nat.* 137:S29–S49.

Ortega, C.P., J.C. Ortega, F.B. Sforza and P.M. Sforza. 2002. Methods for determining concealment of arboreal bird nests. *Wildl. Soc. Bull.* 30:1050–1056.

Osborne, P.E. and B.J. Tigar. 1992. Interpreting bird atlas data using models: an example from Lesotho, Southern Africa. *J. Appl. Ecol.* 29:55–62.

Pearce, J. and S. Ferrier. 2001. The practical value of relative abundance of species for regional conservation planning: a case study. *Biol. Conserv.* 98:33–43.

Penteriani, V. 2002. Goshawk nesting habitat in Europe and North America: a review. *Ornis Fenn.* 79:149–163.

———— and B. Faivre. 1997. Breeding density and nest site selection in a Goshawk *Accipiter gentilis* population of Central Apennines (Abruzzo, Italy). *Bird Study* 44:136–145.

Plumpton, D.L. and D.E. Andersen. 1997. Habitat use and time budgeting by wintering Ferruginous Hawks. *Condor* 98:888–893.

Poirazidis, K., V. Goutner, T. Skartsi and G. Stamou. 2004. Modeling nesting habitat as a conservation tool for the Eurasian Black Vulture (*Aegypius monachus*) in Dadia Nature Reserve, northeastern Greece. *Biol. Conserv.* 111:235–248.

Preston, C. 1980. Differential perch site selection by color morphs of the Red-tailed Hawk. *Auk* 97:782–789.

————. 1990. Distribution of raptor foraging in relation to prey biomass and habitat structure. *Condor* 92:107–112.

Pribil, S. and J. Picman. 1997. The importance of using the proper methodology and spatial scale in the study of habitat selection by birds. *Can. J. Zool.* 75:1835–1844.

Pulliam, H.R. 1988. Sources, sinks, and population regulation. *Am. Nat.* 132:652–661.

Quinn, G.P. and M.J. Keough. 2002. Experimental design and data analysis for biologists. Cambridge University Press, Cambridge, United Kingdom.

Ratcliffe, D. 1962. Breeding density in the Peregrine Falcon *Falco peregrinus* and Raven *Corvus corax*. *Ibis* 104:13–39.

————. 1993. The Peregrine Falcon. T. & A.D. Poyser, London, United Kingdom.

Reich, R.M., S.M. Joy and R.T. Reynolds. 2004. Predicting the location of northern goshawk nests: modeling the spatial dependency between nest locations and forest structure. *Ecol. Model.* 176:109–133.

Reynolds, R.T., E.C. Meslow and H.M. Wight. 1982. Nesting habitat of coexisting Accipiter in Oregon. *J. Wildl. Manage.* 46:124–138.

————, R.T. Graham, M.H. Reiser, R.L. Bassett, P.L. Kennedy, D.A. Boyce, G. Goodwin, R. Smith and E.L. Fisher. 1992. Management recommendations for the Northern Goshawk in the southwestern United States. USDA Forest Service General Technical Report RM-217, Rocky Mountain Forest and Range Experiment Station, Ft. Collins, CO U.S.A.

Rico-Alcázar, L., J.A. Martínez, S. Morán, J.R. Navarro and D. Rico. 2001. Preferencias de hábitat del Águila-azor perdicera (*Hieraaetus fasciatus*) en Alicante (E de España) a dos escalas espaciales. *Ardeola* 48:55–62.

Rolstad, J., E. Rolstad and Ø. Seteren. 2000. Black Woodpecker

nest sites: characteristics, selection and reproductive success. *J. Wildl. Manage.* 64:1053–1066.

Romesburg, H.C. 1981. Wildlife science: gaining reliable knowledge. *J. Wildl. Manage.* 45:293–313.

Rosenfield, R.N., J. Bielefeldt, D.R. Trexel and T.C.J. Doolittle. 1998. Breeding distribution and nest-site habitat of Northern Goshawks in Wisconsin. *J. Raptor Res.* 32:189–194.

Rotenberry, J.T. and S.T. Knick. 1999. Multiscale habitat associations of the Sage Sparrow: implications for conservation biology. *Stud. Avian Biol.* 19:95–103.

Roth, R.R. 1976. Spatial heterogeneity and bird species diversity. *Ecology* 57:773–782.

Rottemborn, S. 2000. Nest-site selection and reproductive success of urban Red-shouldered Hawks in central California. *J. Raptor Res.* 34:18–25.

Rushton, S.P., S.J. Ormerod and G. Kerby. 2004. New paradigms for modeling species distributions? *J. Appl. Ecol.* 41:193–200.

Rutz, C. 2003. Assessing the breeding season diet of Goshawk *Accipiter gentilis*: biases of plucking analysis quantified by means of continuous radio-monitoring. *J. Zool., Lond.* 159:209–217.

Salamolard, M. 1997. Utilisation de l'espace par le busard cendré *Circus pygargus*. *Alauda* 65:307–320.

Sánchez-Zapata, J.A. and J.F. Calvo. 1999. Raptor distribution in relation to landscape composition in semi-arid Mediterranean habitats. *J. Appl. Ecol.* 36:254–262.

Schmutz, J.K., S.M. Schmutz and D.A. Boag. 1980. Coexistence of three species of hawks (*Buteo spp.*) in the prairie-parkland ecotone. *Can. J. Zool.* 58:1075–1089.

Seavy, N.E. and C.K. Apodaca. 2002. Raptor abundance and habitat use in a highly disturbed forest landscape in western Uganda. *J. Raptor Res.* 36:51–57.

Selas, V. 1996. Selection and reuse of nest stands by Sparrowhawks *Accipiter nisus* in relation to natural and manipulated variation in tree density. *J. Avian Biol.* 27:56–62.

————. 1997a. Nest-site selection by four sympatric forest raptors in southern Norway. *J. Raptor Res.* 31:16–25.

————. 1997b. Influence of prey availability on re-establishment of Goshawk *Accipiter gentilis* nesting territories. *Ornis Fenn.* 74:113–120.

———— and C. Steel. 1998. Large brood sizes of Pied Flycatcher, Sparrowhawk and Goshawk in peak microtine years: support for the mast depression hypothesis. *Oecologia* 116:449–455.

———— and T. Rafoss. 1999. Ranging behaviour and foraging habitats of breeding Sparrowhawks *Accipiter nisus* in a continuous forested area in Norway. *Ibis* 141:269–276.

Sergio, F., P. Pedrini and L. Marchesi. 2003. Adaptive selection of foraging and nesting habitat by Black Kites (*Milvus migrans*) and its implications for conservation: a multi-scale approach. *Biol. Conserv.* 112:351–362.

Siders, M.S. and P.L. Kennedy. 1996. Forest structural characteristics of Accipiter nesting habitat: is there an allometric relationship? *Condor* 98:123–132.

Simmons, R. and P.C. Smith. 1985. Do Northern Harriers (*Circus cyaneus*) choose nest site adaptively? *Can. J. Zool.* 63:494–498.

Squibb, R.C. and V.P.W. Hunt. 1983. A comparison of nesting ledges used by seabirds on St. George Island. *Ecology* 64:727–734.

Squires, J. and P.L. Kennedy. 2006. Northern Goshawk ecology: an assessment of current knowledge and information needs for conservation and management. *Stud. Avian Biol.* 31:8–62.

SPEISER, R. AND T. BOSAKOWSKI. 1987. Nest site selection by Northern Goshawks in northern New Jersey and southeastern New York. *Condor* 89:387–394.

STARFIELD, A.M. 1997. A pragmatic approach to modeling for wildlife management. *J. Wildl. Manage.* 61:261–270.

STEENHOF, K., S.S. BERLINGER AND L.H. FREDRICKSON. 1980. Habitat use by wintering Bald Eagles in South Dakota. *J. Wildl. Manage.* 44:798–805.

SUTHERLAND, W.J. 2000. The conservation handbook: research, management & policy. Blackwell Science, London, United Kingdom.

—— AND R.E. GREEN. 2004. Habitat assessment. Pages 251–268 *in* W.J. Sutherland, I. Newton, and R.S. Green [EDS.], Bird ecology and conservation: a handbook of techniques. Oxford University Press, Oxford, United Kingdom.

TAPIA, L., J. DOMÍNGUEZ AND L. RODRÍGUEZ. 2004. Modeling habitat selection and distribution of Hen Harrier (*Circus cyaneus*) and Montagu's Harrier (*Circus pygargus*) in a mountainous area in Galicia (NW-Spain). *J. Raptor Res.* 38:133–140.

——, J. DOMÍNGUEZ AND J. RODRÍGUEZ. In Press. Modeling habitat use and distribution of Golden Eagle *Aquila chrysaetos* in a low-density area of the Iberian Peninsula. *Biodivers. Conserv.*

THIOLLAY, J.M. 1994. Family Accipitridae. Pages 52–205 *in* J. del Hoyo, A. Elliot and J. Sargatal [EDS.], Handbook of the birds of the world, Vol. 2. New World vultures to guineafowl, Lynx Edicions, Barcelona, Spain.

THIRGOOD, S.J., S.M. REDPATH, S. CAMPBELL AND A. SMITH. 2002. Do habitat characteristics influence predation on Red Grouse? *J. Appl. Ecol.* 39:217–225.

——, S.M. REDPATH AND I.M. GRAHAN. 2003. What determines the foraging distribution of raptors on heather moorland? *Oikos* 100:15–24.

THOMPSON, W.L., R.H. YAHNER AND G.L. STORM. 1990. Winter use and habitat characteristics of vulture communal roosts. *J. Wildl. Manage.* 54:77–83.

TITUS, K. 1990. Statistical considerations in the design of raptor population surveys. Pages 195–202 *in* B.G. Pendleton, M.N. LeFranc, Jr., B.A. Millsap, D.L. Krahe, M.A. Madsen, and M.A. Knighton [EDS.], Proceeding from the midwestern raptor management symposium and workshop. National Wildlife Federation, Washington, DC U.S.A.

—— AND J.A. MOSHER. 1981. Nest site habitat selected by woodland hawks in the central Appalachians. *Auk* 98:270–281.

TJENBERG, M. 1983. Habitat and nest site features of Golden Eagles (*Aquila chrysaetos*) (L.), in Sweden. *Swedish Wild. Res.* 12:131–163.

TOBALSKE, C. AND B.W. TOBALSKE. 1999. Using atlas data to model the distribution of woodpecker species in the Jura, France. *Condor* 101:472–483.

VAN HORNE, B. 1983. Density as a misleading indicator of habitat quality. *J. Wildl. Manage.* 47:893–901.

VENIER, L.A. AND L. FAHRIG. 1996. Habitat availability causes the species abundance-distribution relationship. *Oikos* 76:564–570.

WIENS, J.A. 1989a. The ecology of bird communities. Cambridge University Press, Cambridge, United Kingdom.

——. 1989b. Spatial scaling in ecology. *Funct. Ecol.* 3:385–397.

——, J.T. ROTENBERRY AND B. VAN HORNE. 1987. Habitat occupancy patterns of North American shrubsteppe birds: the effects of spatial scale. *Oikos* 48:132–147.

WILLIAMS, B.K., J.D. NICHOLS AND M.J. CONROY. 2002. Analysis and management of animal populations. Academic Press, San Diego, CA U.S.A.

WITHEY, J.C., T.D. BLOXTON AND J.M. MARZLUFF. 2001. Effects of tagging and location error in wildlife telemetry studies. Pages 43–75 *in* J.J. Millspaugh and J.M. Marzluff [EDS.], Radio tracking and animal populations. Academic Press, San Diego, CA U.S.A.

Accessing Nests

10

Joel E. Pagel

Santa Cruz Predatory Bird Research Group,
100 Shaffer Road, Santa Cruz, CA 95060 U.S.A.

Russell K. Thorstrom

The Peregrine Fund,
5668 West Flying Hawk Lane, Boise, ID 83709 U.S.A.

CLIFF- AND TREE-ENTRY TECHNIQUES

Raptors nesting on trees, cliffs or cliff-like structures (bridges, buildings, towers, etc.) create unique circumstances for safe access to nests, eggs, and young. Entries should be limited to biologists who are (1) comfortable with heights, (2) have direct knowledge and handling experience with the species in question, and (3) are thoroughly familiar with safe climbing and rappelling techniques.

Entry to nests, as well as to hunting perches for diet studies, should be undertaken only with sufficient knowledge of the nest or ledge location and the current state of the nesting chronology (see Chapter 19). Searches for nests during climbs or rappels are potentially dangerous, both to the birds and the climber. Noting the exact location of the nest with a photograph taken at an appropriate scale, and recording the azimuth from the observation point to the nest, or having a ground spotter will help the climber locate the best route to the nest.

Equipment

Ropes. **Static**, **semi-static**, and **dynamic** ropes each have their place for tree and cliff research. Static lines have limited stretch, are extremely durable, and are suitable for very long rappels (70 m or more) and tree work, but may be less convenient on smaller cliffs. Static lines are bulky and inflexible, making them more difficult to use on short-distance nest entries. They should never be used for lead climbing where short or long falls are possible.

Dynamic ropes are used for standard rappels, and climbs up to cliff nests, and may be used on nest entries of varying lengths (up to the length of your rope). These ropes may stretch up to 7–10% of rope length, making long rappels "bouncy" and prone to dislodging rocks from above and onto the climber and study species. It is best to use 10.5–11 mm ropes for most nest entry work; thinner, 8.0–9.5 mm, ropes should be avoided even if doubled.

Ropes come in standard lengths of 50, 60, and 70 m, or spools of up to 200 m. Longer ropes provide greater utility for raptor work on cliffs, but are heavier and bulkier. Sometimes, shorter ropes are more appropriate for smaller trees and cliffs for weight, management ease, and swift ascents or descents.

Ropes can be purchased pre-treated to be more water-repellent. These are called dry ropes. Dry treatment lengthens rope life, eases rope handling, and reduces water retained in the rope under wet conditions. Dry dynamic ropes work best for most raptor cliff work, and could be the rope of choice for their versatility. Rope bags are useful for tree climbing, rappelling over brushy or sloping terrain, and when ropes need to be

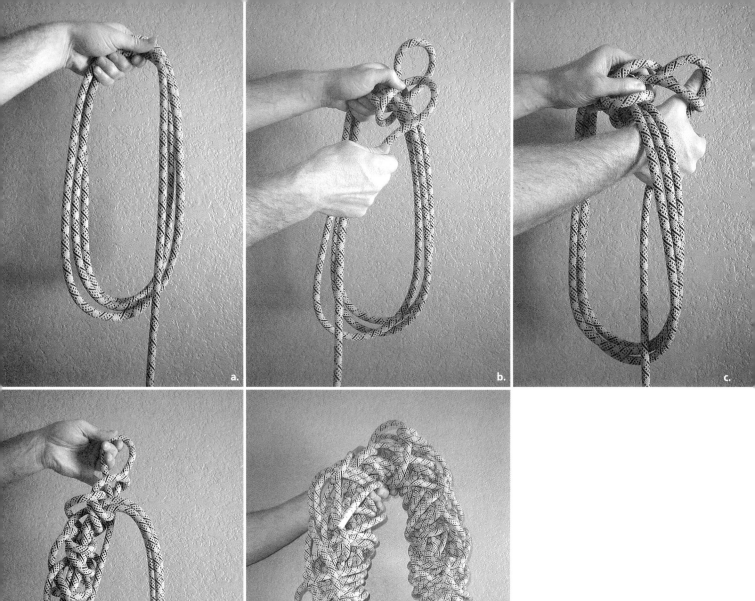

Figure 1. Quickcoil technique (a.– e.). Make three small coils, make loop, pass rope under upper portion of coil making a loop that goes through upper loop, continue over and under with loops for entirety of rope. When finished, the ropes looks like it is a mess, but will easily feed out rope when thrown from the top of a cliff, or when pulled up for lead climbing.

stored while the climber is at the nest. Traditional coils often tangle when thrown from a cliff or dropped through a forest canopy, or when they are "laid" out prior to use, lengthening the duration of the climb and the stress on the study species. The "quick-coil" method can greatly expedite nest entries (see Fig. 1). Ropes, as well as all other climbing gear, should be used exclusively for climbing (i.e., not towing your car), stored and cared for appropriately, inspected frequently for wear or damage (before and after a climb), and replaced as often as necessary.

Harnesses and gear bags. Rock-climbing harnesses are suitable for most raptor work. Specialized harnesses for tree work have metal D-ring attachments to allow the climber to lean back on lanyards while in the tree. Harnesses should fit snugly, have leg loops and double buckles, and not have excessive wear. Indeed, all gear should be checked before and after each climb and discarded if necessary. Tree-climbing harnesses may be more practical when spurring up a tree. Nest entry gear bags should be accessible even while hanging on the rope, have numerous loops for clipping onto the rope or harness, and have double closures. Backpacks worn during climbs can change your balance point; bandolier bags, large fanny packs with a lanyard to attach bag to harness, or closeable nylon climbing buckets attached to the harness directly or hanging from it work best.

Helmets and other personal protective equipment. Climbing helmets should be standard equipment for biologists entering nests. Rock climbing helmets are inexpensive and comfortable. Hockey helmets with plastic or wire face guards have been used for tree entries where the climber faces the dual hazards of tree branches and raptors capable of hitting the climber in the face. Neck guards should be considered for visits to the nests of larger accipiters, and some eagles if the adults have not been captured prior to climbing.

Gloves are recommended for tree climbing, but are not recommended for free climbing on rocks. Fingerless gloves may be helpful for rappelling. Gloves should be removed for processing chicks as rough handling of the chicks may cause damage to developing feathers. All loose clothing as well as long hair should be secured to prevent potential entanglement of tree branches or the rappel device. Prior to rappelling, braid your hair if longer than shoulder length, and tuck it under your shirt (braided hair can serve as extra cushion when a bird strikes your head or neck). Eye protection is helpful when falling sand or other debris is present.

Sturdy hiking boots or shoes are suitable for most raptor nest climbing. Specialized sticky-sole, rock-climbing shoes are helpful when free-climbing up to cliff nests, and can be useful on cliff nest entries where overhangs or some lateral movement is required while entering the nest.

Rappelling and ascending equipment. Ease, simplicity, and familiarity of use is important when selecting rappelling and ascending equipment. There are many descending devices such as figure 8s, belay plates, belay controllers, and more mechanical rappel devices, all of which work well. Caving descenders can be used for very long descents or with wet and dirty ropes, but they are heavy and take considerable practice to use properly. Figure 8s and belay controllers are preferred, as they are easy to use under low-light condi-

tions or when the climber is tired; the only mistake you can make is to not clip the device to your harness. All descenders should be clipped to the climbing harness using a large pear-shaped locking carabiner; however, even locking carabiners can open unexpectedly. The use of two locking carabiners with their gates opposite works best.

Ascenders should be inspected before and after each climb. They should fit your hand comfortably, and be rigged on the top and bottom attachment to weight the device properly. A prusik is a 1.4-m length of 5–6-mm cord tied as a sling and wrapped around the rope three to four times into a prusik knot, with the loop clipped into your harness. It is necessary to use these in concert with your descender. The prusik knot is kept loose around the rope by your free hand during a descent; if you are hit by a branch, rock, bird, or otherwise lose your grip on your descent line, the knot will tighten to reduce the likelihood of an injurious fall. Ascenders can be used for this purpose, but are not recommended as they can cut or damage the rope on severe falls, or can break due to stress fractures of the metal. A prusik cord also should be used as a safety device during fixed-rope ascents on rock or trees. The climber places the prusik knot on the rope, either above the top ascender to be pushed up during ascent, or below the bottom ascender. The prusik knot will tighten on the rope to catch the climber if the ascenders slip or fail. You also may put the prusik on your secondary safety rope and, when pausing to catch your breath during the ascent, pull the trailing rope through the prusik to keep it current with your height. This negates the need to tie into the rope at intervals, which adds weight to the ascending climber. The climber should become adept with rappelling and ascending so as to not require belay.

Anchor points. Preferred anchors for cliff nest entries include natural features such as large boulders, trees, and deep-rooted bushes. Vehicle frames, highway guardrails, and beams may be used as anchors when climbing on bridges and other structures. Where natural anchors are not available, slings, camming devices or chocks can be used. If the area has no natural anchors or available cracks, three or more rebar or concrete form stakes (7 mm x 1.5 m) can be pounded 1 m into firm soil at least 2 m apart to create an anchor for a self-equalizing rappel point. Few cliff nest entries require bolts or pitons to secure the climber; pitons scar rocks during removal, and should be used only if no alternative is available. If the cliff nest will be entered yearly, removable or inconspicuous permanent bolts may be appropriate to expedite nest entry to secure the climber either at the top tie-off point or strategic locations on the climb. Permanent placements of protection, or practice climbs should be done outside of the nest season when nest disturbance is unlikely.

A climbing course taught by qualified instructors at a gym or controlled outdoor situation is often a good start in learning how to access nests safely. However, nothing takes the place of climbing real rocks and trees under diverse conditions to help the biologist understand their physical and mental capabilities when aloft. Advanced knowledge of the limitations of knots, slings, self-equalizing anchors, camming devices, chocks, pitons, and rebar are vital both for the climber and raptor's safety.

See Long (1993) for a thorough explanation of climbing anchors.

Techniques

Descent or ascent to tree and cliff nest sites is inherently dangerous (Fig 2.). Falling debris, unstable rocks, rotten branches, stinging insects, aggressive raptors, and inappropriate technique on the part of the climber contribute to increased risk. Northern Goshawks (*Accipiter gentilis*), Swainson's Hawks (*Buteo swainsoni*), Red-tailed Hawks (*B. jamaicensis*), Harpy Eagles (*Harpia harpyja*), American Kestrels (*Falco sparverius*), and some owls may hit biologists in trees. Bald Eagles (*Haliaeetus leucocephalus*), Red-tailed Hawks, Verreaux's Eagles (*Aquila verreauxi*), Peregrine

Falcons (*F. peregrinus*), and some owls may hit climbers while they are climbing on cliffs near active nest sites. Raptors nesting in urban areas are especially prone to making contact with climbers as they are acclimated to seeing humans and may have lost their fear or "respect." Accipiters and eagles are aggressive, and will strike the climber in the back, neck, or back of head; a light backpack and helmet are sometimes necessary for protection. Owls tend to go for the face and eyes. Swainson's Hawks and Red-tailed Hawks also will hit the climber in the face if given a chance. Climbers may choose protective glasses, or a hockey facemask attached to their helmet to protect themselves from potential facial injuries. Golden Eagles (*A. chrysaetos*), Prairie Falcons (*F. mexicanus*) and Barn Owls (*Tyto alba*) may soar high above the site, or disappear during the nest entry. California Condors (*Gymnogyps californianus*) and Peregrine Falcons may return to the nest ledge during the entry to watch the climber. The latter can be hand-grabbed for banding if done carefully.

Before entry, researchers should consider the nesting chronology of the study species including age of the young, timing of nest entry, exposure of young to the elements, approach and entry disturbance to raptors and other proximal species (nesting passerines, seabirds, mountain goats [*Oreamnos americanus*], sheep, snakes, and stinging insects), fragility of the nest, tree and rock type, moss, weather conditions, falling debris and the presence of waterfalls. Additional considerations include potential rescue options for young raptors, should they flush from the nest; and climbers, should

Figure 2. Climbers should remain clipped into the rope at all times when on nests in cliffs or trees. *(Photo by David Pitkin)*

Figure 3. Climbing trees to enter raptor nests requires skill, knowledge of the species, knowledge of branch strength, and a bit of nerve.

Figure 4. Free climbing large trees is possible, and can be accomplished safely. Use of spurs and a helmet with a facemask augments the safety of the climber.

they fall or otherwise become incapacitated. Expedited entries and departures (within reason, considering safety factors) are goals that should reduce excessive disturbance at nests. In addition to Chapter 19, the following references have stood the test of time on considerations regarding disturbance to raptors during nest entries: Olendorff (1971), Fyfe and Olendorff (1976), Olsen and Olsen (1978), and Grier and Fyfe (1987). Supplemental information on climbing techniques includes Robbins (1970), Dial and Tobin (1994), Benge and Raleigh (1995), Jepson (2000), and Dial et al. (2004).

Tree-nest entries. Tree-nesting raptors pose special problems for raptor biologists attempting to study them. Their nests are limited by the availability of nesting habitat, but a good nest site should provide protection for the eggs, nestlings, and adults from predators (Newton 1979). Many raptors nest in trees because they offer support for platform stick nests or provide natural cavities limiting predator access and at the same time are easily accessible from the air. Raptors nesting in trees are common in both temperate and tropical environments. Accessing raptor nests in trees, and especially those high in the canopy, can be a significant challenge for researchers (Fig. 3). When entering trees, care must

be taken with snakes, bees, wasps, scorpions, other biting or stinging insects, and the potential for rotten and falling limbs.

Methods used for entering nest trees and reaching nests involve a combination of experience with tree and technical rock climbing techniques and equipment. Depending on nest site or cavity location within the tree the techniques used to ascend vary and include: free-climbing by hand from branch to branch, hugging the trunk, free-climbing a vine, using an extendable ladder, ascending a rope with technical climbing equipment, or using climbing spurs (spikes) with a specific harnessed belt with D-rings and lanyard (flip rope). Assistants or spotters should wear protective helmets in case of falling objects (e.g., tree debris, branches, and equipment) and need to stay clear of the area where climbers are working.

Free-climbing trees. Extreme caution and knowing your own ability is important when free-climbing trees. That said, free-climbing to the nest can be done safely on trees, limbs, or vines that offer support for your hands and feet, and that will hold your full body weight (Fig. 4). A good rule to follow is to always have three points of secure contact with climbing substrate (e.g., two hands and one foot, etc.). Carry and use webbing or rope to attach the climber to the tree for added security when at or below the nest, or when resting. Extreme caution is needed in all situations including, but not limited to, wet conditions, trees with biting and stinging insects, aggressive raptors, thorns, and weak or dead branches, vines, and tree trunks.

Using extendable ladders, and tree bicycles. Sectional arborist tree ladders can be used to get the climber up to 20 m into a tree with a straight bole, however tree ladders take considerable practice to use correctly, and time to erect. As such they are best used after the nesting season. Ladders can be difficult to transport and carry into remote field sites and are expensive. Sectional ladders are limited by height, a maximum diameter of the bole of the tree, and trees with few branches on the main trunk. Non-specialty ladders may be used for short climbs, but need to be used cautiously, angled into the branch or trunk, and "footed" or held by an assistant to keep the base from moving. Swiss tree-climbing bicycles (*baumvelo*), which cause no damage to the tree, can be used on trees with straight boles such as those in tree plantations (Seal et al. 1965, Yeatman and Nieman 1978). Even so, the climber must free-climb once the tree crown is reached. Tree-climbing bicycles, although easier to carry than ladders, are more awkward than climbing spurs.

Climbing a rope. Fixed rope ascents often are necessary to get quickly and safely into a tree after a line has been thrown or shot to a solid (live) tree limb. Fixed rope ascents also are used to get out of a cliff nest site. Ascenders should be rigged appropriately, with webbing adjusted to the individual climber. One technique (the frog) allows use of both legs for ascending. Each ascender has a hanging foot loop and security webbing attached to the climbers' harness. Ascending is accomplished by standing in the stirrup and pushing up on the opposite ascender, while simultaneously lifting the corresponding foot. The second technique (single foot) has the climber attach the lower ascender to the harness with about 40–50 cm of 1-cm wide webbing and no leg loop; the other ascender is used with your strongest leg. This method, with practice, can be quicker, and allows the climber to maintain one foot on the cliff or tree to reduce swinging or spinning. The latter also decreases bounce and permits ascent rates of up to 25–30 m per minute. Physical fitness should not be understated; biologists should achieve a minimum of 15 m per minute ascent rate for nest entries or departures on cliffs for nest site manipulations or banding. Ascending fixed

ropes requires significant practice prior to its first use at a nest site, and the climber requires experience with technical rock climbing equipment, including static ropes, ascenders, descenders, webbing, knots, and harnesses.

Nests in large trees require biologists to throw or shoot a weighted line 10–50 m up into the tree, or across the lowest branch. Setting throw lines (monofilament or cord) vary from throwing lines with weights attached to the end, slingshots with and without fishing reels, or crossbows or compound bows with and without reels, dog training shooting devices, and free or spur climbing (Tucker and Powell 1991, Ness 1997, Jepson 2000). The weighted monofilament line is shot over the branch, tied to a 3 mm cord and pulled up, and then tied to a climbing rope that is pulled over the branch and tied off on the non-climbing side. Electricians' tape can be used to taper the union of monofilament and cord, and cord and climbing rope to help negotiate the knot over limbs. Protection of the limb is possible by pulling a rope cover up to the branch via a slipknot, and when positioned, pulling the slipknot out of the rope. This climbing technique is detailed in Jepson (2000), and also has been used to trap raptors in tropical forests (Thorstrom 1996).

Climbing spurs. Biologists should acquire a set of quality climbing spurs, spikes or gaffs and practice prior to accessing raptor nests. Tree spurs must be weighed against legitimate concerns of damaging the tree. Considerable practice prior to entries is necessary to allow the climber to get accustomed to overall balance, and the feeling of leaning back into the harness while attached to a lanyard around the tree. Climbing spurs vary in spur length (pole or tree), and most new models are adjustable to leg length and padding comfort. It is best to wear stiff rubber-soled boots for foot support when using climbing spurs. Spur length should match the thickness of tree bark, and have straps adjustable to fit the climber. Climbers also should use gloves to protect their hands when flipping the lanyard, rope or cord up, down and around trees and branches. This climbing technique is detailed in Jepson (2000). Care should be given to trees with thin bark because of the possibility of introducing holes and scarring that might lead to insect damage and potential disease. The climber needs to practice spur-climbing trees in all conditions and situations, including vines wrapped around trees and limb changes. For limb or branch movement within the tree, a second lanyard, rope or webbing with a locking carabineer works well for security while changing the main climbing line above and below limbs and branches. Dial et al. (2004) developed tree-transfer techniques using specialized rope grabbers shot from crossbows that may be used effectively in large canopy forests where direct ascent of the nest tree may be precluded. Climbing spurs, harness and a lanyard are easy to carry into remote field sites. This method is quick if the climb to the nest site is free of limbs and vines.

Care must be taken to avoid disturbing ants, scorpions, centipedes, wasps and bees that may be under bark and in cavities or attached to limbs on the ascent because of the movement in the tree while climbing. The climber can ascend the tree with a rope looped over a secure branch and tied off by an assistant (spotter) on the ground for a quick descent; or the climber can carry a cord to have an assistant pull up a static line. This method can be used to set lines or ropes for static line ascents in an optimal location to access a nest or nest cavity (Thorstrom 1996).

Biologists should be cautious when ascending to the nest or nest cavity. Having a ground spotter(s) helps to alert the climber if the nestlings are about to fall or jump from the nest and to run down and retrieve any young that do. If a nestling jumps, the climber should stop climbing, mentally note where the bird landed, and direct the spotter to recover it. Talking or humming softly during the climb appears to frighten nestlings less than suddenly arriving unexpected at the level of the nest. Spotters also may help the climber prepare for diving parental birds. The climber should judge whether to continue the approach if nestlings are in a precarious position.

Upon reaching the nest, a small hand-held mirror or telescoping pole mirror is useful in ascertaining the contents of a nest from below the rim. A small flashlight is useful to view inside a nest cavity or cave. The climber should get comfortable at a safe location within reach of the nest site, and should use suitable anchors such as monkeytails or limb loops for protection while in the tree to prevent long falls if a branch breaks, or spur slips. An etrier (i.e., a short, webbed ladder) or looped webbing can be thrown over and fixed on a limb above the nest to assist with nest entry, especially with large nests overhung with nest sticks. Climbers should access nestlings over the lip of the nest either by hand, or with a metal hook that can be placed around the nestling's leg to pull them toward the climber. Processing nestlings (i.e., banding, measuring, and drawing blood) can occur while in the tree if the climber is in a stable position. In many instances, it is possible to lower nestlings to the ground so assistants can process the young or adults. Rounded, padded and ventilated chick bags or lightweight wooden or plastic box inserts for bags with a line to the ground and to a pulley or carabiner near the climber work best to maneuver the bag through limbs, or better yet, out away from the trunk of the tree. Climbers should not sit in or on the nests of even large raptors.

Descent from nest sites can involve climbing or rappelling, or both. When rappelling, the climber should ensure that the rope has free movement over a solid branch so that they may pull the rope down when on the ground. Rope sleeves over the branch should be used to avoid damaging the tree.

Cliff-nest entries. A single rope rappel usually is not the safest way to enter nests on cliffs. Most recreational and professional climbing accidents occur during rappelling; the first author has encountered instances where a weighted rope has been sliced by a falling rock, or the sheath has been cut while passing over sharp rock. Biologists rappelling down cliffs with potential for falling debris, sharp rocks, or where the raptor involved may strike the climber should use two ropes, one for rappelling and the second as the safety line. The use of two different colored ropes works best.

Rappelling is the most efficient method to reach a cliff nest site if the top of the cliff is accessible. Ideally the line of the descent from the tie-off (anchor) to the nest should be as straight down as possible with minimal lateral movement. A spotter in visual or radio contact with the climber can be very helpful. Many nest cliffs have loose rock and other debris that may be dislodged by a careless climber or the ropes. Great care should be taken during the descent to clean the route of loose material and attention paid to where the ropes will contact the cliff above the climber to avoid knocking loose rocks down upon the climber, nestlings, or eggs. Biologists should not descend directly onto an active nest, but rather should rappel about 1 m or so to the side of the ledge or nest and continue past the nest quickly, but in a controlled manner 1–8 m below and out of sight of the chicks. At this point, the biologist can put their ascenders on the rope, and get the banding bag ready before entering the nest. If, on the way past the nest, the behavior of the chicks suggests that they may jump, or are too young or too old for banding, the climber should quickly ascend back up past the nest, or rappel to the ground if possible.

The entry to the nest ledge or the level of the nest where the climber can corner the nestlings should be one swift and controlled movement, allowing the biologist to trap the young in the nest to prevent escape. If a stick nest is present on the cliff, the movement to the level of the nest is still rapid, however the climber usually hangs in front of the nest ledge with both hands available to prevent the nestlings from leaving the nest prematurely. This technique requires considerable practice and balance. Entries for most raptor nests should be timed so they occur after onset of chick thermoregulation, but before loss of most down on the eyases (to prevent premature fledging; see Chapter 19).

Pendulum descents may be necessary near overhangs, or to transfer to different sections of the cliff or trees. The climber descends below the nest, and then swings several to many meters to the ledge, or to a crack below the nest, or the target tree. Normal lead climbing occurs from that point upward to the nest site. The climber should take extra caution on cliffs, as they will need to swing back laterally or out from the nest to the original fall line when done at the nest.

Although climbers sometimes are belayed, or lowered to the nest ledge, we do not recommend either practice under any circumstance, as it often causes debris to fall on the climber and study species, increases rope wear, and does not allow for self-rescue or quick ascents should the timing of the nest entry be inappropriate.

Climbing up to cliff nests may be accomplished with a ladder if access to the cliff is relatively easy, and the nest is not more than 8 m above the ground. Care should be taken to ensure the ladder is solidly "footed" or held by an assistant.

Lead climbing may be used for certain cliff nests. This method of climbing upward to the nest via cracks or holding on to small ledges or protrusions of the rock can be slow unless the route has been climbed during the off-season to locate difficult moves or pitches, or unless it is an easy ascent. Suitable protection devices (e.g., chocks and cam devices) often are placed on the way up the cliff and are necessary throughout the climb to prevent long, injurious falls. Protection placed directly below the nest ledge sometimes allows the climber to stand in an etrier to access the ledge, and allows an extra measure of safety should the climber fall out of the nest. Protection, including a fixed bolt, chock, or a sling wrapped around a rock or vegetation, will be left at the nest ledge to facilitate rappelling back down to the ground.

Sometimes eagles and Osprey (*Pandion haliaetus*) nest on the tops of natural rock pillars, either on shore-lines or in the water. In such situations climbing ropes or conventional ladders may not work. Chubbs et al. (2005) describe a portable anchorbolt ladder to access such nests.

Buildings, towers, and bridges. During most nest entries on man-made structures, building or bridge managers are available for discussion regarding tie-off points, local hazards, as well as for access policy and permits. Climbers should inform local authorities prior to nest entry, so that they are aware of your activities and you are not mistaken for a terrorist or vandal, or believed to be attempting suicide.

Bridges have inherent hazards that should be considered; sudden gusts of winds and their patterns moving through the superstructure can be unpredictable; rain and morning dew can make bridge or building surfaces especially slippery. Biologists always should be tied in to firm metal objects that are securely bolted or welded to the superstructure of the bridge, window washing supports, or the tower ladder or gangway. Rusted metal and some beam edges are sharp or have rough welds that can easily slice a rope or webbing when weighted suddenly during a fall or slip; multiple points of connection to the structure are recommended. Raptors defending their nests may fly through the superstructure of bridges or around corners on power plant towers and suddenly hit the climber. When this is likely, one or more spotters should be used.

Although urban nests often are suited for media coverage, the climber should take into consideration the complexity of the climb prior to inviting the media. If the media are covering the entry, the climber should ensure that proper safety techniques are followed for all involved and that the safety of the raptors is a primary concern. It is best to discuss the needs of the camera operator, and to establish enforceable ground rules, before the climb. Most media people are responsible and want to show your project in the best possible light; however some can be careless and may neglect the safety of the birds and themselves. Mishaps are your responsibility, and accidents involving the birds or media personnel can result in loss of scientific permits, the loss of access to the site, and the permanent alteration or cancellation of your project.

Bridges, buildings and towers may have people, boats or automobiles below. All equipment should be tied with lanyards to the climber, and extreme caution should be used when stepping on ledges or beams to avoid dislodging potential falling objects such as rocks, gravel, loose bolts, and detritus.

Skill and competence with the equipment and safe climbing technique are "musts" prior to any nest entry. Nest entries into active sites are not the place to practice newly learned skills. Physical fitness, and mental clarity are vital in conducting any high-angle nest entries in trees or cliffs, and both aspects require considerable training and preparation to allow the biologist the stamina, strength, agility, and lucidity to make safe decisions before and during the climb. If tree or cliff nest entries are new to you, or you are entering nests of a species you have not previously studied, seek out one or more knowledgeable and experienced biologists, arborists, botanists, silviculturists, etc., who use climbing skills during their work, before preparing to enter a nest. A conversation with an expert may save you hours of grief, and reduce the risk of adversely disturbing the birds. It also may prevent your sudden, albeit accidental, death.

ACKNOWLEDGMENTS

We thank Janet Linthicum, Joe Papp, and Pete Bloom for input and comments on earlier versions of this chapter. Additional helpful comments from Keith Bildstein, David Bird, Ron Jackman, Brian Latta, Allan Mee, Randy Waugh, Amira Ainis, and Shale Pagel helped improve the manuscript.

LITERATURE CITED

BENGE, M. AND D. RALEIGH. 1995. Rock: tools and techniques. Elk Mountain Press, Carbondale, CO U.S.A.

CHUBBS, T.E., M.J. SOLENSKY, D.K. LAING, D.M. BIRD AND G. GOODYEAR. 2005. Using a portable, anchorbolt ladder to access rock-nesting Osprey. *J. Raptor Res.* 39:105–107.

DIAL, R. AND S.C. TOBIN. 1994. Description of arborist methods for forest canopy access and movement. *Selbyana* 15:24–37.

———, S.C. SILLETT, M.E. ANTOINE AND J.C. SPICKLER. 2004. Methods for horizontal movement through forest canopies. *Selbyana* 25:151–163.

FYFE, R.W. AND R.R. OLENDORFF. 1976. Minimizing the dangers of studies to raptors and other sensitive species. *Can. Wildl. Serv. Occ. Paper* 23.

GRIER, J.W. AND R.W. FYFE. 1987. Preventing research and management disturbance. Pages 173–182 *in* B.A. Giron Pendleton, B.A. Millsap, K.W. Cline, and D.M. Bird [EDS.], Raptor management techniques manual. National Wildlife Federation, Washington, DC U.S.A.

JEPSON, J. 2000. The tree climber's companion, 2nd Ed. Beaver Tree Pub., Longville, MN U.S.A.

LONG, J. 1993. How to rock climb: climbing anchors. Chockstone Press, Inc., Evergreen, CO U.S.A.

NESS, T. 1997. Bow and arrow tree entry. www.newtribe.com/technical.html (last accessed 8 August 2006).

NEWTON, I. 1979. Population ecology of raptors. Buteo Books, Vermillion, SD U.S.A.

OLENDORFF, R.R. 1971. Falconiform reproduction: a review. Part 1. The pre-nestling period. Raptor Research Foundation Report No. 1.

OLSEN, P. AND J. OLSEN. 1978. Alleviating the impact of human disturbance on the breeding Peregrine Falcon. 1. Ornithologists. *Corella* 2:1–7.

ROBBINS, R. 1970. Basic rockcraft. La Siesta Press, Glendale, CA U.S.A.

SEAL, D.T., J.D. MATTHEWS AND R.T. WHEELER. 1965. Collection of cones from standing trees. Forestry. Record no. 39. Forestry Commission, London, United Kingdom.

THORSTROM, R.K. 1996. Methods for capturing tropical forest birds of prey. *Wildl. Soc. Bull.* 24:516–520.

TUCKER, G.F. AND J.R. POWELL. 1991. An improved canopy access technique. *North. J. Appl. For.* 8:29–32.

YEATMAN, C.W. AND T.C. NIEMAN. 1978. Safe tree climbing in forest management. Canada Department of Fisheries and Environment, Forestry Service, Ottawa, Canada.

Assessing Nesting Success and Productivity

11

KAREN STEENHOF

U.S. Geological Survey,

Forest and Rangeland Ecosystem Science Center,

Snake River Field Station, 970 Lusk Street, Boise, ID 83706 U.S.A.

IAN NEWTON

Centre for Ecology and Hydrology,

Monks Wood, Abbots Ripton, Huntingdon,

Cambridgeshire PE28 2LS, United Kingdom

INTRODUCTION

Studies of reproductive rates in raptors can be valuable in assessing the status of raptor populations and the factors that influence them. Estimates of nesting success and productivity provide insight into only one component of the demography of a raptor population. Individuals are added to local populations through reproduction, and they are subtracted through mortality. Together with immigration and emigration, these two demographic parameters determine the year-to-year trends in local populations. Reproductive rates usually are easier to evaluate than other aspects of demography, and properly designed studies will allow inferences to be made about relationships between the status of raptor populations and a variety of environmental influences. Unbiased data on reproductive rates allow comparisons among populations in different areas and different years that may reflect differences in land use, contaminant levels, human activity, or variations in natural phenomena, such as weather or prey supply. Such studies may

be essential for identifying effective conservation measures for threatened and declining species. Data on reproduction can help predict the effects of land use changes on raptor nesting populations (U.S. Department of Interior 1979), document effects of contaminants (Newton 1979, Grier 1982), or measure whether a population is reproducing well enough to sustain itself, given existing rates of survival (Henny and Wight 1972). Information on reproductive rates can be useful in deciding whether to list or reclassify an endangered raptor species or whether to allow harvest of a more common species for falconry purposes. Investigations have limited value, however, if objectives are not considered when the study is designed and initiated. Year-to-year fluctuations in nest success and productivity are common in raptors, and short-term decreases in productivity need not affect the long-term stability of populations.

The main objectives of this chapter are to (1) establish standard definitions that will facilitate comparisons of data over time and space, (2) identify the types of information needed to estimate raptor nesting success and productivity, (3) evaluate the advantages and disadvantages of various field techniques, and (4) offer suggestions for procedural and analytical approaches that will minimize bias. We include a glossary of technical terms for reference (Table 1).

CONCEPTS AND DEFINITIONS

To produce young, a raptor must pass successfully through a number of stages. It must first settle in a particular area, establish a **nesting territory** (terms in **bold** are defined in Appendix 1), and acquire a mate. It must then proceed through **nest** building, egg laying, and then to hatching and rearing of young. In this sequential process, birds can fail at any stage.

For the purpose of analyzing reproductive data, a nesting territory is an area that contains, or historically contained, one or more nests (or **scrapes**) within the home range of a mated pair. The term nesting territory should not be confused with the more restricted etho-logical definition of a territory as any defended area. A raptor nesting territory can be thought of as a confined area where nests are found, usually in successive years, and where no more than one pair is known to have bred at one time (Newton and Marquiss 1982). The concept holds even in colonial species, in which the same nest sites tend to be used year after year with the occupants often defending only a small area around their nest.

Individuals that are unable to secure a nesting terri-tory are known as **floaters**. They are usually unpaired and do not reproduce (Postupalsky 1983). Because of the difficulty in counting non-territorial raptors, and their greater mobility, they usually are excluded from analyses of **nesting success** and **productivity**. Howev-er, it may sometimes become possible to consider these birds in analyses of population dynamics (e.g., Ken-ward et al. 1999, Newton and Rothery 2001).

Some individuals are able to secure a nesting terri-tory but not a mate. Postupalsky (1983) recommended that lone territorial birds be excluded from tallies of nesting pairs, but this is seldom practical. Territories that truly have only one adult are difficult to distinguish from those in which the second adult was absent at the time of the nest check, perhaps hunting some distance away. They also often represent only a temporary situa-tion, as a lone bird may soon acquire a mate.

Certain pairs may occupy a territory for only a few days or a few weeks, or may even build a nest, but the process stops here. Not all raptor pairs occupying nest-ing territories lay eggs every year. A major factor influ-encing egg laying is food supply and in poor food years, many territorial pairs in some populations fail to lay eggs (Newton 2002). The proportion of pairs that pro-duce eggs in different years, therefore, can be an impor-tant measure of a population's response to changing food supplies (Steenhof et al. 1997).

Still other territory holders may lay and then desert their eggs or lose them to predation, weather, or other causes. Others may produce eggs that hatch, but then their young die due to a variety of causes and at a vari-ety of ages. Pairs that raise at least one young that is

nearly old enough to fly are usually considered **success-ful**. Of course, additional offspring mortality might occur after this stage (Marzluff and McFadzen 1996) when the young are free-flying, but still fed by their par-ents. Their death at this stage could be measured by a separate detailed study, or accounted for in estimates of juvenile survival, which is usually calculated as starting when the young are banded.

The proportions of pairs that reach these various stages can form a useful basis for comparing different raptor populations or subsets within populations. The most useful comparisons are based on the proportions of territorial pairs (or occupied territories) that produce young, but for practical reasons many studies can only obtain information on the proportion of laying pairs that produce young. Researchers who have good historical information on species that show strong fidelity to well-defined nesting territories (e.g., eagles, Ospreys [*Pan-dion haliaetus*]) can report nesting success and produc-tivity on the basis of territorial pairs or occupied territo-ries in a particular year (Brown 1974, Postupalsky 1974). In short-term investigations or studies of more nomadic raptors, it may be necessary to report success and productivity on the basis of laying pairs. For polyg-ynous or polyandrous species (e.g., harriers, Harris's Hawks [*Parabuteo unicinctus*], etc.), success and pro-ductivity are best reported per mated territorial female or per mated male.

Estimates of productivity based solely on the num-ber of young produced per successful pair can be mis-leading because successful pairs often produce average numbers of young even in years when most pairs fail (Steenhof et al. 1997, 1999). However, **brood size at fledging** can be a useful measure in some calculations (Steenhof and Kochert 1982: see below), depending on the purpose of the study.

CRITERIA FOR CLASSIFYING REPRODUCTIVE EFFORTS

Measurement error occurs when investigators incor-rectly interpret the status of a particular pair or nesting territory, or incorrectly count the number of eggs or young. The ability to determine correctly the status of nests and to count the number of young varies with many factors, including the field situation, observer experience, and weather. Because these factors cannot be held constant, it is sometimes difficult to determine whether differences in estimates reflect measurement error or true differences in productivity. Fraser et al. (1984) analyzed the problem of measurement error in aerial surveys of Bald Eagles (*Haliaeetus leucocephalus*) in the Chippewa National Forest. By running three simulated two-stage surveys in the same year, they were able to compute an error rate caused by mistakes in counts of occupied territories, laying pairs, and fledglings. Using this information, they calculated an estimated standard error that allowed them to test for true differences in productivity among years. The use of simulated surveys to obtain an estimate of variability due to measurement error is a site-specific procedure that must be repeated for each study area and each population. It is most valuable in situations where all territorial pairs have been found.

Territory Occupancy

Evidence that a territory is occupied can be based on observation of two birds that appear to be paired or one or more adults engaged in territorial defense, nest affinity, or other reproductive-related activity. Any indications that eggs were laid or young were reared constitute clear evidence for territorial occupancy. In some species, the presence of a nest that has been recently built, repaired, or decorated may constitute evidence for territorial occupancy, providing that these activities can be ascribed to the species of interest unequivocally. Caution must be used in applying this criterion because of the occasional difficulty in distinguishing old and new nest material. Fresh greenery, several sticks with fresh breaks, or a distinct layer of new material on top of older, weathered sticks usually suggest recent nest repair.

Individuals of some species may occupy territories for short periods only (perhaps less than one day), before moving on to another territory or reverting to a "floating" lifestyle. Some birds can thus easily be missed during a survey, or double-counted if they move from one territory to another in the same study area. Harriers are particularly problematic in this regard, because different individuals may "sky-dance" on different days over the same piece of nesting habitat during migration (e.g., Hamerstrom 1969). Fortunately, this seems not to be an issue for most species, and once a territory is occupied, it seems to remain so at least until the nest fails or the young reach independence.

For long-lived species that re-use the same territories year after year, such as Golden Eagles (*Aquila chrysaetos*) (Watson 1957) and Peregrine Falcons

(*Falco peregrinus*) (Mearns and Newton 1984), an estimate of the proportion of traditional territories occupied by pairs in any given year can be a useful index to the size and status of the nesting population. In species that show less fidelity to particular nesting territories among years, this measure can be misleading because it can grossly underestimate the status of species that normally use nesting territories intermittently or only once, such as Burrowing Owls (*Athene cunicularia*) (Rich 1984), Northern Hawk-Owls (*Surnia ulula*) (Sonerud 1997), Short-eared Owls (*Asio flammeus*) (Village 1987), and Ferruginous Hawks (*Buteo regalis*) (Lehman et al. 1998). For these and similar species, studies should be designed to sample all potential nesting habitat within a study area each year and not only previously occupied territories.

In many species, it is unusual to find all previously known territories occupied in any given year. Over a period of years, some territories may be used every year (or almost every year), whereas others are used irregularly, or very infrequently. In other words, certain territories are used much more often than expected by chance at the population levels found, and others are used much less often. This has led some long-term researchers to distinguish categories of territories, such as "regular and irregular." Typically, occupants of "regular" territories are more often successful than are occupants of less used territories, giving a correlation between occupancy and nest success (Newton 1991, Sergio and Newton 2003). It seems that many raptors are capable of selecting those particular territories where their chances of raising young are high.

Egg Laying

Not all raptor pairs occupying nesting territories lay eggs every year (see above). Evidence of laying may be based on observations of eggs, young, an incubating adult, fresh eggshell fragments, or any other field sign that indicates eggs were laid. However, be aware that some species, such as the Bald Eagle, may assume incubation posture without actually having laid an egg (Fraser et al. 1983).

Laying Date

The laying date of the first egg usually is taken as a measure of the timing of breeding in birds. Laying date is useful because it often correlates with nest success; birds laying earliest in the season usually are the most successful. Laying date also is a critical data element required for some **nest survival** models (Dinsmore et al. 2002). As nests are seldom visited on the very day that the first egg is laid, laying date is usually calculated indirectly, by backdating from some later stage in the cycle. Allowances are then made for the intervals between laying of successive eggs (two days in most raptor species), the **incubation period**, and, in the case of nests found during the **nestling period**, age of the young. Ages of nestlings can be estimated from weights or measurements in some species (e.g., Petersen and Thompson 1977, Bortolotti 1984). Photographic aging keys (e.g., Hoechlin 1976, Moritsch 1983a,b, 1985; Griggs and Steenhof 1993, Boal 1994, Priest 1997, Gossett and Makela 2005) also are useful tools for aging young. Repeated checks during the laying period can help to estimate the date of onset of incubation (Millsap et al. 2004). Otherwise, it is usually difficult to estimate laying date for pairs that fail during incubation. Investigators often assume that nest failure occurred at some specific stage, most typically in mid-incubation, or midway between successive nest checks, the latter check being the one in which failure was discovered. If deserted eggs are present, their stage of development sometimes can be estimated by candling (Weller 1956) to determine the stage of embryo growth, but the observer may still not know how long the eggs have lain unincubated in the nest.

Clutch Size

The number of eggs laid by each pair is useful, but not crucial, in assessments of overall productivity (Brown 1974). Because many raptor species nest on cliffs or in trees, not all nests are readily accessible, and clutch sizes may be difficult or impossible to record. In addition, some raptors are affected adversely by visits to nests during incubation. Because of this, counts of eggs at close range are sometimes associated with increased failure rates (Luttich et al. 1971, Steenhof and Kochert 1982, White and Thurow 1985, Chapter 19). For these reasons, a traditional measure of avian nesting success, the proportion of eggs that hatch and ultimately develop into fledglings, often is not attainable. Data on clutch sizes, however, can provide further insight into the mechanisms of a population's response to food supply or other environmental influences.

Nesting Success and Productivity

Nesting success is defined as the proportion of nesting or laying pairs that raise young to the age of **fledging** (i.e., the age when a fully-feathered offspring voluntarily leaves the nest for the first time). The difference between success per territorial pair and success per laying pair can be large in species that have relatively high rates of non-laying, including Golden Eagles and Tawny Owls (*Strix aluco*) (Southern 1970, Steenhof et al. 1997). It is less important for species in which all or most territorial pairs lay eggs (Steenhof and Kochert 1982).

In many studies, it is impossible to visit each nest on the exact day that young take their first flight; and after young have left the nest, they may be difficult to locate. Once young approach fledging age they become liable to flee from the nest prematurely if approached too closely. As they cannot fly at this stage, they usually flutter to the ground, and unless retrieved, could be vulnerable to predation or drowning. For this reason, it is sensible to check nests a week or more before young are likely to fledge. Most studies of raptors, therefore, consider pairs to be successful when well-grown young are observed in the nest at some point prior to fledging. Studies that consider nests with young of any age to be successful will overestimate nest success because they fail to consider mortality that may occur late in the brood-rearing period. Researchers should consider nest survival models (see below) when it is impossible to check an adequate number of nests at or near fledging.

If investigators wish to compare nest success among years, areas, or treatments, they should establish a standard minimum nestling age at which they consider nests to be successful. This age should be when young are well grown but not old enough to fly and at a stage when nests can be entered safely and after which mortality is minimal until actual fledging. Steenhof (1987) recommended that nests of diurnal raptors be considered successful only if at least one nestling has reached 80% of the average age at first flight. Mortality after this age until first flight is usually minimal (Millsap 1981). Furthermore, young are usually large enough to count from a distance at this stage. For Prairie Falcons (*F. mexicanus*), Golden Eagles, and Red-tailed Hawks (*B. jamaicensis*) nesting in the Snake River Canyon, 80% of fledging age corresponds with the age at which most young are banded (Steenhof and Kochert 1982). The 80% of first-flight age criterion has been used to determine nesting success in studies of several additional raptors, including Ferruginous Hawks, Northern Harriers (*Circus cyaneus*) (Lehman et al.

1998), Snail Kites (*Rostrhamus sociabilis*) (Bennetts et al. 1998), and Northern Goshawks (*Accipiter gentilis*) (Boal et al. 2005). A lower criterion for evaluating nest success (70 or 75% of the age at which young first leave the nest) might be more appropriate for species in which age at fledging varies considerably (i.e., highly sexually dimorphic raptors such as Cooper's Hawks [*A. cooperii*]) or for species that are more likely to leave the nest prematurely when checked. Millsap et al. (2004) considered Bald Eagle nests to be successful if young reached eight weeks of age or approximately 70% of first flight age, and the U.S. Fish and Wildlife Service (2003) considers Peregrine Falcon pairs to be successful when their young are at least 28 days old, or approximately 65% of first flight age. Information about fledging ages of most North American raptors can be found online at the Birds of North America website (http://bna.birds.cornell.edu/BNA/) (Poole 2004). Data on fledging ages of raptors from other parts of the world are in Newton (1979; Table 18) and Cramp et al. (1980). Investigators should consult more recent sources about their study species and use the best available information about variation in fledging ages and susceptibility to disturbance when they define and adopt a minimum age to evaluate success.

Productivity, which refers to the number of young that reach the minimum acceptable age for evaluating success, is usually reported on a per pair basis. In situations with a juvenile sex ratio of 1:1, the number of young per pair is equivalent to **fecundity** (number of females produced per female), a measure that can be incorporated into broader evaluations of a population's demography (e.g., Blakesley et al. 2001, Seamans et al. 2001). After leaving the nest, young normally continue to depend upon their parents (or one parent) for several weeks or months, before becoming independent and dispersing away from the nest vicinity. During the **post-fledging period**, young are sometimes difficult to locate (Fraser 1978). Counts after young have left the nest are unreliable because they tend to miss birds and underestimate the number of young produced. Owls present a special challenge in this regard because the young of many species leave the nest long before they can fly (Forsman et al. 1984) and often at staggered intervals (Newton 2002). Investigators should be aware that the number of young that leave the nest does not always correlate with the number of young that survive to disperse from the nesting territory (Marzluff and McFadzen 1996).

Nest failures. Evidence found at the nest may be helpful in determining the proximate cause of a nest failure. Such signs might include intact, cold eggs, broken eggs, shell fragments, dead nestlings, nestling body parts, or hairs and feathers from likely nest predators. Unhatched eggs can be used for analyses of fertility or contaminant levels. Although a cause of failure often can be assigned in this way, it is important to remember that it may only be the proximate, and not the ultimate, cause. Thus, a female may be short of food, so desert her clutch, which might then be eaten by a predator, leaving shell fragments behind. In this case, the ultimate cause of failure was food shortage, but the proximate cause may be recorded as desertion or predation, depending on whether the observer happened to visit the nest before or after the predator. Nevertheless, assessing proximate causes of nest failure often has proved useful in defining conservation problems, including pesticide-induced shell thinning and egg-breakage (Ratcliffe 1980).

Repeat and double layings. In raptor species that have relatively short breeding cycles and long nesting seasons, pairs that fail early in the breeding cycle (during laying or early incubation) sometimes recycle, and lay another clutch. This usually occurs in a different nest within the same territory. The observer should be aware of this possibility, and check for repeat layings in likely circumstances. Repeat laying does not normally occur in pairs that fail at the nestling stage, presumably because by that stage in the season, pairs would not have time to raise the resulting young before the season ended. However, in at least 15 temperate zone species (Curtis et al. 2005), including Harris's Hawks (Bednarz 1995), American Kestrels (*F. sparverius*) (Steenhof and Peterson 1997), Barn Owls (*Tyto alba*) (Marti 1992), and Long-eared Owls (*A. otus*) (Marks and Perkins 1999), pairs sometimes produce more than one brood in a year. Snail Kites do not necessarily remain paired for successive nestings, but one partner remains to raise the young, while the other moves on, sometimes to re-pair and nest elsewhere (Beissinger and Snyder 1987). Each of these situations requires special attention and interpretation.

FIELD TECHNIQUES

Surveys for raptors may be conducted on foot or from ground vehicles, fixed-wing aircraft, helicopters, or boats (see Chapter 5). The value and accuracy of each of these techniques for locating breeding raptors and their nests depends on the species being surveyed, the nesting substrate, observer experience, the topography and vegetation of the survey area, and the objective of the study. A combination of survey techniques may be most appropriate for specific situations.

Once found, nests on cliffs or trees can be checked from the ground in one of three ways: (1) remote observation, using telescopes or binoculars, (2) close inspection, accessing the nest using ropes or ladders, or (3) inspecting the nest from a short distance, perhaps using a mirror on a telescopic pole (Parker 1972). Mirrors mounted on 15-m poles proved useful in examining the contents of woodland raptor nests (Millsap 1981). Shorter mirror poles (up to 5 m) were used effectively to assess reproductive success of Ospreys nesting on navigational posts (Wiemeyer 1977). Binoculars or telescopes are ideal for cliff situations, but are not as useful where topography or dense vegetation prevents looking down into the nest from above. Observations from a distance may be adequate to confirm the presence of an incubating bird or of young, but they may be less useful in counting young, especially if the full contents of a nest are not visible.

Counts of nestlings from a distance can be particularly difficult if adults stay on the nest to brood or shade young. Climbing to nests is the best way to reduce error in counting young, but it also can be time-consuming and hazardous (see Chapter 10). Climbing requires special training, and the act of climbing to nests sometimes affects the birds adversely (Ellis 1973, Kochert et al. 2002, Chapter 19). Aerial surveys to assess reproduction are most appropriate for large raptors that build large nests in exposed locations. Aerial surveys of productivity have been effective for Ospreys (Carrier and Melquist 1976), Bald Eagles (Postupalsky 1974, Fraser et al. 1983), and Golden Eagles (Boeker 1970, Hickman 1972). In certain situations, helicopter surveys of Osprey reproductive success and productivity can be more cost-effective than ground surveys (Carrier and Melquist 1976), and fixed-wing aerial surveys of Osprey breeding pairs and numbers of fledged young can be as accurate as ground counts (Poole 1981). Both fixed-wing and helicopter surveys of nesting Golden Eagles may be more efficient and cost-effective than ground assessments (Boeker 1970, Hickman 1972, Kochert 1986).

It is easier to age and count young accurately from a slow-flying aircraft than from a fast, fixed-wing airplane (Hickman 1972, Carrier and Melquist 1976). For surveying Golden Eagle productivity, for example,

slow-flying aircraft, such as the Piper Super-Cub, which can travel at speeds of 70 to 120 kmph, are more economical than faster aircraft such as the Cessna 180 series (which travels 110 to 180 kmph) (Hickman 1972). Watson (1993) recommended quieter turbine-engine helicopters to minimize disturbance to Bald Eagles. Even with helicopters, investigators may not always be able to obtain complete brood counts, and ground-based surveys may be necessary to supplement aerial surveys. Most small fixed-wing or rotor-winged aircraft are acceptable for locating nesting pairs early in the season, but slow-flying Super-Cubs or helicopters are preferable during surveys conducted to count young. The accuracy of data can be increased if flights are scheduled for times when low winds improve maneuverability (Carrier and Melquist 1976). To minimize disturbance to Bald Eagles and to maximize safety and data reliability, Watson (1993) recommended conducting helicopter flights on calm, dry days, spending <10 seconds at each nest, staying at least 60 m from the nest, and using binoculars when necessary.

Artificial Nest Sites

Many raptor species breed in areas where a shortage of nesting sites limits nesting density. Provision of artificial sites (boxes or platforms, depending on species) can increase density, and also allow data on nesting success and productivity to be collected in an efficient manner. This is because the locations of all artificial sites are known, and they can be placed in accessible situations, so that nest contents can be easily inspected at every visit. Artificial nest sites, therefore, provide an extremely efficient means of data collection (for a study of more than 100 pairs of Common Kestrels (*F. tinnunculus*) nesting in boxes, see Cavé 1968). However, nesting success in artificial sites may not be the same as that in natural sites, which may be less secure or less sheltered, or vice versa.

Timing of Data Collection

Visits to raptor nests can yield useful information at any stage of the nesting cycle, but for adequate information on numbers and productivity, at least two visits are needed, one at the start of the nesting cycle (ideally around the time of egg-laying) and a second in the late nestling period (ideally just before young fledge). Because not all pairs start nesting at the same time, and, therefore, are out of phase with one another, the ideal time for a survey is a compromise. When surveys of nesting raptors are conducted from aircraft, all pairs can be checked in a short period, but with ground-based surveys, nest checking may have to occur throughout much of the **breeding season**. The objective of the first series of checks is to count the number of pairs associated with nesting territories and (if conducted after laying) the number of pairs with eggs. Some researchers have made these checks after the last clutch has been laid, but before the first brood hatches (Fraser et al. 1983) and before many failures have occurred. In deciduous woodlands, initial surveys made before leaf-out allow nests to be seen more easily (Fuller and Mosher 1981).

The goal of the second set of observations is to count the number of successful pairs and the number of well-grown young. Timing is again a compromise — in this case, between the date that the last brood reaches the minimum acceptable age for success and the date that the earliest brood leaves the nest. In checks that involve close-range observation, care is needed so that frightened young do not leave the nest prematurely. Checks from aircraft or distant vantage points should be scheduled just prior to fledging so that young are large enough to be counted accurately.

Information on the nesting chronology of local raptor populations must be considered when scheduling all nest checks. Some species show wide variations in laying dates within populations, particularly in regions with warmer climates and extended breeding seasons. When there is considerable variation in nesting chronology, more than two surveys may be necessary (Postupalsky 1974). Similarly, when several species are being inventoried, more than two surveys may be needed to accommodate their separate chronologies. When nesting chronology is unknown or highly variable within a species, an intermediate survey after the young hatch, but before they leave the nest, may be necessary to age nestlings and determine when to schedule the final survey.

ANALYTICAL TECHNIQUES TO AVOID BIASED ESTIMATES

In many studies, estimates of nesting success and productivity are based on a sample of pairs rather than the entire nesting population in a defined area. **Sampling error** is the error that occurs when the pairs observed are not representative of the entire population. Obtaining a sample large enough to yield an unbiased estimate

of the parameters of interest is the researcher's greatest challenge. Because nests of most raptor species are relatively inaccessible and widely spaced, there has been a tendency to base productivity estimates on all pairs detected regardless of when or how they were found. The problem with this approach is that the probability of finding a pair is often related, directly or indirectly, to its position or reproductive status. For example, nests low in trees or near roads and openings may be easier to find (Titus and Mosher 1981), but their productivity may be affected by factors related to nest height (e.g., accessibility to predators) or proximity to roads (e.g., availability of road-killed prey).

A more serious problem, common to all studies designed to assess avian reproduction, is that non-laying or early-failing pairs are less likely to be detected than successful pairs (Newton 1979, p. 129). Non-layers spend less time near their nest sites than laying pairs, and unsuccessful pairs spend less time near their nests as the breeding season progresses (Fraser 1978). Nonnesters and unsuccessful pairs have larger home ranges (Marzluff et al. 1997), and unsuccessful pairs may even leave the area altogether soon after failure, especially in migratory populations. Nests with young are usually easier to locate because of audible vocalizations from the young and defending adults, or because of conspicuous "whitewash" or fecal matter around the nest. Because surveys that begin late in the nesting season tend to miss pairs that fail early, they may overestimate nesting success and productivity. Similarly, surveys that simply pool data from nests found at any stage throughout the nesting season also overestimate nest success (Mayfield 1961, 1975; Miller and Johnson 1978). In these situations, the ratio of the number of successful pairs to the total number of all pairs found is clearly of limited value and is equivalent to **apparent nest success** (Jehle et al. 2004).

One approach to minimize bias is to restrict analysis to pairs found prior to the nesting season, or if enough background data are available, to a set of pairs randomly selected prior to the nesting season (Steenhof and Kochert 1982). This approach requires that the success of all selected pairs be determined, but it is not necessary to distinguish non-laying pairs from unsuccessful laying pairs. It is practical only in situations where there is enough historical information on a species that tends to re-use traditional nesting territories (e.g., Golden Eagles). It is inappropriate for many other species of raptors and for most short-term investigations that lack previous information on territories. Some investigators have tried to minimize bias by estimating nesting success only from laying pairs found early in the nesting season (Steenhof and Kochert 1982). However, this approach may greatly reduce sample size. When it is not possible to find all pairs before laying, researchers should consider using nest survival models to estimate the success of laying pairs.

Mayfield (1961) developed an approach to estimate nest success that incorporates data from nests found at various (and sometimes unknown) stages of the nesting cycle. By calculating **daily nest survival** during the time that a nest is under observation and by assuming a constant daily survival rate for all nests, Mayfield's model estimates the probability that all nests will survive over an entire **nesting period**. Several raptor studies have incorporated the Mayfield approach into their assessments of nesting success (e.g., Percival 1992, Bennetts and Kitchens 1997, Barber et al. 1998, Griffin et al. 1998, Lehman et al. 1998). Recently, more sophisticated models of nest survival have been developed that do not require Mayfield's assumption of constant daily survival throughout the nesting period (Dinsmore et al. 2002, Rotella et al. 2004, Shaffer 2004). Unlike Mayfield's original model, the newer models can include many categorical and continuous covariates that allow researchers to evaluate the importance of a variety of spatial and temporal factors that might affect nest survival. The new methods also allow competing models to be assessed via likelihood-based information-theoretic methods (Akaike 1973, Burnham and Anderson 2002). Nest survival models can be implemented in Program MARK (White and Burnham 1999) and in SAS (Rotella et al. 2004).

Nest-survival models allow data to be used from nests found at various times during the nesting season so long as the status of the nest was determined on at least two separate dates within the nesting period. If possible, nest checks should collectively span all stages of the nesting cycle. To use nest survival models, investigators need at least the following information: (1) the date the nest was found and its status on that date, (2) the last date the nest was checked and its status on that date, and, (3) the date the nest was last known to be viable if it had failed by the last check. Investigators also need to know the duration of the "nesting period" for their study species, which can be defined as the time from the laying of the first egg until the first young reaches the **minimum acceptable age for assessing success**. To calculate an appropriate nesting period for a given species, researchers should consider the length of

the laying and incubation periods in addition to the average age at first fledging. Information on each of these parameters is available in Newton (1979; Table 18), Cramp et al. (1980), and Poole (2004). The newer nest survival models have been used mainly for waterfowl, shorebirds, and passerines that nest on or near the ground (Dinsmore et al. 2002, Jehle et al. 2004, Rotella et al. 2004, Shaffer 2004). Raptor studies involving tree and cliff-nesting species differ from studies of ground-nesting birds in that many nests are observed remotely. Nest contents are not always inspected and there often is no way to estimate the age of nests that fail during incubation. In addition, many raptors have a longer nesting season, and many offspring continue to stay at or near the nest after they have made their first flight. Typically, investigators check raptor nests less often (sometimes only 2–3 times each season), and intervals between nest checks are usually longer than in studies of passerines, shorebirds, and waterfowl. For these reasons, adapting the new nest-survival models to raptors can be challenging, and nest survival models that require investigators to know the age of the nest when it is first found (e.g., Dinsmore et al. 2002) may not be useful for raptors. Moreover, studies with long intervals between nest checks may be limited in their ability to evaluate the effects of time-specific variables, including weather. Finally, nest survival models should only be used to estimate nesting success of laying pairs, because it is difficult to define when the nesting period begins for non-laying pairs.

Nest survival models currently available do not estimate survival of individual eggs or young. Therefore, estimates of productivity must be calculated differently. To estimate productivity, the estimate of nesting success must be combined with average brood size at fledging. To estimate productivity per territorial pair, this result must be combined with an independent estimate of the percentage of pairs laying eggs (Steenhof and Kochert 1982). Variances of productivity estimates obtained as products can be calculated using formulas available in Goodman (1960).

ACKNOWLEDGMENTS

This chapter is a contribution from the Snake River Field Station, Forest and Rangeland Ecosystem Science Center, Boise, Idaho and the Centre for Ecology and Hydrology, Huntingdon, United Kingdom. Discussions with Jon Bart, Mike Kochert, and Matthias Leu were helpful in developing ideas presented in this chapter. We thank them and David E. Andersen, Robert E. Bennetts, Robert N. Rosenfield, William E. Stout, and S. Postupalsky for reviewing and commenting on an earlier draft of the manuscript.

LITERATURE CITED

AKAIKE, H. 1973. Information theory and an extension of the maximum likelihood principle. Pages 267–281 in B. Theory, N. Petrov, and F. Csaki, [EDS.], Second International Symposium on Information. Akademiai Kiado, Budapest, Hungary.

BARBER, J.D., E.P. WIGGERS AND R.B. RENKEN. 1998. Nest-site characterization and reproductive success of Mississippi Kites in the Mississippi River floodplains. *J. Wildl. Manage.* 62:1373–1378.

BEDNARZ, J.C. 1995. Harris' Hawk (*Parabuteo unicinctus*). No. 146 in A. Poole and F. Gill [EDS.], The Birds of North America. The Birds of North America, Inc., Philadelphia, PA U.S.A.

BEISSINGER, S.R. AND N.F.R. SNYDER. 1987. Mate desertion in the Snail Kite. *Anim. Behav.* 35:477–487.

BENNETTS, R.E. AND W.M. KITCHENS. 1997. The demography and movements of Snail Kites in Florida. USGS/BRD Tech. Rep. No. 56, Florida Cooperative Fish and Wildlife Research Unit, University of Florida, Gainesville, FL U.S.A.

———, K. GOLDEN, V.J. DREITZ AND W.M. KITCHENS. 1998. The proportion of Snail Kites attempting to breed and the number of breeding attempts per year in Florida. *Fla. Field Nat.* 26:77–83.

BLAKESLEY, J.A., B.R. NOON AND D.W.H. SHAW. 2001. Demography of the California Spotted Owl in northeastern California. *Condor* 103:667–677.

BOAL, C.W. 1994. A photographic and behavioral guide to aging nestling Northern Goshawks. *Stud. Avian Biol.* 16:32–40.

———, D.E. ANDERSEN AND P.L. KENNEDY. 2005. Productivity and mortality of Northern Goshawks in Minnesota. *J. Raptor Res.* 39:222–228.

BOEKER, E.L. 1970. Use of aircraft to determine Golden Eagle, *Aquila chrysaetos*, nesting activity. *Southwest. Nat.* 15:136–137.

BORTOLOTTI, G.R. 1984. Criteria for determining age and sex of nestling Bald Eagles. *J. Field Ornithol.* 55:467–481.

BROWN, L. 1974. Data required for effective study of raptor populations. Pages 9–20 in F.N. Hamerstrom, Jr., B.E. Harrell, and R.R. Olendorff [EDS.], Management of raptors. Raptor Research Report No. 2, Raptor Research Foundation, Inc., Vermillion, SD U.S.A.

BURNHAM, K.P. AND D.R. ANDERSON. 2002. Model selection and multimodel inference: an information-theoretic approach, 2nd Ed. Springer-Verlag, New York, NY U.S.A.

CARRIER, W.D. AND W.E. MELQUIST. 1976. The use of a rotor-winged aircraft in conducting nesting surveys of Ospreys in northern Idaho. *J. Raptor Res.* 10:77–83.

CAVÉ, A.J. 1968. The breeding of the Kestrel, *Falco tinnunculus* L., in the reclaimed area Oostelijk Flevoland. *Neth. J. Zool.* 18:313–407.

CRAMP, S., K.E.L. SIMMONS, R. GILLMOR, P.A.D. HOLLOM, R. HUDSON, E.M. NICHOLSON, M.A. OGILVIE, P.J.S. OLNEY, C.S. ROSE-

LAAR, K.H. VOOUS, D.I.M. WALLACE AND J. WATTEL. 1980. Handbook of the birds of Europe, the Middle East and North Africa: the birds of the Western Palearctic. Oxford University Press, New York, NY U.S.A.

CURTIS, O., G. MALAN, A. JENKINS AND N. MYBURGH. 2005. Multiple-brooding in birds of prey: South African Black Sparrowhawks (*Accipiter melanoleucus*) extend the boundaries. *Ibis* 147:11–16.

DINSMORE, S.J., G.C. WHITE AND F.L. KNOPF. 2002. Advanced techniques for modeling avian nest survival. *Ecology* 83:3476–3488.

ELLIS, D.H. 1973. Behavior of the Golden Eagle: an ontogenic study. Ph.D. dissertation, University of Montana, Missoula, MT U.S.A.

FORSMAN, E.D., E.C. MESLOW AND H.M. WIGHT. 1984. Distribution and biology of the Spotted Owl in Oregon. *Wildl. Monogr.* 87:1–64.

FRASER, J.D. 1978. Bald Eagle reproductive surveys: accuracy, precision, and timing. M.S. thesis, University of Minnesota, St. Paul, MN U.S.A.

———, L.D. FRENZEL, J.E. MATHISEN, F. MARTIN AND M.E. SHOUGH. 1983. Scheduling Bald Eagle reproduction surveys. *Wildl. Soc. Bull.* 11:13–16.

———, F. MARTIN, L.D. FRENZEL AND J.E. MATHISEN. 1984. Accounting for measurement errors in Bald Eagle reproduction surveys. *J. Wildl. Manage.* 48:595–598.

FULLER, M.R. AND J.A. MOSHER. 1981. Methods of detecting and counting raptors: a review. *Stud. Avian Biol.* 6:235–246.

GOODMAN, L.A. 1960. On the exact variance of products. *J. Am. Stat. Assoc.* 55:708–713.

GOSSETT, D.N. AND P.D. MAKELA. 2005. Photographic guide for aging nestling Swainson's Hawks. BLM Idaho Tech. Bull 2005-01. Burley, ID U.S.A.

GRIER, J.W. 1982. Ban of DDT and subsequent recovery of reproduction in Bald Eagles. *Science* 218:1232–1235.

GRIFFIN, C.R., P.W.C. PATON AND T.S. BASKETT. 1998. Breeding ecology and behavior of the Hawaiian Hawk. *Condor* 100:654–662.

GRIGGS, G.R. AND K. STEENHOF. 1993. Photographic guide for aging nestling American Kestrels. USDI Bureau of Land Management. Raptor Research Technical Assistance Center, Boise, ID U.S.A.

HAMERSTROM, F. 1969. A harrier population study. Pages 367–383 in J.J. Hickey [ED.], Peregrine Falcon populations: their biology and decline. University of Wisconsin Press, Madison, WI U.S.A.

HENNY, C.J. AND H.M. WIGHT. 1972. Population ecology and environmental pollution: Red-tailed and Cooper's hawks. Population ecology of migratory birds: a symposium. *USDI Wildl. Res. Rep.* 2:229–250.

HICKMAN, G.L. 1972. Aerial determination of Golden Eagle nesting status. *J. Wildl. Manage.* 36:1289–1292.

HOECHLIN, D.R. 1976. Development of golden eaglets in southern California. *West. Birds* 7:137–152.

JEHLE, G., A.A. YACKEL ADAMS, J.A. SAVIDGE AND S.K. SKAGEN. 2004. Nest survival estimation: a review of alternatives to the Mayfield estimator. *Condor* 106:472–484.

KENWARD, R.E, V. MARCSTRÖM AND M. KARLBOM. 1999. Demographic estimates from radio-tagging: models of age-specific survival and breeding in the Goshawk. *J. Anim. Ecol.* 68:1020–1033.

KOCHERT, M.N. 1986. Raptors. Pages 313–349 in A.L. Cooperrider, R.J. Boyd, and H.R. Stuart [EDS.], Inventory and monitoring of wildlife habitat. Chapter 16. USDI Bureau of Land Management Service Center, Denver, CO U.S.A.

———, K. STEENHOF, C.L. MCINTYRE AND E.H. CRAIG. 2002. Golden Eagle (Aquila chrysaetos). No. 684 in A. Poole, and F. Gill [EDS.], The Birds of North America. The Birds of North America, Inc., Philadelphia, PA U.S.A.

LEHMAN, R.N., L.B. CARPENTER, K. STEENHOF AND M.N. KOCHERT. 1998. Assessing relative abundance and reproductive success of shrubsteppe raptors. J. Field Ornithol. 69:244–256.

LUTTICH, S.N., L.B. KEITH AND J.D. STEPHENSON. 1971. Population dynamics of the Red-tailed Hawk (Buteo jamaicensis) at Rochester, Alberta. Auk 88:75–87.

MARKS, J. S. AND A.E.H. PERKINS. 1999. Double brooding in the Long-eared Owl. Wilson Bull. 111:273–276.

MARTI, C.D. 1992. Barn Owl (Tyto alba). No. 1 in A. Poole, and F. Gill [EDS.], The Birds of North America. The Birds of North America, Inc., Philadelphia, PA U.S.A.

MARZLUFF, J.M. AND M. MCFADZEN. 1996. Do standardized brood counts accurately measure productivity? Wilson Bull. 108:151–153.

———, B.A. KIMSEY, L.S. SCHUECK, M.E. MCFADZEN, M.S. VEKASY AND J.C. BEDNARZ. 1997. The influence of habitat, prey abundance, sex, and breeding success on the ranging behavior of Prairie Falcons. Condor 99:567–584.

MAYFIELD, H.F. 1961. Nesting success calculated from exposure. Wilson Bull. 73:255–261.

———. 1975. Suggestions for calculating nest success. Wilson Bull. 87:456–466.

MEARNS, R. AND I. NEWTON. 1984. Turnover and dispersal in a peregrine Falco peregrinus population. Ibis 126:347–355.

MILLER, H.W. AND D.H. JOHNSON. 1978. Interpreting the results of nesting studies. J. Wildl. Manage. 42:471–476.

MILLSAP, B.A. 1981. Distributional status of falconiformes in west central Arizona - with notes on ecology, reproductive success and management. Technical Note 355. USDI Bureau of Land Management, Phoenix District Office, Phoenix, AZ U.S.A.

———, T. BREEN, E. MCCONNELL, T. STEFFER, L. PHILLIPS, N. DOUGLASS AND S. TAYLOR. 2004. Comparative fecundity and survival of Bald Eagles fledged from suburban and rural natal areas in Florida. J. Wild. Manage. 68:1018–1031.

MORITSCH, M.Q. 1983a. Photographic guide for aging nestling Prairie Falcons. USDI Bureau of Land Management, Boise, ID U.S.A.

———. 1983b. Photographic guide for aging nestling Red-tailed Hawks. USDI Bureau of Land Management, Boise, ID U.S.A.

———. 1985. Photographic guide for aging nestling Ferruginous Hawks. USDI Bureau of Land Management, Boise, ID U.S.A.

NEWTON, I. 1979. Population ecology of raptors. Buteo Books, Vermillion, SD U.S.A.

———. 1991. Habitat variation and population regulation in Sparrowhawks. Ibis 133 suppl. 1:76–88.

———. 2002. Population limitation in Holarctic owls. Pages 3–29 in I. Newton, R. Kavanagh, J. Olsen, and I. Taylor [EDS.], Ecology and conservation of owls. CSIRO Publishing, Collingwood, Victoria, Australia.

——— AND M. MARQUISS. 1982. Fidelity to breeding area and mate in Sparrowhawks Accipiter nisus. J. Anim. Ecol. 51:327–341.

——— AND P. ROTHERY. 2001. Estimation and limitation of numbers of floaters in a Eurasian Sparrowhawk population. Ibis 143:442–449.

PARKER, J.W. 1972. A mirror and pole device for examining high nests. Bird-Banding 43:216–218.

PERCIVAL, S. 1992. Methods of studying the long-term dynamics of owl populations in Britain. Pages 39–48 in C.A. Galbraith, I.R. Taylor, and S. Percival [EDS.], The ecology and conservation of European owls; proceedings of a symposium held at Edinburgh University. UK Nature Conservation; No. 5. Joint Nature Conservation Committee, Peterborough, United Kingdom.

PETERSEN, L.R. AND D.R. THOMPSON. 1977. Aging nestling raptors by 4th-primary measurements. J. Wildl. Manage. 41:587–590.

POOLE, A. 1981. The effects of human disturbance on Osprey reproductive success. Colonial Waterbirds 4:20–27.

——— [ED.]. 2004. The Birds of North America Online. Retrieved January 15, 2005 from The Birds of North American Online database: http://bna.birds.cornell.edu/ Cornell Laboratory of Ornithology, Ithaca, NY U.S.A.

POSTUPALSKY, S. 1974. Raptor reproductive success: some problems with methods, criteria, and terminology. Pages. 21–31 in F.N. Hamerstrom Jr., B.E. Harrell, and R.R. Olendorff [EDS.], Raptor Research Report No. 2. Management of raptors. Raptor Research Foundation, Inc., Vermillion, SD U.S.A.

———. 1983. Techniques and terminology for surveys of nesting Bald Eagles. Appendix D in J.W. Grier et al. [EDS.], Northern states Bald Eagle recovery plan. USDI, Fish and Wildlife Service, Twin Cities, MN U.S.A.

PRIEST, J.E. 1997. Age identification of nestling Burrowing Owls. Pages 127–127 in J.L. Lincer, and K. Steenhof [EDS.], The Burrowing Owl, its biology and management including the proceedings of the first international burrowing owl symposium. Raptor Research Report No. 9, Raptor Research Foundation, Allen Press, Lawrence, KS U.S.A.

RATCLIFFE, D.A. 1980. The Peregrine Falcon. Buteo Books, Vermillion, SD U.S.A.

RICH, T. 1984. Monitoring Burrowing Owl populations: implications of burrow re-use. Wildl. Soc. Bull. 12:178–180.

ROTELLA, J.J., S.J. DINSMORE AND T.L. SHAFFER. 2004. Modeling nest-survival data: a comparison of recently developed methods that can be implemented in MARK and SAS. Anim. Biodiversity Conserv. 27:1–19.

SEAMANS, M.E., R.J. GUTIÉRREZ, C.A. MOEN AND M.Z. PEERY. 2001. Spotted Owl demography in the central Sierra Nevada. J. Wildl. Manage. 65:425–431.

SERGIO, F. AND I. NEWTON. 2003. Occupancy as a measure of territory quality. J. Anim. Ecol. 72:857–865.

SHAFFER, T.L. 2004. A unified approach to analyzing nest success. Auk 121:526–540.

SONERUD, G. 1997. Hawk Owls in Fennoscandia: population fluctuations, effects of modern forestry, and recommendations on improving foraging habitats. J. Raptor Res. 31:167–174.

SOUTHERN, H.N. 1970. The natural control of a population of Tawny Owls (Strix aluco). J. Zool. Lond. 162:197–285.

STEENHOF, K. 1987. Assessing raptor reproductive success and productivity. Pages 157–170 in B.A. Giron Pendleton, B.A. Millsap, K.W. Cline, and D.M. Bird [EDS.], Raptor management techniques manual. National Wildlife Federation, Washington, DC U.S.A.

——— AND M.N. KOCHERT. 1982. An evaluation of methods used to estimate raptor nesting success. J. Wildl. Manage. 46:885–893.

———— AND B. PETERSON. 1997. Double brooding by American Kestrels in Idaho. *J. Raptor Res.* 31:274–276.

————, M. N. KOCHERT, L.B. CARPENTER AND R.N. LEHMAN. 1999. Long-term Prairie Falcon population changes in relation to prey abundance, weather, land uses, and habitat conditions. *Condor* 101:28–41.

————, M.N. KOCHERT AND T.L. MCDONALD. 1997. Interactive effects of prey and weather on Golden Eagle reproduction. *J. Anim. Ecol.* 66:350–362.

TITUS, K. AND J.A. MOSHER. 1981. Nest-site habitat selected by woodland hawks in the central Appalachians. *Auk* 98:270–281.

U.S. DEPARTMENT OF THE INTERIOR. 1979. Snake River Birds of Prey Special Research Report to the Secretary of the Interior. USDI Bureau of Land Management, Boise District, Boise, ID U.S.A.

U.S. FISH AND WILDLIFE SERVICE. 2003. Monitoring plan for the American Peregrine Falcon, a species recovered under the Endangered Species Act. U.S. Fish and Wildlife Service, Divisions of Endangered Species and Migratory Birds and State Programs, Pacific Region, Portland, OR U.S.A.

VILLAGE, A. 1987. Numbers, territory size and turnover of Short-eared Owls *Asio flammeus* in relation to vole abundance. *Ornis Scand.* 18:198–204.

WATSON, A. 1957. The breeding success of Golden Eagles in the northeast highlands. *Scott. Nat.* 69:153–169.

WATSON, J.W. 1993. Responses of nesting Bald Eagles to helicopter surveys. *Wildl. Soc. Bull.* 21:171–178.

WELLER, M.W. 1956. A simple field candler for wildfowl eggs. *J. Wildl. Manage.* 20:111–113.

WHITE, C.M. AND T.L. THUROW. 1985. Reproduction of Ferruginous Hawks exposed to controlled disturbance. *Condor* 87:14–22.

WHITE, G.C. AND K.P. BURNHAM. 1999. Program MARK - estimation from populations of marked animals. *Bird Study* 46 Supplement:120–138.

WIEMEYER, S.N. 1977. Reproductive success of Potomac River Ospreys, 1971. Pages 115–119 in J.C. Ogden [ED.], Transactions of the North American Osprey Research Conference. Transactions and Proceedings Series No. 2. USDI National Park Service, Washington DC U.S.A.

Appendix 1. Glossary of terms frequently used in assessing nesting success.*

Active. An ambiguous term, originally defined by Postupalsky (1974) to describe nests where pairs laid eggs, but used subsequently in many different ways by other authors. The term is now best avoided (S. Postupalsky, pers. comm.), unless clearly defined.

Apparent Nest Success. The ratio of number of successful pairs to the total number of known pairs in a population.

Breeding Season. The period from the start of nest building (refurbishment) or courtship to independence of young.

Brood Size at Fledging. The number of young produced by successful pairs.

Clutch Size. The number of eggs laid in a nest.

Daily Nest Survival. The probability that at least one young or egg in a nest will survive a single day.

Fecundity. The number of female young produced per female. Equivalent to number of young produced per pair, assuming a 1:1 sex ratio among offspring.

Fledging. A fully-feathered young voluntarily leaving the nest for the first time.

Floaters. Birds in either subadult or adult plumage that are not associated with specific nesting territories and do not reproduce. Floaters may be physiologically capable of breeding, but are prevented from doing so by lack of a territory or nesting site. They are usually unpaired.

Incubation Period. The time between the start of incubation and the hatching of an egg, during which the egg is kept at or near body temperature by the parent.

Irregular Territory. Known nesting location occupied only in certain years out of many.

Measurement Error. Misclassification of the status of a particular pair or nesting territory or an inaccurate count of the number of eggs or young.

Minimum Acceptable Age for Assessing Success. A standard nestling age at which a nest can be considered successful. An age when young are well grown but not old enough to fly and at a stage when nests can be entered safely and after which mortality is minimal until actual fledging: 80% of the age that young of a species normally leave the nest of their own volition for many species, but lower (65–75%) for species in which age at fledging varies considerably or for species that are more likely to leave the nest prematurely when checked. Often the same as age at banding.

Nest. The structure made or the place used by birds for laying their eggs and sheltering their young.

Nesting Period. The time from laying of the first egg to the time when at least one young reaches the minimum acceptable age for evaluating success in a given species. This interval can be used to calculate nesting success from estimates of daily survival rates. It can be calculated as the sum of the minimum acceptable age for assessing success, the mean incubation period, and the mean time between laying of the first egg and the onset of incubation.

Nestling Period. The time between hatching of the first egg and the time the first young leaves the nest of its own accord.

Nesting Success. The proportion of pairs that raise at least one young to the minimum acceptable age for assessing success (see above) in a given season, even if it takes >1 attempt. Usually reported per territorial pair or per laying pair.

Nesting Territory. An area that contains, or historically contained, one or more nests (or scrapes) within the home range of a mated pair: a confined locality where nests are found, usually in successive years, and where no more than one pair is known to have bred at one time.

Nest Survival. The probability that a nesting attempt survives from initiation (laying of the first egg) to completion and has at least one offspring that reaches the minimum acceptable age for assessing success.

Nonbreeders. A collective term to describe both floaters and territorial pairs that do not produce eggs.

Post-fledging Period. The time between when young leave the nest (i.e., fledge) and their becoming independent of parental care. Sometimes measured from the time young are banded or are old enough for nests to be considered successful.

Pre-incubation Period. The time between laying of the first egg and onset of incubation.

Productivity. The number of young that reach the minimum acceptable age for assessing success; usually reported as the number of young produced per territorial pair or per occupied territory in a particular year.

Regular Territory. Known nesting territory, in use every, or almost every, year.

Sampling Error. Error that occurs when the pairs observed are not representative of the entire population.

Scrape. A site where falcons, owls, and New World vultures (species that do not construct nests) lay eggs; the depression in substrate (rotting wood chips, old pellets, dust, sand, or gravel) where eggs are deposited.

Successful (nest or pair). One in which at least one young reaches minimum acceptable age for assessing success.

---

* Although definitions in this Glossary are widely accepted among raptor researchers, not everyone uses particular terms in exactly the same way. Therefore, care is needed in making comparisons among studies. It is important to avoid using a familiar term in a different context, and it is equally important to define your terms carefully in your methods section. Doing so will make it easier for others to assess your findings, and to compare them with those of other researchers.

Capture Techniques 12

PETER H. BLOOM
Western Foundation of Vertebrate Zoology,
439 Calle San Pablo, Camarillo, CA 93012 U.S.A.

WILLIAM S. CLARK
2301 S. Whitehouse Crescent, Harlingen, TX 78550 U.S.A.

JEFF W. KIDD
Western Foundation of Vertebrate Zoology,
439 Calle San Pablo, Camarillo, CA 93012 U.S.A.

INTRODUCTION

Many raptor studies, including those involving migration, dispersal, home range use, anatomy, and toxicant ingestion, require that birds be captured for examination, marking, or both. This chapter describes a variety of field-tested techniques for capturing birds of prey. Since the first edition of the Raptor Management Techniques Manual in 1987 (Giron Pendleton et al. 1987) almost no new, radically different raptor trap designs have been invented, but several have undergone design improvements and, importantly, several papers dealing with capture success and raptor trapping outside North America have been published.

As Joseph J. Hickey remarked in the foreword to *Birding with a Purpose* (Hamerstrom 1984:vii), raptor trappers are generally different from the rest of the populace: *"You've heard of wolf trappers, fox trappers, muskrat trappers, and the like. Raptor trappers are different. Officially, they want to band birds to learn about their weight and moult, their later movements, their longevity, and all that. Underneath, they are unabashed admirers of the wildness, magnificent strength, and awesome flight of creatures at the top of the animal pyramid. I wouldn't call them childlike; but they do have a youthful zest, and they will endure any hardship and go to any length to catch their birds."*

While passion and enthusiasm for raptor trapping are contributing factors, trapping success depends upon a number of other factors too. Some species such as Common Kestrel (*Falco tinnunculus*), Black-shouldered Kite (*Elanus caeruleus*), chanting-goshawks (*Melierax* spp.), and Red-shouldered Hawk (*Buteo lineatus*) are easy to capture, whereas others, including many eagles, kites, and condors, require more sophisticated techniques. Some raptors are more easily trapped on migration, whereas others are more easily trapped when nesting. The age of the bird also can be a factor. In general, juvenile raptors are easier to trap than adults, and hungry raptors always are more responsive to traps. Consistently successful trapping comes with experience. Successful trappers not only can trap birds on migration, but also can trap specific individuals during the nesting season, and almost always without injuring them in any way or causing nest failures.

Sometimes easy-to-catch raptors can be difficult to capture. WC noted that winged ants, emerging in droves after rains in Israel, were readily eaten by Steppe Buzzards (*B. buteo vulpinus*), which then completely ignored traps baited with mice and sparrows that had worked well at other times. On the other hand, hard-to-catch raptors often can be trapped when they are especially hungry.

Trapping success often shifts seasonally. For example, Swainson's Hawks (*B. swainsoni*) are relatively easy to trap when breeding in North America, but are difficult to trap during migration, when they are nomadic. On the other hand, Sharp-shinned Hawks (*Accipiter striatus*) are readily captured along well-established migration corridors while migrating but are more difficult to catch during the breeding season.

Knowing what trap to use, when to use it, and what kind of lure to place in it requires familiarity with the target species' ecology and behavior. For example, Merlins (*F. columbarius*) are easily captured with House Sparrows (*Passer domesticus*) as lures in a small dho-gaza trap, but Red-tailed Hawks (*B. jamaicensis*) are more often caught using the same lure in a bal-chatri. This is because Merlins are adapted to aerial pursuit of avian prey, whereas the Red-tailed Hawk is adapted to capturing mammalian prey on the ground. Fuller and Christenson (1976) and Hertog (1987) evaluated several different trap types and quantified the effectiveness of different trapping techniques.

Trapping success may be limited if raptors become "trap-shy." Trap-shyness may be a serious problem if one needs to recapture specific individuals to replace transmitters or to study seasonal weight change or molt, etc. We believe that raptors become trap-shy as a result of negative or unrewarding experiences. Things that affect trap shyness, including trap shape, location, lure used, and proximity to people, should be considered prior to any attempt to recapture a trap-shy bird.

In some situations where capturing many individuals of several species is required, a combination of several strategically placed mist nets, bal-chatris and verbails, etc., with motion detector transmitters, can be extremely productive. Such set-ups typically require two to five people.

Although being caught in a trap can stress a raptor, trapped birds rarely are injured physically. The most frequent causes of trap-related injury or death are predation and weather. Raptors may succumb to temperature extremes or predation if allowed to struggle in traps for long periods. Used properly, the traps discussed in this chapter should rarely result in severe injury or death.

While being handled, raptors may struggle, bite, grab, or vocalize depending upon the species and "personalities" of the birds involved. Buteos and many owls, for example, struggle very little while being handled, and rarely vocalize, whereas accipiters, particularly Northern Goshawks (*A. gentilis*), struggle and vocalize most of the time, and readily attack. Most raptors are capable of inflicting painful wounds with their beaks and talons, and large hawks and owls, eagles, and condors, can cause serious injury. In most species, particularly eagles, the talons are most dangerous. In general most species do not bite. Exceptions include vultures, California Condors (*Gymnogyps californianus*), fish eagles (*Haliaeetus* spp.), White-tailed Hawks (*B. albicaudatus*), and falcons.

Researchers capturing and handling raptors must have the proper permits, which vary among countries. In the U.S., these permits may include U.S. Geological Service bird-marking and salvage permits and endangered-species permits, as well as state permits. See Chapter 25 for details.

CAPTURE TECHNIQUES AND THEIR APPLICATION

The first bible for avian trapping techniques was a small booklet titled *Manual for Bird Banders* by Lincoln and Baldwin (1929). Of about 35 traps and capture techniques described therein, only one, the Number 1 leghold trap, was used for the capture of birds of prey. Stewart et al. (1945) provided one of the earliest comparisons of different raptor traps. Beebe (1964) focused on raptors trapped for use in falconry. McClure (1984) and Bub (1995) have written comprehensive overviews of most avian trapping techniques known worldwide. Today, raptor biologists have the option of using 20 different basic trap designs and numerous variations, as well as trap monitors, which allow many traps to be monitored from a distance.

Some countries prohibit the use of live lures in scientific research, and some researchers avoid live lures for personal reasons. Raptor researchers should ensure that the use of live lures is legal in the country they are trapping and should treat lures humanely and not subject them to undue harm and stress. In the U.S. many, if not all, universities have Animal Care and Use Committees that must approve research using live lures and in Canada at least, approval from an ethics sub-committee is now required (D. Bird, pers. obs.). *Guidelines to the Use of Wild Birds in Research*, a special publication of The Ornithological Council (Gaunt and Oring 1999), includes recommendations on wild-bird care in research.

Trap Monitors

The use of motion-detector transmitters and scanning receivers in raptor trapping has greatly advanced raptor trapping success and the ease with which traps are monitored. Any trap with moving parts, as well as traps that are moved when raptors strike them, can be monitored with these devices. This includes line trapping where as many as 20 bal-chatris, verbails, or Swedish goshawk traps, etc. are deployed over large areas (Bloom 1987). Trap monitors are available at Communications Specialists, Inc. (www.com-spec.com).

Prior to the use of trap monitors, each and every trap had to be under continuous visual surveillance or checked at least hourly. This often meant that some birds were caught and escaped in the interim between checks, and that traps sprung by birds that were not caught were rendered non-functional for the remainder of that hour. Trap monitors also can reduce mortality as the sprung trap is visited immediately after having being triggered.

Trap monitors work when a magnet attached to the transmitter is moved when the trap is sprung, initiating a signal that is detected by the receiver. Depending upon the terrain, traps can be monitored continually from a distance of 2–3 km or more. Two of us (PHB, JFK) that have used trap monitors since 1988 have found them useful in oak woodlands and grasslands of California when trapping resident Red-shouldered Hawks, Cooper's Hawks (*A. cooperii*), Barn Owls (*Tyto alba*); in agricultural areas in Argentina when trapping wintering flocks of Swainson's Hawks; and in tropical forests in India when trapping Crested Serpent Eagles (*Spilornis cheela*) and Crested Goshawks (*A. trivirgatus*). In India, despite not having seen Crested Goshawks for almost a month, we caught six birds in two days after line trapping with trap monitors.

Audio Lures

The use of tape-recorded playbacks of vocalizations can be used to bring raptors closer to field workers for identification and surveying. Playbacks also can be used to attract owls to mist nets. This approach is now used to capture migrating Northern Saw-whet Owls (*Aegolius acadicus*) and Flammulated Owls (*Otus flammeolus*) at most migration banding stations where owls are trapped (e.g., Erdman and Brinker 1997, Evans 1997, Whalen and Watts 1999, Delong 2003).

Playbacks also have been used with excellent success during the breeding season to trap Barn Owls,

Western Screech Owls (*Megascops kennicottii*), Long-eared Owls (*Asio otus*), Spotted Owls (*Strix occidentalis*), Great Horned Owls (*Bubo virginianus*), Red-shouldered Hawks, Northern Goshawks, Cooper's Hawks (PHB, JWK) and other species by drawing the birds into mist nets or other traps in the playback area. It is important to limit playback use during the breeding season as adult behavior and nest success may be influenced by excessive use of this technique.

The Bal-chatri

Because this trap is one of the most successful devices used to date to catch raptors, considerable attention will be paid to it. The bal-chatri, which roughly translates to "noosed umbrella" (Clark 1992), is a wire cage with monofilament nooses tied to the top, sides, or both, with a lure animal placed inside (Fig. 1; Berger and Mueller 1959). The size and shape of the trap depend upon the species being trapped. Typical lures include the house mouse (*Mus musculus*), black rat (*Rattus rattus*), gerbil (*Gerbillus* spp.), House Sparrow, Common Starling (*Sturnus vulgaris*), and Common Pigeon (*Columba livia*). In countries or remote areas where standard such lures are not available, Red Junglefowl (domestic chickens, *Gallus gallus*) or ducks (*Anas* spp.) can be used. Several designs are in widespread use, including (1) quonset (Berger and Hamerstrom 1962, Ward and Martin 1968, Mersereau 1975), (2) cone (Kirsher 1958, Mersereau 1975), (3) octagonal (Erickson and Hoppe 1979), and (4) box with apron (Clark 1967). One modification involves the use of a Plexiglass top for trapping screech owls (Smith and Walsh 1981).

As with many traps involving live lures, trappers should consider the placement of the trap carefully, as traps may be stepped on by farm animals, run over by cars, or moved by people. In some instances ants can kill the lure.

Construction. Mesh and cage size are determined by the species to be captured, and size of the bait animal. The cage must be large enough for the bait animals to move (run or fly) within it, so that the raptor can detect them. The more space the bait animal has to move about, the more likely the raptor is to notice it quickly.

For American Kestrels (*F. sparverius*) we recommend 0.6-cm hardware cloth rather than 1.3-cm cloth, as small- to moderately-sized house mice can escape through or become caught in the large mesh. For most other species up to the size of eagles 1.3-cm hardware

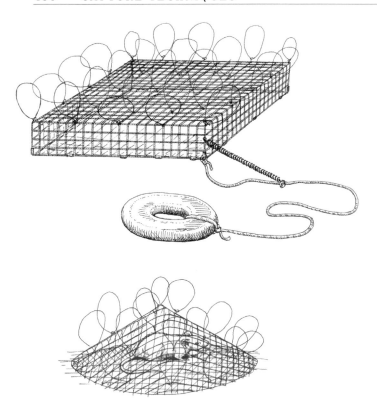

Figure 1. Bal-chatris can be made in a variety of shapes. The box-shaped bal-chatri works well on accipiters, buteos, and owls, whereas the cone-shaped trap works best on kestrels and Burrowing Owls (*Athene cunicularia*).

cloth is best. When using quonset-shaped bal-chatris on large accipiters, buteos, large owls, and Harris's Hawks (*Parabuteo unicinctus*), either 1.3-cm aviary cloth or 2.5-cm chicken wire can be used, especially if the intent is to use a relatively large lure animal (e.g., Common Pigeon). Larger mesh makes the lure animal more visible, particularly at great distances. In addition, the aviary cloth and chicken wire mesh are more flexible and more easily manipulated than hardware cloth, and are easier to work with. On the other hand, aviary- and chicken-wire traps are more easily dented or crushed than are hardware-cloth traps, and nooses cannot be attached as firmly to aviary cloth and chicken mesh as to hardware cloth. Plastic (modeling) cement applied to the area of attachment can alleviate much of this problem.

A single section of hardware cloth or several sections in which the sides, bottom, and top are fastened together can be used to construct the cage. The different flaps or sections of the cage are most efficiently fastened with ring clips (Wiseman 1979) or with wire. The door for the lure animal should be on the bottom of the trap where it will not interfere with nooses.

Camouflaging traps is important. Traps should be spray-painted before the nooses are attached. Good background colors are light green (live vegetation), tan (dead vegetation) and white (snow). Flat colors are preferable to glossy.

A 15- to 25-cm apron of hardware cloth with nooses extending out from the cage increases the potential of capturing individuals that are shy of standing on the cage. This is particularly important for the capture of Burrowing Owls (*Athene cunicularia*).

Some house mice used as lures will pull the nooses inside the bal-chatri, and will gnaw on them or, sometimes, hang themselves. A mouse with a taste for monofilament can ruin hours of work and should be replaced. Placing short lengths of monofilament in the cage sometimes reduces chewing on nooses.

Several knot-tying techniques have been described. Jenkins (1979) found that a running slipknot, which remains closed on the bird's foot or toe once it has tightened, increased trapping success. Unfortunately, the use of such "one-time" knots as part of the noose (the knot also can be used to anchor the noose to the cage) means that all nooses that are closed accidentally must be replaced or retied. We use the traditional technique of an overhand knot for the noose (Collister 1967) and a square knot for the anchor point. This employs simpler knots that can be tied quickly. Unfortunately it may allow more birds to escape than use of the running slipknot. *The North American Bird Banding Manual* illustrates another variation in which an overhand knot is used for the noose but a clinch knot is used at the anchor point (Environment Canada and U.S. Fish and Wildlife Service 1977).

We agree with Jenkins (1979) that the best way to attach nooses to the wire mesh is diagonally across the junction of two wires. Vertical nooses are critical to trapping success. Most nooses can be made to stand vertically by several firm upward tugs on a thick pen tightened in the noose. The knot at the junction of the two wires should be rotated until the noose is in the most vertical position. Plastic cement also may be used to help maintain the erect posture of the noose. When using cement, be certain that it does not weaken the monofilament. See Figure 1 in Berger and Mueller (1959) for tips on attaching nooses so that they stand erect.

The height and spacing of nooses also is important. For small raptors, we make nooses 4-cm tall, spaced at 3 cm. For medium to large raptors we make nooses 5–6.5-cm tall, spaced at 5 cm. Adjacent rows of nooses are staggered. When the trap is complete, we examine it for spaces and add more nooses as needed. Nooses become brittle with age, particularly when traps are rolled in dust or mud or are exposed to long hours in the sun. Depending upon the amount of use that a trap sustains, we replace all nooses one to five times annually.

Bal-chatris are relatively light traps that must be weighted to prevent birds from taking off with them. For traps under constant surveillance we recommend either a 0.7- or 1-kg barbell weight attached to the trap via a 1-m length of nylon cord. Many researchers attach the weight directly to the floor or sides of the bal-chatri, rather than by a nylon cord with a "shock-absorbing" spring. A problem with the first approach is that there is no opportunity to place a spring between the trap and weight and, as a result, both the weight and noose attachments break more readily. Also, we believe that weights attached to the trap make the trap more conspicuous than a weight on a 1-m line, and that this reduces the likelihood of birds striking the trap. Lighter weights can be dragged short distances by the bird, but are less stressful on the nooses and on the bird's toes than are heavier, stationary weights. The only drawback with having the weight separately rather than directly attached to the trap is that the vehicle must be going slower for the drop. When traps are checked hourly, we recommend either tying the trap to a stationary object or using a heavy weight so that the trap is less likely to be dragged off. A spring or other "shock absorber" must be used in situations where the trap is tied to a stationary object. The shock absorber has two functions. First, it reduces the stress on the nooses, which may snap if tugged sufficiently, and second, it reduces the risk to raptors, which may hemorrhage and die if allowed to struggle against a stationary object for more than 10 minutes (pers. obs.). Barn Owls in particular are prone to the latter. The spring should be securely attached on a nylon cord between the trap and the weight to prevent the bird from escaping with a bal-chatri.

Application. The bal-chatri is an extremely effective, versatile, and portable trap. It can be used during all seasons, and has a success rate of up to 85% for most species that are attracted to it. Most North American raptor species have been captured on bal-chatris. In Guatemala, Thorstrom (1996) captured 12 species, including hawks, hawk-eagles, falcons, and owls, with this technique, some of them in trees. In India, Kenya, South Africa, and Israel, PHB, WSC, or both, have used bal-chatris to catch a wide variety of species of kites, accipiters, buzzards, harriers, small to large eagles, fal-

cons and owls. One of the more difficult species to catch with bal-chatris is the Black Kite (*Milvus migrans*).

There are two principal applications. The first is road trapping, in which traps are placed on the roadside from a vehicle in the immediate vicinity of perched raptors. The second is line trapping, in which 10–15 bal-chatris are placed out before the target bird(s) arrive in known use areas. Line trapping is particularly effective on owls and woodland raptors. In both procedures, traps usually are placed to capture perched raptors. Flying accipiters are the occasional exceptions. Road trapping is best attempted from roads with minimal vehicle use, as cars and trucks tend to frighten the birds. That said, it is indeed possible to catch birds along busy highways. Trapping involves driving on country roads while scanning for raptors perched near roads, often on power poles or utility lines. Once a raptor is identified, a weighted bal-chatri is placed on the shoulder of the road. Time spent dropping the trap from the vehicle should be minimal; on the other hand, merely throwing the bal-chatri from a moving vehicle is construed by some as cruel to the lure prey inside. To improve trapping success and avoid injury to the bird and the lure prey: (1) the vehicle should not be stopped when dropping the trap, as this frequently frightens the bird, (2) the door of the vehicle should be closed quietly, (3) the weight and trap should be placed as far from the edge of the road as possible to ensure that the captured hawk or owl does not drag the trap onto the road, and (4) the trap should be placed from the side of the vehicle away from the perched bird, and on roadside opposite to it such that the bird cannot see the trap being set. If it is necessary to stop, do not step out of the vehicle. In all trapping it is best to disassociate yourself from the trap as much as possible. In most instances this can be achieved by not letting the bird see you with the trap. Many birds appear initially suspicious of the "gift-wrapped" food and will not come down to the trap, and will hesitate for a long time and, eventually, leave the area, even when these procedures are followed. Sometimes, birds repeatedly fly down to traps but do not touch it or do not become entangled, and eventually lose interest. Placing a second trap in a new site frequently results in renewed interest and success.

Line trapping differs from road trapping in that 10–15 bal-chatris are placed (1) in a specific territory in an effort to catch a targeted bird or pair of birds or (2) in appropriate habitat across several square kilometers to trap as many birds as possible. As with road trapping, traps should not be placed directly in sight of a bird, but instead near known or suspected hunting and resting perches (e.g., perches near nests where the male or female may spend time while not on the nest). The entire trap line is checked hourly or is monitored via trap monitors. As traps are checked and birds removed, they are reset (i.e., closed nooses are reopened), and captured birds are retained until the entire series of traps is checked.

Line trapping has three advantages over road trapping, including (1) high trapping rates, (2) more appropriate habitat selection for trap placement, and (3) more effective trapping of troublesome or trap-shy individuals through placement of several camouflaged traps. Three disadvantages are that (1) predators occasionally kill birds caught on the trap, (2) birds have more time to free themselves compared with road trapping, and (3) trapping is not as selective in that individuals of non-target species may be caught. The first two disadvantages above can be reduced by using trap monitors that signal the trapper immediately when the trap has been moved or sprung.

When line trapping, there is time to conceal the bal-chatris and tack them to the ground to prevent flipping. Trappers should push surrounding leaves or grass around the trap and place leaves or grass between the nooses on the top, making sure that nothing becomes entangled in the nooses. If you use enough grass or leaves to make the traps difficult to re-locate, the hawk or owl will find them for you. If a bird is trap-shy of bal-chatris, use two to three traps at each perch.

Proper lure animals are important. Pigeons, starlings, and House Sparrows are best for attracting accipiters. Occasionally a house mouse or gerbil can be used. Most falcons do not respond to lures in bal-chatris. One exception is the American Kestrel, which is easily caught on bal-chatris baited with either house mice or House Sparrows. Prairie Falcons (*F. mexicanus*) and Aplomado Falcons (*F. femoralis*) are easily caught on bal-chatris baited with house mice, gerbils, and House Sparrows or combinations thereof. Peregrine Falcons (*F. peregrinus*) and Gyrfalcons (*F. rusticolus*) (B. Anderson, pers. comm.) are rarely if ever caught on bal-chatris.

In our opinion, the best bait animals for capturing buteos are paired combinations of domestic house mice and gerbils, house mice and House Sparrows, or wild and domestic mice. B. Millsap (pers. comm.) has had excellent success trapping Cooper's Hawks using a gerbil-House Sparrow combination. Common Pigeons can be used to lure Cooper's Hawks, Northern Goshawks, Harris's Hawks, and large buteos. Most wild rodents, including *Microtus*, *Peromyscus*, and *Neotoma* spp., are poor attractants because they tend to remain motionless. Northern Harriers (*Circus cyaneus*) and White-tailed Kites (*E. leucurus*) have been caught on bal-chatris with house mice, gerbils and House Sparrows. Medium to large owls are most easily caught on house mice and gerbils. Great Horned Owls can be lured with Common Pigeons. Although Strigidae, which seem to hunt much more by sight than by sound to Tytonidae, readily attack animals in bal-chatris with or without dry leaves in the trap, the latter rarely strike a bal-chatri unless they can hear the sound of the lure animal rustling the leaves.

Although pigeons, starlings, House Sparrows, gerbils, and house mice are the "standard" lures, other species can be used in emergencies (e.g., chipmunks, squirrels, and rabbits). Most native vertebrates are protected by various state and federal laws however, and must not be used without appropriate permits.

The Bartos Trap

A relatively new kind of raptor trap, which blends the concepts of the bow net and the box trap, is the Bartos trap (Bartos et al. 1989). Although this trap has not been used widely to date, it has been used to capture Collared Sparrowhawks (*A. cirrhocephalus*) and Moreporks (*Ninox novaeseelandiae*) in Australia. The trap, which shows considerable potential for capturing small to medium-sized forest raptors lured with small birds, can be suspended at almost any height, in a building, or in a tree near a nest, and does not require attendance, is collapsible for easy transport.

Bow Nets

Several variations of this trap have been used to trap many species, including owls, eagles, falcons, harriers, buteos, and accipiters. The trap consists of two semicircular bows of light metal with gill netting strung loosely between them (Fig. 2). Hinges and springs connect the two semicircles at their bases, the lower one of which is fixed to the ground. When setting the trap, the upper bow is pulled over the lower stationary bow and latched into position. A lure animal, usually a bird, is placed in the center of radius of the trap. When a raptor grabs and holds the lure bird, the trap is triggered, either by a person in a blind pulling a trigger line (Meredith 1943, Mattox and Graham 1968, Clark 1970, Field

Figure 2. Bow nets may be triggered automatically by the raptor itself, or manually from a blind. The above design is triggered manually and uses garage-door springs to power the bow.

1970), by remote control (Meng 1963, Jackman et al. 1994), or by the action of the attacking raptor itself (Tordoff 1954). Bow nets are sometimes superior to bal-chatris because a tethered lure mouse walking about with only a single trigger wire above it appears less intimidating to a raptor than an animal in a cage.

Application. One of the better automatic bow nets is that of Tordoff (1954). The trap, originally designed for small raptors, can be modified for larger species as well. It is easily concealed and has been used successfully on American Kestrels, screech owls, Great Horned Owls, and Long-eared Owls. Kenward and Marcstrom (1983) describe an automatic bow net for use on Northern Goshawks that uses the partially eaten carcass of a raptor kill or a stuffed bird as bait. This trap is usually quite useful unless disturbed by corvids or rendered immobile by ice.

Clark (1970, 1976) describes a typical manually operated bow net that has seen considerable use at a banding station in Cape May, New Jersey, U.S.A., where three bow nets are managed simultaneously. One attribute of the manually operated bow net is its selectivity. In North America, manual bow nets have been used to capture a variety of eagles, accipiters, buteos, and falcons, as well as Burrowing Owls.

According to Clark (1970), the trapper is stationed

in a blind. A pigeon is harnessed in a leather jacket with two lines attached, one of which comes to the blind after going through a bow trap and the other returns to the blind after passing through two guides located at the top and bottom of the lure pole. The two lines are joined at the blind, which allows the trapper to "fly" the pigeon when a hawk is seen in the air, simply by pulling on the first lure line. The fluttering pigeon thus appears to be "injured" and easy prey to the hawk.

If the hawk decides to attack or "stoop" on the pigeon, it is brought back to the center of the bow trap by pulling on the second lure line. Should the hawk continue his stoop and bind to the pigeon, it is captured by triggering the bow trap from the blind.

Another type of bow net useful for capturing large raptors, including eagles (Clark 1970, Field 1970), is a radio-controlled device powered by garage-door springs (Jackman et al. 1994). This type of spring is easily obtained, extremely powerful, and can be adjusted to control speed. Strong springs that will move the bow rapidly are particularly important when trapping eagles because the bow is necessarily large, is concealed by earth and vegetation, and, as a result, is relatively heavy. Proper training in the use of bow nets is mandatory since a bow set off prematurely can strike the bird and cause serious injury or death. Nevertheless, 16 of 19 Bald Eagles (*Haliaeetus leucocephalus*) and 26 of 30 Golden Eagles (*Aquila chrysaetos*) were successfully captured without injury using remote-controlled bownets (Jackman et al. 1994) (Fig. 2). A more recent improved collapsible version of this design has been used to capture hundreds of Bald Eagles and Golden Eagles in California and Arizona (Jackman et al., pers. comm.).

Finally, Q-nets, which are large bow nets that are powered by large bungee cords, have been used to capture carrion-feeding raptors including Southern Crested Caracaras (*Caracara plancus*) (Morrison and McGhee 1996).

Box Traps

Box traps are compartment traps that contain a lower bait cage and an upper section that captures and holds the raptor (Fig. 3). The Swedish goshawk trap originally described by Meredith (1953) has since been improved and made more portable (Meng 1971). The trap's effectiveness in trapping Northern Goshawks is described and compared with similar compartment traps by Karblom (1981) and Kenward and Marcstrom (1983). The Chardoneret trap, which uses a live owl to

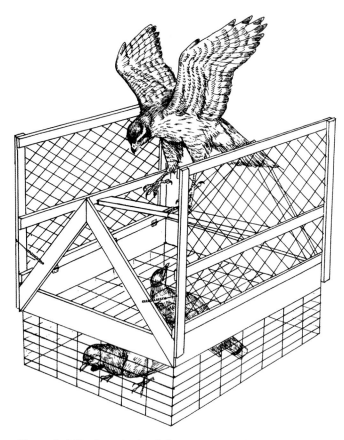

Figure 3. A Northern Goshawk (*Accipiter gentilis*) lands on the trigger stick of a Swedish goshawk trap containing Common Pigeons (*Columba livia*) as bait.

lure raptors during the breeding season (Redpath and Wyllie 1994), is another successful box trap, and Kenward and Marcstrom (1983) provide detailed descriptions of similar compartment traps that may be more useful than the Swedish goshawk in trapping in certain situations.

Application. Meng (1971) provides an excellent description of the materials used in the construction of the Swedish goshawk trap. The Swedish goshawk trap has been used to capture most large hawks (and Great Horned Owls) in North America, as well as Great Black Hawks (*Buteogallus urubitinga*) and Ornate Hawk-Eagles (*Spizaetus ornatus*) in Guatemala (Thorstrom 1996). This trap is particularly useful for falcons because they tend to walk around the trap and do not enter the compartment from above (Meredith 1943). Lure animals typically include Common Pigeons or Common Starlings. Two or more lures should be placed in each trap to increase their movement and visibility, and the trap should be placed in a highly visible location. Although trapping with this method is relatively slow, most birds that enter the trap are caught and do not

escape. The best attribute of the trap is that it only needs to be checked every three hours or so and can be monitored using trap monitors.

Trapping during the nesting season involves the placement of two or three traps between 50 and 200 m from the nest. Trapping during migration entails the placement of 5 to 10 traps spaced 0.5 to 1.0 km apart along pole or fence lines in valleys where birds perch to hunt or, less desirably, on ridges where many birds are moving.

One of us (WC) used from 5 to 10 box traps to capture migrating raptors on return migration in Eilat, Israel. In one season, 45 of 653 captured raptors were caught in them, including 41 Steppe Buzzards, 2 Eurasian Sparrowhawks (*A. nisus*), a Levant Sparrowhawk (*A. brevipes*), and a Black Kite. The next season, using fewer traps, 10 of 445 raptors were caught, including 7 Steppe Buzzards, 2 Levant Sparrowhawks, and a Eurasian Sparrowhawk. The traps were baited with domestic house mice, House Sparrows, or both.

Cannon and Rocket Nets

Cannon and rocket nets are used to capture vultures, eagles, and condors. They are similar, relatively expensive traps, but are very effective and can be used to capture many individuals at a single firing. The technique consists of three to four cannons or rockets that propel a large net over the birds (Mundy and Choate 1973). Animal carcasses are used as bait.

Application. Because they involve explosives, cannon and rocket nets are more dangerous than other traps. The cannon net is less dangerous than the rocket net, and we limit our discussion to it, although much of what follows also applies to rocket nets. Because of the difficulties involved in building this trap, we recommend purchasing it from a manufacturer (e.g., Wildlife Materials Inc., www.wild lifematerials.com). One of the safest and simplest designs is described in Mundy and Choate (1973) (J. Ogden and N. Snyder, pers. comm.). Other designs are in Dill and Thornsberry (1950), Grieb and Sheldon (1956), Marquardt (1960a,b), Thompson and DeLong (1967), and Arnold and Coon (1972). Nets and mesh vary in size depending upon the target species. A 15.2 × 15.2-m net of 10.2 × 20.3-cm mesh is good for trapping eagles. Smaller mesh causes the net to remain airborne too long, allowing birds to escape. Because 10.2 × 20.3-cm mesh is not standard, it must be created by cutting the 10.2 × 10.2-cm mesh to create larger holes. Permits, which must be acquired before detonators (blasting caps) can be purchased, should be applied for one year in advance of the proposed trapping effort.

Cannon nets have been used to capture many species including waterfowl, seabirds, shorebirds, passerines, cranes, and grouse (Dill and Thornsberry 1950, Thompson and DeLong 1967, Arnold and Coon 1972). More recently, raptors, including Black Vultures (*Coragyps atratus*) and Turkey Vultures (*Cathartes aura*), Andean Condors (*Vultur gryphus*), California Condors, Bald Eagles, White-bellied Sea Eagles (*H. leucogaster*) (Hertog 1987) and Golden Eagles have been captured using these traps. The cannon net is one of the best traps available for gregarious species where several individuals accumulate simultaneously. Cannon nets also allow the trapper to capture birds selectively. Non-target species pose little problem when trapping with this method because non-target species feeding at them often attract target species.

To achieve maximum success with eagles, vultures, and condors, lure carcasses should be staked down to render them immoveable. Ideally, the site should be baited at least one week prior to capture attempts, and should be baited continually until the project is finished. Where large nocturnal scavengers such as bears or wild dogs are present, the carcass may need to be removed at the end of each day or be replenished more frequently. Observations should be made from a nearby blind or from about 0.8 km away. Typical bait animals include virtually any medium to large carcass. Fresh rather than rotten carcasses seem to be more attractive to raptors. Stillborn calves are frequently available from dairies.

Between 1982 and 1987, 10 California Condors were captured with cannon nets using the carcasses of stillborn calves as bait (PHB, unpubl. data). Most condors were recaptured using the cannon net at both the same and different locations. During the same period, 43 Golden Eagles were captured, four of which were recaptured. The trap was 100% effective in that none of the targeted condors or eagles escaped capture. On several occasions two to four Golden Eagles were captured with one firing, and Golden Eagles and condors were captured together in the same firing. No injuries or mortalities occurred to either species. In Israel, trappers caught 35 Black Kites with one firing of a cannon net. In Israel, an air-powered cannon net captured European Honey Buzzards (*Pernis apivorus*) when they came to drink at a small pond (WC).

Although selective and efficient, cannon nets are labor-intensive. Initial installation and site preparation requires about four hours. Each firing or preparation for firing takes about one hour. Four people are needed to stretch and fold the net after each firing. Test-firings are needed to determine whether cannons are wired and angled correctly, and to be certain that the net deploys properly.

Selecting a good trap site is important as considerable effort goes into its preparation. A well-camouflaged blind of suitable size should be placed in cover 30–60 m from the trap. Clumps of grass, branches, or both should be placed around each cannon, and the net should be covered lightly with grass.

Cannon and rocket nets set fires easily if dry fuel is available. If wildfires are a strong possibility or the habitat consists of dry vegetation, ridge tops rather than valleys should be used. Regardless of where the trap is set, cut all dry grass from within 5 m of the cannons, and strip this area to near bare earth. Grass within the net landing area, but further than 5 m from the cannons, should be cut to 2 cm. Green grass need not be cut.

Once the trap is ready and desirable raptors are coming to the trap, the blind should be entered about an hour before sunrise. Silence in the blind is important, as eagles, vultures, and condors are highly suspicious and frequently watch for hours before finally settling in to feed. The slightest noise or movement can alert them to the presence of the trappers.

The detonator button on the firing box can be pushed when desired raptors are in position, preferably with their heads down feeding on the carcass. Be certain that no birds are standing where the four projectiles will land, and that no birds are airborne. Once the net has landed and birds secured, they can be taken from under the net and processed. Birds under the net must be separated from each other to avoid having them bite and claw each other. When many raptors are captured with a single firing, each bird should be placed in a restraining device or covered with a light blanket while still under the net.

Cast Lures and Hand Nets

This technique has been used effectively on Great Grey Owls (*S. nebulosa*) and Spotted Owls, but should be successful on other approachable *Strix* spp. The cast lure and hand net consists of a stuffed lure or live rodent attached to a nylon line that is pulled across the terrain, or simply a live rodent placed at the foot of the trapper. When the owl comes into the lure, a fish-landing net is quickly flipped over the bird.

Application. Equipment includes a fish landing net (0.6 × 0.8 m, with a 1.5-m handle), a casting rod, reel and 10-lb test line (Nero 1980). The lure animal is a stuffed or live mouse.

The cast lure and hand net technique described by Nero (1980) to capture Great Gray Owls has since been used to catch Spotted Owls. Essentially, a stuffed lure mouse attached to monofilament fishing line is cast toward the owl and reeled in. The owl is captured with the fish landing net when the lure is close and the owl flies to it or pounces on it. The technique works particularly well with snow on the ground.

The Dho-gaza

The dho-gaza is a mist or gill net suspended between two poles. The nearly invisible net falls with the poles (Harting 1898), detaches from the poles (Meredith 1943, Mavrogordato 1960, Hamerstrom 1963), or slides down the poles when struck by the raptor (Clark 1981). Depending upon the season, a Great Horned Owl or small bird or rodent is used as bait (see below). There are two applications: one involves a large net, used primarily during the nesting season to trap territorial adults (Fig. 4). The other involves a small net and a small lure bird, used primarily during raptor migration or on wintering grounds (Fig. 5).

Elevated small dho-gazas described by Rosenfield and Bielefeldt (1993) and Jacobs and Proudfoot (2002) have been used to increase the recapture rates of "trap shy" individuals previously captured on other types of traps. At close range, taxidermy raptors can elicit strong responses from territorial accipiters.

Application. We use aluminum 1.9–3.7-m or 2.6–4.7-m extension poles. Two 1-m sections of 1-cm reinforcement bar are cut for each net. The two sections of reinforcement bar are pounded into the ground and function as anchor-mounts for each extension pole. The net poles should be painted (mottled green and brown) to blend with the surroundings. A rectangular, 2.1 × 5.5 m, 10.2-cm mesh net is used for large raptors, a 0.8 × 1.2–1.5-m, 10.2-cm mesh net for medium-sized raptors, and a 0.8 × 1.2–1.5-m, 6-cm mesh net for small raptors. Nets are available to licensed banders from sources advertising in *North American Bird Bander* and *Journal of Field Ornithology* (www.avinet.com; AFO Mist Net Sales, Manomet Bird Observatory, P.O. Box 936, Manomet, MA 02345, U.S.A; EBBA Mist Nets, EBBA Net Committee c/o Gale Smith, 8861 Kings Highway, Kempton, PA 19529, U.S.A.).

The net is suspended from each pole by wrapping duct tape around three spring-closure clothespins on each pole (Fig. 4); one clothespin each around top, middle, and 0.6-m from the bottom of each pole. Each side of the net has five loops that would normally slide over the poles for use as a mist net. Since only three loops are necessary for the dho-gaza, we cut off the strings creating the second and fourth loops. To the ends of the three remaining loops, we wrap a 2.5-cm tab of duct tape, rounding off the corners with a pair of scissors. The tabs can then be slipped into the corresponding clothespin on each pole. Care should be taken to insert only the tape tab in the clothespin and no portion of the net loop, because if the loop is inserted in the clothespin, the net will not detach when the raptor flies into it. For small dho-gazas (Fig. 5), we use paper clips rather than clothespins. Each clip is tied to the rod with a rubber band. Tape tabs 1.3-cm long are placed at all four corners of the net and are inserted into the paper clips. For a variation of tab attachment, see Knittle and Pavelka (1994).

One of the bottom corners of the net is attached to a 5-m, 20-lb test monofilament line with a shock absorber and approximately 100-g weight or drag at the other end. The size of the drag depends upon the size of the raptor being trapped. A drag on a long line can be relatively light compared with the size of the bird. If the ensnared raptor becomes airborne the drag pulls it and the net to the ground.

Dho-gazas made from mist nets can be mended when torn; however, nets should be discarded after five major (15-cm hole) repairs. Dho-gazas made from mist nets usually can capture as many as 15 medium-sized or large raptors before they need to be replaced. Dho-gazas made from gill nets last longer, but are more visible.

Small dho-gazas often are used together with bow nets at banding stations along migration corridors. In such instances, gill nets often are used in place of mist nets as hundreds of hawks may be caught weekly. The design most frequently used at stations consists of two or three nets with two poles each and a bow net between them (Clark 1981). The nets are on rings that slide down the poles when a raptor strikes the net. See Clark (1981) for a complete description.

During the nesting season, a large dho-gaza with a Great Horned Owl placed near a nest is probably the most effective trap to use on most small to medium-sized raptors. An array of North American raptors have been captured using large dho-gazas during the breeding season. For a review of this trap's effectiveness, see Bloom et al. (1992).

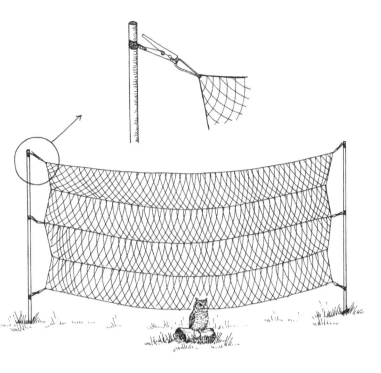

Figure 4. A large dho-gaza with a Great Horned Owl (*Bubo virginianus*) as bait may be used on territorial adult raptors. The inset shows a clothespin attachment to a tape tab on a mist net loop.

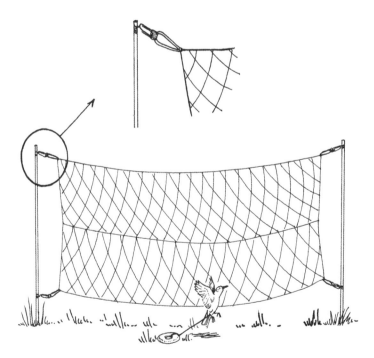

Figure 5. A small dho-gaza with a Common Starling (*Sturnus vulgaris*) as bait may be used for capturing falcons. The inset shows a paper clip attachment on the tape tab.

Great Horned Owls and Eurasian Eagle-Owls (*B. bubo*) make particularly effective lures when used near the nest of raptors because of their predatory potential. A raptor that proves impossible to capture with conventional traps using prey animals as lures may respond readily to a live Great Horned Owl at the nest site. A breeding pair of Greater Spotted Eagles (*A. clanga*) was captured in Poland with a large dho-gaza and live Eurasian Eagle-Owl (Meyburg et al. 2005). Nesting raptors also respond to mounted and mechanical owls, but less vigorously (Gard et al. 1989, Jacobs 1996). Even so, the risk of injury or death either to the lure or to the target is reduced when mechanical or mounted owls are used (McCloskey and Dewey 1999).

Conspecifics are another particularly strong stimulus for many nesting raptors (Elody and Sloan 1984), especially if they are carrying prey within another bird's territory. This relatively rare natural occurrence can be simulated by tethering a conspecific in the nesting territory of the target bird next to a net, and tying a 10-cm line from the bird's leg to a small taxidermy prey. The large dho-gaza has been modified to trap Spanish Imperial Eagles (*A. adalberti*) in Spain, and Golden Eagles in California using a conspecific as the lure (V. Matarranz, pers. comm.). In the latter case, the nets were hand-made using Spiderwire® line (www.spiderwire.com). Three-piece, telescopic, and camouflaged poles were used to elevate the nets. Eighteen of 20 attempts were successful, with the efficiency of the trap highly related to conditions at each site. We recommend using a Great Horned Owl without prey first. If a pair of raptors becomes trap-shy and must be captured again, move the dho-gaza to a new location and use a conspecific with prey. Not all raptors will respond to conspecifics; Red-tailed Hawks, for example, rarely respond.

Another effective variation of this technique, at least for Spotted Owls, is to hand-capture a branching owlet, set it at the middle of the dho-gaza and then handhold a live Great Horned Owl near the chick. The adults usually attack the Great Horned Owl immediately and fly into the net. If the remaining adult is slow to attack, imitations of Great Horned Owl calls and movement by the nestling usually attract the other member of the pair.

For woodland species such as Northern Goshawks, we recommend placing the net and the owl <50 m from the nest tree. The net should be placed so that the lure is between the nest tree or likely perch sites, and the net itself. The owl should be jessed and tied with a 0.6–1-m tether to a short, portable tree stump or log measuring 0.3 × 0.6 m and weighing about 4.5 kg. The owl is positioned halfway across the span of the net, 1.5 m from it. Whenever possible, place at least one of the supporting poles next to a tree, and stretch the net out in the shade. Place the owl about 1.8 m from the tree (slightly inside from the center of the net). This provides the owl with some structural protection from the attacking raptor and will help force the latter to go through, rather than around the net. If feasible, the second pole also should be placed next to a tree as well. In this way, the poles are somewhat hidden, and, if in the shade, the net is less visible. Try to predict the direction of the attack so the net is pulled away from the owl instead of over it. To protect the hawk, set the dho-gaza so that no large trees, logs, or rocks are within 9 m of the front or back of the net. To reduce the amount of time involved in cleaning the net after a capture, remove as many branches, pine cones, etc. from the ground around the set, or have fresh nets available as replacements. Branches on trees adjacent to the net or "landing area" also should be removed.

Using two nets, rather than one, increases the success rate by about 20%. Once the hawk begins stooping at the owl, a capture is virtually assured if the net is in a good, inconspicuous location. The primary benefit of a second net is to facilitate more rapid capture, particularly in wooded areas. A drawback is that two nets and four poles are more conspicuous and require nearly twice as much time to set and occasionally the hawk becomes entangled in both nets.

Injured Great Horned Owls or Red-tailed Hawks that cannot be rehabilitated and would normally be euthanized make ideal lure birds. For humane reasons, we caution against using blind or partially blind birds, or birds with only one leg. Birds with wing injuries that have healed improperly function well, as do imprints. Taxidermy owls with loose feathers blowing in the wind work better than plastic owls. Taking good care of one's lure birds is not only humane, but also makes good sense because they are not readily available. For example, putting a lure bird out for extended periods in the hot sun can lead to dehydration and death.

Contact between lure owls and stooping raptors is rare. Of 1,400 raptors captured using a dho-gaza with a live Great Horned Owl, Bloom et al. (1992) reported one lure-owl death from a raptor strike, one death from a bobcat (*Lynx rufous*), and two deaths from dehydration. In most cases, hawks are attempting to chase the owl from its territory. Exceptions include Northern Goshawks, Great Horned Owls and Great Gray Owls,

which have locked talons with lure owls. When trapping Northern Goshawks, we recommend using a large female Great Horned Owl. We also recommend that the dho-gaza be watched continuously from 5–100 m depending upon the situation.

Trapping during the incubation period may cause nests to fail, and most species are best caught during the nestling and early fledgling stages. One way to reduce or eliminate nest abandonment during incubation is to set the trap >100 m from the nest so that the foraging mate and not the incubating bird is caught. Capturing both members of a pair of accipiters, particularly goshawks, can be difficult and time-consuming. Females usually can be caught within 15 minutes during the nestling period. Males, which are more difficult to trap, are most vulnerable when nestlings are two weeks old.

Using a dho-gaza in open habitats presents its own set of problems, including both increased visibility and wind. In windy situations the net should be placed at right angles to the wind, as the bird will usually stoop into the wind. In some situations it is best to trap at sunrise, when winds are lower and the net is less visible. The hottest part of the day in summer or in arid areas also should be avoided.

As a rule of thumb, we leave the trap out as long as the individual to be captured is still responding (intently looking, stooping, or screaming), or up to three hours. If the bird still shows interest after three hours, we return later and select a different site for trapping. Expending more time at a single set is probably a waste of time, and overly disruptive to the birds.

Small dho-gazas are most effective during migration and on the wintering grounds, particularly on small raptors. Clark (1981) reported that between 1971 and 1979, 6,568 migrants were caught on small dho-gazas in New Jersey. Dho-gazas used at raptor-migration banding stations generally function as back-ups to bow nets (Clark 1981). Lure birds in the bow net attract the raptors, which are subsequently caught in the dho-gaza as they pass the bow net. House Sparrows, starlings, and pigeons are most commonly used as lures.

A wide variety of migrants, including harriers, accipiters, buteos, falcons and small owls have been captured with small dho-gazas in North America (Jacobs and Proudfoot 2002). In Israel, trapped migrants include smaller accipiters and falcons.

Road trapping with small dho-gazas is particularly effective on Prairie Falcons and Merlins. In the case of Prairie Falcons, the trap is placed 60–120 m from the bird. The activities of the trapper are made less obvious by parking the vehicle and stepping out the side of the vehicle opposite the bird. Placing poles in the ground, attaching the net, and dropping the lure bird or mouse should be completed within three minutes. In the winter when the ground is frozen, 5-cm anchor platforms are useful for keeping the poles upright.

Another method that allows rapid placement of one or two dho-gazas is to have the poles mounted on a 1.2 × 1.2-m plywood base with the net or nets in place between them (B. Millsap, pers. comm.). Spreading soil over the base makes the trap less conspicuous.

Mist Nets

As mentioned earlier, mist nets are large, nearly invisible nets that are secured between poles and placed in appropriate habitat to entangle birds. Although used in Europe for centuries, their use in North America was not recognized until 1925 (Grinnell 1925). MacArthur and MacArthur (1974) later expanded on their use in population studies of birds. Mist nets have been used to trap a number of raptors, principally small to medium-sized owls and migrating hawks (O'Neill and Graves 1977, Weske and Terborgh 1981). Walkimshaw (1965) and Mueller and Berger (1967) have both described work in which mist nets were used to capture migrating Northern Saw-whet Owls. Smith et al. (1983) and Reynolds and Linkhart (1984) have both used them to capture Flammulated Owls and Long-eared Owls, respectively.

Similar in function to large dho-gazas, non-detachable mist nets have been used with live Great Horned Owls lures to catch breeding American Kestrels (Steenhof et al. 1994). While mist nets have been used with success on large buteos, these birds are likely to tear them and may be injured when entangled in them. The use of mist nets in Canada and the U.S. requires government permits.

Application. In North America, relatively inexpensive mist nets are available to licensed banders from several sources (www.avinet.com; AFO Mist Net Sales, Manomet Bird Observatory, P.O. Box 936, Manomet, MA 02345, U.S.A; EBBA Mist Nets, EBBA Net Committee c/o Gale Smith, 8861 Kings Highway, Kempton, PA 19529, U.S.A.). We recommend two types of aluminum extension poles as supports. The first, which is used in painting, is 1.9–3.7 m. The second, which is used to clean swimming pools, is 2.6–4.7 m. Both are lightweight and tall. Poles are placed directly over a

0.6–1-m length of reinforcement bar that has been pounded partway into the ground. Nylon cord tied to trees or to stakes can be used to keep the poles upright. Nets should not be so taut as to prevent their horizontally aligned pockets from forming. If the net is too taut, the birds will bounce off of it.

The most common sizes for nets are 2.1 × 12.8 m and 2.1 × 18.3 m. Mesh sizes of 5.8 cm and 10.2 cm (diagonally stretched) are most useful for raptors (Bleitz 1970).

We use a line of 22 12.8-m long, 5.8-cm mesh nets to catch small owls migrating through forests. Nets are checked at 1–3-hour intervals depending upon the temperature, as cold nights tend to stress small owls. During the nesting season, nets can be placed in corridors between known nest site and hunting areas.

We use 12.8-m long, 10.2-cm mesh nets to catch medium-sized and large owls. Nets placed in marshes are particularly effective for Barn Owls, Long-eared Owls, and Short-eared Owls (*A. flammeus*). Owls can be attracted to the net with lure animals in cages or bal-chatris with dry leaves so that the owls can hear the lures moving within. Taped recordings of mouse squeaks played at low volume below the nets also attract owls.

Portions of mist nets can be used to capture raptors that roost or nest in cavities. Barn Owls are easily captured by quietly approaching the cavity entrance and placing a net over it. The nest tree or nest box is then rapped and the owl becomes entangled in the net as it attempts to leave.

In North America, mist nets have been used to catch migrating accipiters, buteos, and falcons (Clark 1970). The nets are used as back-ups to bow nets. Lure birds in the bow net attract the raptors, which subsequently are caught in the mist nets as they pass over the lure. American Kestrels and Sharp-shinned Hawks are best caught with 5.8-cm mesh nets; whereas as larger species including Cooper's Hawks and Red-tailed Hawks are best caught with 10.2-cm mesh nets.

In India we captured several Montagu's Harriers (*C. pygargus*) in a grassland nocturnal roost by placing six 3 × 18-m, 10-cm mesh mist nets at the roost during the daytime. The nets were spaced about 10 m from each other. The birds were flushed into the nets when we walked through the roost two hours after sunset. In spite of our actions the roost remained active with several hundred occupants for several more weeks before the birds left the area.

Although mist nets are not particularly effective at

capturing raptors in rainforests (e.g., six raptors captured in Guatemala during 15,360 net hours), they are one of the best ways to capture tropical species including Plumbeous Kite (*Ictinia plumbea*) and Great Black Hawk (Thorstrom 1996).

Hand Capture and Spot Lighting

Although Spotted Owls are commonly captured by hand, particularly when provided live mice, in most instances capturing raptors by hand is not a technique that one plans to use. Occasionally however, researchers may find themselves in a position where a raptor can be grabbed. This can occur when the bird is unusually agitated, asleep, distracted, in a nest or roost cavity, sick, or fully gorged. A. Harmata (pers. comm.) has hand-captured several eagles with full gorges during windless conditions.

Capturing raptors by hand in open stick nests as they incubate their eggs or brood their young should be avoided. Doing so often results in nest failure, particularly when the bird is on eggs. The risk of damage to the nest contents by the adult's talons is also high. Capturing cavity-nesting raptors on eggs can usually be accomplished safely if the birds are inside the cavity and the hole is temporarily blocked until the bird calms down. A long pole equipped with a blocking device (e.g. a bunched cloth or paddle) on one end is useful for covering nest holes of American Kestrels (D. Bird, pers. obs.).

Spot lighting, which has been used to capture many types of birds, essentially involves shining a light in the target bird's eyes at night, and either grabbing the bird by hand or with a dip net.

Helicopters and Four-wheel-drive Vehicles

The helicopter-capture technique for Golden Eagles was developed by Ellis (1975) and expanded to include use of a net-gun by O'Hara (1986) (see below). The eagle is pursued with the helicopter until it lands on the ground and crouches with its head down while being intimidated by the helicopter (Ellis 1973). The helicopter then lands 75 m from the bird, drops off the biologist, and returns to hover 10–15 m above the eagle. The eagle is then hand-grabbed from behind by the biologist. This technique, which can be extremely effective (four eagles captured in two hours in Montana) (Ellis 1975), works best and most safely in flat terrain with little if any wind. Perched eagles are easier targets than soaring birds (Ellis 1975). The technique is quite useful when eagles are concentrated in livestock areas, and it is desirable to remove them in a short-time period.

The use of a 4-wheel-drive vehicle to pursue and capture raptors can be effective in certain circumstances. This technique was used in flat to undulating terrain in Saudi Arabia to capture Steppe Eagles (*A. nipalensis*) on 48 of 52 approaches. All successful attempts were completed in <15 minutes, with a mean pursuit time of <9 minutes (Ostrowski et al. 2001).

Ground-burrow Traps

This approach involves the use of several different types of traps including small-mammal live traps to capture Burrowing Owls and, occasionally, Barn Owls. The trap is placed in the entrance to a Burrowing Owl nest burrow (Martin 1971, Ferguson and Jorgensen 1981), along with cages (Winchell 1999) or noose carpets (Bloom 1987) placed outside of burrows (Fig. 6). Ferguson and Jorgensen (1981:149) recommended 23 × 23 × 66-cm live traps with single or double doors (www.havahart.com, www.livetrap.com). Winchell (1999) provides useful illustrations of this technique.

Live traps need to be checked periodically throughout the day; noose carpets every 15–60 minutes unless they are attached to trap monitors. Both techniques can be highly efficient; Ferguson and Jorgensen (1981) reported 49 owls captured with live-traps in 150 manhours, and noose carpets, which are less cumbersome, have caught 20 owls in 10 man-hours.

Nest Traps

Several variations of nest traps exist for natural cavities including a wire hoop trap that has been used to catch Barred Forest Falcons (*Micrastur ruficollis*) and Collared Forest Falcons (*M. semitorquatus*) in natural nest cavities (Thorstrom 1996) and American Kestrels in nest boxes (Plice and Balgooyen 1999). As mentioned earlier, small sections of mist nets placed over the openings of nest cavities work well on Barn Owls nesting in trees and on bluffs.

We have used long kitchen tongs and short barbeque tongs to extend our reach into cavities to "capture" young, and, sometimes, adults. Thorstrom (1996) has used a noose pole trap (i.e., a wire rod with nooses attached to leather on the end that the raptor grabs onto), to extract raptors from cavities.

Net Guns

The net gun is a relatively expensive but effective device consisting of a hand-held, 3-barrel gun that uses explosives to propel a net. A number of barrel subsystems and net and mesh sizes are available from Coda Enterprises (www.codaenterprises.com). The net gun can project a variety of triangular or square nets, ranging from lightweight, mist nets to heavy-tensile nets. Depending upon the size and mesh, the net can be projected 15–22 m. The gun, which usually is hand-held when fired, can be fired by remote control.

Although net guns have been used to capture a number of large mammals and birds, eagles are the only raptors that have been captured using this technique (O'Hara 1986). Net guns may be particularly useful with unusually approachable species, and those that roost communally.

Noose Carpets

A noose carpet (Fig. 6) is a weighted piece of hardware cloth festooned with monofilament nooses that is strategically placed on a high-use perch or other surface (Anderson and Hamerstrom 1967, Collister 1967, Kahn and Millsap 1978). A shock-absorbing spring (e.g., a 10–15 cm metal spring or rubber surgical tubing) is placed on the line near the weight to reduce stress on the bird's toes, or on the monofilament.

Application. As for bal-chatris, after which they are modeled, materials for noose carpets include either 0.6- or 1.3-cm hardware cloth or aviary wire, 10- to 40-lb test monofilament, nylon cord, and a metal spring or elastic tubing. Nooses 2.5–5-cm tall should be used for small raptors such as American Kestrels or screech-owls; 5–7.5-cm nooses should be used for medium-sized raptors such as Barn Owls or Red-tailed Hawks; and 10-cm or taller nooses should be used for larger raptors. Ten-, 20-, and 40-lb test monofilament should be used on small, medium-sized, and large raptors, respectively. The size of the perch site to which the noose carpet is to be attached dictates the size of the carpet. Careful positioning is critical.

Noose carpets, which can be used during nesting and migration, have been used to trap vultures, kites, harriers, accipiters, buteos, eagles, falcons, and owls in North America. Thorstrom (1996) used noose carpets in Guatemala to capture several species including kites, hawks, hawk-eagles, falcons, and owls.

Like the Verbail trap (see below), noose carpets do not require bait, although a lure animal can be used.

Figure 6. Noose carpets may be applied to branches and around Burrowing Owl (*Athene cunincularia*) nests.

One simply needs to know the location of the target bird's most frequently used hunting perches. Erecting artificial perch sites in strategic locations, such as a meadow, also works. A 10 × 10-cm by 2.5-m fence post with a 0.6-m, 2.5–5-cm diameter branch or dowel attached perpendicularly on top is ideal for trapping many perch-hunting raptors.

For vultures we use four sections of 1.3-cm hardware cloth cut into 10-cm by 0.6-m lengths. The nooses are attached to the hardware cloth using square knots and are either twisted or glued into a vertical position. Each section of hardware cloth is attached to a 1-m nylon cord with a 1-kg weight and a shock absorber. The four carpets are then placed 0.5-1 m out from the carcass. All four sections of hardware cloth, along with the weight, shock absorber, and nylon cord are lightly but completely covered with soil or grass. Only the nooses remain erect and exposed.

Noose carpets are effective for Turkey Vultures, Bateleurs (*Terathopius ecaudatus*) (Watson and Watson 1985) and, presumably, other carrion feeders. Even

American Kestrels have been captured on noose-carpets baited with carrion (Wegner 1981). A noose carpet wrapped around a dead rabbit has been used to capture Roughlegs (*B. lagopus*) (Watson 1985). A. R. Harmata (pers. comm.) found that pushing meat partially through the mesh makes a carrion set more attractive to buteos and Golden Eagles. For kestrels, a rodent carcass is used as bait, and the nooses are attached to two single strands of wire that are nailed in position over the rodent. Wegner (1981) captured two adult Red-tailed Hawks using this technique. Karblom (1981:140) used leather straps with attached nooses wrapped around a fresh kill or a carcass set out like a fresh kill to capture goshawks at regular plucking sites. Noose carpets also can be used when the target bird has killed and partially eaten a harnessed pigeon without being caught. At such times the raptor is flushed from the pigeon, and the carcass is then covered with a noose carpet (Mavrogordato 1960, Webster 1976, pers. obs.).

Burrowing Owls are most easily captured using a 5–15-cm noose carpet made from 0.6-cm mesh hardware cloth. Approximately 10 4-cm nooses (10-pound test) are tied in a staggered arrangement on each strip of hardware cloth. The carpet is then attached to a weight or stationary object by a 0.6-m length of nylon cord, surgical tubing, or inner tube, and placed 3–30 cm inside the entrance of a burrow. The trap functions best during the nesting season, when young begin to venture from the burrow, but are not capable of sustained flight. When attempting to capture nestlings, the burrow should be checked at 15–30-minute intervals, unless a trap monitor is in place. The process can be repeated until all owls are captured. It is best not to release any of the nestlings until all of them have been caught, but in no case should they be held for more than three hours. Larger carpets can be placed on the apron of the burrow with a dead mouse as an attractant. We have captured more than 500 Burrowing Owls using these procedures and found that they work well on both adults and fledglings.

When using elevated noose carpets, the same safety measures apply as for the Verbail trap (see below) (i.e., the anchor line should reach the ground, and a shock absorber is needed).

Noosed Fish

Noosed fish were first successfully employed by Robards (1967) for Bald Eagles in Alaska, and subsequently modified by Cain and Hodges (1989) and Jackman et al. (1993). Bald Eagles, other fish eagles, Greater Spotted Eagles (*A. clanga*), Western Marsh Harriers (*C. aeruginosus*), and Ospreys have been trapped on noosed fish.

Application. The fish should be the size and species normally taken by the targeted raptor. The entrails are removed and replaced with a block of shaped Styrofoam, placed so the fish will float belly up (see Frenzel and Anthony 1982 and Jackman et al. 1993). Two monofilament lines approximately 1-m long enter the fish through the mouth, pass through a segment of the Styrofoam plug for friction, and exit through the anus. A slipknot is tied at the end of each line to create a noose. The nooses should be 12 cm in diameter with the knots lying at the anus of the fish. The nooses are held to the pectoral region of the fish with a breakaway attachment, one on each side of the fish, with the nooses lying flat on the surface of the water. The ends of the lines that exit the mouth are tied to a 30-lb test monofilament line on a fishing rod and reel or to a 4.5 kg anchor.

Tolerance levels of perched Bald Eagles approached by boats vary. Many will flush if approached closer than 0.4 km, whereas some will allow a boat to pass directly underneath them. Once a target bird is found, the noosed fish is dropped from a slow-moving boat, preferably on the side of the boat opposite to the bird. When the raptor attacks the baitfish and attempts to carry it, one or both nooses close around the bird's toes and the bird falls into the water. The bird is then approached by boat as the line is reeled in.

A variation of this technique is to attach the line from the floating fish to a 4.5-kg anchor, which is then lowered to the river or lake bottom. Anchors lighter than 4.5 kg should not be used in deep water since the eagle's forward momentum may pull the anchor into deeper water, causing the bird to drown. A shock absorber must be placed between the weight and the bait or the noose will snap when the eagle grabs the fish. The shock absorber can be attached to a floating log (drift set).

This trap has been used with success on both Ospreys and Bald Eagles. Frenzel and Anthony (1982) reported near 100% effectiveness on Bald Eagles striking the bait. Having nooses and lines placed correctly is critical as others have experienced lower success using slightly different versions of this trap (Harmata 1985, Jackman et al. 1993). Misses result when either the birds fail to put a toe through a noose, or escape after being temporarily snagged. A variation of this technique used on White-bellied Sea Eagles is provided by Wiersma et al. (2001).

Harnessed Pigeons

The pigeon harness (Webster 1976) is a modified noose trap. Originally designed to capture large falcons, it also is effective in capturing medium-sized to large accipiters, buteos, eagles, and owls. Nylon monofilament nooses are tied or cemented to a leather harness that is attached to a pigeon. Openings in the harness for the legs and wings allow the pigeon to walk or fly about. A 1.5- to 10-m line with a weight or drag at the opposite end is attached to the harness. Modifications include harnessed House Sparrows and starlings used to catch small falcons (Toland 1985), and harnessed ducks, pheasants, and rabbits used to catch eagles.

Application. A pattern is used to outline the harness on 0.3-cm thick leather. The diameter of leg- and wing-holes are modified to fit the individual bird being harnessed so that it can walk and fly. Webster (1976) suggested constructing and using several harnesses of different sizes to overcome this problem. A small shock-absorbing spring attached to the harness behind the legs reduces the number of raptors that escape as a result of broken nooses.

Nooses, similar to those described for the bal-chatri, are attached to the harness after first making the noose itself, and then threading the loose end of it through pinholes in the harness. Several overhand knots are tied together into one large knot that cannot slip back through the hole. The knot is glued to the leather on the inner side of the harness with the noose in the desired position. An alternative technique is to punch two holes in the leather 0.5 cm apart and tie the noose as is done on bal-chatris. Harnesses usually are festooned with 25 4-cm nooses made from 20-lb test monofilament. A 0.2 kg, movable drag weight, such as a stick on a 6-m line, works well in plowed fields. An immovable weight should be used in situations near water.

In our experience, the harnessed pigeons have a success rate of 75–85%. Harnessed pigeons have been used worldwide to capture a variety of raptors, including harriers, accipiters, buteos, falcons, eagles, and owls.

The most common application of this technique involves driving on secondary roads or beaches looking for raptors, and tossing the harnessed pigeon out of the window when a target bird is found. A long, 15-m dragline permits the pigeon to fly some distance. A short, 1.5-m line with an immovable weight can be used in areas with heavy shrub cover. Because buteos and Great Horned Owls usually capture their prey on the ground, a 1.5-meter line is best for capturing these species. Alternatively, because falcons prefer to strike their prey in the air, the trap is more attractive when the pigeon flies. That said, if cover is available and a long line is used, the lure bird will hide in cover once it is aware of the raptor. Long lines also tend to become tangled in shrubs. The more vulnerable the pigeon appears, the higher the probability of successfully trapping.

Another effective use is to place 10–15 harnessed pigeons under buteos and falcons migrating through small, upper-elevation valleys. In such situations the pigeons, which are set 0.5 to 2 km apart on short lines with 1-kg weights with shock absorbers, should be checked hourly or monitored with electronic trap monitors.

As discussed by Webster (1976), the success or failure of a trapping operation depends upon the action of the harnessed pigeon. Wild-caught feral pigeons make the best lures, as they tend to be stronger and attempt to fly more frequently than those raised in captivity and rarely flown.

Noose Poles

A noose pole can be a fishing rod (Zwickel and Bendell 1967, Catling 1972), telescoping pole (Reynolds and Linkhart 1984), or sections of rod or pipe that mount on top of each other to extend high enough to reach the targeted owl (Environment Canada and U.S. Fish and Wildlife Service 1977, Nero 1980). A large nylon monofilament or wire noose at the top of the pole is closed by pulling the line or wire that runs up the side or center of the pole.

Noose poles work best on unusually tame species of raptors including Galapagos Hawks (*B. galapagoensis*) (Faaborg et al. 1980), and several species of owls. Noose poles also have been used to retrieve nestling Barn Owls and Burrowing Owls from deep nest cavities (B. Millsap, pers. comm.).

The Phai trap

The phai, or padam (Mavrogordato 1966, Carnie 1969, Webster 1976), is another infrequently used trap with considerable potential. The phai consists of a live lure surrounded by a small ring of relatively tall nooses connected to a rope or hose (Fig. 7). Large falcons have been trapped with this technique, but other raptors also can be caught. A more portable version designed by C.H. Channing consists of a cage containing a lure bird with large nooses suspended along the top (Fig. 8). This trap is most effective for trapping accipiters.

Application. The ring version of the phai consists of about 25 25-lb test, 15–20-cm diameter monofilament nooses. The ring, which is about 1-m in diameter, is formed by a nylon or rubber hose or a piece of rope. The nooses, tied through perforations, are spaced along the hose so that each overlaps the adjacent noose by about 1.3 cm. Because the nooses are tall and flexible, small 15-cm lengths of wire or twigs are sometimes used to support them.

The portable cage version (Fig. 8) differs from the ring of nooses described above, and shares characteristics of the bal-chatri and noose carpet. The cage consists of 1.3-cm mesh or larger hardware cloth measuring 20 × 20 × 10 cm mounted on a 25 × 25-cm piece of 1.9-cm plywood. Four 20-lb test monofilament nooses are used. Four flexible, but relatively stiff 20-cm rubber rods are attached vertically in each corner of the cage. Ten centimeters of each rod stand above the top of the cage. Two 2-cm lengths of solder (metal alloy used for patch-ing) are wrapped around the top and bottom of each rod. A large monofilament noose is tied to the base of each rod and opened so that the noose is supported by fold-ing the solder once at both the top and bottom of each of the four rods. The solder keeps the nooses open and erect, yet releases from them when a hawk pulls on the noose.

The phai has been used to capture Peregrine Fal-cons and Saker Falcons (*F. cherrug*) (Carnie 1969) using feral pigeons and starlings as lures. We suspect that this method could be used much like a noose carpet for vultures and eagles with carcasses used as bait. Once a perched falcon has been found, the trap is placed at a safe distance using a vehicle to conceal activity. A small quantity of standing grass helps to hide the nooses, and soil can be used to cover the hose.

Recently, the phai was tested while road-trapping Golden Eagles with a rabbit as a lure. Eagles responded within 15 to 90 minutes. The eagles were caught on two of three attempts (Latta et al., pers. comm.).

The cage version of the phai, which can be placed from a vehicle that has stopped momentarily, has been used to trap Cooper's Hawks and should be effective on Northern Goshawks, large buteos, harriers, and Harris's Hawks as well.

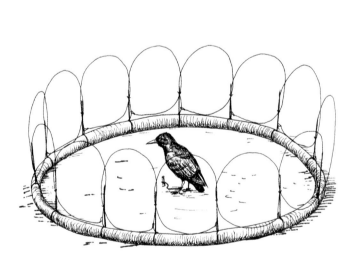

Figure 7. A ring version of the phai trap with a Common Starling (*Sturnus vulgaris*) as bait and a garden hose as the base for the noose attachment.

Figure 8. A cage phai trap, with four large nooses, commonly is used to trap accipiters.

Padded Leg-hold Traps

This trapping technique consists of placing several leg-hold traps with weakened springs and padded jaws in the immediate vicinity of an animal carcass (Fig. 9) to catch scavenging raptors. Today this technique is used almost exclusively to capture vultures and eagles. Initially however, it also was used to capture smaller raptors, including Northern Harriers and buteos (Lincoln and Baldwin 1929, Imler 1937).

Application. The most frequently used trap is a Number 3 double long spring leg-hold (Harmata 1985). Because the springs on new traps are powerful enough to break the toes of eagles, the jaws must be weakened or padded, or, preferably, both. Weakening is accomplished by striking both springs near the bend twice with a heavy hammer. When using traps with two coil springs, one of the springs can be removed to weaken the closure. Each jaw of the trap is wrapped with neoprene or leather and again with friction tape until a thickness of about 0.6 cm is obtained (Imler 1937, Stewart et al. 1945). A. Harmata (pers. comm.) has used this technique extensively and emphasizes the need to provide adequate padding. With adequate padding it may not be necessary to weaken the springs as the latter tends to slow closure.

Leg-hold traps have been used successfully to capture both Bald Eagles and Golden Eagles (Niemeyer 1975, Adkins 1977, Harmata and Stahlecker 1977, Harmata 1985, M. Lockhart, pers. comm.). In India, we used padded leg-hold traps to catch Greater Spotted Eagles and Asian Imperial Eagles (*A. heliaca*) in marshes using dead goats and fish, and Egyptian Vultures (*Neophron percnopterus*) and White-rumped Vultures (*G. bengalensis*) using sheep. Number 3 traps are not recommended for smaller species because of leg bone breakage, but modified smaller traps (e.g., Number 1 and Number 2) may be used on medium-sized to large hawks without causing injury. Imler (1937) trapped Red-tailed Hawks, Roughlegs, Prairie Falcons, and Great Horned Owls using a Number 1 trap placed on poles or around rabbit carcasses, but found that Northern Harriers, Swainson's Hawks, and American Kestrels sustained severe injuries during the process. Stewart et al. (1945) also tried padded leg-hold traps placed on the tops of poles for raptors, but also found them unsatisfactory because several trapped individuals received leg injuries, possibly because the traps being used did not have weakened springs. Padded leg-hold traps should be monitored continually or with electronic trap monitors.

Figure 9. Padded leg-hold traps arranged around a desert cottontail (*Sylvilagus audubonii*) carcass function well in capturing eagles. (*Photo by J. Kidd*)

Entire animal carcasses are used to attract eagles, but parts of animals, including the legs of rabbits and deer, also work. Fish can be used when attempting to trap Bald Eagles. Lures should be staked in sites frequented by eagles, and can be sliced open to make them appear previously fed upon and more attractive. Plucked fur or feathers spread around the carcass also enhance the effectiveness of the lure.

Depending upon the size of the lure, two to six traps are placed around it. A depression is dug for each trap, and the traps are set 2.5–50 cm from the lure. It is important that the trigger is not "haired" but is set for maximum pressure (A. Harmata, pers. comm.).

Before covering the traps, the trigger pan is covered with a 12 × 12-cm piece of cloth that extends under the jaws to prevent soil used to camouflage the trap from filling in under the trigger pan. All parts of the trap are then covered with loose soil or grass. Do not use snow as it may freeze. Each trap is attached to a 1-kg weight via a 1.5-m line with a shock absorber, which also is covered. Some researchers (e.g., Harmata 1985) connect two traps together by their stake-down chains in place of a weight. The technique should not be used during high winds or in hilly or mountainous terrain, and should be used with caution near water.

One major drawback with this technique is that smaller, non-target species also may be attracted to carrion, and if caught, may sustain serious injuries to their legs. The number of non-target individuals captured in leg-hold traps can be reduced by placing 7.6-cm fiberglass building insulation under the pan. Doing so allows for heavier trigger pressure (Harmata 1985).

Pit Traps

There are two basic types of pit traps. Both essentially involve placing a person in a hole with bait nearby. The pit traps, or "dig-ins" (Webster 1976), used to capture Peregrine Falcons on beaches during migration, are usually shallow and temporary while those used to capture eagles are deep and used repeatedly. A feral pigeon is used to lure falcons, whereas a large carcass is used to lure eagles, vultures, and condors. The leg or legs of the target raptors are grabbed by hand by the person inside the pit.

Application. A shallow pit deep enough to hide a person lying down is dug on a beach. A person then lies on their stomach or back in the hole and is buried except for his head and shoulders, which are covered and hidden by an approximately 35 × 50-cm basket (M.A. Jenkins, pers. comm.).

Trapping eagles and vultures with a pit trap requires more material and effort (Fig. 10). A 0.9 × 1.8-m, 1-m deep hole is dug in firm soil. A 2 × 2.4 m sheet of 1.9-cm marine varnished, outdoor plywood is used to cover the pit. The walls of the pit are supported by 1-cm outdoor plywood, and braced with 5.1 × 10.2-cm by 0.9-m wooden beams nailed vertically in the corners to prevent dirt slippage. One beam is jammed and nailed between the vertical beams at the front and back of the pit and a pair of 10.2 × 10.2-cm cross beams should be used to strengthen the roof when used in cattle country or where vehicles might be driven over it. A door the width of the pit is cut at one end of the plywood cover. A rim of beams is then nailed around the edge of the door to prevent debris from falling in when the door is opened. The door is hinged at the rear and a 25-cm diameter hole is cut in the middle of the door. A 30-cm high, 35-cm in diameter basket is secured with wire upside down over the hole. A 15–20-cm by 0.8-m opening between the door and the top of the forward wall of plywood serves as the capture space through which the birds are grabbed. The lure carcass lies in a shallow depression about 0.3 m in front of the capture space. This construction allows a person to kneel or sit in the pit with their head in the basket and peer through the capture space with their hands positioned inside of the opening ready to grab a bird. When left unattended, the door can be protected by screwing a piece of plywood about 12 cm wider and longer than the door onto the framing beams.

When construction is complete, a 10-cm layer of earth is applied to the roof. Great care should be taken in camouflaging the door, basket, and capture space with a thick layer of vegetation, preferably grass. A perch log also can be placed on the roof. Eagles frequently land on the log, thereby alerting the trapper inside of their presence. With the addition of some bushes and more grass on the roof, the trap should virtually "disappear" into the landscape. The vegetation resulting from spring rains will add the finishing touches to the camouflage work. Unlike many other traps, when constructed correctly, the pit trap is completely inconspicuous.

Pit traps are commonly used to capture falcons during coastal migration, and can also be used to trap adults or fledglings on the breeding grounds as well. A pigeon is tied on a 1–5-m line held by the trapper. As the pigeon wanders about on the sand, it is occasionally jerked to attract passing, or preferably, perched falcons. The pigeon also can be tossed from the basket if a falcon is observed flying by. When the falcon begins stooping on the pigeon it is pulled toward the trapper concealed in the basket. When the falcon grasps the pigeon, the trapper grabs the falcon with his hands.

Pit traps are not as effective as other methods of capturing falcons (e.g., harnessed pigeons, dho-gazas), but they do have the advantage of being selective.

Pit-trapping Bald Eagles and Golden Eagles is an ancient and well-tested technique that was used by at least 16 tribes of Native Americans (Wilson 1928). Prebaiting ensures faster results. Rabbits, deer, calf, and waterfowl carcasses can be used as lures. Each carcass should be sliced open and partially plucked or skinned to expose the meat and make it appear as though it has already been fed upon. Large carcasses such as adult deer should be frozen and then cut in half to ensure that the eagle stands on the carcass in a location where it can be reached. The carcass should be placed about 15–30 cm from the basket for eagles, 170 cm for vultures. It typically requires between 12 hours and 4 days for eagles to find and begin feeding on a carcass.

The trap should be entered one hour before sunrise so that no human activity is visible to eagles that might be roosting nearby. At no time during the day should the person inside the trap leave unless a bird has been captured, or the trapping attempt has ended. If possible, the observer should not leave the trap while eagles are present. Rather than waiting in position with arms outstretched and head in the basket, it is easier to wait below the basket in a more comfortable position. Eagles make enough noise with their wings when landing to warn the trapper of their approach. Special precautions should be taken to eliminate any possible sound inside the pit that might frighten target birds.

Figure 10. Construction of pit trap for use on eagles. When finished, cut grass is added to the basket, door, grabbing slot, and foreground. A large animal carcass is used as bait. *(Photos by P. Bloom)*

When an eagle or vulture approaches the carcass, the trapper should get into capture position on his knees at the forward end of the pit, with their head in the basket and hands placed on the edge of the space through which the bird's legs will be grabbed. All movements within the pit must be made very slowly. It may take several minutes for the eagle to rotate into the correct position. The trapper can peer through small holes in the basket or through the capture space in order to properly size up the situation. When the eagle or vulture begins feeding and is in range, the trapper should grab one or both tarsi. When the bird attempts to move forward to fly, it will fall on its breast. At this stage the trapper has two options: to stand up and step out of the pit in order to process the bird or, if several eagles are standing in the vicinity and more birds are needed, to slowly and carefully pull the eagle into the pit. This must be done without frightening the remaining eagles, and can be accomplished by placing both of the bird's legs in one hand, pulling its wings together and holding them with the other hand, and lifting the door slightly with your elbows while pulling the bird inside. Immediately replace the vegetation in front of the capture space. The bird's talons and feet should be wrapped with duct tape and the wings folded together. The bird is then placed in a restraining device (Evans and Kear 1972, Passmore 1979). We have caught two Golden Eagles in five minutes using this technique, which works well when the trap has been pre-baited and several eagles have been feeding there for several days.

The pit trap is rarely used today, probably because of the ease and effectiveness of trapping with leg-hold traps or radio-controlled bow-nets. We believe that the pit trap is probably the safest and most effective method for trapping eagles, vultures, and condors. If the trapper is competent, the success rate is virtually 100%. In many respects pit-trapping eagles is better than using cannon or rocket nets. Although it is possible to capture more than one eagle with each firing of a cannon or rocket net, these traps may start fires or cause fatalities. Once made, a pit trap is operational when a carcass is positioned in front of the opening and the pit is manned.

From 1985 to 1987, 125 Golden Eagles and a Bald Eagle were captured at five pit traps in southern California. Eighteen were recaptured two to three times. Five California Condors were captured in pit traps from 1984 to 1986 (PHB). Whereas some eagles and condors occasionally were missed due to noise in the pit, or impatience and noticeable hand movements of the trapper, 31 consecutive Golden Eagles that made contact with the carcass were captured without a miss. Generally speaking, the principal reason for missing eagles at a pit trap is the inexperience of the trapper or impatience. More often than not, time is on the side of the trapper if the bird is important enough.

Greater Spotted Eagles came regularly and naively to pit traps baited with goats in India, and White-rumped Vultures and Red-headed Vultures (*Sarcogyps calvus*) also fed at the carcasses with a person inside the trap centimeters away (PHB).

Some dangers do exist when using pit traps. Considerable care should be taken, particularly on beaches, to be certain that no vehicles are being driven in the area. Positioning the person so that he can see up and down the beach reduces this potential danger. Likewise, in the case of a deep pit trap, vehicles, horses, or other large ungulates wandering into the area can break through the roof. Trappers also should be aware of any poisonous snakes and invertebrates in the area. In India, large wild dogs attempted to remove the carcass several times and would have succeeded if one of the authors did not hold on to it.

The Verbail and Other Pole Traps

Vernon Bailey invented this trap (Stewart et al. 1945), one of several that are sometimes called pole traps. The Verbail trap consists of two sections: a stand mounted on a perch site or fence post, and a carefully bent length of spring steel shaped into a spring (Fig. 11). When a

Figure 11. The Verbail trap is used most effectively on owls in places where perches are limited, such as marshes, estuaries and prairies. *(Photo by J. Kidd)*

raptor lands on the trigger, the steel spring is released, which closes a 10-cm diameter loop of nylon cord around the bird's leg or legs. The spring is tied to the perch with a nylon line that is usually 1–2.5 m long, depending on perch height, which allows the bird to rest on the ground.

The Verbail is one of the few traps that we recommend be purchased rather than built. We know of no current manufacturer; however, used Verbails sometimes are available on the Internet.

The Verbail can be used effectively on many raptors both during the nesting season and migration. One of its advantages is that no lure is required, although one can be used if needed. Verbails are most effective in areas where perch sites are limited. If natural perch sites are not available, 10.2 × 10.2-cm by 2.5-m posts, erected in strategic locations about 0.4 km apart, work well. Habitats with limited perch availability such as marshes, deserts and grasslands are excellent locations for erecting posts, but a fence post on a prominent hill also can be effective. Most studies during the nesting season require only one or two traps, but during migration or on wintering grounds it is best to have 10 to 20 traps operating simultaneously. Traps should be monitored continuously in extreme cold, and checked hourly in more moderate temperatures, or equipped with trap monitors.

During the nesting season or when attempting to trap specific individuals, a lure animal such as a pigeon or small rodent can be effective, particularly when a Verbail is used in conjunction with another trapping technique, such as a harnessed pigeon or a bal-chatri.

When a hawk or owl flies toward the lure, it often will land first on a nearby perch for a closer inspection. A Verbail on a perch close to other traps works well in such situations. It also is common for a raptor to attack a trap such as a bal-chatri, but veer away just before making contact. A Verbail on a nearby perch provides an opportunity for a second chance at capture.

Verbail traps have been used to capture harriers, accipiters, buteos, and falcons, and are most effective on medium-sized to large owls (JWK, PHB).

Misses tend to increase with larger raptors because the spread of the bird's legs are wide enough, such that when they land, a part of one or both feet cover the loop and cause it to misfire. Increasing the size of all of the parts on the Verbail trap proportionately should increase its effectiveness on Red-tailed Hawks and larger species. Small passerines occasionally set off the trap without being caught due to their small size.

Trapped raptors seldom are killed as a result of being captured by Verbails, but trapped birds can be preyed upon. Four American Kestrels were killed by Red-tailed Hawks, and a Barn Owl and Great Horned Owl were killed by coyotes (*Canis latrans*) (PHB). One way to limit this is to check each trap at 30-minute intervals or, preferably, to use electronic trap monitors. In areas with heavy dew, birds can become soaked in wet grass beneath Verbails. The same can happen in a sudden rain shower with any trap. Wet birds should be dried before release as they are likely to be deeply chilled and hypothermic. Finally, the spring on a Verbail has considerable tension in a closed position. Individuals setting this trap should be aware of this danger to their eyes from a misfire, and should avoid putting their face close to the spring. When traps are set, no one but the person setting them should be as close as 1 m to them.

Another manually operated pole trap described by Dunk (1991) caught 39 White-tailed Kites during 48 attempts, but failed to catch a number of other raptor species that landed on it.

Perch snares, which are powered by rubber strips, are similar to the Verbail trap in terms of how they operate (Prevost and Baker 1984). Although rarely used, this trap would seem to have great potential for trapping a variety of diurnal and nocturnal raptors. No less than 120 Ospreys were captured in West Africa using this device, 17 of which were caught twice; one bird was captured five times (Prevost and Baker 1984).

Power Snares

Power snares, which were first described by Hertog (1987) for use in trapping territorial White-bellied Sea Eagles, are either triggered by the bird itself, or are set off manually or triggered remotely. Remote-controlled power snares have been described by Jackman et al. (1994) and McGrady and Grant (1996).

A remote-controlled trap used at bait sites captured five of seven Bald Eagles (Jackman et al. 1994). Seventy-five percent of attempts were successful at capturing adult Golden Eagles on the nest (McGrady and Grant 1996). Power snares that have been modified for use on the top of wooden fence posts have been used to capture Golden Eagles (Jackman et al., pers. comm.). Potentially this trap could be used on a number of diurnal raptors, particularly scavengers such as eagles, vultures, and, perhaps, caracaras.

Walk-in Traps

Walk-in traps are large cage traps designed for the simultaneous capture of several vultures (McIlhenny 1937, Parmalee 1954; Fig. 12). Such traps are easy to construct and can be very effective when baited with carrion.

Application. Cage size varies and depends upon the number of birds living in the area and the number of birds to be captured. Cages are usually circular, but can be square or rectangular. Cage diameter can vary from 3–12 m with a height of 1.2–1.8 m (Parmalee 1954). Henckel (1982) built a successful trap 3 × 3 m, 1.8-m high that captured as many as 12 vultures daily. Ten-cm

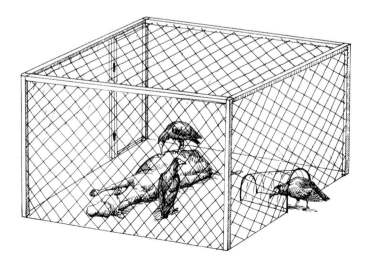

Figure 12. Walk-in trap for capturing Black Vultures (*Coragyps atratus*) and Turkey Vultures (*Cathartes aura*).

braided nylon netting is used to cover the top and sides of the trap. The bottom is left floorless. A door at one end of the cage allows the birds to enter the trap. Vultures enter through a single narrowing funnel that is difficult to exit.

This trap has been used to capture both Black Vultures and Turkey Vultures, with the former being more vulnerable (Parmalee 1954). The trap should be predator-proof, and checked every one to two days depending upon the quantity of food and water remaining. The number of birds caught in some instances is proportional to the size of the trap. In one case, approximately 210 vultures were caught simultaneously in one trap (Parmalee 1954).

We set a walk-in trap for Turkey Vultures in California three times for more than one week each and received only close fly-bys. Success was not obtained until we placed a live, non-releasable Turkey Vulture in the trap, upon which several vultures were captured the next day.

Scaled-down and baited with a tethered lure bird that cannot be killed through the side, this trap also can catch migrating Peregrine Falcons and Merlins (Meredith 1943), as well as accipiters (M. A. Jenkins, pers. comm.) and Northern Harriers. With live lures, ground predators can be a problem.

SUMMARY

Selecting the proper trap and lure for any given situation can be challenging, particularly for the novice. Quantitative studies of trap effectiveness are rare and considerable work remains to be done on this subject (Fuller and Christenson 1976, Bloom 1987, Bloom et al. 1992). The adults of many species that have never been live-trapped await experimentation with different trapping techniques.

Although we have attempted to bring together the important literature sources on trapping raptors, those who delve into the field-studies literature of specific raptors are likely to find descriptions of additional trapping techniques. Those interested in additional details should consult the references cited herein and, most importantly, seek out and gain experience with knowledgeable trappers.

ACKNOWLEDGMENTS

For 35 years a small army of raptor enthusiasts and good friends have spent considerable time assisting us with the capture of over 60,000 raptors representing 107 species of diurnal and nocturnal birds of prey. These people were tremendous in their dedication and passion to studying and conserving raptors and we thank all of you; unfortunately space does not allow proper individual recognition, but we know and you know who you are. R. Thorstrom is thanked for his insights as a trapper of tropical raptor species and for his careful review of the latest edition. J. Nagata provided the superb artwork of traps and birds.

PHB thanks Rebecca Morales for sharing in things natural and for her long-term support and encouragement of his research and conservation efforts.

PHB and WSC thank David Ferguson and the Division of International Conservation of the U.S. Fish and Wildlife Service for sponsoring our expeditions in India and Vibhu and Nikita Prakash and the Bombay Natural History Society for hosting us during our trips. WSC thanks the National Aviary in Pittsburgh, Pennsylvania for travel grants for African field work. S. Porter of Communications Specialists, Inc. is gratefully acknowledged for his support of our research and for the development of the trap transmitter and receivers we use in our studies.

Finally, we thank the many falconers, banders, and raptor biologists who, over the years, have given generously of their ideas. We dedicate this chapter to the memory of Richard R. "Butch" Olendorff who had such a strong influence on so many of us in the raptor world.

LITERATURE CITED

ADKINS, J. 1977. Bald Eagle capture and marking program. Pages 290–294 in Small game management report: 1976–1977. Washington Department of Game, Olympia, WA U.S.A.

ANDERSON, K.A. AND F. HAMERSTROM. 1967. Hen decoys aid in trapping cock Prairie Chickens with bow nets and noose carpets. *J. Wildl. Manage.* 31:829–832.

ARNOLD, K.A. AND D.W. COON. 1972. Modifications of the cannon net for use with cowbird studies. *J. Wildl. Manage.* 36:153–155.

BARTOS, R., P. OLSEN AND J. OLSEN. 1989. The Bartos trap: a new raptor trap. *J. Raptor. Res.* 23:117–120.

BEEBE, F.L. 1964. North American falconry and hunting hawks. North American Falconry and Hunting Hawks, Denver, CO U.S.A.

BERGER, D.D. AND H.C. MUELLER. 1959. The bal-chatri: a trap for the birds of prey. *Bird-Banding* 30:18–26.

————— AND F. HAMERSTROM. 1962. Protecting a trapping station from raptor predation. *J. Wildl. Manage.* 26:203–206.

BLEITZ, D. 1970. Mist nets and their use. *Inl. Bird-Banding News* 42:43–56.

BLOOM, P.H. 1987. Capturing and handling raptors. Pages 99–123 *in* B. A. Giron Pendleton, B.A. Millsap, K.W. Cline, and D.M. Bird [EDS.], Raptor management techniques manual. National Wildlife Federation, Washington DC U.S.A.

—————, J. L. HENCKEL, E.H. HENCKEL, J.K. SCHMUTZ, B. WOODBRIDGE, J.R. BRYAN, R.L. ANDERSON, P.J. DETRICH, T.L. MAECHTLE, J.O. MCKINLEY, M.D. MCCRARY, K. TITUS AND P.F. SCHEMPF. 1992. The Dho-gaza with Great Horned Owl lure: an analysis of its effectiveness in capturing raptors. *J. Raptor Res.* 26:167–178.

BUB, H. 1995. Bird trapping and bird banding. (Translation from German by Hamerstom, F. and K. Wuertz-Schaefer.) Cornell University Press, Ithaca NY U.S.A.

CAIN, S.L. AND J.I. HODGES. 1989. A floating fish snare for capturing Bald Eagles. *J. Raptor Res.* 23:10–13.

CARNIE, S.K. 1969. A Middle Eastern hawking album. *J.N. Am. Falconers Assoc.* 8:30–44.

CATLING, P.M. 1972. An improved technique for capturing Saw-whet Owls. *Ont. Bird Bander* 8:5–7.

CLARK, W.S. 1967. Modification of the bal-chatri trap for shrikes. *EBBA News* 30:147–149.

—————. 1970. Migration trapping of hawks (and owls) at Cape May, N.J. - third year. *EBBA News* 33:181–189.

—————. 1976. Cape May Point raptor banding station - 1974 results. *N. Am. Bird Bander* 1:5–13.

—————. 1981. A modified dho-gaza trap for use at a raptor banding station. *J. Wildl. Manage.* 45:1043–1044.

—————. 1992. On the etymology of the name *Bal-Chatri*. *J. Raptor Res.* 26:196.

COLLISTER, A. 1967. Simple noose trap. *West. Bird Bander* 42:4.

DELONG, J.P. 2003. Flammulated Owl migration project Manzano Mountains, New Mexico - 2003 report. HawkWatch International, Inc., Salt Lake City, UT U.S.A.

DILL, H.H. AND W.H. THORNSBERRY. 1950. A cannon-projected net trap for capturing waterfowl. *J. Wildl. Manage.* 14:132–137.

DUNK, J.R. 1991. A selective pole trap for raptors. *Wildl. Soc. Bull.* 19:208–210.

ELLIS, D.H. 1973. Behavior of the Golden Eagle: an ontogenic study. Ph.D. dissertation, University of Montana, Missoula, MT U.S.A.

—————. 1975. First experiments with capturing Golden Eagles by helicopter. *Bird Bander* 46:217–219.

ELODY, B.I. AND N.F. SLOAN. 1984. A mist net technique useful for capturing Barred Owls. *N. Am. Bird Bander* 9:13–14.

ENVIRONMENT CANADA AND UNITED STATES FISH AND WILDLIFE SERVICE. 1977. North American bird banding manual, Vol. II. Environment Canada, Canadian Wildlife Service, Ottawa, Ontario Canada.

ERDMAN, T.C. AND D.F. BRINKER. 1997. Increasing mist net captures of migrant Northern Saw-whet Owls (*Aegolius acadicus*) with an audiolure. Pages 533–539 *in* R. S. Duncan, D.H. Johnson, and T.H. Nicholls [EDS.], Biology and conservation of owls in the northern hemisphere. USDA Forest Service General Technical Report NC-190, North Central Forest Experiment Station, St. Paul, MN U.S.A.

ERICKSON, M.G. AND D.M. HOOPPE. 1979. An octagonal bal-chatri trap for small raptors. *Raptor Res.* 13:36–38.

EVANS, D.L. 1997. The influence of broadcast tape-recorded calls on captures of fall migrant Northern Saw-whet Owls (*Aegolius acadicus*) and Long-eared Owls (*Asio otus*). Pages 173–174 in R.S. Duncan, D.H. Johnson, and T.H. Nicholls [EDS.], Biology and conservation of owls in the northern hemisphere. USDA Forest Service General Technical Report NC-190, North Central Forest Experiment Station, St. Paul, MN U.S.A.

EVANS, M. AND J. KEAR. 1972. A jacket for holding large birds for banding. *J. Wildl. Manage.* 36:1265–1267.

FAABORG, J., T.J. DE VRIES, C.B. PATTERSON AND C.R. GRIFFIN. 1980. Preliminary observations on the occurrence and evolution of polyandry in the Galapagos Hawk (*Buteo galapagoensis*). *Auk* 97:581–590.

FERGUSON, H.L. AND P.D. JORGENSEN. 1981. An efficient trapping technique for Burrowing Owls. *N. Am. Bird Bander* 6:149–150.

FIELD, M. 1970. Hawk-banding on the northern shore of Lake Erie. *Ont. Bird Bander* 6:52–69.

FRENZEL, R.W. AND R.G. ANTHONY. 1982. Method for live-capturing Bald Eagles and Osprey over open water. *U.S. Fish Wildl. Serv. Res. Infor. Bull.* 82–13.

FULLER, M. R. AND G. S. CHRISTENSON. 1976. An evaluation of techniques for capturing raptors in east-central Minnesota. *Raptor Res.* 10:9–19.

GARD, N.W., D.M. BIRD, R. DENSMORE AND D.M. HAMEL. 1989. Responses of breeding American Kestrels to live and mounted Great Horned Owls. *J. Raptor Res.* 23:99–102.

GAUNT, A.S. AND L.W. ORING [EDS.]. 1999. Guidelines to the use of wild birds in research, 2nd Ed. The Ornithological Council, Washington, DC U.S.A.

GIRON PENDLETON, B. A., B.A. MILLSAP, K.W. CLINE AND D.M. BIRD [EDS.]. 1987. Raptor management techniques manual. National Wildlife Federation, Washington, DC U.S.A.

GRIEB, J.R. AND M.G. SHELDON. 1956. Radio-controlled firing device for the cannon-net trap. *J. Wildl. Manage.* 20:203–205.

GRINNELL, J. 1925. Bird netting as a method in ornithology. *Auk* 42:245–251.

HAMERSTROM, F. 1963. The use of Great Horned Owls in catching Marsh Hawks. *Proc. XIII Int. Ornithol. Congr.* 13:866–869.

—————. 1984. Birding with a purpose. The Iowa State University Press, Ames, IA U.S.A.

HARMATA, A.R. 1985. Capture of wintering and nesting Bald Eagles. Pages 139–159 in J.M. Gerrard and T.N. Ingram [EDS.], The Bald Eagle in Canada: proceedings of Bald Eagle Days, 1983. White Horse Plains Publ., Headingly, Manitoba, Canada.

————— AND D.W. STAHLECKER. 1977. Trapping and colormarking wintering Bald Eagles in the San Luis Valley of Colorado. Unpublished report.

HARTING, J.E. 1898. Hints on the management of hawks to which is added practical falconry. Horace Cox, London, United Kingdom.

HENCKEL, E.H. 1982. Turkey vulture study project. *N. Am. Bird Bander* 7:114.

HERTOG, A.L. 1987. A new method to selectively capture adult territorial eagles. *J. Raptor Res.* 21:157–159.

IMLER, R.H. 1937. Methods for taking birds of prey for banding. *Bird-Banding* 8:156–161.

JACKMAN, R.E., W.G. HUNT, D.E. DRISCOLL AND J.M. JENKINS. 1993. A modified floating-fish snare for capture of inland Bald Eagles. *N. Am. Bird Bander* 18:98–101.

————, W.G. HUNT, D.E. DRISCOLL AND F.J. LAPANSKY. 1994. Refinements to selective trapping techniques: a radio-controlled bow net and power snare for Bald and Golden Eagles. *J. Raptor Res.* 28:268–273.

JACOBS, E.A. 1996. A mechanical owl as a trapping lure for raptors. *J. Raptor Res.* 30:31–32.

———— AND G. A. PROUDFOOT. 2002. An elevated net assembly to capture nesting raptors. *J. Raptor Res.* 36:320–323.

JENKINS, M.A. 1979. Tips on constructing monofilament nylon nooses for raptor traps. *N. Am. Bird Bander* 4:108–109.

KAHN, R.H. AND B.A. MILLSAP. 1978. An inexpensive method for capturing Short-eared Owls. *N. Am. Bird Bander* 3:54.

KARBLOM, M. 1981. Techniques for trapping goshawks. Pages 138–144 in R.E. Kenward and I. Lindsay [EDS.], Understanding the goshawk. International Assocication for Falconry and Conservation of Birds of Prey, Oxford, United Kingdom.

KENWARD, R.E. AND V. MARCSTROM. 1983. The price of success in goshawk trapping. *Raptor Res.* 17:84–91.

KIRSHER, W.K. 1958. Bal-chatri trap for sparrow hawks. *News From Bird Banders* 33:41.

KNITTLE, C.E. AND M.A. PAVELKA. 1994. Hook and loop tabs for attaching a dho-gaza. *J. Raptor Res.* 28:197–198.

LINCOLN, F.C. AND S.P. BALDWIN. 1929. Manual for Bird Banders. United States Department of Agriculture. Misc. Pub. No. 58.

MACARTHUR, R.H. AND A.T. MACARTHUR. 1974. On the use of mist nets for population studies of birds. *Proc. Nat. Acad. Sci.* 71:3230–3233.

MARQUARDT, R.E. 1960a. Smokeless powder cannon with light-weight netting for trapping geese. *J. Wildl. Manage.* 24:425–427.

————. 1960b. Investigations into high intensity projectile equipment for net trapping geese. *Proc. Okla. Acad. Sci.* 41:218–223.

MARTIN, D.J. 1971. A trapping technique for Burrowing Owls. *Bird-Banding* 42:46.

MATTOX, W.G. AND R.A. GRAHAM. 1968. On banding Gyrfalcons. *J.N. Am. Falconers Assoc.* 7:76–90.

MAVROGORDATO, J.G. 1960. A hawk for the bush. Charles T. Branford Co., Newton, MA U.S.A.

————. 1966. A falcon in the field: a treatise on the training and flying of falcons. Knightly Vernon Ltd., London, United Kingdom.

MCCLOSKEY, J.T. AND S.R. DEWEY. 1999. Improving the success of a mounted Great Horned Owl lure for trapping Northern Goshawks. *J. Raptor Res.* 33:168–169.

MCCLURE, E. 1984. Bird Banding. The Boxwood Press, Pacific Grove, CA U.S.A.

MCGRADY, M.J. AND J.R. GRANT. 1996. The use of the power snare to capture breeding Golden Eagles. *J. Raptor Res.* 30:28–31.

MCILHENNY, E.A. 1937. A hybrid between Turkey Vulture and Black Vulture. *Auk* 54:384.

MENG, H. 1963. Radio controlled hawk trap. *EBBA News* 26:185–188.

————. 1971. The Swedish goshawk trap. *J. Wildl. Manage.* 35:832–835.

MEREDITH, R.L. 1943. Methods, ancient, medieval, and modern, for the capture of falcons and other birds of prey. Pages 433–449 in C.A. Wood and F.M. Fyfe [EDS.], The art of falconry. Stanford University Press, Stanford, CA U.S.A.

————. 1953. Trapping goshawks. *J. Falconers Club Am.* 1:12–14.

MERSERAU, G.S. 1975. Modifying the small raptor bal-chatri trap. *EEBA News* 38:88–89.

MEYBURG, B.-U., C. MEYBURG, T. MIZERA, G. MACIOROWSKI AND J. KOWALSKI. 2005. Family break up, departure, and autumn migration in Europe of a family of Greater Spotted Eagles (*Aquila clanga*) as reported by satellite telemetry. *J. Raptor Res.* 39:462–466.

MORRISON, J.L. AND S.M. MCGHEE. 1996. Capture methods for Crested Caracaras. *J. Field Ornithol.* 67:630–636.

MUELLER, H.C. AND D.D. BERGER. 1967. Observations on migrating Saw-whet Owls. *Bird-Banding* 38:120–125.

MUNDY, P.J. AND T.S. CHOATE. 1973. A detonator-propelled cannon net and its use to capture vultures. *Arnoldia* 6:1–6.

NERO, R.W. 1980. The Great Gray Owl. Smithsonian Institution Press, Washington, DC U.S.A.

NIEMEYER, C. 1975. Montana Golden Eagle removal and translocation project. USDI Fish and Wildlife Service, Animal Damage Control. Unpublished Report.

O'HARA, B.W. 1986. Capturing Golden Eagles using a helicopter and net gun. *Wildl. Soc. Bull.* 14:400–402.

O'NEILL, J.P. AND G. GRAVES. 1977. A new genus and species of owl (Aves: Strigidae) from Peru. *Auk* 94:409–416.

OSTROWSKI, S., E. FROMONT AND B.-U. MEYBURG. 2001. A capture technique for wintering and migrating Steppe Eagles in southwestern Saudi Arabia. *Wildl. Soc. Bull.* 29:265–268.

PARMALEE, P.W. 1954. The vultures: their movements, economic status, and control in Texas. *Auk* 71:443–453.

PASSMORE, M.F. 1979. Use of Velcro for handling birds. *Bird-Banding* 50:369.

PLICE, L. AND T. BALGOOYEN. 1999. A remotely operated trap for American Kestrels using nestboxes. *J. Field Ornithol.* 70:158–162.

PREVOST, Y.A. AND J.M. BAKER. 1984. A perch snare for catching Ospreys. *J. Wildl. Manage.* 48:991–993.

REDPATH, S.M. AND I. WYLLIE. 1994. Traps for capturing territorial owls. *J. Raptor Res.* 28:115–117.

REYNOLDS, R.T. AND B.D. LINKHART. 1984. Methods and materials for capturing and monitoring Flammulated Owls. *Great Basin Nat.* 44:49–51.

ROBARDS, F.C. 1967. Capture, handling, and banding of Bald Eagles. Unpublished report submitted to USDI Bureau Sport Fishing and Wildlife, Juneau, AK U.S.A.

ROSENFIELD, R.N. AND J. BIELEFELDT. 1993. Trapping techniques for breeding Cooper's Hawks: two modifications. *J. Raptor Res.* 27:171–172.

SMITH, D.G. AND D.T. WALSH. 1981. A modified bal-chatri trap for capturing screech owls. *N. Am. Bird Bander* 6:14–15.

SMITH, J.C., M.J. SMITH, B.L. HILLIARD AND L.R. POWERS. 1983. Trapping techniques, handling methods, and equipment use in biotelemetry studies of Long-eared Owls. *N. Am. Bird Bander* 8:46–47.

STEENHOF, K., G.P. CARPENTER AND J.C. BEDNARZ. 1994. Use of mist nets and a live Great Horned Owl to capture breeding American Kestrels. *J. Raptor Res.* 28:194–196.

STEWART, R.E., J.B. COPE AND C.S. ROBBINS. 1945. Live trapping hawks and owls. *J. Wildl. Manage.* 9:99–105.

THOMPSON, M.C. AND R.L. DELONG. 1967. The use of cannon and rocket-projected nets for trapping shorebirds. *Bird-Banding* 38:2124–2128.

THORSTROM, R.K. 1996. Methods for capturing tropical forest birds of prey. *Wildl. Soc. Bull.* 24:516–520.

TOLAND, B. 1985. A trapping technique for trap wary American Kestrels. *N. Am. Bird Bander* 10:11.

TORDOFF, H.B. 1954. An automatic live-trap for raptorial birds. *J. Wildl. Manage.* 18:281–284.

WALKIMSHAW, L.H. 1965. Mist-netting Saw-whet Owls. *Bird-Banding* 36:116–118.

WARD, F.P. AND D.P. MARTIN. 1968. An improved cage trap for birds of prey. *Bird-Banding* 39:310–313.

WATSON, J.W. 1985. Trapping, marking, and radio monitoring Rough-legged Hawks. *N. Am. Bird Bander* 10:9–10.

WATSON, R.T. AND C.R.B. WATSON. 1985. A trap to capture Bateleur Eagles and other scavenging birds. *S.-Afr. Tydskr. Natuurnav.* 15:63–66.

WEBSTER, H.M. 1976. The Prairie Falcon: trapping the wild birds. Pages 153–167 *in* A. J. Burdett [ED.], North American falconry and hunting hawks. North American Falconry and Hunting Hawks, Denver, CO U.S.A.

WEGNER, W.A. 1981. A carrion-baited noose trap for American Kestrels. *J. Wildl. Manage.* 45:248–250.

WESKE, J.S. AND J.W. TERBORGH. 1981. *Otus marshalli*, a new species of screech-owl from Perú. *Auk* 98:1–7.

WHALEN, D.M. AND B.D. WATTS. 1999. The influence of audio-lures on capture patterns of migrant Northern Saw-whet Owls. *J. Field Ornithol.* 70:163–168.

WIERSMA, J.M., W. NERMUT AND J.M. SHEPARD. 2001. A variation on the 'noosed fish' method and its suitability for trapping the White-bellied Sea-eagle (*Haliaeetus leucogaster*). *Corella* 25:97–99.

WILSON, G.L. 1928. Hidatsa eagle trapping. *Anthropol. Pap. Am. Mus. Nat. Hist.* 30:101–245.

WINCHELL, C.S. 1999. An efficient technique to capture complete broods of Burrowing Owls. *Wildl. Soc. Bull.* 27:193–196.

WISEMAN, A.J. 1979. On building a better bird trap. *Bird-Banding* 51:30–41.

ZWICKEL, F.C. AND J.F. BENDELL. 1967. A snare for capturing Blue Grouse. *J. Wildl. Manage.* 31:202–204.

Marking Techniques

13

DANIEL E. VARLAND
Rayonier, 3033 Ingram Street, Hoquiam, WA 98550 U.S.A.

JOHN A. SMALLWOOD
Department of Biology & Molecular Biology,
Montclair State University, Montclair, NJ 07043 U.S.A.

LEONARD S. YOUNG
1640 Oriole Lane Northwest, Olympia, WA 98502-4342 U.S.A.

MICHAEL N. KOCHERT
USGS Forest and Rangeland Ecosystem Science Center,
Snake River Field Station, 970 Lusk Street, Boise, ID 83706 U.S.A.

INTRODUCTION

In this chapter we describe techniques for marking raptors for visual identification beginning with a discussion of considerations involved in designing and conducting a marking program. We identify and describe permanent markers that can be used safely and effectively on raptors, including conventional leg bands, colored leg bands, leg markers, and wing markers. We then discuss temporary marking techniques (e.g., paints, dyes, feather imping). Avian marking techniques unsuitable for raptors are not addressed in this chapter but are described by Young and Kochert (1987). These include, but are not limited to, neck collars, nasal saddles and discs, and grafting feathers to the skin.

CONSIDERATIONS IN DESIGNING AND CONDUCTING A MARKING PROGRAM

Selecting Markers

Careful planning is imperative before applying markers, and biologists need to consider many important points before selecting a marker type (Marion and Shamis 1977, Ferner 1979, Barclay and Bell 1988, Nietfeld et al. 1996, Silvy et al. 2005). These include: (1) marker effect on the individual (Will affixing the marker cause pain and stress? Will the marker influence behavior? Will it decrease survival? Will it affect breeding?), (2) marker durability and longevity (Will the marker chosen last for the duration of the study, given both the subject bird's ability to remove or damage it and environment wear and tear?), (3) distance at which marked individuals may be identified and ease of identification (How close can the subject birds be approached for marker identification and to what extent will vegetation impede identification?), (4) need for identifying individuals versus a group, (5) ease in obtaining and, if required, assembling the marker, (6) ease of applying the marker, (7) marker cost, (8) the likelihood that the marker will interfere with other studies or raise public concerns, and (9) the likelihood that the marker will be approved for use by regulatory authorities.

Biologists should be fully aware of the effects that marking may have on the birds they intend to capture and mark (Murray and Fuller 2000). When there is doubt about the effects or effectiveness of a marking technique, trials with captive birds may be in order. Captive studies allow researchers to observe markers and birds at close range, and performance of the marker and its effects on marked individuals can be evaluated.

Developing a Marking Protocol

Careful planning will help ensure that marking objectives are accomplished. Planning a marking program necessarily involves developing a marking protocol or adopting one that already is in place. Although protocols will differ depending on the needs of different species, several basic guidelines should be followed for any protocol to be effective. (1) The protocol should be as simple as possible; usefulness of the marking technique should not be diminished by too complicated a scheme. (2) The protocol should meet the needs of all aspects of the study. (3) The protocol should be effective over the lifetime of the study. (4) The protocol should take into account the species' entire range. (5) Species that appear similar (e.g., Golden Eagles [*Aquila chrysaetos*] and subadult Bald Eagles [*Haliaeetus leucocephalus*]) should be treated as one marking "unit" and should be governed by a common protocol. (6) Techniques that are confused easily (e.g., wing markers and wing streamers) should be treated similarly and should be governed by a common protocol. (7) The Bird Banding Offices that have oversight in the region where the work will occur must approve the protocol.

We recommend accessing Internet web sites of Bird Banding Offices and other organizations that provide information about ongoing avian marking programs.

Bird Banding Offices and Marking Permits

North America. In North America, permits are issued through the North American Bird Banding Program, which is administered jointly by the U.S. Geological Survey (USGS) and the Canadian Wildlife Service (CWS). The Bird Banding Laboratory (BBL; USGS Patuxent Wildlife Research Center, Laurel, Maryland [www.pwrc.usgs.gov/bbl/]) manages the USGS banding program in the U.S. and the Bird Banding Office (BBO; National Wildlife Research Centre, Ottawa, Canada) manages the CWS banding program in Canada. Both the BBL and BBO require the principal investigator to possess an active Federal bird banding permit, and all personnel assisting with marking to have current subpermits if they plan to work independently. Subpermits authorize individuals to mark birds as directed by the principal investigator. Occasionally, individuals are authorized to conduct a marking program under the aus-

pices of a "station" permit held by an individual working for an organization on behalf of its employees. BBL and BBO permits only authorize attachment of conventional bands, which are provided by them at no charge. Authorization for the use of any other type of marker must be requested separately; the BBL and BBO do not supply or underwrite the costs of these markers. State and provincial permit requirements vary. Information on permit requirements may be obtained from the appropriate state or provincial wildlife agencies where the marking is planned. Permits and special authorizations should be carried in the field during marking.

Other geographic areas. Government and privately sponsored marking programs exist around the world. Where permits are required, as in the U.S. and Canada, they must be obtained in advance of fieldwork. North American bird bands and approved markers may be used off the continent with written authorization from the BBL or BBO, however as a rule, their use typically is allowed only in Mexico and Central and South American countries where North American birds migrate and winter. Bird banding, which is called ringing in Great Britain and Europe, is organized and coordinated by the British Trust for Ornithology (BTO; www.bto.org) in Britain and Ireland. The European Union for Bird Ringing (EURING; www.euring.org) is particularly helpful in providing information on ringing schemes both in Europe and elsewhere in the world.

Coordination

Biologists using similar marking schemes on the same or similar species should coordinate their work to reduce possible confusion. Coordinating with and alerting others of your activities also increases the likelihood that marked birds will be observed by other biologists. The Bird Banding Office responsible for oversight and permit approval in the region where the work is planned is an excellent place to gather information on similar marking schemes.

Gathering a Sample of Marked Individuals

The most basic and important assumption underlying studies using marking is that the sample of marked birds is representative of the entire population (Brownie et al. 1985, Williams et al. 2002). Ideally, all individuals in the study have the same probability of capture; however, capture probability often is influenced by factors such as capture methods, intraspecific differences (i.e., age, sex, social status), and behavioral response to trapping (Thompson et al. 1998). Because of this, sampling (marking) is seldom random in studies of raptors. In studies involving recapture of marked individuals, captured birds may become trap-happy (the likelihood of recapture is higher postcapture) or trap-shy (less likely to be caught after initial capture) (Thompson et al. 1998), leading to biases in the data. Most raptors are wary by nature and are far more prone to being trap-shy than trap-happy. Alternative trapping methods may be needed to recapture trap-shy individuals.

To ensure a truly representative sample, a random or stratified random sample of birds should be marked over the entire area or period of interest, and a coordinated survey should be conducted to define the population (Thompson et al. 1998, Chapter 5). Although this may not always be possible, steps to increase the likelihood that the marked sample is representative can be taken in any study. For example, when the entire cohort cannot be marked, nests at which young are marked can be selected randomly. Capture effort can be allocated evenly across an entire migration season, and capture sites can be varied. The steps that should be taken will vary depending on the situation and the purpose of marking, but the guiding principle of selecting a representative sample from an accurately defined population remains the same.

It is easier to accurately define a population and mark a representative sample when the population is sedentary (Brownie et al. 1985). During migration it is difficult, if not impossible, to determine the nature and size of the population the marked sample represents. Biologists typically focus their capture efforts during migration at banding stations, which often are linked with watchsites (see Chapter 6); these generally are on migration corridors and take access logistics into account (e.g., proximity to roads). Data collection, including those involving marking efforts, at banding stations may be described as "cluster sampling" of the larger population (Williams et al. 2002).

Collecting Resighting Data on Marked Individuals

A sound sampling design for accumulating sightings is just as important as a representative sample of marked birds. It is best if observation effort is consistent across the entire area in which marked birds may be resighted. This assumption almost always will be violated when the area is large or parts of the area have limited access.

It certainly is violated if sightings of marked birds by other biologists, amateur ornithologists, and the general public furnish substantial amounts of data, as resightings will be biased in favor of those areas that are populated or frequented by people who will make and report observations of marked birds.

We recommend that marking be used in conjunction with other sources of data in studies of raptor movements and resource use. Spatial tracking by conventional or satellite telemetry, which allows the marked sample to be observed systematically, is an important additional technique in such studies (see Chapter 14). When marking must be used as the primary technique, we recommend that sample sizes be large, that potential observation areas be searched systematically for marked birds, that the marking program be well publicized in the region where the study takes place, and that inferences on movements and resource use be restricted to general patterns and trends.

Because many observations of marked animals go unreported (Williams et al. 2002), announcements that describe the marking program and procedures for reporting sightings of marked birds are useful. Incomplete distribution of announcements will bias reports in favor of areas in which the marking program was publicized. Announcements should continue through the duration of the project to facilitate a similar resighting probability by the public across time.

Special Considerations in Studies of Population Dynamics

Survival (or apparent survival if mortality and emigration are confounded) can be estimated from marked individuals by analyzing data from recoveries of marked or banded birds that are found dead (band-recovery models), or from recaptures or resightings of marked individuals that are alive (mark-recapture models). These two data sources also can be used in combination. Band-recovery models seldom are used with raptors, mainly because of the large sample sizes required to obtain reliable results. For survival estimates, mark-recapture models (e.g., Gould and Fuller 1995, Morrison 2003, Anthony et al. 2006) and combined dead recovery-live recapture models (Kaufman et al. 2003, Craig et al. 2005) often are employed. Using data on nearly 20,000 Tawny Owls (*Strix aluco*) banded over 19 years, Frances and Saurola (2002) calculated age-specific survival rates using all three approaches, and concluded that for birds banded as nestlings the combined dead recovery-live recapture models are best.

The need for particularly durable, visible markers is paramount for mark-recapture studies. Although different types of markers can be handled in the analysis if marking technologies change over time, such analyses can be more complicated. Marker loss can be a serious problem. If that potential exists, a double- or triple-marking scheme should be used (McCollough 1990).

Within the mark-recapture framework, other parameters of interest can be estimated as well, such as population size (Gould and Fuller 1995), rate of population change (Kaufman et al. 2003, Craig et al. 2005, Anthony et al. 2006), and resighting rate (D. Varland, unpubl. data). Software packages, many of which can be downloaded for free from the Internet, are available for analyzing mark-recapture data (see www.phidot.org /software/). Program MARK (White and Burnham 1999) is a leading software package for mark-recapture analyses. Useful references on the mark-recapture literature include Thompson et al. (1998), Williams et al. (2002), and Dinsmore and Johnson (2005).

Marker Characteristics

Before we discuss marker characteristics, we want to point out that individual raptors may have unique plumage or soft-body-part characteristics, such as carbuncles in condors and vultures, which can be used to identify individuals in the field (natural markers). The female Peregrine Falcon (*Falco peregrinus*) that nested for many years on the Sun Life building in Montreal had an unusual "dimple" in her breast (Hall 1970), making her recognizable; R. Wayne Nelson recognized adult peregrines at nest sites by malar stripes and other physical features (Nelson 1988). Today, the ultimate method of identifying individuals is DNA fingerprinting.

Markers employ colors, often with a combination of numbers and letters (alphanumeric code) or, less frequently, symbols. Below we suggest some general guidelines pertaining to the use of colors and characters that will minimize resighting ambiguity and confusion.

Colors. Use as few colors as possible. The more colors used, the greater is the chance of observer error. In studies where markers must be sighted at long distances under adverse conditions, we recommend that, when possible, only three contrasting colors (e.g., red, white, and blue) be used. Use of additional colors may lead to color confusion. Additional colors may be used cautiously if marked birds are observed at close range

under favorable conditions by trained personnel. However, additional colors should be used only if essential to accomplishing study objectives, and alternate means of encoding data are not feasible. Certain pairs of colors should not be used in the same marking program under any circumstances. They include red and orange, yellow and white, dark blue and dark green, and purple and blue. Colors should be bright and bold; pale or pastel colors should not be used. Dark colors may be difficult to distinguish under poor light conditions or against dark earth and vegetation tones (Lokemoen and Sharp 1985). We recommend avoiding use of bicolor markers due to the possibility of seeing only one color (Kochert et al. 1983) and the tendency for colors to "merge" at long distances (Anderson 1963). Red markers on nestlings should be avoided, as these may increase pecking by siblings. Colors should contrast with the birds' coloration (e.g., yellow leg bands will not show well against the yellow legs of a Bald Eagle). When possible, colors present in the plumage or soft body parts of raptors should not be used on those species.

Characters. Characters (letters, numbers, or symbols) provide greater opportunity to identify individual birds. However, characters can be relied upon only if observers will be close enough to marked birds to read them consistently. Trials are useful for determining the ranges at which characters can be identified with different optics. As with colors, as few characters as possible should be used; data that are not essential to study objectives should not be encoded. Characters that are easily confused should be used cautiously in the same program (e.g., a 3 and an 8 or a C and an O can easily be confused if part of the character was obscured). Alphanumerics should be avoided if the general public is likely to be important in reporting birds. In such instances numerical sequences alone are preferred.

Characters can be printed in such a manner so as to reduce similarities. Distinct symbols (e.g., circles, triangles) may be more easily discerned than alphanumeric characters (Lokemoen and Sharp 1985). Colors of characters should contrast well with the background color of the marker; either black or white characters are best. Durable, colorfast paints and inks that adhere well to the marker material should be used to print characters. Marking pens and writing inks should not be used; the marks they produce will fade, blur, or deteriorate relatively quickly. Clear finishes, such as acrylic lacquers, can be used to protect characters, however they may increase the marker's reflectivity, causing glare and making identification difficult under certain light conditions.

Decisions regarding colors and characters should be thought through carefully, especially when initiating long-term studies.

PERMANENT MARKERS

Conventional Bands

Three types of bands, or rings, are used in field studies of raptors (Fig. 1). The butt-end band, or split-ring band, is used for smaller species whose bills are not powerful enough to loosen or remove the band. The butt-end band is placed around the tarsus and closed with banding pliers until the ends meet snugly and evenly. To facilitate a proper fit, specially drilled banding pliers are available commercially, but only for smaller-size, butt-end bands. Lock-on bands, also known as locking tab bands, are used with medium-sized to large raptors (except eagles) where overlapped closure is required to keep the band on the bird's leg. The lock-on bands have two flanges of metal, one longer than the other. The longer flange is folded over the shorter and pressed securely against the latter with pliers, locking the band in place. Rivet bands are used with eagles, whose bills are strong enough to remove butt-end (Berger and Mueller 1960) and, sometimes, even lock-on bands (C. Niemeyer and R. Phillips, pers. comm.). The band is closed snugly by hand, the flanges are pressed together with small vice grips or needle-nose pliers, and then they are riveted together using a pop-riveter.

Bands issued by the BBL and BBO are made of aluminum or a hard-metal alloy. They are inscribed on the outer surface with a unique number and with two means by which individuals who recover a band may report their findings, including a toll-free telephone number and a website address (www.reportband.gov) that replaces the mailing address, beginning in 2007. In Europe, EURING has adopted use of a website address (www.ring.ac) on a trial basis for reporting observations, which is inscribed on the ring in addition to a standard mailing address.

Conventional bands have been used almost exclusively on raptors to mark individuals in the event of recapture and to gradually accumulate information on migration, dispersal, longevity, and causes of death. Band recoveries generally occur by happenstance when individuals not connected with the banding research find and report dead or injured birds, resulting in low data yield (e.g., Broley 1947, Kochert et al. 1983). Because of this, we recommend that raptors be banded only as part of a well-planned and coordinated effort in which large numbers of birds are banded. Casual banding should be done only when raptors are captured or handled for other reasons or, in the case of nestlings, when a biologist has entered the nest for other purposes. Except for New World vultures, a bird receiving a marker or a radio transmitter always should be banded with a conventional band. Occasionally, conventional bands may be appropriate for identifying individual raptors at a distance.

Bands should not be used on Cathartid (New World) vultures because these raptors excrete feces on their legs, presumably for thermoregulation (del Hoyo et al. 1994). Consequently, bands may become impacted with fecal material, causing constriction of the leg and loss of circulation (Henckel 1976). This may result in swelling of the leg and foot below the band and, eventually, the loss of the leg (Henckel 1976) and, possibly, death. Houston and Bloom (2005) documented a shift from the use of leg bands to wing markers in Turkey Vulture (*Cathartes aura*) studies to avoid the problem posed by these species wearing conventional bands.

Bands come in a number of sizes and those of the correct size should always be used. Bands that are too loose may impede proper movement of the foot or become entangled with other objects, and bands that are too tight constrict and injure the bird's leg. Due to the

Figure 1. (Top, from left) Three conventional bands: butt-end, lock-on, and rivet. (Bottom, from left) Two color bands: color metal and color nonmetal. *(Photo by D. Varland)*

high degree of sexual size dimorphism in many raptors, males and females often require different size bands. Size differences among individuals within the same sex also may require the use of different size bands. Banders should measure the leg with a leg gauge to determine the correct band size (see the BBL web site [www.pwrc.usgs.gov/bbl/] for suppliers of leg gauges and other banding equipment, including pliers). Bands that are too loose, too tight, or overlapped must be removed. It also may be necessary to remove a band damaged or forced out of round during the banding process. Great care must be taken that the bird's leg is not injured when removing a band. Pressure must never be exerted on the leg during this process. Larger bands may be removed using two pairs of small vice-grip pliers. This technique is described and illustrated in Hull and Bloom (2001). Bands also may be removed by threading two pieces of wire, such as those used to "string" bands together, between the band and the tarsus on either side of the band's butt end. The free ends of each wire are wrapped around an easy-to-grip object, such as banding pliers, so that the opposing wires may be pulled with sufficient force to open the band.

Banders who find that the bands made available to them by the BBL or another banding organization are a poor fit should notify their supplier and provide advice on band-size improvement. The BBL continues to work on updating band sizing to ensure bird safety (M. Gustafson, pers. comm.).

Ideally, nestling raptors should be banded between one-half and two-thirds of the way through the nestling period (Fyfe and Olendorff 1976). At this time the tarsi are sufficiently developed to hold the appropriate size band, yet the birds are not mobile enough to jump from the nest and fledge inappropriately early (Fyfe and Olendorff 1976). If a goal is to assess productivity, the later the productivity estimate is made, the more accurate it will be (see Chapter 11). However, it is paramount that young in the nest not be disturbed so late in the nesting season as to cause premature fledging. Thus, if it appears that young will fledge prematurely when the nest is approached to band nestlings, banding should be avoided.

Color Bands

For studies involving color bands, each bird should receive a conventional band in addition to at least one color band (Fig. 1), and no more than four bands (two per leg) should be applied altogether. Metal bands should not be stacked; they can flare with time and damage the leg. All birds in a study should receive the same number of color bands, and each leg should receive a consistent number of bands; this allows for immediate identification of birds that have lost bands. When only two bands are attached, bands should be placed on both legs so that observers quickly identify banded birds. Conventional aluminum bands, which are silver, may serve as a "color" band. Adjacent bands should not be the same color as this eliminates confusion as to whether one or two bands were seen (Howitz 1981). General information (e.g., age, sex) should be encoded into the color scheme (Howitz 1981), and the scheme should be designed so that this information will not be compromised by band loss. Birds that frequently are seen together (e.g., members of a pair) should have dissimilar color combinations so that they may be distinguished easily (Howitz 1981). Color combinations should be used in a systematic order to facilitate organization of data and reduce the chances of accidentally using the same combination twice.

Here we identify four types of color bands: metal bands, nonmetal bands, painted bands, and bands wrapped with colored tape. A list of suppliers and manufacturers of metal and nonmetal color bands is maintained on the BBL web site (www.pwrc.usgs.gov/bbl/).

Metal bands. Colored pigments are affixed to metal bands by anodizing, an electrolyzing process in which the band functions as the anode. Anodizing was improved in the early 1990s, making anodized bands less prone to fading (D. Cowen, pers. comm.). As such, anodized bands are now a better choice for raptor studies than were those available when Young and Kochert (1987) reported on marking techniques.

Color metal bands, plain or engraved with alphanumerics or symbols (Fig. 1), are commercially available in North America through ACRAFT Sign and Nameplate Co. Ltd. of Edmonton, Canada. ACRAFT carefully monitors band codes issued to avoid duplication. According to reports to the BBL by field researchers, these bands are mostly durable and colorfast (M. Gustafson, pers. comm.). The only known exception to this among raptors wearing these bands occurred with Galapagos Hawks (*Buteo galapagoensis*) on Santiago Island. In this situation, the bands were so abraded by lava rocks the alphanumerics were unreadable within 4 to 5 years (K. Levenstein, pers. comm.). Color metal bands also can become difficult to read if dirt builds up on them (D. Varland, pers. obs). When this occurs, it may be necessary to recapture individuals and clean their bands.

Metal color bands can be made by anodizing conventional aluminum bands from the BBL or BBO. However, approval first must be granted by the BBL or BBO. These bands will fade somewhat over time. The extent to which fading occurs is not predictable and depends upon factors such as exposure to sunlight, abrasive rocks, and salt water. D. Varland detected little to no fading of blue or red anodized conventional bands worn by Peregrine Falcons in coastal Washington. On the other hand, fading did occur on anodized conventional bands worn by peregrines in the Midwest within four years; purple bands appeared pink and gold bands appeared silver (H. Tordoff, pers. comm.).

Non-metal bands. The plastics, celluloid and Reoplex, and the nonplastic polyvinylchloride Darvic, are common materials used in the manufacture of nonmetal color leg bands (Fig. 1). When wrap-around, or overlap bands are used, special adhesives are employed to ensure bonding between the sides of the band. Nonmetal band suppliers usually offer the adhesive needed with their product or will offer advice on purchase.

Laminated bands consist of two layers of plastic: a colored surface layer bonded to a contrasting white or black base layer. Alphanumerics or symbols may be inscribed on laminated bands by routing the surface layer to expose the contrasting base color. Whereas laminated bands demonstrated durability and retention in use on Spotted Owls (*S. occidentalis*) (Forsman et al. 1996), McCollough (1990) reported poor retention on the stronger-billed Bald Eagle; all 118 attached were lost within four years. In California, some Red-tailed Hawks (*B. jamaicensis*) lost their laminated bands within six years of application (P. Bloom, pers. comm.). These observations suggest that laminated plastic bands should not be used in long-term studies of large raptors capable of exerting substantial force on their bands, or in studies where band loss will bias the data.

Painted bands. Painted bands can be made with chip-free nail polish, which adheres well to bands (M. Gustafson, pers. comm.). Paint, however, wears off with time and is impractical for banding large numbers of birds. Pliers used for attaching painted bands should be wrapped in masking tape to avoid chipping painted surfaces.

Bands wrapped in tape. Wrapping bands with colored cloth tape offers a short-term means of identification, as most raptors quickly tear the tape. Bands such as these sometimes are used to identify raptors released after rehabilitation from injury or sickness.

Summary. Maximum distance for reading alphanumeric codes on bands varies with band size, lighting conditions, behavior of subject birds, and habitat. Alphanumeric codes on bands have been read with a spotting scope at distances of up to 150 m in observations of Galapagos Hawks (K. Levenstein, pers comm.) and Peregrine Falcons (D. Varland, pers. obs.), and up to 190 m with Bald Eagles (McCollough 1990). That said, because of their small size and relative inconspicuousness, color bands should be used as a primary marking technique only in studies where birds can be observed routinely with a spotting scope or with binoculars from relatively short distances.

Additionally, for reasons mentioned above, colored bands should not be used with Cathartid vultures. Color bands also are unsuitable for species whose tarsal feathering is likely to obscure the band, and they should not be used with species that spend large amounts of time standing in tall vegetation, or perching in locations where their tarsi are out of view.

Leg Markers

Leg markers are suitable for the same applications as color bands. In studies employing leg markers, birds also should receive a conventional band. There are two types of leg markers: leg flags, which are fastened around the leg itself, and leg band tags, which are attached to the conventional band (Fig. 2). Durability is necessary with leg markers because they are situated where good leverage can be brought to bear on them by the bird with both the bill and the feet.

Leg flags. Materials used to make leg flags (Fig. 2) include virgin vinyl (i.e., vinyl with no recycled material) (Bednarz 1987, Varland and Loughin 1992), Herculite® (Platt 1980, Warkentin et al. 1990) and Darvic (Fig. 2). Leg flags extending beyond the leg have the potential to interfere with behavior associated with hunting and incubation. Trained falcons wearing long jesses often are pursued by other raptors that mistake the jesses for a prey item and attempt prey robbery (Platt 1980). Herculite® leg flags that extended about 10 cm from the leg had no discernable impact on the hunting success of Merlins (*F. columbarius*) (Warkentin et al. 1990, I. Warkentin, pers. comm.). Far shorter leg flags have been used on Merlins (Fig. 2), Prairie Falcons (*F. mexicanus*) (ca. 1 cm) (Platt 1980); American Kestrels (*F. sparverius*) (3.5 cm) (Varland and Loughin 1992) and Harris's Hawks (*Parabuteo unicinctus*) (2.5 cm) (Bednarz 1987).

Leg flags should be restricted to short-term studies

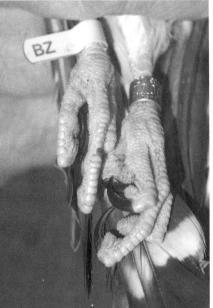

Figure 2. Leg band tag (above) attached to a conventional band on a Bald Eagle (*Haliaeetus leucocephalus*) (worn for 384 days) and leg flag (left) on a Merlin (*Falco columbarius*). *(Photos courtesy of J. Watson [above] and by D. Varland [left])*

because long-term retention of these markers by raptors often is poor (Picozzi and Weir 1976, Platt 1980, I. Warkentin, pers. comm.). Leg flags should be of the correct diameter. Leg flags that are too loose may slip off the foot or become entangled in the toes or on branches, wire, and other objects. Loose markers also are easier for a bird to tear. Markers that are too tight may cause abrasion or restrict circulation.

Leg-band tags. McCollough (1990) attached leg-band tags (Fig. 2) to Bald Eagles in Maine; the markers were retained better than laminated bands (0.6% annual loss rate for leg band tags vs. 35% for laminated bands). Alphanumerics on these tags were readable as far away as 220 m. Leg-band tags made of the vinyl Herculite® were retained well on Bald Eagles marked in Washington, where no marker loss was known in 59 deployments through seven years of monitoring (J. Watson, pers. comm.).

Wing Markers

Wing markers consist of two basic types: wrap-around and piercing, depending upon how the marker is secured to the wing. Markers made of various materials have been attached to or around the patagium. Wrap-around markers are wrapped around the wing and the ends are fastened between a natural break in the feathers, most often between the tertials and scapulars (Kochert et al. 1983; Fig. 3). Piercing markers usually consist of a tag or streamer attached to the wing by a pin or clip that pierces the patagium. Piercing markers are of three general types: a single tag attached to the dorsal surface of the wing (Smallwood and Natale 1998; Fig. 4), two separate tags attached to the dorsal and ventral surfaces of the wing, and a single tag that folds over the leading edge of the wing and is secured both above and below the patagium (Wallace et al. 1980; Fig. 3). In some instances, no pin or clip is employed, and the marker itself pierces the patagium (Sweeney et al. 1985). Sizes and shapes of wing markers vary.

Wing markers have been one of the most commonly used color markers in studies of raptors. They are relatively large and conspicuous, facilitating identification at long distances. Many species have been marked successfully with wing markers, including California Condors (*Gymnogyps californianus*) (Meretsky and Snyder 1992), Black Kites (*Milvus migrans*), Red Kites (*M. milvus*) (Viñuela and Bustamante 1992), Northern Harriers (*Circus cyaneus*) (Picozzi 1984), Common Buzzards (*B. buteo*) (Picozzi and Weir 1976), and Spanish Imperial Eagles (*A. adalberti*) (Gonzalez et al. 1989).

Success of wing markers for falcons has been variable. Wrap-around markers caused substantial feather wear and skin abrasion on Peregrine Falcons and Prairie Falcons to the extent of producing open sores (Sherrod et al. 1981, Kochert et al. 1983). American Kestrels wearing wrap-around markers showed no sign of injury, and they hovered, captured prey, and bred normally (Mills 1975). Pierced wing markers had no observed adverse effects on Common Kestrels (*F. tinnunculus*) (Village 1982) and American Kestrels (Smallwood and Natale 1998). Marker design and attachment methods are particularly important with falcons due to their long, narrow wings and rapid wing beats.

Wing markers are best suited for applications in which the observation area is known and marked birds are likely to be viewed from long distances at which smaller, less conspicuous markers would not be discernible. Examples include studies of nest-site fidelity (Picozzi 1984), dispersal (Miller and Smallwood 1997),

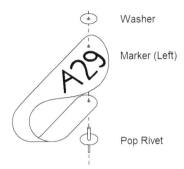

Figure 3. Adult Bald Eagle (*Haliaeetus leucocephalus*) wearing wrap-around wing markers (left) (photo courtesy of H. Allen). The marker is wrapped around the leading edge of the wing and the ends are fastened to the patagium between the tertials and scapulars (right) (Kochert et al. 1983; drawing courtesy of N. Smallwood).

winter-site fidelity (Harmata and Stahlecker 1993), and social relationships (Mossman 1976). Wing markers are highly effective in identifying individuals in behavioral studies (Mendelsohn 1982) and can provide much supplemental information on movements in studies in which conventional radio telemetry also is used (Meretsky and Snyder 1992).

Most wing markers are made from one of three types of materials: vinyl-coated nylon fabrics, upholstery plastics, and semi-rigid plastics. Vinyl-coated nylon fabrics consist of a vinyl coating over a meshlike nylon matrix. This material has been used extensively and is available in a variety of colors and weights. As a group, vinyl-coated nylons are durable and colorfast; wing markers of this material have been worn by Bald Eagles for as long as 22 years (McClelland et al. 2006). Vinyl-coated nylons vary in these characteristics (Nesbitt 1979, Kochert et al. 1983) however, even to the extent that colors of the same material from the same manufacturer may perform quite differently. Materials used with generally good results include Herculite®, Stamoid PE, Suncote, TXN, and Weym-O-Seal (Furrer 1979, Nesbitt 1979, Kochert et al. 1983). On the other hand, Coverlite, Dantex, and Facilon have been known to fade or deteriorate relatively quickly (Guarino 1968, Nesbitt 1979, Kochert et al. 1983). Variable results have been reported for Saflag, the most commonly used vinyl-coated nylon. Saflag has been observed to fade rapidly and dramatically, and to deteriorate relatively quickly (Nesbitt 1979; J. Smallwood, pers. observ.). In contrast, Saflag markers performed well for up to two years in studies of American Woodcock (*Scolopax minor*) (Morgenweck and Marshall 1977) and Band-tailed Pigeons (*Columba fasciata*) (Curtis et al. 1983).

Upholstery plastics such as Masland Duran and Naugahyde are much less durable than vinyl-coated nylon fabrics. Markers cut from these materials have a tendency to curl (Labisky and Mann 1962). For this reason we do not recommend using upholstery plastic for this purpose.

A few studies (e.g., Picozzi 1971, Mudge and Ferns 1978) used markers made of semi-rigid laminated plastics. These materials are very durable and have excellent color retention but sometimes crack if stressed. Common Buzzards occasionally lost markers because the plastic broke between the hole through which the retaining pin passed and the edge of the marker (Picozzi 1971). Laminated plastic markers may not be suitable for use on falcons because of possible severe wing abrasion.

Wing markers should be sized and fitted properly. Markers that are too small are difficult to observe, and markers that are too large may hamper flight. Wallace et al. (1980) equipped nestling Black Vultures (*Coragyps atratus*) and Turkey Vultures with two sizes of marker of the same design. Nestlings with the larger marker (1.5 times as long and 1.9 times the surface area of the smaller marker) fledged on average two weeks later than unmarked vultures. The larger markers caused asynchrony in wing beat, affected soaring ability, and fluttered during flight. Fledging dates of nestlings with the smaller marker were similar to those of unmarked nestlings, and their flight appeared to be unimpaired. If the fit is too loose, markers can cause irritation, and become caught on sticks or other objects, or lost. If the fit is too tight, markers may cause excessive feather wear and abrade the wing.

Mudge and Ferns (1978) developed an equation, T = 5.6L - 411, to estimate suitable marker size (for a single tag on the upper surface of the wing) where T = tag area in mm^2, L = wing length in mm, and width to length ratio of the marker = 3:7. Vinyl-coated nylon fabric should be cut shiny side facing out, and such that the completed marker follows the natural contour of the wing. Markers cut with the shiny side facing in are more likely to curl up when attached to the bird, making them a poor fit and difficult to read. Laminated plastic markers may be curved to the contour of the wing by heating the marker and bending it to the desired shape (Picozzi 1971).

Holes for pins, rivets, and other fasteners should be punched in vinyl-coated nylon markers using a leather punch, awl, or dissecting needle appropriate for the desired size of the hole. Holes in laminated plastic markers should be drilled with a fine bit. The area around the hole may be reinforced with a washer or other material to counteract wear and prevent tearing or breaking that can lead to marker loss.

Unless the number of birds marked is small, colors alone cannot identify individuals. Furthermore, use of colors may be governed by regional, national, or international protocols that restrict the available colors. Thus, colors usually should be used to denote general informa-tion and, if necessary, characters should be used to iden-tify individuals. Wings without a marker should not be part of a marking scheme; this prevents birds that have lost a marker from being misidentified. Marking schemes requiring observation of two wing markers to obtain a single type of data should be avoided so that at least some data still can be gathered if one wing marker is lost or unseen. If marker color is used to denote only one type of information, then left and right markers should be the same. If marker color is used to encode two types of data, then each wing should provide a sep-arate type of information. Characters identifying indi-viduals always should be the same on both markers.

Characters should be as large as possible, extending over the entire exposed portion of the marker. Wrap-around markers typically are preened such that a portion of the marker is obscured, and characters should not be printed there. Characters made from permanent mark-ing pens tend to fade quickly; paint (from paint sticks available at arts and crafts stores) is longer lasting (J. Smallwood, pers. obs.). Buckley (1998) reported that the numbers on cattle ear tags used as wing markers sometimes fade. Printing the band number and address of the researcher on the reverse side of each marker enables identification and reporting if the marker alone is retrieved.

Wing markers have been attached in a number of ways. Wrap-around markers have been fastened with metal eyelets (Southern 1971), staples (Curtis et al. 1983), grommets (Serbheen and English 1979), and pop rivets (Kochert et al. 1983). Pop rivets should be stain-less steel rather than aluminum or copper. Various fas-

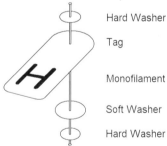

Figure 4. American Kestrel (*Falco sparverius*) wearing piercing marker (left) (photo courtesy of C. Meyer). The single marker is attached to the upper surface of the wing by a monofilament pin that pierces the patagium. The monofilament pin is depicted longer than is need-ed for clarity; actual distance between hard plastic washers is 4 mm (right) (Smallwood and Natale 1998; drawing courtesy of N. Smallwood).

teners have been used to secure piercing markers to the patagium, including metal pins and washers (Mudge and Ferns 1978), nylon pins and washers (Village 1982), plastic cattle-ear tags (Wallace et al. 1980), and pop rivets (Stiehl 1983). Materials used as fasteners should be smooth and round in cross-section, such as the 80-lb test monofilament fishing line used by Smallwood and Natale (1989; Fig. 4), so that any rotation does not injure the tissue surrounding the hole in the patagium.

When applying a piercing wing marker great care must be taken to avoid bones, muscles, tendons, and blood vessels. Isopropyl alcohol can be used to wet feathers to afford better visibility during fastening (Sweeney et al. 1985), as well as to cleanse the skin that is pierced. If a minor blood vessel is ruptured, bleeding usually is limited to one or two drops. A pinch of powdered alum or other coagulant usually stops the bleeding immediately. The fastener should pierce an area slightly distal to the elbow joint at a point about 1/3rd the distance from the biceps to the leading edge of the spread patagium (i.e., a little closer to the biceps) (Smallwood and Natale 1989; Fig. 5). If the fastener is not sufficiently sharp to puncture the patagium easily, a tool such as a dissecting needle may be used to make the smallest hole through which the fastener can pass. Fasteners must hold the marker snugly and securely in place, but not so tightly as to restrict circulation or damage tissue. Pop rivets should be crimped by hand; a rivet tool should not be used because it crimps pop rivets much too tightly, and will crush the patagium (Seel et al. 1982). Persons should consider how the marker contacts the wing when folded. If folding the wing results in the sharp edge of a washer or other fastener rubbing against the biceps, a piece of wing marker fabric (a soft washer) (Fig. 4) may be used to reduce the likelihood of injury (Smallwood and Natale 1989).

During the first few days or weeks following attachment, birds may preen and tug at the markers, presumably in attempts to remove them (Mills 1975, Sweeney et al. 1985). An adult Prairie Falcon removed a marker within 10 minutes of application, and an adult Swainson's Hawk (*B. swainsoni*) removed its wing marker within one week of marking (Fitzner 1980, Kochert et al. 1983). After this initial period, however, most marked birds accept wing markers and do not preen them excessively (Sweeney et al. 1985, Watson 1985).

Although wrap-around and piercing markers both work well with raptors, the latter have certain advantages. Piercing markers can be attached to nestlings at a younger age, whereas wrap-around markers require considerable feather development to hold the marker in place. Piercing markers, other than the fold-over type, do not involve the leading edge of the wing. This is an area where tissue irritation commonly is noted with wrap-around markers and is a critical part of the wing with respect to aerodynamics. The piercing fastener also prevents the marker from rotating around the wing, which has been observed with wrap-around markers (Watson 1985; R. McClelland, pers. comm.).

Minor feather wear and callusing of the patagium have been the most commonly reported effects of wing markers on raptors (Kochert et al. 1983). These effects are caused by chafing of the marker against the wing and in most cases are probably of little consequence. Many workers have observed no feather wear or tissue irritation (e.g., Hewitt and Austin-Smith 1966, Wallace et al. 1980). Indeed, consistent severe abrasion has been noted only with falcons wearing wrap-around markers (Sherrod et al. 1981, Kochert et al. 1983). That said, abrasion occasionally can be severe with some individuals of other species (Harmata 1984). Design, materials, fit, and attachment all influence feather wear and tissue irrita-

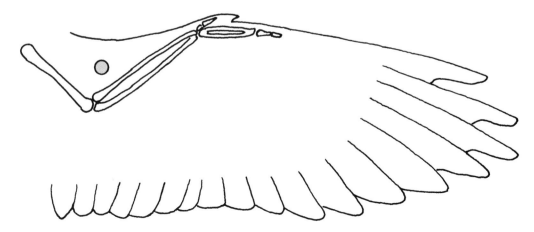

Figure 5. Location on the patagium for piercing fastener (shaded). See text for position relative to biceps and leading edge of patagium. Great care must be taken to avoid muscles, tendons, and blood vessels (redrawn by J. Smallwood from Smallwood and Natale 1998).

tion. A properly fitted and attached marker of a suitable material minimizes the chance of severe abrasion.

Many biologists have found that properly fitted wing markers did not affect flight (Hewitt and Austin-Smith 1966, Mills 1975, Wallace et al. 1980, Kochert et al. 1983). On the other hand, Howe (1980) suggested that wrap-around wing markers worn by Willets (*Catoptrophorus semipalmatus*) increased aerodynamic drag and caused abnormal feather replacement during molts. Marked Willets frequently shook their bodies during flight, suggesting some discomfort.

Wing markers did not affect survival of marked birds in several studies (e.g., Hewitt and Austin-Smith 1966, Kochert et al. 1983). In contrast, marked Willets apparently were more susceptible to predation and nutritional stress during migration, and wing markers may have reduced survival of Ring-billed Gulls (*Larus delawarensis*) and Band-tailed Pigeons (Howe 1980, Curtis et al. 1983, Southern and Southern 1985). Only one of 17 (6%) American Kestrels fitted with wing markers returned to a wintering area the year after capture, whereas 21–27% of banded kestrels returned (Bolen and Derden 1980). In each of the above studies, birds wore wrap-around markers on wings that were relatively long or had rapid wing beats, or both.

Occasionally, an accident involving a wing marker results in the death of a marked bird. A fledgling Bald Eagle that apparently had jumped from its nest and caught one of its markers on a branch died as a result (Gerrard et al. 1978). Non-lethal harm also is possible. Sherrod et al. (1981) suggested that wing markers might make foraging Peregrine Falcons more conspicuous to their potential prey, resulting in lower foraging success and presumably higher mortality. On the other hand, wing-marked Prairie Falcons examined by Kochert et al. (1983) appeared in good nutritional condition.

Studies examining effects of wing markers on breeding behavior and reproduction of raptors suggest that the effects usually are negligible. Marking did not affect nest-site fidelity of Black Vultures, and five of six adult Swainson's Hawks captured and wing-marked on their territories returned to their nests the following spring (Fitzner 1980, Wallace et al. 1980). Breeding success of Red-tailed Hawks, Golden Eagles, American Kestrels, Prairie Falcons, and Common Ravens (*Corvus corax*) for which at least one member of the pair was marked did not differ significantly from that of unmarked pairs (Kochert et al. 1983, Phillips et al. 1991, Smallwood and Natale 1998). Wallace et al. (1980) observed that all young were fledged from all

eight Turkey Vulture and two of three Black Vulture nests where at least one adult was marked.

Harmata (1984:177) suggested that wing markers disrupted relationships between members of Bald Eagle pairs captured at their wintering area. In contrast, wing-marked Golden Eagles appeared to be accepted by other eagles and participated in all the normal breeding behaviors (Phillips et al. 1991).

Wing coloration is important to social relationships in some birds including, for example, Red-winged Blackbirds (*Agelaius phoeniceus*) (Smith 1972). In raptors without natural, colored signal patches on the wings, wing coloration probably is not as important in determining social relationships. However, some raptors, including some harriers and kestrels, exhibit marked sexual dichromatism, so body color likely plays an important role in social behavior in general and breeding behavior in particular. Such species could be especially vulnerable to negative effects of colored wing markers. Although marked American Kestrels appear to behave normally (Mills 1975, Smallwood and Natale 1989), quantitative data are lacking. Until potential behavioral effects are evaluated more fully, wing markers should be used cautiously, with full awareness of unintended consequences. Studies in which wing markers are used should be designed so that quantitative analyses of marker effects are possible. Particular attention should be directed toward discriminating the effects of capture and handling from those of marking *per se*, and evaluating the influences of age, sex, and social status of the marked bird and time of marker application.

TEMPORARY MARKERS

Dyes, Paints and Inks

Dyes and paints have been used to mark a variety of raptors including Bald Eagles (Southern 1963), Golden Eagles (Ellis and Ellis 1975), Verreaux's Eagles (*A. verreauxii*) (Gargett 1973), and Snowy Owls (*Bubo scandiaca*) (Keith 1964). A principal advantage is that no materials are attached to the bird; the color of the plumage is simply altered. This makes the technique suitable for use with almost any species. A chief disadvantage is that even the most permanent dyes, paints, or inks will color the bird only until the next molt. This restricts the technique to short-term studies or applications where birds can periodically be re-marked. Suitable applications include studies of individual development of nestlings from hatching (too young to band) to fledging, studies of nestling behavior, and studies of birds at seasonal concentrations if observations during other times of the year or subsequent years are not of interest. Knowledge of the molt sequence of the subject species is important in predicting marker life; temporary markers have been employed to study molt sequence in Northern Saw-whet Owls (*Aegolius acadicus*) (Evans and Rosenfield 1987, E. Jacobs, pers. comm.)

To effectively mark a bird's plumage, a dye must penetrate feathers well (i.e., be readily absorbed), produce a bright color, and resist fading or washing for several months. Three dyes, Malachite Green, Rhodamine B Extra (bright pink), and picric acid (yellow), consistently display good penetration and brilliance of color, and have the best color retention of dyes tried (Wadkins 1948, Bendell and Fowle 1950, Kozlik et al. 1959). Picric acid, however, sometimes explodes if allowed to crystallize during long-term storage. Therefore, it must be used with great care. Other dyes have been used with variable or poor success. Jones (1950) noted identification problems caused by differential fading of component dyes; the least permanent dye faded first, changing the color of the mark to that of the more permanent dye. Because of this, dyes should not be mixed to produce a third color. Dyes are most effective when applied in a 33% alcohol/66% water solution (Wadkins 1948). Dyes can be applied by spraying, brushing, or dipping the appropriate feathers. Complete saturation is necessary for best results. Dyed feathers should be completely dry before the bird is released. Dyes are most effective with species with light plumage (Kozlik et al. 1959). The BBL recommends that dyes not be used on primary feathers because of the potential for feathers to wear more rapidly (M. Gustafson, pers. comm.).

Both brush-on and spray paints have been used to color plumage, with model airplane paints and bright fluorescent spray paints being used most frequently. Swank (1952) recommended that only flight feathers be painted. Paint should not be applied so heavily that feathers are matted together. Petersen (1979) used a cardboard template to produce marks of a certain shape and to prevent spray paint from drifting onto other body parts. Paint always should be allowed to dry before a bird is released.

Non-toxic blue ink from markers has been applied to more than 8,500 Northern Saw-whet Owls in Wisconsin to study molt sequence (E. Jacobs, pers. comm.). The ink was visible on recaptured birds for up to two years.

Feather Imping

Imping is a technique commonly used in falconry in which a broken flight feather is repaired by replacing the missing distal end of the feather with a corresponding piece from a previously molted feather (Fig. 6). The shaft of the replacement piece is held in place against the shaft of the broken feather with a pin or fine dowel that fits snugly inside both shafts, and may be further secured with glue. Birds may be marked by clipping a natural feather, typically a tail feather, and imping a dyed or brightly colored feather of another species (Wright 1939, Hamerstrom 1942). To increase conspicuousness, the marker feather may be longer than the other natural feathers (Fig. 6). Individual marks are produced by a combination of marker color, alphanumeric or symbol applied to the replacement feather and position in tail (left, center, right).

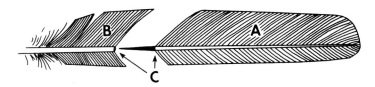

Figure 6. Imping. American Kestrel *(Falco sparverius)* with imped tail feather (top) (photo courtesy of J. Smallwood). The needle protruding from the marker feather (A) is pushed into the natural feather (B) until the cut shafts (C) meet (drawing from Wright 1939, in Young and Kochert 1987).

Feather Clipping

Feather clipping was used to mark large African raptors in the 1970s (Snelling 1970, Gargett 1973, Kemp 1977) but has not been used since, due to its limitations (D. Oschadleus, pers. comm.). The technique involves cutting "windows" in the wings or tail by clipping the vanes from part of the shafts of several adjacent feathers. Individuals are identified by varying the shape and position of the mark(s). Clipping should be done judiciously so that flight is not hampered. The principal advantage of the technique is its simplicity; no materials are used, and plumage is not colored. Clipping is unlikely to affect behavior (Harmata 1984). The chief disadvantage is that marks are inconspicuous when a bird is perched (Snelling 1970, Gargett 1973); this renders the technique of limited use in species that do not fly regularly. Also, the number of shape and position combinations that can be used effectively is limited (Gargett 1973) and the pattern is lost with molting, making this a short-term marking technique. The technique has received little attention in North America.

ACKNOWLEDGMENTS

We thank Mary Gustafson, Tracy Fleming, and Paul Doherty for comments and suggestions on earlier drafts of this manuscript, and Nathan Smallwood for enhancing photographs in Figures 3 and 4.

LITERATURE CITED

ANDERSON, A. 1963. Patagial tags for waterfowl. *J. Wildl. Manage.* 27:284–288.

ANTHONY, R.G., E.D. FORSMAN, A.B. FRANKLIN, D.R. ANDERSON, K.P. BURNHAM, G.C. WHITE, C.J. SCHWARZ, J. NICHOLS, J.E. HINES, G.S. OLSON, S.H. ACKERS, S. ANDREWS, B.L. BISWELL, P.C. CARLSON, L.V. DILLER, K.M. DUGGER, K.E. FEHRING, T.L. FLEMING, R. P. GERHARDT, S.A. GREMEL, R.J. GUTIERREZ, P.J. HAPPE, D.R. HERTER, J.M. HIGLEY, R.B. HORN, L.R. IRWIN, P. . LOSCHL, J.A. REID AND S.G. SOVERN. 2006. Status and trends in demography of Northern Spotted Owls, 1985–2003. *Wildl. Monogr.* 163:1–48.

BARCLAY, R.M.R. AND G.P. BELL. 1988. Marking and observational techniques. Pages 59–79 *in* T.H. Kunz [ED.], Ecological and behavioral methods for the study of bats. Smithsonian Institution Press, Washington, DC U.S.A.

BEDNARZ, J.C. 1987. Successive nesting and autumnal breeding in Harris' Hawks. *Auk* 104:85–96.

BENDELL, I.F.S. AND C.D. FOWLE. 1950. Some methods for trapping and marking Ruffed Grouse. *J. Wildl. Manage.* 14:480–482.

BERGER, D.D. AND H.C. MUELLER. 1960. Band retention. *Bird-Banding* 31:90–91.

BOLEN, E.G. AND D.S. DERDEN. 1980. Winter returns of American Kestrels. *J. Field Ornithol.* 51:174–175.

BROLEY, C.L. 1947. Migration and nesting of Florida Bald Eagles. *Wilson Bull.* 59:3–20.

BROWNIE, C., D.R. ANDERSON, K.P. BURNHAM AND D.S. ROBSON. 1985. Statistical inference from band recovery data - a handbook. USDI Fish and Wildlife Service Resource Publication No. 156, 2nd Ed. Washington, DC U.S.A.

BUCKLEY, N.J. 1998. Fading of numbers from patagial tags: a potential problem for long-term studies of vultures. *J. Field Ornithol.* 69:536–539.

CRAIG, G.R., G.C. WHITE AND J.H. ENDERSON. 2005. Survival, recruitment, and rate of population change of the Peregrine Falcon population in Colorado. *J. Wildl. Manage.* 68:1032–1038.

CURTIS, P.D., C.E. BRAUN AND R.A. RYDER. 1983. Wing markers: visibility, wear, and effects on survival of Band-tailed Pigeons. *J. Field Ornithol.* 54:381–386.

DEL HOYO, J., A. ELLIOTT AND J. SARGATAL [EDS.]. 1994. Handbook of the birds of the world, Vol. 2. New World Vultures to Guineafowl. Lynx Edicions, Barcelona, Spain.

DINSMORE, S.J. AND D.H. JOHNSON. 2005. Population analysis in wildlife biology. Pages 154–184 *in* C. E. Braun [ED.], Research and management techniques for wildlife and habitats, 6th Ed. The Wildlife Society, Bethesda, MD U.S.A.

ELLIS, D.H. AND C.H. ELLIS. 1975. Color marking Golden Eagles with human hair dyes. *J. Wildl. Manage.* 39:445–447.

EVANS, D.L. AND R.N. ROSENFIELD. 1987. Remegial molt in fall migrant Long-eared and Northern Saw-whet owls. Pages 209–214 *in* R.W. Nero, R.J. Clark, R.J. Knapton, and R.H. Hamre [EDS.], Biology and conservation of northern forest owls. USDA Forest Service General Technical Report RM-142, Rocky Mountain Research Station, Ft. Collins, CO U.S.A.

FERNER, J.W. 1979. A review of marking techniques for amphibians and reptiles. *Soc. Study Amphib. Reptiles Herpetol. Circ.* 9.

FITZNER, R.E. 1980. Behavioral ecology of the Swainson's Hawk (*Buteo swainsoni*) in Washington. Pacific Northwest Laboratory, Richland, WA U.S.A.

FORSMAN, E.D., A.B. FRANKLIN, F.M. OLIVER AND J.P. WARD. 1996. A color band for Spotted Owls. *J. Field Ornithol.* 67:507–510.

FRANCES, C.M. AND P. SAUROLA. 2002. Estimating age-specific survival rates of Tawny Owls-recaptures versus recoveries. *J. Appl. Stat.* 29:637–647.

FYFE, R.W. AND R.R. OLENDORFF. 1976. Minimizing the dangers of nesting studies to raptors and other sensitive species. *Can. Wildl. Serv. Occas. Pap.* 23.

FURRER, R.K. 1979. Experiences with a new back-tag for open-nesting passerines. *J. Wildl. Manage.* 43:245–249.

GARGETT, V. 1973. Marking Black Eagles in the Matopos. *Honeyguide* 76:26–31.

GERRARD, I.M., D.W.A. WHITFIELD, P. GERRARD, P.N. GERRARD AND W.I. MAHER. 1978. Migratory movements and plumage of subadult Saskatchewan Bald Eagles. *Can. Field-Nat.* 92:375–382.

GONZALEZ, L.M., B. HEREDIA, J.L. GONZALEZ AND J.C. ALONSO. 1989. Juvenile dispersal of Spanish Imperial Eagles. *J. Field Ornithol.* 60:369–379.

GOULD, W.R. AND M.R. FULLER. 1995. Survival and population size estimation in raptor studies: a comparison of two methods. *J. Raptor Res.* 29:256–264.

GUARINO, I.L. 1968. Evaluation of a colored leg tag for starlings and blackbirds. *Bird-Banding* 39:6–13.

HALL, G.H. 1970. Great moments in action: the story of the Sun Life falcons. *Can. Field-Nat.* 84: 209–230.

HAMERSTROM, F. 1942. Dominance in winter flocks of chickadees. *Wilson Bull.* 54:32–42.

HARMATA, A.R. 1984. Bald Eagles of the San Luis Valley, Colorado: their winter ecology and spring migration. Ph.D. dissertation, Montana State University, Bozeman, MT U.S.A.

——— AND D.W. STAHLECKER. 1993. Fidelity of migrant Bald Eagles to wintering grounds in southern Colorado and northern New Mexico. *J. Field Ornithol.* 64:129–134.

HENCKEL, R.E. 1976. Turkey Vulture banding problem. *N. Am. Bird Bander* 1:126.

HEWITT, O.H. AND P.J. AUSTIN-SMITH. 1966. A simple wing tag for field-marking birds. *J. Wildl. Manage.* 30:625–627.

HOUSTON, C.S. AND P. BLOOM. 2005. Turkey Vulture history: the switch from leg bands to patagial tags. *N. Am. Bird Bander* 30: 59–64.

HOWE, M.A. 1980. Problems with wing tags: evidence of harm to Willets. *J. Field Ornithol.* 51:72–73.

HOWITZ, J.L. 1981. Determination of total color band combination. *J. Field Ornithol.* 52:317–324.

HULL, B. AND P. BLOOM. 2001. The North American bander's manual for raptor banding techniques. The North American Banding Council, Point Reyes, CA U.S.A.

JONES, G.F. 1950. Observations of color-dyed pheasants. *J. Wildl. Manage.* 14:81–83.

KAUFFMAN, M.J., W.F. FRICK AND J. LINTHICUM. 2003. Estimation of habitat-specific demography and population growth for Peregrine Falcons in California. *Ecol. Applic.* 13:1802–1816.

KEITH, L.B. 1964. Territoriality among wintering Snowy Owls. *Can. Field-Nat.* 78:17–24.

KEMP, A.C. 1977. Some marking methods used on a variety of southern African raptors. *Safring News* 6:38–43.

KOCHERT, M.N., K. STEENHOF AND M.Q. MORITSCH. 1983. Evaluation of patagial markers for raptors and ravens. *Wildl. Soc. Bull.* 11:271–281.

KOZLIK, F.M., A.W. MILLER AND W.C. RIENECKER. 1959. Color-marking white geese for determining migration routes. *Calif. Fish Game* 45:69–82.

LABISKY, R.F. AND S.H. MANN. 1962. Backtag markers for pheasants. *J. Wildl. Manage.* 26:393–399.

LOKEMOEN, J.T. AND D.E. SHARP. 1985. Assessment of nasal marker materials and designs used on dabbling ducks. *Wildl. Soc. Bull.* 13:53–56.

MARION, W.R. AND J.D. SHAMIS. 1977. An annotated bibliography of bird marking techniques. *Bird-Banding* 48:42–61.

MCCLELLAND, B.R., P.T. MCCLELLAND AND M.E. MCFADZEN. 2006. Longevity of Bald Eagles from autumn concentrations in Glacier National Park, Montana, and assessment of wing-marker durability. *J. Raptor Res.* 40: *in press.*

MCCOLLOUGH, M.A. 1990. Evaluation of leg markers for Bald Eagles. *Wildl. Soc. Bull.* 18:298–303.

MENDELSOHN, J. 1982. The feeding ecology of the Black-shouldered Kite *Elanus caeruleus* (Aves: Accipitridae). *Durban Mus. Novit.* 13:75–116.

MERETSKY, V.J. AND N.F.R. SNYDER. 1992. Range use and movements of California Condors. *Condor* 94:313–335.

MILLER, K.E. AND J.A. SMALLWOOD. 1997. Natal dispersal and philopatry of Southeastern American Kestrels in Florida. *Wilson Bull.* 109:226–232.

MILLS, G.S. 1975. Winter population study of the American Kestrel in central Ohio. *Wilson Bull.* 87:241–247.

MORGENWECK, R.O. AND W.H. MARSHALL. 1977. Wing marker for American Woodcock. *Bird-Banding* 48:224–227.

MORRISON, J.L. 2003. Age-specific survival of Florida's Crested Caracaras. *J. Field Ornithol.* 74:321–330.

MOSSMAN, M. 1976. Turkey Vultures in the Baraboo Hills, Sauk County, Wisconsin. *Passenger Pigeon* 38:93–99.

MUDGE, G.P. AND P.N. FERNS. 1978. Durability of patagial tags on Herring Gulls. *Ringing & Migr.* 2:42–45.

MURRAY, D.L. AND M.R. FULLER. 2000. A critical review of effects of marking on the biology of vertebrates. Pages 15–64 *in* L. Boitani and T.F. Fuller [EDS.], Research techniques in animal ecology. Columbia University Press, New York, NY U.S.A.

NELSON, R.W. 1988. Do natural large broods increase mortality of parent Peregrine Falcons? Pages 719-728 *in* T.J. Cade, J.H. Enderson, C.G. Thelander, and C. White [EDS.], Peregrine Falcon populations: their management and recovery. The Peregrine Fund, Inc., Boise, ID U.S.A.

NESBITT, S.A. 1979. An evaluation of four wildlife marking materials. *Bird-Banding* 50:159.

NIETFELD, M.T., M.W. BARRETT, AND N. SILVY. 1996. Wildlife marking techniques. Pages 140–168 *in* T. A. Bookhout [ED.], Research and management techniques for wildlife and habitats, 5th Ed. The Wildlife Society, Bethesda, MD U.S.A.

PETERSEN, L. 1979. Ecology of Great Horned Owls and Red-tailed Hawks in southeastern Wisconsin. *Wis. Dep. Nat. Resour. Tech. Bull.* 111.

PHILLIPS, R.L., J.L. CUMMINGS AND J.D. BERRY. 1991. Effects of patagial markers on the nesting success of Golden Eagles. *Wildl. Soc. Bull.* 19:434–436.

PICOZZI, N. 1971. Wing tags for raptors. *The Ring* 68–69:169–170.

———. 1984. Breeding biology of polygynous Hen Harriers *Circus c. cyaneus* in Orkney. *Ornis Scand.* 15:1–10.

——— AND D. WEIR. 1976. Dispersal and cause of death of buzzards. *British Birds* 69:193–201.

PLATT, S.W. 1980. Longevity of Herculite leg jess color markers on the Prairie Falcon (*Falco mexicanus*). *J. Field Ornithol.* 51:281–282.

SEEL, D.C., A.G. THOMSON AND G.H. OWEN. 1982. A wing-tagging system for marking larger passerine birds. *Bangor Res. Station Occas. Pap.* 14.

SERVHEEN, C. AND W. ENGLISH. 1979. Movements of rehabilitated Bald Eagles and proposed seasonal movement patterns of Bald Eagles in the Pacific Northwest. *Raptor Res.* 13:79–88.

SHERROD, S.K., W.R. HEINRICH, W.A. BURNHAM, J.H. BARCLAY AND T.J. CADE. 1981. Hacking: a method for releasing Peregrine Falcons and other birds of prey. The Peregrine Fund, Inc., Ft. Collins, CO U.S.A.

SILVY, N.J., R.R. LOPEZ AND M.J. PETERSON. 2005. Wildlife marking techniques. Pages 339–376 *in* C.E. Braun [ED.], Research and management techniques for wildlife and habitats, 6th Ed. The Wildlife Society, Inc., Bethesda, MD U.S.A.

SMALLWOOD, J.A. AND C. NATALE. 1998. The effect of patagial tags on breeding success in American Kestrels. *N. Am. Bird-Bander* 23:73–78.

SMITH, D.G. 1972. The role of epaulets in the Red-winged Blackbird (*Agelaius phoeniceus*) social system. *Behaviour* 41:251–268.

SNELLING, J.C. 1970. Some information obtained from marking large raptors in the Kruger National Park, Republic of South Africa. *Ostrich* 8:415–427.

SOUTHERN, W.E. 1963. Winter populations, behavior, and seasonal dispersal of Bald Eagles in northwestern Illinois. *Wilson Bull.* 75:42–55.

———. 1971. Evaluation of a plastic wing marker for gull studies. *Bird-Banding* 42:88–91.

SOUTHERN, L.K. AND W.E. SOUTHERN. 1985. Some effects of wing tags on breeding Ring-billed Gulls. *Auk* 102:38–42.

STIEHL, R.B. 1983. A new attachment for patagial tags. *J. Field Ornithol.* 54:326–328.

SWANK, W.G. 1952. Trapping and marking of adult nesting doves. *J. Wildl. Manage.* 16:87–90.

SWEENEY, T.M., J.D. FRASER AND J.S. COLEMAN. 1985. Further evaluation of marking methods for Black and Turkey vultures. *J. Field Ornithol.* 56:251–257.

THOMPSON, W.L., G.C. WHITE AND C. GOWAN. 1998. Monitoring vertebrate populations. Academic Press, San Diego, CA U.S.A.

VARLAND, D.E. AND T.M. LOUGHIN. 1992. Social hunting in broods of two and five American Kestrels after fledging. *J. Raptor Res.* 26:74–80.

VILLAGE, A. 1982. The home range and density of kestrels in relation to vole abundance. *J. Anim. Ecol.* 51:413–428.

VIÑUELA, J. AND J. BUSTAMANTE. 1992. Effect of growth and hatching asynchrony on the fledging age of Black and Red Kites. *Auk* 109:748–757.

WADKINS, L.A. 1948. Dyeing birds for identification. *J. Wildl. Manage.* 12:388–391.

WALLACE, M.P., P.G. PARKER AND S.A. TEMPLE. 1980. An evaluation of patagial markers for cathartid vultures. *J. Field Ornithol.* 51:309–314.

WARKENTIN, I.G., P.C. JAMES AND L.W. OLIPHANT. 1990. Body morphometrics, age structure, and partial migration of urban Merlins. *Auk* 107:25–34.

WATSON, J.W. 1985. Trapping, marking, and radio-monitoring Rough-legged Hawks. *N. Am. Bird Bander* 10:9–10.

WHITE, G.C. AND K.P. BURNHAM. 1999. Program MARK: survival estimation from populations of marked animals. *Bird Study* 46 Supplement:120–139.

WILLIAMS, B.K., J.D. NICHOLS AND M.J. CONROY. 2002. Analysis and management of vertebrate populations. Academic Press, San Diego, CA U.S.A.

WRIGHT, E.G. 1939. Marking birds by imping feathers. *J. Wildl. Manage.* 3:238–239.

YOUNG, L.S. AND M.N. KOCHERT. 1987. Marking techniques. Pages 125–156 *in* B.A. Giron Pendleton, B.A. Millsap, K.W. Cline, and D.M. Bird [EDS.], Raptor management techniques manual. National Wildlife Federation, Washington, DC U.S.A.

Spatial Tracking

14

A. Radio Tracking

SEAN S. WALLS

Biotrack, 52 Furzebrook Road, Wareham, Dorset BH20 5AX,
United Kingdom

ROBERT E. KENWARD

Centre for Ecology and Hydrology, Winfrith Technology Center,
Dorchester
Dorset DT2 8ZD, United Kingdom

Radio-tracking has proved to be an essential tool for raptor studies. This is because it can record individual behavior systematically, not just at the nest or on a particular wintering area, but throughout the year. Radio-tracking can provide geo-specific data on foraging, roosting, and interactions with conspecifics or different species with little of the bias associated with observer location in other types of studies. Records of all tagged individuals, not merely those found at nests or dead, can be used to gain relatively unbiased estimates of breeding rates, survival and the proportionality of mortality agents. Radio-tracking often is the only way to reveal timing, routes and destinations of long-distance dispersal and migration. Such data can be crucial for assessing the impact of change in land use, checking the success of release programs, quantifying the effects of raptors on game, and investigating many other things of interest in wildlife management. Finally, radio-tracking is often the most practical method of getting data on experimental treatments and to parameterize biological models.

Most radio-tracking of raptors, which started about 40 years ago (Southern 1964), has been based on VHF (Very High Frequency) equipment. The last 20 years, however, have seen the maturing of Ultra High Frequency (UHF) technology that uses satellites, either to track tags directly or through Global Positioning Systems (GPS). Such systems can substitute for or complement VHF tracking.

VHF tags cost about $200 (U.S.), can be small (a 2.5-g tag can transmit for four months, and a 20-g tag can last 2 to 3 years) and can be located accurately (typically to within 10–100 m) by manual tracking from distances of 100 to 5000 m. UHF tags for tracking by satellite cost more than $1,000, and require additional payments for each location (typically $12–24 per day). The automated tracking saves labor costs, but there is relatively low accuracy for non-GPS units (e.g., 200–2000 m) and only about 60 transmission days for the smallest, 15-g tags. With intermittent transmission, these tags are uniquely suited for providing information on migration routes. GPS tags have the advantages of both automatic data collection and high accuracy (e.g., 10 m). Until recently, lightweight GPS tags were short-lived and had to be retrieved for downloading locations, but now a combination of solar-powered GPS units and a satellite link has created 30-g tags that supply accurate locations for longer periods, depending on the frequency of positions. That said VHF tracking remains the most successful technique for detailed tracking of small to medium-sized raptors in a local area over a long period.

Equipment, field methods and analysis techniques have been extensively reviewed (Kenward 2001, Millspaugh and Marzluff 2001, Fuller et al. 2005). Here we assume that there is a precise biological question to answer, that one or more of the references above will be consulted, and that experienced radio-trackers will be contacted for help with field techniques. We therefore concentrate on general-planning guidance.

PLANNING

The planning needed to ensure adequately tagged animals and useful data is detailed in White and Garrott (1990) and subsequent reviews. One additional planning consideration is the scope for collecting ancillary information. For example, when collecting locations to estimate home ranges and habitat use, information also can be collected on activity patterns and interactions. If tags are used to monitor whether individuals breed or die, it also is possible to test whether birds that were more active or had larger home ranges or foraged in particular areas were more likely to die or have reduced fecundity. Such a holistic approach leads to understanding of the mechanisms underlying population processes. To maximize the value from an investment in radio-tagging, it is worth considering from the outset what ancillary questions might be investigated.

Movements

The most important point to remember when collecting radio-tracking data is that the number of individuals tracked is a far more important component of sample size than is the number of locations. Simply put, it is better to get adequate samples of locations from many individuals than to get excessive detail on too few individuals. Unless standardized data-collection protocols from previous studies are available, pilot work is needed to assess how often to record locations and check whether individuals have emigrated or died.

If range areas or habitat use is required, is it for an annual or seasonal estimate or a series of snap-shots? If the former, locations should be recorded one or two times a week, at different times of day to avoid timetabling bias. If the latter, analysis of autocorrelation can help to decide how often locations can be recorded without spatio-temporal redundancy. In all cases, incremental analysis helps to decide how many locations make a practical standard range (Kenward 2001). If great detail is required from range outlines and cores, then more locations will be needed (Robertson et al. 1998). After a pilot study to establish standards, locations collected at the same rate over the same period enable robust tests for differences among individuals, populations, sites or seasons.

Studies of static interactions between individuals are based on overlap of home range cores or other territory estimators. Studies of dynamic interactions are more appropriate for finding if related individuals or individuals from a communal roost tend to aggregate. Such analyses require standardized recording of locations from different individuals in rapid succession, with careful planning so that no data are missing (Kenward 2001).

Radio-tracking has revolutionized the study of dispersal, by showing when, how and in what social or environmental contexts individuals make long distance movements beyond a study area. It is wise to check the locations of individuals often at the start of a project on dispersal to establish when they leave. This can be time-consuming, however tracking can be less frequent after pilot work has established the main dispersal periods. Subsequent reduced tracking for each individual allows more birds to be tracked in the same period, with intensive fieldwork restricted to short dispersal periods. When searching for dispersed raptors, the tracker needs to find topographical high-points and to have conviction in following faint signals, even when they are undetectable for 20 km or more after leaving a hilltop. Ground-based searches are easiest if a vehicle can be fitted with a pneumatic mast to raise an antenna 5–10 m, but the most cost-effective searching for birds lost during dispersal may involve mounting antennas onto aircraft wing-struts and conducting aerial surveys.

Survival, Forensics and Breeding

Researchers need not search often to estimate the survival and breeding rates of large, sedentary raptors whose tags will last for several years. Three checks per year, one each during winter, incubation, and rearing, are sufficient. Pre-breeders need more frequent checks to minimize losses during dispersal periods. More frequent checks also are needed to study causes of death, as carcasses can decompose quickly and be scavenged. That said infrequent checks may enable division of deaths into those (a) caused by humans (e.g., using sensitive analyses for poisons and X-rays for traces of lead in bones) (Cooper 1978), (b) associated with human artefacts (e.g., elevated wires, roads, wells, etc.), or (c) due to natural causes. Mortality sensors can speed checking, especially if all tags can be detected from topographical high points, so that only those indicating a death need to be found. When monitoring reintroductions or rehabilitated birds, checks can highlight solvable problems. In such instances, the more frequent the checks, the quicker the remedial action and the higher the likelihood of success.

In all cases, it is imperative to find all birds possible on each survey. Not doing so risks over-estimating shorter movements, as well as survival if birds are lost because their death has produced an undetectable signal. Survival data will be most robust if tags and searching are highly reliable, and if visual or other markers are used for re-sighting checks on the fate of birds with lost signals, to provide a correction for bias.

Analysis

Data analysis should be planned at the start of a study, and suitable software then used in pilot work to optimize data collection (see Planning). Software not only should display data but also should make it quick and easy to repeat analyses on many animals. The software ought to (1) provide all analyses needed, (2) handle the volume of data required (which may be large for GPS tags), and (3) input data and export results of analyses easily. It also should have adequate user-support, including integral or e-mail help. Good software is updated regularly, and it is worth keeping in touch with manufacturers to monitor developments (Larson 2001). Software defines the most efficient way in which to record data, which can help avoid too much re-processing from notebooks or palm-top computers.

Incremental analysis is essential for planning home-range studies, and autocorrelation analysis is a convenience for snapshot estimates (see Movements). These help in the efficient collection of locations from many individuals and in avoiding redundant and pseudo-replicated data from too few birds to enable robust statistical tests. Density-based home-range estimators such as ellipses and, to a lesser extent, contours, require the least locations, but their smoothing can be less suitable for species inhabiting coarse-grained (e.g., blocky or managed environments) than are linkage-based estimators such as mononuclear and cluster polygons (Kenward 2001). Once there are standard ranges from many birds, it is possible to quantify habitat association by comparing where birds were found with what is available to them. Availability should be individual-based (home range outlines or within a circle around a center of activity) rather than map-based, because map limits are set arbitrarily. Those interested in habitat analysis should investigate both compositional analysis (Aebischer et al. 1993) and distance-based analysis (Conner et al. 2003). For survival

analyses, software needs to handle staggered-entry, censored exit, and the inclusion of covariates such as age, sex and habitat (see, for example, White and Garrott 1990 and references in Millspaugh and Marzluff 2001).

EQUIPMENT

Radio-tracking equipment should be specified carefully before they are manufactured because it has to operate on the correct frequency and must be designed specifically, both for the species in question and the aims of the project. Above all, careful consideration should be given to the welfare of each raptor fitted with a tag. Trapping and tagging often is seasonal. As a result most researchers want tags at the same time of year and, consequently, manufacturers become booked at such times for months in advance.

Receiving Equipment

To receive VHF signals a receiver and an antenna are needed, both of which cover the appropriate frequency band to comply with national laws regarding wildlife telemetry. Receivers also must have enough bandwidth to cover all the tags, typically at 10 kHz intervals. The next most important feature is sensitivity (i.e., the ability to pick up weak signals). In addition to sensitivity, weight, waterproofing, and ability to store and scan through pre-set frequencies all are significant practical considerations. Receivers designed specifically for wildlife research cost $500 to $2,500, which is more than similar-looking alternatives intended for other markets, but they will last for many years and are much easier to use. For example, most commercial "scanning" receivers are designed to "modulate" a signal, keeping the same volume even if the signal is changing, which conflicts with the need to use variation in volume for direction finding. When buying a receiver, both tag manufacturers and receiver manufacturers should be consulted.

The antenna that best combines directional accuracy and gain for tracking raptors on the ground is the 3-element Yagi. Flexible elements are less awkward in thick vegetation and when putting them into vehicles. Yagis attached to aircraft should have solid elements. Additional elements can improve reception and directionality, but are cumbersome to use unless attached to a mast. Vehicles need very good suppression or diesel engines to avoid interference with weak signals when on the move.

Signals to indicate behavior and a bird's presence at feeding stations or nests can be logged without mobile tracking if the tags have sensors. The same is true for physiology. It is simpler and less expensive to record from a receiver tuned to one frequency, but for sampling several tagged individuals a programmable logging system is needed. Loggers usually search (via a connected receiver) through the frequencies of several birds, and record pulse characteristics received on each frequency. Although logging can save labor in the long run, neither set-up nor data analyses are simple, and it is important not to underestimate the time required.

Tag Types and Attachment

Tags should transmit on a frequency compatible with the receiving system and about 10 kHz apart from other tags. Tag manufacturers need to know the frequency bands of receivers available to the researcher and the frequencies of any working tags to avoid. Interference in the study area should be checked before specifying

frequencies. Around cities there may be many loud extraneous signals that can damage the hearing of researchers in long-term studies.

Table 1 shows the most common tag attachments for raptors. Researchers should talk with experienced trackers and tag manufacturers about the best technique for the species and project. Minimizing the impact of tags on tagged individuals will contribute to robust and, hence, publishable results, as well as to the welfare of the bird (Murray and Fuller 2000). Tags should be comfortable and entirely humane. One should check that manufacturers have sufficient knowledge of biology or species requirements to produce transmitters without sharp edges or surfaces that may interfere with thermoregulation in cold climates. Tag and harness mass near the upper limit allowable should be avoided for each attachment technique. The allowable mass depends upon the mass and wing-loading of the bird as affected by species, sex, and race. The mass that birds can carry safely determines the battery that can be used, and therefore the life (i.e., the time that it will be active) and range of the tag. A tag that pulses faster is easier to track and a stronger pulse will produce a signal that can

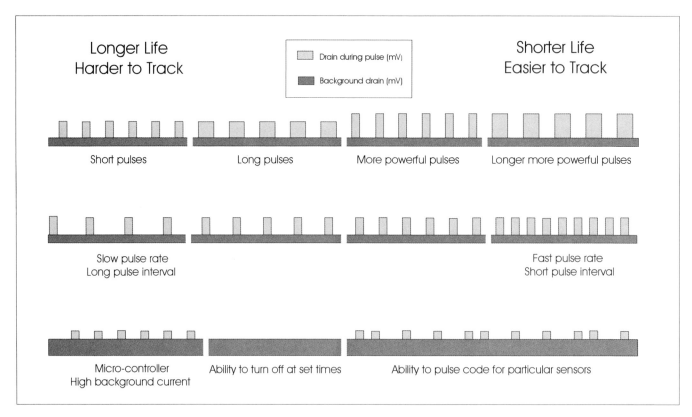

Figure 1. The effects of pulse length, pulse rate and micro-controllers on the life-time and ease of tracking of radio-tags.

Table 1. Techniques for attaching radio-tags to raptors.

Technique	Safety	Considerations
Tailmount	Probably safe if load is less than 2% body mass and attached to two or more feathers.	Feathers must be "hard penned" (i.e., fully grown), therefore one must trap fledglings when out of nest. The tracking stops when the feathers to which the tags are mounted molt.
Backpack	Harness is risky unless carefully fitted.	Can fit to all in the nest just before fledging. Can track for many years and through molts. Tagging at center of lift is best for high-tag mass.
Legmount	No published adverse affects, but might impact hunting.	Tag needs additional protection and a shorter antenna; therefore, life and range for mass of tag are reduced. Can tag all fledglings and track through molt.
Patagial	Only on large raptors with slow wing beat.	Used successfully on condors and large vultures.

be heard from a greater range. However, both requirements draw on battery capacity. To extend the life of the tag, the pulse rate and strength can be reduced to a level at which tracking is more difficult but still practical (Fig. 1). Micro-controllers also can be used to turn tags off during times when there is no need to track, such as during darkness or in winter for migrants. If such controllers are used it is important to ensure that the increased background current of a micro-controller (Fig. 1) does not offset savings from switching off the signal.

Attachment methods must minimize the possibility of entanglement and, where possible, should detach the tag when it stops transmitting. Knowledge of the species is more important than inflexible guidelines or advice from manufacturers. Where possible, potential tag effects should be tested (e.g., by comparison with independent re-sighting data from visual markers on tagged and untagged individuals). If this is not possible, one should consider testing against a low-mass alternative attachment that has little risk of impact, ideally by comparing groups of birds marked in the same season. Doing so is particularly important when using methods that are new or that have known risks. "Tests" also can be based on conservative assumptions. For example, if survival is better than that found with other methods (e.g., banding), effects of tags are probably negligible. Finally, it is worth remembering that males of size-dimorphic raptors may compare best with females if fitted with lighter tags.

LITERATURE CITED

AEBISCHER, A.E., P.A. ROBERTSON AND R.E. KENWARD. 1993. Compositional analysis of habitat use from animal radio-tracking data. *Ecology* 74:1313–1325.

CONNER, L.M., M.D. SMITH AND L.W. BURGER. 2003. A comparison of distance-based and classification-based analyses of habitat use. *Ecology* 84:526–531.

COOPER, J.C. 1978. Veterinary aspects of captive birds of prey. The Steadfast Press, Saul, United Kingdom.

FULLER, M.R, J.J. MILLSPAUGH C.E. CHURCH AND R.E. KENWARD. 2005. Wildlife radiotelemetry. Pages 377–417 in C.E. Braun [ED.], Techniques for wildlife investigations and management, 6th Ed. The Wildlife Society, Bethesda, MD U.S.A.

KENWARD, R.E. 2001. A manual for wildlife radio-tagging. Academic Press, London, United Kingdom.

LARSON, M.A. 2001. A catalog of software to analyze radiotelemetry data. Pages 398–422 in J.J. Millspaugh and J.M. Marzluff [EDS.], Radio tracking and animal populations. Academic Press, San Diego, CA U.S.A.

MILLSPAUGH, J.J. AND J.M. MARZLUFF [EDS.]. 2001. Radio tracking and animal populations. Academic Press, San Diego, CA U.S.A.

MURRAY, D.L. AND M.R. FULLER. 2000. Effects of marking on the life history patterns of vertebrates. Pages 15–64 in L. Boitani and T. Fuller [EDS.], Research techniques in ethology and animal ecology. Columbia University Press, New York, NY U.S.A.

ROBERTSON, P.A., N.J. AEBISCHER, R.E. KENWARD, I.K. HANSKI AND N.P. WILLIAMS. 1998. Simulation and jack-knifing assessment of home-range indices based on underlying trajectories. *J. Appl. Ecol.* 35:928–940.

SOUTHERN, W.E. 1964. Additional observations on winter Bald Eagle populations: including remarks on biotelemetry techniques and immature plumages. *Wilson Bull.* 76:222–237.

WHITE, G.C. AND R.A. GARROTT. 1990. Analysis of wildlife radio-tracking data. Academic Press, New York, NY U.S.A.

B. Satellite Tracking

BERND-U. MEYBURG

World Working Group on Birds of Prey,
Wangenheimstrasse 32, D-14193 Berlin, Germany

MARK R. FULLER

U.S. Geological Survey, Forest Rangeland Ecosystem
Science Center, Snake River Field Station, and
Boise State University Raptor Research Center,
970 Lusk Street, Boise, ID 83706 U.S.A.

INTRODUCTION

Satellite telemetry has revolutionized the study of raptor migration and life histories and will continue to do so in the future (Table 1). This is because tracking systems used in satellite telemetry can regularly estimate and record an individual's location worldwide for several years. Satellite telemetry with birds started in the 1980s (Strikwerda et al. 1986). Since then, satellite telemetry has been based on Ultra High Frequency (UHF) technologies such as the Argos system, that includes the Collecte Localisation Satellites (CLS). More recently, transmitters and Global Positioning Systems (GPS) receivers have become small enough to use on birds. In some cases GPS satellite telemetry will soon supersede land-based VHF tracking.

The Argos System

Satellite telemetry for raptor studies has used the Argos system. Individual birds must be able to carry transmitters, called Platform Transmitter Terminals (PTTs), weighing about 5 g or more. The Argos system provides location estimates and sensor data (e.g., battery voltage, activity, temperature, pressure) from PTTs anywhere around the world. The basics of operation are described in the Argos User Manual (www.argosinc .com/system_overview.htm). Additional recent information is available in the *Proceedings of the Argos Animal Tracking Symposium, 24–26 March 2003* (CLS America 2003), which is available on CD from CLS America, 1441 McCormick Drive, Suite 1050, Largo MD 20774.

Location Estimates of Transmitters by Argos

PTTs are located using the Doppler phenomenon. Polar-orbiting satellites carry Argos receivers. As a satellite approaches the PTT, the frequency received will be higher than the nominal transmitted frequency (401.650 MHz), whereas frequencies lower than 401.650 MHz will be received at the satellite as it moves away from the PTT. At the point of inflection of the Doppler curve, that is, when the received and transmitted frequencies are equal, the position of the transmitter will be perpendicular to the satellite ground track. The system estimates two possible PTT locations, which are symmetrical on each side of the satellite ground track. Argos selects one of these as plausible, but biologists should confirm the validity of the location selected by Argos.

Location estimates based on PTT transmissions and the Argos satellite system are assigned to location classes (LC). *"Location accuracy varies with the geometri-*

Table 1. Topics and questions regarding raptors for which data from satellite telemetry have or are expected to provide information. Some references are provided, and more can be found at the U.S. Geological Survey Raptor Information System (http://ris.wr.usgs.gov/). The keywords below and others can be used to find citations to publications listed in the Raptor Information System.

Annual Movements	• Annual movements (Brodeur et al. 1996, Fuller et al. 2003, Meyburg et al. 2004b, Laing et al. 2005, Steenhof et al. 2005) • Differences among years (Alerstam et al. 2006)
Migration	• Mapping routes of migrating raptors (Meyburg et al. 1995a, 1995b; Brodeur et al. 1996, Fuller et al. 1998, Ellis et al. 2001) • Individual variation (Alerstam et al. 2006) • Ecological barriers, leading lines (sea, mountains, deserts) (Meyburg et al. 2002, 2003) • Bottlenecks; do all individuals pass a narrow area, at what time? (Fuller et al. 1998) • Navigation and orientation (Hake et al. 2001, Thorup et al. 2003a, 2003b, 2006b) • Migration period and timing (Schmutz et al. 1996, Kjellen et al. 2001, Meyburg et al. 2004b) • Age and sex differences, breeding status (Ueta et al. 2000, Ueta and Higuchi 2002, Hake et al. 2003, McGrady et al. 2003, Meyburg et al. 2005, 2006, Soutullo et al. 2006b) • Speed and altitude of migration (Hedenström 1997, Kjellen et al. 2001) • Variation throughout migration (Meyburg et al. 2006) • Daily distances, travel rates (Fuller et al. 1998, Meyburg et al. 1998, Soutullo et al. 2006a) • Daily behavior, stopovers (time of starting and stopping), hunting (Meyburg et al. 1998) • Weather conditions (Meyburg et al. 1998, Thorup et al. 2003b, 2006a) • Ecological conditions along migration routes
Winter or Austral Summer	• Geographical situations of wintering grounds (Woodbridge et al. 1995, Martell et al. 2001, Haines et al. 2003, Higuchi et al. 2005, Steenhof et al. 2005) • Discovery of unknown wintering grounds (Meyburg et al. 1998) • Ranges on wintering grounds (McGrady et al. 2002) • Fidelity to the same area in successive years (Fuller et al. 2003)
Nesting Season	• Home range size, habitat use, and territorial behavior (Meyburg et al. 2006) • Dispersal, philopatry (Rafanomezantsoa et al. 2002, Steenhof et al. 2005) • What accounts for later or earlier arrival in spring at the nest site (influence of weather during migration, later or earlier departure to wintering grounds) (Meyburg et al. 2007b) • Pair continuity over a number of years (Meyburg 2007a) • Behavior of nonbreeding adults, floaters (arrival, fidelity to nest site after failed nesting attempt, possible nomadism) (Meyburg 2007b)
Movements during Immature Stage	• Return to breeding area or remain on the "wintering grounds" (Meyburg et al. 2004a) • Ranging behavior (Meyburg et al. 2004a)
Survival, Mortality, Threats	• Human activity (Eastham et al. 2000) • Other causes (Goldstein et al. 1999, Hooper et al. 1999, Henny et al. 2000, Millsap et al. 2004, Steenhof et al. 2006) • Fate of release birds (Rose et al. 1993, Launay and Muller 2003, Dooley et al. 2004)

cal conditions of the satellite passes, the stability of the transmitter oscillator, the number of messages collected and their distribution in the pass. This means in particular that a given transmitter can have locations distributed over several classes during its lifetime. Classes for which accuracy is estimated and their related values: Class 3: better than 150 m on both axes, 250 m radius, Class 2: better than 350 m, 500 m radius, Class 1: better than 1000 m, 1500 m radius, Class 0: over 1000 m, 1500 m radius. These are estimations at one sigma." (www.cls.fr/html/argos/general/faq_en.html).

Argos location methods are based on three major assumptions: (1) transmission frequency is stable during the satellite pass, (2) the PTT is motionless during the satellite pass, and (3) the altitude of the PTT is known. The LC assigned by Argos usually underestimates the error associated with wildlife applications largely because these assumptions often are violated to some extent when the PTT is on an animal (e.g., Britten et al. 1999, Craighead and Smith 2003). Usually, the accuracy given by Argos is better for the latitude than for the longitude. The given accuracy (e.g., 1 km for LC 1) does not mean that all of the calculated locations (and attributed to LC 1) fall within 1 km, but that about one sigma (one standard deviation) of all estimates are in the nominal accuracy range.

It is important to remember that the best two LCs (LC 2 and LC 3) usually are achieved only 10% to 15% of the time from birds. This occurs for numerous reasons, not the least of which is that many wildlife PTTs do not transmit 1 W of power, upon which the Argos system was designed. Power often is programmed to 0.15 to 0.25 W to conserve energy for prolonged PTT operation. Power output in solar-powered PTTs is adjustable (e.g., from 0.1 to 0.5 W). Reduced radiated power can result in fewer location estimates, and consequently fewer data with which Argos can estimate locations most accurately.

Argos routinely provides Standard LCs (LC 3, LC 2, LC 1, see above), but also can provide Auxiliary LCs (LC 0 > 1000 m, LC A and LC B = no estimate of location accuracy, and LC Z = invalid locations). The Auxiliary LCs are especially important because often there are few Standard LCs from wildlife tracking. Furthermore, the best LC classes do not always include the most accurate location estimates. Thus, wildlife researchers, especially those tracking birds, will want as many location estimates as possible from which to select appropriate data.

Location-estimate error from a given project can vary dramatically depending on the speed of the animal and its behavior, including changes in elevation or altitude (www.cls.fr/manual/; see Appendix 2, Argos location), environmental variables (topography, vegetative cover, marine, atmospheric conditions), and data acquisition and analysis options. Users may specify to Argos values for some factors (e.g., PTT velocity, altitude) and discuss options (e.g., use of digital elevation model, multi-satellite service), and Argos will incorporate these in the estimation procedures. Users also should consult with equipment manufacturers to maximize performance (e.g., PTT power, transmission repetition rate) for the circumstances and objectives of the study. Biologists must determine if the Argos system is appropriate for their objectives, especially if they require regular location accuracy of less than 1 km.

Reduced Argos Performance

A significant difference in actual receptions of PTT transmissions exists in the European region and in Asia (Mongolia, China, Japan), and thus can reduce receptions to less than 10% of the expected data. The affected area is about the size of the satellite footprint (5,000 km in diameter) and seems to be centered in the region of southern Italy (Howey 2005). The cause is ambient broadband noise of significant amplitude around the Argos operating frequencies, which causes interference and affects all PTTs, including GPS models. It essentially limits the number of signals that are received by the satellite (Gros and Malardé 2006). We recommend that users contact CLS to discuss their specific requirements and take advantage of ways to optimize Argos system performance.

Argos Data-validation Procedures

Researchers should examine and carefully filter location estimates before selecting those for analyses. Filtering or data validation procedures usually involve establishing criteria based on animal movement capabilities and behavior (e.g., maximum speed, local versus migration movement; Hays et al. 2001) and inspecting the Argos data for time and distance relationships among location estimates. Many LC 0, LC A, and LC B class points might need to be discarded by filtering, but so might some LC 1, LC 2, and even LC 3 class points. Careful screening also might reveal that some LC 0, LC A, and LC B locations are well within the distance that an animal could have traveled during the period between location estimates, and within a direction that is logical.

Raptor researchers must remember that locations from Argos are estimates and that accuracy and precision vary with animal and environmental factors that are largely unknown. In our experience, the proportion of higher quality LCs (LC 2 and LC 3) varies among PTT-marked animals. Therefore, we recommend that each person establish criteria for the study objectives, species, and environment and then apply those criteria when selecting the location estimates to be used in analyses.

Data Transmission through the Argos System

PTTs transmit a coded identification and data from up to 32 sensors. The signals are digitally encoded on a pulse width of ~ 0.36 seconds and a pulse interval usually between 40 and 90 seconds. The transmitting schedule (i.e., the duty cycle) can be programmed for more transmissions during different periods (e.g., seasons), which can prolong the operational life of battery-powered PTTs.

Transmissions from PTTs are received on polar orbiting satellites and are relayed to processing centers in France and the United States. Records of processed data can be distributed to users in a variety of formats, including Internet access to data received about four hours previously. The cost of data acquisition from Argos varies according to the different agreements between countries and Argos. Costs are assessed as a fee for use of each active platform, for hours of use per day, automatic data distribution service (data via email), fax, telnet, data acquired from the Argos website, and monthly compact discs (CD).

GPS Location of Transmitters

The GPS provides location accuracy to within a few meters. A GPS receiver can be integrated with an Argos PTT. A GPS receiver collects transmissions from at least four satellites, enabling computing of position (in three dimensions), velocity, and time. GPS units can be programmed to collect data at pre-set intervals. Data can be logged in memory and downloaded from the unit (usually requiring recapture), or they can be coded in PTT messages and relayed to users via the Argos system. The GPS estimates are transmitted to Argos during the "on time" of a PTT duty cycle.

The GPS receiver requires considerable energy. Thus, there are radio-tag size and longevity constraints that come into play when using battery power for bird studies. Alternatively, solar-powered GPS-PTTs weigh as little as 22 g. These units include sensors and a 12-channel GPS receiver.

Selection of the PTT

A crucial consideration when choosing a unit is how the PTT size, weight, and attachment might affect the bird (Murray and Fuller 2000). The energy requirements for satellite telemetry limit the minimum mass of units to about 5 g. The mass of the transmitter increases the energy the bird must expend for locomotion. Battery mass and surface areas of solar arrays also are limiting factors for unit size.

Deciding whether to use battery- or solar-powered tags must be made early in study planning. Battery-powered PTTs offer generally reliable performance, but have the disadvantage of a rather short operating life, thus long-term studies (more than three years) normally are not possible. Using 30- to 90-g battery-powered PTTs we regularly received locations from 6 to 18 months, depending on radiated power and duty cycle. Solar-powered transmitters can provide locations for up to several years, and the regularity of data is dependent on enough light on the solar array to charge a battery or capacitor with energy for transmission of the radio signal. Solar-powered GPS-PTT tags need more energy than PTTs. Thus, the problem of recharging these tags is even more acute. One must be sure the feathers do not occlude the solar array to the extent that there is insufficient exposure to light for minimal PTT function. Bird habitat use, such as under-canopy or cave nesting, also can affect solar charging.

The decision of whether to use solar or battery-powered PTTs depends not only on the geography and expected movements of the species to be studied, but also on other factors such as budget, lifestyle of the species, aim of the study (long- versus short-term), etc. In 2007 the price of a PTT was about $3000 (U.S.), and that of a GPS-PTT was about $4000. Costs of delivering data (see above) for several years can be as much or even more than the tag price, depending on how tags are programmed and what Argos services are used.

Attachment of Transmitters

Radio tags can be mounted on tail-feathers, legs, and

wings, but in most studies they are attached to the bird's back using a harness (Fuller et al. 2005). These "backpacks" have the advantage of being fixed near the center of lift which is best for high tag mass. Tags can be fitted to nestlings just before fledging and can be tracked for several years. Most researchers use Teflon® ribbon as harness material, but we found that some raptors (e.g. Asian Imperial Eagles [*Aquila heliaca*] and Lesser Spotted Eagles [*A. pomarina*], Prairie Falcons [*Falco mexicanus*]) remove tags by pulling and cutting through the Teflon® strips with their beaks (Steenhof et al. 2006). The potential complication of feathers over the solar panels on backpacks might be overcome by incorporating a feather guard (Snyder et al. 1989) or thick neoprene rubber on the bottom of the transmitter to elevate the solar array. These modifications might create additional aerodynamic drag and thus, energy needed for flight.

What Causes Termination of Transmissions?

Manufacturers can program a unit to stop transmitting, but most researchers probably would like to receive transmissions for as long as possible. Battery-powered units transmit information about battery voltage so that one can predict depletion of the battery energy. Often however, failure to receive transmissions occurs earlier than expected, raising a question as to what has happened. The causes of failure to receive data are sometimes difficult to determine.

Juvenile and immature birds often die from "natural causes," or perish from persecution. Adults also are subject to heavy persecution in many parts of the world or are killed by electrocution, collisions, etc. Nevertheless, based on observing the bird, recapturing it, or finding it dead much later, we confirmed that several solar-powered PTTs had failed while the birds were alive. In some cases we, or the manufacturer, were unable to determine a reason for the failure. Study planning should account for death of radio-marked birds and the failure of some transmitters.

Our record for long-term tracking is an adult female Greater Spotted Eagle (*A. clanga*). The bird was fitted with a PTT in July 1999 that was still transmitting data in August of 2007. An adult male Lesser Spotted Eagle was tracked as far as Israel on its way back to the breeding grounds almost 6 years after having been marked. When it arrived one month later in Germany we observed the bird with its PTT without an antenna. An

Osprey also lost or removed the antenna after only a few months. It is much easier to find the reasons for tag failure in breeding adults that return to their nest site year after year. There are methods for locating PTTs that are transmitting from a dead bird or detached from the bird (Howey 2002, Bates et al. 2003, Peske and McGrady 2005). Finding the PTT can provide valuable biological information and be cost-effective because most units can be refurbished for about $300 to $500, and used again.

Tracking Options

Finally, satellite telemetry is one of many options for marking raptors. Before deciding to use telemetry we encourage persons to consider carefully (1) their objectives and (2) the possible effects of marking on the birds and their implications for the results. The literature provides many examples of studies in which satellite telemetry has provided valuable information (Table 1). Consultation with manufacturers about options can be very useful, and is especially important for programming the function of transmitters and receivers to maximize performance.

ACKNOWLEDGMENTS

We thank A. Aebischer, P. Howey, B.D. Chancellor, and K. Bates for their helpful comments on drafts of our manuscript.

LITERATURE CITED

ALERSTAM, T., M. HAKE AND N. KJELLÉN. 2006. Temporal and spatial patterns of repeated migratory journeys by Ospreys. *Anim. Behav.* 71:555–566.

BATES, K., K. STEENHOF AND M.R. FULLER. 2003. Recommendations for finding PTTs on the ground without VHF telemetry. Proceedings of the Argos Animal Tracking Symposium, 24–26 March 2003, Annapolis, Maryland. Available on a CD from Service Argos, Inc., Largo MD U.S.A.

BRITTEN, M.W., P. L. KENNEDY, AND S. AMBROSE. 1999. Performance and accuracy evaluation of small satellite transmitters. *J. Wildl. Manage.* 63:1349–1358.

BRODEUR, S., R. DÉCARIE, D.M. BIRD AND M.R. FULLER. 1996. Complete migration cycle of Golden Eagles breeding in northern Quebec. *Condor* 98:293–299.

CLS AMERICA. 2003. Proceedings of the Argos Animal Tracking Symposium, 24–26 March 2003, Annapolis, Maryland. CLS America, Largo, MD U.S.A.

CRAIGHEAD, D. AND R. SMITH. 2003. The implications of PTT location accuracy on a study of Red-tailed Hawks. Proceedings of the Argos Animal Tracking Symposium, 24–26 March 2003, Annapolis, Maryland. Available on a CD from Service Argos, Inc., Largo MD U.S.A.

DOOLEY, J.A., P.B. SHARPE AND D.K. GARCELON. 2004. Abstract: Monitoring movements and foraging of Bald Eagles (*Haliaeetus leucocephalus*) reintroduced on the northern Channel Islands, California using solar Argos/GPS PTTs and VHF telemetry. Pages 26–27 in Program and abstracts: Raptor Research Foundation-California Hawking Club annual meeting, Bakersfield, California, 10–13 November 2004. Raptor Research Foundation, Clovis, CA U.S.A.

EASTHAM, C.P., J.L. QUINN AND N.C. FOX. 2000. Saker *Falco cherrug* and Peregrine *Falco peregrinus* falcons in Asia; determining migration routes and trapping pressure. Pages 247–258 in R.D. Chancellor and B.-U. Meyburg [EDS.], Raptors at risk. World Working Group on Birds of Prey and Owls, Berlin, Germany and Hancock House, London, United Kingdom.

ELLIS, D.H., S.L. MOON AND J.W. ROBINSON. 2001. Annual movements of a Steppe Eagle (*Aquila nipalensis*) summering in Mongolia and wintering in Tibet. *Bombay Nat. Hist. Soc.* 98:335–340.

FULLER, M.R., W.S. SEEGAR AND L.S. SCHUECK. 1998. Routes and travel rates of migrating Peregrine Falcons *Falco peregrinus* and Swainson's Hawks *Buteo swainsoni* in the western hemisphere. *J. Avian Biol.* 29:433–440.

———, D. HOLT AND L.S. SCHUECK. 2003. Snowy Owl movements: variation on the migration theme. Pages 359–366 in P. Berthold, E. Gwinner, and E. Sonnenschein [EDS.], Avian migration. Springer-Verlag, Berlin, Germany.

———, J.J. MILLSPAUGH, K.E. CHURCH AND R.E. KENWARD. 2005. Wildlife radio telemetry. Pages 377–417 in C.E. Braun [ED.], Techniques for wildlife investigations and management, 6th Ed. The Wildlife Society, Bethesda, MD U.S.A.

GOLDSTEIN, M.I., T.E. LACHER, JR., B. WOODBRIDGE, M.J. BECHARD, S.B. CANAVELLI, M.E. ZACCAGNINI, G.P. COBB, E.J. SCOLLON, R. TRIBOLET AND M.J. COOPER. 1999. Monocrotophos induced mass mortality of Swainson's Hawks in Argentina, 1995–96. *Ecotoxicology* 8:201–214.

GROS, P. AND J.-P. MALARDÉ. 2006. Argos performance in Europe. *Tracker News* 7:8.

HAINES, A.M., M.J. MCGRADY, M.S. MARTELL, B.J. DAYTON, M.B. HENKE AND W.S. SEEGAR. 2003. Migration routes and wintering locations of Broad-winged Hawks tracked by satellite telemetry. *Wilson Bull.* 115:166–169.

HAKE, M., N. KJELLÉN AND T. ALERSTAM. 2001. Satellite tracking of Swedish Ospreys *Pandion haliaetus*: autumn migration routes and orientation. *J. Avian Biol.* 32:47–56.

———, N. KJELLÉN AND T. ALERSTAM. 2003. Age-dependent migration strategy in Honey Buzzards *Pernis apivorus* tracked by satellite. *Oikos* 103:385–396.

HAYS, G.C., S. AKESSON, B.J. GODLEY, P. LUSCHI AND P. SANTIDRIAN. 2001. The implications of location accuracy for the interpretation of satellite-tracking data. *Anim. Behav.* 61:1035–1040.

HEDENSTRÖM, A. 1997. Predicted and observed migration speed in Lesser Spotted Eagle *Aquila pomarina*. *Ardea* 85:29–36.

HENNY, C.J., W.S. SEEGAR, M.A. YATES, T.L. MAECHETLE, S.A. GANUSEVICH AND M.R. FULLER. 2000. Contaminants and wintering areas of Peregrine Falcons, *Falco peregrinus*, from the Kola Peninsula, Russia. Pages 871–878 in R.D. Chancellor and B.-U. Meyburg [EDS.], Raptors at risk. World Working Group

on Birds of Prey and Owls, Berlin, Germany and Hancock House, London, United Kingdom.

HIGUCHI, H., H. SHIU, H. NAKAMURA, A. UEMATSU, K. KUNO, M. SAEKI, M. HOTTA, K. TOKITA, E. MORIYA, E. MORISHITA AND M. TAMURA. 2005. Migration of Honey-buzzards *Pernis apivorus* based on satellite tracking. *Ornithol. Science* 4:109–115.

HOOPER, M.J., P. MINEAU, M.E. ZACCAGNINI, G.W. WINEGRAD AND B. WOODBRIDGE. 1999. Monocrotophos and the Swainson's Hawk. *Pesticide Outlook* 10: 97–102.

HOWEY, P. 2002. Useful tips - finding a lost PTT (part 2). Microwave Telemetry, Inc., Newsletter Winter 2000:4. (http://microwave telemetry.com/newsletters.winter00_page4.pdf) (last accessed 7/6/2006).

———. 2005. Argos performance in Europe. *Tracker News* 6:8.

KJELLÉN, N., M. HAKE AND T. ALERSTAM. 2001. Timing and speed of migration in male, female and juvenile Ospreys *Pandion haliaetus* between Sweden and Africa as revealed by field observations, radar and satellite tracking. *J. Avian Biol.* 32:57–67.

LAING, D.K., D.M. BIRD AND T.E. CHUBBS. 2005. First complete migration cycles for juvenile Bald Eagles (*Haliaeetus leucocephalus*) from Labrador. *J. Raptor Res.* 39:11–18.

LAUNAY, F. AND M.G. MULLER. 2003. Falcon release and migration. *Falco* 22:22–23.

MARTELL, M.S., C.J. HENNY, P.E. NYE AND M.J. SOLENSKY. 2001. Fall migration routes, timing, and wintering sites of North American Ospreys as determined by satellite telemetry. *Condor* 103:715–724.

MCGRADY, M.J., T.L. MAECHTLE, J.J. VARGAS, W.S. SEEGAR AND M.C. PORRAS PEÑA. 2002. Migration and ranging of Peregrine Falcons wintering on the Gulf of Mexico coast, Tamaulipas, Mexico. *Condor* 104:39–48.

———, M. UETA, E.R. POTAPOV, I. UTEKHINA, V. MASTEROV, A. LADYGUINE, V. ZYKOV, J. CIBOR, M. FULLER AND W.S. SEEGAR. 2003. Movements by juvenile and immature Steller's Sea Eagles *Haliaeetus pelagicus* tracked by satellite. *Ibis* 145:318–328.

MEYBURG, B.-U., W. SCHELLER AND C. MEYBURG. 1995a. Migration and wintering of the Lesser Spotted Eagle *Aquila pomarina*: a study by means of satellite telemetry. *J. Ornithol.* 136:401–422.

———, J.M. MENDELSON, D.H. ELLIS, D.G. SMITH, C. MEYBURG AND A.C. KEMP. 1995b. Year-round movements of a Wahlberg's Eagle *Aquila wahlbergi* tracked by satellite. *Ostrich* 66:135–140.

———, C. MEYBURG AND J.-C. BARBRAUD. 1998. Migration strategies of an adult Short-toed Eagle *Circaetus gallicus* tracked by satellite. *Alauda* 66:39–48.

———, J. MATTHES AND C. MEYBURG. 2002. Satellite-tracked Lesser Spotted Eagle avoids crossing water at the Gulf of Suez. *Br. Birds* 95:372–376.

———, P. PAILLAT AND C. MEYBURG. 2003. Migration routes of Steppe Eagles between Asia and Africa: a study by means of satellite telemetry. *Condor* 105:219–227.

———, M. GALLARDO, C. MEYBURG AND E. DIMITROVA. 2004a. Migrations and sojourn in Africa of Egyptian Vultures (*Neophron percnopterus*) tracked by satellite. *J. Ornithol.* 145:273–280.

———, C. MEYBURG, T. BELKA, O. SREIBR AND J. VRANA. 2004b. Migration, wintering and breeding of a Lesser Spotted Eagle (*Aquila pomarina*) from Slovakia tracked by satellite. *J. Ornithol.* 145:1–7.

————, C. MEYBURG, T. MIZERA, G. MACIOROWSKI AND J. KOWALSKI. 2005. Family break up, departure, and autumn migration in Europe of a family of Greater Spotted Eagles (*Aquila clanga*) as reported by satellite telemetry. *J. Raptor Res.* 39:462–466.

————, C. MEYBURG, J. MATTHES AND H. MATTHES. 2006. GPS satellite tracking of Lesser Spotted Eagles (*Aquila pomarina*): home range and territorial behaviour. *Vogelwelt* 127:127–144.

————, C. MEYBURG AND F. FRANCK-NEUMANN. 2007a. Why do female Lesser Spotted Eagles (*Aquila pomarina*) visit strange nests remote from their own? *J. Ornithol.* 148:157–166.

————, C. MEYBURG, J. MATTHES AND H. MATTHES. 2007b. Spring migration, late arrival, temporary mate change and breeding success in the Lesser Spotted Eagle *Aquila pomarina*. *Vogelwelt* 128:21–31.

MILLSAP, B., T. BREEN, E. MCCONNELL, T. STEFFER, L. PHILLIPS, N. DOUGLASS AND S. TAYLOR. 2004. Comparative fecundity and survival of Bald Eagles fledged from suburban and rural natal areas in Florida. *J. Wildl. Manage.* 68:1018–1031.

MURRAY, D.L. AND M.R. FULLER. 2000. Effects of marking on the life history patterns of vertebrates. Pages 15–64 in L. Boitani and T. Fuller [EDS.], Research techniques in ethology and animal ecology. Columbia University Press, New York, NY U.S.A.

PESKE, L. AND M.J. MCGRADY. 2005. From the field: a system for locating satellite-received transmitters (PTTs) in the field. *Wildl. Soc. Bull.* 33:307–312.

RAFANOMEZANTSOA, S., R.T. WATSON AND R. THORSTROM. 2002. Juvenile dispersal of Madagascar Fish-eagles tracked by satellite telemetry. *J. Raptor Res.* 36:309–314.

ROSE, E.F., W. ENGLISH AND A. HAMILTON. 1993. Abstract: Paired use of satellite and VHF telemetry on rehabilitated Bald Eagles. *J. Raptor Res.* 27:92.

SCHMUTZ, J.K., C.S. HOUSTON AND G.L. HOLROYD. 1996. Southward migration of Swainson's Hawks: over 10,000 km in 54 days. *Blue Jay* 54:70–76.

SNYDER, N.F.R., S.R. BEISSINGER AND M.R. FULLER. 1989. Solar radio transmitters on Snail Kites in Florida. *J. Field Ornithol.* 60:171–177.

SOUTULLO, A., V. URIOS AND M. FERRER. 2006a. How far away in an hour? Daily movements of juvenile Golden Eagles (*Aquila chrysaetos*) tracked with satellite telemetry. *J. Ornithol.* 147:69–72.

————, V. URIOS, M. FERRER AND S.G. PEÑARRUBIA. 2006b. Post-fledging behaviour in Golden Eagles *Aquila chrysaetos*: onset of juvenile dispersal and progressive distancing from the nest. *Ibis* 148:307–312.

STEENHOF, K., M.R. FULLER, M.N. KOCHERT AND K.K. BATES. 2005. Long-range movements and breeding dispersal of Prairie Falcons from southwest Idaho. *Condor* 107:481–496.

————, K.K. BATES, M.R. FULLER, M.N. KOCHERT, J.O. MCKINLEY AND P.M. LUKACS. 2006. Effects of radio marking on Prairie Falcons: attachment failures provide insights about survival. *Wildl. Soc. Bull.* 34:116–126.

STRIKWERDA, T.E., M.R. FULLER, W.S. SEEGAR, P.W. HOWEY AND H.D. BLACK. 1986. Bird born satellite transmitter and location program. *Johns Hopkins APL Tech. Digest* 7:203–208.

THORUP, K., T. ALERSTAM, M. HAKE AND N. KJELLÉN. 2003a. Can vector summation describe the orientation system of juvenile Ospreys and Honey Buzzards? An analysis of ring recoveries and satellite tracking. *Oikos* 103:350–359.

————, T. ALERSTAM, M. HAKE AND N. KJELLÉN. 2003b. Bird orientation: compensation for wind drift in migrating raptors is age dependent. *Proc. Royal Soc. London, B* 270:S8–S11.

————, T. ALERSTAM, M. HAKE AND N. KJELLÉN. 2006a. Traveling or stopping of migrating birds in relation to wind: an illustration for the Osprey. *Behav. Ecol* 17:497–502.

————, M. FULLER, T. ALERSTAM, M. HAKE, N. KJELLÉN AND R. STRANDBERG. 2006b. Do migratory flight paths of raptors follow constant geographical or geomagnetic courses? *Anim. Behav.* 72:875–880.

UETA, M. AND H. HIGUCHI. 2002. Difference in migration pattern between adult and immature birds using satellites. *Auk* 119:832–835.

————, F. SATO, J. NAKAGAWA AND N. MITA. 2000. Migration routes and differences of migration schedule between adult and young Steller's Sea Eagles *Haliaeetus pelagicus*. *Ibis* 142:35–39.

WOODBRIDGE, B., K.K. FINLEY AND S.T. SEAGER. 1995. An investigation of the Swainson's Hawk in Argentina. *J. Raptor Res.* 29:202–204.

C. Stable Isotopes and Trace Elements

KEITH A. HOBSON

Environment Canada, 11 Innovation Blvd.,

Saskatoon, Saskatchewan, S7N 3H5, Canada

INTRODUCTION

Raptors are relatively large migratory birds and as such are amenable to being equipped with both radio and satellite transmitters. Radio and satellite tracking provide the very best information on individual movements and making connections between breeding, wintering, and stopover sites (Webster et al. 2001). However, in addition to expense and limitations of battery life, like all mark-and-recapture techniques where "recapture" is a locational fix, tracking is limited by the initial marked sample, which is not necessarily representative of the population of interest. This also applies to the use of leg bands and other external markers. Such concerns can be overcome to some degree by the use of endogenous markers, which, because initial marking is not required, rely only on the recaptured population (Rubsenstein and Hobson 2004). Endogenous markers of interest include naturally occurring stable-isotope and trace-element profiles as well as genetic and other molecular markers. Here, I focus on the use of stable isotopes and trace elements to track spatial movements of raptors. Interestingly, raptors have figured prominently in the development of these techniques.

STABLE ISOTOPES

Isotopes are forms of an element that differ only in atomic mass due to a differential number of neutrons in the nucleus. Typically, they have identical chemical properties, but their mass difference confers different kinetic properties on molecules that include them. Stable-isotope abundance of any element is usually expressed as a ratio of the more rare, heavy form to that of the more common, lighter form. Stable-isotope ratios of light elements of greatest interest to ecological applications are those of carbon ($^{13}C/^{12}C$), nitrogen ($^{15}N/^{14}N$), sulfur ($^{34}S/^{32}S$), hydrogen ($^{2}H/^{1}H$) and oxygen ($^{18}O/^{16}O$). Isotopes of heavier elements such as strontium (^{87}Sr) and lead (^{210}Pb) also are particularly useful but require more involved analytical procedures. Stable-isotope ratios of the light elements are measured with isotope-ratio mass spectrometry (IRMS) and are expressed in abundance relative to international standards in delta (δ) notation, and are reported as parts-per-thousand deviation from those standards. This is an extremely well-established field in analytical chemistry and highly accurate measurements are routinely achieved in most laboratories. Fortunately, various biogeochemical processes in nature result in materials that differ in their stable-isotope abundance and these differences can be exploited to infer origins of organisms that come into equilibrium with local food webs.

The basic premise of all stable-isotope applications to animal studies is that isotopic abundance in diet is related directly to isotopic abundance in the consumer. In many cases, consumer tissues differ in their isotopic composition relative to diet by a relatively constant discrimination factor. This simple relationship brings up two important principles in applying stable isotope measurements to food webs in general and to migratory tracking in particular. First, the diet-tissue isotopic discrimination factor can be tissue-specific and these specifications may need to be established experimentally

(Hobson and Clark 1992a). Second, for metabolically active tissues, this relationship is not static but is based on equilibrium time constants related to elemental turnover rates in the tissue (Hobson and Clark 1992b). Thus, choice of tissue is of fundamental importance when deciphering isotopic information. For example, Duxbury et al. (2003) provided experimental evidence that juvenal down or juvenal plumage, but not natal down, of nestling Peregrine Falcons (*Falco peregrinus*) accurately reflects their local diet. Three key components must be considered when inferring origins of migrant birds: (1) the isotopic signature of the source and how this varies spatially, temporally, or both, (2) the isotopic discrimination associated with the tissue being used to reflect that source, and (3) the isotopic turnover rate of that tissue.

CHOICE OF TISSUE

Tissues for isotopic measurement can be metabolically active or inactive. Metabolically active tissues provide a "moving window" of past origins and the width of that window depends on the elemental turnover rate associated with that tissue. For fast-metabolic-rate tissues like liver or blood plasma, the window is in the order of a week (Hobson and Clark 1992a). Muscle and whole blood have slower turnover rates and information can be derived for a period of the order of up to six weeks. Bone collagen has an exceptionally slow turnover rate and so can provide dietary information averaged over years. The problem facing researchers who wish to use metabolically active tissues to infer origins of migratory birds is that precise metabolic turnover rates for wild, migrating birds essentially are unknown (Hobson 2005a).

Metabolically inactive tissues including keratin of feathers and talons present information on origins typical of the period of growth (assuming no endogenous reserves are used in their formation). In cases involving raptors whose molt schedules are well known, the isotopic measurement of a single feather can be a powerful tool in determining migratory connectivity. The disadvantage to using feathers is that if they are lost they can be replaced at locations other than those where they first grew. In addition, we still do not understand molt schedules of several species well enough, and it is possible, although difficult to corroborate, that failed breeders might leave the breeding grounds early and molt en route. The good news is that stable-isotope methods can be used to determine molt patterns as well as breeding origins. Wassenaar and Hobson (2001) confirmed that adult Swainson's Thrushes (*Catharus ustulatus*) molted flight feathers south of their actual breeding grounds. Talons of birds arriving in the spring may give good isotopic information on environments occupied on the wintering grounds because they grow relatively slowly (Bearhop et al. 2003, Mazerolle and Hobson 2005) and so will represent diet on the order of the previous weeks to months.

ISOTOPIC LANDSCAPES

Fortunately, several isotopic patterns known in nature can be exploited to infer origins of migratory birds and other organisms. These patterns vary according to individual isotopes, and how they behave in various biogeochemical reactions. For our purposes, these patterns can be grouped into dietary signals that are related to local biome or climatic conditions and "isoscapes," or to those related to larger-scale isotopic patterns based on underlying geology or continental patterns in precipitation.

The most studied and well known stable isotopic pattern in nature is that of stable carbon isotope signatures associated with photosynthetic pathways. This process is based on fundamentally different molecular fixation of carbon during photosynthesis that results either in a three- (C-3) or four- (C-4) carbon molecular substrate and corresponding different behavior of ^{13}C and ^{12}C in each case. Plants with a C-3 photosynthetic pathway have tissues that are more depleted, or lower in their $\delta^{13}C$ values, than those with a C-4 or CAM pathway. C-3 plants also show remarkable variation in $\delta^{13}C$ signature based on mechanisms associated with water-use efficiency (reviewed by Lajtha and Marshall 1994). The net result is that C-3 plants generally become more enriched in ^{13}C under more xeric conditions than under cooler or more mesic conditions (e.g., Marra et al. 1998). Hobson and Wassenaar (2001) demonstrated that wintering Loggerhead Shrikes (*Lanius ludovicianus*) in the southern United States and northern Mexico originated from areas with food webs ranging from pure C-3 to pure C-4 photosynthetic composition. However, because we do not have useful spatial resolution of the distribution of C-3 versus C-4 biomes throughout much of the range for most species, such information will be quite limited in inferring origins of birds such as shrikes (but see Still et al. 2003). CAM plants are relatively rare in North America but are well represented in dry areas by cacti. Wolf and Martinez del Rio (2000) and Wolf et al. (2002) have examined the dependence of White-winged Doves (*Zenaida asiatica*) and Mourning Doves (*Z. macroura*) on saguaro cactus (*Carnegiea gigantea*) and are currently using this as a marker for populations of doves originating in the American Southwest.

The stable isotopes of several elements including C, N, H, O, S, differ in marine versus terrestrial and freshwater food webs due to isotopic differences in inorganic nutrients available to primary production and, as a result, marine inputs to raptor diets can be traced isotopically. Lott and Smith (2006) were able to correct deuterium isotope (δD) values of feathers from nine different raptor species (see below) to account for links with marine food webs using δ^{34}S measurements. Certainly, dietary reconstructions based on raptor ingestion of seabirds or marine fish, or scavenging on marine-mammal carcasses should be relatively routine using the isotope approach, although there are cases where some terrestrial food webs overlap isotopically with marine food webs (e.g., terrestrial evaporative deposits can have similar δ^{34}S values as marine systems).

Stable-nitrogen isotope ratios (δ^{15}N) are extremely useful as indicators of trophic position (Kelly 2000). However, within terrestrial systems, land-use practices can influence stable-isotope abundance in food webs. Most notably, agricultural practices can alter δ^{15}N values in both upland and wetland systems. Soil nitrogen can vary isotopically within and among sites, but two processes can result in agricultural soils being more enriched in ^{15}N than temperate forest soils. These are the presence of animal-based fertilizers and the greater volatilization of isotopically lighter nitrogenous compounds such as ammonia from agricultural soils as a result of tillage and their lower acidity (Nadelhoffer and Fry 1994).

Deuterium

Without question, the single isotope that has shown the greatest potential for helping to elucidate origins of migratory birds in North America is deuterium. Its usefulness is based on the fact that stable-hydrogen isotope ratios in precipitation show a continent-wide pattern with a general gradient from enriched values in the southeast to more depleted values in the northwest (Fig. 1). This phenomenon is due to the fact that evaporation and precipitation are processes that can discriminate against or favor heavier, deuterium-containing water molecules and are, in turn, influenced by temperature,

relative humidity, distance from oceans and elevation (see Bowen et al. 2005). Following the first avian applications by Chamberlain et al. (1997) and Hobson and Wassenaar (1997), several studies have confirmed the strong association between growing-season average δD values in precipitation and those in feathers of birds grown at those locations (Bowen et al. 2005). Meehan et al. (2001) conducted the first deuterium study on raptors using feathers of Cooper's Hawk (*Accipiter cooperii*) and confirmed the continent-wide pattern could be used to estimate natal origins of birds migrating through Florida. The growing-season deuterium precipitation map was recently constructed for Europe (Hobson 2003). Duxbury (2004) conducted an isotopic baseline study on feathers of Burrowing Owls (*Athene cunicularia*) and Peregrine Falcons with the intent of ultimately tracking migrants to natal or molt origin. However, the most comprehensive feather deuterium map for North American raptors was constructed by Lott and Smith (2006). These authors measured feather δD values from museum specimens of raptors originating from sites across North America and provide a convenient digital isotopic surface amenable to geographic information systems (GIS) queries.

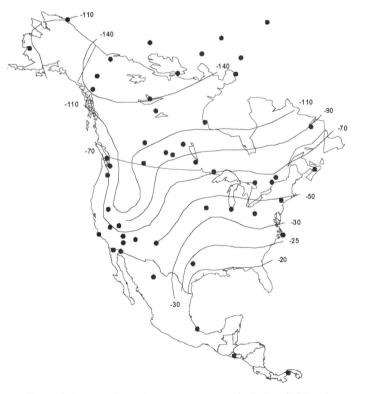

Figure 1. Pattern of growing-season average deuterium (‰) in rainfall for North America (after Hobson and Wassenaar 1997). Dots indicate long-term sampling stations. Note that feathers will be depleted relative to these contours due to isotopic discrimination.

A common question arising from the application of growing-season average precipitation contour maps for deuterium, based on the International Atomic Energy Agency (IAEA) database, is the robustness of the kriged (i.e., geographic) relationships. In any given year, how much variation in these patterns might be expected? This is not an easy question to answer because the geographical and temporal coverage in sampling sites for this database are variable. The patterns depicted by Hobson and Wassenaar (1997) and Hobson (2003) are based on about a 35-year IAEA record. However, several considerations increase our confidence in these relationships, at least qualitatively. The first is that short-term variation in precipitation signals will, to some extent, be smoothed out by the longer-term averaging of the growing season itself. Thus, in many areas, each feather measurement will, in effect, represent the average of many rainfall events and so will tend to smooth short-term fluctuations. This will not necessarily be the case in areas or times of lower precipitation or in areas that are subject to single or synoptic rainfall events. Nor will it apply to areas where groundwater or reservoirs form a significant source of hydrogen for local food webs. For a group of European sites where long-term data are available, several showed extremely small inter-year variation in average growing-season δD in precipitation, of the order of measurement error, whereas others, notably coastal sites, showed variation at least three to four times as high (Hobson 2005b). However, despite numerous potential sources of error, it is remarkable how well long-term average values of δD in precipitation are correlated with δD values of feathers grown in any given year, a relationship now demonstrated independently by several research groups. How well this relationship holds in future given climate change scenarios, of course, is unknown and will be an important area of additional research (Hobson 2005a).

Alternatives to using the long-term average contour maps include the direct measurement of isotopic patterns of interest for a particular year of interest (e.g., Hobson et al. 1999), and the creation of feather isotopic basemaps for each species or taxonomic group of interest (Duxbury 2004, Lott and Smith 2006). Meehan et al. (2003) determined that feathers grown by nestling Cooper's Hawks were more depleted in deuterium than those of attending adults grown at the same site. There are a number of possible explanations for this result including the possibility of dietary differences between age groups. Another possibility is that adult breeding raptors become relatively enriched in deuterium due to evaporative cooling throughout the extended nestling period (Meehan et al. 2003). Experiments with captive birds are needed to confirm if special consideration needs to be given to raptors when associating tissue δD values to origin.

Recently, Smith and Dufty (2005) examined feather δD values of adult and nestling Northern Goshawk (*A. gentilis*) feathers representing breeding territories across western North America. As expected, these authors found a general depletion in feather isotope δD values with latitude and distance from the coast. As with Meehan et al. (2003), these authors found that nestlings had lower δD values than adults at the same location. After controlling for location and local temperature, they also found considerable inter-individual variation in feather isotope profiles related to sex. Adult females had considerably higher δD values than males. Support was found for the hypothesis that such patterns arise from differences in evaporative cooling in those raptors that "work" during feather growth while provisioning young. These authors recommend that future studies using feathers to delineate origin should consider different isotopic basemaps for adults and juveniles.

Trace Elements

Patterns of trace elements in feathers ultimately are derived from diet, which, in turn, is influenced strongly by surficial geology, and as such are expected to provide spatial information. The use of trace elements was a comparatively early approach to using endogenous signatures in avian-migration tracking (early reviews by Means and Peterle 1982, Kelsall 1984). The method has great intuitive appeal because it is possible to measure relative abundance of numerous elements in feathers and so the chances of acquiring a unique signature for an individual or population are increased. Recent developments in analytical techniques allow the routine measurement of concentrations in feathers of numerous elements, including As, Cd, Mg, Mn, Mo, Se, Sr, Co, Fe, Zn, Li, P, Ti, V, Ag, Cr, Ba, Hg, Pb, S, Ni, and Cu. Despite the potential of this technique, the field was largely abandoned a decade ago owing to several concerns over its reliability. Some of these criticisms have since been addressed through improvement in sample-preparation and measurement techniques that made elemental measurements much more reliable, but the stigma remains.

The first attempts to use trace-element analysis to

infer geographical origin were in waterfowl (e.g., Devine and Peterle 1968, Kelsall and Calaprice 1972, Kelsall et al. 1975, Hanson and Jones 1976, Kelsall and Burton 1979). These studies met with variable success but were followed by an excellent study by Parrish et al. (1983), which clearly distinguished three natal populations of Peregrine Falcons by measuring as few as five trace elements in feathers (Fig. 2; see also Barlow and Bortolotti 1988). However, several studies presented evidence of considerable intrapopulation variation in feather elemental profiles related to age (Hanson and Jones 1976) and sex (Hanson and Jones 1974, Kelsall and Burton 1979, Bortolotti and Barlow 1988). The causes of such differences are poorly understood but are likely related to hormonal and metabolic mechanisms influencing secretion of trace elements into feathers. Such variation has been problematic because it usually makes discrimination among populations difficult or may create results that are artifacts of sampling biases (Bortolotti et al. 1990).

In addition to doubts raised over intrapopulational variation in elemental profiles, a more fundamental issue that has not been addressed adequately is how such profiles change among disparate populations. For example, Bortolotti et al. (1989) found that Spruce Grouse (*Falcipennis canadensis*) from similar forest types hundreds of kilometers apart had similar feather elemental compositions, whereas those from adjacent populations occupying different forest types were quite different. Similarly, Szép et al. (2003) determined that feathers from populations of Sand Martins (*Riparia riparia*) grown at locations across Europe varied with age within colonies, and they also showed that similarity and differences in elemental profiles were not related to distance separating colonies. The value of trace-element profiles in making connections between breeding, wintering, and stopover sites, therefore, will depend on the case in question and on how spatially discrete the populations of interest are.

For highly colonial or aggregated species, it may well be possible to characterize the different colonies or breeding regions according to trace-element composition in feathers. If we are fortunate, such areas may have useful elemental fingerprints. For more dispersed species it simply may simply be impossible to describe the trace element profile patterns across the range well enough to reach unambiguous conclusions about origins. This is not to suggest that this field of research will not prove fruitful. Rather, in contrast to the use of continental deuterium precipitation maps, it will simply be

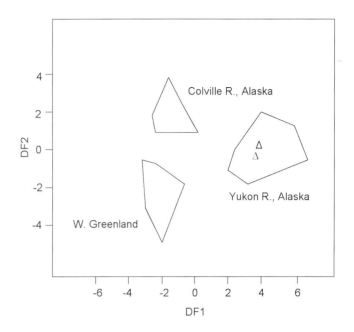

Figure 2. Trace element segregation of three populations of Peregrine Falcons (*Falco peregrinus*). Discriminant function scores based on 14 nestling feather trace-element concentrations as predictors. Polygons represent total outer boundaries of each nestling sample and △ represents two adult birds captured at South Padre Island, Texas (after Parrish et al. 1983).

difficult to make *a priori* predictions about expected trace element profiles, especially at regional scales, without detailed geological information. Trace element profiles in bird feathers may well be useful for less traditional applications. Szép et al. (2003), for example, suggested that because trace element analysis is sensitive to micro-geographical differences among individuals, this approach might be better suited to elucidating migration or wintering behavior at the level of the individual or small group. Bortolotti et al. (1990) suggested that if the effects of age and sex on trace-element profiles were well known, then population demographic information might be gleaned from elemental patterns within study populations.

Measurement techniques for establishing trace-element profiles in tissues have advanced tremendously over the last several decades. Some approaches such as Inductively Coupled Plasma (ICP) techniques require the dissolution of the sample to a liquid form prior to spectral analysis, whereas others such as the Neutron Activation Technique require that the sample be irradiated but not destroyed. Both approaches have advantages and disadvantages. The recent development of ICP-MS technology, which interfaces a mass spectrometer with an ICP machine to provide isotopic measurements of a suite of elements, certainly holds great promise for migration-tracking studies. By increasing the number of elements and species of isotopes that can be examined, it presumably allows for much greater resolution and for tracing isotope signatures hitherto impossible by more conventional MS techniques.

FUTURE DIRECTIONS

The use of stable isotopes and trace elements to track diet and geographical origins of raptors in North America and elsewhere shows significant promise and several programs are now underway that routinely collect feathers for this purpose. Clearly, an understanding of precise molt patterns for various feather tracks will be invaluable for all species of interest. Ideally, obtaining feathers that represent breeding and wintering grounds would allow analysis of two temporal and spatial samples from the same individual. Raptors can be raised in captivity and the continued investigation of isotopic and trace element behavior in experimental birds is highly encouraged.

A number of important areas require continued research (Hobson 2005a, Smith and Dufty 2005, Lott and Smith 2006). For raptors we need to know if feather growth during breeding results in increases in feather δD values and if so, how we might produce appropriate isotopic basemaps for these birds (Lott and Smith 2006). Second, we must better understand factors contributing to variance in precipitation and feather δD values and incorporate a more rigorous statistical approach to how we assign individual birds to origins. Certainly, the advent of GIS tools and Bayesian statistical techniques will be incorporated increasingly into isotopic studies involving raptors (e.g. Mazerolle et al. 2005, Wunder et al. 2005, Lott and Smith 2006). Raptor biologists and enthusiasts are uniquely positioned as a group to assist in the necessary controlled studies involving birds raised on known, isotopically homogenous diets and water sources to answer some fundamental questions related to isotope and trace element techniques. Apart from issues surrounding the evaporative cooling enrichment of raptor tissues during work, more basic information related to elemental turnover and patterns of isotopic distributions among feathers and other tissues within and between individuals are now needed.

ACKNOWLEDGMENTS

I thank David Bird for encouraging this review and an anonymous reviewer who improved an earlier draft of the manuscript.

LITERATURE CITED

BARLOW, J.C. AND G.R. BORTOLOTTI. 1988. Adaptive divergence in morphology and behavior in some New World island birds: with special reference to *Vireo altiloquus*. Pages 1535–1549 in H. Ouellet [ED.], Acta XIX Congressus Internationalis Ornithologici, Ottawa, Ontario, 1986. National Museum of Natural Sciences, Ottawa, Ontario Canada.

BEARHOP, S., R.W. FURNESS, G.H. HILTON, S.C. VOTIER AND S. WALDRON. 2003. A forensic approach to understanding diet and habitat use from stable isotope analysis of (avian) claw material. *Functional Ecol.* 17:270–275.

BORTOLOTTI, G.R. AND J.C. BARLOW. 1988. Some sources of variation in the elemental composition of Bald Eagle feathers. *Can. J. Zool.* 63:2707–2718.

————, K.J. SZUBA, B.J. NAYLOR AND J.F. BENDELL. 1989. Mineral profiles of Spruce Grouse feathers show habitat affinities. *J. Wildl. Manage.* 48:853–866.

————, K.J. SZUBA, B.J. NAYLOR AND J.F. BENDELL. 1990. Intra-population variation in mineral profiles of feathers of Spruce Grouse. *Can. J. Zool.* 68:585–590.

BOWEN, G.J., L.I. WASSENAAR AND K.A. HOBSON. 2005. Application of stable hydrogen and oxygen isotopes to wildlife forensic investigations at global scales. *Oecologia* 143:337–348.

CHAMBERLAIN, C.P., J.D. BLUM, R.T. HOLMES, X. FENG, T.W. SHERRY AND G.R. GRAVES. 1997. The use of isotope tracers for identifying populations of migratory birds. *Oecologia* 109:132–141.

DEVINE, T. AND T.J. PETERLE. 1968. Possible differentiation of natal areas of North American waterfowl by neutron activation analysis. *J. Wildl. Manage.* 32:274–279.

DUXBURY, J.M. 2004. Stable isotope analysis and the investigation of the migration and dispersal of Peregrine Falcons (*Falco peregrinus*) and Burrowing Owls (*Athene cunicularia hypugaea*). Ph.D. thesis. University of Alberta, Edmonton, Alberta, Canada.

————, G.L. HOLROYD AND K. MUEHLENBACHS. 2003. Changes in hydrogen isotope ratios in sequential plumage stages: an implication for the creation of isotope base-maps for tracking migratory birds. *Isot. Environ. Health Stud.* 39:179–189.

HANSON, H.C. AND R.L. JONES. 1974. An inferred sex differential in copper metabolism in Ross' Geese (*Anser rossi*): biogeochemical and physiological considerations. *Arctic* 27:111–120.

———— AND R.L. JONES. 1976. The biogeochemistry of Blue, Snow, and Ross' geese. *Ill. Nat. Hist. Surv. Spec. Publ.* 1.

HOBSON, K.A. 2003. Making migratory connections with stable isotopes. Pages 379–391 in P. Berthold, P. Gwinner, and E. Sannenschein [EDS.], Avian migration. Springer-Verlag, Berlin, Germany.

————. 2005a. Stable isotopes and the determination of avian migratory connectivity and seasonal interactions. *Auk* 122:1037–1048.

————. 2005b. Flying fingerprints: making connections with stable isotopes and trace elements. Pages 235–248 in R. Greenberg and P.P. Marra [EDS.], Birds of two worlds: the ecology and evolution of migratory birds. Johns Hopkins University Press, Baltimore, MD U.S.A.

——— AND R.W. CLARK. 1992a. Assessing avian diets using stable isotopes. II: factors influencing diet-tissue fractionation. *Condor* 94:189–197.

——— AND R.W. CLARK. 1992b. Assessing avian diets using stable isotopes. I: turnover of carbon-13 in tissues. *Condor* 94:181–188.

——— AND L.I. WASSENAAR. 1997. Linking breeding and wintering grounds of neotropical migrant songbirds using stable hydrogen isotopic analysis of feathers. *Oecologia* 109:142–148.

——— AND L.I. WASSENAAR. 2001. A stable isotope approach to delineating population structure in migratory wildlife in North America: an example using the Loggerhead Shrike. *Ecol. Applic.* 11:1545–1553.

———, L.I. WASSENAAR AND O.R. TAYLOR. 1999. Stable isotopes (δD and δ^{13}C) are geographic indicators of natal origins of monarch butterflies in eastern North America. *Oecologia* 120:397–404.

KELLY, J.F. 2000. Stable isotopes of carbon and nitrogen in the study of avian and mammalian trophic ecology. *Can. J. Zool.* 78:1–27.

KELSALL, J.P. 1984. The use of chemical profiles from feathers to determine the origins of birds. Pages 501–515 in J. Ledger [ED.]. Proceedings of the 5th Pan-African Ornithological Congress, Lilongwe, Malawi 1980. South African Ornithological Society, Johannesburg, South Africa.

——— AND R. BURTON. 1979. Some problems in identification of origins of Lesser Snow Geese by chemical profiles. *Can. J. Zool.* 57:2292–2302.

——— AND J. R. CALAPRICE. 1972. Chemical content of waterfowl plumage as a potential diagnostic tool. *J. Wildl. Manage.* 36:1088–1097.

———, W.J. PANNEKOEK AND R. BURTON. 1975. Chemical variability in plumage of wild Lesser Snow Geese. *Can. J. Zool.* 53:1369–1375.

LAJTHA, K. AND J.D. MARSHALL. 1994. Sources of variation in the stable isotopic composition of plants. Pages 1–21 in K. Lajtha and R.H. Michener [EDS.], Stable isotopes in ecology and environmental sciences. Blackwell Scientific, Oxford, United Kingdom.

LOTT, C.A. AND J.P. SMITH. 2006. A GIS approach to estimating the origins of migratory raptors in North America using hydrogen stable isotope ratios in feathers. *Auk* 123:822–835.

MARRA, P.P., K.A. HOBSON AND R.T. HOLMES. 1998. Linking winter and summer events in a migratory bird using stable carbon isotopes. *Science* 282:1884–1886.

MAZEROLLE, D. AND K.A. HOBSON. 2005. Estimating origins of short-distance migrant songbirds in North America: contrasting inferences from hydrogen isotope measurements of feathers, claws, and blood. *Condor* 107:280–288.

———, K.A. HOBSON AND L.I. WASSENAAR. 2005. Combining stable isotope and band-encounter analyses to delineate migratory patterns and catchment areas of White-throated Sparrows at a migration monitoring station. *Oecologia* 144:541–549.

MEANS, J.W. AND T.J. PETERLE. 1982. X-ray microanalysis of feathers for obtaining population data. Pages 465–473 in Transactions of 14th International Congress of Game Biologists.

MEEHAN, T.D., C.A. LOTT, Z.D. SHARP, R.B. SMITH, R.N. ROSENFIELD, A.C. STEWART AND R.K. MURPHY. 2001. Using hydrogen isotope geochemistry to estimate the natal latitudes of immature Cooper's Hawks migrating through the Florida Keys. *Condor* 103:11–20.

———, R.N. ROSENFIELD, V.N. ATUDOREI, J. BIELFELDT, L. ROSENFIELD, A.C. STEWART, W.E. STOUT AND M.A. BOZEK. 2003. Variation in hydrogen stable-isotope ratios between adult and nestling Cooper's Hawks. *Condor* 105:567–572.

NADELHOFFER, K J. AND B. FRY. 1994. Nitrogen isotope studies in forest ecosystems. Pages 22–44 in K. Lajtha and R.H. Michener [EDS.], Stable isotopes in ecology and environmental science. Blackwell Scientific, Oxford, United Kingdom.

PARRISH, J.R., D.T. ROGERS, JR. AND F.P. WARD. 1983. Identification of natal locales of Peregrine Falcons (*Falco peregrinus*) by trace element analysis of feathers. *Auk* 100:560–567.

RUBENSTEIN, D.R. AND K.A. HOBSON. 2004. From birds to butterflies: animal movement patterns and stable isotopes. *Trends Ecol. Evol.* 19:256–263.

SMITH, A.D. AND A.M. DUFTY, JR. 2005. Variation in the stable-hydrogen isotope composition of Northern Goshawk feathers: relevance to the study of migratory origins. *Condor* 107:547–558.

STILL, C.J., J.A. BERRY, G.J. COLLATZ AND R.S. DEFRIES. 2003. The global distribution of C3 and C4 vegetation: carbon cycle implications. *Global Biogeochemical Cycles* 17:1006–1029.

SZÉP, T., A.P. MØLLER, J. VALLNER, B. KOVACS AND D. NORMAN. 2003. Use of trace elements in feathers of Sand Martins *Riparia riparia* to identify molting areas. *J. Avian Biol.* 34:307–320.

WASSENAAR, L.I. AND K.A. HOBSON. 2001. A stable-isotope approach to delineate geographical catchment areas of avian migration monitoring stations in North America. *Environ. Sci. & Technol.* 35:1845–1850.

WEBSTER, M.S., P.P. MARRA, S.M. HAIG, S. BENSCH AND R.T. HOLMES. 2001. Links between worlds: unraveling migratory connectivity. *Trends Ecol. Evol.* 17:76–83.

WOLF, B.O. AND C. MARTINEZ DEL RIO. 2000. Use of saguaro fruit by White-winged Doves: isotopic evidence of a tight ecological association. *Oecologia* 124:536–543.

———, C. MARTINEZ DEL RIO AND J. BABSON. 2002. Stable isotopes reveal that saguaro fruit provides different resources to two desert dove species. *Ecology* 83:1286–1293.

WUNDER, M.B., C.L. KESTER, F.L. KNOPF AND R.O. RYE. 2005. A test of geographic assignment using isotope tracers in feathers of known origin. *Oecologia* 144:607–617.

Energetics

15

C‍HARLES R. B‍LEM

Department of Biology, Virginia Commonwealth University,
Richmond, Virginia 23284 U.S.A.

Studies of raptor energetics are extremely important in understanding the natural history of birds of prey. As terminal predators and sometimes keystone organisms, raptors are important components of many ecosystems. Because they are at the apex of the food pyramid, potentially they are the final repository of heavy metals, pesticides, and other stable compounds, and thus are important indicators of the general health of the system. Early publications describing methods of analyses of ecological energetics (Grodzinski et al. 1975) fail to mention raptors, but their importance has since become obvious.

Gessaman (1987) provided the single previous review of the terms, techniques, and equipment employed in studies of raptor energetics. His excellent survey remains the starting point for anyone attempting to begin work in this field. Much of that paper remains relevant today and anyone beginning a project involving energy analyses should consult it. Gessaman clearly pointed out the methods available at the time for measuring energy metabolism and how these measurements might be applied to studies of the activities of hawks, owls, eagles, and other raptors. Because some of these techniques have become readily available in user-friendly form or have not changed since Gessaman's review, I will not attempt to elaborate upon them, other than to present some of the basic terminology. For reviews of the literature on general avian energetics, see Gessaman (1973), Calder and King (1974), Kendeigh et al. (1977), Walsberg (1983), Blem (1990, 2000, 2004) and Dawson and Whittow (2000).

Compared with energetic studies in other avian taxa, there have been few studies of energy use by raptors. This may be due to the difficulties of maintaining sufficient numbers of relatively large, carnivorous birds in captivity, compared with smaller seed-eating birds. Likewise, caging a large bird that has been accustomed to ranging over a wide area is fraught with more difficulty than is caging a small passerine. Furthermore, because carnivorous birds sometimes egest pellets of undigested materials and drop parts of prey while preparing them for consumption, measuring energy ingestion by raptors may be a bit more difficult than in other groups of birds.

TERMINOLOGY AND CONCEPTS

This chapter describes, in a generic way, the methods that have been applied to measurements of raptor metabolism, and very briefly summarizes results of the few studies that have appeared since Gessaman (1987). Those who need more detail should see Gessaman's paper, or the specific references given below.

There are numerous components of total daily energy expenditure to be considered in studies of energy balance. Historically, terms identifying each of these items have varied from study to study. The words and concepts used here are those most often applied and are in general agreement with recent, significant reviews of avian energetics (e.g., Gessaman 1973, 1987; Karasov 1990, Blem 2000, Dawson and Whittow 2000).

Fundamentally, energy use by birds may be summarized as: **gross energy intake** (GEI) = **metabolized energy** (ME) + energy in egesta (**excretory energy** = EXE; feces, urine, egested pellets; Fig. 1). The proportion of GEI, which becomes ME, is called the **metabolizable energy coefficient** (MEC; Kendeigh et al. 1977, Karasov 1990). Units of metabolism should be expressed as kJ per unit time, or watts, but in many older papers energy units are given in kcal/unit time (1 kJ = 4.184 kcal). ME is the total of the costs of: (1) **basal metabolic processes** ("**basal metabolic rate**" = BMR), (2) thermoregulation (T), (3) **specific dynamic action** (SDA, see below), (4) work (W), and (5) production (P) (Fig. 2). Gross energy intake, excretory energy, and production typically are measured by means of bomb calorimetry in food consumption studies (see below). Components of metabolized energy such as BMR, T, SDA, and W are measured by indirect calorimetry in which metabolism is determined from oxygen consumption or carbon dioxide production (see **Indirect colorimetry**). MEC values also can be used to characterize the relative energy values of different food items. For example, different food items have different MEC values when consumed by the same avian species. Also, the MEC of individual food items may differ among bird species consuming the item.

BMR is the rate of oxygen consumption or carbon dioxide production by a normothermic (normal body temperature) organism: (1) held at ambient temperatures that are not stressful (i.e., within the zone of thermal neutrality; see below), (2) in the inactive phase of their daily activity cycle (i.e., in the dark for some owls, during daylight for all others), and (3) in a post-absorptive condition (not recently fed and without food in the gastrointestinal tract). No major productive processes can be occurring, including molt, fattening, or reproduction. The bird cannot be in hypothermia (i.e., its body temperature [T_b] cannot be below normal levels). BMR is assumed to be the minimal amount of energy expenditure by an endothermic animal under normal, nonstressful conditions. **Standard metabolism** (SM) is the metabolic rate of a bird measured in the same conditions as BMR measurements, except that the effects of thermal conditions are included. Thus, ambient temperature (T_a) may be so low that additional metabolic heat must be generated by the bird to maintain its body temperature, or conversely, ambient temperature is so high that the heat load begins to increase T_b and metab-

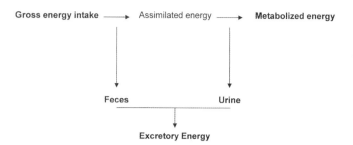

Figure 1. General avian energy balance scheme (modified from Blem 2000).

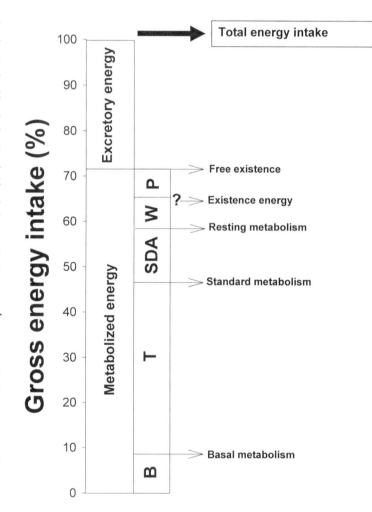

Figure 2. Components of energy use by raptors under differing environmental conditions. P = tissue production energy, W = energy expended in activities involving work, SDA = energy expended when food is being digested (specific dynamic action), T = cost of thermoregulation, and B = basal metabolism (modified from Blem 2000). The model assumes that there is a fixed maximum level of energy use which may vary seasonally and among individual birds.

olism increases along with it. SM changes with insulation, but apparently does not differ significantly among avian classes (Dawson and Whittow 2000). The cost of thermoregulation (T) is a function of the difference between T_b and T_a and how well the bird is insulated. **Fasting metabolic rate** (FMR) is BMR plus the metabolic costs of activity in the respirometry chamber. **Resting metabolic rate** (RMR) is usually measured over a range of ambient temperatures when the animal is relatively inactive, but may have recently eaten (i.e., is not post-absorptive; Kennedy and Gessaman 1991). Specific dynamic action (SDA) is the additional energy generated by digestion of food. This is a function of the exothermic digestive reactions and varies with composition of the food. **Existence energy** (Kendeigh et al. 1977) is the rate of energy use by a bird that is feeding, and subject to varying T_a, but restrained in a cage so that costs of locomotion are minimal (Stalmaster and Gessaman 1982, Hamilton 1985a). Existence energy usually is measured by food-consumption studies (see below) with tests extending for one to several days. Metabolized energy in free-living birds includes existence energy plus the costs of activity plus the costs of various productive processes such as molt, fat, deposition, and reproduction.

METHODS

The major methods for measuring avian metabolism include: (1) indirect calorimetry, (2) food-consumption studies, (3) doubly labeled water studies, (4) applications of telemetry interfaced with methods (1) or (2), and (5) time-energy budgets.

Indirect calorimetry and food consumption studies (by bomb calorimetry) remain the two most common methods for measuring the rate of energy metabolism of birds. Indirect calorimetry is a method by which oxygen consumption and carbon-dioxide production, or both, are measured by special gas analyzers. The specific techniques are complex and computation of energy use depends upon the method used (Gessaman 1987). Food consumption studies are less common but provide a means for quantification of energy metabolism and costs of production by measuring energy content of food, egested materials, and any associated productive processes (production of eggs, changes in biomass, and molt). The energy content of biological materials is commonly measured by means of bomb calorimetry. Total energy balance is a compromise between energy intake and all of the costs of existence: (1) thermoregulation, (2) kinetic energy of locomotion, (3) expenditures in production of body tissues such as reproductive tissues, new plumage, muscle mass, and energy storage as fat, and (4) maintenance. Note that energy storage can be a source or sink of energy, depending upon changes in body-tissue mass.

Laboratory Measurements

Indirect calorimetry. Indirect calorimetry is a method in which oxygen consumption or carbon-dioxide production is quantified, usually by means of open-flow respirometry. In this technique, a stream of air is drawn through a chamber housing the test subject in the dark or, alternatively, air is drawn through a mask fastened on the bird's head in such a fashion that all expired air is captured by the system. The chamber or masked bird is either held within a constant-temperature cabinet, or T_a, is monitored. The general configuration (Fig. 3) usually includes absorbers for carbon dioxide and water for the incoming air stream, and similar absorbers for outgoing air leaving the chamber but prior to going into the oxygen analyzer. Special oxygen analyzers, carbon dioxide detectors, or both, permit quantification of gas concentrations and, ultimately, respiration rates. Gessaman (1987) provides several photographs and diagrams, which illustrate variations in chambers and masks.

There are many ways in which a respirometry system may be configured. In the most generic arrangement (Fig. 3), a pump pulls ambient air through or controls gas mixtures within the respirometry chamber. Pulling air may eliminate pressure problems, which

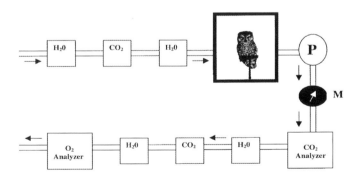

Figure 3. Generic arrangement of respirometry apparatus in a typical open-flow respirometry system for measuring oxygen consumption, carbon dioxide production, or both. Squares labeled H_2O and CO_2 represent tubes containing materials for removal of water and carbon dioxide from the air, respectively. P is a pump and M is a flowmeter.

may affect measurements of oxygen consumption. Gessaman (1987) provides numerous diagrams illustrating potential variations on this theme, and also presents a table that illustrates how such variations may affect the methods of computation. Although the basic equipment remains available, recent computer hardware and software options eliminate many of the problems inherent in the system at the time of Gessaman's paper, thus eliminating the need for the researcher to become a plumber, electrician, and computer-software guru. Presently, complete equipment choices that can be applied without much trouble or knowledge of electronics or computer science are commercially available (e.g., Sable Systems©).

Metabolic measurements made by respirometry include BMR, **standard metabolic rate** (SMR), FMR, and RMR. The rate of energy expenditure can be calculated from the volume of oxygen consumed or carbon dioxide produced. This requires a measurement or estimate of **respiratory quotient** (RQ). RQ is the volume of carbon dioxide produced/volume of oxygen consumed. As RQ increases, the energy equivalent of oxygen consumption increases and carbon dioxide decreases (see Gessaman 1987 for a conversion table).

Open-flow respirometry studies of captive birds enclosed in chambers produce data such as those represented in Fig. 4, where basal metabolism is measured within the **zone of thermal neutrality** (aka, thermal neutral zone; TNZ), and the costs of thermoregulation

are measured above the upper and below the lower critical temperatures. Within the TNZ, heat loss remains constant and, therefore, so does BMR. The balance is maintained by changes in insulation brought about by changes in posture, shunting of blood to and from skin and appendages, and by adjusting the thickness of plumage by fluffing or smoothing feathers. **Upper critical temperature** (UCT) represents the upper limit of effective thermoregulation. The T_a below which metabolism increases (as a result of the onset of shivering) is the **lower critical temperature** (LCT). At higher T_a's above UCT, ineffective heat dissipation results in hyperthermia, which rapidly drives metabolic rates upward as a result of increases in the T_b. Below the lower critical temperature, the metabolic rate increases as an inverse function of T_a and the conductance of the bird's plumage. Laboratory respirometry studies typically do not consider the costs or benefits of radiation and convection, or both. These factors are important to consider in free-living birds because wind movement may cause relatively large increases in SM, whereas basking may decrease SM levels by augmenting body heat due to absorption of solar radiation.

Because SMR and BMR are functions of body mass, metabolism rates typically are expressed on a weight-specific basis. This may present some computational difficulties because metabolic ratios typically do not fit normal distributions, and division of metabolism by body mass may not eliminate the effects of mass (i.e., make measurements of birds of different sizes equal, Blem 1984). Investigators who are not aware of such problems should check methods of covariance analysis as a possible solution.

Conductance (C) is the reciprocal of insulation. Birds with heavy insulation have small C values. At ambient temperatures below LCT, thermal conductance (and hence the reciprocal of insulation) can be calculated as $C = SM/(T_b - T_a)$, but a correction must be made for heat lost from lung and skin surfaces through evaporation. Individuals in torpor (various forms of hypothermia) do not follow these rules. However, notwithstanding some New and Old World vultures (see Bahat et al. 1998, Heath 1962), there is little evidence of adaptive hypothermia in any raptor with the exception of some ephemeral periodic decline of T_b in a few species (see Gessaman 1972).

Measurements of respiration provide insight into a great variety of physiological and ecological factors important in the life of raptors. For example, measurements of BMR and SMR can be used to compare ther-

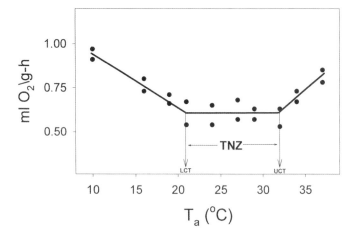

Figure 4. Respiratory metabolism as measured by means of open-flow respirometry for the Barn Owl (*Tyto alba*), redrawn from data in Edwards (1987). TNZ = zone of thermal neutrality = thermal neutral zone. LCT = lower critical temperature. UCT = upper critical temperature.

moregulatory capacity of different raptors (Graber 1962). This is particularly useful in comparisons of differences in insulation among species, during different times of the year, and at different geographic locations (Wasser 1986, Blem 2000). SM changes with differences in insulation, magnitude, and duration of previous exposure to temperature extreme, and biochemical shifts within the bird. Changes in conductance are generally caused by increases or decreases in plumage thickness, but deposition of subcutaneous fat also may cause small changes in insulation (Blem 1990).

In addition, there may be ephemeral physiological adjustments in response to temperature. **Acclimation** involves compensatory physiological changes in response to maintained deviations in ambient temperature, generally under laboratory conditions. **Acclimatization** is a similar change under natural conditions, which may include multiple environmental changes such as seasonal adaptations.

Measurements of SM often have been applied to studies of body-temperature regulation (e.g., Chaplin et al. 1984), but respirometry also has been applied to studies of metabolism of eggs (Hamilton 1985b), roosting (Keister et al. 1985), the costs of flight (Gessaman 1980, Masman and Klaassen 1987), and development of thermoregulation in nestlings (Kirkley and Gessaman 1990).

Measurements of the energetic costs of specific activities occasionally have been combined with amount of time expended in each activity. The resulting time-energy budgets (Goldstein 1990) can be used to address ecological questions about reproduction (Meijer et al. 1989), migration (Smith et al. 1986), foraging (Tarboton 1978, Stalmaster and Gessaman 1984), nesting (Wakely 1978, Brodin and Jonsson 2003), general energy balance (Koplin et al. 1980, Wijnandts 1984, Riedstra et al. 1998), or other life-history phenomena.

Food-consumption studies. In food-consumption studies, the energy content of food ingested and of egesta produced during a defined period is used to compute metabolic rates. These sometimes are referred to as **bioenergetic studies** (see Duke et al. 1973, Gessaman 1978, Kirkwood 1979, Collopy 1986). In such studies, the energy content of items such as feces, pellets, food, and body components is measured by bomb calorimetry. There are several versions of bomb calorimeters, but the most commonly encountered apparatus is the Parr adiabatic calorimeter. Basically, the technique involves the combustion of a known quantity of material in a vessel containing an atmosphere of pure oxygen.

The result is the caloric equivalent (heat of combustion) of the material in kcal/g or J/g. The total energy content of biological substances can be computed by multiplying the total dry mass by its caloric equivalent. The results can be converted from kcal to kJ and vice versa by use of appropriate conversion terms. The most important step in bomb calorimetry is the method used to dry the study materials. If the substance to be analyzed is not fully dry, energy measurements will be low. If the material is exposed to excessive heat during drying, then volatile materials other than water will be driven off or the chemical composition of the material may be changed. Several studies have addressed this problem with slightly different results (e.g., Blem 1968). It appears that freeze-drying (lyophilization) is the best choice for drying substances containing fat. If there is a risk of losing energy from oven-drying materials because a freeze-drier is not available, one can perform determinations in bomb calorimeters with the addition of combustion stimulants (Blem 1968). Such determinations must be corrected for water content of the material and the addition of the combustion stimulant.

Bomb calorimeters usually can analyze only small aliquots (1 g or less) of material, so an unbiased means of sampling large samples of food or tissue is necessary. For example, one could calculate the gross energy intake of a raptor that fully consumes small rodents by converting live mass of the mammal to calories (e.g., Collopy 1986). This is done by drying the whole mammal carcass, thoroughly homogenizing it using a Wiley mill or powerful blender, and testing aliquots of the powdered specimen in the bomb calorimeter. The total energy content of the prey item then can be computed as total dry mass (g) × heat of combustion (per g). The fresh weight of food must be corrected for moisture content; water contributes to mass but not to caloric content. A similar process can be used to measure energy content of excrement and pellets and the difference between these and gross energy intake (GEI - EXE) produces a measurement of metabolized energy (Fig. 1). Energy-use efficiency (**metabolizable energy coefficient** = MEC) is defined as the percent of GEI actually extracted through assimilation after energy losses due to EXE. Measuring MEC in birds is complicated by the fact that avian feces is mixed in the cloaca with urine. Thus, excretory energy represents the energy remaining in feces and pellets (unassimilated) combined with energy lost as urine (assimilated). The difference between energy intake and excretory energy loss is properly termed **apparent assimilated** or **metabolized**

fraction. Division of the metabolized fraction by GEI produces the apparent metabolized coefficient (Kendeigh et al. 1977; also see Karasov 1990). This technique can be used to evaluate digestive efficiency under a variety of environmental conditions (Tollan 1988) or to quantify the bird's ability to extract energy from different foods (Blem 1976a). Conversely, it also can be used to evaluate **use efficiencies** (UC) of different foods (Karasov 1990). UC values are highest in nectarivores (~ 98%) and seed-eaters (80%), but raptors eating arthropods (77%) or vertebrates (75%) also are very efficient (Karasov 1990). Herbivores feeding on grass or conifer needles have low efficiencies, often 40% or less.

Using calorimetric techniques, variation in body composition can be interfaced with energetics to quantify the energetics of lipid deposition (Blem 1976b, 1990), predation (Barrett and Mackey 1975, Wallick and Barrett 1976, Tabaka et al. 1996), development of young (Kirkley and Gessaman 1990, Lee 1998), molt, egg formation, and other life-history phenomena (e.g., Pietiainen and Kolunen 1993, Weathers et al. 2001). Knowledge about energy content of prey can be used to assess hypotheses about prey selection (e.g., Wallick and Barrett 1976, Postler and Barrett 1982, Kirkley and Gessaman 1990, Blem et al. 1993).

Energy storage, particularly lipid reserves, has been quantified in many bird species, but there are few studies involving raptors (but see Smith et al. 1986, Massemin et al. 1997). The lipid depots are composed of triglycerides (triacylglycerols) consisting of three fatty acid molecules attached to a glycerol "backbone." The fatty acids may be of various sizes and caloric contents, and the efficiency of use of them may vary with length of their carbon chain (see Blem 1976b, 1990, for reviews). There has been little, if any, work on the composition of triglycerides in raptors (Blem 1990). Lack of knowledge about triglyceride dynamics in raptors is unfortunate because the pattern of lipid storage and usage in a large carnivorous bird may present unusual clues to important adaptations to stress (Massemin and Handrich 1997). The ability to accumulate fat reserves, which promote survival over extended periods of prey scarcity or during migration, could well be a most significant adaptation in raptors. Lipid provides a rich energy store without great wing loading because of its high heat of combustion (9.0–9.5 kcal/g = 37.7–39.7 kJ/g), and because lipid storage is not accompanied by deposition of much water. Carbohydrate energy stores, such as glycogen, have about one half of the energy content of lipid stores and are accompanied by accumu-lation of about 3 g of water for every g of glycogen reserve (Blem 2000, 2004).

Energy expenditure has been measured in adult raptors (e.g., Gessaman 1972, Koplin et al. 1980, Hamilton 1985a), young raptors (e.g., Hamilton and Neill 1981, Collopy 1986, Kirkley and Gessaman 1990), and in eggs (e.g., Hamilton 1985b, Meijer et al. 1989). Many studies have focused on basic variations of metabolism (Hayes and Gessaman 1980, 1982; Daan et al. 1989, Pakpahan et al. 1989), effects of body size on metabolic rate (Mosher and Matroy 1974), and comparison of the metabolism of different species (Graber 1962, Ligon 1969, Gatehouse and Markham 1970, Ganey et al. 1993). The energetic costs of flight (Masman and Klaassen 1987), growth (Lee 1998), thermoregulation (Arad and Bernstein 1988, Weathers et al. 2001), reproduction (Meijer et al. 1989, Brodin and Jonsson 2003), incubation (Gessaman and Findell 1979), savings during roosting (Keister et al. 1985, McCafferty et al. 2001) and foraging (Wallick and Barrett 1976, Tarboton 1978, Postler and Barrett 1982, Beissinger 1983) also have been measured. I can find no studies of the energetic costs of molt or lipid deposition in raptors. In a few instances, several of the above techniques have been combined to construct energy budgets (e.g., Wakely 1978, Kirkwood 1979, Stalmaster and Gessaman 1982, Wijnandts 1984, Higuchi and Abe 2001), to compare components of energy use (e.g., Graber 1962), or to evaluate ability to survive starvation, harsh winter conditions, or both (Koplin et al. 1980, Handrich et al. 1993a,b, Hohtola et al. 1994, Thouzeau et al. 1999).

Field Measurements

Most of the studies mentioned above employed captive birds. Field studies are more difficult and require special techniques such as those described below.

Doubly-labeled water studies. A less common, more expensive technique for measuring total energy expended during a specific period of time under unrestrained conditions (i.e., the costs of free existence) involves the use of so-called doubly labeled water. This method measures the disappearance rates of isotopes of H* and O* (typically ^{18}O and ^{3}H), which are injected into the test subject. The hydrogen isotope is lost through breathing, urination, and evaporation across the skin. The oxygen isotope is lost in water and in carbon dioxide produced during respiratory metabolism. The loss rate for labeled oxygen is greater than that for labeled hydrogen. As a result, there is a greater difference between slopes of labeled hydrogen and oxygen when greater amounts of carbon dioxide are produced (see Ricklefs et al. 1986, Goldstein 1990).

The method typically used under field conditions is to inject a known amount of doubly labeled water into the bird. The labeled water is then allowed to equilibrate throughout the bird's body fluids. Subsequently, a blood sample is drawn to establish a baseline. At a later time, a second sample of blood is taken and new isotope levels are measured. The difference between the rates of loss of different isotopes between the two sample times can be used to estimate the rate of carbon dioxide creation. The metabolic rate then can be calculated from the rate of CO_2 production. This technique is expensive, but provides a means of measuring energy use by birds involved in natural behavior, such as flight or reproduction. In some instances, the method has been used to measure daily energy expenditure (Masman et al. 1983) and to compare energetics of different sexes (Riedstra et al. 1998).

Telemetric methods. Use of radiotransmitters which can monitor heart rates, electrocardiograms, or breathing rates have been available for quite some time (e.g., Owen 1969, Johnson and Gessaman 1973). Under well-defined circumstances these devices can provide reliable indexes to rates of avian oxygen consumption (Goldstein 1990). Early studies involved relatively large transmitters attached externally that probably contributed to energy demands of flight because of increased friction and wing loading (see Gessaman et al. 1991). Modern devices can be implanted within the body cavity along with small data loggers that can store extensive amounts of information. Under carefully controlled conditions, heart rate can be used reliably to estimate oxygen uptake, although one must be certain to consider a variety of confounding problems (Gessaman 1980, Gessaman et al. 1991).

Time-energy budgets. Time-energy budgets are constructed from extended observations of avian activity interfaced with measurements or estimates of the energetic costs of specific activities. Daily behavior is divided into categories for which energy measurements have been established or estimated, and total energy use is then calculated by adding the products of activity time and energy use for each activity (e.g., Soltz 1984, Craig et al. 1988). In addition to less complex models, comprehensive energy models have been assembled for some birds by combining data from several sources including estimates of energy intake, thermoregulation, and the like. These usually take the form of time-ener-

gy budgets to which measurements of productive energy (P) and physiological costs of thermoregulation (T) have been added.

CONCLUSIONS

Methods for measuring metabolism in raptors have changed relatively little since Gessaman's (1987) review. Exceptions include the fact that instrumentation is more reliable now and that computer software is greatly improved. Techniques are now user-friendly, and a novice investigator does not have to deal with many of the problems encountered earlier. Other than the refinement of stable-isotope techniques, little has been added to the researcher's arsenal. The literature in this area has developed slowly and numerous aspects of raptor life history, physiology, and energetics remain uninvestigated.

LITERATURE CITED

ARAD, Z. AND M.H. BERNSTEIN. 1988. Temperature regulation in Turkey Vultures. *Condor* 90:913–919.

BAHAT, O., I. CHOSHNIAK AND D.C. HOUSTON. 1998. Nocturnal variation in body temperature of Griffon Vultures. *Condor* 100:168–171.

BARRETT, G.W. AND C.V. MACKEY. 1975. Prey selection and caloric ingestion rate of captive American Kestrels. *Wilson Bull.* 87:574–579.

BEISSINGER, S.R. 1983. Hunting behavior, prey selection, and energetics of Snail Kites in Guyana: consumer choice by a specialist. *Auk* 100:84–92.

BLEM, C.R. 1968. Determination of caloric and nitrogen content of excreta voided by birds. *Poult. Sci.* 47:1205–1208.

———. 1976a. Efficiency of energy utilization of the House Sparrow, *Passer domesticus. Oecologia* 25:257–264.

———. 1976b. Patterns of lipid storage and utilization in birds. *Am. Zool.* 16:671–684.

———. 1984. Ratios in avian physiology. *Auk* 101:153–155.

———. 1990. Avian energy storage. Pages 59–114 in D.M. Power [ED.], Current ornithology, Vol. 7. Plenum Press, New York, NY U.S.A.

———. 2000. Energy metabolism. Pages 327–341 in G.C. Whittow [ED.], Avian physiology. Academic Press, New York, NY U.S.A.

———. 2004. Energetics of migration. Pages 31–39 in C.J. Cleveland [ED.], Encyclopedia of energy. Elsevier, New York, NY U.S.A.

———, J.H. FELIX, D.W. HOLT AND L.B. BLEM. 1993. Estimation of body mass of voles from crania in Short-eared Owl pellets. *Am. Midl. Nat.* 129:282–287.

BRODIN, A. AND K.I. JONSSON. 2003. Optimal energy allocation and behaviour in female raptorial birds during the nestling period.

Ecoscience 10:140–150.

CALDER, W.A. AND J.R. KING. 1974. Thermal and caloric relations of birds. Pages 259–413 in D.S. Farner and J.R. King [EDS.], Avian biology, Vol. 4. Academic Press, New York, NY U.S.A.

CHAPLIN, S.B., D.A. DIESEL AND J.A. KASPARIE. 1984. Body temperature regulation in Red-tailed Hawks and Great Horned Owls: responses to air temperature and food deprivation. *Condor* 86:175–181.

COLLOPY, M.W. 1986. Food consumption and growth energetics of nestling Golden Eagles. *Wilson Bull.* 98:445–458.

CRAIG, R.J., E.S. MITCHELL AND J.E. MITCHELL. 1988. Time and energy budgets of Bald Eagles wintering along the Connecticut River. *J. Field Ornithol.* 59:22–32.

DAAN, S., D. MASMAN, S. STRIJKSTRA AND S. VERHULST. 1989. Intraspecific allometry of basal metabolic rate: relations with body size, temperature, composition, and circadian phase in the Kestrel, *Falco tinnunculus. J. Biol. Rhythms* 4:267–283.

DAWSON, W.R. AND G.C. WHITTOW. 2000. Regulation of body temperature. Pages 343–390 in G.C. Whittow [ED.], Avian physiology. Academic Press, New York, NY U.S.A.

DUKE, G.E., J.G. CIGANEK AND O.A. EVANSON. 1973. Food consumption and energy, water and nitrogen budgets in captured Great Horned Owls (*Bubo virginianus*). *Comp. Biochem. Physiol.* 44A:283–292.

EDWARDS, T.C., JR. 1987. Standard rate of metabolism in the Barn Owl (*Tyto alba*). *Wilson Bull.* 99:704–706.

GANEY, J.L., R.P. BALDA AND R.M. KING. 1993. Metabolic rate and evaporative water loss of Mexican Spotted and Great Horned owls. *Wilson Bull.* 105:645–656.

GATEHOUSE, S.N. AND B.J. MARKHAM. 1970. Respiratory metabolism of three species of raptors. *Auk* 87:738–741.

GESSAMAN, J.A. 1972. Bioenergetics of the Snowy Owl (*Nyctea scandiaca*). *Arct. Alp. Res.* 4:223–238.

——— [ED.]. 1973. Ecological energetics of homeotherms. Utah State University Monograph Series 20. Logan, UT U.S.A.

———. 1978. Body temperature and heart rate of the Snowy Owl (*Nyctea scandiaca*). *Condor* 80:243–245.

———. 1980. An evaluation of heart rate as an indirect measure of daily energy metabolism of the American Kestrel. *Comp. Biochem. Physiol.* 65A:273–289.

———. 1987. Energetics. Pages 289–320 in B.A. Giron Pendleton, B.A. Millsap, K.W. Cline, and D.M. Bird [EDS.], Raptor management techniques manual. National Wildlife Federation, Washington, DC U.S.A.

——— AND P. R. FINDELL. 1979. Energy cost of incubation in the American Kestrel. *Comp. Biochem. Physiol.* 63A:57–62.

———, M.R. FULLER, P.J. PEKINS AND G.E. DUKE. 1991. Resting metabolic rate of Golden Eagles, Bald Eagles, and Barred Owls with a tracking transmitter or an equivalent load. *Wilson Bull.* 103:261–265.

GOLDSTEIN, D.L. 1990. Energetics of activity and free living in birds. *Stud. Avian Biol.* 13:423–426.

GRABER, R.R. 1962. Food and oxygen consumption in three species of owls (Strigidae). *Condor* 64:473–487.

GRODZINSKI, W., R.Z. KLEKOWSKI AND A. DUNCAN. 1975. Methods for ecological bioenergetics. IBP handbook No. 24. J.B. Lippincott, Philadelphia, PA U.S.A.

HAMILTON, K.L. 1985a. Food and energy requirements of captive Barn Owls. *Comp. Biochem. Physiol.* 80A:355–358.

———. 1985b. Metabolism of Barn Owl eggs. *Am. Midl. Nat.* 114:209–215.

——— AND A. L. NEILL. 1981. Food habits and bioenergetics of a pair of Barn Owls and owlets. *Am. Midl. Nat.* 106:1–9.

HANDRICH, Y., L. NICOLAS AND Y. LE MAHO. 1993a. Winter starvation in captive Common Barn-Owls: physiological states and reversible limits. *Auk* 110:458–469.

———, L. NICOLAS AND Y. LE MAHO. 1993b. Winter starvation in captive Common Barn-Owls: bioenergetics during refeeding. *Auk* 110:470–480.

HAYES, S.R. AND J.A. GESSAMAN. 1980. The combined effects of air temperature, wind and radiation on the resting metabolism of avian raptors. *J. Therm. Biol.* 5:119–125.

——— AND J.A. GESSAMAN. 1982. Prediction of raptor resting metabolism: comparison of measured values with statistical and biophysical estimates. *J. Therm. Biol.* 7:45–50.

HEATH, J.E. 1962. Temperature fluctuations in the Turkey Vulture. *Condor* 64:234–235.

HIGUCHI, A. AND M.T. ABE. 2001. Studies on the energy budget of captive Ural Owls *Strix uralensis. JPN J. Ornithol.* 50:25–30.

HOHTOLA, E., A. PYORNILA AND H. RINTAMAKI. 1994. Fasting endurance and cold resistance without hypothermia in a small predatory bird: the metabolic strategy of Tengmalm's owl, *Aegolius funereus. J. Comp. Physiol. B. Biochem. Syst. Environ. Physiol.* 164:430–437.

JOHNSON, S.F. AND J.A. GESSAMAN. 1973. An evaluation of heart rate as an indirect monitor of free-living energy metabolism. Pages 44–54 in J.A. Gessaman [ED.], Ecological energetics of homeotherms. Utah University Press, Logan, UT U.S.A.

KARASOV, W.H. 1990. Digestion in birds: chemical and physiological determinants and ecological implications. *Stud. Avian Biol.* 13:391–415.

KEISTER, G.P., JR., R.G. ANTHONY AND H.R. HOLBO. 1985. A model of energy consumption in Bald Eagles: an evaluation of night communal roosting. *Wilson Bull.* 97:148–160.

KENDEIGH, S.C., V.R. DOLNIK AND V.M. GAVRILOV. 1977. Avian energetics. Pages 127–204 in J. Pinowski and S.C. Kendeigh [EDS.], Granivorous birds in ecosystems. Cambridge University Press, Cambridge, United Kingdom.

KENNEDY, P.J. AND J.A. GESSAMAN. 1991. Diurnal resting metabolic rates of Accipiters. *Wilson Bull.* 103:101–105.

KIRKLEY, J.S. AND J.A. GESSAMAN. 1990. Water economy of nestling Swainson's Hawks. *Condor* 92:29–44.

KIRKWOOD, J.K. 1979. The partition of food energy for existence in the Kestrel (*Falco tinnunculus*) and Barn Owl (*Tyto alba*). *Comp. Biochem. Physiol.* 63A:495–498.

KOPLIN, J.R., M.W. COLLOPY, A.R. BAMMANN AND H. LEVENSON. 1980. Energetics of two wintering raptors. *Auk* 97:795–806.

LEE, C.H. 1998. Barn Owl: development and food utilization. *J. Trop. Agric. Food Sci.* 26:151–157.

LIGON, J.D. 1969. Some aspects of temperature relations in small owls. *Auk* 86:458–472.

MASMAN, D. AND M. KLAASSEN. 1987. Energy expenditure during free flight in trained and free-living Kestrels (*Falco tinnunculus*). *Auk* 104:603–616.

———, S. DANN AND H.J.A. BELDHUIS. 1983. Ecological energetics of the Kestrel: daily energy expenditure throughout the year based on time-energy budget, food intake and doubly labeled water methods. *Ardea* 76:64–81.

MASSEMIN, S. AND Y. HANDRICH. 1997. Higher winter mortality of the Barn Owl compared to the Long-eared Owl and the Tawny

Owl: influence of lipid reserves and insulation. *Condor* 99:969–971.

———, R. GROSCOLAS AND Y. HANDRICH. 1997. Body composition of the European Barn Owl during the nonbreeding period. *Condor* 99:789–797.

MCCAFFERTY, D.J., J.B. MONCRIEFF AND I.R. TAYLOR. 2001. How much energy do Barn Owls (*Tyto alba*) save by roosting? *J. Therm. Biol.* 26:193–203.

MEIJER, T., D. MASMAN AND S. DAAN. 1989. Energetics of reproduction in female Kestrels. *Auk* 106:549–559.

MOSHER, J.A. AND P.F. MATROY. 1974. Size dimorphism: a factor in energy savings for Broad-winged Hawks. *Auk* 91:325–341.

OWEN, R.B., JR. 1969. Heart rate, a measure of metabolism in Blue-winged Teal. *Comp. Biochem. Physiol.* 31:431–436.

PAKPAHAN, A.M., J.B. HAUFLER AND H.H. PRINCE. 1989. Metabolic rates of Red-tailed Hawks and Great Horned Owls. *Condor* 91:1000–1002.

PIETIAINEN, H. AND H. KOLUNEN. 1993. Female body condition and breeding of the Ural Owl *Strix uralensis*. *Funct. Ecol.* 7:726–735.

POSTLER, J.L. AND G.W. BARRETT. 1982. Prey selection and bioenergetics of captive screech owls. *Ohio J. Sci.* 82:55–58.

RICKLEFS, R.E., D.D. ROBY AND J.B. WILLIAMS. 1986. Daily energy expenditure by adult Leach's Storm Petrels during the nesting cycle. *Physiol. Zool.* 59:649–660.

RIEDSTRA, B., C. DIJKSTRA AND S. DAAN. 1998. Daily energy expenditure of male and female Marsh Harrier nestlings. *Auk* 115:635–641.

SMITH, N.G., D.L. GOLDSTEIN AND G.A. BARTHOLOMEW. 1986. Is long-distance migration possible for soaring hawks using only stored fat? *Auk* 103:607–611.

SOLTZ, R.L. 1984. Time and energy budgets of the Red-tailed Hawk *Buteo jamaicensis* in southern California. *Southwest. Nat.* 29:149–156.

STALMASTER, M.V. AND J.A. GESSAMAN. 1982. Food consumption and energy requirements of captive Bald Eagles. *J. Wildl. Manage.* 46:646–654.

——— AND J. A. GESSAMAN. 1984. Food consumption and foraging behavior of overwintering Bald Eagles. *Ecol. Monogr.* 54:407–428.

TABAKA, C.S., D.E. ULLREY, J.G. SIKARSKIE, S.R. DEBAR AND P.K. KU. 1996. Diet, cast composition, and energy and nutrient intake of Red-tailed Hawks (*Buteo jamaicensis*), Great Horned Owls (*Bubo virginianus*), and Turkey Vultures (*Cathartes aura*). *J. Zoo Wildl. Med.* 27:187–196.

TARBOTON, W.R. 1978. Hunting and the energy budget of the Black-shouldered Kite. *Condor* 80:88–91.

THOUZEAU, C., C. DUCHAMP AND Y. HANDRICH. 1999. Energy metabolism and body temperature of Barn Owls fasting in the cold. *Physiol. Biochem. Zool.* 72:170–178.

TOLLAN, A.M. 1988. Maintenance energy requirements and energy assimilation efficiency of the Australian Harrier. *Ardea* 76:181–186.

WAKELY, J.S. 1978. Activity budgets, energy expenditures and energy intakes of nesting Ferruginous Hawks. *Auk* 95:667–676.

WALLICK, L.G. AND G.W. BARRETT. 1976. Bioenergetics and prey selection of captive Barn Owls. *Condor* 78:139–141.

WALSBERG, G.E. 1983. Avian ecological energetics. Pages 161–220 in D.S. Farner and J.R. King [EDS.], Avian biology, Vol. 7. Academic Press, New York, NY U.S.A.

WASSER, J.S. 1986. The relationship of energetics of falconiform birds to body mass and climate. *Condor* 88:57–62.

WEATHERS, W.W., P.J. HODUM AND J.A. BLAKESLEY. 2001. Thermal ecology and ecological energetics of California Spotted Owls. *Condor* 103:678–690.

WIJNANDTS, H. 1984. Ecological energetics of the Long-eared Owl *Asio otus*. *Ardea* 72:1–92.

Physiology

16

A. Gastrointestinal

DAVID C. HOUSTON

Division of Environmental and Evolutionary Biology,
Institute of Biomedical and Life Sciences
University of Glasgow, Glasgow G12 8QQ, United Kingdom

GARY E. DUKE

Department of Veterinary Pathobiology,
College of Veterinary Medicine,
University of Minnesota, St. Paul, MN 55126 U.S.A

GASTROINTESTINAL PHYSIOLOGY AND NUTRITION

Most studies of nutrition and gastrointestinal (aka GI) physiology in birds have been conducted on domestic fowl. Birds of prey provide an interesting contrast to domestic fowl because of their carnivorous diets. This part of Chapter 16 summarizes our knowledge of anatomy, gastric secretion and motility, pellet formation and egestion, and the techniques available to study these aspects of raptor biology.

Gastrointestinal Physiology

Anatomical considerations. It is useful to have some notion of anatomy in order to better understand function. The GI tracts of raptors differ significantly from those of domestic fowl, with which most biologists are familiar (Fig.1; Duke 1978). Whereas turkeys have a well-developed crop, that of many raptors is poorly developed, and owls have no crop at all, only a simple enlargement of the esophagus. The crop is largely a food-storage area with little secretory activity, and is exceptionally well developed only in some vultures, whose crop allows them to consume up to 20% of their body weight in a single meal (Houston 1976). The stomach of turkeys, and virtually all other avian species except raptors and Ardeidae, consists of two pairs of alternately contracting muscles that grind food. The meat diet of raptors does not require strong mechanical grinding, and birds of prey have a simpler muscular stomach in which acid secretion and enzyme action start to break down the food. Digestion is continued in the small intestine, which also is the site of absorption. The pancreas fills the entire duodenal loop in turkeys, but occupies only half of the loop in owls, and is even smaller in hawks. There seems to be considerable variation in the total length of the small intestine between species of both raptors and owls. After correction for body-size differences, species such as falcons, which use a method of prey capture that requires extreme acceleration in flight, have a small intestine length about 50% shorter than that found in species such as eagles, buzzards, and kites that have less need for speed and agility when hunting (Barton and Houston 1994a). This may be an adaptation to reduce the overall weight of the digestive tract in those species which have an extremely active hunting strategy, and it does have the consequence of giving such species a reduced digestive efficiency and restricted prey selection (see later). Ceca in birds are highly variable in size, and usually are only conspicuous in certain plant-eating birds, where they are the sites of microbial fermentation of plant-cell walls that cannot otherwise be digested (Klasing 1998). Thus, it is not surprising that they are absent in hawks. They are, however, well developed in owls (Fig. 1). It is not clear why Great Horned Owls (*Bubo virginianus*), which eat almost the same diet as Red-tailed Hawks (*Buteo jamaicensis*), have such a different cecal morphology. Perhaps because owls generally swallow their prey whole, the ceca are used to break down the plant material found in the gut contents of their prey. Cecal droppings of owls are readily distinguished from rectal excreta. In Great Horned Owls on a mouse diet, these droppings occur about once every three days (G. Duke, unpubl. data). This information might be used to determine how long an owl has been roosting at a particular site.

Gastric secretions and motility. Digestive secretions and intestinal absorption have received little investigation in raptors. Gastric secretions have been found to be more acidic (Duke et al. 1975) and to contain more pepsin (Herpol 1964, 1967; Duke et al. 1975) than gastric secretions of granivorous and omnivorous birds; and the pH of the gastric juice of hawks was found to be lower than that of owls (i.e., 1.7 versus 2.4, respectively) (Duke et al. 1975). In an extreme case, this strongly acidic environment enables the Bearded Vulture (*Gypaetus barbatus*) to feed mainly on bones — the only vertebrate known to be able to digest this unpromising diet (Houston and Copsey 1994).

GI motility (i.e., contractile activity) has received considerable attention (Duke et al. 1976b,c; Rhoades and Duke 1977). In more recent years, captive American Kestrels have been used to learn more about this subject (Duke et al. 1997).

Several methods may be used to study GI motility in raptors: (1) tiny strain-gauge transducers (SGT) surgically sutured to the outside surface of the GI tract (called the serosal surface) to monitor smooth muscle contractile activity (Duke et al. 1976b,c), (2) silver bipolar electrodes also sewn onto the serosa to detect electrical potential changes associated with depolarization (contraction) of smooth muscle (Duke et al. 1976c), and (3) radiography using image intensification (a modern type of fluoroscope) and viewing GI contractions on a video monitor or recording observations on video tape (Duke et al. 1976c, Rhoades and Duke 1977). Bioinformation detected by these devices can be recorded on a physiological recorder.

Swallowed foods collect in the crop of hawks and are slowly passed into the stomach. In owls, swallowed food items immediately fill the stomach and lower

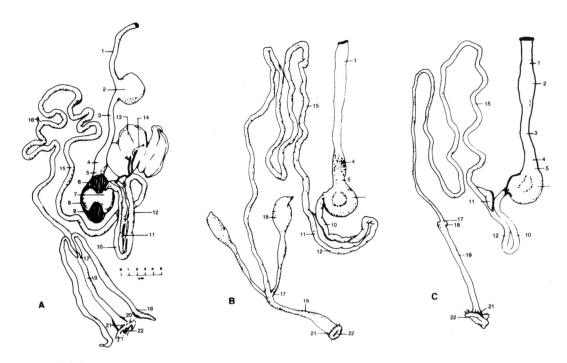

Figure 1. GI tracts of (A) domestic turkey, (B) Great Horned Owl, (C) Red-tailed Hawk. Included are (1) pre-crop esophagus, (2) crop, (3) post-crop esophagus, (4) glandular stomach, (5) isthmus, (6) thin craniodorsal muscle, (6a) muscular stomach of raptor, (7) thick cranioventral muscle, (8) thick caudodorsal muscle, (9) thin caudoventral muscle, (10) proximal duodenum, (11) pancreas, (12) distal duodenum, (13) liver, (14) gall bladder, (15) ileum, (16) Meckel's diverticulum, (17) ileocecocolic junction, (18) cecum, (19) colon, (20) bursa of Fabricus, (21) cloaca, (22) vent, greater curvature. From Duke (1978).

esophagus, and after 20 to 30 minutes the entire meal has been moved into the muscular stomach (Rhoades and Duke 1977). In Great Horned Owls, the motilities of the stomach and duodenum are coordinated and the gastroduodenal contraction sequence involves a contraction wave (called peristalsis) that moves first through the stomach, then on into the duodenum (Kostuch and Duke 1975). The peristaltic contraction is more apparent in the muscular stomach as a flattening or indentation moving around the greater curvature (Kostuch and Duke 1975, Rhoades and Duke 1977).

Pellet formation and egestion. The formation and egestion of pellets is a unique gastrointestinal phenomenon in birds, and is particularly well developed in raptors and especially owls (Rea 1973). Analysis of food remains in pellets is a major aspect of many raptor studies (Mikkola 1983, Yalden 2003). Pellets are formed in the stomach from the indigestible bones, hair or feathers of prey (Reed and Reed 1928, Grimm and Whitehouse 1963, Kostuch and Duke 1975, Rhoades and Duke 1977). The prey remains in owl pellets reflect exactly the prey species eaten (Mikkola 1983). But pellet size varies considerably, and curiously has no correlation with the amount of food eaten (Erkinaro 1973). Raczynski and Ruprecht (1974) showed that some prey bones are digested, some skeletal parts more than others, and that food intake estimates based on pellet remains will underestimate the number of prey items swallowed (see also Chapter 8). Duke et al. (1996) also found considerable variability in parts of food items eaten, pellet size, and pellet egestion frequency in captive American Kestrels. Egestion involves both gastric activity and esophageal antiperistalsis (Duke et al.

1976c), and is considerably different from the mechanisms of vomiting in mammals with a simple stomach, or regurgitation of cud in ruminants (Duke et al. 1976c).

Monitoring of gastric motility in owls shows that food intake, or even the sight of food in hungry owls (Duke et al. 1976b), immediately causes a two- to threefold increase in gastric contractile activity. The first mechanical-digestion phase, with relatively rapid and vigorous motility, moves the entire meal into the muscular stomach, crushes or "macerates" it, and thoroughly mixes it with digestive secretions. The second, or chemical-digestion phase, has low amplitude and low frequency contractions that continue to mix gently ingesta with digestive secretions; most digestion is completed during this phase. During the third phase, fluid is evacuated from the stomach, and pellet formation and egestion occur (Fuller and Duke 1978). The length of these phases and the overall meal-to-pellet interval (MPI) varies directly with the amount eaten by an owl, and thus may be used to estimate meal size.

In order to learn more about other factors that regulate pellet egestion and thus alter the lengths of the three phases and influencing MPI, owls were jessed and attached to perches suspended over a sloping chute within a $1 \times 1 \times 2$-m chamber. Pellets rolled down chutes into wire collecting baskets; a pellet landing in a basket depressed a micro-switch directly under the basket, thereby completing a circuit and activating a marker on a recorder located in another room. The exact time of the event was thus recorded.

Using this technique, six species of owls (Table 1) were fed as many laboratory mice as they wanted during a 30-minute period at two hours after dawn (0900)

Table 1. Mean meal-to-pellet intervals (MPI) in owls.[a]

Species	Number of Birds	Mean MPI ± SE (hour)	Number of Pellets
Eastern Screech Owl *(Megascops asio)*	2	11.86 ± 0.22	29
Great Horned Owl *(Bubo virginianus)*	4	13.25 ± 0.29	36
Snowy Owl *(Bubo scandiaca)*	2	12.02 ± 0.72	35
Barred Owl *(Strix varia)*	2	9.85 ± 0.44	25
Short-eared Owl *(Asio flammeus)*	1	10.22 ± 0.12	132
Northern Saw-whet Owl *(Aegolius acadicus)*	1	10.04 ± 0.32	4

[a] Data modified from Duke et al. (1976a)

daily. The length of the MPI was shorter in smaller-sized owls, but, more significantly, the MPI was directly related to meal size, indicating that the state of ingestion of the meal is important in regulating pellet egestion (Table 2; Duke et al. 1976b).

Experiments involving feeding Great Horned Owls on foods of different composition suggest that the presence of undigested food (proteins or fat) in the stomach seems to inhibit pellet egestion, which will not occur until digestion is complete (Table 3; Duke and Rhoades 1977). There also may be a stimulating effect of undigested material on the gastric mucosa, which contributes to pellet ejection. However, other factors also may be involved. Barred Owls (*Strix varia*) were found to have lengthened MPIs and smaller pellets when fed at a sub-maintenance level until they had lost 10% of their body weight. Analysis of the pellets disclosed that digestion of the meal was more complete in the hungry owls, indicating that the state of hunger may affect MPI (Duke et al. 1980). The constant sight of food may shorten MPI in Short-eared Owls (*Asio flammeus*) (Chitty 1938).

MPI in owls also may be influenced by environmental stimuli. When Great Horned Owls were fed as many mice as they wanted during a 30-minute period at either dawn or dusk, it was found that MPIs were directly related to meal size but that MPI's were longer for meals eaten at dusk than at dawn regardless of the size of the meal (Duke and Rhoades 1977). This is true for Short-eared Owls, too (Chitty 1938). Thus, the portion of the daily cycle during which gastric digestion and pellet formation occur may affect the MPI.

Kuechle et al. (1987) performed a field study using all of the basic information described above and adapting the techniques used therein for telemetry. In free-flying Barred Owls, movements were monitored via a tail-mounted transmitter and gastric motility was monitored via telemetry of signals from an implanted SGT to determine (1) time of ingestion, (2) time of egestion, (3) measurement of the lengths of phases in gastric digestion and thus, (4) estimation of the quantity consumed. Being able to distinguish movements associated with hunting and feeding from other types of movements is significant in understanding owl behavior, and an estimate of daily food consumption in a free-flying owl is very useful in understanding owl energetics.

In owls the MPI is directly correlated with the quantity eaten, but in hawks the major stimulus for pellet egestion is dawn, regardless of the quantity eaten (Balgooyen 1971, Duke et al. 1976b; Table 4). In a light-timed room with dawn set at 0700, the MPIs of hawks were 1 to 2 hours shorter when they were fed at 1100 than when they were fed at 0900. In another study involving Red-tailed Hawks in a room with dawn at 0700, feeding time was shifted from 0800 to 1600, and MPI changed from approximately 2200 to approximate-

Table 2. Mean meal-to-pellet intervals (MPI) as related to food consumption (grams DM/kg) in Great Horned Owls and Eastern Screech-Owls fed at 0900 daily.[a]

Species	Number of Birds	Meal Size	Mean MPI ± SE (hour)	Number of Pellets
Great Horned Owl *(Bubo virginianus)*	4	10	11.76 ± 0.46	4
		11 – 15	12.49 ± 0.35	11
		16 – 20	13.35 ± 0.51	12
		21 – 25	14.71 ± 0.52	9
Eastern Screech Owl *(Megascops asio)*	2	30 – 40	10.92 ± 0.25	9
		41 – 50	11.88 ± 0.28	13
		51 – 60	12.92 ± 0.41	6
		61 – 70	13.75	1

[a] Data modified from Duke et al. (1976a)

Table 3. Mean meal-to-pellet intervals for four Great Horned Owls fed (at 1500) two mice, two mouse skins, or two skins stuffed with various diets.[a]

Diet	Mean Mass of Meal (g)	Mean MPI ± SE (hour)	Number of Pellets
Two 25 g mice	50	15.52 ± 0.45	45
Two mouse skins (with skull)	15	15.26 ± 0.20	8
Two mouse skins plus two pellets [b]	25	8.19 ± 0.26	11
Two pellets only [c]	10	2.75 ± 0.29	5
Two mouse skins plus 35 g of horse meat	50	24.34 ± 1.02	10
Two mouse skins plus 9 g of suet [b]	24	33.74 ± 2.28	11

[a] Table modified from Duke and Rhoades (1977).
[b] Pellets, horse meat, and suet were sewn into the mouse skins with silk suture.
[c] Pellets were force-fed.

Table 4. Mean meal to pellet intervals (MPI) in hawks with dawn (lights on in the holding room) at 0700.[a]

Species	Number of Birds	MPI (hour)			
		Fed at 0900 Mean MPI ± SE	N	Fed at 1100 Mean MPI ± SE	N
Bald Eagle *(Haliaeetus leucocephalus)*	3	21.7 ± 0.4	10	20.9 ± 0.38	10
Northern Goshawk *(Accipiter gentilis)*	4	21.6 ± 0.83	9	20.6 ± 0.17	65
Broad-winged Hawk *(Buteo platypterus)*	2	21.7 ± 0.14	13	20.8 ± 0.13	5
Red-tailed Hawk *(B. jamaicensis)*	6	22.5 ± 0.09	72	20.4 ± 0.14	59
Roughleg *(B. lagopus)*	3	21.7 ± 0.08	79	-	-
Northern Crested Caracara *(Caracara cheriway)*	1	-	-	19.6 ± 0.08	14
American Kestrel [b] *(Falco sparverius)*	1	23.6 ± 0.06	10	-	-

[a] Data from Duke et al. (1976a).
[b] Dawn was approximately 0800.

ly 1800, respectively, a delay of only 4 hours, suggesting that the birds were "attempting" to egest as early in the day as possible (Fuller et al. 1978). It is theorized that whereas owls may hunt either at night or during the daytime, hawks require daylight for hunting (Fuller et al. 1978). Thus, hawks would benefit by egesting a pellet (i.e., emptying the stomach) early in the day, leaving the rest of the day for capturing and ingesting new prey. Hawks conditioned to eating late in the afternoon respond by shifting egestion time to just prior to the anticipated feeding time (Fuller et al. 1978).

Durham (1983) showed that in Red-tailed Hawks pellet egestion occurred at dawn each day even if the hawks had not eaten the day before or if they had eaten only meat without feathers, fur or bone. Thus, in hawks, egestion motility is not just the end result of having ingested, but is apparently an expression of a circadian rhythm. There are other differences between hawks and owls. Owls normally egest a pellet for each meal, while hawks may eat one to three meals before egesting a pellet (Duke et al. 1975, 1976b). The bones of prey receive little digestion in the stomachs of adult owls, whereas bones are virtually entirely digested in the falconiform stomach (Errington 1930, Sumner 1933, Glading et al. 1943, Clark 1972, Duke et al. 1975, 1976b). This is due to the lower pH in the stomach of hawks (Cummings et al. 1976). Nestling owls also digest bones.

The mechanism of pellet egestion in Red-tailed Hawks follows gastric and esophageal contractile activity very similar to that of Great Horned Owls (Durham 1983), with three clear phases of ingestion motility, chemical digestion and pellet formation, and egestion motility. It is likely that a telemetry study, as performed with Barred Owls, using Red-tailed Hawks or other hawks could provide very useful management information.

Ion and water balances. Little is known about ion and water balances in raptors, but the topic is relevant to management of captive birds. For birds weighing 60 g or more, which includes virtually all raptors, evaporative water loss from the respiratory surfaces and the skin in unstressed individuals can be offset by water produced via oxidative metabolism (Bartholomew and Cade 1963). The moisture in freshly killed prey thus can be used to meet (or partially meet) water loss associated with thermal stress, exercise, or both. Most raptors can be maintained in captivity, and even mate and lay eggs, in the absence of drinking water (Bartholomew and Cade 1957, 1963). Captive Great Horned Owls require 4.4–5.3% of their body weight per day as water (Duke et

al. 1973). This intake is lower than that of all but one of 21 species tested by Bartholomew and Cade (1963), including roadrunners (*Geococcyx* spp.), a species adapted to life in an arid environment. Evaporative water loss amounted to approximately 45% of the water ingested with prey in Great Horned Owls (Duke et al. 1973).

Like many other birds, raptors are able to regulate salt and water losses via both the kidney-cloaca system and the nasal salt glands. Urine volumes in Red-tailed Hawks fed beef hearts averaged 30.2 ml/day with sodium and potassium concentrations of 38 and 61 mM/l, respectively. The nasal gland secretions of these birds contained 272 mM/l of sodium and 8 mM/l of potassium (Johnson 1969). Other studies of Red-tailed Hawks have indicated higher sodium and potassium concentrations in both urine (206 and 76 mM/l, respectively) and nasal secretions (380 and 20 mM/l, respectively); similar data were found for eight other falconiform species (Cade and Greenwald 1966). Although functional nasal salt glands are apparently present in all Falconiformes, they have not been reported in Strigiformes.

Nutrition and Food Metabolizability

Nutritional requirements. Small mammals and birds form the bulk of the diet in most raptors. The natural diets (qualitative requirements) of most birds of prey have been studied extensively; some examples are provided in Table 5. The biomass eaten is most important in understanding the energetics of the predator and its impact on the environment. Thus, not only the species of prey and the frequency it occurs in the diet, but also the weight of that prey item must be known. An extensive compilation of prey weights for 35 mammalian and 81 avian prey items was prepared by Steenhof (1983). This includes mean values, determined from a large number of samples in many cases, and separate means for adults (male versus female frequently) and juveniles.

Amounts that must be consumed to maintain a constant body weight under both field and laboratory conditions (quantitative requirements) are known for a few species (Table 5). Food consumption of an individual varies according to level of activity and ambient temperature. Activity is influenced by factors such as day length, prey availability, breeding and nesting, and disturbance. In general, consumption varies inversely with ambient temperature within species and with body size among species (Table 6), as well as directly with activity.

Unfortunately, little is known regarding daily or seasonal requirements for specific nutrients for raptors.

Table 5. Natural foods of some common North American raptors.[a]

Species	Ref.[b]	Percent of Diet				
		Small Rodents	Larger Mammals	Birds	Insects	Other
Northern Harrier (*Circus cyaneus*)	1	98.4	0.3	1.0	-	0.3
Red-shouldered Hawk (*Buteo lineatus*)	1	97.0	-	3.0	-	-
Red-tailed Hawk (*B. jamaicensis*)	1	95.5	1.4	3.1	-	-
Roughleg (*B. lagopus*)	1	98.1	-	1.9	-	-
American Kestrel (*Falco sparverius*)	1	90.3	-	9.9	-	-
Barn Owl (*Tyto alba*)	2	81.6	16.4	2.0	-	-
Eastern Screech Owl (*Megascops asio*)	1	3.4	-	6.3	0.3	-
Great Horned Owl (*Bubo virginianus*)	1	92.3	3.7	3.5	-	0.7
Burrowing Owl (*Athene cunicularia*)	2	12.1	0.7	1.3	85.9	-
Barred Owl (*Strix varia*)	3	53.2	7.8	24.2	4.8	10.0
Long-eared Owl (*Asio otus*)	1	100.0	-	-	-	-
Short-eared Owl (*A. flammeus*)	1	99.3	-	0.7	-	-

[a] Foods were determined by pellet analysis. Foods such as meat from a carcass and insect parts are thoroughly digested in falconiform stomachs and do not appear in pellets.

[b] References: 1 = Craighead and Craighead (1956), 2 = Marti (1969), 3 = Errington (1932).

However, the caloric and nutrient value of some wild and domestic rodents and birds are known (Bird and Ho 1976, Bird et al. 1982; Table 7). These data are useful in assessing the relative nutritive and energy value of wild prey.

The nutrient composition of vertebrate tissues is relatively constant, and as a food source their nutrient balance closely matches that required by birds (Klasing 1998), thus it is unlikely that any macro- or micro-nutrients are limiting in the diet for most species, although a few nutritional disorders have been described in raptors (Cooper 1978). The major difference between prey species is in the relative proportion of fat present, which varies not only between prey species, but also among individuals and between seasons within species. For example, some small passerines can store up to 50% of their body mass as fat prior to migration, making them energetically, high-quality prey.

Almost all raptors eat meat, which is relatively eas-ily digested, and it might be assumed that all species would show similar digestive efficiencies. This, however, seems not to be the case (Barton and Houston 1994b). Digestive efficiency varies from about 75% to 82%, and this is correlated with the length of the digestive tract. Species with short guts tend to digest their food less efficiently than species with long guts, and consequently need to capture proportionately more prey each day. This may be associated with hunting strategy, for the species with short guts and poor digestive efficiency tend to be species which take a high proportion of birds in flight and need the ability to accelerate rapidly (Barton and Houston 1994a). For such species it may be advantageous to have a lightweight, low-volume gut, even if it results in poor digestive efficiency, because by being more agile they can capture more prey. It does, however, have the consequence that short-gut species are forced to feed on prey items with a high energy content (high body fat), and are unable to main-

Table 6. Food consumption at several ambient temperatures for some adult North American raptors kept outside for one year in Ogden, Utah U.S.A.

Species	Ref.	Diet	Body Mass (g)	Amount Eaten per Day		Ambient Temperature (°C)
				Grams	Percent of Body Mass	
Bald Eagle (*Haliaeetus leucocephalus*)	2	mice	3870	219.8	5.6	27
Bald Eagle	5	mixed[c]	3922	344.8	8.8	−10
Bald Eagle	5	mixed[c]	3922	294.5	7.5	5
Bald Eagle	5	mixed[c]	3922	265.2	6.8	20
Northern Goshawk (*Accipiter gentilis*)	2	mice	1100	80.2	7.3	27
Broad-winged Hawk (*Buteo platypterus*)	2	mice	470	29.4	6.3	27
Red-tailed Hawk (*B. jamaicensis*)	2	mice	1320	75.5	5.5	27
Roughleg (*B. lagopus*)	2	mice	1020	48.0	4.7	27
American Kestrel (*Falco sparverius*)	2	chick	105	14.6	13.9	27
Common Kestrel (*F. tinnunculus*)	6	mice	204	24.3	11.9	14
Peregrine Falcon (*F. peregrinus*)	1	mice	680	60.6	8.9	27
Gyrfalcon (*F. rusticolus*)	1	mice	880	70.3	8	27
Barn Owl (*Tyto alba*)	3	mice	603	60.5	10	_[b]
Barn Owl	6	chick	262	28.3	10.8	14
Eastern Screech Owl (*Megascops asio*)	4	mixed	153	39.0	25.4	6
Eastern Screech Owl	2	mice	149	17.1	11.5	27
Great Horned Owl (*Bubo virginianus*)	2	mice	1770	71.2	4.0	27
Great Horned Owl	3	mice	1336	62.6	4.7	_[b]
Snowy Owl (*B. scandiaca*)	1	mice	1900	93.1	4.9	27
Burrowing Owl (*Athene cunicularia*)	3	mice	166	26.4	15.9	_[b]
Barred Owl (*Strix varia*)	2	mice	741	42.9	5.8	27
Barred Owl	4	mixed	625	67.0	11.8	4
Great Gray Owl (*S. nebulosa*)	4	mixed	1045	77.0	7.4	−10
Long-eared Owl (*Asio otus*)	3	mice	291	37.5	12.9	_[b]
Short-eared Owl (*A. flammeus*)	2	mice	432	50.0	11.6	27
Northern Saw-whet Owl (*Aegolius acadicus*)	2	mice	96	12.9	13.4	27

[a] References: 1= Duke et al. (1975), 2 = Duke et al. (1976a), 3 = Marti (1973), 4 = Craighead and Craighead (1956), 5 = Stalmaster and Gessaman (1982), and 6 = Kirkwood (1979).

[b] Data are mean values for birds kept outside for one year in Ogden, Utah.

[c] Chum salmon (*Oncorhynchus keta*), black-tailed jackrabbit (*Lepus californicus*), and Mallard (*Anas platyrynchos*).

Table 7. Partial analysis of nutrient levels in wild and domestic rodents, birds, and an insect.

	Rat[a]	Mouse[a]	Chicken[a]	Day-old Chick[a]	Sparrow[b]	Vole[b]	Grasshopper[b]
Number of animals	10	30	10	30	11	13	89
Average mass (g)	325.7	26.7	386.7	41.2	27	32	0.21
Dry matter % (freeze dried)	34.4	35.4	33.5	27.6	31.6	35.7	31.9
Crude fat (% DM)	22.1	24.9	26.9	24.2	15.9	6.01	6.03
Crude protein (N x 6.25% DM)	62.8	56.1	56.7	62.2	64.9	57.3	75.7
Ash (% DM)	10.0	10.4	9.5	7.4	10.6	10.1	4.8
Crude fiber (% DM)	2.4	1.7	2.0	0.8	0.43	3.85	-
Gross energy (kcal/g DM)	5.78	5.84	5.93	6.02	5.39	4.15	5.02
Calcium (%) DM	2.06	2.38	1.94	1.36	2.94	2.85	0.31
wet mass	0.69	0.84	0.65	0.38	0.94	1.02	0.098
Phosphorus (%) DM	1.48	1.72	1.40	1.00	2.35	2.66	1.27
wet mass	0.51	0.61	0.47	0.28	0.74	0.95	0.41
Ca:P ratio	1.39	1.38	1.39	1.36	1.3	1.1	0.2
Zinc (mg/kg) DM	129.2	134.6	158.0	106.9	109.8	105.5	200.2
wet mass	13.3	47.7	52.8	29.9	34.7	37.7	63.9
Copper (mg/kg) DM	4.5	8.0	4.5	3.2	12.6	13.7	50.3
wet mass	1.5	2.8	1.5	0.9	3.98	4.89	16.1
Manganese (mg/kg) DM	7.5	11.7	9.0	3.0	11.4	14.9	25.1
wet mass	2.5	4.1	3.0	0.8	3.6	5.32	8.01
Iron (mg/kg) DM	175.7	239.1	146.8	121.8	592.0	332.3	331.4
wet mass	58.9	84.6	49.1	34.0	187.2	118.7	105.8
Thiamine (mg/kg) DM	13.3	-	8.5	16.0	-	-	-

[a] From Bird and Ho (1976)
[b] From Bird et al. (1982); House Sparrow (*Passer domesticus*), meadow vole (*Microtus pennsylvanicus*), red-legged grasshopper *(Melanoplus femurrubrum)*

tain their body weight if fed on prey with low fat levels (Taylor et al. 1991). This may explain why many falcons specialize on small passerines and are rarely found feeding on carrion or low-energy prey.

The ceca of owls apparently make little contribution to food digestion since metabolizability of a mouse diet was not significantly different between cecectomized and intact Great Horned Owls (Duke et al. 1981). Water balance also was unaffected by cecectomy.

Kirkwood (1981) calculated maintenance metabolizable energy (ME) based on food intake for several diets at several ambient temperatures for nine strigiform and 22 falconiform species using the linear regression equations ME = 110 $W^{0.679}$, where ME is expressed in kcal/day and W (weight) in kg. Data for Falconiformes and Strigiformes were pooled as separate regressions and were not significantly different. Wijnandts (1984) made similar calculations for 13 strigiforms and 26 falconiforms under caged conditions eating either mice or rats. Metabolizable energy also was calculated from published data on food consumption using a caloric value of 8.4 kJ/g for mice or rats and assumed metabolizability of 76%. Linear regression equations derived for falconiforms and strigiforms were ME = 9.722 $W^{0.577}$ (r = 0.918) and ME = 8.63 $W^{0.578}$ (r = 0.958), respectively, where ME is in kJ/bird/day and W is in g.

SUMMARY

We still have much to learn about the gastrointestinal physiology of raptorial birds. Prey availability (both population size and vulnerability), the nutritive value of the prey, and its metabolizability by raptors all must be considered in evaluating raptor energetics. In these birds with such uniquely carnivorous food habits, further research in this field should prove most fruitful. However, with the tragic passing of co-author Gary Duke, who led the world in the field of avian gastrointestinal physiology in 2006, and no one on the immediate horizon appearing to follow in his footsteps, it may be some time before significant advances in this field are again achieved.

ACKNOWLEDGMENTS

This chapter is dedicated to the memory of my co-author Dr. Gary Duke, the world's foremost authority on avian gastrointestinal physiology. The chapter would

be all too brief without his significant contributions. Gary was a warm, caring human being who loved birds of prey and who was an excellent mentor to many of us in raptor studies. I thank Nigel Barton for his review of an earlier draft of this chapter.

LITERATURE CITED

BALGOOYEN, T.G. 1971. Pellet regurgitation by captive sparrow hawks (*Falco sparverius*). *Condor* 73:382–385.

BARTHOLOMEW, G.A. AND T.J. CADE. 1957. The body temperature of the American Kestrel, *Falco sparverius*. *Wilson Bull.* 69:149–154.

———— AND T.J. CADE. 1963. The water economy of land birds. *Auk* 80:504–511.

BARTON, N.W.H. AND D. C. HOUSTON. 1994a. Morphological adaptation of the digestive tract in relation to feeding ecology of raptors. *J. Zool. Lond.* 232:133–150.

———— AND D.C. HOUSTON. 1994b. A comparison of digestive efficiency in birds of prey. *Ibis* 135:363–371.

BIRD, D.M. AND S.K. HO. 1976. Nutritive values of whole-animal diets for captive birds of prey. *Raptor Res.* 10:45–49.

————, S.K. HO AND D PARÉ. 1982. Nutritive values of three common prey items of the American Kestrel. *Comp. Biochem. Physiol.* 73A:13–515.

CADE, T.J. AND L. GREENWALD. 1966. Nasal salt secretion in falconiform birds. *Condor* 68:338–343.

CHITTY, D. 1938. Pellet formation in Short-eared Owls, *Asio flammeus*. *Proc. Zool. Soc. Lond.* 108A:267–287.

CLARK, R.J. 1972. Pellets of the Short-eared Owl and Marsh Hawk compared. *J. Wildl. Manage.* 36:962–964.

COOPER, J.E. 1978. Veterinary aspects of captive birds of prey. The Standfast Press, Gloucestershire, United Kingdom.

CRAIGHEAD, J.J. AND F.C. CRAIGHEAD, JR. 1956. Hawks, owls and wildlife. Dover Publications, Inc., New York, NY U.S.A.

CUMMINGS, J.H., G.E. DUKE AND A.A. JEGERS. 1976. Corrosion of bone by solutions simulating raptor gastric juice. *Raptor Res.* 10:55–57.

DUKE, G.E. 1978. Raptor physiology. Pages 225–231 in M.E. Fowler [ED.], Zoo and wild animal medicine, 1st Ed. W.G. Saunders Co., Philadelphia, PA U.S.A.

———— AND D.D. RHOADES. 1977. Factors affecting meal to pellet intervals in Great Horned Owls (*Bubo virginianus*). *Comp. Biochem. Physiol.* 56A:283–286.

————, J.C. CIGANEK AND O.A. EVANSON. 1973. Food consumption and energy, water, and nitrogen balances in Great Horned Owls (*Bubo virginianus*). *Comp. Biochem. Physiol.* 44A:283–292.

————, A.A. JEGERS, G. LOFF AND O.A. EVANSON. 1975. Gastric digestion in some raptors. *Comp. Biochem. Physiol.* 50A:649–654.

————, O.A. EVANSON AND A.A. JEGERS. 1976a. Meal to pellet intervals in 14 species of captive raptors. *Comp. Biochem. Physiol.* 53A:1–6.

————, O.A. EVANSON AND P.T. REDIG. 1976b. A cephalic influence on gastric motility upon seeing food in domestic turkeys, Great Horned Owls (*Bubo virginianus*) and Red-tailed Hawks (*Buteo jamaicensis*). *Poult. Sci.* 55:2155–2165.

————, J.C. CIGANEK, P.T. REDIG AND D.D. RHOADES. 1976c. Mechanism of pellet ingestion in Great Horned Owls (*Bubo virginianus*). *Am. J. Physiol.* 231:1824–1830.

————, M.R. FULLER AND B.J. HUBERTY. 1980. The influence of hunger on meal to pellet intervals in Barred Owls. *Comp. Biochem. Physiol.* 66A:203–207.

————, J.E. BIRD, K.A. DANIELS AND R.W. BERTOY. 1981. Food metabolizability and water balance in intact and cecectomized Great Horned Owls. *Comp. Biochem. Physiol.* 68A:237–240.

————, A.L. TEREICK, J.K. REYNHOUT, D.M. BIRD, AND A.E. PLACE. 1996. Variability among individual American Kestrels (*Falco sparverius*) in parts of day-old chicks eaten, pellet size, and pellet egestion frequency. *J. Raptor Res.* 30:213–218.

————, J. REYNHOUT, A.L. TEREICK, A.E. PLACE, AND D.M. BIRD. 1997. Gastrointestinal morphology and motility in American Kestrels (*Falco sparverius*) receiving high or low fat diets. *Condor* 99:123–131.

DURHAM, K. 1983. The mechanism and regulation of pellet egestion in the Red-tailed Hawk (*Buteo jamaicensis*) and related gastrointestinal contractile activity. M.S. thesis, University of Minnesota, St. Paul, MN U.S.A.

ERKINARO, E. 1973. Seasonal variation of the dimensions of pellets in Tengmalm's Owl *Aegolius funereus* and the Short-eared Owl *Asio flammeus*. *Aquilo Ser. Zool.* 14:84–88.

ERRINGTON, P.L. 1930. The pellet analysis method of raptor food habit study. *Condor* 32:292–296.

————. 1932. Food habits of southern Wisconsin raptors, part I: owls. *Condor* 34:176–186.

FULLER, M.R. AND G.E. DUKE. 1978. Regulations of pellet egestion: the effects of multiple feedings on meal to pellet intervals in Great Horned Owls. *Comp. Biochem. Physiol.* 62A:439–444.

————, G.E. DUKE AND D.L. ESKEDAHL. 1978. Regulations of pellet egestion: the influence of feeding time and soundproof conditions on meal to pellet intervals of Red-tailed Hawks. *Comp. Biochem. Physiol.* 62A:433–438.

GLADING, B., D.F. TILLOTSON AND D.M. SELLECK. 1943. Raptor pellets as indicators of food habits. *Calif. Fish Game* 29:92–121.

GRIMM, R.J. AND W.M. WHITEHOUSE. 1963. Pellet formation in a Great Horned Owl: a roentgenographic study. *Auk* 80:301–306.

HERPOL, C. 1964. Activité proteolytique de l'appareil gastric d'oiseaux granivores et carnivores. *Ann. Biol. Anim. Biochem. Biophys.* 4:239–244.

————. 1967. Etude de l'activité proteolytique des divers organes du système digestif de quelques espèces d'oiseaux en rapport avec leur régime alimentaire. *Zeitschrift fur ver gleichended Physiologie* 57:209–217.

HOUSTON, D.C. 1976. Breeding of the White-backed and Ruppell's Griffon Vultures. *Ibis* 118:14–39.

———— AND J. COPSEY. 1994. Bone digestion and intestinal morphology of the Bearded Vulture. *J. Raptor Res.* 28:73–78.

JOHNSON, I.M. 1969. Electrolyte and water balance in the Red-tailed Hawk, *Buteo jamaicensis*. [Abstr.] *Am. Zool.* 9:587.

KIRKWOOD, J.K. 1979. The partitioning of food energy for existence in the Kestrel (*Falco tinnunculus*) and the Barn Owl (*Tyto alba*). *Comp. Biochem. Physiol.* 63A:495–498.

————. 1981. Maintenance energy requirements and rate of weight loss during starvation in birds of prey. Pages 153–157 in J.E. Cooper and A.G. Greenwood [EDS.], Recent advances in the study of raptor diseases. Chiron Publishers, Keighley, West Yorkshire, United Kingdom.

KLASING, K.C. 1998. Comparative avian nutrition. CAB International, Wallingford & New York, NY U.S.A.

KOSTUCH, T.E. AND G.E. DUKE. 1975. Gastric motility in Great Horned Owls. *Comp. Biochem. Physiol.* 51A:201–205.

KUECHLE, V.B., M.R. FULLER, R.A. REICHLE, R.J. SCHUSTER AND G.E. DUKE. 1987. Telemetry of gastric motility data from owls. Proceedings of the 9th international symposium on biotelemetry, Dubrovnik, Yugoslavia.

MARTI, C.D. 1969. Some comparison of the feeding ecology of four owls in north central Colorado. *Southwest Nat.* 14:163–170.

MIKKOLA, H. 1983. Owls of Europe. T. & A.D. Poyser, London, United Kingdom.

RACZYNCKI, J. AND A.L. RUPRECHT. 1974. The effect of digestion on the osteological composition of owl pellets. *Acta Ornithologica* 14:25–37.

REA, A.M. 1973. Turkey Vultures casting pellets. *Auk* 90:209–210.

REED, C.I. AND B.P. REED. 1928. The mechanism of pellet formation in the Great Horned Owl (*Bubo virginianus*). *Science* 68:359–360.

RHOADES, D.D. AND G.E. DUKE. 1977. Cineradiographic studies of gastric motility in Great Horned Owls (*Bubo virginianus*). *Condor* 79:328–334.

STALMASTER, M.V. AND J.A. GESSAMAN. 1982. Food consumption and energy requirements of captive Bald Eagles. *J. Wildl. Manage.* 46:646–654.

STEENHOF, K. 1983. Prey weights for computing percent biomass in raptor diets. *Raptor Res.* 17:15–27.

SUMNER, E.L., JR. 1933. The growth of some young raptorial birds. *Univ. Calif. Publ. Zool.* 40:277–282.

TAYLOR, R., S. TEMPLE, AND D. M. BIRD. 1991. Nutritional and energetic implications for raptors consuming starving prey. *Auk* 108:716–718.

WIJNANDTS, H. 1984. Ecological energetics of the Long-eared Owl *Asio otus*. *Ardea* 72:1–92.

YALDEN, D.W. 2003. The analysis of owl pellets. The Mammal Society, London, United Kingdom.

B. Hematological

DEBORAH J. MONKS
Brisbane Bird and Exotics Veterinary Service
Cnr. Kessels Rd. and Springfield St.
Macgregor, QLD 4109, Australia

NEIL A. FORBES
Great Western Referrals,
Unit 10 Berkshire House, Country Park Business Park
Shrivenham Rd., Swindon, Wiltshire, SN1 2NR, United Kingdom

INTRODUCTION

Disease diagnosis and assessment of therapeutic efficacy often relies on hematological analysis (Howlett 2000, Cooper 2002). Rehabilitation centers can use hematological changes to help detect sub-clinical disease, in pre-release assessment of individuals, and as a prognostic indicator for new admissions. The health of raptor populations can be monitored similarly. Nutritional status, disease, or food supply differences among populations or immune suppression due to various stressors all can be detected using hematology (van Wyk et al. 1998, Cooper 2002).

Because raptors are at the top of many food chains, their health can reflect the health of entire ecosystems (Cooper 2002). Hematological alterations can indicate changing habitat quality and food availability or may imply exposure to pollutants or toxins (Hoffman et al. 1985, Mauro 1987, Bowerman et al. 2000, Seiser et al. 2000). Habitat loss and fragmentation have resulted in increased exposure of some raptor populations to parasites, which can alter host-parasite balances. Increased parasite pathogenicity may be implied by hematological alterations (Loye and Carroll 1995).

Recently, research has focused on determining the reference ranges that distinguish among species of raptors. A number of published references provide parameters by sex and age (Rehder et al. 1982, Ferrer et al. 1987, van Wyck et al. 1998, Bowerman et al. 2000). Computerized databases also are available, including ISIS (www.ISIS.org) and LYNX (Bennett et al. 1991).

The discussion of biochemical parameters is beyond the scope of this chapter. Several comprehensive references provide additional details in this area (Campbell 1994, Joseph 1999, Fudge 2000, Cooper 2002).

Sampling

Physiological variables affecting hematological testing. Blood sampling should be performed as soon as possible after capture and prior to other procedures, as the stress of capture and restraint can alter the leucogram (Wingfield and Farner 1982, Sockman and Schwabl 2001) and result in leucocytosis, heterophilia, lymphocytosis, or lymphopenia (Fudge 2000). Parga et al. (2001) advocated the use of the heterophil/lymphocyte ratio, rather than the absolute number of heterophils or lymphocytes, as a more sensitive indicator of stress in raptors, although the actual ratio may vary among species. Leucocyte numbers also may be altered with concurrent diseases (Howlett 2000, Parga et al. 2001).

Researchers should be aware that other physiological factors might affect hematological parameters. The following variables should be considered when planning hematological testing, so that efforts can be made to minimize their impact: (1) erythrocyte production may decrease with increasing ambient temperatures and may vary with season (Hunter and Powers 1980, Rehder et al. 1982); (2) hematocrit may be increased by high androgen levels and decreased by high estrogen levels; (3) molt decreases hematocrit in both sexes (Sturkie 1976, Rehder et al. 1982); (4) up until fledging, hematocrit and hemoglobin levels increase with age (Rehder et al. 1982, Bowerman et al. 2000); (5) some studies have reported differences in hematocrit between sexes, whereas others have found no correlation (Sturkie 1976, Rehder et al. 1982, Dawson and Bortolotti 1997); (6) Rehder and Bird (1983) demonstrated diurnal variation in hematocrit and red-blood-cell (RBC) count of Red-tailed Hawks (*Buteo jamaicensis*); and (7) certain sedative and anesthetic drugs also can cause leucogram or hemogram changes (Mauro 1987, Fudge 2000). All of this leads us to recommend consistency in sampling.

Venipuncture procedure. Birds that weigh less than 500 g usually are sampled with a 25-gauge hypodermic needle and a 1- or 2-ml syringe, whereas a 23-gauge hypodermic needle is best used for birds of more than 500 g (Cooper 2002). Veins of very small birds can be nicked with a scalpel blade and blood collected in a capillary tube (Dawson and Bortolotti 1997). Butterfly catheters may lessen the effect of bird movements during sampling (Cooper 2002). Smaller-gauge needles increase the risk of hemolysis. The use of larger needles increases the risk of hematomas (Fudge 2000). Excessive negative pressure may collapse veins (Jennings 1996). Adding anticoagulant to the syringe prior to venipuncture may dilute the blood sample, although some authors advocate this if clotting of samples is a problem (Rehder et al. 1982, Cooper 2002). Because of inconsistent results, we do not recommend sampling from talon clipping (Campbell 1994). Regardless of technique, the phlebotomy site must be prepared aseptically to prevent bacterial contamination (Fudge 2000).

Avian blood volume ranges from 6 to 12% of body mass and no more than 5–10% of the total blood volume should be removed. This equates to approximately 0.5–1% of total body mass (Campbell 1988, Fudge 2000). Smaller volumes should be removed from unhealthy or stressed birds (Cooper 2002).

When repeatedly sampling the same bird, allow for sufficient time for erythrocyte replenishment between sampling (Mauro 1987). The average life span of an avian erythrocyte is 28 to 45 days (Rodnan et al. 1957).

That said, American Kestrels (*Falco sparverius*) bled at 10% of blood volume weekly for 20 weeks showed no decrease in hematocrit (Rehder et al. 1982).

Inappropriate restraint can result in blood-vessel laceration and prolonged bleeding. Hematoma formation may considerably increase the total blood volume removed. To reduce this risk, pressure should be applied to the phlebotomy site for at least 1 to 2 minutes after venipuncture, and the bird should not be released until hemostasis is complete.

Normal venipuncture sites. The jugular vein is the largest accessible vein, although hematoma formation there can be a problem in inexperienced hands (see below). The right jugular generally is larger than the left and also has an overlying apteria (featherless area). A popular alternative is the basilic vein, which crosses the elbow ventrally. If the bird is struggling vigorously, it can be difficult to sample, and wing trauma (including fractures) can occur. It is not always easy to find in smaller raptors. Large hematomas can develop after iatrogenic tissue trauma or insufficient post-sampling pressure. A third location for blood sampling is the median metatarsal vein, which is found proximal to the tarsometatarsal joint. There is less chance of hematoma formation in this vein due to the anatomy of the surrounding soft tissue (Fudge 2000).

After 25 years of blood-sampling American Kestrels at the Avian Science and Conservation Centre at McGill University in Montreal, the large jugular vein has become the preferred sampling site. With one person holding the bird's head stable and in an appropriate position to expose the vein, hematomas are rarely encountered using this method (I. Ritchie and D. M. Bird, pers. comm.).

Sample preparation. One or two blood smears should be made at the time of collection using non-anticoagulated blood. Avian blood cells are fragile and rough smear techniques can result in large numbers of unidentifiable (smudge) cells (Jennings 1996, Fudge 2000). If protected from moisture, air-dried, unfixed blood smears may last up to 72 hours (Howlett 2000).

Human pediatric blood tubes, which are available commercially, are quite suitable for raptor blood. The tubes are available as ethylenediaminetetraacetic acid (EDTA), heparin, and plain-gel, and when filled to the line, provide the correct sample to anticoagulant ratio. Hemolysis can be reduced by precise sampling and removing the needle from the syringe prior to filling the sample pots with blood (Fudge 2000). Heparinized capillary tubes may be capped with plasticine and stored.

Choice of anticoagulant. The laboratory where the analyses will be performed should be contacted prior to sample collection for preferred anticoagulant, storage, and other processing information.

Blood for hematological analysis usually is collected into EDTA. Note that erythrocytes may lyze within 24 to 48 hours of exposure to EDTA. This is particularly marked in some non-raptors (Campbell 1994). Heparin may affect the affinity of blood cells to Romanowksy stains and cause clumping of leucocytes and thrombocytes (Jennings 1996, Howlett 2000). Fudge (2000) found that citrated blood provided better cell integrity for automated analysis.

Sample Storage and Processing

Time from sampling to processing should be as short as practical, and samples that are not analyzed immediately should be kept cool (approximately 4°C). One person should do the blood smear staining and interpretation to minimize variability. Wright's, Giemsa, and modified Wright's-Giemsa stains all provide good cell morphology, although new staining techniques also seem to do well (Campbell 1988, Fudge 2000, Samour et al. 2001, Cooper 2002, Kass et al. 2002).

Factors affecting analysis. Sample clotting can occur due to slow sample collection, tissue trauma, inadequate sample mixing and overfilling of anticoagulant sample pots (Jennings 1996). Hemolysis and lipemia can alter a number of hematological and biochemical parameters, including total protein levels (Joseph 1999, Cooper 2002). Inaccurate cell identification can occur with blood smears made from old or anticoagulated blood, exposed to formalin fumes or stained with expired stains (Fudge 2000). Failure to detect hemoparasites can result from poor-quality smearing or staining techniques, operator inexperience, or from the use of poor-quality microscopes (Cooper 2002).

Relevant international agreements, including CITES (Convention on International Trade in Endangered Species of Wild Fauna and Flora), and country and local laws must be considered before transporting samples internationally (Cooper 2002).

Hematological Parameters

Listing "normal" hematological values for each species is beyond the scope of this chapter. Researchers should consult relevant peer-reviewed publications and databases for specific information. General information can be found in Samour (2000), Cooper (2002), and Redig (2003).

Total plasma protein. Although sometimes considered a biochemical parameter, analysis of total plasma protein (TPP) is required for complete interpretation of the erythron, especially in instances of anemia. Protein electrophoresis and fibrinogen determination also may be performed.

The erythron. Evaluation of the erythron involves determining hematocrit or packed cell volume (Hct or PCV - l/l), hemoglobin (Hb - g/l) and the RBC count (x 10^{12}/l) followed by calculation of mean corpuscular volume (MCV), mean cell hemoglobin (MCH) and mean corpuscular hemoglobin concentration (MCHC). For additional information consult Howlett (2000) and Fudge (2000).

The accepted hematocrit range for raptors is 35–55 l/l (0.35–0.55) (Fudge 2000, Cooper 2002). Lower values have been obtained from apparently healthy birds (Rehder et al. 1982). Abnormalities must be interpreted in conjunction with TPP and fibrinogen levels (see Table 1). A reticulocyte count of 5–10% is considered physiologically normal.

Anemia can be characterized as regenerative or non-regenerative, based on the numbers of reticulocytes. Some hematologists maintain that there is no sat-

Table 1. Changes in the Erythron. (Based on Joseph [1999] and Fudge [2000].)

Condition	PCV	TPP	Fibrinogen	TPP:Fibrinogen
Dehydration	Increased	Increased	Increased	>5
Polycythemia	Increased	Normal	Normal	1.5–5.0
Anemia	Decreased	Normal	Normal or increased	Depends on cause
Infection or Inflammation	Normal	Normal	Increased	<1.5

isfactory method of obtaining an objective reticulocyte count in raptors; instead they rely on the subjective analysis of stained blood smears, noting the degree of polychromasia and the numbers of rubricytes, prorubricytes, and rubriblasts, if present (M. Hart, pers. comm.). Others are comfortable using a vital stain such as new methylene blue, or Wright's stain, to preferentially stain reticulocytes (Fudge 2000). In the presence of anemia, a reticulocyte count of <5% indicates a poor regenerative response, whereas a count of >10% indicates a good regenerative response (Cooper 2002). Polychromasia greater than 1–5% also can indicate an appropriate regenerative response. Anemia can be classified according to etiology or erythrocyte morphology (see Tables 2 and 3). Concurrent dehydration may mask the signs of anemia.

Hematocrit and TPP will decrease with chronic undernourishment (Ferrer et al. 1987, Cooper 2002). Unfortunately, the severity and duration of food deprivation required to cause these changes are uncertain. In Common Buzzards (*B. buteo*) starved for 13 days, these parameters changed only when feeding was resumed (Garcia-Rodriguez et al. 1987). Conversely, American Kestrels typically die after five days of starvation (Shapiro and Weathers 1981). Dawson and Bortolotti (1997) found that hematocrit was not accurate in predicting nestling survival in American Kestrels. Body size, species ecology and developmental stage also influence an individual's ability to withstand sub-optimal nutrition.

The leucogram. Detailed anatomy and function of leucocytes is not discussed here. The reader should con-

Table 2. Classification of anemia according to erythrocyte morphology. (Based on Fudge [2000].)

	Type of anemia		
	Normocytic normochromic	Hypochromic microcytic	Hypochromic macrocytic
PCV	Decreased	Decreased	Decreased
MCHC	Normal	Decreased	Decreased
MCV	Normal	Decreased	Increased
Polychromasia	None to slight	Increased	Increased
Anisocytosis	None to slight	Normal to increased	Normal to increased
Possible causes	Generally non-regenerative, reduced RBC production	Iron deficiency, chronic blood loss, chronic disease	Acute blood loss, early stages of lead toxicity

Table 3. Classification of anemia according to etiology. (Based on Campbell [1994, 2000], Fudge [2000], and Howlett [2000].)

Insufficient erythrocyte production	Acute or chronic blood loss	Increased erythrocyte destruction
Malnutrition	Blood-sucking ectoparasites	Hemoparasitism
Chronic disease including mycobacteriosis and aspergillosis	Gastrointestinal parasitism	Bacterial septicemia
Chemicals (lead and aflatoxicosis)	Trauma	Acute aflatoxicosis
Iron and folic acid deficiencies	Rupture of organs or neoplasms	Toxemia
Some neoplasms		

sult Campbell (1988, 1994) and Fudge (2000) for additional information.

Although reference ranges should be established for individual species, it is generally true that vultures and eagles tend to have higher white blood cell (e.g., WBC x 10^9/l) counts than hawks, falcons and owls (N. Forbes, pers. comm.). Both the total and the differential leucocyte count should be obtained. The differential leucocyte count should be expressed both as an absolute count and as a percentage. Consult species-specific reference ranges for normal values. As a guide, in most owl species, the lymphocyte percentage ranges from 40–70%, while in most other raptors, the heterophil is the most common cell (Joseph 2000, Cooper 2002). It is believed that the eosinophil differential can range from 10 to 35% in healthy raptors (Joseph 2000). Conversely, Samour et al. (1996) found that eosinophils were closely associated with parasitism and were not present in such proportions in "normal" individuals. Table 4 lists some leucocyte abnormalities and potential etiologies.

Hemoparasites. Blood parasites are found in many raptors, with incidences varying geographically and among parasite and host species (Joseph 1999). Hemoparasites can cause increased TPP levels, leucocytosis, anemia or death (Garvin et al. 2003, Redig 2003). Table 5 lists hemoparasites in raptors and details regarding pathogenicity, as well as vectors and diagnoses. Pierce (1989) provides a color reference to hemoparasites.

Table 4. Changes in the leucogram of raptors. (Based on Campbell [1994, 2000], Fudge [2000], and Howlett [2000].)

Leucogram changes	Potential etiologies
Leucocytosis	Bacterial infections, including mycobacteriosis, stress, trauma, toxicity, fungal infections including aspergillosis, leukemia
Leucopenia	Overwhelming bacterial infection causing depletion of bone marrow, viremia, depression of bone marrow
Heterophilia	Bacterial infections, including mycobacteriosis, stress, fungal infections including aspergillosis, toxemia
Herteropenia	Overwhelming bacterial infection, viremia, bone marrow suppression, deficiency diseases
Toxic heterophil changes (cytoplasmic basophilia, vacuolization and degranulation, karyorhexis, karyolysis)	Septicemia, viraemia, toxemia, severe infection
Monocytosis	Infection, including mycobacteriosis and aspergillosis, chronic disease, tissue necrosis
Lymphocytosis	Some infectious and metabolic disorders, some neoplasms
Lymphopenia	Stress, uremia, immune suppression, some neoplasms, viremia
Reactive lymphocytes	Infection, including salmonellosis and aspergillosis
Eosinophilia	Parasitism, including hemoparasites, tissue damage, hypersensitivity (questionable)
Eosinopenia	Corticosteroids, stress
Basophilia	Tissue damage, parasites (inconsistent), hypersensitivity (questionable), chronic disease
Fibrinogen (increased)	Infection, inflammation, hemorrhage
Fibrinogen (decreased)	Liver failure
Thrombocytosis	Rebound response to hemorrhage, response to excessive thrombocyte demand (including phagocytosis)
Thrombocytopenia	Excessive peripheral demand or depression of production (e.g., severe septicemia)

Table 5. Hemoparasites of raptors. Based on Cooper (2002), Gutierrez (1989), Joseph (1999), Lacina and Bird (2000), Redig (2003), Remple (2003), Samour and Peirce (1996), Samour and Silvanose (2000).

Species	Site of infection and incidence	Extent pathogenic	Vector and transmission	Diagnosis
Leukocytozoon spp. (Hemosporidia)	Peripheral RBC and WBC — relatively common in a number of species, seasonal incidence	Generally non-pathogenic. May cause illness and occasional deaths due to anemia in young, debilitated or heavily infested birds	Simuliid black flies	Blood smears — non-pigmented gametocytes in RBC cytoplasm. Occasionally found in muscle, heart, spleen, kidney and liver tissues on histology
Hemoproteus spp. (Hemosporidia)	Peripheral RBC — more common in Strigiforms	Generally non-pathogenic. May cause illness and occasional deaths due to anemia in young, debilitated or heavily infested birds	Hippoboscid flies and Culicoides midges	Blood smears — pigmented gametocytes occupying >50% RBC cytoplasm
Plasmodium spp. (Hemosporidia — 34 spp.)	RBC, WBC, thrombocytes and reticulo-endothelial cells. Disease reported in falcons, especially Gyrfalcons (*Falco rusticolus*) and gyr-hybrids	Pathogenicity varies. Clinical signs: anemia, thrombosis, dyspnea, acute death	Culicine, Aedes occasionally Anopheles mosquitoes	Blood smears — pigmented gametocytes, trophozoites and schizonts in RBC, WBC and thrombocytes. May displace nucleus from central position. Unfixed spleen and liver sections
Microfilaria	Free in plasma — sporadic reports in variety of species	Uncertain	Uncertain	Blood smears
Babesia spp. (piroplasm)	Peripheral RBC — a few reports only in Prairie Falcon (*Falco mexicanus*), Saker Falcon (*F. cherrug*), Barn Owl (*Tyto alba*), Bearded Vulture (*Gypaetus barbatus*)	Pathogenicity controversial. Poor performance, possible death	Ticks, including *Ornithodorus concanensis*	Blood smears
Atoxoplasma spp. (coccidia)	Mononuclear leucocytes — reported in Spotted Owl (*Strix occidentalis*)	Uncommonly reported	Ingestion of sporulated oocysts	Reddish intracytoplasmic inclusions indenting leucocyte nucleus
Trypanosomes (flagellated protozoa)	Free in plasma — reported in a variety of species	Not known to be pathogenic	Blood-sucking arthropods	Blood smears. Examination of buffy coat

Management Considerations

Hematological parameters respond to physiological or environmental alterations within hours to weeks. Determining these parameters is easy and inexpensive and can indicate perturbation of the individual, or the population and, in some instances, the ecosystem. If abnormalities are identified, more specific tests (including biochemical, serological and toxicological analysis and polymerase chain reaction [PCR]) should be performed.

The correct interpretation of hematological values requires well-established "normals." Unfortunately, there are gaps in understanding here. Many species have poorly defined "normal" ranges and information regarding basic physiological changes accompanying undernourishment is lacking for many raptors. It also should be noted that databases, published reference ranges, and laboratory reference ranges may have been obtained by different methods and from different numbers of animals in various clinical states and, therefore, may not be directly comparable.

CONCLUSIONS

Although raptor hematology is now a crucial part of clinical veterinary medicine, its use as a management tool in wild populations remains limited. Application to population and conservation medicine has been hampered to date by a scarcity of "normal" values, incorporating age, sex and physiological variables. As more research is conducted, the use of hematological techniques will increase.

LITERATURE CITED

BENNETT, P.M., S.C. GASCOYNE, M.G. HART, J.K. KIRKWOOD AND C.M. HAWKEY. 1991. Development of LYNX: a computer application for disease diagnosis and health monitoring in wild animals, birds and reptiles. *Vet. Rec.* 128:496–499.

BOWERMAN, W.W., J.E. STICKLE, J.G. SIKARSKIE AND J.P. GIESY. 2000. Hematology and serum chemistries of nestling Bald Eagles (*Haliaeetus leucocephalus*) in the lower peninsula of MI, USA. *Chemosphere* 41:1575–1579.

CAMPBELL, T.W. 1988. Avian hematology and cytology. Iowa State University Press, Ames, IA U.S.A.

———. 1994. Hematology. Pages 176–198 in B. Ritchie, G. Harrison, and L. Harrison [EDS.], Avian medicine: principles and application. Wingers Publishing, Lake Worth, FL U.S.A.

COOPER, J.E. 2002. Methods of investigation and treatment. Pages 28–70 in J.E. Cooper [ED.], Birds of prey: health and disease, 3rd Ed. Blackwell Publishing, Oxford, United Kingdom.

DAWSON, R.D. AND G.R. BORTOLOTTI. 1997. Variation in hematocrit and total plasma proteins of nestling American Kestrels (*Falco sparverius*) in the wild. *Comp. Biochem. Physiol.* 117A:383–390.

FERRER, M., T. GARCIA-RODRIGUEZ, J.C. CARRILLO AND J. CASTRO-VIEJO. 1987. Hematocrit and blood chemistry values in captive raptors (*Gyps fulvus, Buteo buteo, Milvus migrans, Aquila heliaca*). *Comp. Biochem. Physiol.* 87A:1123–1127.

FUDGE, A.M. 2000. Laboratory medicine: avian and exotic pets. W.B. Saunders, Philadelphia, PA U.S.A.

GARCIA-RODRIGUEZ, T., M. FERRER, J.C. CARRILLO AND J. CASTRO-VIEJO. 1987. Metabolic responses of *Buteo buteo* to long-term fasting and refeeding. *Comp. Biochem. Physiol.* 87A:381–386.

GARVIN, M.C., B.L. HOMER AND E.C. GREINER. 2003. Pathogenicity of *Hemoproteus danilewskyi*, Kruse, 1890, in Blue Jays (*Cyanocitta cristata*). *J. Wildl. Dis.* 39:161–169.

GUTIERREZ, R.J. 1989. Hematozoa from the Spotted Owl. *J. Wildl. Dis.* 25:614–618.

HOFFMAN, D.J., J.C. FRANSON, O.H. PATTEE, C.M. BUNCK AND H.C. MURRAY. 1985. Biochemical and hematological effects of lead ingestion in nestling American Kestrels (*Falco sparverius*). *Comp. Biochem. Physiol.* 80C:431–439.

HOWLETT, J.C. 2000. Clinical and diagnostic procedures. Pages 28–42 in J.H. Samour [ED.]. Avian medicine. C.V. Mosby, London, United Kingdom.

HUNTER, S.R. AND L.R. POWERS. 1980. Raptor hematocrit values. *Condor* 82:226–227.

JENNINGS, I.B. 1996. Hematology. Pages 68–78 in P.H. Beynon, N.A. Forbes, and N.H. Harcourt-Brown [EDS.], BSAVA manual of pigeons, raptors and waterfowl. BSAVA Publishing, Cheltenham, United Kingdom.

JOSEPH, V. 1999. Raptor hematology and chemistry evaluation. *Vet. Clin. N. Am. Exotic Anim. Pract.* 2:689–699.

———. 2000. Disorders of leukocytes in the raptor. Pages 26–28 in A.M. Fudge [ED.], Laboratory medicine: avian and exotic pets. W.B. Saunders, Philadelphia, PA U.S.A.

KASS, L., G.J. HARRISON AND C. LINDHEIMER. 2002. A new stain for identification of avian leukocytes. *Biotechnic Histochemistry* 77:201–206.

LACINA, D. AND D.M. BIRD. 2000. Endoparasites of raptors: a review and an update. Pages 65–100 in J.T. Lumeij, J.D. Remple, P.T. Redig, M. Lierz, and J.E. Cooper [EDS.], Raptor biomedicine III. Zoological Education Network Inc., Lake Worth, FL U.S.A.

LOYE, J. AND S. CARROLL. 1995. Birds, bugs and blood: avian parasitism and conservation. *Trends Ecol. Evol.* 10:232–235.

MAURO, L. 1987. Hematology and blood chemistry. Pages 269–276 in B.A. Giron Pendleton, B.A. Millsap, K.W. Cline, and D.M. Bird [EDS.], Raptor management techniques manual. National Wildlife Federation, Washington DC U.S.A.

PARGA, M.L., H. PENDL AND N.A. FORBES. 2001. The effect of transport on hematologic parameters in trained and untrained Harris's Hawks (*Parabuteo unicinctus*) and Peregrine Falcons (*Falco peregrinus*). *J. Avian Med. Surg.* 15:162–169.

PEIRCE, M.A. 1989. Blood parasites. Pages 148–169 in C.M. Hawkey and T.B. Dennett [EDS.], A colour atlas of comparative veterinary hematology: normal and abnormal blood cells in mammals, birds and reptiles. Wolfe Medical Publications, Ipswich, United Kingdom.

REDIG, P.T. 2003. Falconiformes (vultures, hawks, falcons, Secretary Bird). Pages 150–160 in M.E. Fowler and R.E. Miller [EDS.]. Zoo and wild animal medicine 5th Ed. Elsevier, St Louis, MI U.S.A.

REHDER, N.B. AND D.M. BIRD. 1983. Annual profiles of blood packed cell volumes of captive American Kestrels. *Can. J. Zool.* 61:2550–2555.

———, D.M. BIRD AND P.C. LAGUË. 1982. Variations in blood packed cell volume of captive American Kestrels. *Comp. Biochem. Physiol.* 72A:105–109.

REMPLE, J.D. 2003. Blood sporozoa of raptors: a review and update. Proceedings of European association of avian veterinarians, 22–26 April 2003, Tenerife, Spain.

RODNAN, G.P., F.G. EBAUGH, JR. AND M.R.S. FOX. 1957. Life span of red blood cell volume in the chicken, pigeon, duck as estimated by the use of Na(2)D,51 (04) with observation on red cell turnover rate in mammal, bird and reptile blood. *Blood* 12:355.

SAMOUR, J.H. [ED.]. 2000. Avian medicine. Mosby, London, United Kingdom.

——— AND M.A. PEIRCE. 1996. Babesia shortii infection in a Saker Falcon (*Falco cherrug*). *Vet. Rec.* 139:167–168.

——— AND C. SILVANOSE. 2000. Parasitological findings in captive falcons in the United Arab Emirates. Pages 117–126 in J.T. Lumeij, J.D. Remple, P.T. Redig, M. Lierz, and J.E. Cooper [EDS.], Raptor biomedicine III. Zoological Education Network, Lake Worth, FL U.S.A.

———, M. A. D'ALOIA, AND J. C. HOWLETT. 1996. Normal hematology of captive Saker Falcons (*Falco cherrug*). *Comp. Hematology Inter.* 6:50–52.

————, J.L. NALDO AND S.K. JOHN. 2001. Staining characteristics of the eosinophil in the Saker Falcon. *Exotic DVM* 3:10.

SEISER, P.E., L.K. DUFFY, A.D. MCGUIRE, D.D. ROBY, G.H. GOLET AND M.A. LITZOW. 2000. Comparison of Pigeon Guillemot, *Cepphus columba*, blood parameters from oiled and unoiled areas of Alaska eight years after the Exxon Valdez oil spill. *Mar. Poll. Bull.* 40:152–164.

SHAPIRO, C.J. AND W.W. WEATHERS. 1981. Metabolic and behavioural responses of American Kestrels to food deprivation. *Comp. Biochem. Physiol.* 68A:111–114.

SOCKMAN, K.W. AND H. SCHWABL. 2001. Plasma corticosterone in nestling American Kestrels: effects of age, handling stress, yolk androgens, and body condition. *Gen. Comp. Endocrinol.* 122:205–212.

STURKIE, P.D. 1976. Body fluids: blood. Pages 102–121 in P.D. Sturkie [ED.]. Avian physiology, 4th Ed. Springer-Verlag. New York, NY U.S.A.

VAN WYK, E., H. VAN DER BANK AND G.H. VERDOORM. 1998. Dynamics of hematology and blood biochemistry in free-living African White-backed Vulture (*Pseudogyps africanus*) nestlings. *Comp. Biochem. Physiol.* 120A:495–508.

WINGFIELD, J.C. AND D.S. FARNER. 1982. Endocrine responses of White-crowned Sparrows to environmental stress. *Condor* 84:399–409.

C. Reproductive

JUAN BLANCO

Centro de Estudios de Rapaces Ibéricas, Junta de Comunidades de Castilla-La Mancha Sevelleja de la Jara, 45671 Toledo, Spain

DAVID M. BIRD

Avian Science and Conservation Centre, McGill University, 21,111 Lakeshore Road, Ste. Anne de Bellevue, Quebec, Canada H9X 3V9

JAMIE H. SAMOUR

Fahad bin Sultan Falcon Center, PO Box 55, Riyadh 11322, Kingdom of Saudi Arabia

INTRODUCTION

The reproductive anatomy and function of raptors have attracted little attention to date. Basic information, such as the presence and location of sperm-storage tubules and the duration of the fertile period, is still unknown for most species. This paucity of knowledge not only acts as a limiting factor for improved reproductive success in captive breeding programs, but also renders it more difficult to understand the reproductive ecology of wild raptors.

The increasing number of endangered raptor species is being accompanied by a growing interest in their biomedicine and captive propagation. More raptors are coming under the scrutiny of microscopes and modern laboratory techniques, and we hope that this will spark a greater interest in resources and research

dedicated to studying reproductive physiology of birds of prey. Meanwhile, readers should consult the limited available studies undertaken in various species used as models, such as the American Kestrel (*Falco sparverius*) (Bird and Buckland 1976, Bakst and Bird 1987), the general review on raptor physiology by Duke (1986), as well as those on the reproductive systems of domestic birds (Johnson 2000, Kirby and Froman 2000) and wild birds (Gee et al. 2004, Samour 2004).

FEMALE REPRODUCTIVE SYSTEM

Reproductive Tract

Ovaries, follicular growth, and ovulation. Unlike the majority of birds, raptors commonly have two functional ovaries (Domm 1939). The phenomenon has been recorded in many species of raptors (Venning 1913, Wood 1932, Boehm 1943, Snyder 1948), but seems more prevalent in accipiters than in Strigiformes (Fitzpatrick 1934).

When growing follicles, females experience a significant increase in body weight. Inability to accomplish this gain may prevent full ovarian growth and egg-laying (Newton 1979, Hardy et al. 1981). Recent studies of Barn Owls (*Tyto alba*) indicate that the onset of reproduction is not triggered by body condition (i.e., an increase in body fat). In fact, the perceived increase in body weight prior to breeding is more likely due to water accumulation as a result of changes in protein metabolism (Durant et al. 2000). Interpretation of the "need" to put on extra body fat as an energy-safe strategy ought to be reconsidered. The rapid growth phase of follicles usually takes 5 to 14 days, during which follicles highest in the growth hierarchy incorporate vitellogenin and low-density lipoprotein in an estrogen-receptor mediated event. In the Golden Eagle (*Aquila chrysaetos*), total fecal estrogen levels progressively increase during the rapid growth phase (Staley 2003), presumably in relation to the increasing activity of the external theca cells of prehierarchal follicles.

Similar to other birds (Wingfield and Farner 1978, Johnson 2000), ovulation of the first egg in Peregrine Falcons (*F. peregrinus*), Golden Eagles, and Asian Imperial Eagles (*A. heliaca*) takes place soon after the estrogen maximum, and coincides with a peak of progesterone and cortisol (J. Blanco, unpubl. data). Both serum luteinizing hormone (LH) and progesterone also peak early in ovulation.

Oviduct. Similar to what has been described for other birds (Gee et al. 2004), the raptor oviduct consists of five distinguishable regions: infundibulum, magnum, isthmus, shell gland, and vagina. The size and mass of the oviduct increase parallel to the ovary early in the breeding season, as regulated by steroid hormones. The presence of sperm storage tubules at the uterovaginal region is poorly documented in raptors. These microscopic structures in the folds of the cervix mucosa have been observed in the American Kestrel (Bakst and Bird 1987; Fig. 1) and in other raptor species (Blanco 2002). These tubules determine the fertile period by maintaining sperm viability and continual release to the site of fertilization.

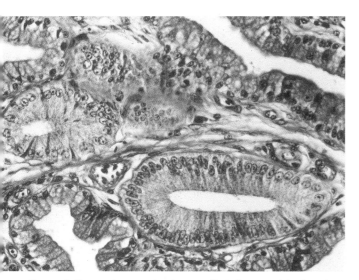

Figure 1. Sperm storage glands were first discovered in American Kestrels (*Falco sparverius*) and are likely found in most other raptors.

Eggs

Egg physiology and variation in eggshell and membrane characteristics. After ovulation the ovum is engulfed by the infundibulum (the site of fertilization), and next descends through the oviduct. The process usually lasts for two or more days, depending on the size of the birds, and involves the addition of numerous layers that conform to the egg. Cuticle, crystallization layers (external, palisade, and mammilary) and eggshell membranes can be differentiated easily in the eggs of raptors. The morphology and size of eggshell pores vary among species (Blanco 2001) and, together with the outer crystallization layer, may be of taxonomic interest.

Falconiformes have been reported to produce more massive eggs than Strigiformes of similar body size (Saunders et al. 1984). Interestingly, body mass is positively correlated with egg width in free-ranging Eleonora's Falcon (*F. eleonorae*) (Wink et al. 1985) and Black Kites (*Milvus migrans*) (Viñuela 1997), but not in captive American Kestrels (Bird and Laguë 1982a). Both inter-annual and intra-seasonal variation in egg-laying dates have been recorded in both captive and free-ranging populations of Peregrine Falcons (Burnham et al. 1984) and Golden Eagles (Blanco 2001), with a significant decrease in length, breadth and initial mass with time.

Certain external factors including stress (Hughes et al. 1986), ambient electromagnetic fields (Fernie et al. 2000a), organochlorine compounds and metabolites (for review see Hickey and Anderson 1968, Ratcliffe 1970, Cooke 1979, Wiemeyer et al. 2001, Chapter 18), heavy metals (Ohlendorf 1989, Blanco 2001), and PCBs (Lowe and Stendell 1991, Fernie et al. 2000b) can induce shifts in eggshell thickness and ultrastructure, as well as in ultrastructure and fiber organization and pattern of the shell membrane.

Clutch size and replacement. Clutch size often is influenced by phylogeny and individual factors including size and age (Brommer et al. 2002). From a global perspective, the number of eggs laid varies latitudinally in some falcons in Australasia (Blanco 2001), as well as longitudinally in several eagles and *Milvus* kites in that region (Olsen and Marples 1993).

The ability to replace clutches has been used as a management tool (see Chapter 23) to augment both captive and wild populations of raptors (Bird and Laguë 1982a). In captive and wild American Kestrels, replacement clutches had fewer eggs than first clutches, but did not differ in fertility, hatchability, and fledging success (Bird and Laguë 1982a,b; Bowman and Bird 1985).

MALE REPRODUCTIVE SYSTEM

Paired reproductive tracts in male birds of prey lie along the dorsal body wall and consist of a testis, epididymis, and a straight ductus deferens, which differs from the highly convoluted version found in some domestic species (J. Blanco, unpubl. data). Spermatogenesis depends on follicle-stimulating hormone (FSH), testosterone, the activity of Sertoli cells and their interaction with the spermatogonial stem cells. Seasonal testicular growth usually takes up to 45 days in the majority of raptor species, a period longer than ovarian growth in the female. FSH and LH, as well as testosterone, are essential for spermatogenesis. The process of spermiogenesis, and the duration of the transport through the excurrent ducts are unknown, but it is clear that fluid is absorbed to concentrate sperm and to become seminal plasma. Seminal plasma differs from blood plasma in electrolyte and protein composition (J. Blanco, unpubl. data). The importance of this process is not well understood, but is likely related to sperm motility more than fertilizing ability, since testicular sperm are able to penetrate the inner periviteline membrane *in vitro*.

Male Gametes

Semen production period, seminal quality, and factors of influence. Semen production period varies among species and individuals, but usually last for nearly three months. Bird and Laguë (1977) described an average period of 74 days for captive American Kestrels with a maximum of 103 days. Longer periods were found for Peregrine Falcons (95 days; Hoolihan and Burnham 1985) and eagles (up to 110 days; Blanco 2002).

Semen production in American Kestrels held in Montreal, Canada begins at about 12 hours and 45 minutes of daylight, and declines considerably at about 15 hours and 45 minutes (Bird and Buckland 1976).

Ejaculate characteristics vary greatly among species and individuals, and with collection method (Bird and Laguë 1976, Boyd et al. 1977, Weaver 1983), male reproductive condition, nutrition (Randal 1994), certain pollutants (Bird et al. 1983) and climate (Bird and Laguë 1977). Concentrations ranging from 31,000 to 40,000 spermatozoa per mm^3 and volumes between 3 and 14.6 µl have been reported for the American Kestrel (Bird and Buckland 1976, Bird and Laguë 1977, Brock 1986). Expectedly, ejaculate volume increases with species size. Semen volume in Peregrine Falcons can be as high as 95 µl (Hoolihan and Burnham 1985), with cell concentrations ranging from 26,000 to 81,000 sperm per µl.

Sperm production varies seasonally; sperm concentration increases early during the breeding season, peaks in mid-season, and declines thereafter. This pattern varies longitudinally. Numbers of spermatogonia, spermatids and abnormal spermatozoa are more likely to be present in the ejaculate both early and late in the season when testes are no longer at their maximum size and when testosterone levels are lower than normal. This is related to the need to ensure maximum sperm quality at the time of maximal frequency of copulation prior to egg laying (Blanco et al. 2002).

Urine contamination of semen and subsequent sperm damage is frequent during collection using forced-massage techniques (Bird and Laguë 1977). Fox (1995) provides a useful description, including an illustration of the various contaminants in raptor semen. The use of modified diluents may help reduce deleterious effects (Blanco et al. 2002). *Escherichia coli* is the most prevalent bacteria contaminating raptor semen. Samples need to be evaluated with caution before artificial insemination to avoid the risk of ascendant salpingitis (Blanco and Höfle 2004).

Artificial insemination. Artificial insemination with fresh semen has been successful in a variety of non-domestic avian species including raptors. This technique has been used as a management tool in several captive breeding projects using fresh diluted semen (Temple 1974, Samour 1986). In the American Kestrel, fertility rates using artificial insemination are similar to those achieved by natural mating (Bird et al. 1976).

Sperm cryopreservation. Semen collected by massage techniques has been cryopreserved and progeny obtained in several species (Gee 1983, Gee et al. 1985, Brock 1986, Parks et al. 1986, Samour 1988, Gee and Sexton 1990, Brock and Bird 1991, Knowles-Brown and Wishart 2001, Wishart 2001). Comparative studies on sperm tolerance to different osmotic conditions, cryoprotectant concentrations and cooling rates indicate considerable variation, even between closely related raptor species (Blanco et al. 2000). Different freezing rates and protocols are described in Brock et al. (1983) and Knowles-Brown and Wishart (2001).

Glycerol and the alternatives, dimethyl sulphoxide (DMSO) and dimethyl acetamide (DMA), have been used widely in sperm cryopreservation in non-domestic species. Sperm from the falcon type (Brock and Bird 1991, Gee et al. 1993) have been successfully cryopreserved using either 13.6% glycerol; 6%, 8%, or 10%

DMSO; or 13.6% DMA. Evaluation of fertilizing ability has been mostly based on progressive motility and fertility after artificial insemination (Gee et al. 1985, Brock and Bird 1991, Gee et al. 1993). In the American Kestrel, motility after thawing averaged 41% and 13% using glycerol and DMA, respectively (Brock and Bird 1991). Post-thaw fertility rates have been obtained following artificial insemination using glycerol as cryoprotectant in the Peregrine Falcon (33.3%) (Parks et al. 1986) and the American Kestrel (11.8%) (Brock and Bird 1991).

Photoperiodism, Reproductive Hormones, and Endocrine Disruptors

The influence of photoperiodism on levels of gonadal hormones in birds generally is well understood, but most of our knowledge of this phenomenon in raptors is based on the use of artificial lighting to induce reproductive activity in captive pairs (Willoughby and Cade 1964, Bird et al. 1980). We know nothing about natural circadian rhythms in raptors, but data collected on other bird families are likely relevant and applicable.

Nelson (1972) and Swartz (1972) were among the first to elucidate the need for photoperiodic stimulation to induce northern-nesting raptors like Gyrfalcons (*F. rusticolus*) and Peregrine Falcons to breed in captivity (i.e. the farther north they originate from, the longer the photoperiod they require). If extra day-length in the form of artificial lighting is to be used, the changes in day-length should be made as gradually as possible to reduce physiological shocks (Bird 1987). At least one successful attempt using artificial photoperiodic changes has been made to induce American Kestrels to undergo an out-of-season breeding period between two consecutive successful spring breeding periods (Bird et al. 1980). An attempt to hasten sexual maturity in kestrels using photoperiod encountered mixed success (Ditto 1996). Such procedures could be used to increase the output of offspring in endangered species breeding programs or to accelerate the turnover of data in experimental research involving captive raptors.

The vast majority of our knowledge about raptor reproductive endocrinology has relied upon blood sampling and plasma-hormone determinations. In the female, plasma corticosterone, progesterone, estradiol 17β and estrone are highest during courtship and egg laying (Rehder et al. 1984, 1986), whereas high levels of androgens, including testosterone, were associated with aggression, territoriality, courtship, nest-building, testicular development, and spermatogenesis in the male (Temple 1974; see also Rehder et al. 1988). Information on plasma levels of lutenizing hormone in American Kestrels can be found in Ditto (1996). More recently, fecal steroid monitoring, which has been used to study seasonality in hormone levels (Bercovitz et al. 1982), the effects of human disturbance (Wasser et al. 1996), steroid excretion lag time (Wasser et al. 1996), and sex determination (Bercovitz and Sarver 1988), shows potential as a safe non-invasive source of information regarding hormone levels.

Exposure to extreme temperatures can limit avian reproduction (Mirande et al. 1996). Drastic temperature fluctuations often reduce semen production (Kundu and Panda 1990), as well as egg-laying and copulation frequency (Bluhm 1985).

The impacts of organochlorine chemicals on reproduction of birds of prey have been well documented (see Chapter 18). Studies indicate that these chemicals also act as endocrine disruptors. For instance, preliminary data by Bowerman et al. (2003) suggest that hormone disruptors, not necessarily estrogen or androgen mimics and their antagonists, are associated with reproductive and teratogenic effects in Bald Eagle (*Haliaeetus leucocephalus*) populations in the Great Lakes Basin. Alterations in reproductive behavior in captive breeding American Kestrels were induced by exposure to Dicofol, one of the last organochlorine pesticides to be banned from use in the U.S. (MacLellan et al. 1996). Other organochlorine chemicals that impact upon reproduction in birds of prey through hormone disruption come in the form of industrial by-products and include polychlorinated biphenyl ethers (PCBs). Captive American Kestrels exposed to PCBs developed more frequent aggressive courtship interactions and experienced clutch abandonment (Fernie et al. 2003); alterations in brood patches also have been observed in PCB-exposed kestrels (Fisher et al. 2006). Most recently, attention has focused on the alarming increase in residue levels of polybrominated diphenyl ethers (PBDEs) in food chains world-wide, arising from the use of brominated flame retardants applied to many household products (Chapter 18). Using the American Kestrel as a model test species, a number of reproductive effects have been documented thus far (cf. Fernie et al. 2006).

SUMMARY

Captive-propagation programs have been extremely useful in maintaining genetic diversity and restoring wild populations of endangered raptors. However, captive breeding success requires knowledge of a species' reproductive behavior, physiology and endocrinology. In addition, species-specific differences in anatomy, gamete or physiological parameters may complicate the task of maintaining captive breeding populations of raptors. Further research is needed to unravel some of the major questions including the spatial requirements and factors involved in the control of reproduction in endangered raptors. Finally, an improved knowledge of the reproductive physiology of raptors will help us better understand the impacts of chemicals released into their environment on their reproduction and, ultimately, their survival.

LITERATURE CITED

BAKST, M.R. AND D.M. BIRD. 1987. Localization of oviductal sperm storage tubules in the American Kestrel (*Falco sparverius*). *Auk* 104:321–324.

BERCOVITZ, A.B. AND P.L. SARVER. 1988. Comparative sex-related differences of excretory sex steroids from day-old Andean Condors (*Vultur gryphus*) and Peregrine Falcons (*Falco peregrinus*): non-invasive monitoring of neonatal endocrinology. *Zoo Biol.* 7:147–153.

———, J. COLLINS, P. PRICE AND D. TUTTLE. 1982. Noninvasive assessment of seasonal hormone profile in captive Bald Eagles (*Haliaeetus leucocephalus*). *Zoo Biol.* 1:111–117.

BIRD, D.M. 1987. Reproductive physiology. Pages 276–282 in B.A. Giron Pendleton, B.A. Millsap, K.W. Cline, and D.M. Bird [EDS.], Raptor management techniques manual. National Wildlife Federation, Washington, DC U.S.A.

——— AND R.B. BUCKLAND. 1976. The onset and duration of fertility in the American Kestrel. *Can. J. Zool.* 54:1595–1597.

——— AND P.C. LAGUË. 1976. Management practices for captive kestrels used as semen donors for artificial insemination. *Raptor Res.* 10:92–96.

———AND P.C. LAGUË. 1977. Semen production of the American Kestrel. *Can. J. Zool.* 55:1351–1358.

——— AND P.C. LAGUË. 1982a. Forced renesting, seasonal date of laying and female characteristics on clutch size and egg traits in captive American Kestrels. *Can. J. Zool.* 60:71–79.

——— AND P.C. LAGUË. 1982b. Fertility, egg weight loss, hatchability, and fledging success in replacement clutches of captives kestrels. *Can. J. Zool.* 60:80–88.

———, P.G. WEIL AND P.C. LAGUË. 1980. Photoperiodic induction of multiple breeding seasons in captive American kestrels. *Can. J. Zool.* 58:1022–1026.

———, P.H. TUCKER, G.A. FOX AND P.C. LAGUË. 1983. Synergistic effects of Aroclor 1254 and Mirex on the semen characteristics of American Kestrels. *Arch. Environ. Contam. Toxicol.* 12:633–640.

BLANCO, J.M. 2001. Ultrastructural alterations of the egg shell and membranes in the Bonelli's Eagle (*Hieraaetus fasciatus*). Thesis, Universidad Complutense de Madrid, Facultad de Veterinaria.

———. 2002. Reproductive system in Mediterranean raptors, science and reproductive techniques. Informe Centro de Estudios de Rapaces ibéricas, JCCM.

——— AND U. HÖFLE. 2004. Bacterial and fungal contaminants in raptor ejaculates and their survival to sperm cryopreservation protocols. Proceedings of the 6th Conference of the European Wildlife Disease Association.

———, G.F. GEE, D.E. WILDT AND A.M. DONOGHUE. 2000. Species variation in osmotic, cryoprotectant and cooling rate tolerance in poultry, eagle and Peregrine Falcon spermatozoa. *Biol. Reprod.* 63:1164–1171.

———, G.F. GEE, D.E.WILDT AND A.M. DONOGHUE. 2002. Producing progeny from endangered birds of prey: treatment of urine contaminated semen and a novel intramagnal insemination approach. *J. Zoo Wildl. Med.* 33:1–7.

BLUHM, C.K. 1985. Social factors regulating avian endocrinology and reproduction. Page 247 in B.K. Follett, S. Ishii, and A. Chandola [EDS.], The endocrine system and the environment. Japan Scientific Societies Press, Tokyo, Japan and Springer-Verlag, Berlin, Germany.

BOEHM, E.F. 1943. Bilateral ovaries in Australian hawks. *Emu* 42:251.

BOWERMAN, W.W., D.A. BEST, O.P. GIESY, M.C. SHIELDCASTLE, M.W. MEYER, S. POSTUPALSKY AND J.G. SIKARSKIE. 2003. Associations between regional differences in polychlorinated biphenyls and dichlorodiphenyldichloroethylene in blood of nestling Bald Eagles and reproductive productivity. *Environ. Toxicol. Chem.* 22:371–376.

BOWMAN, R. AND D.M. BIRD. 1985. Reproductive performance of American Kestrels laying replacement clutches. *Can. J. Zool.* 63:2590–2593.

BOYD, L.L., N.S. BOYD AND F.C. DOBLER. 1977. Reproduction of Prairie Falcons by artificial insemination. *J. Wildl. Manage.* 41:266–271.

BROCK, M.K. 1986. Cryopreservation of semen of the American Kestrel (*Falco sparverius*). M.S. thesis, McGill University, Montreal, Quebec, Canada.

——— AND D.M. BIRD. 1991. Prefreeze and postthaw effects of glycerol and dimethylacetamide on motilty and fertilizing ability of American Kestrel (*Falco sparverius*) spermatozoa. *J. Zoo Wildl. Med.* 22:453–459.

———, D.M. BIRD AND G.A. ANSAH. 1983. Cryogenic preservation of spermatozoa of the American Kestrel *Falco sparverius*. *Int. Zoo Yearb.* 23:67–71.

BROMMER, J.E., H. PIETIAINEN AND H. KOKKO. 2002. Cyclic variation in seasonal recruitment and the evolution of the seasonal decline in Ural Owl clutch size. *Proc. R. Soc. Lond. B Biol. Sci.* 269:647–654.

BURNHAM, W.A., J.H. ENDERSON AND T.J. BOARDMAN. 1984. Variation in Peregrine Falcon eggs. *Auk* 101:578–583.

COOKE, A.S. 1979. Changes in eggshell characteristics of the Sparrowhawk (*Accipiter nisus*) and Peregrine (*Falco peregrinus*) associated with exposure to environmental pollutants during recent decades. *J. Zool.* 187:245–263.

DITTO, M.M. 1996. Lutenizing hormone and the hastening of sexual maturity in the American Kestrel, *Falco sparverius*. Ph.D. Thesis, McGill University, Montreal, Quebec, Canada.

DOMM, L.V. 1939. Modifications in sex and secondary sexual characters in birds. Pages 227–327 in E. Allen [ED.], Sex and internal secretions, 2nd Ed. Williams and Wilkins, Baltimore, MD U.S.A.

DUKE, G.E. 1986. Raptor physiology. Pages 365–375 in M.E. Fowler and E. Murray [EDS.], Zoo and wild animal medicine, 2nd Ed. W.B. Saunders, Philadelphia, PA U.S.A.

DURANT, J.M., S. MASSEMIN, C. THOUZEAU AND Y. HANDRICH. 2000. Body reserves and nutritional needs during laying preparation in Barn Owls. *J. Comp. Physiol. B.* 170:253–260.

FERNIE, K.J., D.M. BIRD, R.D. DAWSON AND P.C. LAGUË. 2000a. Effects of electromagnetic fields on the reproductive success of American Kestrels. *Physiol. Biochem. Zool.* 73:60–65.

———, G.R. BORTOLOTTI, J.E. SMITS, J. WILSON, K.G. DROUILLARD AND D.M. BIRD. 2000b. Changes in egg composition of American Kestrels exposed to dietary polychlorinated biphenyls. *J. Toxicol. Environ. Health* 60:291–303.

———, G. BORTOLOTTI AND J. SMITS. 2003. Reproductive abnormalities, teratogenicity, and developmental problems in American Kestrels (*Falco sparverius*) exposed to polychlorinated biphenyls. *J. Toxicol. Environ. Health* 66:2089–2103.

———, L.J. SHUTT, R.J. LETCHER, I. RITCHIE AND D.M. BIRD. 2006. Changes in the growth, but not the survival, of American Kestrels (*Falco sparverius*) exposed to environmentally relevant levels of polybrominated diphenyl ethers. *J. Toxicol. Environ. Health, Pt. A.* 69:1541–1554.

FISHER, S.A., G.R. BORTOLOTTI, K.J. FERNIE, D.M. BIRD AND J.D. SMITS. 2006. Brood patches of American Kestrels altered by experimental exposure to PCBs. *J. Toxicol. Environ. Health, Pt. A.* 69:1–11.

FITZPATRICK, F.L. 1934. Unilateral and bilateral ovaries in raptorial birds. *Wilson Bull.* 46:19–22.

FOX, N. 1995. Understanding birds of prey. Hancock House, Surrey, British Columbia, Canada.

GEE, G.F. 1983. Avian artificial insemination and semen preservation. Pages 375–398 in J. Delacour [ED.], IFCB symposium on breeding birds in captivity. North Hollywood, CA U.S.A.

——— AND T. J. SEXTON. 1990. Cryogenic preservation of semen from the Aleutian Canada Goose. *Zoo Biol.* 9:361–371.

———, M.R. BAKST AND T.J. SEXTON. 1985. Cryogenic preservation of semen from the Greater Sandhill Crane. *J. Wildl. Manage.* 49:480–484.

———, C.A. MORREL, J.C. FRANSO AND O.H. PATTEE. 1993. Cryopreservation of American Kestrel semen with dimethylsulfoxide. *J. Raptor Res.* 27:21–25.

———, H. BERTSCHINGER, A.M. DONOGHUE, J.M. BLANCO AND J. SOLEY. 2004. Reproduction in nondomestic birds: physiology, semen collection, artificial insemination and cryopreservation. *Avian Poult. Biol. Rev.* 15:47–101.

HARDY, A.R., G.J.M. HIRONS AND P.I. STANLEY. 1981. The relationship of body weight, fat deposit and moult to the reproductive cycle in wild Tawny Owls and Barn Owls. Pages 159–163 in J.E. Cooper and A.G. Greenwood [EDS.], Recent advances in the study of raptor diseases. Chiron Publishers, Keighley, West Yorkshire, United Kingdom.

HICKEY, J.J. AND D.W. ANDERSON. 1968. Chlorinated hydrocarbons and eggshell changes in raptorial and fish-eating birds. *Science* 162:271–273.

HOOLIHAN J. AND W. BURNHAM. 1985. Peregrine Falcon semen: a quantitative and qualitative examination. *Raptor Res.* 19:125–127.

HUGHES, B.O., A.B. GILBERT AND M.F. BROWN. 1986. Categorisation and causes of abnormal egg shells: relationship with stress. *Br. Poult. Sci.* 27:325–337.

JOHNSON, A.L. 2000. Reproduction in the female. Pages 569–596 in P. D. Sturkie [ED.], Avian physiology, 5th Ed. Academic Press, San Diego, CA U.S.A.

KIRBY, J.D. AND D.P. FROMAN. 2000. Reproduction in male birds. Pages 597–615 in P.D. Sturkie [ED.], Avian physiology, 5th Ed. Academic Press, San Diego, CA U.S.A.

KNOWLES-BROWN, A. AND G.J. WHISHART. 2001. Progeny from cryopreserved Golden Eagle spermatozoa. *Avian Poult. Rev.* 12:201–202.

KUNDU, A. AND J.N. PANDA. 1990. Variation on physical characteristics of semen of the white leghorn under hot and humid environment. *Ind. J. Poult. Sci.* 25:195.

LOWE, T.P. AND R.C. STENDELL. 1991. Egg shell modifications in captive American Kestrels resulting from Aroclor 1248 in the diet. *Arch. Chem. Contam. Toxicol.* 20:519–522.

MACLELLAN, K.N.M., D.M. BIRD, D.M. FRY AND J. COWLES. 1996. Reproductive and morphological effects of O,P-Dicofol on two generations of captive American Kestrels. *Arch. Environ. Contam. Toxicol.* 30:364–372.

MIRANDE, C.M., G.F. GEE, A. BURKE AND P. WHITLOCK. 1996. Egg and semen production. Page 45 in D.H. Ellis, G.F. Gee, and C.M. Mirande [EDS.], Cranes: their biology, husbandry and conservation. National Biological Service. Washington, DC and International Crane Foundation, Baraboo, WI U.S.A.

NELSON, R.W. 1972. On photoperiod and captivity breeding of northern peregrines. *Raptor Res.* 6:57–72.

NEWTON, I. 1979. Population ecology of raptors. Buteo Books, Vermillion, SD U.S.A.

OHLENDORF, H.M. 1989. Bioaccumulation and effects of Selenium in wildlife. Pages 133–177 in L.W. Jacobs [ED.], Selenium in agriculture and the environment. SSSA Spec. Publ. No. 23. American Society of Agronomy and Soil Science, Madison, WI U.S.A.

OLSEN, P. AND T.G. MARPLES. 1993. Geographic variation in egg size, clutch size and date of laying of Australian raptors (Falconiformes and Strigiformes). *Emu* 93:167–179.

PARKS, J.E., W.R. HECK AND V. HARDASWICK. 1986. Cryopreservation of spermatozoa from the Peregrine Falcon: post-thaw dialysis of semen to remove glycerol. *J. Raptor Res.* 23:130–136.

RANDAL, N.B. 1994. Nutrition. Pages 63–95 in G.J. Harrison, L.R. Harrison, and B.W. Ritchie [EDS.], Avian medicine: principles and application. Winger Publishing, Inc., Lake Worth, FL U.S.A.

RATCLIFFE, D.A. 1970. Changes attributable to pesticides in egg breakage frequency and eggshell thickness in some British birds. *J. Appl. Ecol.* 7:67–115.

REHDER, N.B., P. C. LAGUË AND D.M. BIRD. 1984. Simultaneous quantification of progesterone, estrone, estradiol 17β and corticosterone in female American Kestrel plasma. *Steroids* 43:371–383.

———, D.M. BIRD AND P.C. LAGUË. 1986. Variations in plasma corticosterone, estrone, estradiol-17β and progesterone concentrations with forced renesting, molt and body weight of captive female American Kestrels. *Gen. Comp. Endocrinol.* 62:386–393.

———, D.M. BIRD AND L. SANFORD. 1988. Plasma androgen levels and body weights for breeding and non-breeding male kestrels. *Condor* 90:555–560.

SAMOUR, J.H. 1986. Recent advances in artificial breeding techniques in birds and reptiles. *Int. Zoo Yearb.* 24/25:143–148.

———. 1988. Semen cryopreservation and artificial insemination in birds of prey. Pages 271–277 in Proceedings of the 5th World Conference on Breeding Endangered Species in Captivity, Cincinnati, OH USA.

———. 2004. Semen collection, spermatozoa cryopreservation, and artificial insemination in non-domestic birds. *J. Avian Med. Surg.* 18:219–223.

SAUNDERS, D.A., G.T. SMITH AND N.A. CAMPBELL. 1984. The relationship between body weight, egg weight, incubation period, nestling period and nest site in the Psittaciformes, Falconiformes, Strigiformes and Columbiformes. *Aust. J. Zool.* 32:57–65.

SNYDER, L.L. 1948. Additional instances of paired ovaries in raptorial birds. *Auk* 65:602.

STALEY, A.M. 2003. Noninvasive fecal steroid measures for assessing gonadal and adrenal function in the Golden Eagle (*Aquila chrysaetos*) and the Peregrine Falcon (*Falco peregrinus*). M.S. thesis, Boise State University, Boise, ID U.S.A.

SWARTZ, L.G. 1972. Experiments on captive breeding and photoperiodism in peregrines and Merlins. *Raptor Res.* 6:73–87.

TEMPLE, S.A. 1974. Plasma testosterone titers during the annual reproductive cycle of Starlings (*Sturnus vulgaris*). *Gen. Comp. Endocrinol.* 22:470–479.

VENNING, F.E.W. 1913. Paired ovaries in the genus *Astur. J. Bombay Nat. Hist. Soc.* 22:199.

VIÑUELA, J. 1997. Adaptation vs. constraint: intraclutch egg-mass variation in birds. *J. Anim. Ecol.* 66:781–792.

WASSER, S.K., K. BEVIS, G. KING AND E. HANSON. 1996. Noninvasive physiological measures of disturbance in the Northern Spotted Owl. *Conserv. Biol.* 11:1019–1022.

WEAVER, J.D. 1983. Artificial insemination. Pages 19–23 in J.D. Weaver and T.J. Cade [EDS.], Falcon propagation: a manual on captive breeding. The Peregrine Fund, Inc. Boise, ID U.S.A.

WIEMEYER, S.N., D.R. CLARK, JR., J.W. SPANN, A.A. BELISLE AND C.M. BUNCK. 2001. Dicofol residues in eggs and carcasses of captive American Kestrels. *Environ. Toxicol. Chem.* 20:2848–2851.

WILLOUGHBY E.J. AND T.J. CADE. 1964. Breeding behavior of the American Kestrel (Sparrow Hawk). *Living Bird* 3:75–96

WINGFIELD, J.C. AND D.S. FARNER. 1978. The endocrinology of a natural breeding population of the White-crowned Sparrow (*Zonotrichia leucophrys pugetensis*). *Physiol. Zool.* 51:188–205.

WINK, M., D. RISTOW AND C. WINK. 1985. Biology of Eleonora's Falcon (*Falco eleonorae*), 7: variability of clutch size, egg dimensions and egg coloring. *Raptor Res.* 19:8–14.

WISHART, G.J. 2001. The cryopreservation of germplasm in domestic and non-domestic birds. Pages 179–200 *in* P.F. Watson and W. V. Holt [EDS.], Cryobanking the genetic resource: wildllife conservation for the future? Taylor and Francis, London, United Kingdom.

WOOD, M. 1932. Paired ovaries in hawks. *Auk* 49:463.

Pathology

A. Disease

17

JOHN E. COOPER
School of Veterinary Medicine, The University of the West Indies
St. Augustine, Trinidad and Tobago

INTRODUCTION

This part of Chapter 17 is concerned with infectious and non-infectious factors that adversely affect the health, well-being and survival of individual birds of prey in the wild or in captivity, and which may influence the conservation status of species in the wild. Toxicology, which is mentioned briefly, is covered primarily in Chapter 18. There are important links between material in this chapter and other aspects of raptor biology that relate to health, including food habits (Chapter 8), reproduction and productivity (Chapter 11), behavior (Chapter 7), physiology (Chapter 16), energetics (Chapter 15) and rehabilitation (Chapter 23). Although ectoparasites and endoparasites are covered elsewhere in Chapter 17, when appropriate, they are mentioned here as well.

I first differentiate "health" and "disease" and define several additional important terms.

Health is a positive concept that is defined by the World Health Organization in relation to humans as *"A state of complete physical, mental and social well-being, not merely the absence of disease or infirmity."* *Disease* (from Old English *dis* = lack of; and *ease*) is taken to mean any impairment of normal physiological function that affects all or part of an organism. As such,

disease can be due to a range of factors, not just infections with pathogens. The causes of disease can be either **infectious**, including viral infections and parasite infestations, or **non-infectious**, including injuries and changes caused by trauma, poisons, genetic factors, or environmental stressors. The causes of disease often are multifactorial. For example, raptors that have been nutritionally deprived (inanition, starvation) more readily succumb to the fungal infection, aspergillosis, than otherwise (Cooper 2002). In this instance, the latter is the proximate (i.e., immediate) cause of death, while the former is the ultimate (i.e., predisposing) cause (Newton 1981). Here, I follow the terminology that is favored by ecologists, rather than medical personnel, in that **macroparasites** include metazoan organisms, such as mites and worms, whereas **microparasites** include single-celled organisms, such as bacteria and protozoa.

The **diagnosis** (detection and recognition) and treatment of disease in birds of prey is primarily the responsibility of the veterinarian but, as will be shown repeatedly in this chapter, those from other disciplines, ranging from anatomists and biochemists to DNA technologists and zoologists, also can and do contribute to this work. **Monitoring of health** of raptors is different from diagnosis. Monitoring of health implies "surveillance of a group or population of birds," and the raptors that are being watched often appear normal. The aim of monitoring in such cases is to compile a health profile of such birds, including understanding which bacteria they carry, whether they have antibodies to certain organisms, their body-condition score, the state of the plumage, etc. The techniques employed in monitoring health often are similar to those used for disease diagnosis. However, the best results are obtained if avian biologists and other non-medical professionals are an integral part of the team (Cooper 1993a).

The two decades that have elapsed since this chapter first appeared in Giron Pendleton et al. (1987) have seen enormous advances in our understanding of the biology of birds of prey and of those diseases that may either cause disease (morbidity) or result in death, and the importance of routine health-monitoring has been widely promoted and put into practice (Cooper 1989).

Health monitoring, essentially, is an early-warning system that can either help to confirm that a population of raptors is free of significant diseases or pathogens or, if these are present, help to ensure that appropriate action is taken without delay. Health monitoring of captive birds of prey is now standard practice in zoos and other establishments, and, increasingly, is the norm in studies on free-living raptors, especially when changes in population numbers or in distribution have been observed or are suspected (Cooper 2002).

The main causes of death and decline in free-living raptors often include environmental factors such as habitat destruction, human persecution, inadvertent human-related injury and poisoning, most of which are well studied in raptors (Newton 1990, Zalles and Bildstein 2000). In contrast, infectious disease as a mortality factor in birds of prey has proved difficult to evaluate, despite the best efforts of various biologists and veterinarians. Important early thinking about the part that might be played by infectious agents in free-living raptors was summarized in Newton (1979) and updated by the same author in 2002 (Newton 2002). Newton discusses the possible impact of infectious agents on raptors and draws attention to the important epidemiological difference between population-dependent and population-independent diseases.

There is increasing evidence from research on other species that when a population of birds becomes isolated and falls below a certain level, infectious (including parasitic) diseases may become relevant factors in demise or survivorship. The effect of infectious disease is likely to be more significant if there is a high inbreeding-coefficient, which can increase susceptibility among individuals. The decline in number of some of the world's birds, and the tendency for many of them to be confined to small islands of suitable habitat, suggests that infectious disease will assume a more pivotal role in the future. Birds of prey occupy a key position at the top of many food chains, and as a result are particularly vulnerable to environmental build-up of infectious

(including parasitic) organisms. Small populations appear to be particularly at risk.

Recently, "wildlife-disease ecology" has evolved as a subject in its own right (Hudson et al. 2002). This has been prompted in part by the recognition of new, emerging infections of domestic livestock and humans, some of them with wild animal reservoirs, and by concerns about the possible adverse effects of micro- and macroparasites on free-living vertebrates. Understanding the dynamics of such diseases often entails the use of mathematical modeling as well as field studies, and, as such, involves scientists from many different backgrounds. As a result, a better understanding of host-parasite relations in wild animal populations is unfolding. This new research is likely to help assess the much-debated role of various organisms in the biology of free-living raptors.

Some people still question the value of health studies on free-living raptors, arguing that other, mainly non-infectious, factors warrant greater attention. Although debatable, the situation is unequivocal for captive birds of prey. Under such circumstances, infectious disease is recognized as presenting a real challenge. Prompt detection is essential and is the focus of any properly formulated health-monitoring program. For many reasons it is desirable that captive raptors remain free of disease. Perhaps even more important is that birds destined for release into the wild are monitored for infectious disease, both to minimize the chances of their disseminating pathogens in their new environment, and to protect them from succumbing to novel organisms that they may encounter there. Such pre-release and pre-translocation health monitoring, or screening, is recommended by the IUCN Reintroductions and Veterinary Specialist Groups (see, for example, Woodford 2001), and is now a standard feature of many conservation programs globally.

Below I discuss the requirements and techniques for investigating diseases and for monitoring free-living and captive birds of prey as part of so-called health studies.

HEALTH-STUDY REQUIREMENTS

Prerequisites for effective health studies include (1) properly trained personnel, (2) appropriate laboratory and field equipment, and (3) effective interdisciplinary collaboration. Each is discussed and commented upon in turn.

The staff and equipment required for health studies

depend upon the degree of investigation planned. For basic health-monitoring studies, where only a representative number of birds are to be examined or a limited series of tests is to be performed ("screening"), a small team and minimal equipment usually are adequate. More extensive and intensive studies, however, usually require specially trained staff and an appropriately equipped laboratory. Interdisciplinary links are especially important in the field, but also are useful in laboratory investigations. It is unlikely that one person or facility will be able to undertake all the tests and analyses required, and some material may need to be sent elsewhere for toxicological analysis or molecular studies, for example.

Personnel

As a general rule, a veterinarian should coordinate clinical or pathological studies, since they will have broad training in animal disease, including a working knowledge of diagnostic and investigative techniques. There also may be legal implications, especially if a diagnosis is being made or if infectious agents are being handled that may present a threat to domestic livestock or human health (see below). If a veterinarian is unavailable in person, although possibly contactable for advice by telephone or e-mail, the biologist should carry out the work alone or with limited assistance. In such cases recruiting individuals who have a background of working in veterinary or medical laboratory technology is recommended, as such people are likely to have knowledge and understanding of appropriate skills in bacteriology, parasitology, and histopathology.

Researchers who regularly conduct health studies without veterinary guidance should be trained to do so. It is preferable to master a limited number of procedures rather than endeavoring to cope with all contingencies. Quality control should be practiced by periodically submitting material to other institutions for independent assessment to check and verify the work.

Laboratory and Field Equipment

Laboratory resources are essential for all health studies on raptors, whether these constitute disease investigations or health monitoring. There is much to be gained if the facilities are part of a larger complex, such as a university department or a veterinary investigation center, as the latter usually provide a range of other disciplines and personnel. If access to a permanent laboratory is not possible (i.e., when working in isolated sites),

laboratory tests may have to be performed in the field. Many clinical kits that can be readily transported and used effectively in difficult terrain, and away from electricity and running water are described in Cooper and Samour (1987). Basic tests can be carried out in the field using equipment and reagents in the kit, whereas others may require material to be transported to a more specialized or better-equipped laboratory.

Whenever and wherever investigations are performed, attention must be paid to the safety of staff and onlookers (see Legal Aspects).

Effective Collaboration with Others

It is important that all those involved in health studies work as a team (Cooper 1993a). From the outset, the raptor biologist should be aware that there are others in disparate disciplines who are likely to provide advice or support. Within a given country, state, or province such collaboration usually is not difficult, but suspicions and jealousy, especially regarding ownership and funding, are possible when things become more regional or international. Researchers should be alert and sensitive to this possibility. Despite closer collaborations among raptor biologists and others recently (Cooper 1993a), a properly coordinated international system for the investigation of morbidity and mortality in birds of prey does not exist (Cooper 1983, 1989, 2002).

TECHNIQUES

Below, I outline some of the methods used to carry out health studies and to sample birds of prey. Details of laboratory and necropsy procedures are given later.

Clinical Methods

Capture techniques are discussed in Chapter 12. The sampling of raptors as part of rehabilitation work is covered in Chapter 23.

Clinical examination and sampling both are part of diagnostic work and health monitoring. This work must be conducted professionally, proficiently, and with a minimum amount of discomfort, pain or stress to the bird. Properly formulated protocols are essential. Detailed information on clinical procedures can be found in several recent texts on raptor medicine and management. Redig (2003) provides an excellent catalog of the veterinary considerations when working with falconiforms or, for that matter, strigiforms, and refers readers who require further information to five authoritative works, including Heidenreich (1997), Lumeij et al. (2000), Redig and Ackermann (2000), Samour (2000), and Cooper (2002).

The principles of clinical investigation include the following sequential stages: (1) history (environmental for free-ranging birds; management for captive birds), (2) observation, (3) clinical examination, (4) taking samples for laboratory investigation, (5) results and diagnosis, and (6) treatment and action. A suggested record sheet for health-monitoring work is in Appendix 1.

Laboratory Investigations

Laboratory investigations are an important part of clinical work, post-mortem examination (see below), and the analysis of environmental samples. Examples of laboratory investigations are depicted in Fig. 1.

Toxicology and chemical analysis are covered in Chapter 18, and are not discussed here. That said, pathologists should work closely with toxicologists and ensure that suitable samples are taken for analysis or stored for later reference. Likewise, carcasses of birds submitted specifically for toxicological examination (e.g., for pesticide analysis) also should be made available for detailed gross and histopathological examination and microbiological studies. Factors other than chemical toxins, including micro- and macroparasites, or underlying renal or hepatic disease, also should be investigated. Other laboratory investigations are discussed and tabulated later in this chapter.

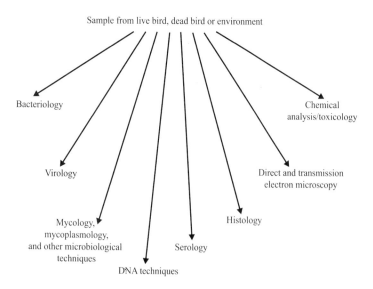

Figure 1. Sample from live bird, dead bird, or the environment.

Special Investigations

Although standard procedures outlined above are applicable to most health studies on raptors, additional laboratory investigations, including microbiological and parasitological monitoring of nests, nest-boxes, aviaries, breeding pens, and incubators, also may prove valuable. Swabs can be taken from such sites and cultured for bacteria and fungi. Food items, likewise, can undergo microbiological or toxicological analysis or both. Ventilation in breeding pens and aviaries can be assessed by smoke tests, and its efficacy calculated by the use of bacteriological "settle plates," or other specific air-sampling methods (Cooper 2002). The laboratory examination of regurgitated pellets is a special feature of raptor health studies that is discussed below.

The Post-mortem or Necropsy Examination

Preparation for a post-mortem examination is all-important. The necessary steps can be summarized as follows:

- Decide why the necropsy is to be carried out. The various categories of examination, each with different objectives, are summarized in Table 1.
- Check that appropriate facilities and equipment are available, including protective clothing and measures aimed at reducing the risk of spread of infectious disease to humans or other animals (see below).
- Be sure that the person carrying out the post-mortem examination is sufficiently knowledgeable about the techniques and precautions that are necessary.
- Be familiar with the normal anatomy of the species (cf. King and McLelland 1984, Harcourt-Brown 2000) as well as its general biology and natural history (Cooper 2003a).

Health and safety. Raptors can present hazards to those who work with them. These include physical dangers when trapping birds on cliffs or retrieving carcasses from marshes or other wetlands, and chemical dangers due to contact with toxic or carcinogenic agents such as formaldehyde. For the purposes of this chapter however, the potential threat of zoonoses, or diseases and infections that are naturally transmissible between vertebrates and humans, is particularly relevant. A review of zoonotic infections that might be acquired from birds, including raptors, was produced some years

Table 1. Categories of post-mortem examination of raptors.

Purpose	Category	Comment
To determine the cause of death.	Diagnostic	Routine diagnostic techniques are followed.
To ascertain the cause of ill-health (not necessarily the cause of death).	Diagnostic-health monitoring	Usually routine, but detailed examinations and laboratory tests may be needed to detect non-lethal changes.
To provide background information on supposedly normal birds on the presence or absence of lesions, parasites, or of other factors, such as fat reserves or carcass composition.	Health monitoring	As above.
To provide information for a legal case or similar investigation, including determining the circumstances of death or the possibility that the bird suffered pain or distress while it was alive.	Forensic-legal	Usually very different from the categories above. The approach depends upon the questions being asked by police or enforcement bodies who requested the necropsy. There must be a proper "chain of custody/ evidence." All material and wrappings should be retained until the case is closed (Cooper and Cooper 2007).
For research purposes, such as collection of tissue samples or studies on organ weight.	Investigative	Depends upon the requirements of the research worker.

ago (Cooper 1990). A number of publications have followed on the heels of new hazards, including West Nile virus. Palmer et al. (1998) provides a useful general reference to zoonoses, including information on both animals and humans.

It is both useful and legally astute for researchers to have an up-to-date list of zoonoses that may be contracted from birds. Infectious agents that once were considered unimportant in humans now are recognized as being potentially pathogenic. Many of these "opportunistic" species take advantage of a debilitated host; in particular, an individual that is immunosuppressed as a result of another infectious disease (e.g., HIV-AIDS, malaria, etc.), malnutrition or on account of medication that is reducing the immune response. It is prudent to assume that any raptor might be a source of organisms that are pathogenic to humans. If this precautionary approach is followed and appropriate safeguards taken, the risks involved in carrying out an examination of a live or dead bird are minimized.

The specific precautions used to restrict the spread of zoonotic infections depend upon the circumstances. In some countries national health and safety legislation may require the employer of those studying wild birds (including handling, post-mortem examinations or sample-taking) to compile a "risk assessment" before the work commences. The researcher, veterinarian, or technician will need to follow prescribed rules and take appropriate precautions. In some countries rules may not exist or may be poorly enforced. Nevertheless, researchers have a responsibility to protect colleagues and assistants, and it is wise to compile a code of practice aimed at minimizing the risk of infection (Cooper 1996).

Necropsy technique. Many methods have been advocated for the post-mortem examination of birds. Some have been devised by veterinarians, usually for the diagnosis of specific diseases (Wobeser 1981, Hunter 1989, Cooper 1993b, 2002, 2004). Others have been devised by ornithologists interested in wild bird mortality or those needing to obtain samples for research (van Riper and van Riper 1980). A basic technique for those working in the field, especially in areas where access to professional advice is limited, is detailed in Cooper (1983). Specific guidance for the necropsy of birds of prey is provided in Cooper (2002).

Necropsy methods should be efficient and reproducible. A post-mortem examination is not simply a matter of "opening up the body." It is a structured operation that involves both external and internal observations and, usually, detailed investigations of organs and tissues. Young birds and embryos require a different approach (Cooper 2004b and below).

A comprehensive necropsy, which encompasses features of both "diagnostic" and "health-monitoring" investigations, including a range of tests and analyses, in addition to collection of biometric data (see Appendix 2), can be time-consuming. Detailed and exhaustive work is vital when rare or threatened species of raptors are involved or deaths have occurred under unusual circumstances. Under more typical circumstances, when time is at a premium and common species are involved, lengthy and detailed investigations of every bird may not be feasible. At such times the abbreviated post-mortem protocol outlined below can be followed, coupled with the appropriate storage of material for subsequent studies:

■ Upon receipt of the specimen, record the history and give the bird a unique reference number. This not only is good practice, but is an essential precaution (to facilitate chain of custody/evidence) if legal action is underway or likely to occur (Cooper and Cooper 2007).

■ Examine the bird externally (including beak, buccal cavity, auditory canal, preen gland, and cloaca). Record (and quantify) any parasites, lesions, or abnormalities. Comment on plumage and molt using standard ornithological protocols.

■ Weigh the bird. Record standard measurements. The body mass of a bird is of limited value without measurement of its linear dimensions (i.e., wing chord [carpus], tarsus, culmen, combined head and bill, and sternum). The body mass is the most important and should form part of every examination.

■ Dissect (open) the bird from the ventral surface by lifting or removing the entire sternum. Examine superficial internal organs. Record any lesions or abnormalities.

■ Remove and set aside in clean (preferably sterile) containers, the heart, liver and gastrointestinal tract, ligating the esophagus and rectum to prevent the spillage of their contents. Examine deeper internal organs. Note any lesions or abnormalities.

■ Fix in 10% formalin small portions of the lung, liver, and kidney, and any organ or tissue that appears to be abnormal (enlarged, unusual color, containing distinct lesions, etc.).

■ Open the proventriculus, gizzard, and portions of intestine. Search with the naked eye and a hand lens for food, other material (e.g., pellets), parasites, or lesions. Examination is facilitated if the material is placed in a Petri dish together with a little saline, and illuminated from below.

Save any interesting contents or parasites and make an effort to quantify them, for example, by estimating the proportion of the intestine examined and counting the number seen.

■ After examination, freeze and save the bird's carcass, (or, if more than one bird is available, some frozen and others fixed in formalin) until a decision can be made as to further tests that may need to be performed (see below).

■ Record how and where the body and samples are saved, and include a reminder that they may need to be processed or discarded at a later date.

Appropriate equipment, including a scalpel with blade, scissors, and two pairs of forceps, must be used when conducting the examination. Small ophthalmological instruments may be needed when necropsying nestlings of small raptors, whereas larger, heavy-duty instruments may prove more serviceable for large raptors, such as eagles. Rat-toothed forceps are ideal for grasping tissues during dissection, but can damage samples destined for the histology laboratory. A hand lens or dissecting loupe is invaluable for the investigation of small birds and detecting tiny lesions.

Key features of any post-mortem examination include (1) recording all that is seen or done, (2) taking of samples, and (3) retaining material for subsequent study. The prime objective of any person who is carrying out a post-mortem examination, regardless of training and experience, is to observe and to record. There is an inherent danger in attempting to interpret findings during the post-mortem examination. Something that may appear significant initially, such as damage to a pectoral muscle or pallor of the liver, subsequently may prove to be of little consequence as other findings cast a different light on the case. Bacteriological examination, which typically does not yield results for 3 to 4 days, may reveal that a bird that died with an injured muscle or pale liver, actually died from an overwhelming bacterial infection. Thus, it is prudent to reserve judgment until all tests are complete. If a provisional diagnosis is essential, this should be issued with the caveat that it is tentative, and may be modified pending further results. Many investigations of raptor mortality have been compromised by premature judgments based on inadequate information.

The assessment of "condition," although controversial, is considered an important index in studies on survival and reproductive success. Methods of assessing condition in birds include:

■ Relating body mass to linear measurements (see above). Unwrapped carcasses undergo gradual evaporation, therefore weight loss should be taken into account.

■ Assessing and scoring the amount of fat, both subcutaneous and internal.

■ Measuring muscle (especially pectoral) size, both macroscopically and histologically.

■ Taking whole-body measurements using, for example, the TOBEC system (Samour 2000).

All of these methods have their own devotees. Which is used depends upon the protocol being followed and the facilities available. However, it is important that some assessment of condition be made in order to relate findings from one bird to another. Thus, measurements of carpus must be routine, as should calculations of body mass. A scoring system should be devised and applied to parameters such as the quantity of fat that is visible or the size of pectoral muscles.

Space does not permit detailed discussion of all systems, but mention is made of the reproductive tract because of its importance in assessing and measuring breeding success (Newton 1998). Careful examination of the genitals is essential. Sexing a dead raptor is generally not difficult. However, if a bird is immature or not yet in breeding condition the gonads may be difficult to see. In some instances, post-mortem change (autolysis) can make detection impossible. The use of a hand lens and strong reflected light often helps, but if this also fails, a portion of the kidney and the presumed gonad can be examined histologically to determine the sex. Notes always should be taken of the appearance of the ovary or testes. In the falconiforms, the presence or absence of a vestigial right ovary should be recorded as part of developing a biomedical database. The color of the testes should be noted as they are sometimes pigmented. Whenever possible, and always when a series of birds is being examined and compared, the size of the gonad(s) should be noted by measuring, weighing or scoring. Assessing follicle development in the ovary also is important.

Other observations on the reproductive tract can provide additional information. A readily visible, well-developed, left oviduct usually indicates that the bird has laid eggs. For many species reliable data on oviduct size and appearance are lacking. The size of the organ should be recorded by measuring, weighing or scoring.

Study of the reproductive system can be supplemented by histological examination. The gonad and tract, or parts of them, should be fixed in buffered 10% formalin, and hematoxylin and eosin-stained sections should be prepared. After measuring and weighing, the reproductive organs can be fixed for study at a later date.

Weighing organs, especially the liver, heart, spleen, kidney and brain, is encouraged whenever possible. Changes in organ to body-mass ratios often occur during infectious and non-infectious diseases.

The retention of material following post-mortem examination, referred to frequently above, is important for several reasons:

■ It may be necessary to go back to the carcass later in order to carry out additional investigations. This may prove necessary, for example, if histopathology suggests a bacterial infection, in which case unfixed samples can be taken and cultured to identify the causal organisms.

■ Carcasses or other material may be required for legal (forensic) purposes, if, for example, a court action relating to the bird's death is to be brought (Cooper and Cooper 2007).

■ Material may be needed for research. This requirement can range from whole bodies, study skins, or skeletons for museums, to the retention of relevant samples for morphometric study of gross or microscopic anatomy. In some cases, the bird's carcass and or tissues may be needed for a reference collection (see below).

The likely fate of carcasses, tissues and specimens should be assessed initially before the examination is conducted. Appropriate containers will be needed, and a decision must be made as to how to dissect the bird and preserve its body and tissues. For example, tissues for histology can be stored in 10% buffered formalin, but this method will destroy most microorganisms and damage DNA. Freezing, on the other hand, will preserve most microorganisms and DNA but will hamper histological and electron microscope work. Plastic and glass containers may influence results if they are used to store samples for certain toxicological analyses.

Facilities for storing carcasses and tissues may be limited, in which case a decision has to be made as to what is retained and for how long. As a general rule, following a post-mortem examination, the bird's carcass and tissues can be kept in a refrigerator at 4°C for up to 5 days, after which, if still needed, they should be frozen at -20°C, or fixed in formalin, ethanol, or a combination of both. Material from threatened, endangered, or endemic species should be retained for future reference or retrospective studies (Cooper and Jones 1986, Cooper et al. 1998). If a specific reference collection for the species exists, the carcass, except for small portions of tissue, including the liver, should be fixed in formalin. The latter should be frozen or fixed in ethanol.

Considerations for necropsying neonates and eggs. The examination of young, neonatal raptors is not as straightforward as it may seem. They are not simply smaller versions of the adult bird. The nestling's immune system is just beginning to develop and respond to antigens in the environment (see below). Its powers of thermoregulation usually are poorly developed, especially in nidicolous species such as raptors. These and other features mean that susceptibility to certain infectious agents, as well as to physical factors such as cold, may be enhanced. Investigation of the young bird should follow standard techniques for "neonates" that were originally developed for domestic poultry (Cooper 2002). An important feature of the necropsy of young birds is the examination, measurement, and sampling of the bursa of Fabricius. This organ, which lies adjacent to the cloaca, is a key component of the immune system and its investigation is imperative if mortality and morbidity of young birds is to be fully investigated. The bursa as well as the thymus, another part of the immune system, should be examined, weighed or measured and fixed in formalin for subsequent examination. If an investigator is in doubt over the examination of young birds, they should seek the advice of an experienced avian pathologist. This also applies to necropsying eggs and embryos (see below).

The comprehensive examination of raptor eggs is highly specialized. Most information in this area comes from studies involving domestic fowl and other galliforms and, more recently, passerines and psittacines (Cooper 2002, 2003a). Unfortunately, the examination of eggs often does not follow a standard protocol. Toxicologists, for example, examine and take samples differently from pathologists, who are particularly interested in infectious diseases, developmental abnormalities, and incubation failures. A detailed description of specific techniques for examining eggs appears in Appendix 3, and a recommended report form is provided in Appendix 4. Measuring eggshell thickness is an important part of assessing eggs, whether or not the eggs are fertile. Various methods can be used. A useful index is described in Ratcliffe (1970). Eggshells should be stored dry for future reference.

Laboratory Investigations

Laboratory investigation of samples is an important component of clinical work, as well as an essential component of necropsy examination and a useful adjunct to environmental studies. An extensive range of tests is available depending upon the situation and resources available. For example, carcasses of raptors found near a chemical spill are likely to undergo toxicological analyses rather than cultured for bacteria, fungi, or viruses. Unfortunately, laboratory procedures are expensive and the cost of some may be prohibitive. Funding may permit only a limited number of tests on a sample of birds, with the remainder being stored for investigation later. When this occurs, researchers should store the carcasses and tissues appropriately (see above). This includes safety concerns. Glutaraldehyde, for example, which must be stored below 40°C if it is not to deteriorate, is toxic to humans and must be handled accordingly. Examples of investigative tests on whole birds (both live and dead) and tissues are given in Tables 2 and 3, respectively. Although a few of the techniques listed can be learned quickly (e.g., detecting of helminth and protozoan parasites, preparing cytological preparations, etc.), others will require technical assistance.

Table 2. Investigative tests on live and dead birds.

Investigative test	Live birds	Dead birds
Clinical examination	+	-
Post-mortem examination	-	+
Radiology	+	+
Hematology	+	+/-[a]
Clinical chemistry	+	+/-
Microbiology	+	+
Toxicology	+/-	+
Histology	+/-	+
Electron microscopy	+/-	+
Chemical analysis of carcass	-	+

[a] of limited value.

Table 3. Laboratory tests on samples from raptors.

Samples	Available from	Comments
Blood in appropriate anticoagulant for hematological and clinical chemical analysis and detection of hemoparasites.	Usually only from live birds; occasionally, small samples can be retrieved from birds that died very recently.	Various blood tests can be conducted, and databases of reference values are being established. The subject is a specialized one and reference should be made to standard texts including Campbell (1995) and Hawkey and Dennett (1989). Blood smears also can be valuable, but experience is needed to produce good preparations and the possibility of error, especially when looking for and quantifying hemoparasites, is high. Consult Cooper and Anwar (2001), Feyndich et al. (1995), and Godfrey et al. (1987).
Blood without anti-coagulant (serum) for serological investigation.	Usually only from live raptors; occasionally small samples can be retrieved from birds that have died very recently.	Serology, usually to detect antibodies to viruses and other organisms, has an important part to play in both disease diagnosis and health monitoring. Various serological tests are available and each demands skill in performance and interpretation. A rise in antibody titer usually is considered indicative of exposure to a specific organism. The increase, however, can take time and may not be apparent in birds that have only recently contracted an infection.
Tissues fixed in 10% formalin (preferably buffered) for histology.	Dead birds; occasionally live biopsies, but usually only from a dermal lesion or one that is surgically accessible.	Fixed tissues can be stored indefinitely and examined at a later stage. The general rule is to take lung, liver, and kidney (LLK) tissue, plus any organs that show abnormalities or which are considered important because they may provide useful information (e.g., bursa of Fabricius and thymus of young birds, which can yield data on immune status). Samples, usually, should not exceed 20 millimeters^2 and fixative volume should be ten times that of the tissue. Small carcasses can be fixed whole, following opening for processing.
Tissues fixed in glutaraldehyde for transmission electronmicroscopy (TEM).	As above.	Generally as above, but only tiny samples are taken. Scanning electronmicroscopy (SEM) employs different techniques and is not considered here.
Cytological preparations.	As above.	Easy to take and inexpensive to process (readily done in any veterinary practice or in the field). Produces results rapidly. Usually consist of touch preparations or impression smears which can give valuable information about tissues within a few minutes. The samples first must be blotted on filter paper to remove excess blood.
Swabs, organ and tissue samples, and other specimens for microbiological and other investigations.	Live or dead birds, dermal lesions, mouth or cloacal swabs, internal organs (carcasses only).	Usually sampled with swabs (in transport medium if they are to be sent elsewhere). Includes portions of tissue as well as exudates and transudates (Hunter 1989, Scullion 1989). If culture is not possible, an impression smear stained with Gram or other stains often provides useful information.
Tissues for toxicological examination.	Dead birds mainly, but some small samples can be taken from live birds as well (e.g., blood or muscle biopsies for pesticide analysis, and feathers for heavy metal and other analyses).	It is important that samples from wild bird casualties are taken and stored routinely for toxicological analysis. Samples for toxicology usually are kept frozen for later analysis. Samples should be taken and stored even when there is no immediate prospect of their being analyzed.
Droppings, including feces and urates as voided, for parasitological and other tests.	Both live (recently voided droppings) and dead birds (removed from the cloaca).	Droppings provide a means of diagnosing some diseases and obtaining health-monitoring data with minimal disturbance to the live bird (Cooper 1998). Droppings often are passed when a raptor is restrained or handled. The fecal component can be used to detect internal parasites, to provide information on other changes in the intestine (e.g., the presence of blood, undigested food, etc.) or to investigate the origin of recently ingested food. Feces also can be used to detect bacteria, fungi and viruses. Molecular techniques, including PCR, now are being used to detect the antigens of pathogenic organisms and to provide other information based on DNA technology. The urate component of feces can be used to investigate kidney function and also may yield parasites associated with the renal system. In all cases, fresh samples provide the most reliable results.
Stomach and crop contents.	Usually from dead birds. Stomach and crop washings can be obtained from live birds or regurgitation can be stimulated by physical or chemical means. Regurgitated pellets can provide valuable information.	As above.
Feathers.	Both live and dead birds.	Can be examined for lesions, analyzed for heavy metals, and used in studies involving mitochondrial DNA (Cooper 2002).

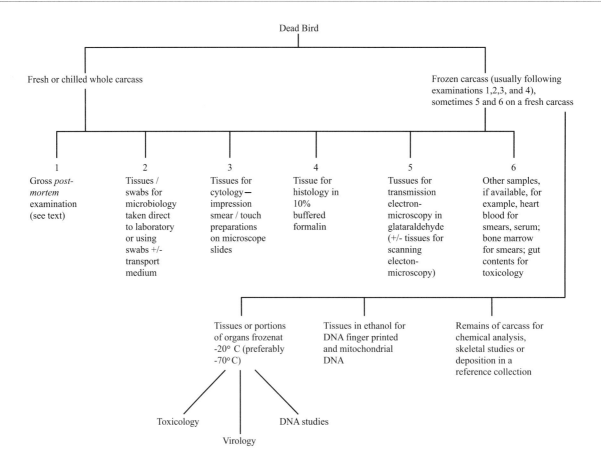

Figure 2. Sample taking during post-mortem examination of birds.

One difficulty often faced is deciding which specimens to keep and how they should be preserved. Figure 2 illustrates the range of possibilities for some post-mortem samples and the various methods used. When material is sparse, a "triage" system may need to be instituted.

Interpretation of Findings

The analysis and interpretation of results can present problems. For example, one cannot assume that a firearm killed a hawk that has lead shot in its body. The shot may be longstanding, related to a previous non-fatal, shooting and of no relevance to the bird's death. One also must distinguish between the cause of death and factors that may have contributed to it (i.e., the "proximate" versus "ultimate" causes). For example, a bird with avian tuberculosis or pox may become so weak that it is unable to hunt and as a result, is killed while scavenging by the roadside; the bird in question will have died of trauma, but the most significant patho-

logical finding would be acid-fast *Mycobacterium* organisms in its internal organs.

Finding micro- or macroparasites on or in a bird also can be misleading. Sometimes parasites are acquired from prey species (e.g., lice from corvids), or from another carcass in the post-mortem room. Even when such organisms are bona fide isolates, their relevance may not be clear. Intestinal worms associated with hemorrhage in a bird's intestine, or bacteria isolated from a hot, swollen foot, clearly are likely to be of some significance, but what if such organisms are found without such lesions? Are they of importance? Much remains to be learned about the biology of pathogens (Reece 1989) and host-parasite relations (Cooper 2001) in free-living birds. Until that happens, it is best to record findings, both qualitatively and quantitatively, and to attempt to relate these to the bird's body condition and its systemic health. In this regard, data from captive raptors can provide useful references for wild bird casualties (Cooper 2003b).

The cause of death is "euthanasia" when the bird

has been killed on humanitarian grounds or to obtain fresh material for examination. In such cases, the aim of any investigation is to detect underlying lesions or factors that may have contributed to the bird's ill health or influenced its behavior.

Interpretation of pathological findings is particularly difficult. Mistakes can be made easily by those who are unfamiliar with the various disciplines involved. Thus, the profuse growth of a potentially pathogenic bacterium from a carcass does not necessarily mean that the organism was the cause of death; if the bird has been dead for some time it may have invaded the tissues post mortem. Likewise, the detection of a distinct pathological lesion, such as an interstitial nephritis, need not indicate that the raptor died of kidney disease; the renal damage may be chronic and not sufficiently severe to have proved fatal. In all cases, careful collation of results is necessary, and diagnosis and conclusions should be made only in the light of all information and findings available. Records are essential and, if possible, should be computerized to facilitate retrieval and analysis. Field and other preliminary data also should be retained. It is important to recall that in health studies on raptors a "diagnosis" is not necessarily the objective. Apparently minor background findings of parasitism or unusual gonads, for example, may be far more relevant, especially when the study is part of a larger, population-monitoring program.

From the above it is clear that care must be taken with regard to terminology. A "diagnosis" is one thing, the "cause of death" another, and underlying health-status yet another. Gross and laboratory findings need to be interpreted in the context of the background, history, and circumstances under which the birds were found and examined, the species and sex and age ratios involved, and other extraneous factors, including weather, that may have played a part.

Interpretation of findings also can be hampered by the lack of reliable reference values. For example, recently there have been great advances in our knowledge of the hematology and blood biochemistry of birds of prey, however the data available largely relate only to species that are kept or bred in captivity, or have been subjected to detailed study in the wild, and for some species little or no information is available (cf. Tryland 2006). Likewise, toxicological investigations can be thwarted because of a paucity of "normal" background values, as well as sub-lethal and lethal values for a given species. Although extrapolation is sometimes possible, it is far from ideal.

The absence of basic data remains a cause for concern. For instance, the normal ranges of organ mass and organ to body-mass ratios of most species of raptors are not known, and yet such information could be gathered easily if proper records were kept and findings freely disseminated. There is a need to involve scientists from all disciplines, including undergraduate and postgraduate students and "amateur" naturalists, in filling such gaps in our knowledge. Comprehensive databases on host-parasite relations of different families of birds also are needed (Cooper 2003b). These should encompass basic biological parameters of raptor hosts, as well as information about the macro- and microparasites associated with particular species, whether or not the latter are considered to be pathogenic. A useful first step is to compile local, national, and regional checklists of parasites together with the names of the hosts with which they are associated.

These caveats aside, several useful references for interpreting laboratory findings do exist. They include Randall and Reece (1996) on histopathology, Hawkey and Dennett (1989) and Campbell (1995) on hematology, and Scullion (1989) and Cooper (in Fudge 2000) on microbiology.

Legal Considerations

In the United Kingdom and several other countries, the making of a formal diagnosis, even as a result of examining a dead bird, is restricted by law to the veterinary profession (Cooper 1987). There are other legal considerations in raptor pathology as well. Health and safety legislation may dictate how and where clinical examination, sample taking or a post-mortem investigation is performed. Where a zoonotic disease is suspected, the legislation may demand a risk assessment and, perhaps, that the necropsy is only performed if appropriate protection (i.e., clothing, equipment, and facilities) is available for all those involved, and that the personnel are appropriately experienced or trained. Laws may restrict the movement of carcasses or specimens (Cooper 1987, 2000). Within countries, such laws usually relate primarily to postal requirements for adequate packing and transport. When moving samples from one country to another, the situation becomes more complex because conservation legislation, especially CITES (Convention on International Trade in Endangered Species), may apply. The Ministry or Department of Agriculture of the receiving country is likely to require documentation describing the type of material that is being transported, particularly its likely pathogenicity. If the raptors in question are covered by CITES, there will be an additional need for permits. In addition, the movement of small specimens, including blood smears, or tissues for DNA study, remains a cause of frustration for those that wish to send samples to colleagues or laboratories in other countries. Even the smallest sample can fall into the category of a "recognized derivative" under CITES and, therefore, require appropriate documentation and authorization. Recently, there have been moves to obtain exemptions for such material, especially if the samples in question are required for important diagnostic or forensic purposes. CITES continues to debate the issue and, at the time of writing, the likely outcome appears to favor introducing a "fast-track" system for small, but urgent, samples (see Chapter 25). Those involved in health studies on birds of prey should be familiar with the relevant legislation and adhere to it.

In many countries, legislation relevant to health studies on raptors is non-existent or is poorly enforced. In such circumstances, it is good practice to work toward "in-house" protocols and to develop and use guidelines that, although not legally binding, help to ensure high standards of work (Cooper 1996). In all instances, the status of raptor biology is not served by breaching the law or broadly established professional protocols, however tedious and inconvenient they may appear.

CONCLUSIONS

Health studies are an important component of raptor management, both in the wild and in captivity. Of particular and increasing significance is health monitoring. Those working with raptors need to be aware of developments in this field, especially the new technology that is now available for the detection of organisms and antibodies.

The value of an interdisciplinary approach to the study of the diseases and health parameters of raptors cannot be over-emphasized. For centuries, in Europe, Arabia, and the Far East, it was the falconers, who kept and flew birds of prey, who knew most about the natural history of raptors and how to detect early signs of ill health in their charges. These people always maintained that keeping a hawk in good health was preferable to treating ailments, and many early texts advised on how this might be achieved through proper management (Cooper 2002). Charles d'Arcussia, the French noble-

man, whose book on falconry was first printed in 1598 (Loft 2003), had a refreshingly positive approach to the question of disease and advocated the following: "If you want to maintain the health of your hawks take as guides those who are experienced and can lead you forward with their advice." This admonition remains relevant today. Raptor biologists have unprecedented access to literature, ranging from field notes and scientific papers to the Internet, and are able to take advantage of the numerous developments in clinical medicine and laboratory investigation that have characterized the past three decades. That said, we must remain wary of working in isolation and instead collaborate with others working in various disciplines that now contribute to our understanding of the health and diseases of birds of prey.

ACKNOWLEDGMENTS

I am grateful to my friend, David Bird, for inviting me to contribute again to this work, and to Oxford University Press for permitting me to reproduce, in part, sections of my chapter in "Bird Ecology and Conservation" edited by Sutherland, Newton, and Green (2004).

LITERATURE CITED

CAMPBELL, T.W. 1995. Avian hematology and cytology. Iowa State University Press, Ames, IA U.S.A.

COOPER, J.E. 1983. Guideline procedures for investigating mortality in endangered birds. International Council for Bird Preservation, Cambridge, United Kingdom.

———— [ED.]. 1989. Disease and threatened birds. Technical Publication No. 10. International Council for Bird Preservation, Cambridge, United Kingdom.

————. 1990. Birds and zoonoses. *Ibis* 132:181–191.

————. 1993a. The need for closer collaboration between biologists and veterinarians in research on raptors. Pages 6–8 in P.T. Redig, J.E. Cooper, J.D. Remple, and D.B. Hunter [EDS.], Raptor biomedicine. University of Minnesota Press, Minneapolis, MN U.S.A.

————. 1993b. Pathological studies on the Barn Owl. Pages 34–37 in P.T. Redig, J.E. Cooper, J.D. Remple, and D.B. Hunter [EDS.], Raptor biomedicine. University of Minnesota Press, Minneapolis, MN U.S.A.

————. 1998. Minimally invasive health monitoring of wildlife. *Animal Welfare* 7:35–44.

————. 2001. Parasites and birds: the need for fresh thinking, new protocols and co-ordinated research in Africa. *Ostrich Suppl.* 15:229–232.

————. 2002. Birds of prey: health & disease. Blackwell Science Ltd., Oxford, United Kingdom.

————. 2003a. Captive birds. World Pheasant Association and Hancock Publishing, London, United Kingdom.

———. 2003b. Jiggers and sticktights: can sessile fleas help us in our understanding of host-parasite responses? Pages 287–294 in Jesus M. Perez Jimenez [ED.], Memoriam of Prof. Dr. Isidoro Ruiz Martinez. University of Jaen, Spain.

———. 2004a. Information from dead and dying birds. Pages 179–210 in W.J. Sutherland, I. Newton, and R.E. Green [EDS.], Bird ecology and conservation. Oxford University Press, Oxford, United Kingdom.

———. 2004b. A practical approach to the post-mortem examination of eggs, embryos and chicks. *J. Caribb. Vet. Med. Assoc.* 4:14–17.

——— AND M.A. ANWAR. 2001. Blood parasites of birds: a plea for more cautious terminology. *Ibis* 143:149–150.

——— AND C.G. JONES. 1986. A reference collection of endangered Mascarene specimens. *The Linnean* 2:32–37.

——— AND M.E. COOPER. 2007. Introduction to veterinary and comparative forensic medicine. Blackwell Publishing Ltd., Oxford, United Kingdom.

——— AND J. SAMOUR. 1987. Portable and field equipment for avian veterinary work. Proceedings of the European Committee of the Association of Avian Veterinarians, 19–24 May 1997, London, United Kingdom.

———, C.J. DUTTON AND A.F. ALLCHURCH. 1998. Reference collections in zoo management and conservation. *Dodo* 34:159–166.

COOPER, M.E. 1987. An introduction to animal law. Academic Press, London, United Kingdom and New York, NY U.S.A.

———. 1996. Community responsibility and legal issues. *Semin. Avian Exotic Pet Med.* 5:37–45.

———. 2000. Legal considerations in the international movement of diagnostic and research samples from raptors - conference resolution. Pages 337–343 in J.T. Lumeij, J.D. Remple, P.T. Redig, M. Lierz, and J.E. Cooper [EDS.], Raptor biomedicine III. Zoological Education Network, Lake Worth, FL U.S.A.

FEYNDICH A.M., D.B. PENCE AND R.D. GODFREY. 1995. Hematozoa in thin blood smears. *J. Wildl. Dis.* 31:436–438.

FUDGE, A.M. [ED.]. 2000. Laboratory medicine: avian and exotic pets. W.B. Saunders, Philadelphia, PA U.S.A.

GIRON PENDLETON, B. A., B.A. MILLSAP, K.W. CLINE AND D.M. BIRD [EDS.]. 1987. Raptor management techniques manual. National Wildlife Federation, Washington, D.C. U.S.A.

GODFREY, R.D., A.M. FEDYNICH AND D.B. PENCE. 1987. Quantification of hematozoa in blood smears. *J. Wildl. Dis.* 23:558–565.

HARCOURT-BROWN, N. 2000. Birds of prey: anatomy, radiology and clinical conditions of the pelvic limb. CD ROM. Zoological Education Network, Lake Worth, FL U.S.A.

HAWKEY, C.M. AND T.B. DENNETT. 1989. A colour atlas of comparative veterinary haematology. Wolfe, London, United Kingdom.

HEIDENREICH, M. 1997. Birds of prey: medicine and management. Blackwell, Wissenschafts-Verlag, Oxford, United Kingdom.

HUDSON, P.J., A. RIZZOLI, B.T. GRENFELL, H. HEESTERBEEK AND A.P. DOBSON [EDS.]. 2002. The ecology of wildlife diseases. Oxford University Press, Oxford, United Kingdom.

HUNTER, D.B. 1989. Detection of pathogens: monitoring and screening programmes. Pages 25–29 in J.E. Cooper [ED.], Disease and threatened birds. Technical Publication No. 10. International Council for Bird Preservation, Cambridge, United Kingdom.

KING, A.S. AND J. MCLELLAND. 1984. Birds, their structure and function. Ballière Tindall, London, United Kingdom.

LOFT, J. 2003. D'Arcussia's falconry: a translation. St. Edmundsberry Press, Suffolk, United Kingdom.

LUMEIJ, S.J., J.D. REMPLE, P.T. REDIG, M. LIERZ AND J.E. COOPER [EDS.]. 2000. Raptor biomedicine III. Zoological Education Network, Lake Worth, FL U.S.A.

NEWTON, I. 1979. Population ecology of raptors. Buteo Books, Vermillion, SD U.S.A.

———. 1981. Mortality factors in wild populations - chairman's introduction. Page 141 in J.E. Cooper, A.G. Greenwood, and P.T. Redig [EDS.], Recent advances in raptor diseases. Chiron, Yorkshire, United Kingdom.

———. [ED.]. 1990. Birds of prey. Facts on File, New York, NY U.S.A.

———. 1998. Population limitation in birds. Academic Press, London, United Kingdom.

———. 2002. Diseases in wild (free-living) bird populations. Pages 217–234 in J.E. Cooper [ED.], Birds of prey: health and disease. Blackwell Science Ltd., Oxford, United Kingdom.

PALMER, S.R., LORD SOULSBY AND D.I.H. SIMPSON [EDS.]. 1998. Zoonoses. Oxford University Press, Oxford, United Kingdom.

RANDALL, C.J. AND R.L. REECE. 1996. Color atlas of avian histopathology. Mosby-Wolfe, London, United Kingdom and Baltimore, MD U.S.A.

RATCLIFFE, D.A. 1970. Changes attributable to pesticides in egg breakage frequency and eggshell thickness in some British birds. *J. Appl. Ecol.* 7:67–113.

REDIG, P.T. 2003. Falconiformes (vultures, hawks, falcons, secretary bird). Pages 150–161 in M.E. Fowler and R.E. Miller [EDS.], Zoo and wild animal medicine, 5th Ed. Saunders, St. Louis, MI U.S.A.

——— AND J. ACKERMANN. 2000. Raptors. Pages 180–214 in T.N. Tully, M.P.C. Lawton, and G.M. Dorrestein [EDS.], Avian medicine. Butterworth-Heinemann, Oxford, United Kingdom.

REECE, R.L. 1989. Avian pathogens: their biology and methods of spread. Pages 1–23 in J.E. Cooper [ED.], Disease and threatened birds. Technical Publication No. 10. International Council for Bird Preservation, Cambridge, United Kingdom.

SAMOUR, J.H. 2000. Avian medicine. Mosby, London, United Kingdom.

SCULLION, F.T. 1989. Microbiological investigation of wild birds. Pages 39–50 in J.E. Cooper [ED.], Disease and threatened birds. Technical Publication No. 10. International Council for Bird Preservation, Cambridge, United Kingdom.

TRYLAND, M. 2006. "Normal" serum chemistry values in wild animals. *Vet. Rec.* 158:211–212.

VAN RIPER, C. AND S.G. VAN RIPER. 1980. A necropsy procedure for sampling disease in wild birds. *Condor* 82:85–98.

WOBESER, G.A. 1981. Necropsy and sample preservation techniques. Pages 227–242 in G.A. Wobeser [ED.], Diseases of wild waterfowl. Plenum Press, New York, NY U.S.A.

WOODFORD, M.H. [ED.]. 2001. Quarantine and health screening protocols for wildlife prior to translocation and release into the wild. OIE, VSG/IUCN, Care for the Wild International and EAZWV, Paris, France.

ZALLES, J.I. AND K.L. BILDSTEIN [EDS.]. 2000. Raptor watch: a global directory of raptor migration sites. Birdlife International, Cambridge, United Kingdom, and Hawk Mountain Sanctuary, Kempton, PA U.S.A.

Appendix 1.   HEALTH MONITORING OF LIVE BIRDS OF PREY

Species: _____   Location: _____   Reference: _____

Relevant history: _____

Circumstances of monitoring

Numbers of birds involved: _____   Details: _____

Personnel involved: _____

Other comments:

OBSERVATION

Behavior: _____

Bird unaware of observer: _____

Bird aware of observer: _____

EXAMINATION

Clinical signs: _____

Injuries or external lesions and distinguishing features: _____

Plumage, molt, and preen gland: _____

Ectoparasites: _____

　　　Species: _____

　　　Numbers: _____

Body mass: _____   Carpal length: _____

Other measurements: _____   Condition score: _____

Samples

　　Feathers:

　　Feces:

　　Swabs:

　　Blood:

　　Others:

Follow-up tests

Reported by: _____   Date: _____   Time: _____

Assisted by: _____

Appendix 2.   POST-MORTEM EXAMINATION (NECROPSY) OF DEAD BIRDS OF PREY

Species: _____   Reference No: _____

Date of submission: _____   Origin: _____

Band (ring) number: _____   Other identification: _____

Relevant history and circumstances of death:

Request (category of necropsy): diagnosis (cause of death or ill-health), health monitoring, forensic investigation, research, or other:

Special requirements regarding techniques to be followed, instructions regarding fate of body or samples:

Submitted by: _____   Date: _____

Received by: _____   Date: _____

MEASUREMENTS   Carpus:_____   Tarsus:_____   Other:_____   Body mass: _____

Condition score: Obese or fat / good / fair or thin / poor

State of preservation: Good / fair / poor / marked autolysis

Storage since death: Refrigerator / ambient temperature / frozen / fixed

EXTERNAL OBSERVATIONS, including preen gland, state of moult, ectoparasites, skin condition, lesions, etc.:

MACROSCOPIC EVALUATION on opening the body, including position and appearance of organs, lesions, etc.:

ALIMENTARY SYSTEM:

MUSCULOSKELETAL:

CARDIOVASCULAR:

RESPIRATORY:

URINARY:

REPRODUCTIVE:

LYMPHOID (including bursa and thymus):

NERVOUS:

Appendix 2. *continued.*

OTHER SAMPLES TAKEN

_____	Bact	Paras	Hist	DNA	Cytology	Other (e.g., serology)
_____	Bact	Paras	Hist	DNA	Cytology	Other (e.g., serology)
_____	Bact	Paras	Hist	DNA	Cytology	Other (e.g., serology)
_____	Bact	Paras	Hist	DNA	Cytology	Other (e.g., serology)
_____	Bact	Paras	Hist	DNA	Cytology	Other (e.g., serology)
_____	Bact	Paras	Hist	DNA	Cytology	Other (e.g., serology)
_____	Bact	Paras	Hist	DNA	Cytology	Other (e.g., serology)

LABORATORY FINDINGS

Date: _____ Initials:_____ Reported to whom: _____

PRELIMINARY REPORT (based on gross findings and immediate laboratory results, e.g., cytology)

Reported to: _____ Date:_____ Time: _____

FINAL REPORT (based on all available information)

FATE OF BODY / TISSUES

Destroyed / frozen / fixed in formalin (other) / retained for Reference Collection / sent elsewhere

FATE OF RING/BAND (if appropriate)

PM examination performed by: _____ Date:_____ Time: _____

Assisted by: _____

Appendix 3.   PROTOCOL FOR EXAMINATION OF UNHATCHED EGGS OF BIRDS OF PREY

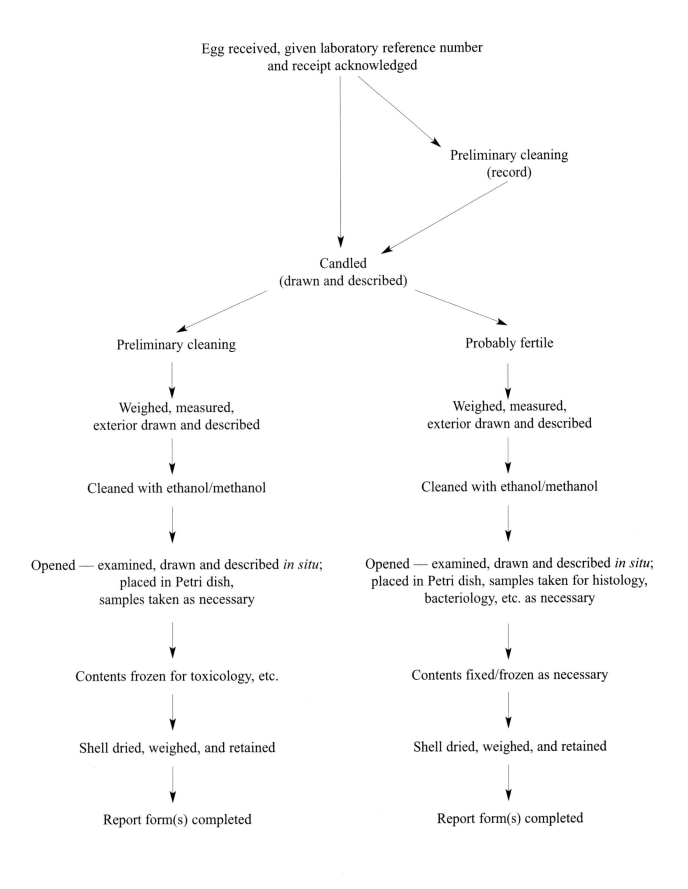

Appendix 4.   EXAMINATION OF EGGS AND EMBRYOS OF BIRDS OF PREY

Reference number: _____

Received (date):_____ (by): _____

Receipt acknowledged by: _____ Date: _____

Method of packing/wrappings:

History:

EGG / EMBRYO EXAMINATION (to be completed for each specimen)

Species: _____

Owner / Origin: _____

Weight of whole unopened egg:_____ Length:_____ Width: _____

External appearance:

Appearance on candling:

    Embryo

    Air cell

    Blood vessels

    Fluids

Appearance when opened:

Contents:

Embryo:

    Length (crown-rump)

    Amniotic cavity

    Allantoic cavity

    Yolk sac

Other comments:

Microbiology:

Histopathology:

Other tests:

Samples sent elsewhere:

Weight of dried eggshell:_____ Thickness (measurement or index): _____

Samples stored:

COMMENTS

Examination performed by:_____ Date:_____ Time: _____

Assisted by:_____

B. Ectoparasites

JAMES R. PHILIPS
Math/Science Division, Babson College,
Babson Park MA 02457-0310 U.S.A.

INTRODUCTION

Fly ectoparasites that feed on blood include biting midges, blackflies, blowflies, louse flies, mosquitoes, and carnid flies. Additional blood-feeding insects that parasitize raptors include cimicid bugs, fleas, and some chewing lice. Other chewing lice feed on feathers. Although usually nonparasitic, scavenging skin beetle (Dermestidae) larvae have even been found in wounds in Snail Kite (*Rostrhamus sociabilis*) nestlings in Florida, and other raptors in Africa and Europe (Snyder et al. 1984). Arachnid ectoparasites of raptors include blood-sucking ticks and mites, mites that feed on feather material, and mites, including chiggers, that feed on tissue. Most mites are external parasites, but some skin mites burrow into and under the skin, and some mites colonize the respiratory tract. In addition to direct pathological effects, raptor ectoparasites can have indirect pathological effects because a weakened host is more vulnerable to infection. Bacterial and fungal infections caused by ectoparasites can occur in wounds, and many flies, ticks, and mites act as disease vectors as well. Philips (2000, 2006a,b) reviewed the parasitic mites of raptors and the author maintains an online checklist of raptor hosts and their mite ectoparasites.

Levels of ectoparasite infestation vary greatly among and within species of raptors. Raptor ectoparasite management involves collection, preservation, and identification of ectoparasites, followed, when necessary, by treatment of affected birds and control measures to reduce the ectoparasite levels in the nest or local environment. Clayton and Walther (1997) reviewed collection and preservation techniques of avian ectoparasites. Beynon et al. (1996) list ingredient formulas for six ectoparasiticides useful in the treatment and control of insects and mites that parasitize raptors.

Raptors and their nests should be surveyed and monitored for ectoparasites as causes of direct pathology and disease transmission. Raptor ectoparasites, such as the lice of the threatened Galapagos Hawk (*Buteo galapagoensis*), can serve as excellent markers of host population differentiation (Whiteman and Parker 2005). Host-specific ectoparasites of endangered raptor species are themselves endangered species.

Insects and blood-filled ticks and mites are much more noticeable to the naked eye than most mites. Feather mites often look like grains of sand, and 0.25-mm chiggers and skin mites as "specks." Species identification often requires ectoparasite dissection, particular ectoparasite clearing techniques, particular slide-mount media, and specialized taxonomic expertise. The mite fauna of raptors is largely unknown, and many new species remain to be discovered. Below I detail the types of flies, cimicid bugs, fleas, chewing lice, ticks, and mites that parasitize birds of prey.

FLIES (DIPTERA)

Biting Midges (Ceratopogonidae)

Boorman (1993) provides an identification key to adults in blood-sucking ceratopogonid genera.

Biting midges, which often are called "no-see-ums," transmit filarial nematodes, blood protozoans *Haemoproteus* and *Leucocytozoon*, and the Thimiri arbovirus to birds (Mullen 2002). After a blood meal, female midges lay eggs in habitats ranging from moist compost and manure to water in tree holes, freshwater marshes, and mangrove swamps. Females can be aspirated from hosts or collected by using light-traps with black-light lamps and carbon dioxide. Midges can be preserved in 1–2% formalin or 70–80% alcohol. Controlling these ectoparasites is difficult. Most screens are not effective and neither is general application of insecticides to kill larvae. Eliminating breeding habitat and general application of insecticides as mists or fogs in early evening when adults are most active can reduce populations.

Blackflies (Simuliidae)

Crosskey and Howard (1997) provide an inventory of the blackflies of the world. Blackflies are the main vectors of *Leucocytozoon* in birds, and also they transmit *Trypanosoma* and filarial nematodes (Adler and McCreadie 2002). Adler et al. (2004) list North American blackfly species, their raptor and other hosts, and the species of Leucocytozoon they transmit.

Blackflies have killed nestling Red-tailed Hawks (*B. jamaicensis*) (Brown and Amadon 1968, Smith et al. 1998), nestling Merlins (*Falco columbarius*) (Trimble 1975), and have weakened nestling Cape Vultures (*Gyps coprotheres*) (Boshoff and Currie 1981). Blackflies tend to feed on the crown, back, and shoulders of raptors. Biting occurs during the day in the open, and adult blackflies can be collected from hosts with an aspirator, or with sticky silhouette or carbon dioxide traps. After a blood meal, females lay their eggs in running water. Fluid preservation destroys important taxonomic features, so adults should be micropinned through the thorax after they dry in a freezer for 5 weeks (Crosskey 1993).

Blackfly control, which mainly targets the larvae, uses the entomopathogenic bacterium, *Bacillus thuringiensis var. israelensis*, applied to bodies of water by hand or from the air. Providing shelters for captive birds helps protect them from blackflies.

Mosquitoes (Culicidae)

The mosquitoes of the world are listed in Knight and Stone (1977) and its supplements (Knight 1978, Ward 1984, Gaffigan and Ward 1985, Ward 1992). There are many regional identification keys, including that of Darsie and Ward (2005) to species of the U.S. and Canada, and a key to world genera by Mattingly (1971).

Mosquitoes transmit many viruses to birds, including encephalomyelitis viruses, West Nile virus, and poxvirus (Foster and Walker 2002). They also are vectors of avian malaria (*Plasmodium*) and filarial nematodes. After a blood meal, female mosquitoes lay eggs on water or wet surfaces under floating vegetation or in the walls of wet tree-holes (Service 1993). Mosquitoes can be collected from hosts and in shaded resting places using aspirators, and with carbon dioxide traps and light-traps. Specimens should not be preserved in liquids but micropinned through the thorax.

Approaches to control of mosquito populations include reducing their breeding habitat; using light mineral oils, organophosphates, insect-growth regulators, or *Bacillus thuringiensis var. israelensis* to kill the aquatic larvae; applying residual insecticides to adult resting surfaces; and direct contact spraying or fogging of organophosphates, carbamates, pyrethrins and synthetic pyrethroids. Screens can protect captive birds.

Louse Flies (Hippoboscidae)

Maa (1963) lists the louse flies of the world and provides genera and species-group identification keys.

Avian louse flies, often called flat flies, tend to remain on their host unless disturbed, and they sometimes bite humans that handle infested birds. Larvae develop in the female and pupate in birds' nests and roosts immediately when born. Louse flies transmit the blood protozoans, *Haemoproteus* and *Trypanosoma*, through biting, and carry lice and the ectoparasitic skin mites, *Strelkoviacarus*, *Microlichus*, and *Myialges*, on their exterior to new bird hosts (Philips 1990, Lloyd 2002). Louse flies have tested positive for West Nile virus, but their role as vectors of this and other viruses is unconfirmed. Infestation of several dozen louse flies does not seem to harm raptors, but when levels exceed 80, raptors become emaciated and too weak to hunt. Louse flies, which range in size from 4 to 7 mm, can be caught with air nets and by hand, and can be pinned or preserved in ethanol. Infested birds can be treated with pyrethroid dust.

Myiasis Flies (Calliphoridae, Muscidae)

Sabrosky et al. (1989) provides a key and a host list for Nearctic *Protocalliphora*, and lists the Palearctic species. Whitworth (2003, 2006) provides a species key to *Protocalliphora* pupae. Furman and Catts (1982) designed a key to a variety of myiasis-causing fly genera.

Nest flies of raptors include the Holarctic and Oriental blow flies *Protocalliphora* (Calliphoridae), the European carrion flies *Lucilia sericata* and *Calliphora* (Calliphoridae), and the tropical flies *Philornis* and *Passeromyia* (Muscidae), all of which lay their eggs in nests or on nestlings. The maggots of these flies cause myiasis by burrowing into host tissues and sucking blood (Baumgartner 1988). Ear cavities, noses, the ventral surface, and feather sheaths are preferred sites. After feeding, larvae drop off the host to digest their blood meal and pupate.

Myiasis is known to kill nestling Northern Harriers (*Circus cyaneus*) (Hamerstrom and Hamerstrom 1954), Sharp-shinned Hawks (*Accipiter striatus*) (Delannoy and Cruz 1991), Verreaux's Eagles (*Aquila verreauxii*) (Gargett 1977), Gyrfalcons (*F. rusticolus*) (Poole and Bromley 1988), and Prairie Falcons (*F. mexicanus*) (White 1963), and to weaken nestling Red-tailed Hawks (Tirrell 1978) and prolong their development (Catts and Mullen 2002). Burrowed larvae will evacuate nestlings if the breathing opening of the larvae is blocked with petroleum jelly or if nestling orifices are flushed with saline solution. Mineral oil can be used to remove them from ear cavities. Maggots should be relaxed before being preserved in ethanol (Hall and Smith 1993). This can be accomplished by placing them into water just below the boiling point, or into acetic alcohol (one part glacial acetic acid to three parts 90% ethanol). Dissecting nest material can yield pupae. Treatment involves removing larvae and applying antibiotics to the wound to prevent infection. Nests can be dusted with pyrethroids.

Carnid Flies (Carnidae)

Carnid flies can be identified using the fly family key of Arnett (2000). Grimaldi (1997) discusses the species, of which the most well known is *Carnus hemapterus*, and lists all avian hosts.

Carnus larvae scavenge in nests. Wingless adults either suck the blood of nestlings or feed on their skin secretions. Infestations are characterized by scabby axillae. Heavy infestations cause reduced pack-cell volumes in Barn Owls (*Tyto alba*) (Schulz 1986), reduced body mass in Common Kestrels (*F. tinnunculus*) (Heddergott 2003), and nestling mortality in Northern Saw-

whet Owls (*Aegolius acadicus*) (Cannings 1986). The fly seems harmless to American Kestrels (*F. sparverius*) (Dawson and Bortolotti 1997). *Carnus* occurs in North America, Europe, Africa, and Malaysia. Specimens can be collected from hosts by hand or from nests by Tullgren funnel extraction of nest material (Mullen and O'Connor 2002), and then preserved in ethanol. Insecticide dusts can be used to treat hosts and control infestations in nests.

CIMICID BUGS (BED BUGS)

Cimicid bugs (Cimicidae) lay eggs where hosts live. Both adults and nymphal stages suck blood. One species in particular regularly attacks raptors. The Mexican chicken bug (*Haematosiphon inodorus*) has killed nestling Bald Eagles (*Haliaeetus leucocephalus*) (Grubb et al. 1986) as well as nestling Red-tailed Hawks and Prairie Falcons (Platt 1975, McFadzen and Marzluff 1996), and has caused nestling California Condors (*Gymnogyps californianus*) to fledge prematurely (Brown and Amadon 1968). The swallow bedbug (*Oeciacus vicarious*) occurs in Prairie Falcon aeries. The bugs hide in nests or cracks near hosts during the day, and feed mainly at night near the eyes and at the base of the host's legs and wings. Cimicid bugs can be collected with forceps, Tullgren funnel extraction, or dissection of nest material, or be forced out of cracks with pyrethroid or kerosene sprays (Schofield and Dolling 1993). Specimens can be preserved in ethanol. Usinger (1966) provides species identification keys for the family and an avian host list. Treatment and control involve spraying hosts, nests, and surfaces near the host with insecticides including pyrethrins.

FLEAS (SIPHONAPTERA)

Regional identification keys with host lists include Holland (1985) for Canada, and Benton and Shatrau (1965) and Lewis et al. (1988) for parts of the U.S. Lewis (1993) provides a key to medically important flea genera globally. Arnett (2000) provides an identification key to families, and Lewis (1993) provides more detailed keys to some of the taxa.

Fleas of adult raptors bite hosts to obtain blood and lay their eggs on their hosts or in nests, where larvae are scavengers. Typically, more fleas are found in nests than on hosts. One exception is the sticktight flea (*Echidno-*

phaga gallinaacea), which remains attached to hosts in unfeathered places around the head. Burrowing Owls (*Athene cunicularia*) in particular seem to be infested with fleas when nesting (Smith and Belthoff 2001). Fleas can be collected from hosts with insecticide dusts, and by dissecting nest material or via extraction in a Tullgren funnel. They can be preserved in 80% ethanol. Treatment and control involve pyrethrin dusts and insect growth regulators (Lewis 1993, Durden and Traub 2002).

CHEWING LICE (MALLOPHAGA)

Price et al. (2003) provide a list of avian lice globally, their hosts, and identification keys to genera by host.

Chewing lice usually are transferred by direct contact and, less frequently, by louse flies. Their feeding can damage feathers, and scratching in response to infestation can cause additional damage. Heavy louse infestations cause anemia, weight loss, and death. Lice can be collected from hosts with forceps, or by ruffling feathers after dusting with insecticidal powder (Clayton and Drown 2001). During necropsy, carcasses can be washed with detergent or skinned, and skin and feathers dissolved using trypsin or potassium hydroxide (Furman and Catts 1982). Detergent washes also will yield mites, whereas dissolving tends to destroy most mites. Resulting solutions are sieved or filtered to collect specimens. Specimens should be preserved in 95% ethanol. Insecticidal dusts and resin strips are useful in treatment and control (Durden 2002).

TICKS (IXODIDA)

Varma (1993) provides an identification key to tick families and genera.

Larval, nymphal and adult ticks all suck blood, often from different hosts. Individuals remain attached to hosts for as long as two days (Sonenshine et al. 2002). Eyelids and the bases of beaks are usual feeding sites. Most ticks are ambush parasites found in litter and soil that latch on to passing hosts. Avian soft ticks (Argasidae — *Argas* and *Ornithodoros*) and some hard ticks (Ixodidae — *Ixodes*) live in nests and burrows. Ticks transmit avian spirochetosis and Lyme disease, and are vectors for *Babesia* spp., an anemia-causing protozoan known to occur in Prairie Falcons (Croft and Kingston 1975). They also transmit viruses and tularemia bacteria to birds. Some species produce a toxin in their saliva that induces paralysis. Ticks have killed nestling Prairie Falcons (Webster 1944, Oliphant et al. 1976) and Peregrine Falcons (*F. peregrinus*) (Schilling et al. 1981), and tick paralysis killed an adult Powerful Boobook (*Ninox strenua*) (Fleay 1968) in Australia. Ticks can be collected directly from hosts by dissecting nest material by extraction with a Tullgren funnel, by dragging a blanket or sheet over vegetation, and with carbon-dioxide traps. Ethanol preserves soft ticks, and Pampel's fluid (2 ml glacial acetic acid, 6 ml 40% formalin, 30 ml distilled water, and 15 ml 95% ethanol) prevents hard tick scutal patterns from fading.

Ticks should be removed carefully from hosts with forceps, making certain to avoid leaving the mouthparts embedded in the skin. A drop of ethanol or oil can be used to detach individuals. Antibiotics should be applied to the point of attachment once the tick has been removed. Pyrethroid dusts are useful in control.

MITES (ACARINA)

Blood-sucking Mites

Varma (1993) provides an identification key to the most important species of *Dermanyssus* and *Ornithonyssus*.

Nidicolous mites in the genera *Dermanyssus* and *Ornithonyssus* and their less common relatives, as well as rhinonyssid nasal-cavity mites, feed on blood. Rhinonyssid nasal-cavity mites that cause rhinitis or sinusitis usually are limited to a few individuals per host (Mullen and O'Connor 2002). *Sternostoma* can clog air sacs, causing wheezing and mortality. *Dermanyssus* and *Ornithonyssus* populations can mass on hosts, causing anemia and weight loss. Tropical fowl mites (*Ornithonyssus bursa*), which usually feed near the vent, have killed nestling Snail Kites (Sykes and Forrester 1983) and a captive adult Eurasian Sparrowhawk (*A. nisus*). *Ornithonyssus* transmits encephalitis viruses, and *Dermanyssus* transmits the white blood cell-infecting protozoan *Lankesterella* (Box 1971). Nasal mites can be collected from live hosts by flushing the nares with water, whereas *Dermanyssus* and *Ornithonyssus* can be obtained by ruffling feathers dusted with insecticide powder, or from nest material by dissection or extraction using a Tullgren funnel. Mites should be preserved in Oudemans' fluid (5 parts glycerine, 8 parts glacial acetic acid, and 87 parts 70% alcohol) to prevent hardening. Treatment and control of external mites involves pyrethroid and other insecticide dusts or sprays. Rhononyssid mites can be controlled with dichlorvos pest strips or pyrethrin-piperonyl butoxide spray (Ritchie et al. 1994).

Skin and Tissue-eating Mites

Skin-mite identification keys are outdated, incomplete, scattered or in some cases, nonexistent. Krantz (1978) provides family keys for mites, overall.

Skin or tissue-eating mites on raptors include *Pneumophagus* in the lungs and air sacs, Ereynetidae in the nasal cavity, Turbinoptidae in the outer nares, Hypoderatidae under thigh and underbody skin, Syringophilidae in quills, and Analgidae, Cheyletiellidae, Epidermoptidae, Harpirhynchidae, Knemidocoptidae, and Trombiculidae (chiggers) on or in the skin. Cheyletiellid mites also feed on blood, and, as with epidermoptid and harpirhynchid mites, can cause edema, hyperkeratosis, and feather loss, with secondary infections in skin lesions. *Knemidocoptes* can cause development of scaly-face and scaly-leg encrustations. Females of *Strelkoviacarus* and *Microlichus* are phoretic on louse flies, whereas *Myialges* females lay their eggs on these flies. Hypoderatid mites reproduce in nests, but their adults are nonfeeding and short-lived. Chiggers, often a cause of dermatitis, are larval mites whose nymphal and adult forms are soil predators. Skin mites can be collected from hosts in skin scrapings, and with detergent washes during necropsies. Hypoderatid mites may be revealed as lumps under the skin. Chiggers can be collected by placing a black disk on the ground below the bird, which will attract them (Mullen and O'Connor 2002). Skin and tissue-entry mites can be preserved in Oudemans' fluid. Ivermectin can be used to treat infestation of nasal, skin and syringophilid quill mites.

Feather-eating Mites

Thirteen families and 22 genera of feather-eating mites parasitize raptors. Gaud and Atyeo (1996) provide keys to genera of the feather mites of the world.

Many mites live on feathers where they scavenge fungi, lipids, bacteria, and feather fragments. A few live in the rachis and quill and eat medulla tissue. Feather and quill mites are most abundant on wing feathers. Feather mites can be collected by ruffling feathers dusted with insecticides. Most quill mites require dissection of shed feathers or quills during necropsy. Oudemans' fluid can be used for preservation. Pyrethrin dusts reduce feather mite populations, while dichlorvos pest strips or ivermectin can be used to treat quill mite infestations (Ritchie et al. 1994).

Feather Microbiology

The microbiology of raptor feathers is poorly known. Hubalek (1974a,b, 1981) surveyed the keratinophilic and other fungi on Common Kestrels and European owls, whereas Rees (1967) found two fungal genera on the feathers of Australian raptors. Pinowski and Pinowska (unpubl. data) have reviewed the feather fungal literature, and concluded that feather fungi are not very important in that they remain mostly dormant and rarely destroy feathers, and do not regulate the numbers of other feather ectoparasites. Bacteria also degrade feathers (Goldstein et al. 2004), but Cristol et al. (2005) found no evidence that they affect feathers on living birds. Although many North American birds have been examined for these bacteria (Burtt and Ichida 1999, Muza et al. 2000), raptors have yet to be studied in this regard.

LITERATURE CITED

ADLER, P.H. AND J.W. McCREADIE. 2002. Black flies (Simuliidae). Pages 163–183 in G.R. Mullen and L. Durden [EDS.], Medical and veterinary entomology. Academic Press, San Diego, CA U.S.A.

———, D.C. CURRIE AND D.M. WOOD. 2004. Black Flies (Simuliidae) of North America. Comstock, Ithaca, NY U.S.A.

ARNETT, R.H., JR. 2000. American Insects, 2nd Ed. CRC Press, Boca Raton, FL U.S.A.

BAUMGARTNER, D.L. 1988. Review of myiasis (Insecta: Diptera: Calliphoridae, Sarcophagidae) of Nearctic wildlife. *Wildl. Rehab.* 8:3–46.

BENTON, A.H. AND SHATRAU, V. 1965. The bird fleas of eastern North America. *Wilson Bull.* 77:76–81.

BEYNON, P.H., N.A. FORBES AND N.H. HARCOURT-BROWN. 1996. Manual of raptors, pigeons and waterfowl. British Small Animal Veterinary Association, Iowa State University Press, Ames, IA U.S.A.

BOORMAN, J. 1993. Biting midges (Ceratopogonidae). Pages 288–309 in R.P. Lane and R.W. Crosskey [EDS.], Medical insects and arachnids. Chapman and Hall, London, United Kingdom.

BOSHOFF, A.F. AND M.H. CURRIE. 1981. Notes on the Cape Vulture colony at Potberg, Bredasdorp. *Ostrich* 52:1–8.

BOX, E.D. 1971. *Lankesterella (Atoxoplasma)*. Pages 309–312 in J.W. Davis, R.C. Anderson, L. Karstad, and D.O. Trainer [EDS.], Infectious and parasitic diseases of wild birds. Iowa State University Press, Ames, IA U.S.A.

BROWN, L. AND D. AMADON. 1968. Eagles, hawks and falcons of the world. Vols. 1, 2. Country Life Books, Feltham, United Kingdom.

BURTT, E.H., JR. AND J.M. ICHIDA. 1999. Occurrence of feather-degrading bacilli in the plumage of birds. *Auk* 116:364–372.

CANNINGS, R.J. 1986. Infestations of *Carnus hemapterus* Nitzsch (Diptera: Carnidae) in Northern Saw-whet Owl nests. *Murrelet* 67:83–84.

CATTS, E.P. AND G.R. MULLEN. 2002. Myiasis (Muscoidea, Oestroidea). Pages 317–348 in G.R. Mullen and L. Durden [EDS.], Medical and veterinary entomology. Academic Press, San Diego, CA U.S.A.

CLAYTON D.H. AND D.M. DROWN. 2001. Critical evaluation of five methods for quantifying chewing lice (Insecta: Phthiraptera). *J. Parasitol.* 87:1291–1300.

——— AND B.A. WALTHER. 1997. Collection and quantification of arthropod parasites of birds. Pages 419–440 in D.H. Clayton and J.E. Moore [EDS.], Host-parasite evolution: general principles and avian models. Oxford University Press, Oxford, United Kingdom.

CRISTOL, D.A., J.L. ARMSTRONG, J.M. WHITAKER AND M.H. FORSYTH. 2005. Feather-degrading bacteria do not affect feathers on captive birds. *Auk* 122:222–230.

CROFT, R.E. AND N. KINGSTON. 1975. *Babesia moshkovskii* (Schurenkova 1938) Laird and Lari, 1957 from the Prairie Falcon, *Falco mexicanus*, in Wyoming; with comments on other parasites found in this host. *J. Wildl. Dis.* 11:230–233.

CROSSKEY, R.W. 1993. Blackflies (Simuliidae). Pages 241–287 in R.P. Lane and R.W. Crosskey [EDS.], Medical insects and arachnids. Chapman and Hall, London, United Kingdom.

——— AND T.M. HOWARD. 1997. A new taxonomic and geographical inventory of world blackflies (Diptera: Simuliidae). Natural History Museum, London, United Kingdom.

DARSIE, M.W. AND R.A. WARD. 2005. Identification and geographical distribution of the mosquitoes of North America, north of Mexico, 2nd Ed. University Press of Florida, Gainesville, FL U.S.A.

DAWSON, R.D. AND G. BORTOLOTTI. 1997. Ecology of parasitism of nestling American Kestrels by *Carnus hemapterus* (Diptera: Carnidae). *Can. J. Zool.* 75:2021–2026.

DELANNOY C.A. AND A. CRUZ. 1991. *Philornis* parasitism and nestling survival of the Puerto Rican Sharp-shinned Hawk (*Accipiter striatus venator*). Pages 93–103 in J.E. Loye and M. Zuk [EDS.], Bird-parasite interactions. Ecology, evolution and behavior. Oxford University Press, Oxford, United Kingdom.

DURDEN, L. 2002. Lice (Phthiraptera). Pages 45–65 in G.R. Mullen and L. Durden [EDS.], Medical and veterinary entomology. Academic Press, San Diego, CA U.S.A.

——— AND R. TRAUB. 2002. Fleas (Siphonaptera). Pages 103–125 in G.R. Mullen and L. Durden [EDS.], Medical and veterinary entomology. Academic Press, San Diego, CA U.S.A.

FLEAY, D. 1968. Night watchmen of bush and plain. Taplinger, New York, NY U.S.A.

FOSTER, W.A. AND E.D. WALKER. 2002. Mosquitoes (Culicidae). Pages 203–262 in G.R. Mullen and L. Durden [EDS.], Medical and veterinary entomology. Academic Press, San Diego, CA U.S.A.

FURMAN, D.P. AND E.P. CATTS. 1982. Manual of medical entomology. Cambridge University Press, Cambridge, United Kingdom.

GAFFIGAN, T.V. AND R.A. WARD. 1985. Index to the second supplement to "A Catalog of the Mosquitoes of the World", with corrections and additions. *Mosquito Syst.* 17:52–63.

GARGETT, V. 1977. A 13-year study of the Black Eagles in the Mato-pos, Rhodesia, 1964–1976. *Ostrich* 48:17–27.

GAUD, J. AND W.T. ATYEO. 1996. Feather mites of the world (Acarina: Astigmata): the supraspecific taxa. *Ann. Mus. Roy. Afr. Centr., Sci. Zoo.* 277, Parts I and II.

GOLDSTEIN, G., K.R. FLORY, B.A. BROWNE, S. MAJID, J.M. ICHIDA AND E.H. BURTT, JR. 2004. Bacterial degradation of black and white feathers. *Auk* 12:656–659.

GRIMALDI, D. 1997. The bird flies, genus Carnus: species revision, generic relationships, and a fossil *Meoneura* in amber (Diptera: Carnidae). *Amer. Mus. Nov.* No. 3190.

GRUBB, T.G., W.L. EAKLE AND B.N. TUGGLE. 1986. *Haematosiphon inodorus* (Hemiptera: Cimicidae) in a nest of a Bald Eagle (*Haliaeetus leucocephalus*) in Arizona. *J. Wildl. Dis.* 22:125–127.

HALL, M.J.R. AND K.G.V. SMITH. 1993. Diptera causing myiasis in man. Pages 429–469 in R.P. Lane and R.W. Crosskey [EDS.], Medical insects and arachnids. Chapman and Hall, London, United Kingdom.

HAMERSTROM, E. AND F. HAMERSTROM. 1954. Myiasis of the ears of hawks. *Falconry News and Notes* 1:4–8.

HEDDERGOTT, M. 2003. Parasitierung nestjunger Turmfalken *Falco t. tinnunculus* durch die Gefiederfliege *Carnus hemapterus* (Insecta: Milichiidae, Diptera). *Vogelwelt* 124:201–205.

HOLLAND, G.P. 1985. The fleas of Canada, Alaska and Greenland. *Mem. Entomol. Soc. Can.* 30.

HUBALEK, Z. 1974a. Dispersal of fungi of the family Chaetomiaceae by free-living birds. I. a survey of records. *Ces. mykologie* 28:65–79.

———. 1974b. The distribution patterns of fungi in free-living birds. *Acta Sci. Nat. Brno* 8:1–51.

———. 1981. Keratinophilic fungi from the feathers of free-living birds. *Folia Parasitol.* 28:179–186.

KNIGHT, K.L. 1978. Supplement to a catalog of the mosquitoes of the world (Diptera: Culicidae). Entomological Society of America, College Park, MD U.S.A.

——— AND A. STONE. 1977. A catalog of the mosquitoes of the world (Diptera: Culicidae). Entomological Society of America, College Park, MD U.S.A.

KRANTZ, G.W. 1978. A manual of acarology, 2nd Ed. Oregon State University Book Stores, Corvallis, OR U.S.A.

LEWIS, R.E. 1993. Fleas (Siphonaptera). Pages 529–575 in R.P. Lane and R.W. Crosskey [EDS.], Medical insects and arachnids. Chapman and Hall, London, United Kingdom.

———, J.H. LEWIS, and C. MASER. 1988. The fleas of the Pacific Northwest. Oregon State University Press, Corvallis, OR U.S.A.

LLOYD, J.E. 2002. Louse flies, keds and related flies (Hippoboscoidea). Pages 349–362 in G.R. Mullen and L. Durden [EDS.], Medical and veterinary entomology. Academic Press, San Diego, CA U.S.A.

MAA, T.C. 1963. Genera and species of Hippoboscidae (Diptera): types, synonymy, habitats and natural groupings. *Pac. Ins. Monogr.* 6.

MATTINGLY, P.F. 1971. Contributions to the mosquito fauna of Southeast Asia XII. Illustrated keys to the genera of mosquitoes (Diptera): Culicidae). *Contrib. Amer. Entomol. Inst.* 7:1–84.

McFADZEN, M.E. AND J.M. MARZLUFF. 1996. Mortality of Prairie Falcons during the fledging-dependence period. *Condor* 98:791–800.

MULLEN, G.R. 2002. Biting midges (Ceratopogonidae). Pages

163–183 *in* G. R. Mullen and L. Durden [EDS.], Medical and veterinary entomology. Academic Press, San Diego, CA U.S.A.

———— AND B.M. O'CONNOR. 2002. Mites (Acari). Pages 449–516 in G.R. Mullen and L. Durden [EDS.], Medical and veterinary entomology. Academic Press, San Diego, CA U.S.A.

MUZA, M.M., E.H. BURTT, Jr. AND J.M. ICHIDA. 2000. Distribution of bacteria on feathers of some eastern North American birds. *Wilson Bull.* 112:432–435.

OLIPHANT, L.W., W.J.P. THOMPSON, T. DONALD AND R. RAFUSE. 1976. Present status of the Prairie Falcon in Saskatchewan. *Can. Field-Nat.* 90:365–367.

PHILIPS, J.R. 1990. What's bugging your birds? Avian parasitic arthropods. *Wildl. Rehab.* 8:155–203.

————. 2000. A review and checklist of the parasitic mites (Acarina) of the Falconiformes and Strigiformes. *J. Raptor Res.* 34:210–231.

————. 2006a. A list of the parasitic mites of the Falconiformes. http://raptormites.babson.edu/falcmitelist.htm (last accessed 8 August 2006).

————. 2006b. A list of the parasitic mites of the Strigiformes. http://raptormites.babson.edu/owlmitelist.htm (last accessed 8 August 2006).

PLATT, S.W. 1975. The Mexican chicken bug as a source of raptor mortality. *Wilson Bull.* 87:557.

POOLE, K.G. AND R.G. BROMLEY. 1988. Natural history of the Gyrfalcon in the central Canadian Arctic. *Arctic* 41:31–38.

PRICE, R W., R.A. HELLENTHAL, R.L. PALMA, K.P. JOHNSON AND D.H. CLAYTON. 2003. The chewing lice: world checklist and biological overview. *Ill. Nat. Hist. Surv. Spec. Publ.* 24.

REES, R.G. 1967. Keratinophilic fungi from Queensland - II. Isolations from feathers of wild birds. *Sabouradia* 6:14–18.

RITCHIE, B.W., G.J. HARRISON AND L.R. HARRISON. 1994. Avian medicine: principles and applications. Wingers, Lake Worth, FL U.S.A.

SABROSKY, C.W., G.F. BENNETT AND T.L. WHITWORTH. 1989. Bird blowflies (Protocalliphora) in North America (Diptera: Calliphoridae) with notes on the Palearctic species. Smithsonian Institution Press, Washington, DC U.S.A.

SCHILLING, F., M. BOTTCHER AND G. WALTER. 1981. Probleme des Zeckenbefalls bei Nestlingen des Wanderfalken (*Falco peregrinus*). *J. Ornithol.* 122:359–367.

SCHOFIELD, C J. AND W.R. DOLLING. 1993. Bedbugs and kissing-bugs (bloodsucking Hemiptera). Pages 483–511 in R.P. Lane and R.W. Crosskey [EDS.], Medical insects and arachnids. Chapman and Hall, London, United Kingdom.

SCHULZ, T.A. 1986. Conservation and rehabilitation of the Common Barn-owl. Pages 146–166 in P. Beaver and D.J. Mackey [EDS.], Wildlife Rehabilitation, Vol. 5. Daniel James Mackey, Coconut Creek, FL U.S.A.

SERVICE, M.W. 1993. Mosquitoes (Culicidae). Pages 120–240 in R.P. Lane and R.W. Crosskey [EDS.], Medical insects and arachnids. Chapman and Hall, London, United Kingdom.

SMITH, B. AND J.R. BELTHOFF. 2001. Identification of ectoparasites on Burrowing Owls in southwestern Idaho. *J. Raptor Res.* 35:159–161.

SMITH, R.N., S.L. CAIN, S.H. ANDERSON, J.R. DUNK AND E.S. WILLIAMS. 1998. Blackfly-induced mortality of nestling Red-tailed Hawks. *Auk* 115:369–375.

SNYDER, N.F.R., J.C. OGDEN, J.D. BITTNER AND G.A. GRAU. 1984. Larval dermestid beetles feeding on nestling Snail Kites, Wood Storks, and Great Blue Herons. *Condor* 86:170–174.

SONENSHINE, D.E., R.S. LANE AND W.I. NICHOLSON. 2002. Ticks (Ixodida). Pages 517–558 in G.R. Mullen and L. Durden [EDS.], Medical and veterinary entomology. Academic Press, San Diego, CA U.S.A.

SYKES, P.W., JR. AND D.J. FORRESTER. 1983. Parasites of the Snail Kite in Florida and summary of those reported for the species. *Fl. Field Nat.* 11:111–116.

TIRRELL, P.B. 1978. *Protocalliphora avium* (Diptera) myiasis in Great-horned Owls, Red-tailed Hawks, and Swainson's Hawks in North Dakota. *Raptor Res.* 12:21–27.

TRIMBLE, S.A. 1975. Habitat management series for unique or endangered species. Report 15. Merlin *Falco columbarius*. USDI Bureau of Land Management Tech Note 271.

USINGER, R.L. 1966. Monograph of Cimicidae. Thomas Say Foundation, Baltimore, MD U.S.A.

VARMA, M.R.G. 1993. Ticks and mites (Acari). Pages 597–658 in R.P. Lane and R.W. Crosskey [EDS.], Medical insects and arachnids. Chapman and Hall, London, United Kingdom.

WARD, R.A. 1984. Second supplement to "A Catalog of the Mosquitoes of the World: (Diptera: Culicidae)". *Mosquito Syst.* 16:227–270.

————. 1992. Third supplement to "A Catalog of the Mosquitoes of the World: (Diptera: Culicidae)". *Mosquito Syst.* 24:177–230.

WEBSTER, H., JR. 1944. A survey of the Prairie Falcon in Colorado. *Auk* 61:609–616.

WHITE, C.M. 1963. Botulism and myiasis as mortality factors in falcons. *Condor* 65:442–443.

WHITEMAN, N.K. AND P.G. PARKER. 2005. Using parasites to infer host population history: a new rationale for parasite conservation. *Anim. Conserv.* 8:175–181.

WHITWORTH, T.L. 2003. A key to the puparia of 27 species of North American *Protocalliphora* Hough (Diptera: Calliphoridae) from bird nests and two new puparial descriptions. *Proc. Entomol. Soc. Wash.* 105:995–1033.

————. 2006. Keys to the genera and species of blow flies (Diptera: Calliphoridae) of America north of Mexico. *Proc. Entomol. Soc. Wash.* 108:689–725.

C. Endoparasites

OLIVER KRONE

Leibniz-Institute for Zoo and Wildlife Research (IZW)
Alfred-Kowalke-Str. 17, D-10315 Berlin, Germany

INTRODUCTION

Endoparasites are organisms that, in their developmental or adult stages, live in animals called hosts. Endoparasites, which include single-celled protozoa, worms (helminths), and arthropods, invade nearly all organs of animals. Protozoans are found in digestive and respiratory systems, muscles, blood, and feces of their hosts. Several endoparasitic worms feed on the ingesta in the intestine of the definitive (final) host or are attached to the mucosal layer within the intestine or the trachea where they suck blood or epithelial cells. Other worms are found in specific organs or only parts of organs. Some worms migrate through different internal organs during their stages of development. Parasitic arthropods, including ticks, mites, flies, mallophages, and fleas, often are found on the skin or feathers of their hosts. Only a few arthropods enter the internal organs of the hosts. Endoparasitic arthropods include mites living in layers of the skin or subcutaneously, and the larval stages of flies (maggots) that burrow through internal organs.

It is not the aim of this chapter to describe all endoparasites and their ways of life but rather to provide information on several relevant examples.

The traditional doctrine in parasitology states that a good parasite does not harm its host in a way that it weakens or kills the host, because this also would affect the parasite itself. And indeed, long-term, well-adapted parasites often are less pathogenic to their traditional hosts, whereas evolutionarily young parasites can harm their hosts severely. That said, the host–parasite relationship is a dynamic evolutionary system that may be compared to an arms race, in which both sides alter their behavior in response to the other in a way that sustains the interaction (Van Valen 1973). Dobson et al. (1992) described parasitic worms as natural enemies causing a permanent drain of energy in their hosts that affects the behavior and reproductive success of them. Depending on age, immune status, and infection pressure, parasites can invade their hosts to different degrees, and probably are a strong selective force on their hosts.

Parasitism as a way of life developed independently in different taxa. Endoparasites are believed to have evolved several million years ago. The oldest parasitic roundworms (nematodes) were found in beetles embedded in amber from the Eocene (Conway Morris 1981). Compared with their hosts, parasites are relatively simple organisms, many of which have "degenerated" during evolution, although parasites have the advantage that processes such as digestion or locomotion are provided by the host. Tapeworms living in a nutritionally rich environment — the intestine of their hosts — have reduced their digestive system and are able to reabsorb their food through their cuticula. Parasites also have developed new abilities in response to their parasitic lifestyles (e.g., host-finding mechanisms, resistance against the host's immune and digestive system, and new organs such as sensoric receptors). As a result of such adaptations, the genome of parasites can be bigger than those of the free-living parasite relatives and, in some cases, bigger than those of the hosts they occupy (Poulin 1998).

As an adaptation to their way of life endoparasites often develop complex life cycles, including stages of sexual and asexual reproduction. Sexual multiplication normally occurs in the definitive host they forage in. The developing stages of endoparasites leave the host actively or passively to move to the next suitable environment. The life cycle of a parasite can be direct or indirect (i.e., via intermediate hosts). Complex life cycles often contain more than one intermediate host required by the parasite to reach the definitive host. Sexual reproduction generally occurs in the definitive host and asexual reproduction occurs in intermediate hosts. On the way to the next host many developmental stages may be lost. To increase the likelihood of host infection, parasite fecundity often has increased during its evolution. As a result, some roundworms can produce about 200,000, and some tapeworms up to 720,000 eggs per day (Crompton and Joyner 1980).

Parasites have developed a diversity of strategies to reach their definitive host. They often try to produce as many eggs as possible of which only a few develop to mature parasites. Sometimes intermediate hosts are manipulated by their parasites to become an easier victim to a predator. Several species of flukes that parasitize piscivorous birds migrate into the eyes of the last intermediate host (fish) reducing their ability to see (Odening 1969). Protozoa of the genus *Sarcocystis* use raptors as definitive hosts and mice or birds as intermediate hosts where they form cysts in the muscles. Infected mice changed their behavior and are twice as likely to be eaten by Common Kestrels (*Falco tinnunculus*) as are non-infected mice (Hoogenboom and Dijkstra 1987). Behavioral changes due to parasitic infections in definitive hosts (birds of prey) also have been described. Experimentally infected American Kestrels (*F. sparverius*) exhibited decreased flight activity during their reproduction and a prolonged courtship behavior compared to control birds (Saumier et al. 1991).

Whether an infection with parasites induces clinical signs depends on the status of the immune system, hormone status, and infection pressure. An infection with a few ascarids, for example, may cause some irritation in the intestinal mucosa but may not necessarily affect the condition of the host. A heavy infection can block the lumen partially or completely, resulting in a perforation or rupture. The metabolic products of worms also can

harm the health of the host. Continued parasitic infection can weaken the host's immune system, thereby enabling additional parasites to enter the host. In such instances the parasitic infection is considered a factorial disease.

ENDOPARASITES OF BIRDS OF PREY

Endoparasites are frequently detected in raptors. Indeed, in some populations of birds of prey, 90% of all individuals have helminths (e.g., Krone 2000, Sanmartin et al. 2004).

Endoparasites found in birds of prey include protozoans, roundworms (nematodes), spiny-headed worms (acanthocephala), flukes (digenetic trematodes), tapeworms (cestodes), and tongue worms (pentastomida).

I provide a broad introduction here. A more extensive review of endoparasites found in birds of prey is in Lacina and Bird (2000). A more general overview of helminths in birds is in Rausch (1983). For information on the biology and treatment of parasites in raptors see Krone and Cooper (2002).

Protozoa

Unicellular parasites usually are very small and only visible by a microscope (Fig. 1). Among the eight classes of protozoa, two are of major interest in raptor parasitology.

Trichomonas. One of the oldest recognized and most significant diseases in raptors, "frounce" or "crop canker," is caused by *Trichomonas gallinae*, a single-celled organism that belongs to the class Zoomastigophorea. These spindle- to pear-shaped small flagellated protozoans reproduce by simple division. Direct transmission via the feeding of nestlings with crop-milk occurs in Columbiformes. The parasite lives on and in the mucosal layers of the oropharynx, oesophagus, and crop. Infection in raptors occurs via the ingestion of contaminated prey. Pigeons and doves are the main reservoirs of *Trichomonas*, but other birds, including Passeriformes, also can be infected. Deep, yellow, crumbly, caseous lesions often are found in advanced infections in the upper digestive tract. These abscesses can grow up to the size of ping-pong balls and, depending on where they are, can mechanically block the passage of food or respiration. Raptors feeding on birds are more likely to be infected. Nestlings of urban goshawks often carry the agent at high prevalence in their pharynx

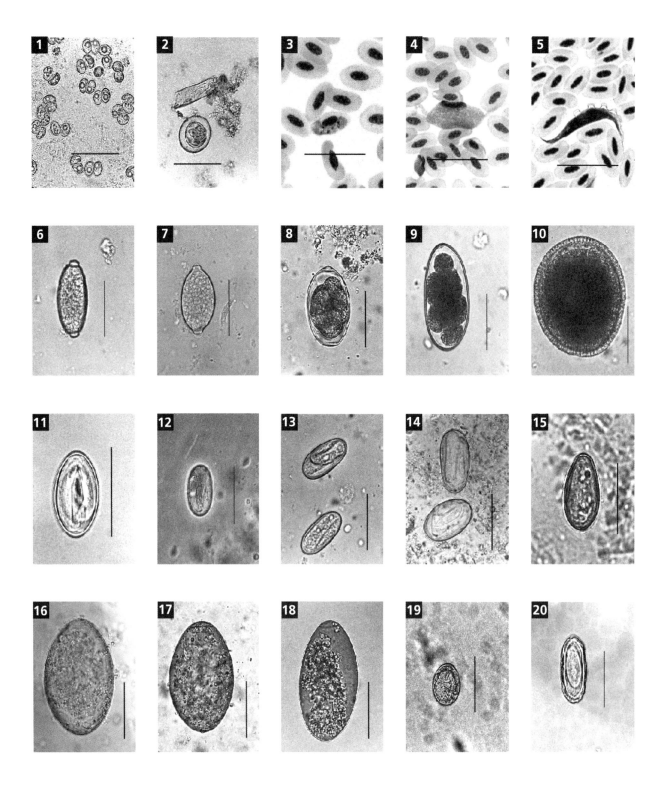

Figure 1. All protozoan parasites in line one are photographed at 1000x magnification (scale 50μm). The helminth eggs in lines two to four are photographed at 400x and 630x magnification (scale 50μm). (1) *Sarcocystis* sp., (2) *Caryospora* sp., (3) *Haemoproteus* sp., (4) *Leucocytozoon* sp., (5) *Trypanosoma avium*, (6) *Capillaria tenuissima*, (7) *Eucoleus dispar*, (8) *Syngamus trachea*, (9) *Hovorkonema variegatum*, (10) *Porrocaecum* sp., (11) *Microtetrameres cloacitectus*, (12) *Synhimantus laticeps*, (13) *Physaloptera alata*, (14) *Serratospiculum tendo*, (15) *Metorchis* sp., (16) *Nematostrigea serpens*, (17) *Strigea falconispalumbi*, (18) *Neodiplostomum attenuatum*, (19) *Cladotaenia globifera*, (20) *Centrorhynchus* sp.

without showing clinical symptoms. Sick birds may develop stomatitis (i.e., inflammation of the mucous membrane of the mouth) and have difficulty in swallowing food items. Birds often dehydrate and starve. Secondary bacterial infections can complicate and speed up the disease process.

Trypanosoma. Flagellated blood parasites of the genus *Trypanosoma* also belong to the class Zoomastigophorea. Their life cycle is indirect with the parasite being transmitted by the bite of hippoboscid flies. The pathogenicity of this genus in birds is unknown. Most diagnoses are made unintentionally by examining blood smears. Although the taxonomy is unclear, Bennett (1970) concluded that *T. avium* is the only valid species occurring in birds. Molecular parasitology should help to resolve this issue.

Sarcocystis and Ferenkelia. Coccidia of the genera *Sarcocystis* and *Frenkelia* belong to the class Sporozoea (subclass: Coccidia). These protozoans live in the mucosal layers of the intestine, where they reproduce sexually. The sporocysts excreted by the feces of the definitive host must be ingested by an intermediate host (mouse, bird). Within the intermediate host, the parasite reproduces asexually several times before cysts are built in the muscle (*Sarcocystis*) or brain (*Frenkelia*). The life cycle of the parasite is completed when a cyst in the mouse or bird is ingested by the raptor. Infections with *Sarcocystis* and *Frenkelia* spp. are seldom pathogenic. Nestlings may develop clinical symptoms such as diarrhea, feces with blood, and emaciation. Odening (1998) listed seven *Sarcocystis* spp. for the Falconiformes and four for the Strigiformes. He also declared the genus *Frenkelia* to be a synonym of *Sarcocystis* not only because of their same morphology, but also because of their developmental features.

Caryospora. The coccidia of the genus *Caryospora* (class: Sporozoea) live in the intestines of raptors. Excreted parasites are set free by the feces but need several days before reaching infectivity. The life cycle is direct, but also can involve an intermediate host. In breeding centers for birds of prey, *Caryospora* infections frequently cause problems, especially in young birds. To date, more than 14 species of *Caryospora* have been described from birds of prey (Böer 1982, Klüh 1994, Upton et al. 1990).

Leucocytozoon, Haemoproteus, Plasmodium. All three of these genera of blood parasites belong to the class Sporozoea (subclass: Coccidia). Blood-feeding insects (mosquitoes, hippoboscid flies, simulids), in which the sexual reproduction of these parasites occurs, are vectors. In the avian host, the parasites reproduce asexually in specific tissues. Only in the last stage do the parasites appear in the blood while waiting for a blood-feeding insect to infect. *Plasmodium* is more pathogenic than *Leucocytozoon* and *Haemoproteus*. *Plasmodium*, in particular, causes problems in translocated birds from areas where birds are not immunologically adapted to these parasites (e.g., Arctic, Antarctic, Himalayas). Six species of *Haemoproteus*, one species of *Leucocytozoon* and eight species of *Plasmodium* occur in falconiforms. Four species of *Haemoproteus*, nine species of *Plasmodium*, and one *Leucocytozoon* are known to occur in Strigiformes (Bennett et al. 1993, 1994; Telford et al. 1997, Valkiunas 1997).

Rare Protozoan Parasites in Raptors

Other blood parasites seldom reported are *Hepatozoon* spp. and *Haemogregarina* spp. (subclass: Coccidia), Babesia spp. (subclass: Piroplasmia), and Rickettsia-like organisms. *Toxoplasma gondii* uses a broad range of vertebrates as intermediate hosts, including, apparently, raptors (Lindsay et al. 1993).

Cawthorn (1993) reported two species of *Eimeria* (subclass: Coccidia) in Falconiformes and four species in Strigiformes, not including two new species described by Upton et al. (1990) in the latter group. A rarely reported protozoan infection of unknown origin is found in the kidneys of owls without causing inflammatory alterations. Burtscher (1966) diagnosed renal coccidiosis in three species of owls in Germany.

Helminths

Parasitic helminths are worms in the phyla Platyhelminthes and Nemathelminthes. Parasitic worms in the phylum Pentastomida are rarely found in raptors. Platyhelminthes are represented in raptors by the classes Trematoda and Cestoda. Among the Trematoda the subclass Digenea, and among the Cestoda the subclass Eucestoda, are of major interest in raptor parasitology. Nematodes belong to the class Nemathelminthes, which includes Acanthocephala.

These metazoan parasites usually are visible with the naked eye. The nematodes have a fully developed digestive system and the trematodes have an incompletely developed digestive tract. Cestodes and acanthocephalans digest material via their tegument.

Most *nematodes* (roundworms) are long, thread-formed worms that are pointed at both ends. Sexes are separate and the females are generally larger than the males. Oviparous as well as viviparous species exist. The life cycle can be very simple (i.e., direct) or complex with intermediate and paratenic (accumulative) hosts, or both (Anderson 2000, Lee 2002).

The *acanthocephalans* (spiny-headed worms) are divided into a body (sometimes with a spinous surface) and a proboscis at the anterior end. The proboscis, which is armed with hooks, serves as an attachment organ. Sexes are separate. Eggs contain a spiny-armed larva. The developmental cycle of acanthocephalans that inhabit birds of prey is often indirect, including intermediate hosts (e.g., locusts). "Paratenic hosts," including amphibians, reptiles, and mammals, feed on locusts and accumulate the larva before the parasite reaches its definitive host.

Digenetic *trematodes* (flukes) usually are oval and dorso-ventrally flattened, with two suckers (i.e., an oral sucker surrounding the mouth, and a ventral sucker). Digenetic trematodes are mainly hermaphroditic. Exceptions include the schistosomes. Some species are capable of self-fertilization. Eggs are relatively large and always have an operculum, or cap. The life cycle of the digenetic trematodes is by far the most complex among Platyhelminthes, and also is among the most complex animals (Cheng 1986).

The *cestodes* (tapeworms) are divided into three regions: the head (scolex), the neck (proliferation zone), and the strobila (chain of proglottids). The scolex, which serves as an attachment organ, generally bears hooks and suckers. The strobila, the largest part of the cestode, is made of proglottids, the single segments which generally contain a complete hermaphrodite set of reproductive organs maturing towards the posterior end of the worm. The last proglottids are gravid (i.e., filled with eggs). The eggs contain larva (oncosphere) with three pairs of hooks. Most cestodes require an intermediate host for their development.

The *pentastomids* (tongue worms) are elongated and often segmented. Four or six rudimentary legs are present on the larvae. Adult pentastomids have two pairs of sclerotized hooks in the mouth region. Females are larger than males. The eggs contain a fully developed larva. Although the life cycle usually includes an intermediate host, the one case of a pentastomid diagnosed in a White-backed Vulture (*Gyps bengalensis*) appeared to be direct (Riley et al. 2003).

SAMPLING TECHNIQUES

Sampling techniques differ between living and dead birds. In living birds, blood, saliva, mucosal scrapings, and feces should be examined fresh and, therefore, are of better quality than those from carcasses. Interpretation of the results can be difficult as several parasites occur only in the peripheral blood or feces at some stages of development or follow a specific seasonal or daily cycle (i.e., a negative blood smear or fecal sample does not mean that the bird is not infected by the parasite). Doaster and Goater (1997) provide a good overview regarding collection and quantification techniques for avian helminths and protozoans.

Protozoa

A wet-cotton swab is used to collect saliva or mucosa from the bird's oropharynx to examine for *Trichomonas gallinae*. This swab expressed into warm water should reveal highly motile flagellate parasites when positive. Flagellates should be stained with Giemsa-solution. A more sensitive technique is to grow the parasite in a culture medium. Allowing the parasite to multiply for 3 days at 38°C and then scanning a drop of medium under a microscope is recommended. It is not possible to collect *Trichomonas* sp. from dead birds because the flagellate, which is temperature-sensitive, dies within few a minutes after the host dies.

Trypanosoma spp. often are randomly detected in classical blood smears (see blood parasites). A more reliable technique is to cultivate blood or bone marrow on blood agar. Kucera (1979) described a simple method for field diagnosis of avian trypanosomes using small penicillin bottles.

Coccidia such as *Sarcocystis* spp. or *Caryospora* spp. are diagnosed in the feces or intestinal mucosa of their definitive host. A direct smear from a fecal sample often is sufficient to find oocysts of coccidia. The standard method is flotation using a solution with a high specific gravity (i.e., saturated sugar- or NaCl-solution). A McMaster chamber can be used (see Appendix 1) to quantify the number of oocysts or helminth eggs.

Fresh fecal samples yield the highest-quality parasite stages. Fresh samples can be obtained by covering with foil the ground where the bird normally defecates. This is easily done in captive birds and also can be done in free-ranging birds with a known roosting site. During collection one should avoid the urinal part of the feces which makes the direct smear difficult to read. Uric-acid crystals are opaque and parasite stages may be hidden. If the sample is sent by mail, an unbreakable sealed container should be used. If the transportation requires more than three days a small amount of isotonic buffered 4% formalin solution (i.e., approximately half of the volume of the sample itself) should be added to reduce bacterial growth. Often a direct smear is sufficient to diagnose parasite developmental stages, but sometimes a concentration of eggs or oocysts is needed. Simple flotation can be used to concentrate samples. It is not necessary to pass raptor fecal samples through a wiremesh filter, as they do not contain large amounts of plant matter. It is important to dissolve a small part of the sample in a saturated sugar or NaCl solution and mix it thoroughly until all large particles are broken down. It should then stand for 30 min, after which the surface film can be removed with a cover glass or a pipette for examination. The suspension also can be centrifuged to concentrate the non-floating material at the bottom, but there are many different flotation methods described in the parasite text books. For the direct-smear method a small drop of isotonic solution (RLS) helps to dilute the material so that eggs or oocysts may become easily visible. A 100 to 400x microscope is adequate to find and identify parasite stages.

Coccidian oocysts are diagnosed by their size and appearance: oocysts contain two sporocysts with four sporozoites each (Sarcocystis-type), or four sporocysts with two sporozoites each (Eimeria-type); in the Sarcocystis spp. the oocyste-membrane is very thin, giving it an appearance of a "double-egg." The genera Sarcocystis and Frenkelia (Fig. 1.1) cannot be differentiated using the oocysts which are excreted sporulated (i.e., sporocycsts and sporozoites are visible). *Caryospora* oocysts (Fig. 1.2) are much larger and are not sporulated at the time of excretion, resembling a fried egg. Their sporulated oocysts contain one sporocyst with eight sporozoites. Blood protozoa are found intracellular in erythrocytes or leucocytes, or both (Fig. 1.3-4) or in the plasma (Fig. 1.5). Identification to species in helminths is possible in only a few cases including the species *Capillaria tenuissima* (Fig. 1.6) and *Eucoleus dispar* (Fig. 1.7), both of which can be differentiated by their egg surfaces: striated in the former and dotted in the latter. Eggs in this helminth family have typical plug-like prominences at each pole. Other eggs can be determined only to the family or genus. Eggs of the family Syngamidae with *Syngamus trachea* (Fig 1.8) and *Hovorkonema variegatum* (Fig. 1.9) contain a number of blastomeres. Ascarid eggs including *Porrocaecum* sp.

(Fig 1.10) have a golf-ball appearance with a dented surface that often attracts debris. Spirurid eggs (Fig. 1.11-14) are asymmetric and often contain a folded embryo. Trematode eggs (Fig. 1.15-18) are characterized by an operculum at the upper tip of the egg through which the larvae (miracidium) hatches. Most cestode eggs (Fig. 1.19) of raptors already contain a larvae with three pairs of hooks. Acanthocephalan eggs (Fig. 1.20) are embryonated with three shells, sometimes with visible hooks. A McMaster chamber (see Appendix 1) can be used to count eggs or protozoan oocysts.

Blood parasites typically are examined by the use of a blood smear. To perform a blood smear, a small amount of blood is taken from the bird, preferably with a syringe and a needle. Insulin needles with a small diameter cause a minimal lesion in the skin and vessel of the bird. The blood must be pulled and pushed slowly from the syringe so that the cells are not ruptured. A small drop of blood is placed on one end of a slide. A second slide can be used to smear the blood across the first slide. The second slide is arranged with its small edge at an angle of 45° to the first horizontal slide to allow the blood to spread along the edge. With the blood attached to the edge, the slide is pushed across the horizontal slide to create a thin blood film. It is important that the thin film tapers off on the slide with a distinct margin. The monolayer near the margin can be used to identify single blood cells and blood parasites. Air-dried blood smears should be fixed in pure methanol for one minute soon after preparation.

Helminths

Endoparasitic worms can be obtained from living birds using antihelmintics that kill or paralyze the worms, which will then be excreted in the feces within 24 hours (cf. Cooper 2002, Heidenreich 1997).

The most reliable method for collecting helminths is to dissect a bird after it has died. Carcasses from rehabilitation centers or wildlife clinics often are available for this purpose. One should have the appropriate background information on the bird (i.e., species, age, sex, circumstances of finding, location, kept in captivity, medically treated, date, name and address of finder, etc.) before starting a necropsy. This information will help to evaluate the biological data obtained from the bird. Standard dissection protocols for birds and raptors are provided by Latimer and Rakich (1994) and Cooper (2002). All internal organs should be examined completely. Most helminths are found in the digestive system. Thus, the oropharynx, oesophagus, proventriculus, gizzard, small intestine, large intestine, bile ducts, pancreatic ducts, and cloaca should be opened longitudinally and examined. Arranging the tract in a spiral in large Petri dishes and examining it carefully at low magnification (6–60x) under a stereo-microscope helps one avoid overlooking even the smallest worms. The trachea, air sacs, and body cavity also should be scanned under a stereo-microscope. Other internal organs such as the lungs, liver with gall bladder, and kidneys should be dissected under the stereo-microscope to look for migrating larva or for parasitic cysts. Impression smears should be taken from the spleen, liver, kidneys, and lungs and stained with Giemsa solution to check for protozoan parasites. Scrapings from the mucosal layer of the intestine should be examined for the presence of oocycsts. Helminths should be handled with care so as not to destroy features important for identification. The worms can be removed gently from the attached side and washed in tap water or normal saline solution. Helminths from fresh birds should be killed in a standardized way. Nematodes and acanthocephala can be heated carefully in a glycerin (5%):ethanol (70%) solution to prevent contraction. Trematodes and cestodes can be relaxed in a refrigerator prior to fixation. Trematodes should be fixed in a Bouin-solution (see Appendix 1) for 24 hours and cestodes can be killed and fixed in a 10% formalin-solution (neutral buffered) and then are stored in a glycerine (5%):ethanol (70%) solution. Use of these solutions helps to identify the parasite using morphological features. They are not appropriate for a genetic analysis. For this purpose, specimens should be conserved in pure ethanol or frozen until DNA analysis is possible.

IDENTIFICATION TECHNIQUES

To identify protozoan parasites it may be necessary to stain them or to allow further development (e.g., sporulation). Methanol fixed-blood smears are stained in Giemsa-solution for 20 to 30 minutes (see Appendix 1). After staining, the remains of the staining solution are washed away with tap water and flushed in aqua dest. After air-drying, the blood smears are examined microscopically at 25, 100, 400, and 1000x magnification. Some coccidian parasites are excreted unsporulated (e.g., *Caryospora*, *Eimeria*). To enable sporulation small fecal sample or a mucosal scrape is given on a slide together with a drop of water and covered with a

cover slip. This preparation is then placed in a Petri dish together with a moistened piece of pulp and kept for 24 to 48 (72) hours at room temperature. The identification is then performed using the literature listed above.

The classical identification of helminths is based on morphological features such as body size, buccal capsule, spicules, ornaments, suckers, testes, cirrus, hooks, probosces, etc. Internal structures needed for identification become visible by passing the nematodes and acanthocephala through a lactophenol-mixture (20 g crystalline phenol, 20 ml lactic acid, 10 ml glycerin, 20 ml aqua dest) or a lactoglycerol-mixture (equal parts of lactic acid, glycerol and distilled water) for some minutes. Trematodes and cestodes need to be stained for identification. After fixing, trematodes should be squeezed between two slides and stained with an alum-carmine-solution (see Appendix 1). To stain the trematodes, picric acid needs to be washed out with 70% ethanol for about 24 hours. The specimens then need to be washed in aqua dest, followed by staining with alum-carmine solution for 10 to 60 minutes, and then washed in aqua dest again. An alternative staining method using Gower's acetic carmine is described by Schell (1970). Dehydrate the specimen sequentially in 60%, 70%, 80%, 96% ethanol for 3 to 10 minutes in each concentration. Then wash the cestodes in pure n- or isopropanol for 3 to 10 minutes. Next clear them in xylene for 10 to 15 minutes and mount them in Canada balsam prior to identification. The cestodes are stained for eight minutes in a hydrochloric acid-carmine solution (See Appendix 1) and then transferred into a 1% hydrochloric-acid ethanol solution. Depending on the quality of the cestodes, color changes occur within 30 minutes. The cestodes are then washed in 60% ethanol and moved through a series of higher concentrated alcohols for dehydration starting with 70% ethanol for 24 hours followed by 96% ethanol for 24 hours. Finally, the specimens are washed in pure propanol for 10 to 15 minutes and cleared in xylol and mounted in Canada balsam. Schmidt (1986) also described a staining method using hematoxylin. All helminths are microscopically examined at 25, 100, 400, and 1000x magnification. Using internal (and external) structures for identification requires some experience.

Useful identification guides for helminths to the family or genus level (very rarely to species level) are uncommon and sometimes appear in languages other than English. The list of references below should be useful. Identifying nematodes can be accomplished using keys provided by Skrjabin (1953, 1957, 1963,

1964, 1965, 1967, 1968), Anderson et al. (1974a,b, 1975a,b, 1976a,b, 1978a,b, 1980a,b, 1982), Hartwich (1975, 1994), and Anderson and Chabaud (1983). Acanthocephala can be identified using Chochlova (1986). Trematodes can be identified using Skrjabin (1950, 1959, 1960, 1971), Dubois (1968, 1970), Gibson et al. (2002), and Jones et al. (2004). Cestodes can be identified using Abuladze (1964), Chertkova and Kosupko (1978), Schmidt (1986), and Khalil et al. (1994). To identify rare endoparasites it often is necessary to read the original or, when available, revised species descriptions. Doing so often requires an extensive literature search.

MOLECULAR PARASITOLOGY

Molecular parasitology is a new and fast evolving discipline. The tools described below represent a small selection of those available. Molecular-biology techniques, including DNA sequencing, can be useful in identifying species as well as in answering questions of systematics (Gasser 2001). To understand the mechanisms of parasite origin, phylogenetic studies are needed and correct identification of specimens is a prerequisite (Blaxter 2001).

Parasitic protozoa may be identified by comparing sequences of the internal transcribed spacer region 1 (ITS-1) of the ribosomal RNA (Marsh et al. 1999) or the 18S small subunit (SSU) of the ribosomal DNA (Jenkins et al. 1999). The PCR protocol of Bensch et al. (2000), as modified by Hellgren et al. (2004) and Waldenström et al. (2004), can be used to amplify sequences of the cytochrome b gene of the avian blood-parasites *Haemoproteus*, *Plasmodium*, and *Leucocytozoon*.

Different molecular markers can be used to study nematodes. Slowly evolving genes such as cytochrome c, globin, RNA II polymerase, and heat shock protein 70, are useful in this regard at higher taxonomic levels (i.e., Order or higher). The ribosomal DNA contains several conservative coding sequences including SSU, 28S or large subunit (LSU) and 5.8S, and highly variable non-coding sequences ITS-1 and ITS-2 (Blaxter 2001). The conservative 5.8S sequence is suitable for phylogenetic studies at the level of Order or higher (Chilton et al. 1997). The ITS sequences are useful for genus or subfamily levels (Chilton et al. 2001, Morales-Hojas et al. 2001).

The cytochrome c oxidase gene I (COI gene) can be used to differentiate some types of trematodes (Wongratanacheewin et al. 2001, Pauly et al. 2003) and cestodes (Bowles and McManus 1994) at genus or species level. The 3' end of the ITS-1 element can be used in elucidating phylogenetic relationships of distinct taxa (Schulenburg et al. 1999), and the full ITS-1 sequence is useful for differentiating trematodes at the species level.

Because they provide more sensitive tools, such as detecting low parasite burdens with specific markers, molecular methods will help achieve deeper insights into parasite diversity by detecting morphologically undistinguishable (i.e., cryptic) species. As a result, some infections of protozoic and metazoic parasites will be more easily diagnosed, because they are often overlooked with classical methods, including blood parasite infections detected with blood smears.

LITERATURE CITED

ABULADZE, K.I. 1964. Essentials of cestodology. Taeniata of animals and humans and diseases caused by them. 4 (Russian). Publisher Nauka, Moscow.

ANDERSON, R.C. 2000. Nematode parasites of vertebrates: their development and transmission. CAB Publishing, CAB International, Wallingford, Oxon, United Kingdom.

———— AND A.G. CHABAUD [EDS.]. 1983. CIH keys to the nematode parasites of vertebrates 10. Commonwealth Agricultural Bureaux, London, United Kingdom.

————, A.G. CHABAUD AND S. WILLMOTT [EDS.]. 1974a. CIH keys to the nematode parasites of vertebrates 1. Commonwealth Agricultural Bureaux, London, United Kingdom.

————, A.G. CHABAUD AND S. WILLMOTT [EDS.]. 1974b. CIH keys to the nematode parasites of vertebrates 2. Commonwealth Agricultural Bureaux, London, United Kingdom.

————, A.G. CHABAUD AND S. WILLMOTT [EDS.]. 1975a. CIH keys to the nematode parasites of vertebrates 3, part 1. Commonwealth Agricultural Bureaux, London, United Kingdom.

————, A.G. CHABAUD AND S. WILLMOTT [EDS.]. 1975b. CIH keys to the nematode parasites of vertebrates 3, part 2. Commonwealth Agricultural Bureaux, London, United Kingdom.

————, A.G. CHABAUD AND S. WILLMOTT [EDS.]. 1976a. CIH keys to the nematode parasites of vertebrates 3, part 3. Commonwealth Agricultural Bureaux, London, United Kingdom.

————, A.G. CHABAUD AND S. WILLMOTT [EDS.]. 1976b. CIH keys to the nematode parasites of vertebrates 4. Commonwealth Agricultural Bureaux, London, United Kingdom.

————, A.G. CHABAUD AND S. WILLMOTT [EDS.]. 1978a. CIH keys to the nematode parasites of vertebrates 5. Commonwealth Agricultural Bureaux, London, United Kingdom.

————, A.G. CHABAUD AND S. WILLMOTT [EDS.]. 1978b. CIH keys to the nematode parasites of vertebrates 6. Commonwealth Agricultural Bureaux, London, United Kingdom.

————, A.G. CHABAUD AND S. WILLMOTT [EDS.]. 1980a. CIH keys to the nematode parasites of vertebrates 7. Commonwealth Agricultural Bureaux, London, United Kingdom.

————, A.G. CHABAUD AND S. WILLMOTT [EDS.]. 1980b. CIH keys to the nematode parasites of vertebrates 8. Commonwealth Agricultural Bureaux, London, United Kingdom.

————, A.G. CHABAUD AND S. WILLMOTT [EDS.]. 1982. CIH keys to the nematode parasites of vertebrates 9. Commonwealth Agricultural Bureaux, London, United Kingdom.

BENNETT, G.F. 1970. *Trypanosoma avium* Danilewsky in the avian host. *Can. J. Zool.* 48:803–807.

————, M.A. BISHOP AND M.A. PEIRCE. 1993. Checklist of the avian species of *Plasmodium* Marchiafava & Celli, 1885 (Apicomplexa) and their distribution by avian family and Wallacean life zones. *Syst. Parasitol.* 26:171–179.

————, M.A. PEIRCE AND R.A. EARLÉ. 1994. An annotated checklist of the valid avian species of *Haemoproteus, Leucocytozoon* (Apicomplexa: Haemosporida) and *Hepatozoon* (Apicomplexa: Haemogregarinidae). *Syst. Parasitol.* 29:61–73.

BENSCH, S., M. STJERNMAN, D. HASSELQUIST, Ú. ÚSTMAN, B. HANSSON, H. WESTERDAH AND R. TORRES PINHEIRO. 2000. Host specificity in avian blood parasites: a study of *Plasmodium* and *Haemoproteus* mitochondrial DNA amplified from birds. *Proc. R. Soc. Lond. B* 267:1583–1589.

BLAXTER, M.L. 2001. Molecular analysis of nematode evolution. Pages 1–24 in M.W. Kennedy and W. Harnett [EDS.], Parasitic nematodes: molecular biology, biochemistry and immunology. CAB International, London, United Kingdom.

BÖER, B. 1982. Untersuchungen über das Vorkommen von Kokzidien bei Greifvögeln und die Entwicklung von zwei *Caryospora*-Arten der Falken (*Caryospora neofalconis* n. sp. und *Caryospora kutzeri* n. sp.). Veterinary Medicine dissertation, Hanover, Germany.

BOWLES, J. AND D.P. MCMANUS. 1994. Genetic characterisation of the Asian *Taenia*, a newly described taeniid cestodes of humans. *Am. J. Trop. Med. Hyg.* 50:33–44.

BURTSCHER, H. 1966. Nieren-Kokzidiose bei Eulen. *Wien. Tierärztl. Monat.* 53:654–666.

CAWTHORN, R.J. 1993. Cyst-forming coccidia of raptors: significant pathogens or not? Pages 14–20 in P.T. Redig, J.E. Cooper, J.D. Remple, and B. Hunter [EDS.], Raptor biomedicine. University of Minnesota Press, Minneapolis, MN U.S.A.

CHENG, T.C. 1986. General parasitology, 2nd. Ed. Academic Press, Inc., London, United Kingdom.

CHERTKOVA, A.N. AND G.A. KOSUPKO. 1978. The Suborder Mesocestoidata SKRJABIN, 1940. Pages 118–229 in K.M. RYZHIKOV [ED.], Essentials of cestodology. Tetrabothriata and Mesocestoidata / Cestodes of birds and mammals (in Russian). Publisher Nauka, Moscow, Russia.

CHILTON, N.B., H. HOSTE, G.C. HUNG, I. BEVERIDGE AND R.B. GASSER. 1997. The 5.8S rDNA sequences of 18 species of bursa nematodes (Order Strongylida): comparison with Rhabditid and Tylenchid nematodes. *Int. J. Parasitol.* 27:119–124.

————, L.A. NEWTON, I. BEVERIDGE AND R.B. GASSER. 2001. Evolutionary relationships of trichostrongylid nematodes (Strongylida) inferred from ribosomal DNA sequence data. *Mol. Phylogenet. Evol.* 19:367–386.

CHOCHLOVA, I.K. 1986. The acanthocephala of the terrestric vertebrates of the USSR (Russian). Publisher Nauka, Moscow, Russia.

CONWAY MORRIS, S. 1981. Parasites and the fossil record. *Parasitol.* 82:489–509.

COOPER, J.E. [ED]. 2002. Birds of prey: health and diseases, 3rd Ed.

Blackwell Science Ltd, Oxford, United Kingdom.

CROMPTON, D.W. AND S.M. JOYNER. 1980. Parasitic worms. Wykeham, Publications Ltd., London, United Kingdom.

DOASTER, G.L. AND C.P. GOATER. 1997. Collection and quantification of avian helminths and protozoa. Pages 396–418 in D.H. Clayton and J. Moore [EDS.], Host-parasite evolution: general principles and avian models. Oxford University Press, Oxford, United Kingdom.

DOBSON, A.P., P.J. HUDSON AND A.M. LYLES. 1992. Macroparasites: worms and others. Pages 329–348 in M.C. Crawley [ED.], Natural enemies. Blackwell Scientific Publications, Oxford, United Kingdom.

DUBOIS, G. 1968. Synopsis des Strigeidae et des Diplostomatidae (Trematoda). *Mém. Soc. Neuchâtel. Sci. Nat.* 10:1–258.

————. 1970. Synopsis des Strigeidae et des Diplostomatidae (Trematoda). *Mém. Soc. Neuchâtel. Sci. Nat.* 10:259–727.

GASSER, R.B. 2001. Identification of parasitic nematodes and study of genetic variability using pcr approaches. Pages 53–82 in M.W. Kennedy and W. Harnett [EDS.], Parasitic nematodes: molecular biology, biochemistry and immunology. CAB International, London, United Kingdom.

GIBSON, D.I., A. JONES AND R.A. BRAY [EDS.]. 2002. Keys to trematoda, Vol. 1. CAB International and The Natural History Museum, London, United Kingdom.

HARTWICH, G. 1975. I. Rhabditida und Ascaridida. Die Tierwelt Deutschlands. 62. Teil. Gustav Fischer Verlag, Jena, Stuttgart, Germany.

————. 1994. II. Strongylida: Strongyloidea und Ancylostomatoidea. Die Tierwelt Deutschlands. 68. Teil. Gustav Fischer Verlag, Jena, Stuttgart, Germany.

HEIDENREICH, M. 1997. Birds of prey: medicine and management. Iowa State Press, Ames, IA U.S.A.

HELLGREN, O., J. WALDENSTRÖM AND S. BENSCH. 2004. A new PCR assay for simultaneous studies of *Leucocytozoon, Plasmodium*, and *Haemoproteus* from avian blood. *J. Parasitol.* 90:797–802.

HOOGENBOOM, I. AND C. DIJKSTRA. 1987. *Sarcocystis cernae:* a parasite increasing the risk of predation of its intermediate host, *Microtus arvalis. Oecologia* 74:86–92.

JENKINS, M.C., J.T. ELLIS, S. LIDDELL, C. RYCE, B.L. MUNDAY, D.A. MORRISON AND J.P DUBEY. 1999. The relationship of *Hammondia hammondi* and *Sarcocystis mucosa* to other heteroxenous cyst-forming coccidia as inferred by phylogenetic analysis of the 18S SSU ribosomal DNA sequence. *Parasitol.* 119:135–42.

JONES, A., R.A. BRAY AND D.I. GIBSON [EDS.]. 2004. Keys to the Trematoda, Vol. 2. CAB International and The Natural History Museum, London, United Kingdom.

KHALIL, L.F., A. JONES AND R.A. BRAY. 1994. Keys to the cestode parasites of vertebrates. CAB International, University Press, Cambridge, United Kingdom.

KLÜH, P.N. 1994. Untersuchungen zur Therapie und Prophylaxe der *Caryospora*-Infektion der Falken (Falconiformes: Falconidae) mit Toltrazuril sowie die Beschreibung von zwei neuen *Caryospora*-Arten der Falken (*C. megafalconis* n. sp. und *C. boeri* n. sp.). Veterinary Medicine dissertation. Hanover, Germany.

KRONE, O. 2000. Endoparasites in free-ranging birds of prey in Germany. Pages 101–116 in J.T. Lumeij, J.D. Remple, P. Redig, M. Lierz, and J.E. Cooper [EDS.], Raptor biomedicine III. Zoological Education Network, Lake Worth, FL U.S.A.

———— AND J.E. COOPER. 2002. Parasitic diseases. Pages 105–120 in

J.E. Cooper [Ed.], Birds of prey: health and diseases, 3rd Ed. Blackwell Science Ltd, Oxford, United Kingdom.

KUCERA, J. 1979. A simple cultivation method for field diagnosis of avian trypanosomes. *Folia Parasitol.* 26:289–293.

LACINA, D. AND D. BIRD. 2000. Endoparasites of raptors: a review and an update. Pages 65–99 in J.T. Lumeij, D. Remple, P.T. Redig, M. Lierz, and J.E. Cooper [EDS.], Raptor biomedicine III. Zoological Education Network, Lake Worth, FL U.S.A.

LATIMER, K.S. AND P.M. RAKICH. 1994. Necropsy examination. Pages 356–381 in B.W. Ritchie, G.J. Harrison, and L.R. Harrison [EDS.], Avian medicine: principles and application. Wingers Publishing, Inc. Lake Worth, Florida.

LEE, D.L. [ED.]. 2002. The biology of nematodes. Taylor and Francis, London, United Kingdom and New York, NY U.S.A.

LINDSAY, D.S., P.C. SMITH, F.J. HOERR AND B.L. BLAGBURN. 1993. Prevalence of encysted *Toxoplasma gondii* in raptors from Alabama. *Parasitol.* 79:870–873.

MARSH, A.E., B.C. BARR, L. TELL, D.D. BOWMANN, P.A. CONRAD, C. KETCHERSIDE AND T. GREEN. 1999. Comparison of the internal transcribed spacer, ITS-1, from *Sarcocystis falcatula* isolates and *Sarcocystis neurona*. *J. Parasitol.* 85:750–757.

MORALES-HOJAS, R., R.J. POST, A.J. SHELLEY, M. MAIA-HERZOG, S. COSCARÓN AND R.A. CHEKE. 2001. Characterisation of nuclear ribosomal DNA sequences from *Onchocerca volvulus* and *Mansonella ozzardi* (Nematoda: Filaroidea) and development of a PCR-based method for their detection in skin biopsies. *Internat. J. Parasitol.* 31:169–177.

ODENING, K. 1969. Entwicklungswege der Schmarotzerwürmer. Akad. Verl. Geest & Portig K.-G., Leipzig, Germany.

———. 1998. The present state of species-systematics in *Sarcocystis* Lankester, 1882 (Protista, Sporozoa, Coccidia). *Syst. Parasitol.* 41:209–233.

PAULY, A., R. SCHUSTER AND S. STEUBER. 2003. Molecular characterisation and differentiation of opistorchiid trematodes of the species *Opisthorchis felineus* (Rivolta, 1884) and *Metorchis bilis* (Braun, 1790) using polymerase chain reaction. *Parasitol. Res.* 90:409–414.

POULIN, R. 1998. Evolutionary ecology of parasitism. Chapman & Hall, London, United Kingdom.

RAUSCH, R.L. 1983. The biology of avian parasites: helminths. Pages 367–442 in D.S. Farner, J.R. King, and K.C. Parkes [EDS.], Avian Biology, Vol. VII. Academic Press, New York, NY U.S.A.

RILEY, J, J.L. OAKS AND M. GILBERT. 2003. *Raillietiella trachea* n. sp., a pentastomid from the trachea of an Oriental White-backed Vulture *Gyps bengalensis* taken in Pakistan, with speculation about its life-cycle. *Syst. Parasitol.* 56:155–161.

SANMARTIN, M.L., F. ÁLVAREZ, G. BARREIRO AND J. LEIRO. 2004. Helminth fauna of falconiform and strigiform birds of prey in Galicia, Northwest Spain. *Parasitol. Res.* 92:255–263.

SAUMIER, M.D., M.E. RAU AND D.M. BIRD. 1991. Behavioural changes in breeding American Kestrels infected with *Trichinella pseudospiralis*. Pages 290–313 in J.E. Loye and M. Zuk [EDS.], Bird-parasite interactions. Oxford University Press, Oxford, United Kingdom.

SCHELL, S. C. 1970. The trematodes. William C. Brown Co., Dubuque, IA U.S.A.

SCHMIDT, G.D. 1986. Handbook of tapeworm identification. CRC Press, Inc., Boca Raton, FL U.S.A.

SCHULENBURG VD, J.H., U. ENGLISCH AND J.W. WAGELE. 1999. Evolution of ITS1 rDNA in the Digena (Platyhelminthes: trematoda): 3' end sequence conservation and its phylogenetic utility. *J. Mol. Evol.* 48:2–12.

SKRJABIN, K.J. [ED.]. 1950. The trematodes of animals and men / Essentials of trematodology 4, (in Russian). Publisher Akademy of Science, Moscow, Russia.

——— [ED.]. 1953. The nematodes of animals and men / Essentials of nematodology 2, part 2, (in Russian). Publisher Akademy of Science, Moscow, Russia.

——— [ED.]. 1957. The nematodes of animals and men / Essentials of nematodology 6, (in Russian). Publisher Akademy of Science, Moscow, Russia.

——— [ED.]. 1959. The trematodes of animals and men / Essentials of trematodology 16, (in Russian). Publisher Akademy of Science, Moscow, Russia.

——— [ED.]. 1960. The trematodes of animals and men / Essentials of trematodology 17, (in Russian). Publisher Akademy of Science, Moscow, Russia.

——— [ED.]. 1963. The nematodes of animals and men / Essentials of nematodology 11, (in Russian). Publisher Akademy of Science, Moscow, Russia.

——— [ED.]. 1964. The nematodes of animals and men / Essentials of nematodology 12, (in Russian). Publisher Akademy of Science, Moscow, Russia.

——— [ED.]. 1965. The nematodes of animals and men / Essentials of nematodology 14, (in Russian). Publisher Akademy of Science, Moscow, Russia.

——— [ED.]. 1967. The nematodes of animals and men / Essentials of nematodology 16, (in Russian). Publisher Akademy of Science, Moscow, Russia.

——— [ED.]. 1968. The nematodes of animals and men / Essentials of nematodology 21, (in Russian). Publisher Akademy of Science, Moscow, Russia.

——— [ED.]. 1971. The trematodes of animals and men / Essentials of trematodology 24, (in Russian). Publisher Akademy of Science, Moscow, Russia.

TELFORD, S.R., JR., J.K. NAYAR, G.W. FORRESTER AND J.W. KNIGHT. 1997. *Plasmodium forresteri* n.sp. from raptors in Florida and southern Georgia: its distinction from *Plasmodium elongatum* morphologically within and among host species and by vector susceptibility. *J. Parasitol.* 83:932–937.

UPTON, S.J., T.W. CAMPBELL, M. WEIGEL AND R.D. MCKNOWN. 1990. The Eimeriidae (Apicomplexa) of raptors: review of the literature and description of new species of the genera *Caryospora* and *Eimeria*. *Can. J. Zool.* 68:1256–1265.

VALKIUNAS, G. 1997. Bird haemosporida. Acta Zool., Lithuanica 3–5, 608pp.

VAN VALEN, L. 1973. A new evolutionary law. *Evol. Theory* 1:1–30.

WALDENSTRÖM, J., S. BENSCH, D. HASSELQUIST AND Ö. ÖSTMAN. 2004. A new nested polymerase chain reaction method very efficient in detecting *Plasmodium* and *Haemoproteus* infections from avian blood. *J. Parasitol.* 90:191–194.

WONGRATANACHEEWIN, S., W. PIMIDONMING, R.W. SERMSWAN AND W. MALEEWONG. 2001. *Opisthorchis viverrini* in Thailand - the life cycle and comparison with *O. felineus*. *J. Parasitol.* 51:207–214.

Appendix 1. Recipes for solutions and the McMaster chamber mentioned in the text.

Acid-carmine solution	Boil 4 g carmine, 15 ml aqua dest, and 1.5 ml concentrate hydrochloric acid using a Liebig cooler. After cooling add 85 ml 95% ethanol.
Alum carmine solution	Boil 5 g potassium-aluminium-sulfate, 2 g carmine, and 100 ml aqua dest for 1 hour. When cool, filter the solution and add some thymol crystals for preservation. Store the solution in a refrigerator.
Bouin-solution	Mix one part 40% formalin plus three parts aqua dest filled with picric acid until saturation. Add one part glacial acetic acid to 10 parts of this stock solution.
Giemsa-solution	Mix 10 ml Giemsa with 190 ml distilled aqua dest buffered to pH 7.2 for 10 minutes at 40°C.
McMaster Chamber	The specific slide made of glass or plastic can be used to count parasite eggs or oocysts of protozoa per gram of feces. This standard method is often described in classical text books of parasitology, but information also can be obtained from the homepage of the Food and Agriculture Organization of the United Nations (FAO): www.fao.org/ag/AGAInfo/resources/documents/Parasitology/EggCount/Purpose.htm (last accessed 17 August 2006).

Toxicology

18

CHARLES J. HENNY
USGS Forest and Rangeland Ecosystem Science Center
3200 SW Jefferson Way, Corvallis, OR 97331 U.S.A.

JOHN E. ELLIOTT
Environment Canada, Pacific Wildlife Research Centre
5421 Robertson Road, RR1, Delta, BC, V4K 3N2 Canada

INTRODUCTION

In earlier years (1947–1985), many contaminant-related problems concerning raptors were related to chlorinated hydrocarbon (CH) insecticides, such as DDT, dieldrin, heptachlor, and chlordane, most of which now have been banned in the U.S. and elsewhere. Other contaminants mentioned in the first edition of this manual (Peakall 1987) included mercury, lead, polychlorinated biphenyls (PCBs), and acid deposition, the latter impacting fish populations in poorly buffered lakes and, therefore, adversely affecting Ospreys (*Pandion haliaetus*) and Bald Eagles (*Haliaeetus leucocephalus*). Secondary poisoning of raptors by anticoagulant rodenticides and organophosphorus (OP) pesticides was beginning to be evaluated. The extirpation of the Peregrine Falcon (*Falco peregrinus*) from the eastern U.S. by 1964, and major reductions in numbers elsewhere around the world, was due primarily to DDT, and, perhaps, other CHs. The recovery of the peregrine in the U.S., following the 1972 ban on the widespread use of DDT and much effort in reintroducing the species, and its eventual delisting in 1999 from being an Endangered

Species, was recently told in *Return of the Peregrine* (Cade and Burnham 2003).

Overall, the relative importance of specific contaminant issues today is not the same as discussed in the first edition of this manual, and new issues have emerged. That said persistent CHs still adversely influence some species at selected locations (e.g., DDE-reduced nesting success and significantly thinned the eggshells of some Ospreys breeding along the lower Columbia River in 1997–98, even though the population was increasing at the time [Henny et al. 2004]).

This chapter is subdivided into different classes of environmental contaminants that may adversely affect raptor populations. For each class of contaminants, we present: (1) structure and chemistry (what they are), (2) sources and use patterns (where and how they are used), (3) fate and transport (how mobile they are in the environment), (4) toxicology (what their basic mode[s] of action are), (5) effects criteria (what residue concentration and biochemical response in which tissues should be investigated; Table 1), and (6) techniques for studying field exposure and effects (Table 2).

As a note of caution, residue concentrations in the literature may be presented in several ways, which can be confusing (e.g., wet weight [ww], dry weight [dw], lipid weight [lw]). Sometimes the methods section of a paper must be read carefully to determine which value was used; it is critical to understand this terminology because reported concentrations vary tremendously depending upon how data are presented, as well as with the percent moisture and percent lipid in the tissue examined. Concentrations (C) readily can be converted (e.g., $C_{dry} = C_{wet}\ 100/100 - \%$ moisture).

Table 1. Selection of estimated toxicity threshold values for contaminants in raptor species[a].

Species	Chemical [b]	Tissue	Effect [c]	Value w.w. (units)	Ref. [d]
Bald Eagle (*Haliaeetus leucocephalus*)	DDE	Egg	15% reduction in shell thickness	16 mg/kg	1
		Egg	Corresponds to 0.7 young/occupied territory	5.9 mg/kg	2
		Egg	Significant reduction in productivity	12 mg/kg	3
		Egg	Embryo lethality	5.5 mg/kg	2
		Plasma	Corresponds to 5.9 mg/kg in eggs	41 µg/kg	2
		Brain	Lowest value poisoned adult	212 mg/kg	4
	Dieldrin	Brain	Lowest value poisoned adult	3.6 mg/kg	5
	ΣPCBs	Egg	Reduced probability of producing young	20 mg/kg	2
		Plasma	Corresponds to 20 mg/kg in eggs	189 µg/kg	2
	TCDD TEQs	Egg	NOAEL hatching	303 ng/kg	2
		Egg	NOAEL CYP1A induction	135 ng/kg	2
		Egg	LOAEL CYP1A induction	400 ng/kg	2
White-tailed Eagle (*H. albicilla*)	DDE	Egg	LOAEL productivity	6.0 mg/kg	6
		Egg	Strong reduction in desiccation index	8.5 mg/kg	6
		Egg	Corresponds to 0.7 young/occupied territory	10.5 mg/kg	6
	ΣPCBs	Egg	LOAEL for productivity	25 mg/kg	6
	TCDD TEQs	Egg	LOAEL for embryo mortality	320 ng/kg	6
Osprey (*Pandion haliaetus*)	DDE	Egg	Corresponds to 0.8 young/occupied nest	4.2 mg/kg	7
	TCDD TEQs	Egg	NOAEL for productivity	162 ng/kg	8
		Egg	NOAEL for hatching	136 ng/kg	9
		Egg	NOAEL for CYP1A induction	36 ng/kg	9
		Egg	LOAEL for CYP1A induction	130 ng/kg	9
American Kestrel (*Falco sparverius*)	PCB 126	Egg	Embryonic LD_{50}	65 µg/kg	10
		Egg	Significant increase in malformations and edema	2.3 µg/kg	10
	PCB 77	Egg	Embryonic LD_{50}	688 µg/kg	10
	ΣPCBs	Egg	Effects on reproductive and endocrine endpoints	34 mg/kg	11,12
	HE	Egg	Reduced productivity	1.5 mg/kg	13
Peregrine Falcon (*F. peregrinus*)	DDE	Egg	Reduced productivity	15–20 mg/kg	14
	ΣPCBs	Egg	Reduced productivity	40 mg/kg	14
Common Kestrel (*F. tinnunculus*)	MeHg	Brain	Mortality	25–33 mg/kg	15
	MeHg	Liver	Mortality	50–120 mg/kg	15
Red-tailed Hawk (*Buteo jamaicensis*)	MeHg	Liver	Mortality	20 mg/kg	16

[a] Sensitivity to most contaminants is species-specific. Much additional information on non-raptorial species is available (see Beyer et al. [1996]).

[b] DDE = p,p'-dichlorodiphenyl-dichloroethylene; ΣPCBs = sum polychlorinated biphenyl congeners; TCDD TEQs = 2,3,7,8-tetrachlorodibenzo-p-dioxin toxic equivalents; PCB 126 = 3,3',4,4',5-penta-CB (one of the most toxic PCB congeners); PCB 77 = 3,3',4,4'-tetra-CB (one of the most toxic PCB congeners); HE = heptachlor epoxide; MeHg = methylmercury.

[c] NOAEL = no-observed-adverse-effect-level; LOAEL = lowest-observed-adverse-effect-level; LD_{50} = acute oral median lethal dosage.

[d] References: 1, Wiemeyer et al. 1993; 2, Elliott and Harris 2002; 3, Nisbet and Risebrough 1994; 4, Garcelon and Thomas 1997; 5, Prouty et al. 1977; 6, Helander et al. 2002; 7, Wiemeyer et al. 1988; 8, Woodford et al. 1998; 9, Elliott et al. 2001; 10, Hoffman et al. 1998; 11, Fernie et al. 2001; 12, Smits et al. 2002; 13, Henny et al. 1983; 14, Peakall et al. 1990; 15, Koeman et al. 1971; 16, Fimreite and Karstad 1971.

Table 2. Examples of studies using recommended techniques in raptor field ecotoxicology.

Chemical(s) of concern	Sampling matrix	Technique(s)	Reference[a]
Persistent organic pollutants	Egg	**Salvage unhatched eggs or fragments for chemistry** (combined with productivity and other measurements)	1,2,3
	Egg	**Sample egg technique** (collection of eggs for chemistry) (combined with productivity and other measurements)	4,5,6
	Egg	**Laboratory incubation of fresh eggs** (chemical analysis of the yolk sacs or of a sibling egg; morphology, histology, biochemistry of organs)	7,8
	Egg	**Egg swap experiments** (can be combined with collection of eggs and behavioral observations; always combined with productivity, and potentially, other measurements)	9,10
	Major organs: liver, kidney, brain	**Mortality monitoring**: collection of dead and moribund birds for necropsy and chemistry, biochemistry, and histology	3,11
	Blood	**Capture**: migrant or breeding birds, residues	12,13
Mercury	Egg	**Sample egg technique** (see above)	14
	Blood	**Nestlings**: residues	14
	Feathers	**Adults and nestlings**: residues	14
	Liver, kidney, brain	**Nestlings**: residues	14
Lead	Blood	**Nestlings, adults**: residues, ALAD, protoporphyrin, hemoglobin	15,16
Anti-cholinesterase insecticides	Brain	**Cholinesterase activity**: dead or moribund birds	17,18
	Blood	**Cholinesterase activity**: dead or moribund birds	19
	Blood	**Cholinesterase activity**: captured birds	20
	Crop contents	**Chemical residues**: dead birds or surgically removed from live,9 poisoned birds	18,21

Note: ALAD = delta-aminolevulinic acid dehydratase.

[a] References: 1, Ratcliffe 1970; 2, Newton and Galbraith 1991; 3, Newton 1988; 4, Blus 1984; 5, Henny et al. 1983; 6, Henny et al. 2004; 7, Elliott et al. 1996a; 8, Elliott et al. 2001; 9, Wiemeyer et al. 1975; 10, Woodford et al. 1998; 11, Prouty et al. 1977; 12, Henny et al. 1996; 13, Court et al. 1990; 14, DesGranges et al. 1998; 15, Henny et al. 1991; 16, Henny et al. 1994; 17, Henny et al. 1987; 18, Elliott et al. 1996b; 19, Elliott et al. 1997b; 20, Hooper et al. 1989; 21, Henny et al. 1985.

CHLORINATED HYDROCARBON (CH) INSECTICIDES

Chemistry and Toxicology

Matsumura (1985) characterized these synthetic organic insecticides by (1) the presence of carbon, chlorine, hydrogen and sometimes oxygen atoms, including C-Cl bonds, (2) the presence of cyclic carbon chains (including benzene rings), (3) whether or not they were preferentially lipid-soluble, and (4) their stability in the environment. These compounds generally persist in the environment and biomagnify in food chains (some more than others), with raptors at the top of food chains and, especially bird-eating and fish-eating species, being particularly vulnerable. Generally, there are three kinds of CH insecticides: (1) DDT and its analogs (including methoxychlor and dicofol [kelthane]), (2) benzene hexachloride (BHC) isomers including lindane, and (3) cyclodiene compounds (including chlordane, heptachlor, aldrin, dieldrin (HEOD), endrin, toxaphene, mirex, kepone, endosulfan and telodrin. All are neuroactive agents whose modes of action include effects on ion permeability (DDT group) or effects as agents for nerve receptors (BHC and cyclodienes).

These compounds were first introduced in the late 1940s and early 1950s, and, with few exceptions, were banned in most industrialized countries during the 1970s or shortly thereafter. Their principal uses were in agriculture and for disease-vector control. Effects of persistent CH insecticides on raptor populations were widely documented and were catastrophic for some species. Continuing concern reflects their persistence, biomagnification and continued public health use for mosquito control in some countries. In theory, raptors may become exposed to CHs at great distances from application sites due to: (1) atmospheric transport (e.g., elevated concentrations in the Canadian arctic [Barrie et al. 1992]), (2) migratory prey species transporting material from distant sources, or (3) migratory raptors themselves transporting material from distant sources (Henny et al. 1982).

Criteria and Techniques

DDT (and its breakdown product DDE), heptachlor, dieldrin, and perhaps other CH insecticides can cause reduced productivity (Lockie et al. 1969, Ratcliffe 1970, Henny et al. 1983). Unhatched (failed) eggs have been and continue to be analyzed to determine the con-taminants causing reduced reproductive success (e.g., Wegner et al. 2005). A nonviable Peregrine Falcon egg analyzed in 1960 represents the earliest study of pollutant-related effects on raptors (Moore and Ratcliffe 1962). As Peakall (1987) pointed out, examining eggs is advantageous because it directly examines the target (i.e., the nonviable egg). CH residue concentrations in the egg are directly related to levels in the adult female (Norstrom et al. 1985), which is not necessarily true for other classes of pollutants. CH concentrations reported from nonviable eggs remaining in the nest, after the expected hatch date (a non-random sample), are usually biased towards higher values, if CHs adversely influenced hatchability. Scientists prefer residues from a randomly collected single "sample egg" (Blus 1984) (1 to 2 weeks into incubation) from a series of nests to evaluate possible effects of CHs on success of eggs remaining in the clutch and to document contaminant levels in populations. Collecting a sample egg can cause nest abandonment for some raptor species, such as Bald Eagles (Grier 1969), which negatively influences productivity, whereas for other species, including Ospreys, nests are rarely abandoned after a short visit for egg collection. The reduction for each Osprey egg collected (usually from a three-egg clutch), for example, was only 0.28 young fledged per active nest (Henny et al. 2004; Fig. 1). The sensitivity of eggs to this group of insecticides is species-specific, and as such, no single diagnostic egg concentration can be used for all species, e.g., DDE adversely influences Osprey reproductive success above 4.2 mg/kg (ww) (Wiemeyer et al. 1988), Bald Eagle above 5.9 mg/kg (ww) (Elliott and Harris 2002), and Peregrine Falcon above 15–20 mg/kg (ww) (Peakall et al. 1990). As expected, the degree of eggshell thinning caused by a given egg concentration of DDE, the only CH insecticide known to thin eggshells except for the structurally similar dicofol (Bennett et al. 1990), also varies among families of raptors and even among species within the same family (Peakall 1975). Usually, shell thickness is compared to pre-DDT era norms based upon eggshells in museums.

CH insecticides, especially the cyclodienes, also kill birds, and when dead raptors are found and these insecticides suspected, the brain should be analyzed and residues compared to diagnostic concentrations based on laboratory studies (see criteria in Beyer et al. 1996). Peakall (1996) reviewed the causes of death of Bald Eagles found dead in the U.S. by a network of federal, state, and private investigators from 1966 to 1983. The

Figure 1. Increasing Osprey (*Pandion haliaetus*) populations have pioneered back into some potentially contaminated industrial sites, e.g., Seattle Harbor, U.S.A., where they are again being used as an indicator species to evaluate numerous contaminants. Note Opsrey nest with two eggs in the foreground *(Photo by J. Kaiser, USGS).*

percentage of deaths attributed to dieldrin decreased following a ban on its use (i.e., 13% in 1966–70, 6.5% in 1971–74, 3.0% in 1975–77, and 1.7% in 1978–83). Decreases in mortality and increases in natality in the late 1970s and 1980s were followed by population increases in those species adversely affected in earlier years. The best evidence of the impact of dieldrin poisoning is the long-term study of Eurasian Sparrowhawks (*Accipiter nisus*) in Britain (Newton et al. 1986). A recent re-analysis of these data shows that at least 29% of the sparrowhawks in the area of high cyclodiene use died directly from dieldrin poisoning, which led to a population decline (Sibley et al. 2000). Comparison of temporal trends in populations of Sharp-shinned Hawks (*A. striatus*) with both egg residues and usage patterns of dieldrin and DDT in North America support the possibility that dieldrin poisoning also may have impacted North American accipiters (Elliott and Martin 1994). Chlordane, persisting from earlier efforts to control turf pests in parks and gardens, recently poisoned songbirds and raptors, particularly Cooper's Hawks (*A. cooperii*) (Stansley and Roscoe 1999).

Blood plasma can be used to monitor long-term CH residue trends in raptor populations and to evaluate local exposure (Henny and Meeker 1981, Court et al. 1990, Elliott and Shutt 1993, Jarman et al. 1994). Migratory species (both raptors and their prey) often are exposed elsewhere during their travels. Based upon DDE measured in blood plasma of migratory Peregrine Falcons captured on the Texas coast as they departed and returned to the U.S. during migration, Henny et al. (1982) concluded DDE at that time was largely accumulated during winter in Latin America. This study continued for long-term monitoring purposes and, with the use of satellite telemetry to locate breeding and wintering localities, documented the decrease of DDE in arctic-breeding peregrines from the late 1970s to 1994 (Henny et al. 1996).

Two general types of CH studies continue: (1) long-term monitoring of the productivity and population sizes of species previously in trouble, often with egg or blood-plasma collections for residue analyses, and (2) evaluations of potentially sensitive species based upon diet (i.e., fish or bird-eaters) or at locations with limited information.

POLYCHLORINATED BIPHENYLS (PCBS), POLYCHLORINATED DIBENZO-P-DIOXINS (PCDDS), AND POLYCHLORINATED DIBENZOFURANS (PCDFS)

Chemistry and Toxicology

These related chemicals are released to the environment mainly from industrial and commercial chemical sources. Being relatively persistent and volatile, they have dispersed throughout the global environment where they biomagnify, particularly in aquatic food chains. Some of the highest PCB concentrations in biota have been reported in eagle and falcon species, and thus have been investigated as potential factors in populations of raptors with chronic low productivity. Exposure and effects on wildlife, including raptors, have been reviewed by Hoffman et al. (1996) and Rice et al. (2003).

PCBs were used for a variety of purposes including manufacture of electrical transformers, and formulation of lubricating and cutting oils, pesticides, plastics, paints, etc. More than a billion kilograms were produced worldwide, with a third having been released into the environment (Tanabe 1988). PCB use has been banned or heavily restricted in most countries since the late 1970s.

Neither PCDDs nor PCDFs are deliberately produced commercially, but they are formed either as by-products during synthesis of other chemicals, such as chlorophenolic herbicides, or during combustion of chlorine-containing materials. Incineration of municipal and industrial wastes is the major global source of dioxins, which can be transported long distances and deposited in soils and sediments (Czuczwa et al. 1984).

The number and position of chlorine atoms determines the chemical and biological attributes of each dioxin, furan, or PCB isomer (Fig. 2). More chlorine atoms generally lead to greater fat solubility and resistance to degradation. The most toxic isomers have chlorines at the 2,3,7,8 (PCDD/Ds) or 3, 3',4,4' (PCBs) positions. Those congeners are more planar in shape and readily bind a cellular protein known as the *Ah* or arylhydrocarbon receptor, which leads to a variety of biological responses.

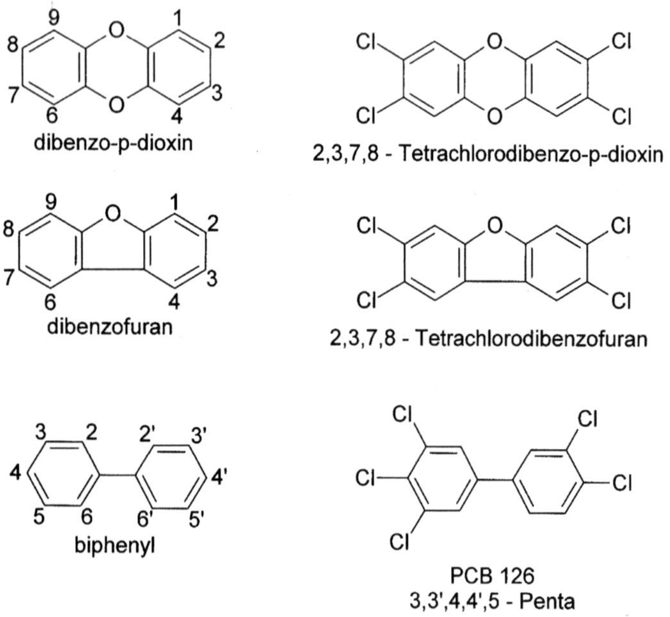

Figure 2. Structure of PCDDs, PCDFs and PCBs.

Toxicity relative to 2,3,7,8-TCDD can be compared using TCDD Toxic Equivalents (TEQs) (Van den Berg et al. 1998). Embryos and growing nestlings are at most risk to TCDD toxicity (Peterson et al. 1993). In laboratory studies, birds of prey including kestrels were less sensitive to PCBs than were quail and chickens (Elliott et al. 1990, 1991, 1997a), but more sensitive than Common Terns (*Sterna hirundo*) (Hoffman et al. 1998). Feeding an environmentally relevant concentration of PCBs to American Kestrels (*F. sparverius*) caused reproductive effects, as well as altered immune and endocrine endpoints (Fernie et al. 2001, Smits et al. 2002). Less persistent congeners, which were less likely encountered in the field, appeared more toxic than the persistent ones.

Toxic effects have been well studied in wild Great Lakes colonial waterbirds. A set of toxic symptoms referred to as GLEMEDS (Great Lakes embryo mortality, edema, and deformities syndrome) has been attributed to exposure to dioxin-like chemicals in gull, tern and cormorant populations (Gilbertson et al. 1991). Bowerman et al. (1994) reported bill deformities in Bald Eagles, but no quantitative relationship between incidence and contaminant exposure. However, kestrel embryos exhibited malformations and edema when eggs were injected with concentrations of PCB-126 at considerably lower levels than measured in Great Lakes Bald Eagles, supporting the contention for dioxin-like chemicals as the cause of observed defects in Bald Eagle chicks (Hoffman et al. 1998).

Despite high concentrations in eggs, it has proved difficult to link PCB concentrations with significant reproductive effects in raptor populations, including peregrines, Ospreys, and accipiters (Newton et al. 1986, Wiemeyer et al. 1988, Peakall et al.1990, Elliott et al. 2001). Statistical associations between productivity and concentrations of PCBs in eggs were found for Bald Eagles. However, the strong intercorrelation with DDE, which showed a greater effect on productivity (Wiemeyer et al. 1984, 1993), was a confounding factor in that and other studies. A more recent analysis of available data for Bald Eagles showed significant associations between productivity and DDE, but not PCBs (Elliott and Harris 2002). In a long-term study of White-tailed Eagles (*H. albicilla*) in Sweden, Helander et al. (2002) found a correlation for PCBs and the incidence of embryo mortality, but not with productivity. Data from Helander et al. (2002), supported by laboratory evidence (Hoffman et al. 1998, Fernie et al. 2001), indicate that PCBs have affected *Haliaeetus* populations in

Figure 3. Osprey (*Pandion haliaetus*) nest on an easily accessible (with U.S. Coast Guard approval) channel marker near a paper mill on lower Columbia River (River Mile 44) downstream of Portland, Oregon. Ospreys nest at regular intervals along some major river systems and bays which can result in strategic or random sampling of eggs or blood for contaminant evaluation *(Photo by J. Kaiser, USGS).*

areas of high exposure; however, effects are difficult to separate not only from DDE, but also from ecological factors such as food supply (Dykstra et al. 1998, Elliott and Norstrom 1998, Elliott et al. 1998, Gill and Elliott 2003, Elliott et al. 2005a).

Sublethal effects of dioxin-like chemicals have been reported in some raptors. In a study of Ospreys breeding on a river in Wisconsin, nestlings grew more slowly at a site contaminated with 2,3,7,8-TCDD from a pulp mill than at uncontaminated sites (Woodford et al. 1998; Fig. 3). Induction of cytochrome P450 liver enzymes (CYP1A) of a type responsive to exposure to 2,3,7,8-TCDD toxic equivalents has been reported in embryos of both Ospreys and Bald Eagles breeding near bleached kraft pulp mills (Elliott et al. 1996a, 2001).

Criteria and Techniques

Collection and analysis of eggs is still the preferred method for investigating exposure of raptors to PCBs and related chemicals. The pros and cons of collecting fresh versus unviable eggs, and the use of the sample egg technique discussed in the section on chlorinated hydrocarbon pesticides apply equally here. Concentrations of PCBs in eggs and related compounds diagnostic of effects, such as embryo survival or overall nest success, have not been defined clearly for most species.

Threshold levels for PCBs in eggs were estimated using older analytical methods, such as 40 mg/kg (ww) for peregrines (Peakall et al. 1990). Based on a review and re-analysis of existing data for Bald Eagles, Elliott and Harris (2002) suggested that the reproductive effect threshold was at least 20 mg/kg (ww) total PCBs for Bald Eagles. Helander et al. (2002) determined the lowest observable effect level of 25 mg/kg (ww) (500 mg/kg [lw]) for PCB effects on productivity of White-tailed Eagles. Combining an egg swap design (for more details see Peakall [1987:325] and the discussion below) with regular measurements of chick growth rates, Woodford et al. (1998) suggested a no-observable-adverse-effect-level (NOAEL) of at least 136 ng/kg (ww) for 2,3,7,8-TCCD for the hatchability of Osprey eggs.

Effects of dioxin-like chemicals in birds also can be studied using the technique of laboratory incubation of wild eggs. This approach separates egg-intrinsic effects from adult behavior (egg-extrinsic) and also permits measurement of biomarkers in hatchlings. Using this approach, Elliott et al. (2001) determined a no-effect level for TEQs in Osprey nestlings and a lowest-observable-adverse-effect level of 130 ng/kg (ww) TEQs for hepatic CYP1A induction. Effects of dioxin-like chemicals were studied in laboratory-incubated Bald Eagle eggs (Elliott et al. 1996a) and critical values subsequently recalculated using updated toxic equivalence factors (Elliott and Harris 2002). The results indicate a NOAEL of 135 ng/kg (ww) and lowest-observed-effect-level (LOAEL) of 400 ng/kg for CYP1A induction, and, for embryo toxicity, a NOAEL of 303 ng/kg.

The possible role of dioxin-like chemicals in instances of chronic low reproductive success can be investigated by experimental manipulation of eggs in the field. The logistics of such experiments are complex, given factors such as potential nest abandonment, and the lack of synchronicity in timing of breeding. Embryonic mortality can be caused not only by toxicants within the egg (an intrinsic factor), but also by inadequate parental care caused by the pollutant load (an extrinsic factor), or by a combination of both. These factors can be separated by an egg-exchange experiment between clean and contaminated sites. Adult:Egg combinations in such an experiment (and expected results) include: clean, clean (normal reproduction); clean, contaminated (intrinsic only); contaminated, clean (extrinsic only); and contaminated, contaminated (both intrinsic and extrinsic). For this type of research to be successful, clean and contaminated sites must be near identical from an ecological perspective, including food availability. Swapping of eggs between treatment and reference sites has provided valuable information in contaminant studies of Ospreys (Wiemeyer et al. 1975), particularly when combined with intensive observation of nesting behavior (Woodford et al. 1998). Nest surveillance, whether directly by an observer or by use of video recording technology, has proved useful in factoring contaminant and ecological variables (Dykstra et al. 1998, Elliott et al. 1998, Gill and Elliott 2003).

Measurement of contaminant levels in blood samples of nestling raptors provides a non-destructive approach, particularly for threatened populations (Elliott and Norstrom 1998, Olsson et al. 2000, Bowerman et al. 2003). Adults also can be trapped either at the nest (Court et al. 1990, Newson et al. 2000), or during migration (Elliott and Shutt 1993), and their blood sampled to assess exposure to PCBs. Diagnostic values for plasma generally are not available, but a value of 189 μg/kg (ww) total PCBs in nestling plasma was suggested as being correlated with 20 mg/kg (ww) in eggs of Bald Eagles (Elliott and Harris 2002).

LEAD

Chemistry and Toxicology

Sources of lead include lead mining, smelting and refining activities, battery-recycling plants, areas of high vehicular traffic, urban and industrial areas, sewage and spoil-disposal areas, dredging sites, and areas with heavy hunting pressure (Eisler 2000). Most of these sources are local, but until recently, lead exposure from spent shotgun pellets and vehicular traffic were much more widespread. Amounts of lead in roadside soils increased as a direct result of the combustion of gasoline containing organo-lead additives. After about a two-decade phase-out, lead additives in gasoline were totally banned in 1996 for on-road vehicles in the U.S. Since 1998, similar regulations were approved in the European Union, progressively restricting and finally banning the use of leaded gasoline in vehicles.

Lead concentrations in livers of Common Kestrels (*F. tinnunculus*) from both rural and city regions of southeastern Spain decreased significantly between 1995–97 and 2001 (Garcia-Fernandez et al. 2005). The U.S. banned the use of lead shot to hunt waterbirds in 1991. Lead shot was similarly banned in the 1990s in Canada, Denmark, Finland, The Netherlands, and Nor-

way, and in portions of Australia and Sweden (see country policies in Miller et al. [2002]). Thus, two widespread sources of lead were eliminated or were in the process of being reduced in many countries, although lead from the earlier use remains in the environment, and lead shot and bullet use continue for other types of hunting in most countries.

Lead modifies the function and structure of kidney, bone, the central nervous system, and the hematopoietic system, and produces adverse biochemical, histopathological, neuropsychological, fetotoxic, teratogenic, and reproductive effects (Eisler 2000). Lead poisoning in raptors has been fairly well documented since the 1970s. Secondary poisoning from consumption of lead-poisoned or shot waterfowl is believed to be the predominant source of lead exposure for wintering Bald Eagles and Golden Eagles (*Aquila chrysaetos*) (Feierabend and Myers 1984). Upland-foraging raptors and scavengers that typically include game birds and mammals in their diet are also at risk for lead poisoning (Kim et al. 1999, Clark and Scheuhammer 2003, Fry 2003, Wayland et al. 2003).

Criteria and Techniques

Depending on its severity, lead poisoning causes specific clinical signs including depression, foul-smelling breath, lime green feces, nonregenerative anemia, vomiting, diarrhea, ataxia, blindness, and epileptiform seizures (Gilsleider and Oehme 1982). Subclinical or chronic lead exposure usually decreases the ability to hunt and predisposes raptors to injury from environmental hazards such as vehicles, power lines, etc., which could partially explain why many raptors were admitted to rehabilitation centers with miscellaneous trauma (Kramer and Redig 1997). Blood-lead concentrations between 0.2–0.6 mg/kg (ww) were classified as subclinical lead exposure and birds with concentrations between 0.61–1.2 mg/kg classified as clinical (treatable) lead poisoning. Blood-lead concentrations >1.2 mg/kg were invariably associated with death (Kramer and Redig 1997). Blood parameters (g-aminolevulinic acid dehydrase [ALAD], hematocrit, proporphyrin, hemoglobin) have been used in field studies. ALAD inhibition of 80% is often associated with decreased hemoglobin and hematocrits (see references in Henny 2003). Lead-poisoning categories in livers based upon Pain (1996) include: <2 mg/kg (ww) (background), 2–5.9 mg/kg (subclinical), 6–15 mg/kg (clinical) and >15 mg/kg (severe clinical).

Lead poisoning has been documented in at least 14 species of raptors that eat or scavenge prey containing lead shot or bullets (including hunter-wounded birds and mammals). These include California Condor (*Gymnogyps californianus*), Andean Condor (*Vultur gryphus*), King Vulture (*Sarcoramphus papa*), European Honey Buzzard (*Pernis apivorus*), Bald Eagle, White-tailed Eagle, Steller's Sea Eagle (*H. pelagicus*), Western Marsh Harrier (*Circus aeruginosus*), Red-tailed Hawk (*Buteo jamaicensis*), Roughleg (*B. lagopus*), Golden Eagle, Prairie Falcon (*F. mexicanus*), Peregrine Falcon and Great Horned Owl (*Bubo virginianu*) (Locke and Friend 1992, Pain et al. 1994, Kim et al. 1999, Eisler 2000, Clark and Scheuhammer 2003). Most available information was reported from U.S., Canada, Europe and Japan. That most raptors regurgitate pellets (i.e., undigested bones, fur, feathers, and often lead shot) definitely reduces their exposure to lead. Shot has been reported in field-collected pellets, and a laboratory study with five Bald Eagles showed that of 196 shot ingested, only 18 were retained at death, with a median retention time of 2 days (Pattee et al. 1981). Based on these and other findings, Henny (1990) concluded that without pellet casting, the Bald Eagle probably would have become extirpated because of lead poisoning in portions of its range a hundred years ago, long before the lead problem was understood. Thus, lead poisoning could have been much more serious for raptors than it has been.

To avoid various threats, particularly lead poisoning, all remaining California Condors were brought into captivity in 1987. Release of captive-propagated condors began in 1992 and, despite extensive efforts to reduce incidental exposure of condors to lead ammunition fragments, they continue to suffer from acute lead poisoning (Meretsky et al. 2000, Fry 2003). California Condors feed mainly on soft tissues, rarely ingesting bones, hair or feathers (Snyder and Snyder 2000), and thus not only reduced the need to cast pellets, but also increased exposure to ingested lead fragments. Lead is a problem not only in the U.S., but also worldwide. During the winter of 1998–1999 in Hokkaido, Japan, 16 Steller's Sea Eagles and 9 White-tailed Eagles died of lead poisoning after consuming sika deer (*Cervus nippon*) remains containing lead-bullet fragments (Kurosawa 2000).

Lead poisoning of raptors from mining sources has been studied at the Coeur d'Alene (CDA) lead mining and smelting complex in northern Idaho, U.S.A. (Henny et al. 1991, 1994; Henny 2003). Waterfowl were most affected, due to their consumption of sediment (Beyer et al. 2000). Raptors do not ingest sediment, and most raptors do not digest bones of prey species (a major storage area in vertebrates for lead), thus it became clear why Ospreys, hawks and owls in the CDA basin were less contaminated with lead from mining sources than were waterfowl.

MERCURY

Chemistry and Toxicology

Toxicity of mercury to birds was reviewed by Scheuhammer (1987). Toxicity depends on whether mercury is in the organic or inorganic form. Only a small percentage of inorganic mercury is absorbed, but almost all organic mercury is absorbed by the intestine. Biotic and abiotic methylation in nature of inorganic mercury produces methylmercury (MeHg), which fish accumulate from water and their diet; nearly all mercury in fish flesh is MeHg. MeHg can adversely affect developing neural tissue in birds, with fish-eating birds being especially vulnerable.

Historically, mercury was used extensively in gold and silver extraction, in the chlor-alkali industry, in the manufacture of electrical instruments, in pharmaceuticals, in agricultural fungicides, in the pulp and paper industry as a slimicide, and in the production of plastics (Eisler 2000). Other activities that contribute significantly to the global input of environmentally available mercury include the combustion of fossil fuels; mining and reprocessing of copper and lead, runoff from abandoned cinnabar mines; wastes from nuclear reactors, pharmaceutical plants, and military ordinance facilities; incineration of municipal solid wastes and medical wastes; and disposal of batteries and fluorescent lamps (Eisler 2000). Long-range atmospheric transport of mercury has resulted in elevated mercury loadings great distances from source sites, including remote lakes in Canada (Lucotte et al. 1995). Since 1985, mercury has accumulated in flooded soils of the Florida Everglades at a much higher rate than decades earlier. The increase was attributed to increased global and regional deposition, and is similar to increases reported in Sweden and the northern U.S. (Rood et al. 1995). Elevated mercury concentrations have resulted in closing many lakes and rivers to fishing because of human health concerns. In general, the number of mercury-contaminated fish and wildlife habitats has increased progressively. Increased mercury concentrations in lakes are attributed to increased atmospheric emissions and to acid rain in poorly buffered systems.

Concerns about mercury exposure of raptors were especially high in Europe and North America during the 1960s and 1970s and are again reaching high levels in more recent years. The earlier interest was associated with the agricultural use of alkyl mercury as a fungicide applied as a seed dressing. This killed many seed-eating birds and secondarily poisoned many raptors (Berg et al. 1966, Jenson et al. 1972). Alkyl mercury was introduced around 1940 and was banned as a seed dressing in Sweden in 1966 (Johnels et al. 1979). Most mercury issues have been associated with aquatic systems and species, but the fungicide use resulted in exposure of upland species, including Eurasian Sparrowhawk, Common Buzzard (*B. buteo*), Merlin (*F. columbarius*), and Common Kestrel.

Contemporary interest in mercury includes: (1) atmospheric deposition from coal-fired power plants worldwide, especially the Arctic and the northeastern U.S. and adjacent Canada, which contaminates fish stocks and exposes fish-eating wildlife, (2) the Amazon Basin where mining operations annually discharge 90–120 tons of mercury into local ecosystems (Nriagu et al. 1992) affecting local breeding populations of birds, including raptors, and perhaps neotropical migrants (e.g., fish-eating Ospreys that nest in eastern North America), and (3) in many parts of the world, localized historic mining sites for mercury, or where mercury was used to extract gold or silver.

Criteria and Techniques

Mercury monitoring procedures have included eggs, liver, kidneys, whole blood and feathers (we recommend that personnel at the analytical chemistry laboratory wash feathers with a metal-free alkaline detergent to remove adhering particulate matter). Shunting MeHg into growing feathers is an important sequestering process in birds. And indeed, essentially all mercury in blood, eggs, and feathers is MeHg. Feathers from museum specimens have been used to provide a long-term evaluation of mercury exposure, although care needs to be taken about consistency in the specific feathers analyzed. Livers and kidneys of many raptors found dead were routinely analyzed only for total mercury (THg). THg concentrations reported in birds "dying of mercury poisoning" showed considerable variation, e.g., White-tailed Eagles (all mg/kg ww): Finland, liver 4.6

to 27.1, kidney 48.6 to 123.1; Germany, liver 48.2, 91, kidney 120; Baltic Sea, liver 30, 11, 33 (see Thompson 1996). This variability may be associated with the presence of differing ratios of inorganic mercury and the more toxic MeHg. It has been known for some time that birds (especially seabirds) demethylate MeHg (the form readily absorbed and usually ingested) and sequester it in the liver and kidneys in the less toxic inorganic form. Forms of mercury present in the liver and kidneys have been analyzed in recent years and provide better insight into mercury toxicity and sequestration. Recent studies of waterbirds along the Carson River in Nevada (a highly-contaminated historic mining site) revealed interesting aspects of mercury toxico-dynamics in birds and evidence of some histologic effects (Henny et al. 2002). The theoretical "effect criterion" of mercury in eggs is ~ 0.80 mg/kg (ww) (Heinz 1979, Newton and Haas 1988), but see Oehme (2003). Thompson (1996) rightfully implies that no single mercury criterion in eggs applies to all species, which is similar to the species-specific findings reported earlier for CHs.

Perhaps the best approach to monitoring mercury in raptors is to sample whole blood (highly correlated and 1:1 ratio with MeHg in the liver [Henny et al. 2002]), or to sample newly grown feathers of young (all grown about the same time), which are highly correlated with blood concentrations in young at the time of feather growth. Feathers from adults are more complicated and reflect blood concentrations when the feather was grown (which may represent mercury exposure at different locations for a migratory species), or different degrees of depuration (via feathers) depending upon when in the molt cycle the collected feather was grown. Heinz and Hoffman (2003) reported that once a bird begins ingesting elevated levels of mercury in the diet, it only takes a few days before depositing high levels of mercury in its eggs. High levels of mercury also should appear rapidly in both blood and growing feathers.

ORGANOPHOSPHORUS (OP) AND CARBAMATE (CB) INSECTICIDES

Chemistry and Toxicology

When many of the CH insecticides were banned, they were largely replaced by shorter-lived but more toxic cholinesterase(ChE)-inhibiting OP and CB insecticides. The agents comprising this type of insecticide, which have a common mechanism of action, arise from two different chemical classes, the esters of phosphoric or phosphorothioic acid (OP) and those of carbonic acid (CB) (Ecobichon 1996). These insecticide classes, primarily developed in the 1950s and 1960s, were generally considered non-persistent and non-bioaccumulative, and, therefore, at low risk for raptor secondary poisoning, occurring through eating intoxicated prey. Many OP and CB compounds have high acute toxicity (low amounts kill vertebrates), especially when compared with the CHs, but they do not bioaccumulate or biomagnify up food chains. Their high acute toxicity results in numerous raptor poisonings and deaths. Secondary poisoning of raptors from these acutely toxic chemicals is most likely from exposure to the unabsorbed compound remaining in the gastro-intestinal tract of the prey (Hill and Mendenhall 1980, Hill 1999), which is in contrast to the importance of residual metabolites accumulated in post-absorptive tissues and fat for CHs. Early reports of OP secondary poisoning involving raptors involved Swamp Harriers (*C. approximans*) in New Zealand killed by parathion and fensulfothion (Mills 1973), and about 400 raptors killed in Israel after eating voles and birds poisoned with monocrotophos (azodrin) (Mendelssohn and Paz 1977).

The principal toxicity of OP and CB pesticides is based on disruption of the nervous system by inhibition of ChE activity in the central nervous system and at neuromuscular junctions with death generally attributed to acute respiratory failure (O'Brien 1967). When an OP or CB binds to ChE, a relatively stable bond is formed and prevents the ChE from deactivating the neurotransmitter, acetylcholine. The clinical signs following an acute exposure include lethargy, labored breathing, excessive bronchial secretion (salivation), vomiting, diarrhea, tremors, and convulsions. These toxic indicators are useful when sick animals are found near an area of recent applications, but the signs are not uniquely different from poisoning by other neurotoxins (Hill 2003).

Criteria and Techniques

OP and CB pesticides have resulted in hundreds of incidents of wildlife mortality from disease vector control and agriculture (including forest and range management). When many dead and moribund animals of mixed species are found in an area of known OP or CB treatment, the casual association may be evident but is not conclusive without biochemical and chemical confirmation (Hill 2003). Proper diagnosis depends upon demonstration of brain ChE inhibition consistent with

levels indicative of toxicity or exposure and chemical detection of residues of the causative agent. Hill (2003) pointed out that the last step is sometimes difficult because neither OP nor CB residues tend to accumulate in tissues, but that a strong inferential diagnosis is possible by demonstrating inhibited brain ChE activity and "detection" of the anti-ChE agent in either ingesta or tissues.

Normal brain ChE values are obtained from raptors (same species, because normal values are species-specific) not exposed to OPs or CBs and used as a basis for comparison. Some published normal values for North American raptors (10 species of vultures, hawks, eagles, falcons, and owls) are available (Hill 1988), but before using them for comparative purposes, it is critical that observed values be based upon the same methodology. The concurrent running of "controls" for normal values on the same instrument is preferred. Another alternative is to use a suitable reactivation technique to determine the degree of inhibition. In cases of OP poisoning, ChE activity can be reactivated *in vitro* by the oxime 2-PAM (Fairbrother 1996), and for carbamalated ChE (which is less stable) simple *in vitro* heat will serve as a rapid indicator of CB exposure (Hill and Fleming 1982).

A conservative threshold of 50% inhibition in whole brain ChE activity of a bird found dead is generally considered diagnostic of death from anti-ChE poisoning. Even so, 70–95% is commonly reported for birds killed in nature by OP insecticides (Hill 2003). In contrast, when birds are killed by CB pesticides, whole brain ChE activity often is not nearly as inhibited (ChE levels may vary from near normal to only 70% inhibition). Lesser degrees of ChE inhibition may reflect spontaneous postmortem reactivation of the enzyme (Hill 1989), or that death occurred as a result of initial inhibition of the peripheral nervous system and its control of vital functions prior to the brain being completely inhibited. If immediate analysis is not available, store carcasses frozen (preferably at –80°C) prior to ChE analyses, especially if CBs are suspected. Freezing, however, will hinder the ability to detect other causes of death (e.g., deaths from infectious diseases).

Toxic consequences to raptors from OP and CB applications usually last only a few days, but exceptions do occur. Treatment of cattle with the OP, famphur (poured directly on back of cattle with ladle), kills warble larvae in the blood stream. Black-billed Magpies (*Pica pica*) died several months following application of famphur, and hawks and owls also died from second-

Figure 4. Bald Eagle (*Haliaeetus leucocephalus*) feeding on duck carcass in the Fraser River Delta of British Columbia, Canada. A single duck carcass can attract many Bald Eagles and other raptors, and is a prime vector of insecticides to birds of prey in that environment (*Photo by S. Lee, CWS*).

Figure 5. Juvenile Bald Eagle (*Haliaeetus leucocephalus*) from Fraser River Delta, British Columbia, Canada, with symptoms of anticholinesterase poisoning. Note dilated pupils, clenched talons and the inability to stand. The bird died 2 days later of phorate poisoning (*Photo by J. Elliott, CWS*).

ary exposure (Henny et al. 1985). Unabsorbed famphur persisted on cattle hair (sampled at weekly intervals) under field conditions for at least 3 months, and magpies that ingested cattle hair died. One Red-tailed Hawk that consumed a magpie died from secondary poisoning 10 days after cattle treatment and another was found incapacitated about 13 days after treatment with blood plasma ChE inhibited 82%.

Plasma ChE, which is more variable than brain ChE, can be used to measure exposure by comparing the observed value to a norm for the species. Caution is indicated for diagnostic use of plasma ChE, because this source of non-specific ChE is more labile and prone to dissociation from inhibitor than is brain ChE (Hill and Fleming 1982). However, acute exposure to potentially lethal levels, at least of OPs, resulted in complete inhibition of plasma ChE activity in many Bald Eagles and other raptors (Elliott et al. 1997b, J. E. Elliott, unpubl. data).

Prior to the famphur study in 1982–83, raptors were not routinely evaluated for OP or CB poisoning. When testing was initiated between March 1984 and March 1985, eight Bald Eagles, two Red-tailed Hawks, and one Great Horned Owl were identified as killed by OP pesticides including fenthion and famphur (Henny et al. 1987). In 1989 and 1990, secondary poisoning of Bald Eagles and Red-tailed Hawks was documented in the Fraser River delta in British Columbia, Canada (Elliott et al. 1996b). Crop contents of the dead raptors, which contained mainly duck parts, included the granular insecticides, carbofuran and fensulfothion (Fig. 4). Elliott et al. (1996b) concluded that enough granular insecticide persists in the low pH conditions of the delta to cause waterfowl kills and secondary poisoning of raptors several months after application, which was supported by subsequent research (Wilson et al. 2002). In 1992–94 additional Bald Eagles and a Red-tailed Hawk in the same area died from phorate, another granular OP insecticide (Elliott et al. 1997b; Fig. 5). Dead eagles usually were found at roost sites rather than in agricultural fields. Persistence of granular formulations causing secondary poisoning is likely not confined to the Fraser River delta, as a similar scenario involving carbofuran poisoning of several hundred waterfowl and some raptors was reported from California (Littrell 1988).

Under laboratory conditions, 14 American Kestrels were presented with House Sparrows (*Passer domesticus*) dermally exposed to Rid-A-Bird (11% fenthion active ingredient). All kestrels died within 3 days (Hunt et al. 1991). In another scenario, Red-tailed Hawks wintering in orchards in central California were dermally exposed to several dormant-season OP sprays (Hooper et al. 1989, Wilson et al. 1991).

Consumption of freshly sprayed insects by raptors can lead to mortality as well. Large numbers of Swainson's Hawks (*B. swainsoni*) from North America died following grasshopper control in Argentina (Woodbridge et al. 1995). During the 1995–96 austral summer, as many as 3,000 individuals were killed in a single incident and at least 18 different incidents were witnessed totaling about 5,000 Swainson's Hawks (Canavelli and Zacagnini 1996). The OP monocrotophos (first associated with raptor deaths in Israel [Mendelssohn and Paz 1977]) was responsible for the Swainson's Hawk deaths. For additional incidents, see the overall compilation of raptor poisonings by OPs and CBs with emphasis on Canada, U.S., and the U.K. (Mineau et al. 1999), and from the U.S. in 1985–94 (Henny et al. 1999). Raptor poisonings have been frequent under current OP and CB use practices (Henny et al. 1999), although only a few products and formulations have been responsible for most of the incidents.

A high proportion of raptor-poisoning cases in the U.K. resulted from deliberate misuse or abuse of OPs and CBs, whereas the proportion of deliberate poisonings was smaller in North America where problems with labeled uses were as frequent as abuse cases (Mineau et al. 1999).

VERTEBRATE-CONTROL CHEMICALS

Chemistry and Toxicology

A variety of chemicals has been used to control mammal, particularly rodent, and bird populations in urban and agricultural situations. The risk of secondary poisoning of raptors can be high, as many raptor species prey on rodents or other targeted small mammals such as ground squirrels, whereas other species are drawn to scavenge carcasses. Secondary poisoning of raptors has been reported for strychnine (Reidinger and Crabtree 1974) and anti-coagulants (Hegdal and Colvin 1988, Newton et al. 1990, Stone et al. 1999). Chemicals such as sodium monofluoroacetate (Compound 1080) are registered in some jurisdictions for control of livestock predators, whereas CH and anticholinesterase insecticides in particular have seen widespread illegal use for predator control in many countries, and have poisoned many raptors directly or secondarily (Mineau et al. 1999).

Strychnine is a convulsant that works by lowering the stimulation threshold of spinal reflexes. It is toxic to birds at low concentrations, with LD_{50}s ranging from 2.0 to 24.0 mg/kg (ww). The Golden Eagle LD_{50} is 4.8 to 8.1 mg/kg (Hudson et al. 1984). Strychnine was widely used in North America in grain baits to control small mammals, including prairie dogs that were considered pests in range and forestlands. Aboveground use was banned in 1983 by the EPA based on secondary poisoning concerns for listed species.

Anti-coagulants now dominate rodent control worldwide. They function by interfering with the action of Vitamin K-dependent clotting factors in the liver, killing the animal via fatal hemorrhaging. The first generation of 4-hydroxy coumarin-based anticoagulants is typified by warfarin, widely used since the 1940s, but to which rodents became resistant in many areas. Second-generation products, such as difenacoum, bromadialone, and brodifacoum, were subsequently developed. These chemicals are used widely around farm buildings, food storage facilities and in urban settings to control commensal rodents. They have greater potency than the first-generation versions, and also are more persistent and toxic to non-target species. Field and forestry use of second-generation anticoagulants has increased and replaced other poisons such as 1080 and zinc phosphide (Eason et al. 2002).

The hazard to wildlife posed by anticoagulants has been known for some time (Mendenhall and Pank 1980, Townsend et al. 1981). Duckett (1984), for example, reported anticoagulants causing a population collapse of Barn Owls (*Tyto alba*) in Malaysia. A field study in Virginia found that attempts to control orchard voles with brodifacoum resulted in the death of at least five radio-tagged Eastern Screech-Owls (*Megascops asio*) (Hegdal and Colvin 1988). Newton et al. (1990) found that 10% of Barn Owls found dead in Britain contained residues of difenacoum or brodifacoum in their livers, and that exposure to those compounds posed a potential threat to populations. Stone et al. (1999) reported 26 raptors that died from hemorrhage with hepatic residues of anticoagulants, principally brodifacoum, but including warfarin, diphacinone and bromadialone. Secondary ingestion was the presumed source, and Great Horned Owls (13 cases) and Red-tailed Hawks (seven cases) were the most often poisoned species, although a variety of other raptor species were affected.

Brodifacoum, in particular, has been used to

remove rats from island seabird colonies in various locations, and thus posed a risk to raptors and scavengers. Bald Eagles were exposed to brodifacoum, but with no evidence of adverse effects during a successful rat-eradication program on Langara Island, British Columbia (Howald et al. 1999). Swamp Harriers were among the wildlife poisoned by secondary ingestion of brodifacoum during rat-control projects on islands in New Zealand (Eason et al. 2002).

Techniques

Programs to routinely monitor raptor debilitation and mortality following "vertebrate-control operations" can provide valuable information on incidence of exposure and poisoning (Newton et al. 1990, Stone et al. 1999). Considerable variation exists in avian sensitivity to different anticoagulants and among species to each chemical, which makes it difficult to determine diagnostic liver residue concentrations. Brodifacoum appears to pose a particular risk to raptors not only due to its greater toxicity in general, and to some owls in particular (Newton et al. 1990), but also because of its greater persistence and widespread use. Finding of any residues of brodifacoum in livers of raptors is cause for concern and an indication of potentially lethal exposure of local populations. More intensive monitoring methods, including live-capture for blood sampling of residues and clotting times, and telemetry of raptor populations at risk may be indicated in specific circumstances (Colvin and Hegdal 1988, Howald et al. 1999).

ROTENONE AND OTHER PISCICIDES

As many as 30 piscicides have been used extensively in fisheries management in the U.S. and Canada since the 1930s. Today, only four are registered for general or selective fish control or sampling (Finlayson et al. 2000). The general piscicides include antimycin and rotenone (most extensively used in the U.S.). Lampricides include lamprecid and bayluscide. Rotenone is a naturally occurring substance derived from the roots of tropical plants in the bean family (Leguminosae). It has been used for centuries to capture fish in areas where these plants are found naturally. Rotenone inhibits a biochemical process at the cellular level, making it impossible for fish to use the oxygen absorbed in the blood and needed for respiration (Oberg 1967).

Fisheries managers in North America began to use rotenone for fisheries management in the 1930s. By 1949, 34 states and several Canadian provinces used rotenone to manage fish populations. The piscicide was applied first to ponds and lakes, and then to streams in the early 1960s (Schnick 1974). Finlayson et al. (2000) reported that rotenone residues in dead fish are generally very low (< 0.1 mg/kg [ww]) and not readily absorbed through the gut of the animal eating the fish. While secondary toxicity of rotenone by a fish-eating bird or mammal does not appear to be an issue, the loss of food supply following rotenone treatment of a lake has been shown to reduce reproductive success of fish-eating raptors and loons. Bowerman (1991) reported significantly lower Bald Eagle production rates in Michigan at inland breeding areas treated within 3.2 km of nests for rough fish removal during the treatment year and 2 years following compared to the same sites in non-treatment years (0.57 vs. 1.30 young per occupied nest). Production was even more reduced when treatment locations were within one km of nesting sites (0.39 vs. 1.31). At most lakes in Michigan, fish were manually removed and not killed with rotenone. California mitigated an impact to nesting Bald Eagles by transferring eggs from a nest to an approved eagle recovery program (California Department of Fish and Game 1991). Similarly, Oregon provided supplemental salmon for a pair of Bald Eagles nesting at Hyatt Reservoir in 1990 following rotenone treatment in the fall of 1989; the pair produced one young (J. L. Kaiser, pers. comm.). Michigan mitigated impacts on loons by delaying treatments until chicks fledged (Finlayson et al. 2000).

Ospreys were studied in Oregon associated with an operational use of rotenone. Nesting populations at Hyatt Reservoir (the treatment) and Howard Prairie Reservoir (the control) were studied for two years before application (Henny and Kaiser 1995). Production rates (young/occupied nest) in 1988 and 1989 were similar at both Hyatt (1.48 and 1.44) and Howard Prairie (1.50 and 1.50). Rotenone was applied in autumn 1989 (after Osprey departure) and nesting numbers did not change appreciably in 1990 at Hyatt (11 nests) with no fish present (not yet restocked) or at Howard Prairie (29 nests). Productivity in 1990 was higher at the control reservoir (2.07), and lower at the treatment reservoir (1.00) (C. J. Henny and J. L.Kaiser, unpubl. data), and correlated with low prey delivery rates at Hyatt. Several young died shortly before fledging at Hyatt in 1990, and more days were required to fledge at Hyatt in 1990, which implies food shortages and a slower growth rate. As in the Michigan Bald

Eagle study, production rates at Hyatt Reservoir were depressed in the second and third years after fish removal (0.55 and 1.09 young/occupied nest in 1991 and 1992).

Magnitude of the rotenone effect seems to be related to two factors: (1) the distance to alternative sources of fish, and (2) the timing of the restocking program. After treatment and restocking with game fish, foraging must change to a different cohort of fish (e.g., trout or bass) that are likely less abundant, and more difficult to capture. Bullheads, suckers and chubs, the usual target species of rotenone operations, are usually abundant, prefer shallow water and are slow-moving (i.e., fish characteristics preferred by Ospreys).

EMERGING CONTAMINANTS

Polybrominated Diphenyl Ethers (PBDEs)

The group of chemicals termed persistent organic pollutants (POPs), which includes "legacy" contaminants such as CH pesticides and PCBs, have certainly posed the most serious threat to raptors, including global population declines. Many POPs-type chemicals are considered important in a variety of commercial applications with large quantities of some compounds continuing to be produced. Polybrominated diphenyl ethers (PBDEs) are widely used as flame-retardants in plastic and textile products. PBDEs can affect thyroid hormone and neuronal systems in laboratory animals (Danerud et al. 2001, Danerud 2003) and persist, bioaccumulate and biomagnify in predatory fish, mammals and birds in many ecosystems (de Wit 2002). PBDE residues were reported in Swedish raptors (Jansson et al. 1993), and a variety of isomers (including supposedly non-accumulative types) have been reported in eggs of Peregrine Falcons from Sweden (Lindberg et al. 2004). The eggs of Little Owls (*Athene noctua*) in Belgium collected in 1998–2000 contained PBDEs (Jespers et al. 2005). PBDEs also were found in Osprey eggs from Maryland and Virginia in 2000 and 2001 (Rattner et al. 2004), and from Washington and Oregon in 2002–2004 (C. J. Henny, unpubl. data). Osprey eggs collected between 1991 and 1997 along major rivers in British Columbia had PBDE concentrations that increased 10-fold over that time period, raising concerns over possible health effects if increases continued (Elliott et al. 2005b). Hydroxylated PBDE metabolites, including known thyroxine mimics, recently were reported in blood samples

Figure 6. Juvenile Bald Eagles (*Haliaeetus leucocephalus*) at a nest in the Fraser Valley of British Columbia, Canada. Pre-fledging birds weigh about 4 kg and can provide a large quantity of blood for measurement of contaminants and biomarkers without any adverse effects. Sampling should be scheduled when young are about 6 weeks of age *(Photo D. Haycock, CWS).*

of Bald Eagle nestlings from British Columbia and California (McKinney et al. 2006; Fig. 6).

Kestrels hatched from eggs injected during incubation with a mixture of PBDEs at a concentration of 1500 ng/g (ww) intended to simulate exposure of Great Lakes Herring Gulls (*Larus argentatus*) exhibited some effects on retinol, thyroid, and oxidative stress parameters (Fernie et al. 2005).

Sulfonated Perfluorochemicals

Perfluoroactane sulfonate (PFOS) was the active ingredient in Scotchguard™ stain and water repellents; perfluoroactanoic acid was used in manufacture of Teflon® and related coatings. In 2000, 3M Corporation committed to eliminate all PFOS use in Scotchguard™ by 2002, while the use of related compounds is undergoing EPA review. These compounds are present as complex mixtures of fluorine atoms substituted on carbon-carbon bonds, which have presented a challenge to the analytical chemist. They have been shown to be persistent and widely transportable in the environment. There is evidence that structurally similar chemicals affect a variety of biological processes including endocrine function. Blood samples of Bald Eagles from various locations in the U.S. had substantial amounts of PFOS, as did livers of White-tailed Eagles from Poland and Germany (Kannan et al. 2001, 2002). PFOS also were found in Osprey eggs from Chesapeake Bay (Rattner et al. 2004). No data are available to determine whether these chemicals are having a significant effect on wild birds.

Diclofenac

In addition to vultures, which do so regularly, many raptor species scavenge dead prey during periods of inclement weather or when normal prey are scarce. Eagles and buteos, in particular, have been lethally exposed to a wide array of contaminants, particularly lead and various pesticides, from scavenging, as documented elsewhere in this report. As obligate scavengers, vultures are at particular risk of exposure to many chemicals. During the 1990s, catastrophic declines in populations of *Gyps* vulture species took place on the Indian subcontinent (Prakesh et al. 2003). A comprehensive investigation of the causes of mortality in the White-rumped Vulture (*Gyps bengalensis*) in Pakistan identified the main factor as renal failure caused by exposure to diclofenac, a non-steroidal anti-inflammatory drug (Oaks et al. 2004). Diclofenac was readily available in the region and widely used to treat hoofed livestock. Vultures appear to consume the drug while feeding on treated livestock, the carcasses of which are typically left for scavengers. There is further evidence that diclofenac also is the major cause of vulture decline in India and probably across the range of the impacted species (Green et al. 2004). Efforts to restrict or alter the use of diclofenac and similar drugs are presently underway, but may be too late to save the white-rumped and possibly the other vulture species in the wild (Green et al. 2004). In May 2006, a letter from the Drug Controller General (India) indicated that diclofenac formulations for veterinary use in India were to be phased out within three months.

Toxins of Biological Origin

We found no reports of raptors poisoned from toxins in algal blooms although sea eagles and Ospreys in particular, could be at risk. Threats from plant toxins are not confined to marine ecosystems. Beginning in the early 1990s in the southeastern U.S., Bald Eagles were found dying from a nervous system condition referred to as avian vacuolar myelinopathy (AVM), thought to originate from feeding on similarly afflicted American Coots (*Fulica americana*) (Thomas et al. 1998). Recent findings point to a toxin present in cyanobacteria, which grow on the common invasive water plant hydrilla, as the cause of AVM (Birrenkott et al. 2004, Wiley et al. 2004).

Such toxic hazards may occur naturally. Halogenated dimethyl bipyrroles, believed to be of natural origin and structurally similar to products of marine chromobacterium, were found to accumulate in tissues of Bald Eagles and seabirds (Tittlemier et al. 1999). A laboratory dosing study with kestrels found evidence of clinical effects, but concluded that those chemicals did not pose an acute reproductive threat to avian populations (Tittlemier et al. 2003). The increasing perturbation and pollution of ecosystems by exotic species, nutrients and contaminants, along with climatic fluctuations, may increase the future likelihood of similar phenomena.

Newly Registered Chemicals

In addition to the thousands of commercial chemicals presently in use, new products are introduced each year. Many jurisdictions require that all pesticides and pharmaceuticals undergo extensive evaluation for toxicity and environmental fate prior to registration for use (www.epa.gov/opptintr/newchems/pubs/expbased.htm). Concern about the development and use of compounds with endocrine-disrupting properties has prompted extensive new screening requirements and requirements to test other types of commercial chemicals (Huet 2000, Gross et al. 2003). Despite those stringent testing protocols, the increased volume and chemical diversity of new products combined with increasing human populations and economic activity almost guarantees that new chemicals or new usage patterns will pose future environmental threats.

Raptors, and, in particular, scavenging species, face increasing and unexpected threats to their survival from the introduction of new commercial chemicals, despite pre-market testing requirements. From the unpredicted effects of DDE on development of eggshells to the exposure and sensitivity of vultures to diclofenac, most of the ecological consequences of those chemicals would not have been identified even by the current relatively rigorous testing procedures.

CONCLUSIONS

With more chemicals registered each year, raptors are exposed to a seemingly endless number of contaminants. At about the time adverse effects of one contaminant or a group of contaminants diminish (usually following much research and a ban or limitation on its use), other contaminants emerge as problems, and the cycle continues. The diversity of raptors inhabiting the planet, with their many feeding strategies and characteristics, place some species in perilous situations. Some traits help raptors cope with selected contaminants. They include pellet-casting by owls and many other raptors which eliminates much ingested lead shot, and demethylation (by many species, especially adults) of toxic MeHg to a less toxic form. However, other traits make entire species or individual populations exceedingly vulnerable to certain contaminants (e.g., flocking behavior of Swainson's Hawks on wintering grounds in Argentina). Scavenging species, including vultures, and many eagles and buteos, particularly are vulnerable to secondary poisoning by feeding on carcasses contaminated by lead shot, pesticides, and veterinary pharmaceuticals. Populations of some species have recovered from DDT. These include the Osprey, which tolerates humans and is now beginning to nest again in many polluted areas, and is being promoted as an indicator species to monitor the health of large rivers, bays, and estuaries, a role the species initially played many years ago. There is an ongoing need to monitor raptor populations, and to investigate reports of poor productivity or unusual mortality and to report it to appropriate authorities.

Readers of this chapter are on the frontline. Many times initial reports of contaminant issues come from field workers who are studying other aspects of raptor biology. We could cite many examples, but space does not permit. The bottom line is that raptor biologists need to remain vigilant.

ACKNOWLEDGMENTS

We appreciate the review of an earlier draft of this chapter by Elwood F. Hill and D. Michael Fry.

LITERATURE CITED

BARRIE, L.A., D. GREGOR, B. HARGRAVE, R. LAKE, D. MUIR, R. SHEARER, B. TRACEY AND T. BIDLEMAN. 1992. Arctic contaminants: sources, occurrence and pathways. *Sci. Total Environ.* 122:1–74.

BENNETT, J.K., S.E. DOMINGUEZ AND W.L. GRIFFIS. 1990. Effects of dicofol on Mallard eggshell quality. *Arch. Environ. Contam. Toxicol.* 19:907–912.

BERG, W., A. JOHNELS, B. SJÖSTRAND AND T. WESTERMARK. 1966. Mercury content in feathers of Swedish birds from the past 100 years. *Oikos* 17:71–83.

BEYER, W.N., G.H. HEINZ AND A.W. REDMON-NORWOOD [EDS.]. 1996. Environmental contaminants in wildlife — interpreting tissue concentrations. Lewis Publishers, Boca Raton, FL U.S.A.

———, D.J. AUDET, G.H. HEINZ, D.J. HOFFMAN AND D. DAY. 2000. Relation of waterfowl poisoning to sediment lead concentrations in the Coeur d'Alene River basin. *Ecotoxicology* 9:207–218.

BIRRENKOTT, H.H., S.B. WILDE, J.J. HAINS, J.R. FISCHER, T.M. MURPHY, C.P. HOPE, P.G. PARNELL AND W.W. BOWERMAN. 2004. Establishing a food-chain link between aquatic plant material and avian vacuolar myelinopothy in Mallards (*Anas platyrhynchos*). *J. Wildl. Dis.* 40:485–492.

BLUS, L.J. 1984. DDE in birds' eggs: comparison of two methods for estimating critical levels. *Wilson Bull.* 96:268–276.

BOWERMAN, W.W., IV. 1991. Factors influencing breeding success of Bald Eagles in upper Michigan. MA thesis, Northern Michigan University, Marquette, MI U.S.A.

———, T.J. KUBIAK, J.B. HOLT, D. EVANS, R.G. ECKSTEIN, C.R. SINDELAR, D.A. BEST AND K.D. KOZIE. 1994. Observed abnormalities in mandibles of nesting Bald Eagles (*Haliaeetus leucocephalus*). *Bull. Environ. Contam. Toxicol.* 53:450–457.

———, D.A. BEST, J.P. GIESY, M.C. SHIELDCASTLE, M.W. MEYER, S. POSTUPALSKY AND J.G. SIKARSKIE. 2003. Associations between regional differences in polychlorinated biphenyls and dichlorodiphenyldichloroethylene in blood of nestling Bald Eagles and reproductive productivity. *Environ. Toxicol. Chem.* 22:371–376.

CADE, T.J. AND W. BURNHAM [EDS.]. 2003. Return of the peregrine: a North American saga of tenacity and teamwork. The Peregrine Fund, Inc., Boise, ID U.S.A.

CALIFORNIA DEPARTMENT OF FISH AND GAME. 1991. Northern pike eradication project - Draft subsequent environmental impact report. Inland Fisheries Division, Sacramento, CA U.S.A.

CANAVELLI, S.B. AND M.E. ZACCAGNINI. 1996. Mortandad de aguilucho langostero (*Buteo swainsoni*) en la region pampeana: primera approximacion al problema. Instituto Nacional de Technologia Agropecuaria, Parana, Entre Rios, Argentina.

CLARK, A.J. AND A.M. SCHEUHAMMER. 2003. Lead poisoning in upland-foraging birds of prey in Canada. *Ecotoxicology* 12:23–30.

COLVIN, B.A. AND P.L. HEGDAL. 1988. Procedures for assessing secondary poisoning hazards of rodenticides to owls. Pages 64–71 in S.A. Shumake and R.W. Bullard [EDS.], Vertebrate pest control and management materials. American Society of Testing and Materials, Philadelphia, PA U.S.A.

COURT, G.S., C.C. GATES, D.A. BOAG, J.D. MACNEIL, D.M. BRADLEY, A.C. FESSER, R.J. PATTERSON, G.B. STENHOUSE AND L.W. OLIPHANT. 1990. A toxicological assessment of Peregrine Falcons *Falco peregrinus tundrius* breeding in the Keewatin District of the Northwest Territories, Canada. *Can. Field-Nat.* 104:255–272.

CZUCZWA, J.M., B.D. MCVEETY AND R.A. HITES. 1984. Polychlorinated dibenzo-p-dioxins and dibenzofurans in sediments from Skskiwit Lake, Isle Royal. *Science* 26:226–227.

DANERUD, P.O. 2003. Toxic effects of brominated flame retardants in man and wildlife. *Environ. Internat.* 29:841–853.

———, G.S. ERIKSEN, T. JOHANNESSON, P.B. LARSEN AND M. VILUK-SELA. 2001. Polybrominated diphenyl ethers: occurrence, dietary exposure and toxicology. *Environ. Health Perspect.* 109 Supplement:49–68.

DESGRANGES, J.-L., J. RODRIQUE, B. TARDIF AND M. LAPERLE. 1998. Mercury accumulation and biomagnification in Ospreys (*Pandion haliaetus*) in the James Bay and Hudson Bay regions of Quebec. *Arch. Environ. Contam. Toxicol.* 35:330–341.

DE WIT, C.A. 2002. An overview of brominated flame retardants in the environment. *Chemosphere* 46:583–624.

DUCKETT, J.E. 1984. Barn owls (*Tyto alba*) and the 'second generation' rat baits utilized in oil palm plantations in Peninsula Malaysia. *Planter* 60:3–11.

DYKSTRA, C.R., M.W. MEYER, D.K. WARNKE, W.H. KARASOV, D.E. ANDERSEN, W.W. BOWERMAN AND J.P. GIESY. 1998. Low reproductive rates of Lake Superior Bald Eagles: low food delivery rates or environmental contaminants? *J. Great Lakes Res.* 24:32–44.

EASON, C.T., E.C. MURPHY, G.R.G. WRIGHT AND E.B. SPURR. 2002. Assessment of risks of brodifacoum to non-target birds and mammals in New Zealand. *Ecotoxicology* 11:35–48.

ECOBICHON, D.J. 1996. Toxic effects of pesticides. Pages 643–689 in C.D. Klaassen [ED.], Casarett and Doull's toxicology: the basic science of poisons, 5th Ed. McGraw-Hill, New York, NY U.S.A.

EISLER, R. 2000. Handbook of chemical risk assessment: health hazards to humans, plants and animals, Vols. 1-3. Lewis Publishers, Boca Raton, FL U.S.A.

ELLIOTT, J.E. AND M.L. HARRIS. 2002. An ecotoxicological assessment of chlorinated hydrocarbon effects on Bald Eagle populations. *Reviews Toxicol.* 4:1–60.

——— AND P.A. MARTIN. 1994. Chlorinated hydrocarbons and shell thinning in eggs of Accipiter hawks in Ontario, 1986–1989. *Environ. Pollut.* 86:189–200.

——— AND R.J. NORSTROM. 1998. Chlorinated hydrocarbon contaminants and productivity of Bald Eagle populations on the Pacific coast of Canada. *Environ. Toxicol. Chem.* 17:1142–1153.

——— AND L. SHUTT. 1993. Monitoring organochlorines in blood of Sharp-shinned Hawks *Accipiter striatus* migrating through the Great Lakes. *Environ. Toxicol. Chem.* 12:241–250.

———, S.W. KENNEDY, D.B. PEAKALL AND H. WON. 1990. Polychlorinated biphenyl (PCB) effects on hepatic mixed function oxidases and porphyria in birds. I. Japanese Quail. *Comp. Biochem. Physiol.* 96C:205–210.

———, S.W. KENNEDY, D. JEFFREY AND L. SHUTT. 1991. Polychlorinated biphenyl (PCB) effects on hepatic mixed function oxidases and porphyria in birds. II. American Kestrel. *Comp. Biochem. Physiol.* 99C:141–145.

———, R.J. NORSTROM, A. LORENZEN, L.E. HART, H. PHILIBERT, S.W. KENNEDY, J.J. STEGEMAN, G.D. BELLWARD AND K.M. CHENG. 1996a. Biological effects of polychlorinated dibenzo-p-dioxins, dibenzofurans, and biphenyls in Bald Eagle (*Haliaeetus leucocephalus*) chicks. *Environ. Toxicol. Chem.* 15:782–793.

———, K.M. LANGELIER, P. MINEAU AND L.K. WILSON. 1996b. Poisoning of Bald Eagles and Red-tailed Hawks by carbofuran and fensulfothion in the Fraser delta of British Columbia, Canada. *J. Wildl. Dis.* 32:486–491.

———, S.W. KENNEDY AND A. LORENZEN. 1997a. Comparative toxicity of polychlorinated biphenyls to Japanese Quail (*Coturnix c. japoniica*) and American Kestrels (*Falco sparverius*). *J. Toxicol. Environ. Health* 51:57–75.

———, L.K. WILSON, K.M. LANGELIER, P. MINEAU AND P.H. SINCLAIR. 1997b. Secondary poisoning of birds of prey by the organophosphorus insecticide, phorate. *Ecotoxicology* 6:219–231.

———, I.E. MOUL AND K.M. CHENG. 1998. Variable reproductive success of Bald Eagles on the British Columbia coast. *J. Wildl. Manage.* 62:518–529.

———, L.K. WILSON, C.J. HENNY, S.F. TRUDEAU, F.A. LEIGHTON, S.W. KENNEDY AND K.M. CHENG. 2001. Assessment of biological effects of chlorinated hydrocarbons in Osprey chicks. *Environ. Toxicol. Chem.* 20:866–879.

———, L.K. WILSON AND B. WAKEFORD. 2005b. Polybrominated diphenyl ether trends in eggs of aquatic and marine birds from British Columbia, Canada, 1979–2002. *Environ. Sci. Technol.* 39:5584–5591.

ELLIOTT, K.H., C.E. GILL AND J.E. ELLIOT. 2005a. The influence of tide and weather on provisioning rates of chick-rearing Bald Eagles in Vancouver Island, British Columbia. *J. Raptor Res.* 39:1–10.

FAIRBROTHER, A. 1996. Cholinesterase-inhibiting pesticides. Pages 52–60 in A. Fairbrother, L.N. Locke, and G.L. Hoft [EDS.], Noninfectious diseases of wildlife, 2nd Ed. University of Iowa Press, Ames, IA U.S.A.

FEIERABEND, J.S. AND O. MYERS. 1984. A national summary of lead poisoning in Bald Eagles and waterfowl. National Wildlife Federation, Washington, DC U.S.A.

FERNIE, K.J., J.E. SMITS, G.R. BORTOLOTTI AND D.M. BIRD. 2001. Reproduction success of American Kestrels exposed to dietary polychlorinated biphenyls. *Environ. Toxicol. Chem.* 20:776–781.

———, J.L. SHUTT, G. MAYNE, D.J. HOFFMAN, R.J. LETCHER, K.G. DROUILLARD AND I.J. RITCHIE. 2005. Exposure to polybrominated diphenyl ethers (PBDEs): changes in thyroid, vitamin A, glutathione homeostasis, and oxidative stress in American Kestrels (*Falco sparverius*). *Toxicol. Sci.* 88:375–383.

FIMREITE, N. AND L. KARSTAD. 1971. Effects of dietary methylmercury on Red-tailed Hawks. *J. Wildl. Manage.* 35:293–300.

FINLAYSON, B.J., R.A. SCHNICK, R.L. CAITTEUX, L. DEMONG, W.D. HORTON, W. MCCLAY, C.W. THOMPSON AND G.J. TICHACEK. 2000. Rotenone use in fisheries management: administrative and technical guidelines manual. American Fisheries Society, Bethesda, MD U.S.A.

FRY, D.M. 2003. Assessment of lead contamination sources exposing California Condors. Final Report submitted to California Department of Fish and Game, Sacramento, CA U.S.A.

GARCIA-FERNANDEZ, A.J., D .ROMERO, E. MARTINEZ-LOPEZ, I. NAVAS, M. PULIDO AND P. MARIA-MOJICA. 2005. Environmental lead exposure in the European Kestrel (*Falco tinnunculus*) from southeastern Spain: the influence of leaded gasoline regulations. *Bull. Environ. Contam. Toxicol.* 74:314–319.

GARCELON, D.K. AND N.J. THOMAS. 1997. DDE poisoning in an adult Bald Eagle. *J. Wildl. Dis.* 33:299–303.

GILBERTSON, M., T.J. KUBIAK, J. LUDWIG AND G.A. FOX. 1991. Great

Lakes embryo mortality, edema, and deformities syndrome (GLEMEDS) in colonial fish-eating birds: similarity to chick edema disease. *J. Toxicol. Environ. Health* 33:455–520.

GILL, C.E. AND J.E. ELLIOTT. 2003. Influence of food supply and chlorinated hydrocarbon contaminants on breeding success of Bald Eagles. *Ecotoxicology* 12:95–111.

GILSLEIDER, E. AND F.W. OEHME. 1982. Some common toxicoses in raptors. *Vet. Hum. Toxicol.* 24:169–170.

GREEN, R.E., I. NEWTON, S. SCHULTZ, A.A. CUNNINGHAM, M. GILBERT, D.J. PAIN, AND V. PRAKESH. 2004. Diclofenac poisoning as a cause of vulture population declines across the Indian subcontinent. *J. Appl. Ecol.* 41:793–800.

GRIER, J.W. 1969. Bald Eagle behavior and productivity responses to climbing to nests. *J. Wildl. Manage.* 33:961–966.

GROSS, T.S., B.S. ARNOLD, M.S. SEPULVEDA AND K. MCDONALD. 2003. Endocrine disrupting chemicals and endocrine active agents. Pages 1033–1098 in D.J. Hoffman, B.A. Rattner, G.A. Burton, Jr. and J. Cairns, Jr. [EDS.], Handbook of ecotoxicology, 2nd Ed. Lewis Publishers, Boca Raton, FL U.S.A.

HEGDAL P.L. AND B.A. COLVIN. 1988. Potential hazard to Eastern Screech-Owls and other raptors of brodifacoum bait used for vole control in orchards. *Environ. Toxicol. Chem.* 7:245–260.

HEINZ, G.H. 1979. Methylmercury: reproductive and behavioral effects on three generations of Mallard ducks. *J. Wildl. Manage.* 43:394–401.

———— AND D.J. HOFFMAN. 2003. Mercury accumulation and loss in Mallard eggs. *Environ. Toxicol. Chem.* 23:222–224.

HELANDER, B.A., A. OLSSON, A. BIGNERT, L. ASPLUND AND K. LITZEN. 2002. The role of DDE, PCB, coplanar PCB and eggshell parameters for reproduction in the White-tailed Sea Eagle (*Haliaeetus albicilla*) in Sweden. *Ambio* 31:386–403.

HENNY, C.J. 1990. Mortality. Pages 140–150 in I. Newton [ED.], Birds of prey: an illustrated encyclopedic survey by international experts. Golden Press Pty. Ltd., Silverwater, Australia.

————. 2003. Effects of mining lead on birds: a case history at Couer d'Alene Basin, Idaho. Pages 755–766 in D.J. Hoffman, B.A. Rattner, G.A. Burton, Jr. and J. Cairns, Jr. [EDS.], Handbook of ecotoxicology, 2nd Ed. Lewis Publishers, Boca Raton, FL U.S.A.

———— AND J.L. KAISER. 1995. Effects of rotenone use to kill "trash" fish on Osprey productivity at a reservoir in Oregon. *J. Raptor Res.* 29:58 (Abstract).

———— AND D.L. MEEKER. 1981. An evaluation of blood plasma for monitoring DDE in birds of prey. *Environ. Pollut.* 25A:291–304.

————, F.P. WARD, K. E. RIDDLE AND R.M. PROUTY. 1982. Migratory Peregrine Falcons *Falco peregrinus* accumulate pesticides in Latin America during winter. *Can. Field-Nat.* 96:333–338.

————, L.J. BLUS AND C.J. STAFFORD. 1983. Effects of heptachlor on American Kestrels in the Columbia Basin, Oregon. *J. Wildl. Manage.* 47:1080–1087.

————, L.J. BLUS, E.J. KOLBE AND R.E. FITZNER. 1985. Organophosphate insecticide (famphur) topically applied to cattle kills magpies and hawks. *J. Wildl. Manage.* 49:648–658.

————, E.J. KOLBE, E.F. HILL AND L.J. BLUS. 1987. Case histories of Bald Eagles and other raptors killed by organophosphorus insecticides topically applied to livestock. *J. Wildl. Dis.* 23:292–295.

————, L.J. BLUS, D.J. HOFFMAN, R.A. GROVE AND J.S. HATFIELD. 1991. Lead accumulation and Osprey production near a mining site on the Coeur d'Alene River, Idaho. *Arch. Environ. Contam. Toxicol.* 21:415–424.

————, L.J. BLUS, D.J. HOFFMAN AND R.A. GROVE. 1994. Lead in hawks, falcons and owls downstream from a mining site on the Coeur d'Alene River, Idaho. *Environ. Monit. Assess.* 29:267–288.

————, W.S. SEEGAR AND T.L. MAECHTLE. 1996. DDE decreases in plasma of spring migrant Peregrine Falcons, 1978–1994. *J. Wildl. Manage.* 60:342–349.

————, P. MINEAU, J.E. ELLIOTT AND B. WOODBRIDGE. 1999. Raptor poisoning and current insecticide use: what do isolated kill reports mean to populations? *Proc. Internat. Ornithol. Congr.* 22:1020–1032.

————, E.F. HILL, D.J. HOFFMAN, M.G. SPALDING AND R.A. GROVE. 2002. Nineteenth century mercury: hazard to wading birds and cormorants of the Carson River, Nevada. *Ecotoxicology* 11:213–231.

————, R.A. GROVE, J.L. KAISER AND V.R. BENTLEY. 2004. An evaluation of Osprey eggs to determine spatial residue patterns and effects of contaminants along the lower Columbia River, USA. Pages 369–388 in R.D. Chancellor and B.-U. Meyburg [EDS.], Raptors Worldwide. World Working Group for Birds of Prey and Owls, Berlin, Germany and MME, Budapest, Hungary.

HILL, E.F. 1988. Brain cholinesterase activity of apparently normal wild birds. *J. Wildl. Dis.* 24:51–61.

————. 1989. Divergent effects of postmortem ambient temperature on organophosphorus and carbamate-inhibited brain cholinesterase in birds. *Pest. Biochem. Physiol.* 33:264–275.

————. 1999. Wildlife toxicology. Pages 1327–1363 in B. Ballantyne, T.C. Marrs, and T. Syversen [EDS.], General and applied toxicology, Vol. 2, 2nd Ed. Macmillan Reference Ltd., London, United Kingdom.

————. 2003. Wildlife toxicology of organophosphorus and carbamate pesticides. Pages 281–312 in D.J. Hoffman, B.A. Rattner, G.A. Burton, Jr. and J. Cairns, Jr. [EDS.], Handbook of ecotoxicology, 2nd Ed. Lewis Publishers, Boca Raton, FL U.S.A.

———— AND W.J. FLEMING. 1982. Anticholinesterase poisoning of birds: field monitoring and diagnosis of acute poisoning. *Environ. Toxicol. Chem.* 1:27–38.

———— AND V.M. MENDENHALL. 1980. Secondary poisoning of Barn Owls with famphur, an organophosphate insecticide. *J. Wildl. Manage.* 44:676–681.

HOFFMAN, D.J., C.P. RICE AND T.J. KUBIAK. 1996. PCBs and dioxins in birds. Pages 165–207 in W.N. Beyer, G.H. Heinz, and A.W. Redman-Norwood [EDS.], Environmental contaminants in wildlife: interpreting tissue concentrations. Lewis Publishers, Boca Raton, FL U.S.A.

————, M. J. MELANCON, P.N. KLEIN, J.D. EISEMANN AND J.W. SPANN. 1998. Comparative developmental toxicity of planar polychlorinated biphenyl congeners in chickens, American Kestrels and Common Terns. *Environ. Toxicol. Chem.* 17:747–757.

HOOPER, M.J., P.J. DETRICH, C.P. WEISSKOPF AND B.W. WILSON. 1989. Organophosphorus insecticide exposure in hawks inhabiting orchards during winter dormant spraying. *Bull. Environ. Contam. Toxicol.* 42:651–659.

HOWALD, G.R., P. MINEAU, J.E. ELLIOTT AND K.M. CHENG. 1999. Brodifacoum poisoning of avian scavengers during rat control on a seabird colony. *Ecotoxicology* 8:431–447.

HUDSON R.H., R.K. TUCKER AND M.A. HAEGLE. 1984. Handbook of toxicity of pesticides to wildlife, 2nd Ed. U.S. Fish and Wildlife

Service, Resour. Publ. 153, Washington, DC U.S.A.

HUET, M.-C. 2000. OECD activity on endocrine disrupter test guideline development. *Ecotoxicology* 9:77–84.

HUNT, K.A., D.M. BIRD, P. MINEAU AND L. SHUTT. 1991. Secondary poisoning hazard of fenthion to American Kestrels. *Arch. Environ. Contamin. Toxicol.* 21:84–90.

JANSSON, B., R. ANDERSSON, L. ASPLUND, K. LITZEN, K. NYLUND, U. SELLSTROM, U.B. UVEMO, C. WAHLBERG, U. WIDEQVST, T. ODSJO AND M. OLSSON. 1993. Chlorinated and brominated persistent organic compounds in biological samples from the environment. *Environ. Toxicol. Chem.* 12:1163–1174.

JARMAN, W.M., S.A. BURNS, W.G. MATTOX AND W.S. SEEGAR. 1994. Organochlorine compounds in the plasma of Peregrine Falcons and Gyrfalcons nesting in Greenland. *Arctic* 47:334–340.

JENSON, S., A.G. JOHNELS, M. OLSSON AND T. WESTERMARK. 1972. The avifauna of Sweden as indicators of environmental contamination with mercury and chlorinated hydrocarbons. *Proc. Internat. Ornithol. Congr.* 15:455–465.

JESPERS, V., A. COVACI, J. MAERVOET, T. DAUWE, S. VOORSPOELS, P. SCHEPENS AND M. EENS. 2005. Brominated flame retardants and organochlorine pollutants in eggs of Little Owls (*Athene noctua*) from Belgium. *Environ. Pollut.* 136:81–88.

JOHNELS, A.G., G. TYLER AND T. WESTERMARK. 1979. A history of mercury levels in Swedish fauna. *Ambio* 8:160–168.

KANNAN, K., J.C. FRANSON, W.W. BOWERMAN, K.J. HANSEN, J.D. JONES AND J.P. GIESY. 2001. Perfluooctane sulfonate in fish-eating water birds including Bald Eagles and albatrosses. *Environ. Sci. Technol.* 35:3065–3070.

———, K., S. CORSOLINI, J. FALANDYSZ, G. OEHME, S. FOCARDI AND J.P. GIESY. 2002. Perflorooctanesulfonate and related florinated hydrocarbons in marine mammals, fishes and birds from coasts of the Baltic and the Mediterranean Seas. *Environ. Sci. Technol.* 36:3210–3216.

KIM, E-Y, R. GATO, H. IWATA, Y. MASUDA, S. TANABE AND S. FUJITA. 1999. Preliminary survey of lead poisoning of Steller's Sea Eagle (*Haliaeetus pelagicus*) and White-tailed Sea Eagle (*Haliaeetus albicilla*) in Hokkaido, Japan. *Environ. Toxicol. Chem.* 18:448–451.

KOEMAN, J.H., J. GARSEEN-HOEKSTRA, E. PELS AND J.J.M. deGOEIJ. 1971. Poisoning of birds of prey by methylmercury compounds. *Meded. Faculteit, Landbouwweton* 36:43–49.

KRAMER, J.L. AND P.T. REDIG. 1997. Sixteen years of lead poisoning in eagles, 1980–95: an epizootiologic view. *J. Raptor Res.* 31:327–332.

KUROSAWA, N. 2000. Lead poisoning in Steller's Sea Eagles and White-tailed Sea Eagles. Pages 107–109 *in* M. Ueta and M. McGrady [EDS.], First Symposium on Steller's and White-tailed sea eagles in east Asia. Wild Bird Society, Tokyo, Japan.

LINDBERG P., U. SELLSTROM, L. HAGGBERG AND C.A. de WIT. 2004. Higher brominated diphenyl ethers and hexabromocyclododecane found in eggs of Peregrine Falcons (*Falco peregrinus*) breeding in Sweden. *Environ. Sci. Technol.* 38:93–96.

LITTRELL, E.E. 1988. Waterfowl mortality in rice fields treated with the carbamate, carbofuran. *Calif. Fish Game* 74:226–231.

LOCKE, L.N. AND M. FRIEND. 1992. Lead poisoning of avian species other than waterfowl, Pages 19–22 in D.J. Pain [ED.], Lead poisoning in waterfowl. IWRB Special Publ. No. 16, Brussels, Belgium.

LOCKIE, J.D., D.A. RATCLIFFE AND R. BALBARRY. 1969. Breeding success and organochlorine residues in Golden Eagles in west Scotland. *J. Appl. Ecol.* 6:381–389.

LUCOTTE, M., A. MUCCI, C. HILLAIRE-MARCEL, P. PICHET AND A. GRONDIN. 1995. Anthropogenic mercury enrichment in remote lakes of northern Quebec (Canada). *Water Air Soil Pollut.* 80:467–476.

MATSUMURA, F. 1985. Toxicology of insecticides, 2nd Ed. Plenum Press, New York NY U.S.A.

McKINNEY, M.A., L.S. CESH, J.E. ELLIOTT, T.D. WILLIAMS, D.K. GARCELON AND R.J. LETCHER. 2006. Novel brominated and chlorinated contaminants and hydroxylated analogues among North American west coast populations of Bald Eagles (*Haliaeetus leucocephalus*). *Environ. Sci. Technol.* 40:6275–6281.

MENDELSSOHN, H. AND U. PAZ. 1977. Mass mortality of birds of prey caused by Azodrin, an organophosphorus insecticide. *Biol. Conserv.* 11:163–170.

MENDENHALL, V.M. AND L.F. PANK. 1980. Secondary poisoning of owls by anticoagulant rodenticides. *Wildl. Soc. Bull.* 8:311–315.

MERETSKY, V.J., N.F.R. SNYDER, S.R. BEISSINGER, D.A. CLENDENEN AND J.W. WILEY. 2000. Demography of the California Condor: implications for reestablishment. *Conserv. Biol.* 14:957–967.

MILLER, M.J.R., M.E. WAYL AND G.R. BORTOLOTTI. 2002. Lead exposure and poisoning in diurnal raptors: a global perspective. Pages 224–245 in R. Yosef, M.L. Miller and D. Pepler [EDS.], Raptors in the new millennium. International Birding and Research Center, Eilat, Israel.

MILLS, J.A. 1973. Some observations on the effects of field applications of fensulfothion and parathion on bird and mammal populations. *Proc. New Zealand Ecol. Soc.* 20:65–71.

MINEAU, P., M.R. FLETCHER, L.C. GLASER, N.J. THOMAS, C. BRASSARD, L.K. WILSON, J.E. ELLIOTT, L.A. LYON, C.J. HENNY, T. BOLLINGER AND S.L. PORTER. 1999. Poisoning of raptors with organophosphorus and carbamate pesticides with emphasis on Canada, U.S. and U.K. *J. Raptor Res.* 33:1–37.

MOORE, N.W. AND D.A. RATCLIFFE. 1962. Chlorinated hydrocarbon residues in the egg of a Peregrine Falcon *Falco peregrinus* from Perthshire. *Bird Study* 9:242–244.

NEWSON, S.C., C.D. SANDAU, J.E. ELLIOTT, S.B. BROWN AND R.J. NORSTROM. 2000. PCBs and hydroxylated metabolites in Bald Eagle plasma: comparison of thyroid hormone and retinol levels. Proc. Internat. Conf. of Environ. Chem., Ottawa, ON, 7–11 May 2000.

NEWTON, I. 1988. Determination of critical pollutant levels in wild populations, with examples from organochlorine insecticides in birds of prey. *Environ. Pollut.* 55:29–40.

——— AND E.A. GALBRAITH. 1991. Organochlorines and mercury in eggs of Golden Eagles *Aquila chrysaetos* from Scotland. *Ibis* 133:115–120.

——— AND M.B. HAAS. 1988. Pollutants in Merlin eggs and their effects on breeding. *Br. Birds* 81:258–269.

———, J.A. BOGAN, AND P. ROTHERY. 1986. Trends and effects of organochlorine compounds in sparrowhawk eggs. *J. Appl. Ecol.* 23:461–478.

———, I. WYLLIE AND P. FREESTONE. 1990. Rodenticides in British Barn Owls. *Environ. Pollut.* 68:101–117.

NISBET, I.C.T. AND R.W. RISEBROUGH. 1994. Relationship of DDE to productivity of Bald Eagles *Haliaeetus leucocephalus* in California and Arizona, USA. Pages 771–773 in B.-U. Meyburg and R.D. Chancellor [EDS.], Raptor Conservation Today. Pica Press, Berlin, Germany.

NORSTROM, R.J., J.P. CLARK, D.A. JEFFREY, H.T. WONAND AND A.P.

GILMAN. 1985. Dynamics of organochlorine compounds in Herring Gulls *Larus argentatus*: distribution of [^{14}C] DDE in free-living Herring Gulls. *Environ. Toxicol. Chem.* 5:41–48.

NRIAGU, J.O., W.C. PFEIFFER, O. MALM, C.M. MAGALHAES DE SOUZA AND G. MIERLE. 1992. Mercury pollution in Brazil. *Nature* 356:389.

OAKS, J.L., M. GILBERT, M.Z. VIRANI, R.T. WATSON, C.U. METEYER, B.A. RIDEOUT, H.L. SHIVAPRASAD, S. AHMED, M.J.I. CHAUDHRY, M. ARSHAD, S. MAHMOOD, A. ALI AND A.A. KHAN. 2004. Diclofenac residues as the cause of vulture population decline in Pakistan. *Nature* 427:630–633.

OBERG, K. 1967. On the principal way of attack of rotenone in fish. *Arch. Zool.* 18:217–220.

O'BRIEN, R.D. 1967. Insecticides action and metabolism. Academic Press, New York, NY U.S.A.

OEHME, G. 2003. On the toxic level of mercury in the eggs of *Haliaeetus*. Pages 247–256 in B. Helander, M. Marquiss, and W. Bowerman [EDS.], Sea Eagle 2000. Proceedings from an international sea eagle conference, Swedish Society for Nature Conservation, Stockholm, Sweden.

OLSSON, A., K. CEDER, A. BERGMAN AND B. HELANDER. 2000. Nestling blood of the White-tailed Sea Eagle (*Haliaeetus albicilla*) as an indicator of territorial exposure to organohalogen compounds: an evaluation. *Environ. Sci. Technol.* 34:2733–2740.

PAIN, D.J. 1996. Lead in waterfowl. Pages 225–264 in W.N. Beyer, G.H. Heinz, and A.W. Redman-Norwood [EDS.], Environmental contaminants in wildlife: interpreting tissue concentrations. Lewis Publishers, Boca Raton, FL U.S.A.

———, J. SEARS AND I. NEWTON. 1994. Lead concentrations in birds of prey from Britain. *Environ. Pollut.* 87:173–180.

PATTEE, O.H., S.N. WIEMEYER, B.M. MULHERN, L. SILEO AND J.W. CARPENTER. 1981. Experimental lead shot poisoning in Bald Eagles. *J. Wild. Manage.* 45:806–810.

PEAKALL, D.B. 1975. Physiological effects of chlorinated hydrocarbons on avian species. Pages 343–360 in R. Hague and V. Freed [EDS.], Environmental dynamics of pesticides. Plenum Press, New York NY U.S.A.

———. 1987. Toxicology. Pages 321–329 in B.A. Giron Pendleton, B.A. Millsap, K.W. Cline, and D.M. Bird [EDS.], Raptor management techniques manual. National Wildlife Federation, Washington, DC U.S.A.

———. 1996. Dieldrin and other cyclodiene pesticides in wildlife. Pages 73–97 in W.N. Beyer, G.H. Heinz, and A.W. Redman-Norwood [EDS.], Environmental contaminants in wildlife: interpreting tissue concentrations. Lewis Publishers, Boca Raton, FL U.S.A.

———, D.G. NOBLE, J.E. ELLIOTT, J.D. SOMERS AND G. ERICKSON. 1990. Environmental contaminants in Canadian Peregrine Falcons, *Falco peregrinus*: a toxicological assessment. *Can. Field-Nat.* 104:244–254.

PETERSON, R.E., H.M. THEOBALD AND G.L. KIMMEL. 1993. Developmental and reproductive toxicity of dioxins and related compounds: cross species comparisons. *Critical Rev. Toxicol.* 23:283–335.

PRAKESH, V., D.J. PAIN, A.A. CUNNINGHAM, P.F. DONALD, N. PRAKESH, A. VERMA, R. GARGI, S. SIVAKUMAR AND A.R. RAHMANI. 2003. Catastrophic collapse of Indian White-backed *Gyps bengalensis* and Long-billed *Gyps indicus* vulture populations. *Biol. Conserv.* 109:381–390.

PROUTY, R.M., W.L. REICHEL, L.N. LOCKE, A.A. BELISLE, E. CRO-

MARTIE, T.E. KAISER, T.G. LAMONT, B.M. MULHERN AND D.M. SWINEFORD. 1977. Residues of organochlorine pesticides and polychlorinated biphenyls and autopsy data for Bald Eagles, 1973–74. *Pestic. Monit. J.* 11:134–137.

RATCLIFFE, D.A. 1970. Changes attributable to pesticides in egg breakage frequency and eggshell thickness in some British birds. *J. Appl. Ecol.* 7:67–115.

RATTNER, B.A., P.C. MCGOWAN, N.H. GOLDEN, J.S. HATFIELD, P.C. TOSCHIK, R.F. LUKEI, JR., R.C. HALE, I. SCHMITZ-ALFONSO AND C.P. RICE. 2004. Contaminant exposure and reproductive success of Ospreys (*Pandion haliaetus*) nesting in Chesapeake Bay regions of concern. *Arch. Environ. Contam. Toxicol.* 47:126–140.

REIDINGER, R.F. AND D.G. CRABTREE. 1974. Organochlorine residues in Golden Eagles, United States: March 1964–July 1971. *Pest. Monit. J.* 8:37–43.

RICE, C.P., P. O'KEEFE AND T. KUBIAK. 2003. Sources, pathways and effects of PCBs, dioxins and dibenzofurans. Pages 501–573 in D.J. Hoffman, B.A. Rattner, G.A. Burton, Jr. and J. Cairns, Jr. [EDS.], Handbook of ecotoxicology, 2nd Ed. Lewis Publishers, Boca Raton, FL U.S.A.

ROOD, B.E., J.F. GOTTGENS, J.J. DELFINO, C.D. EARLE AND T.L. CRISMAN. 1995. Mercury accumulation trends in Florida Everglades and savannas marsh flooded soils. *Water Air Soil Pollut.* 80:981–990.

SCHNICK, S. 1974. A review of the literature on the use of rotenone in fisheries. U.S. Fish and Wildlife Service, LaCrosse, WI U.S.A.

SCHEUHAMMER, A.M. 1987. The chronic toxicity of aluminum, cadmium, mercury and lead in birds: a review. *Environ. Pollut.* 46:263–295.

SIBLEY, R.M., I. NEWTON AND C.H.WALKER. 2000. Effects of dieldrin on population growth rates of sparrowhawks 1963–1986. *J. Appl. Ecol.* 37:540–546.

SMITS J.E., K.J. FERNIE, G.R. BORTOLOTTI AND T.A. MARCHANT. 2002. Thyroid hormone suppression and cell-mediated immunomodulation in American Kestrels (*Falco sparverius*) exposed to PCBs. *Arch. Environ. Contam. Toxicol.* 43:338–344.

SNYDER, N. AND H. SNYDER. 2000. The California Condor, a saga of natural history and conservation. Academic Press, London, United Kingdom.

STANSLEY, W. AND D.E. ROSCOE. 1999. Chlordane poisoning of birds in New Jersey, USA. *Environ. Toxicol. Chem.* 18:2095–2099.

STONE, W.B., J. OKONIEWSKI AND J.R. STEDELIN. 1999. Poisoning of wildlife with anticoagulant rodenticides in New York. *J. Wildl. Dis.* 35:187–193.

TANABE, S. 1988. PCB problems in the future: foresight from current knowledge. *Environ. Pollut.* 50:5–28.

THOMAS, N.J., C.U. METEYER AND L. SILEO. 1998. Epizootic vacuolar myelinopathy of the central nervous system of Bald Eagles (*Haliaeetus leucocephalus*) and American Coots (*Fulica americana*). *Vet. Pathol.* 35:479–487.

THOMPSON, D.R. 1996. Mercury in birds and terrestrial mammals. Pages 341–356 in W.N. Beyer, G. Heinz, and A.W. Redmon-Norwood [EDS.], Environmental contaminants in wildlife: interpreting tissue concentrations. Lewis Publishers, Boca Raton, FL U.S.A.

TITTLEMIER, S.A., R.J. NORSTROM, M. SIMON, W.M. JARMAN AND J.E. ELLIOTT. 1999. Identification and distribution of a novel brominated chlorinated heterocyclic compound in seabird eggs.

Environ. Sci. Technol. 33:26–33.

———., J.A. DUFFE, A.D. DALLAIRE, D.M. BIRD AND R.J. NORSTROM. 2003. Reproductive and morphological effects of halogenated dimethyl bipyrroles on captive American Kestrels (*Falco sparverius*). *Environ. Toxicol. Chem.* 22:1497–1506.

TOWNSEND, M.G., M.R. FLETCHER, E.M. ODAM AND P.I. STANLEY. 1981. An assessment of the secondary poisoning hazard of warfarin to Tawny Owls. *J. Wildl. Manage.* 45:242–248.

VAN DEN BERG, M., L. BIRNBAUM, A.T.C. BOSVELD, B. BRUNSTRÖM, P. COOK, M. FEELEY, J.P. GIESY, A. HANBERG, R. HASEGAWA, S.W. KENNEDY, T. KUBIAK, J.C. LARSEN, F.X.R. VAN LEEUWEN, A.K.D. LIEM, C. NOLT, R.E. PETERSON, L. POELLINGER, S. SAFE, D. SCHRENK, D. TILLITT, M. TYSKLIND, M. YOUNES, F. WÆRN AND T. ZACHAREWSKI. 1998. Toxic equivalency factors (TEFs) for PCBs, PCDDs, and PCDFs for humans and wildlife. *Environ. Health Perspect.* 106:775–792.

WAYLAND, M., L. K. WILSON, J. E. ELLIOTT, M. J. R. MILLER, T. BOLLINGER, M. MCADIE, K. LANGELIER, J. KEATING, AND J. M. W. FROESE. 2003. Mortality, morbidity and lead poisoning of eagles in western Canada, 1986–98. *J. Raptor Res.* 37:8–18.

WEGNER, P., G. KLEINSTAUBER, F. BRAUM, AND F. SCHILLING. 2005. Long-term investigation of the degree of exposure of German Peregrine Falcons (*Falco peregrinus*) to damaging chemicals in the environment. *J. Ornithol.* 146:34–54.

WIEMEYER, S.N., P.R. SPITZER, W.C. KRANTZ, T.G. LAMONT AND E. CROMARTIE. 1975. Effects of environmental pollutants on Connecticut and Maryland Ospreys. *J. Wildl. Manage.* 39:124–139.

———, T.G. LAMONT, C.M. BUNCK, C.R. SINDELAR, F.J. GRAMLICH, J.D. FRASER AND M.A. BYRD. 1984. Organochlorine pesticide, polychlorobiphenyl, and mercury residues in Bald Eagle eggs 1969–79 and their relationships to shell thinning and reproduction. *Arch. Environ. Contam. Toxicol.* 13:529–549.

———, C.M. BUNCK AND A.J. KRYNITZKY. 1988. Organochlorine pesticides, polychlorinated biphenyls and mercury in Osprey eggs 1970–1979 and their relationship to shell thinning and productivity. *Arch. Environ. Contam. Toxicol.* 17:767–787.

———, C.M. BUNCK AND C.J. STAFFORD. 1993. Environmental contaminants in Bald Eagle eggs 1980–1984 and further interpretations of relationships to productivity and shell thickness. *Arch. Environ. Contam. Toxicol.* 24:213–227.

WILEY, F.E., A.H. BIRRENKOTT, S.B. WILDE, T.M. MURPHY, C.P. HOPE, J.J. HAINS AND W.W. BOWERMAN. 2004. Investigating the link between avian vacuolar myelinopathy and a novel species of cyanobacteria. SETAC 2004, 14–18 November, Portland, OR U.S.A. (Abstract).

WILSON, B.W., M.J. HOOPER, E.E. LITTRELL, P.J. DETRICH, M.E. HANSEN, C.P. WEISSKOPF AND J.N. SEIBER. 1991. Orchard dormant sprays and exposure of Red-tailed Hawks to organophosphates. *Bull. Environ. Contam. Toxicol.* 47:717–724.

WILSON, L.K., J.E. ELLIOTT, R.S. VERNON AND S.Y. SZETO. 2002. Retention of the active ingredients in granular phorate, terbufos, fonofos, and carbofuran in soils of the Lower Fraser Valley and their implications for wildlife poisoning. *Environ. Toxicol. Chem.* 21:260–268.

WOODBRIDGE, B., K.K. FINLEY AND S.T. SEAGER. 1995. An investigation of the Swainson's Hawk in Argentina. *J. Raptor Res.* 29:202–204.

WOODFORD J.E., W.H. KRASOV, M.E. MEYER AND L. CHAMBERS. 1998. Impact of 2,3,7,8-TCDD exposure on survival, growth, and behaviour of Ospreys breeding in Wisconsin, USA. *Environ. Toxicol. Chem.* 17:1323–1331.

Reducing Management and Research Disturbance

19

ROBERT N. ROSENFIELD

Department of Biology, University of Wisconsin
Stevens Point, WI 54481 U.S.A.

JAMES W. GRIER

Department of Biological Sciences, North Dakota State University
Fargo, ND 58105 U.S.A.

RICHARD W. FYFE

Box 3263, Fort Saskatchewan, Alberta, T8L 2T2 Canada

INTRODUCTION

Researchers may disturb raptors in several ways during breeding or other seasons, and in so doing skew the results of their fieldwork. For example, disturbance may be a problem in achieving unbiased estimates of reproductive success and other behavior. It is thus desirable to understand and minimize the effects of disturbance on research work and on the birds themselves. Because raptor conservation has received considerable attention, we have much information on the actual or potentially deleterious effects that researchers and managers have had or may inflict on raptors. In this chapter we discuss some of the problems associated with research and management disturbance to raptors and offer possible solutions.

Destructive effects of human activity on raptors are varied and rather well documented in both non-technical and technical publications. The sub-lethal and lethal effects of various toxic chemicals have produced a rich literature (Parker 1976, White et al. 1989, Goldstein et al. 1996, Mineau et al. 1999, Klute et al. 2003, Ratcliffe 2003). Other threats to raptor populations stem from the loss and degradation of habitat due to logging, agriculture, industrial pollution, climate change, recreational activities, weapons-testing noise, and even still, direct persecution through shooting, trapping, and poisoning (Bildstein et al. 1993, White 1994, Fuller 1996, Kirk and Hyslop 1998, Brown et al. 1999, Fletcher et al. 1999, Wood 1999, Noon and Franklin 2002, Klute et al. 2003, Newbrey et al. 2005). Impacts of researcher disturbance on breeding raptors also have been documented, including nesting failures after climbs to nests (Boeker and Ray 1971, Luttich et al. 1971), lowered nesting success (Wiley 1975, Buehler 2000), and displacement of birds from home ranges (Andersen et al. 1986, 1990).

That said, many species of raptors worldwide recently have found ways to co-exist and breed successfully in human-altered and occupied environments, often nesting on man-made structures such as power-line poles, buildings, smoke-stacks and bridges (e.g., Bird et al. 1996). Although raptors as a group often are described as being sensitive to human disturbance, especially when nesting (Newton 1979, Snyder and Snyder 1991, Roberson et al. 2002), recent studies report numerous populations of "forest" raptors nesting successfully in human-dominated landscapes. For example, 70 breeding territories of Northern Goshawks (*Accipiter gentilis*) were found within Berlin, Germany in 1999 (Krone et al. 2005). Cities, in fact, now harbor some of the highest nesting densities yet recorded for some woodland and forest species, including Mississippi Kite (*Ictinia mississippiensis*) (Parker 1996), Red-tailed Hawk (*Buteo jamaicensis*) (Stout et al. 2006) and Cooper's Hawk (*A. cooperii*) (Rosenfield et al. 1995, Boal and Mannan 1999).

Raptor scientists have documented behavioral and demographic differences between raptors nesting in rural areas where disturbance is reduced and those nesting in relatively high-disturbance settings including urban areas where birds are more habituated to human presence and are less wary and, sometimes, more aggressive as well (Götmark 1992, Steidl and Anthony 1996, Bielefeldt et al. 1998, Aradis and Carpaneto 2001, W.E. Stout and A.C. Stewart, pers. comm.; see also Andersen et al. 1989). Northern Goshawks in Britain, central Europe, and Japan nest in close proximity to humans in rural landscapes where some populations are not especially prone to disturbance (Squires and Kennedy 2005). The docile behavior and interactions with humans, for instance, indicate low levels of direct human impact on the Spotted Owl (*Strix occidentalis*) (Gutierrez et al. 1995).

There are several reviews of negative human impacts on raptors (e.g., Stalmaster and Newman 1978, Newton 1979, Keran 1981; see also various species accounts on raptors in the Birds of North America series [Poole 2004]). Much literature suggests that human disturbance is a problem during the nesting period, especially during incubation (e.g., Fyfe and Olendorf 1976, Boal and Mannan 1994, Roberson et al. 2002). Management attempts to lessen such impacts, including buffer zones around nests and timed restrictions on activities, are described by Stalmaster and Newman (1978), Suter and Jones (1981), Grier et al. (1983), Squires and Reynolds (1997), Erdman et al. (1998), Jacobs and Jacobs (2002), and Watson (2004). Attempts to minimize investigator disturbance are varied and include actions such as building tunnels to observation blinds (Nelson 1970, Shugart et al. 1981), limiting the duration of nest visits (Rosenfield and Bielefeldt 1993a, Squires and Kennedy 2006), and using small, silent cameras installed near nests to reduce or eliminate the need for repeated visits to nests by observers (Booms and Fuller 2003, Rogers et al. 2005, Smithers et al. 2005).

The effectiveness of minimizing disturbance associated with research and management activities with nesting raptors is rarely known or reported (Gotmark 1992). This is because disturbance is difficult to measure and, generally, is not directly quantified by raptor researchers (but see Grier 1969, Busch et al. 1978,

White and Thurow 1985, Crocker-Bedford 1990). Also, because raptors tend to nest at relatively low densities, the effects of disturbance may be harder to detect because of difficulties in collecting large enough samples of nests (Gotmark 1992; but see Riffel et al. 1996). Researcher disturbance *per se* is not mentioned in some reviews of management efforts for various raptor species of high conservation profile (Cade et al. 1988, Reynolds et al. 1992, Klute et al. 2003, Andersen et al. 2005). However, one report (United States Fish and Wildlife Service 1998) stated that observations of nests for short periods after the young hatch, or trapping of adults for banding or attaching radio transmitters during nesting, did not cause nest desertion. The report concluded that disturbance usually is not a significant factor affecting the long-term survival of any North American goshawk population.

Grier (1969) found no disturbance effects from a large-scaled, three-year, controlled experimental study of possible effects from climbing to Bald Eagle (*Haliaeetus leucocephalus*) nests in northwestern Ontario, Canada. Similarly, Steenhof (1998) indicated that properly designed field studies have no measurable effect on Prairie Falcon (*Falco mexicanus*) populations; and that during 24 years of research on this species in the Snake River Birds of Prey Natural Conservation Area in Idaho, investigators caused egg or nestling losses at only 11 (0.7%) of 1,555 nesting attempts (Steenhof 1998).

Likewise, during hundreds of thousands of hours of research on and monitoring of Spotted Owls, including more than 2,065 captures with no deaths, there was no clear evidence of significant impact by research activity except for a negative effect on reproduction from backpack radio transmitters (Gutierrez et al. 1995, see below). In a review that compared the effects of investigator disturbance at nests, non-raptors seemed to be more vulnerable to disturbance effects than were raptors (Gotmark 1992). Although sample sizes were small, one possible reason for the disparity may have been that raptor biologists made comparatively fewer visits to nests or employed relatively benign forms of disturbance compared with methods used to study other birds (Gotmark 1992).

Because the problems of general human disturbance are so diverse (Riffel et al. 1996) and are discussed in Chapter 20, we focus on the responsibilities and possible consequences of the actions of researchers and managers. Most literature on disturbance deals with research on breeding raptors. Fyfe and Olendorff (1976) reviewed and provided excellent suggestions for remedying a variety of research and management disturbance problems among nesting raptors. Below we summarize their suggestions and offer modifications of some of their suggestions in light of Gotmark (1992), as well as attempt to coordinate some of our recommendations with those presented elsewhere in this book.

PRELIMINARY CONSIDERATIONS

It is imperative that researchers and managers consult the technical literature, as well as knowledgeable persons during the design of their projects to learn of potential disturbance problems that could arise from field activities, along with ways to minimize such disturbance. They should not rely solely on literature as disturbance effects may not always be mentioned in papers (Gotmark 1992).

Some disturbance problems may be species- or site-specific, or both. For example, White and Thurow (1985) found that Ferruginous Hawks (*B. regalis*) in Utah were quite susceptible to disturbance at their nest. In some areas but not others, Swainson's Hawks (*B. swainsoni*) may desert their nests if they are visited by humans during incubation (Houston 1974, England et al. 1997). Cooper's Hawks in Wisconsin do not desert nests in trees climbed to count eggs (Rosenfield and Bielefeldt 1993a). However, Erdman et al. (1998) cautioned against flushing incubating, congeneric Northern Goshawks there; they do not climb to their nests to count eggs because they believe that such activity will cause the birds to desert their nest, although they did not document this effect.

Within a population, most individuals exhibit variation in behavioral responses to human presence (Grier 1969, Andersen et al. 1989, McGarigal et al. 1991, Gotmark 1992), and some variation relates to the bird's activity at the time of approach. Snail Kites (*Rostrhamus sociabilis*), for example, are approachable on their foraging grounds but tend to be very sensitive to human intrusion around nests (Snyder and Snyder 1991). In response to human presence, breeding Bald Eagles were less likely to flush, and flushed at shorter distances to people than did nonbreeding adults (Steidl and Anthony 1996). Successfully reducing disturbance to raptors may call for close attention to their behavior and a willingness to break off operations if signs of stress become evident, such as prolonged alarm calling, extended absence of adults from a nest (during which time predators may gain access to nests [Craighead and Craighead 1956]), or a shift of activity within home ranges (Andersen et al. 1990).

Raptor scientists should contact the proper agencies for procuring research permits and, when appropriate, seek approval of field procedures from an animal-care entity. Within the U.S. and Canada, wild birds are given legal protection through The Migratory Bird Treaty Act and the Migratory Bird Conservation Act, respectively. Any research that involves disturbing, handling, collecting, or in any way manipulating wild birds requires written approval from the appropriate Federal, State, or Provincial regulatory authorities in North America. Details regarding permit applications and wildlife protection in North America are in Little (1993), and can be obtained directly from the U.S. Fish and Wildlife Service regional offices or the Canadian Wildlife Service, or Provincial wildlife authorities as appropriate. In addition, researchers and managers working with raptors also may need approval for projects including field work from their institution's Animal Care and Use Committee.

When possible, we urge researchers and managers to seek training on the use of techniques by participating in workshops that provide, for example, field instruction by experienced biologists in the natural history of raptors and their usual responsiveness to people, training on how to find and monitor nests, and knowledge of ways to collect reproductive data by observing or climbing to raptor nests (e.g., Jacobs and Jacobs 2002). In Wisconsin, Erdman et al. (1998) indicated that their workshops and field training on the nesting biology of Northern Goshawks generated so much interest and cooperation among U.S. Forest Service personnel that these employees helped double the number of known goshawk territories in a national forest. Workshops often are announced in newsletters published by the Ornithological Societies of North America and by the Raptor Research Foundation. Appropriate government agency offices also are good sources of this information.

An inadvertent and indirect cause of disturbance at nest sites involves public knowledge of the locations and resulting attention. Problems can result if individuals seek to deliberately harm or collect the birds, and when well-intentioned individuals interfere by their presence at the site. Such disturbance from unauthorized falconers, photographers, birders, zoologists, and even wildlife managers is well known. As a result, several concerned wildlife groups have adopted resolutions recommending that such site information be kept confidential, but made available to the appropriate land managers. Problems result not only for the birds, but also for persons working with them, including future access and spending money to guard and protect the sites. The obvious solution to this is maintaining the confidentiality of site-specific information, even in reports, graduate theses, and scientific publications. In case of conflicts with freedom of information laws or regulations, site information may be placed under provisions of special protection or kept in the files of researchers or other agencies not subject to public disclosure. Maps and site-specific information also can be kept at widely dispersed locations, at various offices and in different files, with only statistical results stored in central, public locations. Dispersed information would likely slow access by unauthorized persons, increase inability to find all of the information, and facilitate detection of unauthorized use. Ellis (1982) refers to the treatment of information to ensure protection and privacy as "information management." Information management is extremely important and should be heeded by all persons working with birds of prey. Proper attention to information management not only will reduce disturbance, but also will greatly reduce or eliminate the need for eyrie wardens and other forms of site protection.

BASIC RECOMMENDATIONS FOR REDUCING DISTURBANCE AT NESTS

Nest Desertion

Nest desertion, which is serious, can be unpredictable. In general, the likelihood of desertion varies by nesting stage, among different species, among different individuals within species, and probably gender. Nest desertion due to researcher disturbance is poorly documented in many studies. Nest desertion may be underestimated because of the likelihood that abandoned nests may be preyed upon or scavenged before they are detected (Gotmark 1992). It is generally believed, and some studies show, that nest desertion due to disturbance is more likely to occur early, as opposed to late, in a season. For example, in an intensive 14-year study of Cooper's Hawks in Wisconsin, involving multiple and repeated sources of potential disturbance (e.g., attempts [often successful] to trap adults at *all* stages of nesting, climbs to nests to count eggs and band young), including an estimated cumulative total of more than 3,000 visits over 3–4 months to 330 nests, only four (1.2%) nests were known to have failed due to researcher dis-

turbance. All four were deserted following extended visits of about 1 hour by field workers during the incubation stage (Rosenfield and Bielefeldt 1993a, R. Rosenfield, unpubl. data). In all four instances, only females deserted; males tried unsuccessfully at all four sites to incubate clutches for about 7–10 days following their mates' desertion (R. Rosenfield, unpubl. data).

Human activities near nests with young rarely cause nest abandonment, and then only because of severe disturbance. For example, logging activities including cutting, loading, and skidding within 50–100 m of a goshawk nest can cause abandonment even when 20-day-old nestlings are present (J. Squires, unpubl. data). Intra-seasonal nest desertion due to researcher presence is highly unlikely after young hatch, but Golden Eagle (*Aquila chrysaetos*) pairs whose young were banded in three Rocky Mountain states were more likely to move to alternate nests or not breed the following year than pairs whose young were not banded (Harmata 2002). Nesting Bald Eagles responded similarly in coastal British Columbia, Canada (D. Hancock, unpubl. data), but no such effects have been found for Bald Eagles in Ontario, where a study involving several thousand climbs into nests revealed no difference in nesting success between nests that were climbed into and those that were not (Grier 1969). It also seems likely that little disturbance occurs to other adult raptors, including Eastern Screech Owls (*Megascops asio*) and Barn Owls (*Tyto alba*), that are caught in nestboxes while incubating (Taylor 1991; K. Steenhof, per comm.).

Some species may be quite tolerant of various forms of researcher disturbance during the earlier pre-incubation and incubation periods. For example, using bait birds in traps set out before dawn, Rosenfield et al. (1993b) captured 38 different adult Cooper's Hawks (25 males, 13 females) at 41 nests that were under construction during the pre-incubation period in Wisconsin. Trapping at this time was expeditious because traps were placed precisely where the hawks were expected to appear at dawn. The hawks detected the human-controlled movement of bait birds quickly and were usually caught (or missed) within 0.5 hours. None of the nests were deserted and 98% of the 41 pairs laid eggs; whereas among 127 pairs they discovered at the pre-incubation stage and where trapping was not attempted, 93% laid eggs (Rosenfield and Bielefeldt 1993b).

No nest desertions occurred during 35 years of capturing more than 400 adult Ospreys (*Pandion haliaetus*) in Michigan using a dome-shaped, noose "carpet" trap set over eggs or young (S. Postupalsky, pers. comm.).

Adult Ospreys were attentive and were caught within minutes during incubation, whereas trapping took 1 to 2 hours during the nestling stage. Trapping of Ospreys always was done during rainless periods. Houston and Scott (1992) reported no "adverse effects" with the use of noose carpets to trap adult Ospreys in Saskatchewan. Slip-noose traps also have been used on nests during the incubation period to trap adult Eurasian Sparrowhawks (*A. nisus*) in Scotland and Peregrine Falcons (*F. peregrinus*) in West Greenland, with no known desertions by adults attributed to this disturbance (Newton 1986, W.G. Mattox, unpubl. data). Catching adult American Kestrels (*F. sparverius*) in boxes is a common technique that rarely results in disturbance unless it occurs during egg-laying (K. Steenhof, pers. comm.). It also seems that little disturbance occurs to other adult raptors (e.g., Western Screech Owls [*M. kennicottii*] and Barn Owls) caught in nest boxes when incubating (K. Steenhof, pers. comm.).

Cooper's Hawks in Wisconsin also have been caught during the incubation period using a mist net placed near plucking posts with an owl as a lure (Rosenfield and Bielefeldt 1993a). The limbs and other perches where males pluck and transfer prey to females are usually about 50–100 m from the nest and typically out of view of the nest. Males call immediately to their mates upon arrival at plucking posts, which also alerts the researcher hidden nearby to play a recording of a Cooper's Hawk alarm call to draw attention to the owl. Males, the target of this technique, usually are caught within 15 minutes after they detect the owl. If not, the researchers immediately leave the nest area to minimize disturbance to the incubating female. This type of disturbance has been used at 40 nests and resulted in captures of 35 males and, inadvertently, seven females (the other 33 females remained on their nests), with no nest desertions attributed to this technique (R. Rosenfield and J. Bielefeldt, unpubl. data). Adult male Broad-winged Hawks (*B. platypterus*) and Sharp-shinned Hawks (*A. striatus*) also have been caught at prey-transfer sites in Wisconsin without causing nest desertion (E. Jacobs, pers. comm.).

Raptor biologists often use blinds near the nest to study nesting behavior (e.g., Harris and Clement 1973, Kennedy and Johnson 1986, Bielefeldt et al. 1992, and see Chapter 5 of this book). Blinds are erected either during late incubation or, more often, during the early nestling stage during favorable weather, and often are placed 5–20 m horizontally from and a little above the nest to facilitate observation. Some researchers also have placed blinds within 2 m of Eurasian Spar-

rowhawk nests to allow for more accurate identification of prey (Newton 1978, Geer and Perrins 1981). Such close placement allowed the researchers to extend tongs through a hole in the blind to retrieve some of the songbird prey that were leg-banded; prey were replaced with the tongs after the bands were removed (Geer and Perrins 1981). Initially, the parent birds flew off when tongs were extended, but soon they became so accustomed to the procedure that tugs-of-war developed over prey items that the researchers tried to remove. Adult Prairie Falcons also tolerate observation blinds within 2 m of their nests (Sitter 1983).

Nesting adult raptors appear to habituate to blinds, as well as to people entering and leaving them (e.g., Geer and Perrins 1981, Steenhof 1998, but see Snyder and Snyder 1991 for Snail Kites). Nest abandonment, apparently, is rare, although researchers have generally not detailed their procedures on blind placement and the behavior of adults in response to human activity. Adult females returned to nests within 20 minutes of completion of blind set-up, and no desertions of nests occurred after blinds were installed in about 2 hours and within 5 m of nests when young were about 1 week old at each of three Broad-winged Hawk and five Cooper's Hawk nests in Wisconsin, and at four Peregrine Falcon nests in West Greenland (Rosenfield 1983, Bielefeldt et al. 1992, Rosenfield et al. 1995, R. Rosenfield, pers. obs). On the other hand, at a Gyrfalcon (*F. rusticolus*) nest on Ellesmere Island in 1973, the male, but not the female, abandoned a brood of four young when a wooden blind was relocated from hundreds of meters from the nest to a spot approximately 12 m away (D. Muir and D. M. Bird, pers. comm.). Blinds can be constructed during short work periods (< 2 hours) over a series of days to reduce disturbance of parent raptors (Geer and Perrins 1981, Boal and Mannan 1994).

It is generally assumed that nesting adult raptors will behave normally around blinds, but one adult female Broad-winged Hawk uttered alarm calls and attacked a cloth-covered blind, piercing it with her talons (Rosenfield 1978). In another instance an adult female Peregrine Falcon called and attacked a blind placed near her nest in West Greenland. The female responded to an "apparent" intruding conspecific, herself, because from the nest she could see her image reflected in the blind's one-way glass, which later was angled to prohibit mirroring (Rosenfield et al. 1995a, R. Rosenfield, pers. obs). Both of these females ceased calling within 3 days of blind installation and both fledged all their young. Adult males at these sites seemed disturbed by the blinds. One male Broad-winged Hawk and two male Peregrine Falcons uttered alarm calls when they flew by the blinds and appeared hesitant at times to land on their nests (R. Rosenfield, pers. obs). There was no indication, however, that their hunting activity and prey deliveries were adversely influenced by the presence of blinds (R. Rosenfield, pers. obs). One study has reported that nestling Cooper's Hawks exposed to frequent handling and study from blinds were more likely to die from human causes, especially shooting (Snyder and Snyder 1974).

Compared with observers hidden in blinds, the recent technology of using remote cameras to record nest activities can minimize researcher disturbance at raptor nests (Delaney et al. 1998, Booms and Fuller 2003, Rogers et al. 2005). Cameras are silent, small in size (ca. 12 × 4 × 4 cm, L × W × H), and can be installed on the nest tree or a nearby tree (Delaney et al. 1998), or on rock at a cliff site (e.g., Booms and Fuller 2003, Rogers et al. 2005). In time-lapse cameras, a long (75 m) video cable links the camera to a recording unit and power source, thus allowing researchers to change tapes at locations out of view of adult raptors on nests (Delaney et al. 1998, Booms and Fuller 2003). Responses of nesting birds to camera installation vary by species and individuals, timing of camera placement during the nesting season, and length of time needed for camera installation. Camera set-up time averaged 42 minutes at 20 nests of incubating Mexican Spotted Owls (Delaney et al. 1998), and took an average of about 2 hours at 10 nests with 4–7 day-old Northern Goshawks (Rogers et al. 2005). Camera installation during the mid-incubation to early nestling stage (young = 5 days old) took 2–4 hours at each of three Gyrfalcon cliff nests (Booms and Fuller 2003). Researchers reported no nest abandonment in response to remote cameras used with Mexican Spotted Owls (Delaney et al. 1998), Cooper's Hawks (Estes and Mannan 2003), Gyrfalcons (Booms and Fuller 2003), and Northern Goshawks (Lewis et al. 2004, Rogers et al. 2005). However, Cain (1985) reported abandonment of three Bald Eagle nests after installing cameras during the late incubation and early nestling periods. In some bird studies, miniature remote cameras have attracted predators. Thus, researchers may want to camouflage or hide cameras (Green 2004). Conversely, the use of cameras may repel predators and potentially bias an investigation aimed at documenting nest predation (Green 2004).

Raptor biologists frequently use broadcasts of conspecific vocalizations during population surveys to elicit behavioral responses of woodland raptors, determine their presence, or find nests (e.g., Forsman 1983, Rosenfield et al. 1988, Mosher et al. 1990, McLeod and Andersen 1998; also see Chapter 5). Prolonged playing of calls can lure some adult females repeatedly away from their nests, and broadcast calls also can attract potential avian predators such as American Crows (Corvus brachyrhynchos) (R. Rosenfield, unpubl. data). It is possible that broadcasts of raptor calls could result in nest abandonment or depredation of eggs or nestlings, or both. However, we are not aware of any such reports, or any published recommendations by raptor scientists about minimizing disturbance while using broadcast calls. On the other hand, while conducting experiments to evaluate the probability of detecting nesting Northern Goshawks, researchers did not use broadcast trials during incubation in part because they believed that broadcasts could disturb incubating females and cause egg loss (Roberson et al. 2005). These researchers also ended broadcast trials 2 hours before sunset to reduce the possibility of attracting nocturnal predators (i.e., Great Horned Owls [Bubo virginianus] and fishers [Martes pennanti]) to fledglings.

Lastly, many raptor researchers investigate movement and other behavior of breeding adults through the use of radio marking and associated technology (Fuller et al. 1995, and see Chapter 14 of this manual). Nesting adults sometimes are caught and radiotagged at the incubation stage. For example, across 9 years in West Greenland researchers trapped and radiotagged adult Peregrine Falcons (mostly females) at more than 600 eyries using noose gin traps placed among eggs. They recorded no abandonment at any nests, and did not detect any difference in productivity at nests where adults were radiotagged versus nests that were not disturbed by trapping and radiomarking of adults (W. Mattox, unpubl. data).

Adult raptors are more commonly radiotagged during the nestling stage so as not to compromise the viability of fragile eggs during the time it takes to capture, attach a transmitter, and allow birds to resume nesting activities. Investigators implicitly assume that radiomarked individuals behave and survive normally (Conway and Garcia 2005), especially if radio transmitters are small relative to the animal's mass (Reynolds et al. 2004). Several studies have investigated the effects of radio tagging on the behavior of breeding raptors, and none reported nest abandonment. However, decreased productivity in Golden Eagles, including nesting success, fledglings per occupied territory, and brood size, in one of three breeding seasons was associated with the presence of radio transmitters (Marzluff et al. 1997). Vekasy et al. (1996) reported no effect of radio tagging on Prairie Falcon nesting success and brood size, but indicated that biases may occur in certain years of varying weather and prey availability; they suspected that radio-tagged female Prairie Falcons may have had lower productivity and thus, they tended to quickly release gravid females without attaching radio tags. In related research, Spotted Owls carrying backpack transmitters had lower productivity than leg-banded owls (Foster et al. 1992). Although 25 of 29 radiomarked adult Northern Goshawks successfully fledged young, the annual survival of breeding male goshawks that carried a tailmount was lower than for males that carried backpack-style radio transmitters (Reynolds et al. 2004). Careful selection of an attachment method, practice on captive or wild non-nesting birds and, if required, innovation and testing can minimize potential effects of radio marking raptors and reduce the overall time spent attaching radios to nesting adults (Fuller et al. 1995).

We reiterate that some species may be less tolerant than others of research activity during the nesting season. For instance, breeding adult Gyrfalcons are relatively shy and do not seem to habituate as readily to radio tagging as do other nesting raptors. After being outfitted with a satellite-received, platform transmitter terminal (PTT), one female Gyrfalcon in West Greenland did not feed her young, which eventually died (M. Yates and T. Maechtle, pers. comm.). In another study in Greenland using PTTs, K. Burnham (pers. comm.) has never observed nest abandonment by Gyrfalcons, but he did not radio adults until nestlings were about 20–25 days old and can thus tolerate the several hours that breeding adults, especially females, may take to "accept" transmitters.

Damage to Eggs and Young by Frightened Adults

When incubating or brooding small young, adults often respond to human approach by hunkering down in the nest, presumably to avoid detection. Incubating or brooding adults sometimes also carefully walk to the rim of the nest before flying off. At other times, an adult is disturbed suddenly and bolts so quickly that the eggs or young, which are between or underneath its feet, are catapulted out of the nest cup or scrape onto the nest rim or out of the nest completely. Eggs on the nest rim likely will not be moved back into the nest by an adult, but sometimes young crawl back or are picked up by adults

and returned to the cup (Olsen 1993). It also is possible for an adult to puncture an egg or to trample small young under circumstances of a sudden exit. Fortunately, these types of situations appear to be very rare. Problems are more likely during the days just prior to and after hatching, when adults of most species sit "tighter" (some birds will stay on the nest until a climber is halfway up a tree or down a cliff), making a sudden departure more likely, at a time when small, and weak young are dislodged easily. When researchers cannot see clearly into a tree nest, detect other sign of young (by looking for whitewash on the ground beneath a nest), or otherwise determine that a nest is active, they often tap the nest tree in an attempt to induce detectable movement by a tending adult to confirm occupancy. Tapping also may cause a fast exit by an adult, who accidentally may dislodge ad eject eggs or nestlings from the nest. To reduce this possibility, it is better to seek a distant vantage point and use binoculars or a spotting scope to determine occupancy. If this is not possible, one should slowly approach from a distance in an obvious and visible manner, perhaps even making sounds, so that the adults have an opportunity to detect one's presence and leave the nest in a less frantic manner. Walking tangentially rather than directly toward a nest will help slow the approach and is less threatening to the birds. We recommend tapping trees as a last resort and use only moderately strong repeated strikes, which tend to cause minimal movement in adults. When doing so, one should watch for ejected young. In two cases a nestling was knocked completely from a nest (among the thousands of visits made by the authors to nests of many species of raptors across North America). In one of these instances, R. Rosenfield (unpubl. data) caught a 5 day-old Cooper's Hawk in mid-air and returned the uninjured bird to the nest where it eventually fledged.

Cooling, Overheating, and Loss of Moisture from Eggs or Young

Eggs and small young (less than 7 days old), in particular, are vulnerable to chilling, overheating, and dehydration when the parents are kept away from the nest. The temporary cooling of eggs apparently does not pose a serious problem during normal field procedures, and some species can tolerate adult trapping procedures during incubation (see Nest Desertion above). Researchers that climb raptor nests during the incubation period to determine clutch size generally do not report weather or other conditions at the time they counted eggs

(Reynolds and Wight 1978, Janik and Mosher 1982, Andrusiak and Cheng 1997, Petty and Fawkes 1997). Climbs of < 10 minutes at more than 500 Cooper's Hawk nests in Wisconsin did not appear to result in nest abandonment or egg loss due to cooling (R. Rosenfield and J. Bielefeldt, unpubl. data); nests were never climbed when temperatures were < 18°C, and the estimated maximum time that females were off their nests during such visits for clutch counts was 20 minutes.

Nestling raptors in hot environments, or in nests exposed to direct sunlight, may face extreme thermal- and water-balance problems. When stressed these individuals rely heavily on increased respiratory water loss via panting to combat hyperthermia. Heat-induced death of nestlings has been reported in several species of raptors, including Red-tailed Hawk (Fitch et al. 1946), Galapagos Hawk (*B. galapagoensis*) (deVries 1973), Golden Eagle (Beecham and Kochert 1975) and Peregrine Falcon (Nelson 1969). In most nestling birds of prey, the only source of water (except for metabolic water) is from food provided by the parents (Kirkley and Gessaman 1990), and thus missed feedings due to prolonged researcher presence may, besides diminishing nutrient intake, compromise water balance of chicks. Older nestlings avoid exposure to direct sunlight by moving to shaded parts of the nest, and when the nest is exposed to direct sunlight, attending adults shade their young with outstretched wings and tail. Heat-stressed nestlings tend to position themselves on the perimeter of the nest, presumably to enhance the effectiveness of convective cooling (Kirkley and Gessaman 1990). Some young may be heat-stressed and already near their limits of tolerance even without the added burden of disturbance. The situations vary with location (e.g., cooling is more likely in the Arctic whereas drying occurs in desert or grassland areas, and overheating in lower latitudes), although panting in response to direct sunlight can occur even at seemingly cool temperatures. For instance, a pair of 17 day-old Red-tailed Hawks began to pant in the early-morning sunlight at 08:30 when the air temperature was 13°C (Kirkley and Gessaman 1990). Temperature and humidity are generally most favorable for nestlings in forested areas. Extremes are possible in all places however, and should always be considered. Wind, precipitation, and direct sunlight can exacerbate the situation. The times that nestlings can be exposed to adverse conditions are increased in timid species such as Gyrfalcons, Golden Eagles, and Snowy Owls (*B. scandiaca*), where parental birds stay away from the nest for extended times wait-ing for intruders to leave the area. To avoid such situations, keep visits as brief and unobtrusive as possible and consider weather, position of the sun, and time of day. If possible, do not visit nests at time of hatching, or during periods of extreme weather and avoid visiting unshaded nests during the hottest part of the day. If visits during inclement conditions are necessary and unavoidable, put the eggs or young in a fur-lined glove or protective container, or cover them with a piece of cloth or branches with leaves. Do not conduct adult trapping activities at nests until young can thermoregulate. See Steenhof et al. (1994) and Erdman et al. (1998) for details. And terminate trapping activity about 2 hours before sunset to allow adults ample time to return to their nest and resume normal behavior.

Premature Fledging and Banding Young

Fledging occurs when young first leave the nest. In most species, fledging is a gradual process that includes combinations of climbing, jumping, and flapping before flight feathers of the wings and tail are completely grown and sustained flight is possible (Newton 1986, Rosenfield and Bielefeldt 2006). Approximate fledging dates of selected North American and European raptors can be found online in the Birds of North America accounts (Poole 2004), Newton (1979b, Table 18), and Cramp and Simmons (1980). In most studies of raptor productivity researchers visit nests to count and, at the same time, band young. Fyfe and Olendorff (1976) suggested that the optimum time for banding is when the young are approximately one-half to two-thirds fledging age. This is because until about one-half of fledging age, a nestling's legs and feet are not fully grown, and a band may slip down a leg and encircle the foot. Prior to two-thirds fledging age, nestlings tend to move minimally when researchers are at the nest. Nestlings also tend to struggle less when handled at these ages, and banding at this stage is relatively straightforward and proceeds relatively quickly. Thus, overall time at the nest is reduced. When a researcher reaches a nest with older, unfledged youth, the young often spread their wings, move quickly to the opposite edge of the nest, and lean backward (often precariously) in a defensive posture. At this age, young are easily startled and may fledge or leave the nest prematurely by trying to fly off, or by stepping to the edge of the nest or onto branches or cliff ledges from which they may fall. The results of such falls depend on the bird's age and condition, and where it lands. Finally, there always is the risk of injury, loss, or increased vulnerability to predation due to premature fledging.

When reaching for older nestlings we recommend moving a hand slowly toward and at the level of the individual's feet, and allowing the bird to grab your hand if possible so as to establish reliable contact. If several birds are about to jump from a tree nest, try and capture them one at a time by reaching up and letting them grab a hand without putting your (obtrusive) head and shoulders above the nest. A makeshift poultry hook is useful in some situations (Grier 1969). We also recommend putting older young in a backpack to confine their movements. When the pack is closed, the darkness inside the bag seems to calm the nestlings. If young do jump, take care to mark where they went, retrieve them and, if there are no injuries, place in the nest one at a time. Again keep most of your body below the nest and replace the "jumper" nestlings in the reverse order in which they jumped. Thus, the young that jumped first and, presumably, would be more likely to jump again, will be minimally disturbed. One should then depart slowly and quietly from the nest. Premature fledging is best avoided by visiting nests early in the season. When older nestlings are encountered, they are best left alone or approached slowly and handled with extra caution. If handled carefully and slowly, young can be distracted from fledging and will adjust to the presence of the intruder. We also recommend that observations from blinds be discontinued about 3 to 4 days before the young are due to fledge to avoid causing them to leave the nest prematurely (Geer and Perrins 1981, Rosenfield et al. 1995).

Steenhof (1987) recommended visiting nests when young are about at 80% fledging age to assess nest success and productivity, a somewhat later nest visitation time than the one-half to two-thirds fledging age discussed above (Fyfe and Olendorff 1976). Steenhof and Newton (Chapter 11, this volume) now encourage determining an appropriate standard for timing of nest visits to assess productivity and band young of various raptor species. There are, however, temporal differences in behavioral development among species, and among populations of the same species (i.e., young may develop more slowly or more quickly in some populations [Rosenfield and Bielefeldt 1993a, 2006; Curtis and Rosenfield 2006; S. Postupalsky, pers. comm.). Temporal differences are sometimes accentuated in raptors because of reversed size dimorphism, in which smaller males develop faster and fledge earlier than females. For example, male Ferruginous Hawk nestlings leave nests about 10 days earlier than female nestlings

(Bechard and Schmutz 1995). A universal application of an 80% fledging-age metric may make it difficult for researchers to capture mobile young, lead to unsafe handling of older young, and result in premature fledging, all of which can result in inaccurate productivity estimates. Consequently, we recommend that counts and banding of young should be done when young are at about 70% fledging age for Cooper's Hawks in British Columbia, North Dakota, and Wisconsin (Rosenfield and Bielefeldt 1999, 2006; Stout et al. 2007; A Stewart, unpubl. data); and at about 65% fledging age for Red-shouldered Hawks (*B. lineatus*) in Wisconsin (E. and J. Jacobs, unpubl. data); and 55% fledging age for Sharp-shinned Hawks in Wisconsin (E. Jacobs and R. Rosenfield, unpubl. data). Researchers should be cognizant of the possibility of population-specificity in nestling development when timing their nest visits, and should first learn how to handle nestling raptors in the field by spending time in the field with experienced researchers.

Avian and Mammalian Predation

Fyfe and Olendorff (1976) indicated that avian predators including jaegers, gulls, and corvids often visually cue onto unattended nests, and that after researchers had disturbed nests, predators might raid nests while the adults are away. Although in non-raptorial species avian predators have been shown to respond to or follow field workers and to prey on nests visited by investigators there is no direct evidence of this in the literature concerning raptor nests (Gotmark 1992). Even so, a crow (*Corvus* sp.) has been observed throwing one of two, unattended, small Great Horned Owl nestlings out of a nest about 30 minutes after a researcher climbed to it. Although the researcher returned this unhurt bird to its nest, several days later researchers found both owlets dead at the base of the nest tree and attributed their deaths to attacks by crows due to his presence at the nest (Craighead and Craighead 1956). Adult Cooper's Hawks are frequently mobbed and rarely struck by Northwestern Crows (*C. caurinus*) when researchers are near their nests in British Columbia, but the crows do not visit nests during the presence of researchers and there is no evidence of unattended eggs or nestlings being preyed upon by crows after visits (A. Stewart and R. Rosenfield, pers. obs.). Gotmark (1992) suggested that when avian predation has been documented, the predators responded opportunistically to unattended nests or young rather than to observer presence *per se*.

Some authors assume or emphasize that mammalian predators might find nests by following scent trails left by researchers during their nest visits (e.g., Hamerstrom 1970, Poole 1981, Gawlick et al. 1988), and that this problem is particularly serious for ground-nesting raptors (Fyfe and Olendorff 1976). Mammals also are sometimes thought to follow tracks in the vegetation made by observers. In his review, Gotmark (1992) found no evidence of increased predation by mammals due to researcher presence at nests. He also was unable to locate a study that documented mammalian predators following observers. He noted, however, that if precautions like avoiding the creation of trails in vegetation is effective (as recommended by Hamerstrom [1970]) and were being taken by researchers, such behavior may have influenced his inability to find investigator effects. Finally, to avoid drawing attention to a nesting area, one should withdraw from the site to complete field notes.

Mishandling Birds

Both raptors and handlers can be injured during improper handling. Young birds with growing bones, feathers, and talons are particularly vulnerable (see Chapter 12). How to handle birds correctly is best learned in the field from someone with experience.

Miscellaneous Considerations

A number of precautions can help reduce disturbance to raptors by observers, researchers and managers. These include using teams of two people instead of single individuals and giving special care to banding and marking of raptors. Using two workers enhances safety both for the researchers and birds, and permits greater efficiency in note-taking and carrying equipment, which, in turn, reduces the amount of time spent in the area. In addition, Speiser and Bosakowski (1991) noted that two or more observers elicited milder, less aggressive encounters with nesting adult goshawks (which sometimes strike researchers).

When trapping breeding adult raptors, some researchers advocate using mist nets rather than dho-gazas (see Chapter 12). The use of mist nets probably lowers time spent at the nest because they do not collapse after a strike and, therefore less time is needed to reset the net. Contact between a lure owl and a trapped bird rarely occurs with mist nets, a possibility that often is uncontrollable with the dho-gaza (Steenhof et al.

1994, Erdman et al. 1998). We recommend using broadcasts of conspecific calls while conducting adult trapping activities at nests — especially in wooded areas where visibility is limited — because they often more quickly draw attention of parents to a decoy and can reduce time spent at the nest (Erdman et al. 1998, R. Rosenfield, unpubl. data). Steenhof et al. (1994) reported that broadcasting Great Horned Owl calls did not expedite trapping American Kestrels.

Many species of nesting raptors are surveyed or studied from fixed-wing aircraft or helicopters without adverse disturbance effects (e.g., Grier et al. 1981, Kochert 1986, Andersen et al. 1989, Watson 1993, McLeod and Andersen 1998, Kochert et al. 2002). Knowledge of a species' tolerance to low-level flying aircraft is critical and researchers should use only experienced pilots when surveying raptors (Kochert 1986). In a novel study, White and Nelson (1991) monitored habitat use and the hunting behavior of a male Peregrine Falcon and a female Gyrfalcon by following these nesting adults (even in hunting stoops!), with helicopters at a distance of 30–50 m. They emphasized that despite the potential lethal threat of doing so, both to the birds and the human observers (some Gyrfalcons attack helicopters), the technique produced information almost impossible to collect by more conventional methods (see Chapter 5). The young at one of their study nests, however, were depredated about 3 weeks after the project ended. But either the same adult pair or another used the same eyrie the following year.

ACKNOWLEDGMENTS

Our research has been funded by the National Audubon Society, the National Wildlife Federation, the Canadian Wildlife Service, the U.S. Fish and Wildlife Service, the U.S. Forest Service, the North Dakota Game and Fish Department, the North Dakota, Wisconsin, and Great Lakes Falconers' Association, the Personnel Development Committee and the Letters and Science Foundation of the University of Wisconsin-Stevens Point. We thank K. Burnham, W. Stout, W. Mattox, P. Kennedy, S. Postupalsky, E. Jacobs, and A. Stewart for providing information and helpful discussions regarding researcher disturbance. D. Ellis, J. Bielefeldt, K. Steenhof, A. Stewart, and K. Bildstein improved the manuscript with their editorial suggestions. R. Rosenfield acknowledges sabbatical support of his campus.

LITERATURE CITED

ANDERSEN, D.E., O.J. RONGSTAD AND W.R. MYTTON. 1986. The behavioral response of a Red-tailed Hawk to military training activity. *Raptor Res.* 20:65–68.

———, O.J. RONGSTAD AND W.R. MYTTON. 1989. Response of nesting Red-tailed Hawks to helicopter overflights. *Condor* 91:296–299.

———, O.J. RONGSTAD AND W.R. MYTTON. 1990. Home-range changes in raptors exposed to increased human activity levels in southeastern Colorado. *Wildl. Soc. Bull.* 18:134–142.

———, S. DESTEFANO, M.I. GOLDSTEIN, K. TITUS, C. CROCKER-BEDFORD, J.J. KEANE, R.G. ANTHONY AND R.N. ROSENFIELD. 2005. Technical review of the status of Northern Goshawks in the western United States. *J. Raptor Res.* 39:192–209.

ANDRUSIAK, L.A. AND K.M. CHENG. 1997. Breeding biology of the Barn Owl (*Tyto alba*) in the lower mainland of British Columbia. Pages 38–46 in J.R. Duncan, D.H. Johnson, and T.H. Nicholls [EDS.]. Biology and conservation of owls of the Northern Hemisphere. USDA Forest Service General Technical Report NC-190, North Central Research Station, St. Paul, MN U.S.A.

ARADIS, A. AND G.M. CARPANETO. 2001. A survey of raptors on Rhodes: an example of human impacts on raptor abundance and distribution. *J. Raptor Res.* 35:70–71.

BECHARD, M. J. AND J.F. SCHMUTZ. 1995. Ferruginous Hawk (*Buteo regalis*), No. 172 in A. Poole and F. Gill [EDS.], The birds of North America. The Birds of North America, Inc., Philadelphia, PA U.S.A.

BEECHAM, J.J. AND M.N. KOCHERT. 1975. Breeding biology of the Golden Eagle in southwestern Idaho. *Wilson Bull.* 87:506–513.

BIELEFELDT, J., R.N. ROSENFIELD AND J.M. PAPP. 1992. Unfounded assumptions about diet of the Cooper's Hawk. *Condor* 94:427–436.

———, R.N. ROSENFIELD, W.E. STOUT AND S.M. VOS. 1998. The Cooper's Hawk in Wisconsin: a review of its breeding biology and status. *Passenger Pigeon* 60:111–121.

BILDSTEIN, K.L., J. BRETT, L. GOODRICH AND C. VIVERETTE. 1993. Shooting galleries: migrating raptors in jeopardy. *Am. Birds* 47:38–43.

BIRD, D.M., D.E. VARLAND AND J.J. NEGRO [EDS.]. 1996. Raptors in human landscapes: adaptations to built and cultivated environments. Academic Press, London, United Kingdom.

BOAL, C.W. AND R.W. MANNAN. 1994. Northern Goshawk diets in ponderosa pine forests on the Kaibab Plateau. *Stud. Avian Biol.* 16:97–102.

——— AND R.W. MANNAN. 1999. Comparative breeding ecology of Cooper's Hawks in urban and exurban areas of southeastern Arizona. *J. Wildl. Manage.* 63:77–84.

BOEKER, E.L. AND T.D. RAY. 1971. Golden Eagle populations studies in the southwest. *Condor* 73:463–467.

BOOMS, T.L. AND M.R. FULLER. 2003. Time-lapse video system used to study nesting Gyrfalcons. *J. Field Ornithol.* 74:416–422.

BROWN, B.T., G.S. MILLS, C. POWELS, W.A. RUSSELL, G.D. THERRES AND J.J. POTTIE. 1999. The influence of weapons-testing noise on Bald Eagle behavior. *J. Raptor Res.* 33:227–232.

BUEHLER, D.A. 2000. Bald Eagle (*Haliaeetus leucocephalus*), No. 506 in A. Poole and F. Gill [EDS.], The birds of North America. The Birds of North America, Inc., Philadelphia, PA U.S.A.

BUSCH, D.E., W.A. DEGRAW AND N.C. CLAMPITT. 1978. Effects of handling-disturbance on heart rate in the Ferruginous Hawk (*Buteo regalis*). *Raptor Res.* 12:122–125.

CADE, T.J., J.H. ENDERSON, C.G. THELANDER AND C.M. WHITE. 1988. Peregrine Falcon populations: their management and recovery. The Peregrine Fund, Inc., Boise, ID U.S.A.

CAIN, S.L. 1985. Nesting activity time budgets of Bald Eagles in southeast Alaska. M.S. thesis, University of Montana, Missoula, MT U.S.A.

CONWAY, C.J. AND V. GARCIA. 2005. Effects of radiotransmitters on natal recruitment of Burrowing Owls. *J. Wildl. Manage.* 69:404–408.

CRAMP, S. AND K.E.L. SIMMONS. 1980. Handbook of the birds of Europe the Middle East and North Africa, Vol. I. Oxford University Press, Oxford, United Kingdom.

CRAIGHEAD, J.J. AND F.C. CRAIGHEAD. 1956. Hawks, owls, and wildlife. Stackpole Co., Harrisburg, PA U.S.A.

CROCKER-BEDFORD, D.C. 1990. Goshawk reproduction and forest management. *Wildl. Soc. Bull.* 18:262–269.

CURTIS, O.E. AND R.N. ROSENFIELD. 2006. Cooper's Hawk (*Accipiter cooperii*). The Birds of North America Online. The Birds of North America Online database: http//bna.birds.cornell.edu/. Cornell Laboratory of Ornithology, Ithaca, NY U.S.A.

DELANEY, D.K., T.G. GRUBB AND D.K. GARCELON. 1998. An infrared video camera system for monitoring diurnal and nocturnal raptors. *J. Raptor Res.* 32:290–296.

DE VRIES, T. 1973. The Galapagos Hawk: an eco-geographical study with special reference to its systematic position. Ph.D. dissertation, Free University of Amersterdam, Netherlands.

ELLIS, D.H. 1982. The Peregrine Falcon in Arizona: habitat utilization and management recommendations. Institute for Raptor Studies Research Report I, Oracle, AZ U.S.A.

ENGLAN, A.S., M.J. BECHARD AND C.S. HOUSTON. 1997. Swainson's Hawk (*Buteo swainsoni*), No. 265 in A. Poole and F. Gill [EDS.], The Birds of North America. The Birds of North America, Inc., Philadelphia, PA U.S.A.

ERDMAN, T.C., D.F. BRINKER, J.P. JACOBS, J. WILDE AND T.O. MEYER. 1998. Productivity, population trend, and status of Northern Goshawks, *Accipiter gentilis atricapillus*, in northeastern Wisconsin. *Can. Field-Nat.* 112:17–27.

ESTES, W.A. AND R.W. MANNAN. 2003. Feeding behavior of Cooper's Hawks at urban and rural nests in southeastern Arizona. *Condor* 105:107–116.

FITCH, H.S., F. SWENSON AND D.F. TILLOTSON. 1946. Behavior and food habits of the Red-tailed Hawk. *Condor* 48:205–237.

FLETCHER, R.J., JR., S.T. MCKINNEY AND C.E. BOCK. 1999. Effects of recreational trails on wintering diurnal raptors along riparian corridors in a Colorado grassland. *J. Raptor Res.* 33:233–239.

FORSMAN, E.D. 1983. Methods and materials for locating and studying Spotted Owls in Oregon. USDA Forest Service General Technical Report PNW-GTR 162, Pacific Northwest Research Station, Portland, OR U.S.A.

FOSTER, C.C., E.D. FORSMAN, E.C. MESLOW, G.S. MILLER, J.A. REID, F.F. WAGNER, A.B. CAREY AND J.B. LINT. 1992. Survival and reproduction of radio-marked adult Spotted Owls. *J. Wildl. Manage.* 56:91–95.

FULLER, M.R. 1996. Forest raptor population trends in North America. Pages 167–208 in R.M. Degraff and R.I. Miller [EDS.], Conservation of faunal diversity in forested landscapes. Chapman Hall, NY U.S.A.

———, W.S. SEEGAR, J.M. MARZLUFF AND B.A. HOOVER. 1995.

Raptors, technological tools and conservation. *Trans. N. Am. Wildl. Nat. Resour. Conf.* 61:131–141.

FYFE, R.W. AND R.R. OLENDORFF. 1976. Minimizing the dangers of nesting studies to raptors and other sensitive species. Occas. Paper No. 23. Canadian Wildlife Service, Ottawa, Ontario, Canada.

GAWLICK, D.E., M.E. HOSTETLER AND K.L. BILDSTEIN. 1988. Napthalene mothballs do not deter mammalian predators at Red-winged Blackbird nests. *J. Field Ornithol.* 59:189–191.

GEER, T.A. AND C.M. PERRINS. 1981. Notes on observing nesting accipiters. *J. Raptor Res.* 15:45–48.

GOLDSTEIN, M.I., B. WOODBRIDGE, M.E. ZACCAGNINI AND S.B. CANAVELLI. 1996. An assessment of mortality of Swainson's Hawks on wintering grounds in Argentina. *J. Raptor Res.* 30:106–107.

GÖTMARK, F. 1992. The effects of investigator disturbance on nesting birds. Pages 63–104 in D.M. Power [ED.], Current Ornithology, Vol. 9. Plenum Press, New York, NY U.S.A.

GREEN, R.E. 2004. Breeding biology. Pages 57–83 in W.J. Sutherland, I. Newton and R.E. Green [EDS.], Bird ecology and conservation: a handbook of techniques. Oxford University Press, New York, NY U.S.A.

GRIER, J.W. 1969. Bald Eagle behavior and productivity responses to climbing to nests. *J. Wildl. Manage.* 33:961–966.

———, J.M. GERRARD, G.D. HAMILTON AND P.A. GRAY. 1981. Aerial visibility bias and survey techniques for nesting Bald Eagles in northwestern Ontario. *J. Wildl. Manage.* 45:83–92.

———, F.J. GRAMLICH, J. MATTSSON, J.E. MATHISEN, J.V. KUSSMAN, J.B. ELDER AND N.F. GREEN. 1983. The Bald Eagle in the northern United States. Pages 41–66 in S.A. Temple [ED.], Bird Conservation. University of Wisconsin Press, Madison, WI U.S.A.

GUTTIERREZ, R.J., A.B. FRANKLIN AND W.S. LAHAYE. 1995. Spotted Owl (*Strix occidentalis*), No. 179 in A. Poole and F. Gill [EDS.], The Birds of North America. The Birds of North America, Inc., Philadelphia, PA U.S.A.

HAMERSTORM, F. 1970. Think with a good nose near a nest. *Raptor Res. News* 4:79–80.

HARMATA, A. 2002. Encounters of Golden Eagles banded in the Rocky Mountain West. *J. Field Ornithol.* 73:23–32.

HARRIS, J.T. AND D.M. CLEMENT. 1975. Greenland Peregrines at their eyries. *Medd. Grønl.* 205:1–28.

HOUSTON, C.S. 1974. Mortality in ringing: a personal viewpoint. *Ring* 80:215–220.

——— AND F. SCOTT. 1992. The effect of man-made platforms on Osprey reproduction at Loon Lake, Saskatchewan. *J. Raptor Res.* 26:152–158.

JACOBS, J.P. AND E.A. JACOBS. 2002. Conservation assessment for Red-shouldered Hawk on national forests for north central states. Unpubllished Report, USDA Forest Service, Eastern Region. www.fs.fed.us/r9/wildlife/tes/ca-overview/docs/red-shoulderedhawk_ca_1202final.pdf

JANICK, C.A. AND J.A. MOSHER. 1982. Breeding biology of raptors in the central Appalachians. *Raptor Res.* 16:18–24.

KENNEDY, P.L. AND D.R. JOHNSON. 1986. Prey-size selection in nesting male and female Cooper's Hawks. *Wilson Bull.* 98: 110–115.

KERAN, D. 1981. The incidence of man-caused and natural mortalities to raptors. *Raptor Res.* 13:65–78.

KIRK, D.A. AND C. HYSLOP. 1998. Population status and recent trends in Canadian raptors: a review. *Biol. Conserv.* 83:91–118.

KIRKLEY, J.S. AND J.A. GESSAMAN. 1990. Water economy of nestling Swainson's Hawks. *Condor* 92:29–44.

KLUTE, D.S., L.W. AYRES, M.T. GREEN, W.H. HOWE, S.L. JONES, J.A. SHAFFER, S.R. SHEFFIELD AND T.S. ZIMMERMAN. 2003. Status assessment and conservation plan for the Western Burrowing Owl in the United States. USDI Fish and Wildlife Service, Biol. Tech. Publ. FWS/BTP-R6001-2003. Washington, DC U.S.A.

KOCHERT, M.N. 1986. Raptors. Pages 313–349 in A.L. Cooperrider, R.J. Boyd, and H.R. Stuart [EDS.], Inventory and monitoring of wildlife habitat. Chapter 16. USDI Bureau of Land Management Service Center, Denver, CO U.S.A.

———, K. STEENHOF, C.L. MCINTYRE AND E.H. CRAIG. 2002. Golden Eagle (*Aquila chrysaetos*), No. 684. *In* A. Poole and F. Gill [EDS.], The birds of North America. The Birds of North America, Inc., Philadelphia, PA U.S.A.

KRONE, O., R. ALTENKAMP AND N. KENNTNEN. 2005. Prevalence of *Trichomonas gallinae* in Northern Goshawks from the Berlin area of northeastern Germany. *J. Wildl. Dis.* 41:304–309.

LEWIS, S.B., P. DESIMONE, M.R. FULLER AND K. TITUS. 2004. A video surveillance system for monitoring raptor nests in a temperate rainforest environment. *Northwest Sci.* 78:70–74.

LITTLE, R. 1993. Controlled wildlife. Vol. 1, federal permit procedures; Vol. II, federally protected species; Vol. III, state permit procedures. Association of Systematic Collections, Washington, DC U.S.A.

LUTTICH, S.N., L.B. KEITH AND J.D. STEPHENSON. 1971. Population dynamics of the Red-tailed Hawk (*Buteo jamaicensis*) at Rochester, Alberta. *Auk* 88:75–87.

MARZLUFF, J.M., M.S. VEKASY, M.N. KOCHERT AND K. STEENHOF. 1997. Productivity of Golden Eagles wearing backpack transmitters. *J. Raptor Res.* 31:223–227.

MCGARIGAL, K., R.G. ANTHONY AND F.B. ISAACS. 1991. Interactions of humans and Bald Eagles on the Columbia River Estuary. *Wildl. Monogr.* 115.

MCLEOD, M.A. AND D.E. ANDERSEN. 1998. Red-shouldered Hawk broadcast surveys: factors affecting detection of responses and population trends. *J. Wildl. Manage.* 62:1385–1397.

MINEAU, P., M.R. FLETCHER, L.C. GLASES, N.J. THOMAS, C. BRASSARD, L.K. WILSON, J.E. ELLIOT, L.A. LYON, C.J. HENNY, T. BOLLINGER AND S.L. PORTER. 1999. Poisoning of raptors with organophosphorus and carbamate pesticides with emphasis on Canada, U.S. and U.K. *J. Raptor Res.* 33:1–37.

MOSHER, J.A., M.R. FULLER AND M. KOPENY. 1990. Surveying forest-dwelling hawks by broadcast of conspecific vocalizations. *J. Field Ornithol.* 61:453–461.

NELSON, M.W. 1969. Status of the Peregrine Falcon in the Northwest. Pages 61–72 in J.J. Hickey [ED.], Peregrine Falcon populations: their biology and decline. University of Wisconsin Press, Madison, WI U.S.A.

NELSON, R.W. 1970. Some aspects of the breeding behavior of Peregrine Falcons on Langara Island, B.C. M.S. thesis, Calgary University, Alberta, Canada.

NEWBREY, J.L., M.A. BOZEK AND N.D. NIEMUTH. 2005. Effects of lake characteristics and human disturbance on the presence of piscivorous birds in northern Wisconsin, USA. *Water Birds* 28:478–486.

NEWTON, I. 1978. Feeding and development of Sparrowhawk *Accipiter nisus* nestlings. *J. Zool., Lond.* 184:465–487.

———. 1979. Effects of human persecution on European raptors. *Raptor Res.* 13:65–78.

———. 1986. The Sparrowhawk. T. and A.D. Poyser, Calton, United Kingdom.

NOON, B.R. AND A.B. FRANKLIN. 2002. Scientific research and the Spotted Owl (*Strix occidentalis*): opportunities for major contributions to avian population ecology. *Auk* 119:311–320.

OLSEN, P. 1993. Birds of prey of Australia. Australian Museum and Angus & Robertson, Sydney, Australia.

PARKER, J.W. 1976. Pesticides and eggshell thinning in the Mississippi Kite. *J. Wildl. Manage.* 40:243–248.

———. 1996. Urban ecology of the Mississippi Kite. Pages 45–52 in D.M. Bird, D.E. Varland, and J.J. Negro [EDS.], Raptors in human landscapes: adaptations to built and cultivated environments. Academic Press, London, United Kingdom.

PETTY, S.J. AND B.L. FAWKES. 1997. Clutch size variation in Tawny Owls (*Strix aluco*) from adjacent valley systems: can this be used as a surrogate to investigate temporal and spatial variations in vole density. Pages 315–324 in J.R. Duncan, D.H. Johnson, and T.H. Nicholls [EDS.], Biology and conservation of owls of the Northern Hemisphere. USDA Forest Service General Technical Report NC-190, North Central Research Station, St. Paul, MN U.S.A.

POOLE, A. 1981. The effect of human disturbance on Osprey reproductive success. *Colon. Waterbirds* 4:20–27.

———. [ED.]. 2004. The Birds of North America Online database: http://bna.birds.cornell.edu/. Cornell Laboratory of Ornithology, Ithaca, NY U.S.A.

RATCLIFFE, D. 2003. Discovering the causes of Peregrine decline. Pages 23–33 in T.J. Cade and W. Burnham [EDS.], Return of the Peregrine: a North American saga of tenacity and teamwork. The Peregrine Fund, Inc., Boise, ID U.S.A.

REYNOLDS, R.T. AND H.M. WIGHT. 1978. Distribution, density, and productivity of accipiter hawks in Oregon. *Wilson Bull.* 90:182–196.

———, G.C. WHITE, S.M. JOY AND R.W. MANNAN. 2004. Effects of radiotransmitters on Northern Goshawks: do tailmounts lower survival of breeding males? *J. Wildl. Manage.* 68:25–32.

———, M.H. REISER, R.L. BASSETT, P.L. KENNEDY, D.A. BOYCE, JR., G. GOODWIN, R. SMITH AND E.L. FISHER. 1992. Management recommendations for the Northern Goshawk in the southwestern United States. USDA Forest Service General Technical Report RM-217, Rocky Mountain Forest and Range Experiment Station, Fort Collins, CO U.S.A.

RIFFEL, S.K., K.J. GUTZWILLER AND S.H. ANDERSON. 1996. Does repeated human intrusion cause cumulative declines in avian richness and abundance? *Ecol. Appl.* 6:492–505.

ROBERSON, A.M., D.E. ANDERSEN AND P.L. KENNEDY. 2002. The Northern Goshawk (*Accipiter gentilis*) in the western Great Lakes Region: a technical conservation assessment. Minnesota Cooperative Fish and Wildlife Research Unit, University of Minnesota, St. Paul, MN U.S.A.

———, D.E. ANDERSEN AND P.L. KENNEDY. 2005. Do breeding phase and detection distance influence the effective area surveyed for Northern Goshawks? *J. Wildl. Manage.* 69: 1240–1250.

ROGERS, A.S., S. DESTEFANO AND M.F. INGRALDI. 2005. Quantifying Northern Goshawk diets using remote cameras and observations. *J. Raptor Res.* 39:303–309.

ROSENFIELD, R.N. 1978. Attacks by nesting Broad-winged Hawks. *Passenger Pigeon* 40:419.

———. 1983. Nesting biology of Broad-winged Hawks in Wisconsin. M.S. thesis, University of Wisconsin-Stevens Point, Stevens Point, WI U.S.A.

——— AND J. BIELEFELDT. 1993a. Cooper's Hawk (*Accipiter cooperii*), No. 75. In A. Poole and F. Gill [EDS.], The birds of North America. The Birds of North America, Inc., Philadelphia, PA U.S.A.

——— AND J. BIELEFELDT. 1993b. Trapping techniques for breeding Cooper's Hawks: two modifications. *J. Raptor Res.* 27:170–171.

——— AND J. BIELEFELDT. 1999. Mass, reproductive biology, and nonrandom pairing in Cooper's Hawks. *Auk* 116:830–835.

——— AND J. BIELEFELDT. 2006. Cooper's Hawk (*Accipiter cooperii*). Pages 162-163 in N.J. Cutright, B.R. Harriman, and R.W. Howe [EDS.], Atlas of the breeding birds of Wisconsin. Wisconsin Society for Ornithology, Inc. Madison, WI U.S.A.

———, J. BIELEFELDT AND R.K. ANDERSON. 1988. Effectiveness of broadcast calls for detecting breeding Cooper's Hawks. *Wildl. Soc. Bull.* 16:210–212.

———, J.W. SCHNEIDER, J.M. PAPP AND W.S. SEEGAR. 1995. Prey of Peregrine Falcons breeding in West Greenland. *Condor* 97:763–770.

———, J. BIELEFELDT, J.L. AFFELDT AND D.J. BECKMANN. 1995. Nesting density, nest area reoccupancy, and monitoring implications for Cooper's Hawks in Wisconsin. *J. Raptor Res.* 29:1–4.

RUTZ, C., A. ZINKE, T. BARTELS AND P. WOHLSEIN. 2004. Congenital neuropathy and dilution of feather melanin in nestlings of urban-breeding Northern Goshawks (*Accipiter gentilis*). *J. Zoo Wildl. Med.* 35:97–103.

SHUGART, G.W., M.A. FITCH AND V.M. SHUGART. 1981. Minimizing investigator disturbance in observational studies of colonial birds: access to blinds through tunnels. *Wilson Bull.* 93: 565–569.

SITTER, G. 1983. Feeding activity and behavior of Prairie Falcons in the Snake River Birds of Prey Natural Area in southwestern Idaho. M.S. thesis, University of Idaho, Moscow, ID U.S.A.

SMITHERS, B.L., C.W. BOAL AND D.E. ANDERSEN. 2005. Northern Goshawk diet in Minnesota: an analysis using video recording systems. *J. Raptor Res.* 39:264–273.

SNYDER, N.F.R. AND H.A. SNYDER. 1974. Increased mortality of Cooper's Hawks accustomed to man. *Condor* 76:215–216.

——— AND H.A. SNYDER. 1991. Birds of prey: natural history and conservation of North American raptors. Voyageur Press, Inc., Stillwater, MN U.S.A.

SODHI, N.S., L.W. OLIPHANT, P.C. JAMES AND I.G. WARKENTIN. 1993. Merlin (*Falco columbarius*), No. 44 in A. Poole and F. Gill [EDS.], The birds of North America. The Bird of North America, Inc., Philadelphia, PA U.S.A.

SPEISER, R. AND T. BOSAKOWSKI. 1991. Nesting phenology, site fidelity, and defense behavior of Northern Goshawks in New York and New Jersey. *J. Raptor Res.* 25:132–135.

SQUIRES, J.R. AND R.T. REYNOLDS. 1997. Northern Goshawk (*Accipiter gentilis*), No. 298 in A. Poole and F. Gill [EDS.], The Birds of North America. The Birds of North America, Inc., Philadelphia, PA U.S.A.

SQUIRES, J.A. AND P.L. KENNEDY. 2006. Northern Goshawk ecology: an assessment of current knowledge and information needs for conservation and management. *Stud. Avian Biol.* 31:8–62.

STALMASTER, M.V. AND J.R. NEWMAN. 1978. Behavioral responses of wintering Bald Eagles to human activity. *J. Wildl. Manage.* 42:506–513.

STEENHOF, K. 1987. Assessing raptor reproductive success and productivity. Pages 157–170 in B.A. Giron Pendleton, B.A. Millsap, K.W. Cline, and D.M. Bird [EDS.], Raptor management techniques manual. National Wildlife Federation, Washington, DC U.S.A.

———. 1998. Prairie Falcon (*Falco mexicanus*), No. 346 in A. Poole and F. Gill [EDS.], The birds of North America. The Birds of North America, Inc., Philadelphia, PA U.S.A.

———, G.P. CARPENTER AND J.C. BEDNARZ. 1994. Use of mist nets and a live Great Horned Owl to capture breeding American Kestrels. *J. Raptor Res.* 28:194–196.

STEIDL, R.J. AND R.G. ANTHONY. 1996. Responses of Bald Eagles to human activity during summer in interior Alaska. *Ecol. Appl.* 6:482–491.

STOUT, W.E., R.N. ROSENFIELD, W.G. HOLTON AND J. BIELEFELDT. 2007. Nesting biology of urban Cooper's Hawks in Milwaukee, Wisconsin. *J. Wildl. Manage.* 71:366–375.

———, S.A. TEMPLE AND J.M. PAPP. 2006. Landscape correlates of reproductive success for an urban/suburban Red-tailed Hawk population. *J. Wildl. Manage.* 70:989–997.

SUTER, G.W., II AND J.L. JONES. 1981. Criteria for Golden Eagle, Ferruginous Hawk, and Prairie Falcon nest site protection. *Raptor Res.* 15:12–18.

TAYLOR, I.R. 1991. Effects of nest inspections and radiotagging on Barn Owl breeding success. *J. Wildl. Manage.* 55:312–315.

UNITED STATES FISH AND WILDLIFE SERVICE (USFWS). 1998. Northern Goshawk status review. U.S. Fish and Wildlife Service, Office of Technical Support, Portland, OR U.S.A.

VEKASY, M.S., J.M. MARZLUFF, M.N. KOCHERT, R.N. LEHMAN AND K. STEENHOF. 1996. Influence of radio transmitters on Prairie Falcons. *J. Field Ornithol.* 67:680–690.

WATSON, J.W. 1993. Responses of nesting Bald Eagles to helicopter surveys. *Wildl. Soc. Bull.* 21:171–178.

———. 2004. Responses of nesting Bald Eagles to experimental pedestrian activity. *J. Raptor Res.* 38:295–303.

WHITE, C.M. 1994. Population trends and current status of selected western raptors. *Stud. Avian Biol.* 15:161–172.

——— AND T.L. THUROW. 1985. Reproduction of Ferruginous Hawks exposed to controlled disturbance. *Condor* 87:14–22.

——— AND R.W. NELSON. 1991. Hunting range and strategies in tundra breeding peregrine and Gyrfalcon observed from a helicopter. *J. Raptor Res.* 25:49–62.

———, D.A. BOYCE AND R. STRANECK. 1989. Observations on *Buteo swainsoni* in Argentina, 1984 with comments on food, habitat alteration and agricultural chemicals. Pages 79–87 in B.-U. Meyburg and R.D. Chancellor [EDS.], Raptors in the modern world. World Working Group for Birds of Prey and Owls, London, United Kingdom.

WILEY, J.W. 1975. The nesting and reproductive success of Red-tailed Hawks and Red-shouldered Hawks in Orange County, California, 1973. *Condor* 77:133–139.

WOOD, P.B. 1999. Bald Eagle response to boating activity in north-central Florida. *J. Raptor Res.* 33:97–101.

Mitigation

20

RICHARD E. HARNESS
EDM International, Inc.
4001 Automation Way, Fort Collins, CO 80525-3479 U.S.A.

INTRODUCTION

Throughout history, divergence between human interests and raptors has led to impacts upon birds of prey. Direct impacts include deliberate persecution, illegal trade, and collection. Indirect impacts include human activities that may have an unintentional adverse impact on raptors. Unintended adverse impacts often result from technological advancements, including urbanization and pesticide use.

This chapter presents an overview of a number of human activities that affect raptors, and identifies mitigating measures that have been used to counter negative impacts. Sometimes, several activities collectively create an impact requiring several mitigating techniques. Topics covered include some of the most frequently documented examples of human impacts. This chapter builds on Postovit and Postovit (1987), which should be consulted for additional detail.

DIRECT IMPACTS

Direct impacts to raptors include shooting, trapping, and poisoning. Like natural mortality, raptor deaths due to persecution can be either compensatory or additive (Newton 1979). Compensatory mortality occurs when deliberate persecution replaces natural mortality. Additive mortality occurs when persecution adds to the natural mortality. Persecution tends to be most damaging immediately prior to breeding, when the population is at a seasonal low (Newton 1979). Larger species are more vulnerable to persecution than are smaller species because they occur at lower densities, have lower reproductive rates, and take longer to reach maturity (Newton 1979).

When population declines are due to exploitation or persecution, legal protection and education are the most appropriate conservation methods. Education and public involvement are critical to dispel prejudices against raptors (Postovit and Postovit 1987). Although there are many cases in which persecution has had an adverse impact on raptor populations, raptors typically rebound from this threat once direct impacts are reduced or removed (Newton 1979).

Shooting and Trapping

For centuries, raptors were shot and trapped to protect farm and game animals from depredation (Newton 1979). The phenomenon has been widespread, common, and even encouraged through bounties. Alaska, for example, paid bounties for more than 100,000 Bald Eagles (*Haliaeetus leucocephalus*) from 1917 to 1952 to protect salmon stocks (Robards and King 1966). Initially, only larger raptors were persecuted. However, as game bird and poultry farms became common, smaller raptors increasingly were shot and trapped (Newton 1979). Even today, Ospreys (*Pandion haliaetus*) are killed to protect fish at aquaculture farms, and vultures are shot near airports to prevent bird-aircraft collisions. In addition, raptors are shot for recreation at migration bottlenecks (Xirouchakis 2003), along roadsides from utility poles (Olson 1999), and for the illegal feather trade (Delong 2000).

Large diurnal raptors are more frequently shot because they are conspicuous (Snyder and Snyder 1974). Gregarious species, such as vultures, are particularly susceptible to mass shootings (Newton 1979). Immature birds not wary of humans are at greater risk (Ellis et al. 1969).

Raptors are killed deliberately through the use of traps (Brooker 1990). Traps may be set at nests, around live or dead bait, or on artificial perches. The most common traps are leg-hold traps with spring jaws (Newton 1979). When a raptor lands on a spring trap, the jaws snap together and the bird is held until it dies or is removed and killed (Newton 1979).

Mitigation discussion. Today many countries protect raptors from such indiscriminate killing, minimizing impacts on many long-term raptor populations. Yet, despite protection, some persecution persists, which can affect certain populations (e.g., California Condor [*Gymnogyps californianus*] [Cade et al. 2004]). Although there are cases in which raptor predation creates economic hardship (e.g., goshawks killing game birds at release sites [Newton 1979]), overall, depredation is probably negligible. In cases where raptors do create financial hardship, it is best to compensate the landowner for the losses and to work with the property owner to limit future damages. Education is crucial to eliminating prejudices toward raptors (Postovit and Postovit 1987).

Poisoning

Poison baits are used both legally and illegally to control a variety of animal pests (Newton 1990). These baits can result in both intentional and unintentional raptor kills. Unintentional poisoning occurs when a scavenging raptor either eats poisoned bait set out for another animal or feeds on the poisoned carcass of a target animal (i.e., secondary poisoning) (Newton 1979) (see *Indirect Impacts* and *Pesticides and Contaminants* below for details).

Vultures often are impacted by intentional poisoning (Houston 1996). Because these scavengers are gregarious, it is easy to poison large numbers of birds at a single time (Ledger 1988). Eagles also are sometimes intentionally poisoned with acute toxins, such as strychnine, to prevent them from killing lambs (Brooker 1990, Newton 1990).

Birds may fly away after ingesting poison and die elsewhere. As a result, the cause of death may not be apparent. Stock-tank drowning may occur when birds seek water after ingesting poison (Mundy et al. 1992).

Mitigation discussion. As with shooting and trapping, education is critical to eliminate prejudices toward raptors (Postovit and Postovit 1987). Care should be taken when handling raptor carcasses as they may have had contact with a poison. Some organophosphate pesticides, like monocrotophos, are absorbed through the skin (EXTOXNET 1996); therefore, gloves should be used to handle carcasses to prevent possible contamination. If poisoning is quickly diagnosed, there are antidotes for some organophosphorous, carbamate, and rodenticide compounds (Ontario Ministry of the Environment 1995). However, poisoned birds will require a rapid and precise diagnosis, as well as prolonged rehabilitation.

Traditional and Cultural Practices

The ceremonies of some traditional groups entail hunting and sacrificing animals. Birds of prey usually have little importance as a food source, but they often have great symbolic value giving them a powerful role in traditional ceremonies (White 1913). Raptors are sometimes killed due to these beliefs. For example, members of the Hopi Eagle Clan practice an annual ritual requiring the sacrifice of young Golden Eagles (*Aquila chrysaetos*). The eaglets are reared by hand until July when they are smothered. According to the Hopi belief, the sacrificed eagles carry the prayers of the Hopi back to their spirit home (Williams 2001). In South Africa, traditional folklore holds that vultures have such keen eyesight that they can see into the future. Poachers hunt Bearded Vultures (*Gypaetus barbatus*) for their heads, which are prized by gamblers playing the new national lottery (Marshall 2003). Although many modern cultures support the right of traditional peoples to practice traditional customs, such support often fades when the fate of an endangered species is at stake.

Mitigation discussion. Modern peoples often assume that education is the key to halt customs of killing endangered or threatened species. Members of traditional cultures however, may believe that it is modern people who need to be educated about traditional beliefs (Kaye 2001).

INDIRECT IMPACTS

Indirect impacts are numerous, diverse, and often negative. These impacts can be lethal or sublethal. Sublethal impacts may affect raptors in a variety of ways that are difficult to detect (e.g., decreased reproductive rate, eggshell thinning). Additionally, once indirect impacts are detected, they can be difficult to reverse. Unlike direct impacts, raptors cannot learn to recognize many indirect impacts in order to avoid or habituate to them (Postovit and Postovit 1987).

Habitat Loss, Modification, and Fragmentation

Habitat destruction and alteration due to human-population growth impact raptors (Newton 1979). Habitat alteration occurs in many ways. Habitats can be completely replaced or they can be significantly modified. When this occurs, many animals dependent upon these areas are displaced and may not find suitable habitat in surrounding areas.

Habitat fragmentation may result from incremental and cumulative changes to an area. When habitats are split into smaller units, they may provide less favorable habitat for some species, and the relative carrying capacity of the habitat is likewise reduced, supporting fewer individuals. Habitat also can be degraded slowly over a long period, resulting in the disappearance of suitable prey, perching sites, and nest sites. Although some species can adapt to altered areas, fragmented habitats generally are not as productive for native species as natural areas, and raptor numbers are reduced (Newton 1990).

Habitat destruction is probably the most devastating impact raptors face. Raptors requiring unique habitats or large home ranges are at greatest risk. Habitat changes due to urban, suburban, and rural encroachment are most often permanent. Urbanization tends to favor disturbance-tolerant species (e.g., American Kestrels [*Falco sparverius*], Red-tailed Hawks [*Buteo jamaicensis*], and Great Horned Owls [*Bubo virginianus*]) at the expense of less tolerant species. Many raptors are migratory, so it is important to consider breeding habitat, migratory corridors, and wintering grounds.

Mitigation discussion. Areas must be set aside to preserve animals with large territories. Land can be purchased outright or protected with conservation easements (DeLong 2000). Knowledge of landscape needs is essential in order to understand and preserve the distribution of raptor species. Evaluating the effects of habitat change on a species is complex and must address the particular species' needs (Redpath 1995). The structure and dynamics of populations measured at small spatial scales may not reflect the characteristics of the overall population across the landscape (Kareiva and Wennergren 1995).

Transportation

Vehicle collisions. Raptors are drawn to roadways for many reasons. Roadways can provide a steady supply of carrion from vehicle collisions (Platt 1976) and right-of-way mowing. Utility poles along roadways provide attractive hunting and roosting perches for raptors preying on mice, voles, and other rodents (Robertson 1930, Bevanger 1994). During cold winter months, roads may provide a source of heat and salt, both of which may attract prey (Meade 1942, Dhindsa et al. 1988). Road salt also attracts large animals that may become road casualties, providing carrion (Noss 1990). Raptor species adapted to hunting along roadsides are at particular risk of colliding with vehicles.

In areas where birds regularly feed on the carcasses of road-killed animals, carcasses should be pulled off the road. In persistent problem areas, signs can be posted alerting motorists to slow down due to the possibility of encountering raptors on the road (DeLong 2000). Lower speed limits also can be deployed.

Aircraft collisions. In addition to being a risk to human life, bird–aircraft collisions cost the world's aviation industry millions of dollars annually (Sodhi 2002). Most aircraft strikes involve gulls (Laridae), Common Starlings (*Sturnus vulgaris*), and blackbirds (Icteridae), but raptor strikes also have been documented (Lesham and Bahat 1999).

Israel has one of the most sophisticated programs for managing bird–aircraft strikes. The country is along a major bottleneck of bird migration and twice each year millions of raptors fly through Israel's limited air space (Shirihai et al. 2000). Bird migration patterns are monitored and mapped using motorized gliders, drones, radar, and ground observers (Leshem and Bahat 1999). Maps with "Bird Plagued Zones" are updated yearly and provided to military pilots. Air-space restrictions are developed for lower flight altitudes using real time radar (Leshem and Bahat 1999).

In the U.S., the Bird-Wildlife Aircraft Strike Hazard Team (BASH), an organization committed to reducing wildlife-related hazards with aircraft, has developed a bird-avoidance model (U.S. Air Force 2004).

Vegetation management and habitat modification are important tools used to make airports less desirable for birds and other wildlife (Sodhi 2002). Many airports mow vegetation around airstrips to decrease the cover for rodents and other prey. This makes the area less likely to attract raptors. Birds are sometimes hazed using noisemakers or gunshot, but birds often habituate to loud noises (Sodhi 2002). Denying roosting sites can be important and installing anti-perching devices on airport facilities is sometimes used to discourage raptor use (Transport Canada 2001). However, no single anti-perching device will deter all perch-hunting species (Avery and Genchi 2004).

Energy and Communication Infrastructure

Energy and mineral development. The total world consumption of marketed energy is predicted to expand by 54% between 2001 and 2025 (Energy Information Administration 2004). Developing nations, including Asia, China, and India, are expected to account for the greatest increase in world energy consumption (U.S. Department of Energy 2004). The increasing demand for energy will result in more land being used for the exploration and extraction of petroleum, natural gas, and coal. There also will be an increase in the number of dams, nuclear power plants, and renewable energy sources.

The development of these resources will result in direct, indirect, and cumulative environmental impacts. Because many of these activities occur in remote locations, raptors can be impacted adversely (Murphy 1978). Impacts to raptors include habitat fragmentation and loss, displacement from disturbance, and a reduction in prey. Energy development can impact nesting, roosting, and foraging areas. Indirect impacts may result from road construction, soil and vegetation disturbance, and increased air and water pollution.

As with all large construction projects, it is important to conduct and complete an environmental analysis prior to construction. This analysis should include baseline studies of the land, vegetation, water, air, terrestrial and aquatic resources, and of human interactions. Regulatory agencies and communities should be provided an opportunity to comment on projects in an open forum.

Baseline surveys should be conducted for raptor nests, and projects should consider the life histories of raptors. These data should be incorporated into the project plans and environmental assessment, which would address facility construction, operation, and maintenance. Reclamation plans should be developed, when appropriate, and should include revegetation, off-site habitat enhancement, and the construction of artificial nest sites where applicable. If lands are managed properly, reclaimed areas have the potential to provide breeding habitat for raptors (Yahner and Rohrbaugh 1998).

Post-construction monitoring is a critical component of energy development as it is the only way to assess accurately the validity of pre-construction mitigating measures. Ongoing assessment enables corrective measures to be taken if mitigating efforts are determined to be ineffective.

Power-line electrocution. During the 1970s and early 1980s, electric industry efforts in North America to reduce raptor electrocutions were widespread. Predictions about mitigating the problem were overly optimistic and raptors continue to be electrocuted, possibly in large numbers (Lehman 2001). Raptor electrocutions remain a persistent problem throughout the world and although most power line mortality is probably compensatory, in certain parts of the world power lines are responsible for the decline of some raptors. In Spain, for example, electrocution was responsible for the population decline of the Spanish Imperial Eagle (*A. adalberti*) in Doñana National Park (Ferrer and de la Riva 1987).

Electrocution occurs in many ways, depending on pole design (Janss and Ferrer 1999). In North America, power lines typically are constructed using non-conductive wood power poles and wood crossarms (Fig. 1). In

Figure 1. Golden Eagle (*Aquila chrysaetos*) perched on a wooden electricity distribution structure.

Figure 2 (left). Raptor nest on a concrete electricity pole with steel crossarms.

Figure 3 (above). Swainson's Hawk (*Buteo swainsoni*) on an electric transformer pole.

Europe and in many places elsewhere, conductive steel and concrete poles are more common (Janss and Ferrer 1999). The latter often are fitted with grounded steel crossarms (Fig. 2), resulting in possible wire-to-crossarm, wire-to-pole, and wire-to-wire contacts (Janss and Ferrer 1999). This type of construction affects a broader group of bird species due to the greatly reduced clearances (Janss and Ferrer 1999). The use of steel poles is becoming more prevalent in the U.S., where similar problems have occurred (Harness 1998).

Protecting raptors from electrocution depends on the type of power-line configuration and size of the bird (APLIC 2006). In areas using conductive steel and concrete poles and crossarms, the critical clearance often is body length because a perching bird needs to touch only one energized wire (Janss and Ferrer 1999). For this reason, insulation often is the preferred method to prevent contact with conductive structures. Perch deterrents used to prevent wire-to-wire contacts in North America (APLIC 2006) are less effective on conductive structures (Janss and Ferrer 1999). Equipment, such as transformers (Fig. 3), are universally problematic and should be installed with insulated jumper wires and protective bushing covers (Janss and Ferrer 1999, van Rooyen 2000, Harness and Wilson 2001, Platt 2005).

Because many utility configurations are used even at a regional level, specific construction practices and habitat use must be determined before developing appropriate mitigation measures (Mañosa 2001). Effective retrofitting requires a thorough understanding of the pole configuration, the at-risk birds, and other contributing factors (e.g., bird behavior, size, age, prey species, preferred habitat, season, weather, wind, and topography). Three solid references provide guidance on these issues: *Suggested Practices for Avian Protection on Power Lines: State of the Art in 2006* (APLIC 2006), *Birds and Power Lines — Collision, Electrocution and Breeding* (Ferrer and Janss 1999) and *Suggested Practices for Raptor Protection on Power Lines: The State of the Art in 1996* (APLIC 1996). In Europe, *Caution: Electrocution! Suggested Practices for Bird Protection on Power Lines* is available in German, English, and Russian from the German Society for Nature Conservation (NABU).

Power-line collisions. Although birds of prey spend considerable time in the air, collisions with power lines occur relatively infrequently compared with other species (Bevanger 1994). As discussed in *Mitigating Bird Collisions with Power Lines: The State of the Art in 1994* (APLIC 1994), aerial hunters like raptors possess

Figure 4. Perch deterrents minimize electrocution risks, but potential perch sites remain.

excellent flying abilities along with binocular vision. Furthermore, raptors generally do not fly in restrictive flocks. Although raptors are agile flyers with excellent eyesight, they are likely to be more susceptible to colliding with power lines when preoccupied or distracted (e.g., during territorial defense or prey pursuit) (Olendorff and Lehman 1986, Thompson 1978). Except in the case of a critically endangered species (e.g., California Condor), collisions with power lines are a random, low-level, and biologically inconsequential mortality factor for raptors (Olendorff and Lehman 1986).

Lines with persistent problems or lines that are likely to affect sensitive or rare species should be marked to make them more visible. One reference describing ways to address and mitigate bird collisions is *Mitigating Bird Collisions with Power Lines: The State of the Art in 1994* (APLIC 1994).

Power-line depredation. Raptors regularly use power poles as hunting perches (Benson 1981). For a sensitive species such as the Sage Grouse (*Centrocercus urophasianus*), it is believed that power lines and other artificial structures allow raptors to prey on displaying male grouse, nesting hens, and brooding chicks (Connelly et al. 2004). Additionally, ground-nesting Burrowing Owls (*Athene cunicularia*) can be at risk (Fitzner 1980).

Many products are available to manage raptors perching on utility structures. However, these devices (triangles, spikes, etc.) are specifically designed to pre-

vent bird electrocutions and not to exclude birds from all possible perching locations (EPRI 2001). There are many locations on poles where it is not possible to prevent perching with existing products (Fig. 4). Some researchers have concluded that poles and other perches near critical grouse breeding areas should be eliminated to preclude depredation (Connelly et al. 2000).

Power-line nesting. Power lines can positively impact raptors by providing nesting sites. Steenhof et al. (1993) showed that Ferruginous Hawks (*B. regalis*) nesting on transmission line towers were more successful than those nesting on natural substrates. Ospreys also have benefited from nesting on utility structures (Henny and Kaiser 1996). Significant negative biological impacts from electric and magnetic fields were not documented in birds nesting on power lines by Lee et al. (1979), but see more recent studies on both wild and captive American Kestrels (e.g., Fernie et al. 2000).

Raptors are more likely to nest on poles with double crossarms, which provide a wider platform (EPRI 2001). Where crossarms are required, employing single apitong crossarms rather than double crossarms, eliminates a place for raptors to nest (Fig. 5). When retrofitting existing poles, a stick deflector can be used to deter raptor nesting (Fig. 6). Perch deterrents do not effectively prevent nesting, and may actually facilitate nest construction by providing an anchor to attach sticks (Fig. 7; EPRI 2001).

Raptor nesting on structures should not be discouraged unless it results in operational problems. If a nesting raptor causes outages, the utility company should install a new unenergized pole with a nesting platform on the edge of the right-of-way and relocate the nest to this platform (APLIC 2006). Nearby power poles should be retrofitted to protect fledgling birds from possible electrocution (Dwyer and Mannan 2004).

Wind turbines. Wind-energy facilities (Fig. 8) provide an alternative to fossil fuel electricity production. However, this technology is not without risk to wildlife and may result in an increased threat of collision for raptors (Estep 1989). Factors contributing to increased collision risk include the raptor species present, prey concentrations, turbine design, migration routes, daily movement corridors, topographic features, and the position of a turbine in a string of turbines (Anderson et al. 1999b).

The National Wind Coordinating Committee has published a useful document titled *Studying Wind Energy/Bird Interactions: A Guidance Document* (Anderson et al. 1999b). The United States Fish and Wildlife Ser-

Figure 6. Stick deflector at a double dead-end structure (Kaddas).

Figure 5. Apitong crossarm used to eliminate nesting attempts.

Figure 7. Raptor nest on a structure fitted with anti-perching devices (Tri-State G&T).

Figure 8. Wind turbine with pronghorn antelope (*Antilocapra americana*).

vice (USFWS) also has developed *Interim Guidelines to Avoid and Minimize Wildlife Impacts from Wind Turbines* (USFWS 2003). The primary objectives of these guidelines are to (1) assist developers in deciding whether to proceed with wind-energy development, (2) describe a procedure to determine pre-construction study needs to verify use of potential sites by wildlife, and (3) provide recommendations for monitoring potential sites post-construction to identify, quantify, or verify actual impacts (or lack thereof). The USFWS also developed a "Potential Impact Index" (PII) as part of the guidelines (USFWS 2003). The PII is a tool used to determine potential impacts to birds and bats from proposed wind-energy development at a particular site.

Communication towers. Bird kills at communication tower sites have come under increased scrutiny. Although some diurnal birds are at risk from these structures, very few raptor collisions with communication towers have been recorded (Avatar Environmental et al. 2004).

Dams and Water Management

Dams and associated water-management actions can have positive or negative impacts. Positive impacts include providing additional habitat for fish, eagles, and Ospreys (Henny et al. 1978, Steenhof 1978, Van Daele and Van Daele 1982, Grover 1984, Detrich 1985). In general, piscivorous raptors benefit from the construction of dams (Postovit and Postovit 1987). Creation of riparian woodlands following the impoundment of water also has benefited some species, including the Common Black Hawk (*Buteogallus anthracinus*) in Mexico (Rodríguez-Estrella 1996).

The height (head) and length of the dam, amount of water released, and presence of fish help determine the suitability of dams as feeding sites for Bald Eagles (Brown 1996). When water is released, prey are concentrated in smaller areas, resulting in increased prey availability and predation (Bryan et al. 1996). Discharge during hydroelectric generation is important because it can release dead and injured fish, which are easy prey for eagles (Stalmaster 1987). Furthermore, below-dam waters often are ice-free throughout the winter, attracting waterfowl and making fish available to eagles. The availability of perch trees near reservoirs is another important component for eagles and other raptors (Stalmaster 1987, Brown 1996).

Impoundments are tradeoffs for the habitat displaced by the water, and altered water levels in them can negatively impact raptors associated with nearby riparian habitats (Schnell 1979). Nesting raptors along reservoir shorelines also may be impacted by recreational activities on and near the reservoir (see *Recreational Disturbance*).

As with dam construction, altering wetlands can negatively impact some species while improving habitat for others. Species at greatest risk from wetland manipulation are those associated with or restricted to wetlands. In the U.S., wetland management has adversely impacted Snail Kites (*Rostrhamus sociabilis*), which depend upon apple snails (*Pomacea paludosa*) in naturally flooded wetlands. Manipulating water levels and otherwise changing water flow impacts the snails and, subsequently, the kites (Sykes 1979, Shapiro et al. 1982).

Mitigation discussion. Advanced planning and robust impact analysis is recommended when manipulating or creating wetlands. Species-specific mitigation measures should be developed on a case-by-case basis, depending on the species affected, extent of the anticipated impacts, and location of the project. Long-term mitigation could include off-site habitat enhancement, human-use restrictions near sensitive areas including nest sites, communal roosts, and feeding areas, and changes to the proposed water management regime.

Forestry

Because the degree to which raptors depend on forests varies among species, forest management exerts a wide range of impacts (Postovit and Postovit 1987). Some forest practices result in dramatic changes in forest structure and composition, resulting in either positive or negative effects on raptors. Thiollay (1996), for example, reported that managed forests in western Indonesia preserve no more than a quarter of the original raptor forest community. In contrast, Reynolds et al. (1982) noted that in western Oregon some accipiters benefit from forest harvest and shortened rotation. Newly constructed roads and trails may lead to increased human access, disturbance, and further habitat fragmentation.

Mitigation discussion. A well-managed forest integrates wildlife-habitat and forest-management goals. Critical considerations include the intensity of logging and the length of forest rotation (Postovit and Postovit 1987). Raptors can use managed forests successfully where the forest structure (snags, canopy, layering, etc.) is adequate for the birds' needs (Horton 1996). To accomplish this, the ecological responses to forest management practices must be known. Managing for forest-dwelling raptors might include preserving diverse habitat features. Buchanan et al. (1999), for example, reported that total snags/ha, shrub cover, canopy closure, and coarse woody-debris cover are important to the Northern Spotted Owl (*Strix occidentalis*). Finn et al. (2002) stated that nesting Northern Goshawks (*Accipiter gentilis*) are more likely to occupy historical nest sites with a high overstory depth and low shrub cover. Forests should be managed to preserve prey species and it should be recognized that altered areas will provide opportunities for non-forest species, including raptors, to compete for limited resources (Kenward 1996).

It may be important to protect existing nest sites of sensitive forest-breeding raptors. These nests should be inventoried, mapped, and species-specific buffer zones established around them, as warranted (Mooney and Taylor 1996). Disturbance during nesting can affect raptors in many ways leading to a variety of impacts, including nest desertion (Newton 1979). Accordingly, forestry activities such as tree marking and logging should be avoided during the critical periods of nest building and incubation to reduce the likelihood of nest abandonment.

Agriculture

Farming and ranching primarily affect raptors by altering habitat. Changes brought about by agriculture can be either positive or negative (Postovit and Postovit 1987). Farming, as opposed to ranching, typically causes greater habitat changes and has more adverse impacts on raptors (Postovit and Postovit 1987). Intensive farming can result in lower prey populations for some raptor species such as the Ferruginous Hawk (Gilmer and Stewart 1983), although ranching typically is more compatible for many raptors (Postovit and Postovit 1987). The removal of native shrublands to "improve" pastures (Hamerstrom 1974, Murphy 1978) and the removal of prey species such as burrowing mammals (Zarn 1974) adversely and significantly impact raptors.

Agricultural activities can create problems through the use of pesticides and other chemicals (Postovit and Postovit 1987). Additionally, removing water from natural sources for agriculture negatively impacts some raptors (Gould 1985). Benefits from agriculture include the planting of trees as wind breaks and their subsequent use as nesting and perching sites for raptors (Postovit and Postovit 1987). Agricultural structures such as windmills and dwellings also may provide nest sites (Olendorff 1973). Other agricultural impacts include fence collisions, stock-tank drownings, and carcass-disposal practices (Anderson 1977, Ledger 1979, Newton 1990).

Fence collisions. Wire fences occasionally are responsible for raptor deaths when birds collide with them, become entangled in the wires, or impale themselves on barbed wire (Anderson 1977). Burrowing Owls also have been killed by electric fences (Staff and Wire Reports 1998).

In problem areas, making fences more visible, including the use of commercially available swinging plates, can help reduce collisions (Harness et al. 2003).

Stock-tank drownings. Raptors may use stock tanks for bathing, drinking (Houston 1996), or perching. In addition, raptors may be attracted to tanks if prey is present in the area (Craig and Powers 1976). Both diurnal and nocturnal raptor species have been known to drown in stock-watering tanks (Anderson et al. 1999a).

In South Africa, Cape Vultures (*Gyps coprotheres*) often drown in stock tanks (Ledger 1979), possibly when they respond to thirst caused by strychnine poisoning (Mundy et al. 1992). The gregarious nature of this species may contribute to mass drownings. The flapping wings of a single trapped individual may attract other vultures that mistakenly interpret the behavior as a feeding opportunity (Mundy et al. 1992).

Solutions include installing plastic floats and wooden planks or branches in water tanks (Anderson et al. 1999a). Keeping the reservoir full also helps animals escape. Farmers can be convinced to modify the tanks by informing them of the problem and emphasizing that carcasses will pollute the water and render it unsuitable for humans and livestock.

Carcass-disposal practices. Many Old World and New World vultures feed on large-ungulate carcasses (Houston 1996). With the conversion of native areas to agriculture, domestic livestock have replaced many native ungulates. Vultures have adapted to this transition in areas where livestock carcasses are available (Houston 1996). However, where carcasses are buried or burned to prevent the spread of disease, access to food can be restricted. In some areas, modern farming has supplanted herding with intensive stock farms (Iezekiel et al. 2004), resulting in fewer carcasses to scavenge. Improved sanitation in many African and Asian cities has reduced access to carcasses (Newton 1990). Furthermore, regulations may prohibit traditional carcass-disposal methods due to concerns regarding diseases, such as bovine spongiform encephalopathy (Camina 2004).

One problem associated with domestic carcasses is the use of the systemic painkiller and non-steroidal anti-inflammatory veterinary drug, diclofenac (Chapter 18). Diclofenac use in South Asia has been shown to cause visceral gout in White-rumped Vultures (*G. bengalensis*) and is believed likely to act similarly in other *Gyps* species. In just over a decade of use in livestock, diclofenac is suspected of causing the near extinction of three vulture species (Oaks et al. 2004).

Vulture "restaurants" (i.e., sites where large animal carcasses are provided as an artificial food source for vultures) have been set up in several countries in an attempt to conserve vultures. These sites provide a regular, uncontaminated food supply (Houston 1996).

Vulture restaurants should be placed away from fences and power lines to reduce collision and electrocution risks (Piper 2003). Plastic bags and other non-food waste should not be dumped at these sites (Piper 2003). Carcasses of livestock euthanized with barbiturates and those containing diclofenac should be removed so scavenging raptors are not indirectly poisoned (Piper 2003).

In addition to maintaining or bolstering populations locally, vulture restaurants provide opportunities for the general public to observe and photograph feeding vultures and conservationists with the chance to educate the public on the benefits of vultures. Vulture restaurants typically attract mammalian scavengers that may need to be managed. Finally, restaurants should be designed so the resulting smell does not offend nearby residents (Piper 2003) (see Chapter 22 for details).

Pesticides

Pesticides are a relatively recent threat to raptors (see Chapter 18 for details). As discussed by Newton (1979), widespread pesticide use began in the 1940s after World War II with the development of inexpensive and, at least initially, effective crop pesticides and herbicides. Pesticides can directly, or indirectly, poison birds. Pesticides also may cause indirect impacts by reducing prey species (Rands 1985). Whereas direct impacts to raptors typically affect local populations, pesticides can impact raptors on a regional or even global scale (Newton 1979).

Organochlorines. The term "organochlorine" refers to a chemical containing carbon, chlorine, and, sometimes, other elements. Organochlorine compounds include herbicides, insecticides, fungicides, and industrial chemicals such as PCBs. Additionally, organochlorine compounds include DDT and cyclodienes such as aldrin, dieldrin, endrin, and heptachlor (European Environment Agency 1995).

As discussed in Newton (1979), organochlorine compounds are very stable, fat-soluble, and environmentally persistent. Organochlorines bioaccumulate in fatty tissue in animals, resulting in further concentrations in successive links in the food chain. When used as insecticides, these chemicals are widely dispersed, increasing the potential exposure to birds and other animals even in remote locations. Raptors at greatest risk of bioaccumulating organochlorines are those that eat birds and fish (Newton 1979).

DDT (dichloro-diphenyl-trichloroethane). Animals metabolize DDT into DDE, which is less toxic than DDT (Newton 1979). Within fat, organochlorines are relatively non-toxic unless the fat is suddenly metabolized, such as during a food shortage or on migration. When this happens, death may result if high concentrations are present (Newton 1979). Furthermore, DDE has been shown to significantly impact reproduction through eggshell thinning, resulting in egg breakage (Newton 1979).

An unprecedented decline in Peregrine Falcons (*F. peregrinus*) in Europe and North America occurred with the widespread use of DDT (Ratcliffe 1967). Subsequent restrictions on the use of this synthetic pesticide are credited with a return to increasing eggshell thickness and dramatic increases in Peregrine Falcon reproductive success in these same countries (Newton 1979, Henny et al. 1999). Because of DDT's persistence (a chemical half-life of 15 years), it still remains a problem in some watersheds in the U.S. (Sharpe 2004). Although no longer widely used in North America and Europe, DDT remains in use in developing nations for disease-vector control (Malaria Foundation International 2006, see also Chapter 18).

Organophosphates. Most broad-spectrum insecticides currently in use are organophosphates (Pesticide News 1996, Chapter 18). Organophosphates are relatively inexpensive and mainly are used to "protect" food crops from insects. These chemicals break down more rapidly than organochlorine pesticides and have largely replaced the latter in agricultural pest control (Pesticide News 1996). Unfortunately, organophosphate compounds affect the nervous system of both vertebrates and invertebrates and as such, include some of the most toxic chemicals used in agriculture.

There are numerous reports of avian deaths attributed to organophosphate compounds through secondary poisoning (Henny et al. 1999). In Argentina, an estimated 20,000 Swainson's Hawks (*B. swainsoni*) died in 1996 after farmers applied monocrotophos to alfalfa fields for grasshopper control (Goldstein et al. 1999). These raptors were killed both by direct exposure and by eating contaminated grasshoppers. Raptors exhibiting gregarious behavior and opportunistically feeding on debilitated prey may be at higher risk (Mineau et al. 1999).

Carbamates. Carbamates are synthetic chemicals widely used as pesticides, including herbicides, insecticides, and fungicides. They are less persistent in the environment than organochlorines. One carbamate, the insecticide carbofuran, is highly toxic to birds (EXTOXNET 1996). Birds are susceptible to carbofuran when they ingest granules of it, which resemble grain seeds (Erwin 1991). Although the granular production of carbofuran generally has been phased out based solely on the danger it presents to birds, some granular formulations are still in use today. Raptors are vulnerable to carbamate poisoning when they scavenge prey poisoned by it (Erwin 1991).

Rodenticides. Rodenticides are second only to insecticides in their use in agriculture (Chapter 18). Rodenticides are either acute neurotoxins or anticoagulants that cause internal bleeding and eventual death (Corrigan and Moreland 2001). Strychnine and zinc phosphide are examples of acute toxins (Corrigan and Moreland 2001). Strychnine is used to control mammalian predators such as gray wolves (*Canis lupus*), foxes (Canidae), and coyotes (*C. latrans*) (Newton 1990).

Anticoagulant rodenticides inhibit the enzymes responsible for vitamin K recycling, which ultimately reduces blood clotting. This results in the increased permeability of capillaries throughout the body and widespread internal hemorrhaging. Death usually occurs several days after bait ingestion or after several feedings. Newer anticoagulants can cause death after only a single dose (Corrigan and Moreland 2001).

The use of rodenticides can result in either primary or secondary poisoning of non-target wildlife (Corrigan and Moreland 2001). Secondary raptor poisoning due to both legal and illegal use has been documented (Newton 1990). There may be a greater chance of raptors being poisoned secondarily by anticoagulants than by acute toxins. Anticoagulants are slower to take effect (1 to 10 days) and during this period poisoned rodents are more susceptible to predation because they are disoriented and sluggish (Delong 2000).

1080 (sodium monofluoroacetate). 1080 is a water-soluble salt that is highly toxic to mammals (Green 2004). The compound was developed in the 1940s to control rodents and predators. The bulk of world usage is in New Zealand and, to a lesser extent, Australia (Green 2004), where it is used to control possums, rabbits, foxes, and other introduced vertebrate pests. Although birds have a relatively high tolerance to 1080, kites and eagles have died from secondary poisoning (McIlroy 1984).

Mitigation discussion. Landowners who apply pesticides should do so strategically and with caution. The use of pesticides should be regarded as just one tool in an overall pest management program. The key to good pest-management includes implementing proper sanitation practices, removing food sources, and using appropriate biological control. All pesticides must be used in accordance with their legally binding labels. In the case of rodent control, this includes searching for carcasses and burying or burning them to avoid secondary exposures (Corrigan and Moreland 2001). Granular formulations toxic to avian species should be limited because they are sometimes mistaken as grain by birds (Henny et al. 1999).

Industrial Contaminants

Contaminants such as polychlorinated biphenyls (PCBs), polybrominated diphenyl ethers (PBDEs), and heavy metals are another relatively recent threat to raptors (Chapter 18).

PCBs (polychlorinated biphenyls). PCBs are industrial products used in transformers and capacitors for insulating purposes. They also are used as lubricants and as plasticizers in paints (Eisler and Belisle 1996). Unlike pesticides, these chemicals are not deliberately released into the environment. Organochlorine contaminants persist in the environment and have been found in many organisms (Eisler and Belisle 1996). In birds, PCBs have been correlated with embryo deaths and deformations (Ludwig et al. 1996). They also may enhance the effect of organochlorine pesticides as endocrine disrupters (Lincer 1994, cited in Henny et al. 1999). Today, many industries have restricted or eliminated their use of PCBs. Even so, low levels still are being detected in fish-eating birds (Braune et al. 1999).

PBDEs (polybrominated diphenyl ethers). Chemical fire retardants have become common in many consumer products. One of the most frequently used is a class of bromine-based chemicals known as polybrominated diphenyl ethers, or PBDEs (Chapter 18). PBDEs occur in numerous products and, because PBDEs are not chemically bonded to plastics or foam products, they often leach out into the environment. PBDEs and other brominated fire retardants (BFRs) are similar in chemical structure to PCBs. Like PCBs, they are persistent in the environment, soluble in fats, and can bioaccumulate in the food chain.

While human health impacts of these chemicals are not well studied, they are known to cause neurological damage in laboratory animals. PBDEs have been detected in Peregrine Falcon eggs in Sweden (Lindberg et al. 2001).

Manufacturers are working to develop substitutes for PBDEs, and the European Union (EU) banned all PBDEs in electronic products beginning in 2006. Similar restrictions have been proposed in the U.S.

Mercury. Mercury is a naturally occurring element present throughout terrestrial and aquatic environments (Chapter 18). Human sources of environmentally active mercury include coal-burning power plants, industrial boilers, hazardous-waste incineration, and chlorine production. When mercury enters aquatic environments, microorganisms convert it into methylmercury (Eisler 1987). Methylmercury is a toxic form of mercury, and elevated levels can cause serious reproductive effects and addled eggs (Newton 1979). Methylmercury is the most common form of organic mercury found in the environment and is soluble in water and fat, allowing it to bioaccumulate in organisms (Newton 1979).

Organisms in aquatic ecosystems tend to have the highest levels of methylmercury (Eisler 1987). As a result, fish-eating raptors including Ospreys and fish eagles are particularly vulnerable to this toxin (Stjernberg and Saurola 1983).

Lead. Raptors are exposed to lead when they consume animals that have ingested lead shot or have been shot with lead pellets (Kramer and Redig 1997, Henny et al. 1999, Chapter 18). Secondary lead detection has been documented for 35 raptor species in 18 countries (Miller et al. 2002). Spent gunshot and fishing weights are the primary source of lead in wildlife (Scheuhammer and Norris 1996, Henny et al. 1999). Waterfowl ingest lead from the bottom of ponds while foraging (Kramer and Redig 1997).

Bullet fragments containing lead can adversely impact raptors (Newton 1990). Eagles have been poisoned after eating carrion containing lead fragments. The response to ingested lead varies. One study showed that as few as 10 lead pellets can kill a Bald Eagle (Pat-

tee et al. 1981). In Japan, both Steller's Sea Eagles (*H. pelagicus*) and White-tailed Eagles (*H. albicilla*) have died after eating lead fragments from rifle bullets in deer carcasses (Masterov and Saito 2003). Scavenging raptors, including the California Condor, are especially vulnerable to lead poisoning (Janssen et al. 1986). Unlike eagles, condors do not regularly "cast" indigestible materials such as bones and fur or feathers and this may increase their exposure to this threat (Graham 2000).

In addition to direct mortality, sublethal effects may weaken raptors and leave them unable to hunt (Kramer and Redig 1997). Another sublethal effect is severe visual impairment (Redig 1979).

Canada and the U.S. have banned the use of lead shot for hunting waterfowl, however eagles still are poisoned after ingesting lead from hunter-shot upland game (Craig and Craig 1995, Kramer and Redig 1997). More attention is being paid to lead in all environmental contexts, and today non-lead shot, bullets, and fishing sinkers are available, but unfortunately, are not yet widely accepted (Graham 2000).

Introduced Diseases

The introduction of non-native species and diseases is a global problem. As human mobility increases, many organisms associated with people will continue to be dispersed throughout the world, often with adverse consequences.

A number of diseases (bacterial, viral, and fungal) and parasites (internal and external) afflict raptors (Cooper 1969, see Chapter 17, part 1 for details). It can be difficult to assess the impact of diseases because their presence in raptors may be masked by other conditions, including starvation. The solitary nature of many raptors probably affords them some protection from outbreaks of major diseases. However, diseases and parasites can be spread among raptor and vulture species using communal roosts or during migration (Cooper 1990).

The effects of West Nile virus (WNV) on North American bird populations remain a concern. After the mosquito-born disease first appeared in New York in 1999, the virus spread across the continental U.S. in 5 years (U.S. Geological Survey 2003). Diurnal and nocturnal raptors are reported to have a high incidence of WNV infection (Fitzgerald et al. 2003). In the U.S., Great Horned Owls and Red-tailed Hawks particularly have been affected by WNV (Saito et al. 2004).

The introduction of the aquatic plant, hydrilla (*Hydrilla verticillata*), into the U.S. in the 1960s for use in aquariums has been detrimental to navigation, power generation, water intakes, and water quality. Recent field and feeding studies also have implicated exotic hydrilla and an associated epiphytic cyanobacteria species as a link in an emerging avian disease in waterbirds and eagles feeding on them (Wilde 2004). A neurotoxin produced by the bacterial epiphyte may have caused avian vacuolar myelinopathy reported in Bald Eagle deaths beginning in 1994 (National Wildlife Health Center 2001).

Mitigation discussion. Vaccines have been developed for WNV and are available for use in free-ranging raptors including threatened or endangered species (e.g., Aplomado Falcon [*F. femoralis*], California Condor). However, since the animal must be captured for treatment, vaccines may not be effective for wild populations. Their use in captive raptors, on the other hand, may reduce the risk that these diseases pose for wild raptors.

Recreational Disturbance

The effects of recreational disturbance on raptors vary, depending on the species and disturbance type, magnitude, and duration (Preston and Beane 1996). Nesting birds are particularly susceptible to disturbance, which can result in altered foraging patterns, foraging efficiency, and reproductive success (Steidl 1995). Some raptors readily adapt to a human environment while others do not (Fletcher et al. 1999). Studies indicate that raptors are more sensitive to humans approaching on foot than to humans in vehicles (Skagen 1980, Chapter 19). Birds not subjected to direct human persecution might habituate to human activity (Keller 1989), although the extent to which they do so may depend on a number of factors, including the timing, extent, and the type of activity involved.

Mitigation discussion. For sensitive raptor species, management zones are used to protect raptors from human impacts during nesting (Olendorff et al. 1980). A primary zone is established around a nest to protect critical habitat throughout the year, and to seasonally protect all disturbance during nesting. Disturbance from low-flying aircraft may be included. A larger secondary zone generally is used to provide an additional buffer from extreme disturbances such as logging, land clearing, and construction activities during the nesting season. Minor activities such as hiking, bird-watching, camping, and fishing may be permitted in the secondary zone throughout the year, or there may be temporal or seasonal restrictions.

As urbanization and recreational pressures increase, additional research is needed to determine both the magnitude of human-caused disturbance and the proper way to manage it. Public education is a critical component to explain the reason for recreational restrictions (Steenhof 1978).

Urbanization

The Raptor Research Foundation held a symposium titled "Raptors Adapting to Human-Altered Environments" in 1993 during which papers were presented on both negative and positive impacts of raptors in human-altered landscapes. The book *Raptors in Human Landscapes* (Bird et al. 1996) that resulted from this symposium is an excellent reference for this area of conservation concern.

CONCLUSIONS

As research on birds of prey continues, additional impacts and mitigating measures will come to light. Increasingly, raptor biologists will be confronted not only with determining the source and scope of these impacts, but also will have to define ecological and political processes to reverse them. To do so effectively, researchers will need to use rigorous science and good communication skills to promulgate their conservation efforts among those trying to protect raptors.

ACKNOWLEDGMENTS

I am indebted to a number of people who helped update this chapter. Foremost are Bonnie and Howard Postovit who created the original document. Much of the heavy lifting on this document was done by Lori Nielsen and Joel Hurmence who researched various impacts and mitigating measures. I also am indebted to numerous reviewers including Jerry Craig, Daniel Varland, and Mike Kochert. Chuck Henny provided valuable guidance on industrial contaminants. Finally, I would like to thank Chris van Rooyen, Mark Anderson, Jon Smallie, Albert Froneman, Reuven Yosef, Alvaro Camiña Cardenal and Ivan Demeter for providing a global perspective on raptor issues.

LITERATURE CITED

ANDERSON, H.L. 1977. Barbed wire impales another Great Horned Owl. *Raptor Res.* 11:71–72.

ANDERSON, M.D., A.W.A. MARITZ AND E. OOSTHUYSEN. 1999a. Raptors drowning in farm reservoirs in South Africa. *Ostrich* 70:139–144.

ANDERSON, R., M. MORRISON, K. SINCLAIR, D. STRICKLAND, H. DAVIS, W. KENDALL AND NATIONAL WIND COORDINATING COMMITTEE. 1999b. Studying wind energy/bird interactions: a guidance document. National Wind Coordinating Committee, Washington, DC U.S.A.

AVATAR ENVIRONMENTAL, LLC, EDM INTERNATIONAL, INC., AND PANDION SYSTEMS, INC. 2004. Notice of Inquiry comment review: avian/communication tower collisions. Prepared for the Federal Communications Commission (FCC). 30 September 2004.

AVERY, M.L. AND A.C. GENCHI. 2004. Avian perching deterrents on ultrasonic sensors at airport wind-shear alert systems. *Wildl. Soc. Bull.* 32:718–725.

AVIAN POWER LINE INTERACTION COMMITTEE (APLIC). 1994. Mitigating bird collisions with power lines: the state of the art in 1994. Edison Electric Institute, Washington DC U.S.A.

———. 2006. Suggested practices for avian protection on power lines: state of the art in 2006. APLIC, Edison Electric Institute, and the California Energy Commission, Washington, DC U.S.A. and Sacramento, CA U.S.A.

BENSON, P.C. 1981. Large raptor electrocution and powerpole utilization: a study in six western states. Ph.D. dissertation, Brigham Young University, Provo, UT U.S.A.

BEVANGER, K. 1994. Bird interactions with utility structures: collision and electrocution, causes and mitigating measures. *Ibis* 136:412–425.

BIRD, D.M., D.E. VARLAND AND J.J. NEGRO. 1996. Raptors in human landscapes. Academic Press Inc., San Diego, CA U.S.A.

BRAUNE, B.M., B.J. MALONE, N.M. BURGESS, J.E. ELLIOTT, N. GARRITY, J. HAWKINGS, J. HINES, H. MARSHALL, W.K. MARSHALL, J. RODRIGUE, B. WAKEFORD, M. WAYLAND, D.V. WESELOH AND P.E. WHITEHEAD. 1999. Chemical residues in waterfowl and gamebirds harvested in Canada, 1987–95. Canadian Wildlife Service Technical report series No. 326, Canadian Wildlife Service, Ontario, Canada.

BROOKER, M.G. 1990. Persecution of the Wedge-tailed Eagle. Page 196 in I. Newton [ED.], Birds of Prey. Facts on File, New York, NY U.S.A.

BROWN, R.D. 1996. Attraction of Bald Eagles to habitats just below dams in Piedmont North and South Carolina. Pages 299–306 in D.M. Bird, D.E. Varland, and J.J. Negro [EDS.], Raptors in human landscapes. Academic Press Inc., San Diego, CA U.S.A.

BRYAN, A.L., JR., T.M. MURPHY, K.L. BILDSTEIN, I.L. BRISBIN, JR. AND J.J. MAYER. 1996. Use of reservoirs and other artificial impoundments by Bald Eagles in South Carolina. Pages 285–298 in D.M. Bird, D.E. Varland, and J.J. Negro [EDS.], Raptors in human landscapes. Academic Press Inc., San Diego, CA U.S.A.

BUCHANAN, J.B., J.C. LEWIS, D.J. PIERCE, E.D. FORSMAN AND B.L. BISWELL. 1999. Characteristics of young forests used by Spotted Owls on the western Olympic Peninsula, Washington. *Northwest Sci.* 73:255–263.

CADE, T.J., S.A.H. OSBORN, W.G. HUNT AND C.P. WOODS. 2004. Commentary on released California Condors *Gymnogyps californianus* in Arizona. Pages 11–25 in R.D. Chancellor and B.-U. Meyburg [EDS.], Raptors Worldwide. World Working Group on Birds of Prey and Owls, Berlin, Germany.

CAMINA, A. 2004. Consequences of bovine spongiform encephalopathy (BSE) on breeding success and food availability in Spanish vulture populations. Pages 27–44 in R.D. Chancellor and B.-U. Meyburg [EDS.], Raptors Worldwide. World Working Group on Birds of Prey and Owls, Berlin, Germany.

CONNELLY, J.W., M.A. SCHROEDER, A.R. SANDS AND C.E. BRAUN. 2000. Guidelines to manage Sage Grouse populations and their habitats. *Wildl. Soc. Bull.* 28:967–985.

———, S.T. KNICK, M.A. SCHROEDER AND S.J. STIVER. 2004. Conservation assessment of Greater Sage Grouse and sagebrush habitats. Unpublished report submitted to the Western Association of Fish and Wildlife Agencies, Cheyenne, WY U.S.A.

COOPER, J.E. 1969. Current work on raptor diseases. *Raptor Res. News* 3:94–95.

———. 1990. Infectious and parasitic diseases. Page 144 in I. Newton [ED.], Birds of Prey. Facts on File, New York, NY U.S.A.

CORRIGAN, R.M. AND D. MORELAND. 2001. Rodent control: a practical guide for pest management professionals. GIE Publishing, Cleveland, OH U.S.A.

CRAIG, E.H. AND T.H. CRAIG. 1995. Lead levels in Golden Eagles in southeastern Idaho. *J. Raptor Res.* 29:54–55.

CRAIG, T.H. AND L.R. POWERS. 1976. Raptor mortality due to drowning in a livestock watering tank. *Condor* 78:412.

DELONG, J.P. 2000. HawkWatch International raptor conservation program: issues and priorities. HawkWatch International, Salt Lake City, UT U.S.A.

DETRICH, P.J. 1985. Status of the Bald Eagle in California. Pages 81–83 in N. Venizelos and C. Grijalva [EDS.], Raptors. Proceedings of the 7th Annual Wildlife Conference, 4–6 February 1983. San Francisco Zoological Gardens and the California Academy of Sciences, San Francisco, CA U.S.A.

DHINDSA, M.S., J.S. SANDHU AND H.S. TOOR. 1988. Roadside birds in Punjab (India): relation to mortality from vehicles. *Environ. Conserv.* 15:303–310.

DWYER J.F. AND R.W. MANNAN. 2004. Mitigating raptor electrocution in Tucson, Arizona. Abstract of paper presented at the 2004 Raptor Research Foundation Annual Meeting, Bakersfield, CA U.S.A.

EISLER, R. 1987. Mercury hazards to fish, wildlife, and invertebrates: a synoptic review. U.S. Fish and Wildlife Service Biological Report 8 (1.10), Contaminant Hazard Reviews Report 31.

——— AND A.A. BELISLE. 1996. Planar PCB hazards to fish, wildlife, and invertebrates: a synoptic review. U.S. National Biological Service Biological Report 31, Contaminant Hazard Reviews Report 31.

ELECTRIC POWER RESEARCH INSTITUTE (EPRI). 2001. Distribution Wildlife and Pest Control, Technical Report 1001883, EPRI, Palo Alto, CA U.S.A.

ELLIS, D.H., D.G. SMITH AND J.R. MURPHY. 1969. Studies on raptor mortality in western Utah. *Great Basin Nat.* 29:165–167.

ENERGY INFORMATION ADMINISTRATION. 2004. International Energy Outlook 2004. U.S. Department of Energy, Washington, DC U.S.A. www.eia.doe.gov/oiaf/archive/ieo04/index.html (last accessed 11 May 2006).

ERWIN, N. 1991. Carbofuran and bird kills: regulation at a snail's pace. *J. Pestic. Reform* 11:15–17.

ESTEP, J.A. 1989. Avian mortality at large wind energy facilities in California: identification of a problem. Staff report no. P700-89-001. California Energy Commission, Sacramento, CA U.S.A.

EUROPEAN ENVIRONMENT AGENCY. 1995. Glossary. European Environment Agency, Copenhagen K, Denmark. http://glossary.eea.eu.int/EEAGlossary/O/organochlorines (last accessed 11 May 2006).

EXTENSION TOXICOLOGY NETWORK (EXTOXNET). 1996. Pesticide information profiles. Cooperative Extension Offices of Cornell University, NY, Oregon State University, OR, the University of Idaho, ID, the University of California at Davis, CA, and the Institute for Environmental Toxicology, Michigan State University, MI U.S.A. http://extoxnet.orst.edu/pips/carbofur.htm (last accessed 11 May 2006).

FERNIE K.J., D.M. BIRD, R.D. DAWSON AND P.C. LAGUË. 2000. Effects of electromagnetic fields on reproductive success of American Kestrels. *Physiol. Biochem. Zool.* 73:60–65.

FERRER, M. AND M. DE LA RIVA. 1987. Impact of power lines on the population of birds of prey in the Doñana National Park and its environments. *Ric. Biol. Selvaggina* 12:97-98.

——— AND G.F.E. JANSS. 1999. Birds and power lines; collision, electrocution and breeding, 1st Ed. Quercus, Madrid, Spain.

FINN, S.P., D.E. VARLAND AND J.M. MARZLUFF. 2002. Does Northern Goshawk breeding occupancy vary with nest-stand characteristics on the Olympic Peninsula, Washington? *J. Raptor Res.* 36:265–279.

FITZGERALD, S.D., J.S. PATTERSON, M. KIUPEL, H.A. SIMMONS, S.D. GRIMES, C.F. SARVER, R.M. FULTON, B.A. STEFICEK, T.M. COOLEY, J.P. MASSEY AND J.G. SIKARSKIE. 2003. Clinical and pathologic features of West Nile virus infection in native North American owls (Family Strigidae). *Avian Dis.* 47:602–610.

FITZNER, R.E. 1980. Impacts of a nuclear energy facility on raptorial birds. Pages 9–33 in R.P. Howard and J.F. Gore [EDS.], A workshop on raptors and energy developments. U.S. Fish and Wildlife Service, Boise, ID U.S.A.

FLETCHER, R.J., JR., S.T. MCKINNEY AND C.E. BOCK. 1999. Effects of recreational trails on wintering diurnal raptors along riparian corridors in a Colorado grassland. *J. Raptor Res.* 33:233–239.

GILMER, D.S. AND R.E. STEWART. 1983. Ferruginous Hawk populations and habitat use in North Dakota. *J. Wildl. Manage.* 47:146–157.

GOLDSTEIN, M.I., T.E. LACHER, JR., B. WOODBRIDGE, M.J. BECHARD, S.B. CANAVELLI, M.E. ZACCAGNINI, G.P. COBB, E.J. SCOLLON, R. TRIBOLET AND M.J. HOOPER. 1999. Monocrotophos-induced mass mortality of Swainson's Hawks in Argentina, 1995–96. *Ecotoxicology* 8:201–214.

GOULD, G.I., JR. 1985. A case for owls. Pages 14–21 in N. Venizelos and C. Grijalva [EDS.], Proceeding of the 7th Annual Wildlife Conference, 4–6 February 1983. San Francisco Zoological Gardens and the California Academy of Sciences, San Francisco, CA U.S.A.

GRAHAM, F., JR. 2000. The day of the condor. *Audubon* 102:46.

GREEN, W. 2004. The use of 1080 for pest control - a discussion document. Animal Health Board and the Department of Conservation, Wellington, New Zealand.

GROVER, K.E. 1984. Nesting distribution and reproductive status of Ospreys along the upper Missouri River, Montana. *Wilson Bull.* 96:496–498.

HAMERSTROM, F. 1974. Raptor management. Pages 5–8 in R.N. Hamerstrom, Jr., B.E. Harrell, and R.R. Olendorff [EDS.], Proceedings of the Conference on Raptor Conservation Techniques, 1973. Management of Raptors. Fort Collins, CO U.S.A. Raptor Research Report No. 2.

HARNESS, R. 1998. Steel distribution poles-environmental implications. Pages D1 1–5. 1998 Rural Electric Power Conference, St. Louis, Missouri, 26–28 April. Institute of Electricity and Electronic Engineers, Inc., New York, NY U.S.A. (Paper No. 98 D1).

———, AND K.R. WILSON. 2001. Electric-utility structures associated with raptor electrocutions in rural areas. *Wildl. Soc. Bull.* 29:612–623.

———, MILODRAGOVICH AND J. SCHOMBURG. 2003. Raptors and power line collisions. *Colo. Birds* 37:118–122.

HENNY, C.J., D.J. DUNAWAY, R.D. MALLETTE AND J.R. KOPLIN. 1978. Osprey distribution, abundance, and status in western North America: I. The northern California population. *Northwest Sci.* 52:261–271.

——— AND J.L. KAISER. 1996. Osprey population increases along the Willamette River, Oregon, and the role of utility structures, 1976–93. Pages 97–108 in D.M. Bird, D.E. Varland, and J.J. Negro [EDS.], Raptors in human landscapes. Academic Press Inc., San Diego, CA U.S.A.

———, P. MINEAU, J.E. ELLIOTT AND B. WOODBRIDGE. 1999. Raptor poisonings and current insecticide use: what do isolated kill reports mean to populations? in N.J. Adams and R.H. Slotow [EDS.], Proceedings of the 22nd International Ornithological Congress, Durban, Johannesburg. BirdLife South Africa, Johannesburg, South Africa.

HORTON, S.P. 1996. Spotted Owls in managed forests of western Oregon and Washington. Pages 215–231 in D.M. Bird, D.E. Varland, and J.J. Negro [EDS.], Raptors in human landscapes. Academic Press Inc., San Diego, CA U.S.A.

HOUSTON, D.C. 1996. The effect of altered environments on vultures. Pages 328–335 in D.M. Bird, D.E. Varland, and J.J. Negro [EDS.], Raptors in human landscapes. Academic Press Inc., San Diego, CA U.S.A.

IEZEKIEL, S., D.E. BAKALOUDIS AND C.G. VLACHOS. 2004. The status and conservation of Griffon Vulture *Gyps fulvus* in Cyprus. Pages 67–73 in R.D. Chancellor and B.-U. Meyburg [EDS.], Raptors Worldwide. World Working Group on Birds of Prey and Owls, Berlin, Germany.

JANSS, G.F.E. AND M. FERRER. 1999. Mitigation of raptor electrocution on steel power poles. *Wildl. Soc. Bull.* 27:263–273.

JANSSEN, D.L., J.E. OOSTERHUIS, J.L. ALLEN, M.P. ANDERSON, D.G. KELTS AND S.N. WIEMEYER. 1986. Lead poisoning in free-ranging California Condors. *J. Am. Vet. Med. Assoc.* 189:1115–1117.

KAREIVA, P. AND U. WENNERGREN. 1995. Connecting landscape patterns to ecosystem and population processes. *Nature* 373:299–302.

KAYE, E. 2001. Letter to the Editor. *Audubon.* http://magazine.audubon.org/letter/letter0105.html (last accessed 11 May 2006).

KELLER, V. 1989. Variation in the response of Great Crested Grebes *Podiceps cristatus* to human disturbance - a sign of adaptation? *Biol. Conserv.* 49:31–45.

KENWARD, R.E. 1996. Goshawk adaptation to deforestation: does Europe differ from North America? Pages 233–243 in D.M. Bird, D.E. Varland, and J.J. Negro [EDS.], Raptors in human landscapes. Academic Press Inc., San Diego, CA U.S.A.

KRAMER, J.L. AND P.T. REDIG. 1997. Sixteen years of lead poisoning in eagles, 1980–95: an epizootiologic view. *J. Raptor Res.* 31:327–332.

LEDGER, J. 1979. Drowning of Cape Vultures in circular water tanks in South Africa. Abstract of paper presented at International Symposium on Vultures, 23–26 March 1979, Santa Barbara Museum of Natural History, Santa Barbara, CA U.S.A.

———. 1988. Tackling the problem of vulture poisoning. *Bokmakierie* 40:4–5.

LEE, J.M., JR., T.D. BRACKEN AND L.E. ROGERS. 1979. Electric and magnetic fields as considerations in environmental studies of transmission lines. Pages 55–73 in R.D. Phillips et al. [EDS.], Biological effects of extremely low frequency electromagnetic fields. Conf 78106. National Technical Information Service, Springfield, VA U.S.A.

LEHMAN, R.N. 2001. Raptor electrocution on power lines; current issues and outlook. *Wildl. Soc. Bull.* 29:804–813.

LESHEM, Y. AND O. BAHAT. 1999. Flying with the birds. Yedioth Ahronoth & Chemed Books Ltd., Tel Aviv, Israel.

LINCER, J.L. 1994. A suggestion of synergistic effects of DDE and Aroclor 1254 on reproduction of the American Kestrel *Falco sparverius*. Pages 767–769 in B.-U. Meyburg and R.D. Chancellor [EDS.], Raptor Conservation Today. World Working Group on Birds of Prey and Owls, Berlin, Germany and Pica Press, London, United Kingdom.

LINDBERG, P., U. SELLSTRÖM, L. HÄGGBERG AND C. DE WIT. 2001. Polybrominated flame retardants (PBDEs) found in eggs of Peregrine Falcons *(Falco peregrinus)*. Page 109 in Abstracts from the 4th Eurasian Congress on Raptors, 25–29 September 2001, Seville, Spain.

LUDWIG, J.P., H. KURITA-MATSUBA AND H.J. AUMAN. 1996. Deformities, PCBs, and TCDD-Equivalents in Double-crested Cormorants *(Phalacrocorax auritus)* and Caspian Terns *(Sterna caspia)* of the upper Great Lakes 1986–1991: testing a cause-effect hypothesis. *J. Great Lakes Res.* 22:172–197.

MALARIA FOUNDATION INTERNATIONAL. 2006. Malaria control campaign. www.malaria.org/DDTpage.html (last accessed 15 March 2006).

MAÑOSA, S. 2001. Strategies to identify dangerous electricity pylons for birds. *Biodiversity Conserv.* 10:1997–2012.

MARSHALL, L. 2003. Gamblers fuel trade in "lucky" vulture heads in Africa. *National Geographic News.* 25 February 2003. http://news.nationalgeographic.com/news/2003/02/0225_030225_SAvultures.html (last accessed 11 May 2006).

MASTEROV, V.B. AND K. SAITO. 2003. Problems of conserving Steller's Sea Eagle *Haliaeetus pelagicus* in the southern part of its home range and on its wintering grounds. Abstract from 6th World Conference on Birds of Prey and Owls, 18–23 May 2003, Budapest, Hungary.

MCILROY, J.C. 1984. The sensitivity of Australian animals to 1080 poison: VII. native and introduced birds. *Aust. Wildl. Res.* 11:373–385.

MEADE, G.M. 1942. Calcium chloride - a death lure for crossbills. *Auk* 59:439–440.

MILLER, M.J.R., M.E. WAYLAND AND G.R. BORTOLOTTI. 2002. Lead exposure and poisoning raptors: a global perspective. Pages 224–245 in R. Yosef, M.L. Miller, and D. Pepler [EDS.], Raptors in the New Millennium, Proceedings of the World Conference on Birds of Prey and Owls, International Birding and Research Center - Eilat, Eilat, Israel.

MINEAU, P., M.R. FLETCHER, L.C. GLASER, N.J. THOMAS, C. BRAS-

SARD, L.K. WILSON, J.E. ELLIOTT, L.A. LYON, C.J. HENNY, T. BOLLINGER AND S.L. PORTER. 1999. Poisoning of raptors with organophosphorous and carbamate pesticides with emphasis on Canada, U.S. and U.K. *J. Raptor Res.* 33:1–37.

MOONEY, N.J. AND R.J. TAYLOR. 1996. Value of nest site protection in ameliorating the effects of forestry operations on Wedge-tailed Eagles in Tasmania. Pages 275–282 in D.M. Bird, D.E. Varland, and J.J. Negro [EDS.], Raptors in human landscapes. Academic Press Inc., San Diego, CA U.S.A.

MUNDY, P., D. BUTCHART, J. LEDGER AND S. PIPER. 1992. The vultures of Africa. Academic Press Inc., San Diego, CA U.S.A.

MURPHY, J.R. 1978. Management considerations for some western hawks. *Trans. N. Am. Wildl. Nat. Resour. Conf.* 43:241–251.

NATIONAL WILDLIFE HEALTH CENTER. 2001. Fact sheet: avian vacuolar myelinopathy: an unexplained neurologic disease. USDI Geological Survey, Madison, WI U.S.A.

NEWTON, I. 1979. Population ecology of raptors. Buteo Books, Vermillion, SD U.S.A

———. 1990. Birds of Prey. Facts on File, New York NY U.S.A.

NOSS, R. 1990. Ecological effects of roads. Pages 1–7 in J. Davis, [ED.], Killing roads: a citizen's primer on the effects and removal of roads. Earth First! Biodiversity Project Special Publication, Tucson, AZ U.S.A.

OAKS, J.L., M. GILBERT, M.Z. VIRANI, R.T. WATSON, C.U. METEYER, B.A. RIDEOUT, H.L. SHIVAPRASAD, S. AHMED, M.J.I. CHAUDHRY, M. ARSHAD, S. MAHMOOD, A. ALI AND A.A. KHAN. 2004. Diclofenac residues as the cause of vulture population decline in Pakistan. *Nature* 427:630–633.

OLENDORFF, R.R. 1973. The ecology of the nesting birds of prey of northeastern Colorado. U.S. International Biological Program Grassland Biome Tech. Rep. 211.

———, R.S. MOTRONI AND M.W. CALL. 1980. Raptor management - the state of the art in 1980. USDI Bureau of Land Management, Denver, CO U.S.A.

——— AND R.N. LEHMAN. 1986. Raptor collisions with utility lines: an analysis using subjective field observations. Unpublished Report submitted to Pacific Gas and Electric Company, San Ramon, CA U.S.A.

OLSON, C.V. 1999. Hawk shooting: not just a problem of the past. Page 37 in Raptor Research Foundation Annual Meeting, program and abstracts, 3–7 November 1999, La Paz, Baja California, Sur, Mexico.

ONTARIO MINISTRY OF THE ENVIRONMENT. 1995. Pesticides safety handbook. Sustainable Production Branch, Saskatchewan Agriculture and Food Occupational Health and Safety Division of Saskatchewan Labour, Ontario, Canada.

PATTEE, O.H., S. WIEMEYER, B.M. MULHERN, L. SILEO AND J.W. CARPENTER. 1981. Experimental lead-shot poisoning in Bald Eagles. *J. Wildl. Manage.* 45:806–810.

PESTICIDE NEWS. 1996. Organophosphate insecticides. *Pestic. News* 34:20–21.

PIPER, S.F. 2003. Vulture restaurants; conflict in the midst of plenty. 6th World Conference on Birds of Prey and Owls, 18–23 May 2003, Budapest, Hungary.

PLATT, C.M. 2005. Patterns of raptor electrocution mortality on distribution power lines in southeast Alberta. M.S. thesis, University of Alberta, Edmonton, Alberta, Canada.

PLATT, J.B. 1976. Bald Eagles wintering in a Utah desert. *Am. Birds* 30:783–788.

POSTOVIT, H.R. AND B.C. POSTOVIT. 1987. Impacts and mitigation techniques. Pages 183–213 *in* B. A. Giron Pendleton, B. A. Millsap, K. W. Cline and D. M. Bird. [EDS.], Raptor management techniques manual. National Wildlife Federation, Washington, DC U.S.A.

PRESTON, C.R. AND R.D. BEANE. 1996. Occurrence and distribution of diurnal raptors in relation to human activity and other factors at Rocky Mountain Arsenal, Colorado. Pages 365–374 in D.M. Bird, D.E. Varland, and J.J. Negro [Eds.], Raptors in human landscapes. Academic Press Inc., San Diego, CA U.S.A.

RANDS, M.R.W. 1985. Pesticide use on cereals and the survival of Grey Partridge chicks: a field experiment. *J. Appl. Ecol.* 22:49–54.

RATCLIFFE, D.A. 1967. Decrease in eggshell weight in certain birds of prey. *Nature* 215:208–210.

REDIG, P.T. 1979. Raptor management and rehabilitation. Pages 226–237 *in* Management of northcentral and northeastern forests for nongame birds. USDA Forest Service Gen. Tech. Rep. NC-51, North Central Forest Experiment Station, Minneapolis, MN U.S.A.

REDPATH, S.M. 1995. Impact of habitat fragmentation on activity and hunting behavior in the Tawny Owl, *Strix aluco. Behav. Ecol.* 6:410–415.

REYNOLDS, R.T., E.C. MESLOW AND H.M. WIGHT. 1982. Nesting habitat of coexisting Accipiters in Oregon. *J. Wildl. Manage.* 46:124–138.

ROBARDS, F.C. AND J.G. KING. 1966. Census, nesting, and productivity of Bald Eagles in southeast Alaska, 1966. U.S. Fish and Wildlife Service, Juneau, AK U.S.A.

ROBERTSON, J.M. 1930. Roads and birds. *Condor* 32:142–146.

RODRÍGUEZ-ESTRELLA, R. 1996. Response of Common Black Hawks and Crested Caracaras to human activities in Mexico. Pages 355–363 in D.M. Bird, D.E. Varland and J.J. Negro [EDS.], Raptors in human landscapes. Academic Press Inc., San Diego, CA U.S.A.

SAITO, E.K., L. SILEO, D.E. GREEN, C.U. METEYER, D.E. DOCHERTY, G.S. MCLAUGHLIN, AND K.A. CONVERSE. 2004. Raptor mortality due to West Nile virus in several states, 2002. Page 39 in Raptor Research Foundation Annual Meeting Abstracts. 10–13 November 2004, Bakersfield, CA U.S.A.

SCHEUHAMMER, A.M. AND S.L. NORRIS. 1996. The ecotoxicology of lead shot and lead fishing weights. *Ecotoxicology* 5:279–295.

SCHNELL, J.H. 1979. Behavior and ecology of the Black Hawk *(Buteogallus anthracinus)* in Aravaipa Canyon (Graham/Pinal Counties), Arizona, 4th Progress Report submitted to USDI Bureau of Land Management, Safford, AZ U.S.A.

SHAPIRO, A.E., F. MONTALBANO, III AND D. MAGER. 1982. Implications of construction of a flood control project upon Bald Eagle nesting activity. *Wilson Bull.* 94:55–63.

SHARPE, P.B. 2004. Twenty-five years of Bald Eagle restoration in southern California and the continuing effects of DDT. Page 39 in Raptor Research Foundation Annual Meeting Abstracts, 10–13 November 2004, Bakersfield, CA U.S.A.

SHIRIHAI, H., R. YOSEF, D. ALON, G.M. KIRWAN AND R. SPAAR. 2000. Raptor migration in Israel and the Middle East; a summary of 30 years of field research. International Birding and Research Center - Eilat, Eilat, Israel.

SKAGEN, S.K. 1980. Behavioral responses of wintering Bald Eagles to human activity on the Skagit River, Washington. Pages 231–241 in R.L. Knight, G.T. Allen, M.V. Stalmaster, and C.W. Servheen [EDS.], Proceedings of the Washington Bald Eagle Symposium. The Nature Conservancy, Seattle, WA U.S.A.

SNYDER, H.A. AND N.F.R. SNYDER. 1974. Increased mortality of Cooper's Hawks accustomed to man. *Condor* 76:215–216.

SODHI, N.S. 2002. Competition in the air: birds versus aircraft. *Auk* 75:400–414.

STAFF AND WIRE REPORTS. 1998. Nets to protect birds from prison fences. www.prisonactivist.org/pipermail/prisonact-list/1998-April/001686.html (last accessed 11 May 2006).

STALMASTER, M.V. 1987. The Bald Eagle. Universe Books, New York, NY U.S.A.

STEENHOF, K. 1978. Management of wintering Bald Eagles. U.S. Fish and Wildlife Service Report FWS/OBS-78/79.

———, M.N. KOCHERT AND J.A. ROPPE. 1993. Nesting by raptors and Common Ravens on electrical transmission line towers. *J. Wildl. Manage.* 57:271–281.

STEIDL, R.J. 1995. Human impacts on the ecology of Bald Eagles in interior Alaska. Ph.D. dissertation, Oregon State University, Corvallis, OR U.S.A.

STJERNBERG, T. AND P. SAUROLA. 1983. Population trends and management of the White-tailed Eagle in northwestern Europe. Pages 307–318 in D.M. Bird, N.R. Seymour, and J.M. Gerrard [EDS.], Biology and management of Bald Eagles and Ospreys: Proceedings of 1st International Symposium on Bald Eagles and Ospreys, Montreal, 28–29 October 1981. Macdonald Raptor Research Centre, McGill University and Raptor Research Foundation, Montreal, Canada.

SYKES, P.W. 1979. Status of the Everglade Kite in Florida 1968–1978. *Wilson Bull.* 91:495–511.

THIOLLAY, J.M. 1996. Rain forest raptor communities in Sumatra: the conservation value of traditional agroforests. Pages 245–261 *in* D. M. Bird, D. E. Varland, and J. J. Negro [EDS.], Raptors in human landscapes. Academic Press Inc., San Diego, CA U.S.A.

THOMPSON, L. S. 1978. Transmission line wire strikes: mitigation through engineering design and habitat modification. Pages 27–52 in M.L. Avery [ED.], Impacts of transmission lines on birds in flight: proceedings of a workshop, Oak Ridge Associated Universities, Oak Ridge, Tennessee, 31 January–2 February 1978. USDI Fish and Wildlife Service, Washington, DC U.S.A.

TRANSPORT CANADA. 2001. Sharing the Skies: an aviation industry guide to the management of wildlife hazards (TP 13549E). Transport Canada, Ontario, Canada.

U.S. AIR FORCE. 2004. Bird/Wildlife aircraft strike hazard (BASH) management techniques. Air Force Pamphlet 91–212.

U.S. DEPARTMENT OF ENERGY. 2004. International energy outlook 2004. Report No. DOE/EIA-0484. USDOE Office of Integrated Analysis and Forecasting, Washington, DC U.S.A.

U.S. FISH AND WILDLIFE SERVICE. 2003. Interim guidelines to avoid and minimize wildlife impacts from wind turbines. U.S. Fish and Wildlife Service, Washington, DC U.S.A. www.fws.gov/habitatconservation/wind.pdf (last accessed 11 May 2006).

U.S. GEOLOGICAL SURVEY. 2003. Fact sheet: effects of West Nile Virus. USDI Geological Survey, Washington, DC U.S.A..

VAN DAELE, L.J. AND H.A. VAN DAELE. 1982. Factors affecting the productivity of Ospreys nesting in west-central Idaho. *Condor* 84:292–299.

VAN ROOYEN, C.S. 2000. Raptor mortality on powerlines in South Africa. Pages 739–750 in R.D. Chancellor and B.-U. Meyburg [EDS.], Raptors at risk. World Working Group on Birds of Prey and Owls, Berlin, Germany and Hancock House, Blaine, WA U.S.A.

WHITE, J., [ED.]. 1913. *Handbook of Indians of Canada*, Published as an Appendix to the Tenth Report of the Geographic Board of Canada, Ottawa, Canada.

WILDE, S.B. 2004. Avian vacuolar myelinopathy (AVM) linked to exotic aquatic plants and a novel cyanobacterial species. Abstract at the 2004 Joint Conference of the American Association of Zoo Veterinarians, Wildlife Disease Association, American Association of Wildlife Veterinarians. San Diego, CA U.S.A.

WILLIAMS, T. 2001. Golden Eagles for the Gods: if a species is essential to religious practices of Native Americans, why would they recklessly kill it? And why would the Feds encourage them? http://magazine.audubon.org/incite/incite0103.html (last accessed 11 May 2006).

XIROUCHAKIS, S. 2003. Causes of raptor migration in Crete. Pages 849–859 in R.D. Chancellor and B.-U. Meyburg [EDS.], Raptors Worldwide. World Working Group on Birds of Prey and Owls, Berlin Germany and MME/Birdlife Hungary, Budapest, Hungary.

YAHNER, R.H. AND R.W. ROHRBAUGH, JR. 1998. A comparison of raptor use of reclaimed surface mines and agricultural habitats in Pennsylvania. *J. Raptor Res.* 32:178-180.

ZARN, M. 1974. Habitat management series for unique or endangered species. Report No. 11: Burrowing Owl *(Speotyto cunicularia hypugea)*. USDI Bureau of Land Management Technical Note 250.

Captive Breeding

Joseph B. Platt
PCR Services Corp.
1 Venture Suite 150, Irvine, CA 92692 U.S.A.

David M. Bird and Lina Bardo
Avian Science and Conservation Centre of McGill University
21,111 Lakeshore Road, Ste. Anne de Bellevue
Quebec, H9X 3V9 Canada

INTRODUCTION

Birds of prey have been held in captivity for thousands of years by many cultures. However, it was not until the 20th century that they were bred in captivity and manipulated in the manner of domestic species. In his review of breeding records of aviculturists, zoos, and falconers, Cade (1986) found reports of 15 species that had bred in captivity by the 1950s, and 22 by 1965. None of these occurrences was part of an organized or sustained program.

It was the idea of saving a diminishing species that provided the catalyst to bring together the people and resources needed to overcome the challenges of consistently breeding these highly aggressive birds. Peregrine Falcons (*Falco peregrinus*) were disappearing from breeding sites in North America and Europe because of the contamination of their prey by DDT (Ratcliffe 1980). Western nations were committed to clean up the food chain, but would declining species such as Peregrine Falcons, Ospreys (*Pandion haliaetus*) and Bald Eagles (*Haliaeetus leucocephalus*) recover?

In the early 1960s, Willoughby and Cade (1964) demonstrated that it was feasible to breed American Kestrels (*F. sparverius*) in large numbers for scientific study. The Raptor Research Foundation was formed in 1966 by a group of falconers and biologists mainly focused on saving the Peregrine Falcon. Under its aegis, information and ideas were exchanged between private breeders and institutions. In North America, government and institutional programs were begun; the largest included The Peregrine Fund at Cornell University, the U.S. Fish and Wildlife Service's program at Patuxent, Maryland, the Canadian Wildlife Service's facility in Wainwright, Alberta, the Saskatchewan Co-operative Falcon Project at the University of Saskatchewan, and the Macdonald Raptor Research Centre at McGill University. At the same time Europe saw the creation of the Hawk Trust in the United Kingdom, and various falconry groups in Germany to promote breeding of large raptors for falconry and conservation.

Within 10 years significant progress had been made in understanding the behavior and management of breeding pairs as well as the art of incubation and the care of young. Survey articles by Cade (1986, 2000) documented that hundreds of large falcons were being produced each year and at least 83 raptor species had been bred by 1985.

The successes have continued. Captive breeding and the related manipulation of wild-produced eggs have proved critical in the re-establishment of at least 13 species. These include the California Condor (*Gymnogyps californianus*), Red Kite (*Milvus milvus*) in Britain, Bald Eagle, White-tailed Eagle *(H. albicilla)* in Scotland, Bearded Vulture (*Gypaetus barbatus*), Griffon Vulture (*Gyps fulvus*), Harris's Hawk (*Parabu-*

teo unicinctus), Mauritius Kestrel (*F. punctatus*), Aplomado Falcon (*F. femoralis*), Lanner Falcon (*F. biarmicus*), Peregrine Falcon on two continents, Barn Owl (*Tyto alba*) and Eurasian Eagle-Owl (*Bubo bubo*) in Europe. Another dozen species have been bred and released on a smaller scale (Cade 2000).

This chapter presents a summary of guidelines to the successful breeding of captive birds of prey. Raptors are a diverse group, one to which generalities do not always apply.

Artificial insemination and the use of imprints is one aspect of breeding that came about because of close association between trained raptors and their handlers. These birds, both male and females, accept humans as mates. They court, solicit copulation, and raise fostered young with humans. The mechanics of this specialized aspect of breeding is well presented in the literature (Weaver and Cade 1985, Fox 1995) and is not discussed here. The raptors presented in this chapter are divided into six major categories: large falcons, small falcons, eagles, hawks, owls, and vultures and condors. Since captive propagation of raptors began with the breeding of large falcons, we have placed an initial detailed emphasis on these raptors, followed by descriptions of the variations in breeding practices for the remaining groups of species. Within each raptor group, aspects of cage design, feeding methods, breeding behavior, and natural and artificial incubation and brooding methods are discussed. No matter what group of raptors the reader is interested in breeding, it is worthwhile to peruse the entire chapter for useful tips that are likely applicable across the board.

All birds of prey are protected by government agencies and the importation of exotic species is highly regulated (see Chapter 25). One should always verify which permits are necessary to set up a breeding facility and to acquire and raise raptors before beginning the project. Security to protect the birds from predators, thieves, and vandals also must be considered in the design and operation of a facility.

ORIGINS OF BREEDING STOCK

Acquiring the Birds

Raptors for captive breeding can be acquired from several sources; some are taken from the wild, particularly for species conservation programs (Cox et al. 1993). Special permits are required to remove birds from the wild or to import them from other countries. Birds can be collected as eggs or fledglings and then hand-raised or raised by existing captive pairs. These individuals grow up accustomed to their confinement and are generally well adjusted (Weaver and Cade 1985, Toone and Risser 1988, Jenny et al. 2004). Hand-rearing young to fledging age can cause significant socialization problems due to imprinting and should be avoided if natural pairing is intended. On the other hand, if nestlings are hand-reared in groups of two or more conspecifics, they also will imprint on one another. If these nestlings are placed in groups in flight pens at post-fledging, the human imprint phenomenon can be reversed, at least with American Kestrels (D. Bird, unpubl. obs.). Hand puppets can also be used (see Condors and Vultures section).

Some species will breed in captivity when caught as adults (e.g., California Condors [Wallace 1994, Harvey et al. 2003], American Kestrels [D. Bird, unpubl. obs.]), but this is less likely for larger falcons (Weaver and Cade 1985). Injured, unreleasable wild owls frequently have been used as natural breeders if not too severely compromised (McKeever 1979) and endangered raptors held in rehabilitation programs can be used as semen donors (Blanco et al. 2002). All newly acquired birds should be quarantined and tested for disease and parasites before being used in a breeding program (Toone and Risser 1988).

Female raptors usually are larger and more aggressive than males. To reduce the risk of injury or death to the male, the male should be placed in the breeding pen several days or weeks before the female (Heidenreich 1997). This may allow the territorial male to exert a certain degree of dominance over the newly introduced female. Even so, in some species such as Merlins, the female will suddenly and explicably kill her long-time mate (D. Bird, unpubl. obs.).

SEXING AND PEDIGREES

Many raptors are size- or plumage-dimorphic, and thus can be sexed easily (D'Aloria and Eastham 2000). A few, however, are size- or plumage-monomorphic. For the latter, breeders must resort to collecting blood or excreta in order to perform radioimmunoassays to test for the presence of testosterone or estrogen (Saint Jalme 1999). Birds also can be sexed using standard DNA blood analyses and karyotyping (Saint Jalme 1999, Leupin and Low 2001). Bald Eagles have been sexed using laparoscopy (Mersmann et al. 1992, Parry-Jones 2000). Observing the behavior and vocalizations of interacting birds also can be an indication of their sex (McKeever 1979).

There always is a risk of inbreeding when working with a small population. Stock secured from other captive populations may already be inbred. Severely reduced wild populations also may be highly related. Careful records should be kept and genetic fingerprinting (microsatellite marking) can be used to ascertain the relationships between birds (Toone and Risser 1988). Programs such as KINSHIP have been used to test the pedigree of potential pairs to ensure that inbreeding is reduced (Gautschi et al. 2003).

LARGE FALCONS

In 1983, The Peregrine Fund, Inc. (now based at the World Center for Birds of Prey in Boise, Idaho) produced a publication: *Falcon propagation: a manual on captive breeding*, edited by Jim Weaver and Tom Cade (revised in 1985), that contains sections on "Artificial Incubation of Falcon Eggs" by William Burnham, and "Incubation and Rearing" by Willard Heck and Dan Konkel that are especially useful. These sections are summarized in the earlier version of this manual (Burnham et al. 1987). Both are among the best sources of general information available on the propagation of large falcons and other species of raptors. We draw heavily from these documents and refer the reader to them for greater detail.

Cage Design

The Peregrine Fund's breeding facilities formerly in Ithaca, New York and Fort Collins, Colorado and now in Boise, Idaho were among the most thoroughly researched for the reproduction of large falcons. These facilities serve as a model for those now belonging to a good number of private falcon breeders in many parts of the world (e.g., the Middle East, the United Kingdom, Europe, North America). The chambers are designed primarily for peregrine-sized falcons with other facilities for raptors ranging from kestrels to eagles. The chambers are grouped on either side of a two-story central hallway from which the chambers are viewed and serviced. The buildings are basically "pole barns." The floors of the chambers measure 3 × 6 m. The roof of each breeding barn is sloped, making the chambers 6 m high

on the interior wall and 4.2 m high on the outside wall. The outer wall is open and covered with two layers of wire mesh on the outside and vertical bars of 1.3-cm thin walled electrical conduit placed at 6.2-cm centers on the inside. The PVC bars prevent the birds from coming in contact with the mesh, which is 15 cm beyond the bars. The roof is solid except for a 9 × 3-m panel of mesh and bars to allow light and air flow. The walls of the chambers are painted plywood, which provides a smooth washable surface. The floor and nest ledge are covered with pea-sized smooth gravel, which has smooth edges and does not compact, thus providing a "giving" surface for landing birds. The bottom meter of the outer wall is paneled with metal sheeting to keep out snow. Predator barriers are buried around the buildings to protect the birds from potential predators and rodents (Weaver and Cade 1985). The service corridor runs down the center of the barn on the first and second floor, allowing keepers access to each pen for maintenance and observation of the breeding pairs. The floors of these corridors are soundproofed with carpeting. Strategically placed one-way glass panels permit keepers to observe the birds (Weaver and Cade 1985, Jenny et al. 2004).

In two-story breeding facilities, the upper corridor should provide hatch-door access for removing eggs from the nest ledge without entering the pen. Similarly, each pen should have access ports for food to be slipped into the room and the bath to be removed and replaced without keepers entering the chamber. Raising the bath pan above the floor reduces the amount of feathers and debris in the water. Food should be provided using inclined chutes from the upper and lower corridors (Jenny et al. 2004). Microphones to detect copulatory behavior (J. Weaver, pers. comm.) or even better, closed-circuit television (K. McKeever, pers. comm.) greatly improve the ability to monitor a pair's behavior.

Facilities built after the ones in Ithaca employed several smaller barns offering identical types of pens in order to reduce the risk of spreading of disease (Weaver and Cade 1985). As mentioned earlier, The Peregrine Fund design has been modified by other breeders. In drier climates, most of the roof of each pen can be barred like the walls instead of being fully covered. Parts of the roof should still be covered to provide shelter and shade for the birds. In facilities where disturbance from traffic or people cannot be avoided, the walls can be solid metal or wood sheeting and the roofs meshed or barred. Vents can be added to sidewalls where extra air circulation is needed. Screening can be placed over the wire mesh to protect the birds against mosquito-born diseases such as West Nile virus (Weaver and Cade 1985, K. McKeever, pers. comm.).

The Aplomado Falcon breeding facilities at the World Center for Birds of Prey consist of 3 × 6.1-m breeding pens with roofs sloped from 4.3 to 5.5 m. The structure is solid except for two roof skylights and one wall window with 4.3-cm bar spacing (Jenny et al. 2004). Circular cages also have been successful with large falcons, allowing them to fly in circles for exercise (Heidenreich 1997). Gyrfalcons (*F. rusticolus*) have been bred in circular pens of more than 20 m in diameter and up to 6 m high (Heidenreich 1997).

Nest ledges ranging from 0.75 × 1.25 m to 1.25 × 3 m have been successful for large falcons (Parry-Jones 2000). Nest ledges should be lined with clean aquarium sand or small gravel (Fig. 1). Aplomado Falcons are given a choice between two 0.6-m^2 nest boxes lined with cedar chips (*Thuja* spp.) on a nest ledge (Jenny et al. 2004). Perches in the pens should have at least 1 m of clearance above them, and some branch perches should be higher than the nest ledge to provide lookout posts. Coco fiber doormats or AstroTurf® (some carpet fibers can cause the bird's talons to become entangled) should be placed on flat shelf perches or on the lips of nest ledges to reduce the potential of bruising a bird's foot when landing (Weaver and Cade 1985, Jenny et al. 2004). Large smooth rocks also can serve as perches on the ground. Perches should be placed such that excreta do not foul other perches or the bath pan.

Figure 1. A female white Gyrfalcon (*Falco rusticolus*), arguably one of the more difficult large falcons to breed in captivity, solicits copulation on her gravel-lined nesting ledge.

Pen floors can be covered with coarse gravel overlaid with 10 cm of pea-sized gravel for rapid drying and good drainage. For cases where the photoperiods of the birds must be adjusted, or in case of emergency, light fixtures should be added to the pens in such a way that prevents the birds from perching on or shattering them (Weaver and Cade 1985). Birds also may require radiant heat panels near perches or heat tape on nest ledges if they are being kept at a facility with temperatures below their accustomed range (Jenny et al. 2004).

Food, Feeding, and Watering Procedures

Large falcons at the World Center for Birds of Prey facilities are fed quail, day-old chicks and 5-week-old chickens alternately. Large food items are cut in pieces. Feeding is once per day. During cold weather the daily ration may be offered in two half-portions to prevent it from becoming frozen before being consumed. Vitamin supplements (e.g., Avitron) are added to food in the breeding season. During the breeding season smaller food items are given more frequently to encourage males to begin food transfers with the female as part of the pair bonding process. Water baths consist of large, open, shallow pans and are changed once per week or as needed. Clean pans are used every time. It may be necessary to remove baths in colder months.

Capture of Falcons in the Chamber

To provide clean chambers at the World Center for Birds of Prey, the falcons are captured and moved immediately after the breeding season and in mid-winter. In some cases it is necessary to capture and briefly hold a female while her eggs are removed from the nest ledge. A defensive female can break an egg or the whole clutch; males are less of a problem. Birds are caught with a long-handled net. A slight noise made prior to entering the chamber reduces the chance of a panic flush resulting in injury.

After entering the chamber and netting the bird or birds, an assistant allows the bird to grip his gloved hands to prevent self-inflicted foot punctures. When punctures occur they may cause low-level foot infections (small scab and limited swelling). To minimize stress however, there is no attempt to treat such individuals, as they seldom, if ever, develop a serious chronic foot problem. Rarely, a sprained wing results during capture; it may last from a few hours to several weeks. Appropriate provisioning of food and perches to limit flight should lead to recovery, but persistence of the condition warrants examination by a veterinarian.

Moving the birds to a clean chamber offers the chance to examine their health. Talons are clipped extremely short. The beak also is trimmed if overgrown. The whole operation takes only a few minutes, and the bird may be hooded if necessary (Burnham 1983).

Courtship Behavior

Courtship behavior of large falcons varies among species, but often involves flight displays, vocalizations, and food transfers between the pair. Courtship displays in Peregrine Falcons include fly-bys by the male near the female, scraping of the gravel on the nest ledge by both sexes, food transfers from male to female, and the female calling for food or chasing the male for food. Courtship can progress to displays and vocalizations on the nest ledge, including "hitch-winged" displays by the male, and solicitation by the female (see Fig. 1). For further information on courtship displays, see Weaver and Cade (1985) and Platt (1989).

Management for Production

Large falcons may take 2–3 years before reaching sexual maturity (Parry-Jones 2000). They usually lay 2–6 eggs, depending on the species. Incubation is typically 30–35 days. Chicks usually are fledged in about 6 weeks (Parry-Jones 2000). Ideally, young birds should be placed in communal pens to encourage natural socialization in order to make pairing more successful (Weaver and Cade 1985). The need to propagate certain genetically valuable individuals, especially with small populations of endangered species, may mean that some pairings are decided by keepers.

Birds may have to be paired together for several years before nesting successfully, and some individuals take longer to mature than others. Pairing a younger bird with an experienced breeder can increase the new bird's chance of having a successful first breeding season (Jenny et al. 2004). Pairs that continue to fail to produce young should be separated and offered new mates. Pairs should be monitored for aggression and be separated if necessary. The birds should be disturbed as little as possible, ideally using microphones or cameras in the pens, or one-way observation windows, or both to monitor them (Weaver and Cade 1985).

To increase egg production, pairs can be forced to double-clutch by removing the first clutch or by remov-

ing eggs sequentially and incubating them artificially (Weaver and Cade 1985, Jenny et al. 2004). This should be done only with more experienced pairs. First-time breeders should be given the chance to raise their own first clutch, unless the breeder suspects potential problems. Second clutches can be replaced by the hatched chicks of the first clutch for the adults to raise, whereas the second clutch is incubated artificially. Burnham (1983) and Weaver and Cade (1985) are good sources for information on this subject.

Incubation and Hatching Procedures

Unless otherwise indicated below, most of the details on the procedures of artificial incubation can be taken from this section, as there are similarities in the procedures for all raptors.

Eggs can be incubated naturally unless there is concern that the pair will damage the eggs, or if double-clutching is desired. Males and females often share the task of incubation (Weaver and Cade 1985). Before incubating eggs artificially, several factors must be considered. First, even if artificial incubation is to be used, ideally eggs should be incubated naturally for the first 7–10 days to increase their chance of hatching (Burnham 1983, Weaver and Cade 1985, Jenny et al. 2004). This natural incubation also can be achieved using chickens, but facilities must be built for the chickens and only specific-species and specific individuals can be used (Weaver and Cade 1985). Second, if breeders want artificially incubated eggs to hatch at approximately the same time, eggs can be stored temporarily before incubation at 14–15°C and at 60–80% humidity for up to 5 days while being turned four times per day (Weaver and Cade 1985, Parry-Jones 2000). In the wild, incubation of eggs usually does not begin until the last or second last egg is laid.

The room used for artificial incubation must remain as undisturbed as possible, and without direct sunlight or temperature fluctuations, which might affect the internal temperature of the incubators (Weaver and Cade 1985, Parry-Jones 2000).

The incubators, hatchers, and brooders should be cleaned and disinfected every 2 weeks when in use and every year before the start of the breeding season (Heck and Konkel 1985, Parry-Jones 2000). Eggs can be transferred to an alternative incubator during the process. The machines should be disassembled, cleaned with bactericidal and fungicidal disinfectants (e.g., Hibiscrub, Virkon) and all wiring should be cleared of dust with pressurized air (Weaver and Cade 1985, Parry-Jones 1998). The re-assembled incubator should be fumigated with formaldehyde gas or a similar agent for about 20 minutes (Weaver and Cade 1985). The gas should be allowed to dissipate for several hours before the machine is considered safe to use.

A facility should have a minimum of three incubators, one to act as an incubator, a second to serve as a hatcher, and a third to act as a backup (Weaver and Cade 1985, Parry-Jones 2000). Many types of incubators exist ranging in cost from hundreds to thousands of dollars; The Peregrine Fund uses "Roll-X" counter-top incubators that take up minimum space and are easy to clean (Burnham 1983, Weaver and Cade 1985). All incubators should have a double-temperature control system, with a second thermostat acting as an override system should the primary thermostat fail to keep the temperature in a safe range.

The ideal incubating temperature for Peregrine Falcon eggs appears to be 37.5°C (Heck and Konkel 1985). Humidity, which determines the rate of water loss from within the egg, can be manipulated by placing Petri dishes of distilled water in the incubator, and varying their number to achieve the humidity desired (Weaver and Cade 1985). Artificial incubation usually is begun at approximately 30% humidity (Burnham 1983, Weaver and Cade 1985, Parry-Jones 2000). A dial hygrometer can be used to monitor humidity.

The number of times an egg must be turned may vary depending on the species, but eggs should be turned at regular intervals. Incubators can be programmed to turn eggs, eggs can be turned by hand, or both methods can be used (Burnham 1983, Weaver and Cade 1985, Parry-Jones 2000). Eggs should be rotated between 45° and 90°, and turns should be done in alternate directions. The turning grid that the eggs are placed on must be adjusted for egg size to reduce the risk of breaking the eggs.

Eggs should be tested for fertility even if they are being naturally incubated. Infertile eggs can be immediately removed to encourage pairs to re-lay if it is not too late in the season. Candling can be used to determine if the eggs are fertile (Burnham 1983). Thin-shelled or lightly pigmented eggs can be candled using incandescent lights, whereas thick-shelled or heavily pigmented eggs can be examined with ultra-violet candlers (Weaver and Cade 1985). A good-quality candler will avoid overheating the egg.

An egg must lose an appropriate amount of water to ensure proper hatching (see Burnham 1983, Weaver and Cade 1985). The weight of eggs must be monitored individually and the rate of loss regulated by manipulating the humidity to which it is exposed. On average, Peregrine Falcon eggs lose 18% of their weight before hatching; including 15% before pipping (the first crack in the shell) (Burnham 1983, Parry-Jones 2000). If the eggs are losing weight too rapidly or too slowly, the humidity in the incubator can be adjusted to slow or speed the process (Heck and Konkel 1985). Further information about candling and adjusting weight loss of problematic eggs can be obtained in Burnham (1983), Weaver and Cade (1985) or Parry-Jones (2000).

Forty-eight hours or less before pipping, the air cell in the egg will extend and move down one side of the egg (Burnham 1983, Weaver and Cade 1985). Eggs should not be turned after the expansion begins. When pipping occurs, the eggs should be placed in the hatcher with the pipped end up (Weaver and Cade 1985). Soft padding such as gauze should be placed under each egg and each egg should be surrounded by a ring of metal, a wire mesh corral or plexiglass to prevent other hatching chicks from bumping the egg. Containing the egg in this manner also facilitates keeping pedigree records, should two or more young hatch simultaneously overnight. The hatcher's humidity should range from 55% to 60% and the temperature should be similar to that in the incubator (Burnham 1983, Weaver and Cade 1985, Parry-Jones 2000).

The pip-to-hatch interval averages 50 hours and ranges from 24 to 72 hours (Burnham 1983, Heck and Konkel 1985). Patience is needed to prevent well-intended help from injuring the chick. The yolk sac is outside the body while the chick is in the egg and it must be absorbed before hatching. The chick also must turn within the egg, extending the pip into a line of breakage around the egg. Low humidity may dry the egg, causing the chick to become stuck within the egg. The calling of newly hatched chicks within the hatcher appears to stimulate the chick within the egg. The newly hatched chicks should have their navels swabbed with 1% iodine antibiotic ointment containing Bacitracin, and be placed in a brooder (see below) with a sterile corncob litter base when they have dried (Weaver and Cade 1985, Parry-Jones 2000). Burnham (1983) or Heck and Konkel (1985) should be consulted for problems with hatching such as unretracted yolk sacs or dried-egg membranes.

Brooding and Hand-Rearing

Two approaches to brooding are used. In still-air brooders, the temperature is constant throughout the chick's space and the breeder must modify it for the bird's comfort. K-pads and an infrared bulb suspended over the chick allow the hatchling to move between warmer and cooler portions of its environment. The chicks are kept in shallow aluminum cake pans filled with corncob covered with paper towel that is changed after every feeding (Weaver and Cade 1985). The corncob should be formed into a cup to prevent the chick's legs from splaying. A 25-cm diameter aluminum ring or corral surrounds the pan to catch the young's defecation and the whole fixture is placed on newspaper sheets. This system is easy to clean. Birds can be placed two to four per pan initially (fewer as they grow). Humidity and temperature (36°C) must be monitored and adjusted as needed. Chicks will huddle if they are cold or spread out and pant if they are hot (Weaver and Cade 1985). Chicks can be brooded under infrared lights hung overhead, though they should first be covered with a cloth to protect them from dehydration (Heidenreich 1997). A bottom heater also can be used to warm the chick's abdomens to enhance digestion (Heidenreich 1997).

The K-pad brooder consists of a pad filled with circulating heated water draped in a tent-like fashion over a prop in a pan filled with corncob. The temperature of the pan should be 38°C. Chicks should be placed in the brooder on patches of gauze under the pad and should be covered with towels. As the chicks age, the towels can be removed and the temperature reduced in the K-pad (Weaver and Cade 1985). The chicks in this brooder also should be encircled by a corral of plexiglass or aluminum to contain defecation. The corncob in the brooder should be changed as needed.

Temperature should be dropped daily by 1°C in the brooder until, after approximately 10–13 days, the chicks can be raised at room temperature in pairs in corncob-filled pans (30 cm in diameter) with aluminum corrals (33-cm diameter) around them (Weaver and Cade 1985, Parry-Jones 2000, Jenny et al. 2004). A cup should be formed in the corncob to prevent the legs from splaying out. Exposure to humans should be limited after two weeks of age to avoid imprinting (Jenny et al. 2004).

Chicks are not fed until they are at least 8 hours old. Nestlings younger than 10 days should be fed fresh, adult Coturnix Quail (*Coturnix coturnix japonica*) for the best growth (Heck and Konkel 1985). The quail is

skinned, and the head, neck, digestive tract and limbs are removed. The meat is then finely ground and refrigerated until needed, although fresh food should be prepared daily. Young usually are fed every 3–5 hours except at night. Aplomado Falcons initially are fed five times daily and feeding is reduced to three times daily as they age (Jenny et al. 2004). Eventually feeding is reduced to once a day. Older chicks can be fed a mix of 50% ground six-week-old chicken and 50% ground horsemeat with a vitamin and mineral supplement (especially D3). A probiotic such as Avipro Paediatric (Vetark) can be added to the food every few days as an alternative (Parry-Jones 2000).

The meat should be warmed to room temperature before feeding or it should be freshly killed. The food should be wet with Ringer's solution or 0.9% saline before feeding to make swallowing easier (Weaver and Cade 1985, Heidenreich 1997). The chicks are fed with a pair of blunt forceps. Adult calls may have to be imitated to get the young to accept food. Older chicks should be encouraged to feed themselves directly from a bowl. The chick will continue to beg even if full and should not be overfed. Chicks with food still in their stomachs will have round, firm abdomens and should not be fed. Well-ground bone can be added after a few days. At 10 days the young can eat from a bowl, and the ground meat should include small body feathers to encourage casting. For problem chicks see Heck and Konkel (1985), Weaver and Cade (1985), or Parry-Jones (2000).

SMALL FALCONS

Cage Design

Breeding pens for American Kestrels range in size from 15.2 × 6.1 × 1.8 m to 1.5 × 1.2 × 1.2 m (Bird 1982, 1985, 1987; Parks and Hardaswick 1987). Much larger outdoor designs are used at the Patuxent Wildlife Research Center in Laurel, Maryland (Porter and Wiemeyer 1970, 1972). Aviaries can be built entirely out of wood frames and wire mesh, or with polyethylene or plywood walls with a wire mesh roof and floor that is elevated off the ground (Bird 1985). These small raptors will successfully breed in stove-sized cardboard boxes with a mesh roof and a nest box attached (Fernie et al. 2000).

Solid walls should be used when the facility is in a heavily disturbed area (Bird 1985). Mesh roofs should be partly covered with plywood to provide shelter from sun and rain. Basic necessities include a food port, a nest box, one or two 2-cm rope perches and one 5-cm wide wooden perch for copulations, and a one-way glass window for observation. Nest boxes generally are 25 × 25 × 36 cm high and have an access port for checking on the eggs (Bird 1985). American Kestrels can be wintered in single sex flocks of 20–25 birds in indoor, unheated flight pens of 6.1 × 6.1 × 2.4 m (Bird 1985) with concrete floors equipped with drains. Wood shavings should be placed on the concrete floor to absorb feces.

Red-necked Falcons (*F. chicquera*) have been bred in rectangular pens 3.6 × 3.6 × 2.4 m high, as well as in polygon-shaped pens with a floor area of 17 m² and a height of 2.4 m (Olwagen and Olwagen 1984). The pens are constructed of treated wood frames and lined with plastic sheeting separating the pairs visually. The floor of the breeding pens consists of 0.5-cm concrete stone. The roofs of the pens are covered with metal roof sheeting, with one third of the roof covered with 25 × 50-mm mesh to allow for natural lighting. In this polygon cage design, only the corners are sheltered by metal sheeting. Shade cloth can be added just below the mesh (Olwagen and Olwagen 1984).

Red-necked Falcons use other birds' nests, therefore a selection of man-made and crows' nests are provided under the sheltered roof. The birds have a mesh feeding platform and plastic water trays for easy cleaning.

Food, Feeding, and Watering Procedures

Red-necked Falcons generally are fed 30–50 g of food per bird per day including day-old chicks, small passerines, doves, pigeons, mice and beef. Larger food items have to be defeathered and cut up. Vitamin-mineral supplements such as Beefee (Centaur Laboratories [Pty] Ltd.) can be added every 4 days (Olwagen and Olwagen 1984). Some small falcons also may require supplements of insects such as mealworms and crickets, and parents of some species should be provided with skinned food to feed their chicks, since down or fur can affect their digestive tracts (Parry-Jones 2000). Many breeders feed their small falcons day-old chicks (Heidenreich 1997). American Kestrels sometimes are fed only the latter (Bird and Ho 1976, Surai et al. 2001). In fact, American Kestrels have been maintained and bred in captivity at McGill University in Montreal for 34 years while fed a mono-diet of day-old cockerels without any apparent nutritional problems (D. Bird, unpubl. data). If desired though, small falcons can also be main-tained on laboratory mice or commercial zoo diets (Porter and Wiemeyer 1970, 1972).

During the winter, food quantities may have to be doubled, and in freezing temperatures, birds should be fed twice daily (Bird 1987). While American Kestrels fed on moist day-old cockerels seldom drink, baths should be provided in temperatures above freezing. Alternatively, on a hot day, a garden hose perforated for watering lawns can be placed on the mesh roof of breeding pens to provide the birds with showers (Bird 1987).

Courtship Behavior

Food transfers are common in courtship behavior, as are nest inspections and vocalizations (Olwagen and Olwagen 1984). Pair-bonding can be encouraged in Red-necked Falcons by anchoring large food items to the feeding platform, encouraging the pair to eat together (Olwagen and Olwagen 1984). Generally, males feed females in courtship feeding, but in Red-necked Falcons, the female feeds the male. Successful courtship feeding often is followed by copulation. Feeding may continue even after the bond is formed in order to strengthen it (Olwagen and Olwagen 1984). Courtship behavior for American Kestrels has been documented by Willoughby and Cade (1964), Porter and Wiemeyer (1970, 1972) and others.

Management for Production

Small falcons often will breed in their first year and usually can lay between 2 and 6 eggs (Parry-Jones 2000). Birds can be double-clutched and can recycle in as little as 10–14 days (11 days for the American Kestrel [Bird 1987]). Some pairs of American Kestrels can produce up to 3 or 4 clutches per season (Bird 1987) or as many as 26 eggs if removed as laid (D. Bird, unpubl. obs.). Some pairs of falcon species, such as Merlins, should be separated after the breeding season to prevent injury from aggression (Heidenreich 1997). Antagonistic pairs should be separated and re-paired (Bird 1987).

Incubation and Hatching Procedures

American Kestrel eggs can be stored in a refrigerator safely for up to one week before beginning artificial incubation. Using Marsh Farms Roll-X incubators, eggs are kept at 37.5°C with 55% humidity (Bird 1987). Eggs can be turned hourly by the automatic turning device in the incubators or turned by hand at least 4 times daily. Once the eggs pip, they are placed in another Roll-X incubator serving as a hatcher set at 36.9°C and 55% humidity. The eggs are placed on a wire-mesh floor in a small wire-mesh corral, lined with masking tape to minimize sharp edges, to prevent the hatched chicks from moving around and for identification during hatching (Bird 1987). The eggs are kept in the hatcher for 2 days and then moved to a brooder.

Brooding and Hand-Rearing

Chicks can be kept in brooders separately in wire-mesh corrals or in groups in small bowls with cups formed from soft paper, which is changed after every feeding (Bird 1987). American Kestrels are brooded under heat lamps and can be reared in groups of up to five chicks. Temperature is adjusted in the brooder until 10–14 days, when the birds are comfortable at room temperature (Bird 1987).

Chicks are fed bits of neonatal mice within the first 24 hours after hatching and initially are fed 4 times daily (Bird 1987). Within a few days, they can be fed mashed day-old chicks that have been skinned with beak and legs removed (Bird 1987). Vitamin supplements can be added on occasion. Chicks can feed from bowls at 2 weeks of age and can be fed larger food items at this time.

When capable of flight, American Kestrels (but not Common Kestrels [*F. tinnunculus*]) can be housed and wintered in sex-segregated flocks of 20–30 birds in

Figure 2. American Kestrels (*Falco sparverius*), which are extremely easy to breed in captivity, can be wintered in sex-segregated flocks of 25–30 individuals in large flight pens.

large flight pens measuring 6.6 × 6.6 × 1.3 m and equipped with rope perches (Fig. 2). A concrete floor with drains for cleaning and is otherwise covered in wood shavings to soak up feces, works well with these small falcons (D. Bird, unpubl. obs.).

EAGLES

Cage Design

Eagles have been bred in pens ranging from 1.8 × 2.4 × 2.4 m to 48 × 30 × 33 m. They breed best in tall, elongated pens (Carpenter et al. 1987). Eagle pens generally range from 18 to 34 m² in floor space with heights ranging from 2.5–3 m (Heidenreich 1997). White-tailed Eagles have been bred in pens of 7 × 8 × 5 or 6 m high and pens of 9 × 13 × 5 to 6 m high (Carpenter et al. 1987). According to Parry-Jones (1991), minimum recommended sizes for eagle pens are 9 × 4.5 × 4.8 m or 6 × 3 × 3.6 m, depending on the size of the bird. Bald Eagles have been bred in pens of 22 × 11 × 5.5 m high at the Patuxent Wildlife Research Center (Carpenter et al. 1987). The pen frames are constructed from utility poles with wooden roof beams, and the walls and roof consist of 2.5 × 5-cm or 2.5 × 2.5-cm vinyl-coated mesh wire. Plywood sheets protect at least one corner of each pen from the weather. Aluminum roofing is used to cover the roof above the nest platform, which is 1.2 × 1.2 m in size, 3.7 m above the ground. The sides of the platform are 34 cm high, and the floor of the nest platform is 2.5 × 2.5-cm mesh covered with straw and sticks (Carpenter et al. 1987).

Nest ledges in eagle pens can be 2.4 m wide and 4.5 m long with a 23-cm high lip (Parry-Jones 1991). Nest ledges of 2.5 × 5 m are generally bolted to the back wall of the pens (Parry-Jones 2000).

Hardware cloth is buried underground to a depth of 1 m to keep out potential predators (Carpenter et al. 1987). Wooden stumps serve as feeding platforms serviced through feeding ports. Perches in eagle pens consist of tree-like structures using a vertical 30-cm diameter pine pole with branches attached (Parry-Jones 1991). Perches 6.6 to 10 cm in diameter that range in length from 1.2 to 5.5 m span the pens (Carpenter et al. 1987). Perches should have enough space around them to prevent damage to the birds' wings (Parry-Jones 2000). A shelter should be provided for the birds with a floor space of 4 m² and a height of 2 m (Heidenreich 1997).

Food, Feeding, and Watering Procedures

At Patuxent, eagles are fed 6 days per week in the non-breeding season and 7 days per week in the breeding season (Carpenter et al. 1987). The pairs should be fed twice per day when they have chicks to raise. Food is supplied in quantities so that there always is some left over. Eagles are fed whole animals, including poultry, fish and laboratory mammals, depending on the species of bird (Carpenter et al. 1987). Day-old chicks should be supplemented with vitamins and minerals. If the food, especially fish, is stored frozen, its nutrient quality might be reduced, so vitamins should be added to the food (Carpenter et al. 1987). Eagles always should have access to fresh water for drinking and bathing.

Courtship Behavior

Courtship behavior in eagles includes territorial behavior, nest building, mutual preening, communal roosting and copulation. A breeder should understand the progression of behavioral development in order to ensure its sequence. For more information on eagle courtship behavior, the reader is advised to consult Carpenter et al. (1987), Heidenreich (1997) and Parry-Jones (2000).

Management for Production

Some species of eagles lay only one egg per clutch, while others will lay up to 5 (Parry-Jones 2000). Incubation time can range up to 61 days. Fledging can take up to 6 months in large species. Certain species will raise several chicks at a time, while others will raise only the first chick hatched and let any other young die. Such behavior can be avoided in some species by providing the adults with enough food to sustain several chicks. If the young are aggressive to one another, it is wise to raise one by hand until it is old enough to defend itself (Parry-Jones 2000).

Bald Eagles can be double-clutched if the first clutch is removed early enough (Wood and Collopy 1993), but eagles may take a long time to lay a replacement clutch. On average, it takes Bald Eagles 32 days to lay a second clutch (Heidenreich 1997).

The male should be placed in the pen days or weeks before the female to allow him to become familiar with the territory before the larger female is added, and pairs should be separated if they are incompatible. New breeders should have their eggs removed and replaced with dummy eggs to determine the parent's effectiveness as caregivers before allowing them to raise their own clutch. A new pair should not be double-clutched during their first breeding effort (Carpenter et al. 1987).

Incubation and Hatching Procedures

Eagles may be aggressive towards potential nest threats; therefore, anyone attempting to remove eggs from a nest should take caution (Heidenreich 1997). Generally, adults begin incubation with the first egg and share the task. Eggs can be incubated under bantam chickens to within 2–3 days of hatching, when they are transferred to artificial incubators (Carpenter et al. 1987). Artificial incubation is performed at temperatures of 37.4° to 37.6°C and eggs are turned every 2 hours. The eggs are placed blunt-end elevated in the incubator and laid out flat approximately 5 days before hatching, at which time turning should be stopped (Carpenter et al. 1987). Pipped eggs are kept in a humid hatcher at 36.9°C.

Brooding and Hand-Rearing

A major concern for some eagle species is siblicide (Heidenreich 1997). Often chicks will hatch several days apart, giving the oldest the advantage in size. The risk of fighting is reduced as the nestlings age. If the parents have been permitted to hatch their young, it is advised to remove the younger and weaker chicks, hand-rear them until they are strong enough to defend themselves and then move them to the nest (Heidenreich 1997). The birds should not be hand-reared after 3 weeks of age because of the risk of imprinting (Carpenter et al. 1987).

After artificially incubated eggs hatch, the young are allowed to dry in the hatcher and then they are shifted to a paper towel and straw-filled cardboard box in a humid brooder set at 35°C. Temperature is reduced until the chicks can tolerate room temperature, usually at approximately 3 weeks of age. They are fed minced fish and chicken or minced skinned mammals using blunt forceps. Vitamin supplements and digestive enzymes are added. The young are fed 6 times per day initially, and the feedings are reduced with age (Carpenter et al. 1987).

HAWKS AND HARRIERS

Cage Design

Pen sizes for hawks vary with the size of the birds, as

well as with their temperament. Very nervous, rapid fliers should not be placed in pens large enough to allow them to build up excessive speed and present a collision risk with the cage walls (Heidenreich 1997). Pens with 10–18 m² of floor space that are 2.5 m high have been successful (Heidenreich 1997). Minimum size for hawk pens ranges from 6 × 3 × 3.6 m to 4.5 × 2.4 × 2.4 m. The Falconry Centre has bred various hawk species in pens of 3 × 6 m, with sloped roofs 4 to 6.7 m high (Parry-Jones 1991). For other aviary design considerations for hawks, see Crawford (1987).

Some species of hawks cannot remain together year-round because of the risk of the larger female harming the male (Heidenreich 1997, Parry-Jones 2000). To resolve this, Northern Goshawk (*Accipiter gentilis*) breeding pens are designed as two adjacent pens separated by a sliding, barred window so that the birds can see each other. If the pair begins to show interest in each other (e.g., male offering female food through the bars and the female assuming copulatory postures and vocalizing), the door is opened to permit them to mate. The birds can be separated immediately thereafter if necessary (Heidenreich 1997). Harris's Hawks, on the other hand, are relatively social raptors, and placing several birds together can be beneficial for breeding (Heidenreich 1997). A male also may mate with multiple partners.

For most species, two nests should be provided. For accipiters and buteos, a freestanding metal basket on a pole and the other on a shelf in a pen corner should suffice (Crawford 1987). As an alternative, one could provide one long nest ledge (e.g., 1.2 m wide and 3 m long) with a lip of 23 cm, which serves to give the pair some choice on actual location (Parry-Jones 1991). Northern Harriers (*Circus cyaneus*) require two 1-m² platforms 15 cm off the ground and screened by long grass. All nests should be filled with sticks and, for harriers, grass. Extra twigs and conifer branches should be provided on the ground for the pair to adjust their nest (Crawford 1987). The substrate in the hawk pens at the World Bird Sanctuary in St. Louis, Missouri is composed of 2 cm of gravel covered with 8 cm of pea-sized gravel (Crawford 1987). Perches consist of branches set at various heights, partly covered with AstroTurf® to reduce the chance of bruising to the birds' feet (Parry-Jones 2000). If injured hawks and harriers are being bred, they may require walk-up ramps to reach the perches. Males may require perches out of sight from aggressive mates. Such perches require two exits so that a pursued male cannot become cornered (Crawford 1987). Shelters for

the birds also can be added and can range in size from 2 to 4 m² with a height of 2 m (Heidenreich 1997).

Food, Feeding, and Watering Procedures

Hawks can be fed a variety of rats, mice, chickens, adult quail, and rabbits (Crawford 1987). They also can be given venison, day-old chicks, and guinea pigs at times. Normally fed daily, smaller portions of food are offered during the breeding season several times per day to encourage the male to make food transfers with the female (Crawford 1987).

Courtship Behavior

Courtship behavior includes food transfers, mutual preening and a variety of postures, nest construction, and vocalizations (Parry-Jones 2000). Readers should consult the general literature on species-specific behavior.

Management for Production

Accipiters are known to be nervous by nature and often are more vocal (Crawford 1987). If they are to be bred successfully, they should be kept isolated from human contact as much as possible (Parry-Jones 2000). When first introduced together, the pair should be observed for signs of severe aggression. If aggression does not lessen, they should be re-paired with other individuals.

Incubation and Hatching Procedures

If eggs are to be incubated artificially, they should be removed from the nest 7 days after the last egg is laid, and should be kept in incubators at 37.5°C at less than 50% humidity (Crawford 1987). Readers should consult the large-falcon section above for more information.

Brooding and Hand-Rearing

Accipiter chicks are fed ground quail, and buteos are fed ground rats or mice, but see Crawford (1987) for more details on the various diets used for different aged chicks. Vitamin and mineral supplements usually are added to these diets. Generally, for the first 10 days of life, chicks are fed 4 times daily, then 3 times daily, and after 21 days, twice daily.

Several hawk species such as Common Buzzard (*Buteo buteo*) and Red-tailed Hawk (*B. jamaicensis*) sometimes kill siblings in brooders (Heidenreich 1997), so caution must be taken if they are to be hand-reared. Though most 2-week-old young raised in a brooder can be safely returned to their parents, Northern Goshawk young often are initially afraid of their natural parents, and may try to escape from the nest ledge, so strict observation of their behavior must be made when they are initially returned (Parry-Jones 1991).

OWLS

Cage Design

Most owls are relatively sedentary raptors and require less space in captivity than other birds of prey (McKeever 1979). Cage designs vary depending on the size and habits of the owls they are meant to house. The recommendations of Parry-Jones (1998) are as follows: for large owls such as Great Horned Owls (*B. virginianus*), 3 × 4.8 × 2.4 m high to 3.6 × 4.8 × 2.7 to 4.2 m high; for medium-sized owls, 3 × 3 × 2.4 m high to 3 × 3.6 × 2.7 to 4.2 m high; for smallish owls such as Tawny owls (*Strix aluco*) and Barn Owls, 1.8 × 3 × 1.8 m high to 2.4 × 3 × 2.4 to 3.6 m high, and for tiny owls (owlets, *Otus* spp.): 1.5 to 1.8 × 3 × 2.4 to 3.6 m high. Alternatively, the Owl Research Foundation in Ontario, Canada (McKeever 1979) offers the following minimum cage size requirements for large owls: 9.1 × 3.6 × 3 m high; for medium-sized owls such as Barred Owls (*S. varia*): 7.3 × 3 × 3 m high, and for small owls such as screech owls (*Megascops* spp.) and Northern Saw-whet Owls (*Aegolius acadicus*): 5.5 × 2.4 × 2.4 m high. Barn Owls also have been bred successfully in 5 × 4 × 2.5-m outdoor aviaries with a nest box provided (Durant et al. 2004). They have even been bred in 1.5 × 3 × 4-m pens with wooden nest boxes measuring 0.5 × 0.5 × 0.5 m (Rich and Carr 1999).

One style of owl pen used at the Falconry Centre in the U.K. consists of three solid walls and one mesh wall (Parry-Jones 1998). These pens have completely covered roofs made from Onduline or fiber and concrete and are equipped with ceiling lights to adjust photoperiods if necessary. The base of the walls consists of a low brick wall, and treated tongue and groove cladding is used to build the upper portion of the walls of the pens. The floors are constructed from sloped cement for better water drainage and perch holes can be built directly into the floor. All pens have an access door leading into a closed maintenance passageway large enough to permit passage of a wheelbarrow to facilitate cage cleaning and food deliv-ery. One-way glass observation windows, food ports, and access ports to nest ledges or boxes are standard for each pen (Parry-Jones 1998).

The Owl Research Foundation has had great success with a cage design involving two breeding pens connected to each other by corridors that can be closed once a pair has been successfully established in each pen (McKeever 1979). This has worked well for breeding Northern Hawk-Owls (*Surnia ulula*) (McKeever 1995). The corridors, 1.5 to 6 m in length and fitted with removable gates, connect various breeding pens. The gates are opened in early spring, allowing the birds natural mate selection, and again in fall, allowing young to leave their "natal territory" (McKeever 1995).

Overall, this method leads to better pair formation. Each breeding pen is further divided into hunting and nesting sections. The pen frames are constructed of sealed spruce timber 5 × 10 cm thick and built up on steel stakes driven into the ground below the frost line. The roofs are sloped to shed rain and snow. The edges of the roofs are solid instead of meshed to provide shelter for the birds from the weather. For diurnal owls, transparent fiberglass or opaque Coroplast™ can be used (McKeever 1979). Additional wooden slats are strategically placed on the roof for shade. The rest of the roof consists of wire mesh ranging in size from welded mesh up to chain link depending on the bird's size. The walls of the cages are designed of the same material as the roofs, and the ratio of solid wall to mesh wall depends on the needs of the species in question and on the climate (e.g., more protection in regions with cold winters). A completely sheltered area should be available at all times in each pen. The lower section of the walls at ground level is lined with fiberglass to allow live mice to be inserted into the pens for the purposes of release training. Pens are placed out of line of sight of each other using vegetation to afford the pairs privacy. White and ultraviolet fluorescent lights fixed to the pen's roof can be used to adjust the bird's photoperiod to match that of its natural environment (McKeever 1979).

Burrowing Owls (*Athene cunicularia*) have been bred in 5 × 10-m buildings separated into private burrows for each pair, with outdoor flyways for each pair and a communal 3 × 33-m flyway surrounding the individual breeding areas for the non-breeding season (Leupin and Low 2001). An outdoor aviary of 18 × 18 m divided into three breeding pens also has been used successfully. Tunnels connect each of these pens to an individual underground nest chamber. These pens can be turned into a communal flyway by dropping partitions after the breeding season. Tunnels are built using 15-cm diameter perforated flexible plastic pipes, and artificial burrows can be constructed using three 11- to 19-l plastic buckets joined together (Leupin and Low 2001).

Nests in the pens vary among species, and can range from a nest ledge to a nest box to an open-topped box on the ground (Parry-Jones 1998). Boxes and ledges can have 10 cm of peat over a base of pea-sized gravel or 15 cm of sand in them for the owls to dig a scrape in.

Perches vary in size depending on the species, and can consist of tree stumps, logs, branches, grapevine, rocks or rope (McKeever 1979, Parry-Jones 1998). Males must have a high roost in the nest area as a lookout station. If one or both of the pair are permanently injured birds, perches should be designed to allow them to travel to all the important features in their pen.

If training the young to hunt is necessary, at least two food boxes should be placed in the hunting area of each pen (McKeever 1979). All pens should have a built-in bath with an access port for cleaning (Parry-Jones 1998). Pools range from 30 to 90 cm in size and from 1 to 15 cm deep depending on the owl's size and can be made of cement or brick covered in concrete (McKeever 1979, Parry-Jones 1998).

Ground cover in the pens varies among species and can include small gravel and peat moss, wood chips, leaves and turf (McKeever 1979). Pens with concrete floors should be overlaid with gravel or 10 cm of sand (Parry-Jones 1998). In laboratory conditions, pen floors can be covered with cage bedding (recycled newspaper), which is changed every 2 weeks (Rich and Carr 1999). Rocky gravel also can be used to line the ground outside the pens to discourage potential burrowing predators (Parry-Jones 1998). Some species such as Snowy Owls (*B. scandiaca*) require large clear spaces in their pens for take-off, whereas others can be provided with trees and logs to provide a more forested setting (McKeever 1979).

Food, Feeding, and Watering Procedures

Most owls eat rodents; a few species eat birds, fish, amphibians, or insects. It is best to feed adult mice (20–50 g) to owls, and occasionally to offer rats and rabbits to large owls, which wears down their beaks and talons. Weanling rats and mice do not have the nutrient content of their adult counterparts, so vitamin and mineral supplements may have to be injected into the food several times per week (McKeever 1979). Burrowing Owls have been fed daily with laboratory mice, weanling rats,

Common Starlings (*Sturnus vulgaris*) and House Sparrows (*Passer domesticus*) (Martell et al. 2001). Barn Owls in laboratory settings have been fed laboratory mice daily (Durant et al. 2004). Eagle owls have been successfully fed chicks, quail, rats, mice, parts of rabbits and guinea pigs (Parry-Jones 1998). Eurasian Scops Owls (*Otus scops*) and owlets can be supplemented with mealworms, crickets or locusts (Parry-Jones 1998). Day-old cockerels appear to offer poor nutritional supplements for owls, and their down can serve as an intestinal irritant (McKeever 1979). Two-week-old chickens are more suitable, but only for large owls.

The birds eat much more (sometimes more than twice as much) in winter and during the breeding season than in summer (McKeever 1979).

Courtship Behavior

Courtship behavior includes food transfers, vocalizations, and the digging of a nest scrape (Parry-Jones 1998). Females often will base mate choice on the male's territory or pen size (McKeever 1979).

Management for Production

Ideally, owls should be permitted to select their own mates to have a more successful pair bonding (McKeever 1979). Many owls have perennial pair bonding, and an individual taken from the wild or one whose mate has recently died may not show interest in another mate for several years (McKeever 1979, Parry-Jones 1998).

Incubation and Hatching Procedures

Eggs can be incubated by the adults or, where double clutching is desired, eggs can be incubated artificially. Eggs can be removed for artificial incubation or for surrogate incubation using chickens. Some owls will lay a second clutch approximately 2 weeks after the first is removed. Eggs can be candled after 8–10 days to test for fertility (Parry-Jones 1998). Apart from the differences presented below, details for artificial incubation and hatching are similar to those in the Large Falcon section. Readers are referred to Heck and Konkel (1985) or Parry-Jones (1998).

It is ideal to have several incubators set at a different temperature and humidity to transfer the eggs between as needed to insure proper weight loss (Parry-Jones 1998). Eggs should be cleaned with an egg disinfectant before being placed in the incubator. Incubators for owl eggs are kept between 37.3°C and 37.4°C. The humidity should be adjusted to lose 15% egg weight by the pipping stage (Parry-Jones 1998).

Approximately 70% of Barn Owl eggs will hatch in incubator conditions, though they cannot be stored at low temperatures beforehand (Rich and Carr 1999). For good hatching results, Barn Owl eggs should be turned once every two hours (Rich and Carr 1999).

Brooding and Hand-Rearing

Owl nestlings can be hand-reared in small groups to insure proper socialization (Parry-Jones 1998). The brooder should be set up a week in advance to stabilize the temperature to 35°C or 37°C. Brooder temperatures can be adjusted to suit the chick's comfort. Once the young are dry, their navels should be disinfected and placed inside containers filled with sand and lined with paper towel forming a cup (Parry-Jones 1998). Containers should be cleaned at every feeding.

Hand-reared owls should be fed mashed, freshly killed mice that have been skinned, with teeth, tail, limbs and intestines removed (McKeever 1979). They also can be fed day-old chicks, rabbit and quail prepared the same way (Parry-Jones 1998). Vitamins such as Plex-Sol C (Vet-A-Mix Inc.) can be added to the food (Rich and Carr 1999). Vitamin and mineral supplements such as MVS 30 (Vydex) or Nutrobal (Vetark) and probiotics (e.g., Avipro by Vetark) can be mixed in their food as an alternative (Parry-Jones 1998). The young often have to be encouraged to eat by touching the food to the sides of their beaks and imitating the parents' calls (McKeever 1979, Parry-Jones 1998). The birds should be fed with round-ended forceps.

Owl nestlings should not be fed in the first 24–36 hours after hatching, thus allowing their yolk sacs to be fully absorbed (McKeever 1979). They can be given dextrose in water until that time. Young should not be overfed; this can be judged based on the feel of the stomach (it will be firm if the bird is full). Weight should be monitored. At 2–3 weeks of age, they generally can feed themselves from a bowl of minced food offered 3 times per day (Parry-Jones 1998). At this time they can be returned to the parents if desired; otherwise, imprinting can become a concern. According to McKeever (1979), owls imprint on a parental figure between their second and sixth week of life. Fostering young owls to an adult of their own species works so long as they are about the same age as that of the biological young of the foster parents.

CONDORS AND VULTURES

Cage Design

California Condors and Andean Condors (*Vultur gryphus*) breed in cliff cavities in the wild. Breeding pens should include a flight pen, a connecting catch pen to capture the birds as needed, and roosting and nesting areas (Toone and Risser 1988). Such facilities have been used at the San Diego Zoo, Los Angeles Zoo, and the Patuxent Wildlife Research Center. Pens for pairs of California Condors are 12.2 × 24.4 m in size and 6.1 to 7.3 m high (Toone and Risser 1988, Snyder and Snyder 2000, Harvey et al. 2003). Pens of this size in a breezy area can actually permit condors and vultures to soar briefly (Toone and Risser 1988). Pens are constructed of poles and cable covered by 5.1 × 10.2-cm welded mesh with visual barriers of corrugated metal sheeting on the sides of the pens adjacent to human activity to reduce disturbance (Toone and Risser 1988, Cox et al. 1993). Visual barriers of metal sheets also are placed at ground level between the pens, but birds are permitted to see each other from higher perches (Harvey et al. 2003). California Condor pens of 12.2 × 6.1 × 6.1 m also have been used (Snyder and Snyder 2000). Chain-link pens of 9.1 × 18.2 m in size and 3.6 to 9.1 m in height have been used to breed condors successfully (Cox et al. 1993). Andean Condors have been bred in pens that are 12 × 18 × 6 m high (Toone and Risser 1988), but also in pens of 5.5 × 11 × 5.3 m high (Ricklefs 1978). King Vultures (*Sarcoramphus papa*) have bred in pens one-third that size with a nest consisting of a raised wooden box.

The nesting area for California Condors at the San Diego Zoo consists of an open-fronted roost 1.5 × 1.5 m in size with a perch in it (Toone and Risser 1988). Near this roost is a 1.5 × 1.5 × 1.8-m high box with an entranceway that serves as the nesting area (Harvey et al. 2003). The floor of this box is covered with sand. Simulated rock caves also can be used (Toone and Risser 1988). The nest area should have a small, 30 × 35-cm access door for handlers to have access to eggs or young, and the roost area also should have a door for maintenance purposes (Toone and Risser 1988).

Vultures and condors require a great deal of space to land, as well as wide perch surfaces because their feet are not designed for gripping (Parry-Jones 2000). Perches for California Condors can be 5 × 15-cm thick wooden planks installed with the wide side as the perching surface. Some perches should be far enough from the roost to allow for flight back and forth. For bathing purposes a pool of 1.8 × 2.4 m is suitable (Toone and Risser 1988).

Birds can be monitored with cameras placed inside the nest boxes, and adults can be observed from blinds outside of the pens (Cox et al. 1993). Heating lamps or perch heaters may be necessary if the birds are being bred outside of their normal climate (Parry-Jones 2000).

Food, Feeding, and Watering Procedures

The adult and juvenile California Condor diet at the San Diego Zoo consists of 0.5 kg of cat food (e.g., Nebraska Brand Feline Diet), mackerel, 2-day old chicks, and a rat or rabbit daily (Toone and Risser 1988). All food is fed fresh. The birds also have been fed Nebraska Brand Canine Diet, beef spleen and rainbow trout (Harvey et al. 2003). The birds are fasted twice a week on non-consecutive days.

Adult vultures of various species also have been fed a diet of cow's heads and whole rabbits twice per week (Mundy and Foggin 1981). Other facilities have been successful with fresh whole rabbits, chicken and horsemeat provided daily for the birds in the early morning (Dobado-Berrios et al. 1998). Water always is available for them. For problems with getting wild-caught birds accustomed to the captive diet, see Toone and Risser (1988).

Courtship Behavior

Pair bonds in Lappet-faced Vultures (*Torgos tracheliotus negevensis*) begin forming in their second year of age (Mendelssohn and Marder 1984). Bond formation includes the "head-stretch and turn" display, as well as the passing of nest material to each other. California Condors perform a "wings out and head down display" (Cox et al. 1993). Allopreening and approaching each other also are signs of interest in a potential mate (Ricklefs 1978, Cox et al. 1993). The birds may use the skin on their necks as a display to potential mates (e.g., puffing up their throats to show off colors or to create a drumming noise). They also perform courtship dances. Successful pairings often last until one of the mates dies (Parry-Jones 2000).

Management for Production

Condors and vultures require 5–8 years to reach sexual maturity depending on the species, and young require 3–6 months before fledging (Toone and Risser 1988, Cox et al. 1993, Parry-Jones 2000). They also have a low productivity rate; adults often breed only once every 2 years (Cox et al. 1993). Young California Condors were originally placed in pairs as early as possible to encourage bond formation. Current workers now raise fledglings in groups and place them in pairs at or after sexual maturity. Pair selection should be based on genetic considerations, although this does not always work as planned. Incompatible pairs should be separated and re-paired after 2 years of unsuccessful breeding. Extra birds are housed together in a group, which allows a chance for natural mate selection to occur (Cox et al. 1993). Bond formation can take a year or more to form and more time may be needed before a pair produces a successful fledging (Cox et al. 1993).

The pairs should be kept out of public view to reduce disturbance as much as possible (Cox et al. 1993). Condors and vultures normally lay a one-egg clutch on the ground, on ledges, in tree holes or in the undergrowth (Parry-Jones 2000). Andean and California condors have lain up to 3 eggs in a season after the first egg was removed. If recycling occurs, it is typically after about 30 days. Incubation can range from 40 to 55 days, depending on the species.

Incubation and Hatching Procedures

First-time breeders of an endangered species often are given a dummy egg, or the egg of a less vulnerable species, as a trial experience (Harvey et al. 2003). Both adults will incubate the egg. Breeders should watch for aggressive behavior between adults while exchanging positions in the nest as this can harm the egg (Harvey et al. 2003).

If artificial incubation is to be used, it is best to remove the eggs from the nest after a week of natural incubation (Mendelssohn and Marder 1984, Snyder and Snyder 2000). Lappet-faced Vulture eggs have been incubated successfully at 34.5°C and 40% humidity with 5 turns per day (Mendelssohn and Marder 1984). California Condor eggs have been successfully incubated at 36.3°C to 36.7°C with a humidity allowing for a 12–14% mass loss in the egg (Saint Jalme 1999, Snyder and Snyder 2000). Eggs are turned by machine every hour with an extra turn by hand every 12 hours. Young usually hatch approximately 48–68 hours after pipping (Mendelssohn and Marder 1984, Snyder and Snyder 2000).

Brooding and Hand-Rearing

Hand-rearing vultures and condors can raise imprinting concerns. When parents or surrogates cannot be used to raise the young, a successful alternative is to house young vultures and condors individually in protective cages (1.2 × 1 × 1 m) inside an aviary containing a pair of adults of the same species (Mendelssohn and Marder 1984). Initially the young can be hand-reared in a temperature-controlled room equipped with a mirror to give the hatchling a non-human image to focus on while keepers are feeding it. The use of hand puppets resembling the adults' heads to feed the young also has been successful in raising California Condors and Andean Condors (Mendelssohn and Marder 1984, Toone and Risser 1988, Cox et al. 1993, Wallace 1994). After releasing the puppet-raised young into the wild however, it was discovered that the birds were attracted to human habitations. Those tendencies appear to have lessened with age, and since survivorship is not significantly different, use of puppets to minimize imprinting continues to be one of the main methods of rearing for release.

Nestlings are first fed 24 hours after hatching and initially are fed 3 times per day (Mendelssohn and Marder 1984). Feedings are reduced from 3 times to 2 times to 1 time per day as the chick ages. Young raised by parents are fed a regurgitated diet. The recipes recommended by Heidenreich (1997) also have been successful. Lytren, an electrolyte solution, can be provided to the nestlings several hours after feeding if they are not digesting their food rapidly enough. For further problems with feeding young, consult Toone and Risser (1988).

After several days the young can be weaned onto small, skinned mice, and later on to mice with the fur peeled but still attached to the carcass (Toone and Risser 1988). The young also can be fed pieces of skinned, de-boned rats or small mice warmed in digestive enzymes using tweezers (Mendelssohn and Marder 1984). Nestlings also have been raised on lean meat, liver, lung, spleen, and guinea pigs (Mundy and Foggin 1981). After reaching a couple of weeks of age, supplements of vitamin D3 and bone fragments every few days also have been found to be useful (Mundy and Foggin 1981, Mendelssohn and Marder 1984). After several weeks, large pieces of skinned meat with bone can be fed to the young, and after a month, whole rats or large pieces of meat (Mendelssohn and Marder 1984). Young can feed themselves at approximately 3 months and can then be put on an adult diet.

GENERAL HEALTH CONCERNS

When holding and breeding birds in captivity, especially when those birds are rare or endangered, one must always be aware of the potential threats to their breeding stock. Burrowing Owl breeding facilities in British Columbia are geographically separated from one another to protect the birds from total population loss in the case of an infectious disease outbreak (Leupin and Low 2001). A footbath outside of incubation rooms and each pen also may be necessary to prevent the spread of disease (Giron Pendleton et al. 1987). Unexplained mortality of breeding adults, hatchlings or embryos should be tested for the presence of bacterial, viral, parasitical or fungal infections (Battisti et al. 1998). Blood samples, feces, pellets, unhatched eggs and cloacal swabs should be taken from each bird occasionally and tested as well. Cultures also should be taken periodically from the bird's food source (Battisti et al. 1998).

Illnesses in raptors can result in symptoms such as lethargy or a loss of appetite, and even in death (see Chapters 16, 17, and 23 for more specific information). Disease can result from infections of *Salmonella*, *Chlamydia* and *Mycoplasma* (Battisti et al. 1998). Food sources such as poultry, day-old chicks and mice are common sources of *Salmonella* infections (Battisti et al. 1998, Lany et al. 1999). *Salmonella* also can be transmitted to young or to eggs from contaminated meat deposited in the nest, from fecal contamination, or from direct ovarian transmission (Battisti et al. 1998).

Annual exams are recommended for all birds, if only to gather physiological data on the birds that may prove useful in the future (Ricklefs 1978, Toone and Risser 1988). The exams should be timed to provide the least stress on the birds (e.g., during cage cleaning). Video monitoring, observations via one-way glass, and a perch-weighing system can be used to gather continual information on the birds (Toone and Risser 1988). Minor injuries such as wing sprain can occur when birds are handled. Often the best and safest treatment for these injuries is to leave the birds alone and let them heal (Weaver and Cade 1985).

Diet always is a concern in captive breeding programs for raptors (Clum et al. 1997, Cooper 2002). Because of cost and facility location, breeders may not have access to the bird's natural food. Breeders often have to resort to commercial diets and domestic food sources such as quail, chickens or chicks, rats, mice or guinea pigs (Clum et al. 1997, Cooper 2002). Studies of the domestic species used for food suggest that nutritional content is variable depending on the food sources' diet, age and sex, as well as on the storage method (e.g., frozen or freshly killed; Clum et al. 1997). In general, lipid content of the food is sufficient because captive birds are generally less active, require less energy and are provided with excess food compared to their wild counterparts. We know little about the potential impacts (e.g., atherosclerosis) of long-term feeding of high-lipid foods such as day-old cockerels on raptor health. Vitamin and mineral content also are of concern to captive breeders (Clum et al. 1997). Diet must be optimum before and during egg laying for females to have a successful breeding season (Cooper 2002). Moreover, the eggs of certain species produced in captivity have different fatty acid profiles than those of wild birds of the same species, possibly due to diet (Surai et al. 2001). This may affect hatchability and the survival rate of chicks.

GENERAL FACILITY CARE

Overfeeding should be avoided so that uneaten food does not accumulate. Baths should be placed by a small hatch so that their regular removal for cleaning does not disrupt breeding birds. Nest platforms and boxes should be cleaned before and after the breeding season. Nest grass for harriers should be changed before each nesting season (Giron Pendleton et al. 1987).

Pens should be cleaned once or twice a year with a disinfectant wash and rinse (McKeever 1979, Olwagen and Olwagen 1984, Weaver and Cade 1985, Parry-Jones 1998) and usually at a time when minimum stress will be inflicted on the birds, such as in the autumn after the breeding season. Ideally, birds are captured, examined and moved to an already clean chamber. If the pair must be held until the original chamber is cleaned, they can be placed in individual boxes that are dark but well-ventilated and kept in a cool, quiet location. Nest ledges or boxes, perches and food platforms must be scrubbed well and rinsed. Mats, perches, nest ledges, or boxes should be replaced if needed. Gravel or sand substrate must be raked and replaced if necessary (Weaver and Cade 1985). If an outdoor pen has a grass, the grass and other vegetation should be mowed or pruned regularly (Giron Pendleton et al. 1987). Heavy paint on the walls can prevent insects from destroying wooden walls and facilitates washing.

SUMMARY

Captive breeding of raptors can be a useful tool in reintroduction, research, educational programs, zoos, and falconry. An important consideration to be made before breeding raptors is what will be done with surplus birds. Animals should not be bred in captivity unless the offspring are intended for release, research, or the enhancement of useful captive populations.

Captive breeding of raptors has come a long way. Advances in incubation, artificial insemination, and the hand-rearing of young have increased the success of captive breeding projects. Breeders now are more aware of health concerns, behavioral needs and dietary supplements that can enhance the quality of life of captive birds and thus improve their breeding. However, much has yet to be learned about the behavior and biology of captive and wild raptors, which could further improve captive breeding and conservation projects.

LITERATURE CITED

BATTISTI, A., G. DI GUARDO, U. AGRIMI AND A. I. BOZZANO. 1998. Embryonic and neonatal mortality from Salmonellosis in captive bred raptors. *J. Wildl. Dis*. 34:64–72.

BIRD, D.M. 1982. The American Kestrel as a laboratory research animal. *Nature* 299:300–301.

———. 1985. Evaluation of the American Kestrel (*Falco sparverius*) as a laboratory research animal. Pages 3–9 in J. Archibald, J. Ditchfield, and H.C. Rowsell [EDS.], The contribution of laboratory animal science to the welfare of man and animals. 8th ICLAS/CALAS Symposium, Vancouver, 1983. Verlag, Stuttgart, Germany.

———. 1987. Captive breeding - small falcons. Pages 364–366 in B. A. Giron Pendleton, B.A. Millsap, K.W. Cline, and D.M. Bird [EDS.], Raptor management techniques manual. National Wildlife Federation, Washington, D.C. U.S.A.

——— AND S.K. HO. 1976. Nutritive values of whole-animal diets for captive birds of prey. *Raptor Res*. 10:45–49.

BLANCO, J.M., G.F. GEE, D.E. WILDT AND A.M. DONOGHUE. 2002. Producing progeny from endangered birds of prey: treatment of urine-contaminated semen and a novel intramagnal insemination approach. *J. Zoo Wildl. Med*. 33:1–7.

BURNHAM, W. 1983. Artificial incubation of falcon eggs. *J. Wildl. Manage*. 47:158–168.

———, J.D. WEAVER AND T.J. CADE. 1987. Captive breeding - large falcons. Pages 359–363 in B.A. Giron Pendleton, B.A. Millsap, K.W. Cline, and D.M. Bird [EDS.], Raptor management techniques manual. National Wildlife Federation, Washington, D.C. U.S.A.

CADE, T.J. 1986. Reintroduction as a method of conservation. *Raptor Res. Rep*. 5:72–84.

———. 2000. Progress in translocation of diurnal raptors. Pages 343–372 in R.D. Chancellor and B.-U. Meyburg [EDS.], Raptors at Risk. World Working Group on Birds of Prey and Owls, Berlin, Germany and Hancock House Publishers, Blaine, WA U.S.A.

CARPENTER, J.W., R. GABEL AND S.N. WIEMEYER. 1987. Captive breeding - eagles. Pages 350–355 in B.A. Giron Pendleton, B.A. Millsap, K.W. Cline, and D.M. Bird [EDS.], Raptor management techniques manual. National Wildlife Federation, Washington, D.C. U.S.A.

CLUM, N.J., M.P. FITZPATRICK AND E.S. DIERENFELD. 1997. Nutrient content of five species of domestic animals commonly fed to captive raptors. *J. Raptor Res.* 31:267–272.

COOPER, J.E. [ED.]. 2002. Birds of prey: health and disease, 3rd Ed. Blackwell Science Ltd., Oxford, United Kingdom.

COX, C.R., V.I. GOLDSMITH AND H.R. ENGLEHARDT. 1993. Pair formation in California Condors. *Am. Zool.* 33:126–138.

CRAWFORD, W.C., JR. 1987. Captive breeding - hawks and harriers. Pages 356–358 in B.A. Giron Pendleton, B.A. Millsap, K.W. Cline, and D.M. Bird [EDS.], Raptor management techniques manual. National Wildlife Federation, Washington, D.C. U.S.A.

D'ALORIA, M.A. AND C.P. EASTHAM. 2000. DNA-based sex identification of falcons and its use in wild studies and captive breeding. *Zool. Middle East* 20:25–32.

DOBADO-BERRIOS, P.M., J.L. TELLA, O. CEBALLOS AND J.A. DON-AZAR. 1998. Effects of age and captivity on plasma chemistry values of the Egyptian Vulture. *Condor* 100:719–725.

DURANT, J.M., S. MASSEMIN AND Y. HANDRICH. 2004. More eggs the better: egg formation in captive Barn Owls (*Tyto alba*). *Auk* 121:103–109.

FERNIE K.J., D.M. BIRD, R.D. DAWSON AND P.C. LAGUË. 2000. Effects of electromagnetic fields on reproductive success of American Kestrels. *Physiol. Biochem. Zool.* 73:60–65.

FOX, N. 1995. Understanding birds of prey. Hancock House Publishers, Blaine, WA U.S.A.

GAUTSCHI, B., G. JACOB, J.J. NEGRO, J.A. GODOY, J.P. MULLER AND B. SCHMID. 2003. Analysis of relatedness and determination of the source of founders in the captive Bearded Vulture, *Gypaetus barbatus*, population. *Conserv. Gen.* 4:479–490.

GIRON PENDLETON, B.A., B.A. MILLSAP, K.W. CLINE AND D.M. BIRD [EDS.]. 1987. Raptor management techniques manual. National Wildlife Federation, Washington, D.C. U.S.A.

HARVEY, N.C., S.M. FARABAUGH, C.D. WOODWARD AND K. MCCAF-FREE. 2003. Parental care and aggression during incubation in captive California Condors (*Gymnogyps californianus*). *Bird Behaviour* 15:77–85.

HECK, W.R. AND D. KONKEL. 1985. Incubation and rearing. Pages 34–76 in Falcon propagation: a manual on captive breeding. The Peregrine Fund Inc., Ithaca, NY U.S.A.

HEIDENREICH, M. [ENGLISH TRANSLATION BY Y. OPPENHEIM]. 1997. Birds of prey: medicine and management. Blackwell Science Ltd., Oxford, United Kingdom.

JENNY, J.P., W. HEINRICH, A.B. MONTOYA, B. MUTCH, C. SANDFORT AND W.G. HUNT. 2004. From the field: progress in restoring the Aplomado Falcon to southern Texas. *Wildl. Soc. Bull.* 32:276–285.

LANY, P., I. RYCHLIK, J. BARTA, J. KUNDERA AND I. PAVLIK. 1999. Salmonellae in one falcon breeding facility in the Czech Republic during the period 1989–1993. *Veterinarni Medicina* 44:345–352.

LEUPIN, E.E. AND D.J. LOW. 2001. Burrowing Owl reintroduction efforts in the Thompson-Nicola region of British Columbia. *J. Raptor Res.* 35:392–398.

MARTELL, M.S., J. SCHLADWEILER AND F. CUTHBERT. 2001. Status and attempted reintroduction of Burrowing Owls in Minnesota, U.S.A. *J. Raptor Res.* 35:331–336.

MCKEEVER, K. 1979. Care and rehabilitation of injured owls: a user's guide to the medical treatment of raptorial birds - and the housing, release training and captive breeding of native owls. The Owl Rehabilitation Research Foundation, Ontario, Canada.

———. 1995. Opportunistic response by captive Northern Hawk Owls (*Surnia ulula*) to overhead corridor routes to other enclosures, for purpose of social encounters. *J. Raptor Res.* 29:61–62.

MENDELSSOHN, H. AND U. MARDER. 1984. Hand-rearing Israel's Lappet-faced Vulture *Torgos tracheliotus negevensis* for future captive breeding. *Int. Zoo Yearb.* 23:47–51.

MERSMANN, T.J., D.A. BUEHLER, J.D. FRASER AND J.K.D. SEEGAR. 1992. Assessing bias in studies of Bald Eagle food habits. *J. Wildl. Manage.* 56:73–78.

MUNDY, P.J. AND C.M. FOGGIN. 1981. Epileptiform seizures in captive African vultures. *J. Wildl. Dis.* 17:259–265.

OLWAGEN, C.D. AND K. OLWAGEN. 1984. Propagation of captive Red-necked Falcons *Falco chicquera*. *Koedoe* 27:45–59.

PARKS, J.E. AND V. HARDASWICK. 1987. Fertility and hatchability of falcon eggs after insemination with frozen Peregrine Falcon semen. *J. Raptor Res.* 21:70–72.

PARRY-JONES, J. 1991. Falconry: care, captive breeding and conservation. David & Charles, Devon, United Kingdom.

———. 1998. Understanding owls: biology, management, breeding, training. David & Charles, Devon, United Kingdom.

———. 2000. Management guidelines for the welfare of zoo animals – falconiformes. The Federation of Zoological Gardens of Great Britain and Ireland, London, United Kingdom.

PLATT, J.B. 1989. Gyrfalcon courtship and early breeding behavior on the Yukon North Slope. *Sociobiol.* 15:43–72.

PORTER, R.D. AND S.N. WIEMEYER. 1970. Propagation of captive kestrels. *J. Wildl. Manage.* 34:594–604.

——— AND S.N. WIEMEYER. 1972. DDE in dietary levels in captive kestrels. *Bull. Environ. Contam. Toxicol.* 8:193–199.

RATCLIFFE, D.A. 1980. The Peregrine Falcon. Buteo Books, Vermillion, SD U.S.A.

RICH, V. AND C. CARR. 1999. Husbandry and captive rearing of Barn Owls. *Poult. Avian Biol. Rev.* 10:91–95.

RICKLEFS, R.E. 1978. Report of the Advisory Panel on the California Condor. Audubon Conservation Report No. 6. National Audubon Society, New York, NY U.S.A.

SAINT JALME, M. 1999. Endangered avian species captive propagation: an overview of functions and techniques. *Proc. Int. Cong. Birds Rep., Tours*:187–202.

SNYDER, N. AND H. SNYDER. 2000. The California Condor: a saga of natural history and conservation. Academic Press, San Diego, CA U.S.A.

SURAI, P.F., B.K. SPEAKE, G.R. BORTOLOTTI AND J.J. NEGRO. 2001. Captivity diets alter egg yolk lipids of a bird of prey (the American Kestrel) and of a Galliforme (the Red-legged Partridge). *Physiol. Biochem. Zool.* 74:153–160.

TOONE, W.D. AND A.C. RISSER, JR. 1988. Captive management of the California Condor *Gymnogyps californianus*. *Int. Zoo Yearb.* 27:50–58.

WALLACE, M.P. 1994. The control of behavioral development in the context of reintroduction programs in birds. *Zoo Biol.* 13:491–499.

WEAVER, J.D. AND T.J. CADE [EDS.]. 1985. Falcon propagation: a manual on captive breeding. The Peregrine Fund Inc., Ithaca, NY U.S.A.

WILLOUGHBY, E.J. AND T.J. CADE. 1964. Breeding behavior of the American Kestrel (Sparrow Hawk). *Living Bird* 3:75–96.

WOOD, P.B. AND M.W. COLLOPY. 1993. Effects of egg removal on Bald Eagle productivity in northern Florida. *J. Wildl. Manage.* 57:1–9.

Augmenting Wild Populations and Food Resources

22

JUAN JOSÉ NEGRO AND JOSÉ HERNÁN SARASOLA
Estación Biológica de Doñana CSIC
Avda de María Luisa s/n, Pabellón del Perú 41013 Sevilla, España

JOHN H. BARCLAY
Albion Environmental, Inc.
1414 Soquel Avenue, No. 205, Santa Cruz, CA 95062 U.S.A.

INTRODUCTION

Many populations of raptors, including several in North America, have substantially declined or currently exist at levels that merit population augmentation (Stattersfield and Capper 2000; see below). **Augmenting wild populations** has been defined as *"increasing a population whose numbers have been reduced"* (Barclay 1987). We retain that definition and continue to separate population augmentation into (1) techniques that take advantage of a population's ability to increase by its own reproductive efforts, and (2) those that involve adding individuals from outside of the population.

Management programs must be based on an understanding of the life history of the species in question and a thorough assessment of the conservation status of the population. They should include research to identify factors that have contributed to reducing the population, and an evaluation of whether the population is likely to respond favorably to management attempts. What follows is based on the assumption that there is sufficient information about the life history of a species, including its past and present conservation status, to select and employ effective management techniques; and that any critical limiting factors will not nullify input from population-augmentation techniques.

Raptors are long-lived species that produce relatively few fledglings when they attempt to breed (Newton 1979). In species with this demographic strategy, adult survival is the life-history trait that contributes most to population growth (Lande 1988). In other words, even small changes in adult survival rates may have a larger impact on the persistence of populations over time than, for example, breeding success (Hiraldo et al. 1996). Keeping in mind that the best strategy to augment a raptor population is to enhance adult survival, we have to acknowledge that this parameter may not be amenable to human alteration, and that productivity can become the only parameter susceptible to improvement. As this is often the case, we have focused on management techniques aimed at increasing productivity.

REPRODUCTIVE MANIPULATION

"A population can be increased by manipulating [its] reproductive biology, i.e., increasing the number of young produced by each breeding pair so these individuals will eventually contribute to the breeding segment of the population" (Barclay 1987). Below we discuss various methods for population augmentation in order of their place in the reproductive cycle.

Clutch Manipulation

Consider a situation in which fertile eggs of a nesting population hatch below the normal rate due to eggshell thinning or other causes. Historically, this occurred in some populations of Bald Eagles (*Haliaeetus leucocephalus*), Ospreys (*Pandion haliaetus*), and Peregrine Falcons (*Falco peregrinus*) as a result of pesticide contamination (Hickey and Anderson 1968, Ratcliffe 1970, Anderson and Hickey 1972, Jefferies 1973, Peakall 1976, see Chapter 18 for details). In such situations, vulnerable thin-shelled eggs can be removed shortly after the start of incubation and replaced with artificial eggs, so that incubation continues. The real eggs are incubated artificially and the young produced are returned to the nests (Fyfe and Armbruster 1977, Burnham et al. 1978, Engel and Isaacs 1982). The overall production of young obtained should be higher than if the original eggs had been left with the pairs (Cade 1978, Fyfe et al. 1978, Spitzer 1978). This method has been used successfully with Peregrine Falcons (Burnham et al. 1978, Fyfe et al. 1978, Walton and Thelander 1983, Cade and Burnham 2003) and Bald Eagles (Wiemeyer 1981, Engel and Isaacs 1982).

A variation of clutch manipulation involves transplanting eggs from a population that is producing uncontaminated eggs to one where hatching success is low (Bennett 1974, Armbruster 1978, Burnham et al. 1978). Whole clutches can be relocated or single eggs can be removed from selected pairs. Care should be taken to ensure that transplanted eggs placed in the same nest have had about the same amount of incubation so some degree of hatching synchrony is maintained. Egg relocations have been successful with Ospreys (Spitzer 1978), Prairie Falcons (*F. mexicanus*) and Peregrine Falcons (Walton 1977, Armbruster 1978). Translocations of Bald Eagle eggs have been less successful, particularly when using eggs produced by captive pairs, and overall, egg transplants are not effective for managing Bald Eagle populations (Wiemeyer 1981, Engel and Isaacs 1982).

The technique of forced renesting or "double clutching" also can be used to increase productivity. Initial clutches of eggs are removed early during incubation and are not replaced with artificial eggs. Removal of an entire clutch early in incubation usually results in the production of a replacement clutch, which is left in place for the pair to incubate. The initial clutch is incubated artificially, and the young raised from these eggs are returned to the population by fostering or hacking (techniques discussed in the next section). This tech-

nique has the potential of doubling the productivity of manipulated pairs (Monneret 1974, 1977; Kennedy 1977, Burnham et al. 1978, Fyfe et al. 1978, Cade 1980). This technique requires careful monitoring of the nesting pairs to determine when to remove the first set of eggs. It is used routinely in captive breeding of raptors to increase annual production of young, and from this work it appears that the best time to remove the first eggs is after about one week into incubation. Bird and Laguë (1982a,b,c) provide details of the influence of forced renesting on captive-breeding American Kestrels (*F. sparverius*).

Management programs involving manipulation of incubation behavior should not be considered unless technical resources and expertise in incubating raptor eggs and rearing their young are available. If such a technique seems applicable, we recommend that a small number of pairs, perhaps two or three, be tested first, so that field logistics and other details involved in handling live eggs can be worked out. If the resources are available, this technique offers the greatest potential for increasing the productivity of a nesting population.

If a population of nesting pairs is producing fertile eggs with normal hatching success and the technical resources to incubate eggs and rear young are not available, then it is better to defer from any management involving egg manipulations. The nestling stage is the next part of the reproductive cycle in which techniques can be applied to increase productivity.

Brood Manipulation

The number of young reared to independence can be augmented by increasing brood size to the normal maximum for a species. This has been done with species that experience death of young nestlings due to fratricide (aka siblicide or cainism). Brood size is reduced to one by removing nestlings at an age before sibling rivalry develops. These young are hand-reared and then returned to the nest at an age beyond which fratricide is likely. Another alternative is to place the removed young in a duplicate nest separated physically by a barrier, and allowing the parents to raise both young. This variant, called "siblicide rescue" (Cade 2000), has been applied mostly in large eagles, and allows the weaker young, the so-called biblical "Abel" to be separated from its sibling "Cain," during the period during which sibling attacks are most likely to occur.

When an outside source of young to "foster" to breeding pairs is needed to increase the number of young reared to fledging, additional young can come from other wild populations or from captive breeding projects (Cade 1980, Wiemeyer 1981). Nestlings can be removed selectively from breeding pairs in a population that can withstand such harvest, and then placed in nests in the population to be augmented (Spitzer 1978). The optimum time for these translocations varies with the species, but is usually about mid-way through the nestling stage. Burnham et al. (1978) recommended placing Peregrine Falcon nestlings into nests when they are 2 to 3 weeks of age. Young that no longer require brooding and are at the stage where they start to tear their own food from prey delivered to the nest are the best candidates for such translocations (Fyfe et al. 1978). Care should be taken to ensure that the translocated young are about the same age as the young with which they are placed (Wiemeyer 1981). Close observations should follow the translocations to ensure that the fostered young are accepted by the adults and that enough food is being delivered to provide for the entire brood. Broods should not be increased if there is any indication that local prey availability might be inadequate to enable the adults to provide for additional young. Nor should they be increased above the normal maximum for the species unless "supplemental feeding" is feasible (see below).

In some populations, there will be local differences in prey availability and feeding rates by different nesting pairs, which often is reflected in locally lower brood sizes and production of fledged young (Newton 1979). The technique of translocating and fostering young can be applied in these situations by reducing brood size in areas with low prey availability, and placing removed young into nests in areas where prey availability and feeding rates are higher.

The technique of fostering captive-reared young to nests containing young of the same age also can be used to augment a population. When using captive-reared young for fostering, one should consider how they have been raised and whether they will respond appropriately to their foster parents (Cade 1980, Wiemeyer 1981). Ideally, captive-reared young to be used for fostering should be raised by conspecific parents so they will adjust easily to their eventual "wild" parents. In cases where this is not possible, young should be placed into foster nests at an earlier age. Captive-reared young that have been hand-raised to an advanced age should not be used for fostering in wild areas. It also is advisable to place the young in the surrogate nest early in the day to allow time to thoroughly evaluate the behavior of the

adults. In the event the fostered young are not being fed or accepted, the researcher should remove them, return them to the captive breeding facility or a raptor rehabilitation facility, or, perhaps, place them in another nest.

DEMOGRAPHIC SUPPLEMENTATION

Cross-fostering

Cross-fostering consists of placing young of one species into the nest of another species. Many raptors have been cross-fostered, either in captivity or in the wild (Bird et al. 1985). There always is a risk though, that cross-fostered individuals will become imprinted upon the surrogate parental species. If this occurs, the former choose individuals of the latter species as mates. In captive experiments using American Kestrels and Common Kestrels (*F. tinnunculus*), females made "mistakes" (i.e., chose a mate of the wrong species) about half of the time (D. Bird, unpubl. data). Successful breeding with the correct species though, has been observed in captive-raised Peregrine Falcons fostered into nests of heterospecific raptors in both California (B. Walton, unpubl. data) and Germany (C. Saar, pers. comm.). There are no current raptor management programs based on cross-fostering in the wild so far as we know. However, there have been attempts to use cross-fostering to develop phylopatry of the foster parents to a particular nesting locale. A pair of Ospreys that had lost their clutch and were at risk of leaving the area was kept in place by introducing Black Kite (*Milvus migrans*) nestlings into the nest (M. Ferrer, pers. comm.). In this case, the focus of management was the adoptive parental Ospreys, not the kites. Cross-fostered kites were immediately adopted and, whether they became imprinted to the Ospreys was not of immediate concern, as Black Kites were abundant locally. In Montreal, Canada, American Kestrel nestlings were successfully fostered to a nest of Peregrine Falcons to maintain their breeding interest in the nest site until peregrine nestlings could be exchanged for them (D. Bird, unpubl. data).

Hacking

Hacking, the controlled release of young raptors into the wild, is the most frequently used technique to reintroduce or augment raptor populations (Sherrod et al. 1981). Nestling raptors raised in captivity or in wild nests are translocated alone or in small groups of three to five individuals to the hacking site. The hacking site generally consists of a wooden or metal tower with a large enclosure at the top constructed in such a way as to provide the birds with a view of their surroundings. For some time, individuals are fed in the enclosure, without seeing their handlers. At about the natural fledging time for the species, the front of the enclosure is opened and the birds inside have the opportunity to fly freely and explore the surroundings. Food continues to be provided in the enclosure for some time after it has been opened, and released individuals often stay in the area for weeks or months before dispersing or migrating. This technique has been successful in many situations, including the reintroduction of Ospreys in the U.S. and United Kingdom; Bald Eagles in the U.S.; Bearded Vultures (*Gypaetus barbatus*) in the Alps; Red Kites (*M. milvus*) in the United Kingdom; Peregrine Falcons in the U.S., Sweden, and Spain; and Lesser Kestrels (*F. naumanni*) in Spain (Table 1).

For the migratory and colonial Lesser Kestrel, a species in which colonies seem to grow due to conspecific attraction (Serrano and Tella 2003, Serrano et al. 2004), the hacking procedure has been combined with the use of captive individuals that serve to lure back the hacked individuals at the end of their return migration in the spring. We know of four independent hacking projects in Spain that have created new colonies of Lesser Kestrels using this technique. Intriguingly, a hacking project in the city of Sevilla in southern Spain where no live birds were placed failed to establish a breeding colony at the hacking site after several years and the release of more than 150 hacked individuals, some of which were observed as breeders in nearby colonies.

SUPPLEMENTAL FEEDING

Historically, supplemental feeding has not been particularly successful (Archibald 1978). Early attempts to increase breeding output by supplemental feeding did not work as expected in both White-tailed Eagles (*H. albicilla*) (Helander 1978) and California Condors (*Gymnogyps californianus*) (Wilbur 1978). Clutch and egg sizes and hatching success did not differ between Burrowing Owl (*Athene cunicularia*) pairs supplemented and those not supplemented during incubation (Wellicome 2000). However, Burrowing Owl pairs supplemented during the nestling stage raised more young, suggesting that supplemental feeding was more effective at that stage (Wellicome 1997). On the other hand,

Newton and Marquis (1981) found that supplemental feeding increased clutch size for Eurasian Sparrowhawk (*Accipiter nisus*).

Supplemental feeding may be used to increase the likelihood of raptors breeding in certain locales. It also can be used at critical periods of the breeding season to increase productivity. These efforts are not to be confused with the controversial practice of putting out food to attract raptors for ecotourists. To date, scavenging species have been supplemented with food more often than has been done with predatory raptors (Knight and Anderson 1990). Most of these species in which supplemental feeding has been used as a management technique are highly social and numerous such that large numbers can be fed simultaneously by placing dead animals or meat scraps in designated feeding stations. Often used to enhance populations of carrion-eaters such as vultures, such stations have been referred to as "vulture restaurants" (see below). This management technique also has been used with territorial predatory species such as the endangered Spanish Imperial Eagle (*Aquila adalberti*), in which case both dead prey and live animals, including rabbits (*Oryctolagus cuniculus*) are placed or released in open-top enclosures. In addition, the re-establishment of wild populations of susliks (*Citellus citellus*), a colonial ground squirrel, in mountainous regions in Hungary, has helped support breeding pairs of Saker Falcons (*F. cherrug*) and Asian Imperial Eagles (*A. heliaca*) (Bagyura et al. 1994).

Vulture Restaurants

In many places, traditional sources of food for many populations of vultures and other scavengers have declined dramatically in the last 100 years. Wild ungulates that once provided food for vultures in the western U.S., Africa, and Asia, are now absent or severely diminished in many places (Mundy et al. 1992). Changes in livestock management and traditional stock-raising practices (e.g., pastoralism versus intensive production) also have reduced the availability of domestic livestock carcasses. More recently, outbreaks of mad-cow disease in Europe and the measures adopted by government agencies, such as the incineration of dead livestock from farms, has led to a reduction in food availability for scavenger species (Tella 2001). On the Indian subcontinent, the use of the anti-inflammatory drug, diclofenac, to treat cattle has led to the near extirpation of several species of vultures (Oaks et al. 2004). Supplemental feeding thus has evolved as a common

Table 1. Management techniques employed to restore populations of 24 species of birds of prey in North America, Europe, and Africa (after Cade 2000).

Species	Fostering	Cross-fostering	Hacking	Release
California Condor (*Gymnogyps californianus*)			X	
Andean Condor (*Vultur gryphus*)			X	X
Osprey (*Pandion haliaetus*)			X	
Red Kite (*Milvus milvus*)			X	
Bald Eagle (*Haliaeetus leucocephalus*)	X		X	
White-tailed Eagle (*H. albicilla*)			X	
Bearded Vulture (*Gypaetus barbatus*)			X	
Griffon Vulture (*Gyps fulvus*)				X
Cinereous Vulture (*Aegypius monachus*)			X	X
Montagu's Harrier (*Circus pygargus*)			X	
Northern Goshawk (*Accipiter gentilis*)			X	X
Harris's Hawk (*Parabuteo unicinctus*)	X		X	X
Common Buzzard (*Buteo buteo*)			X	
Harpy Eagle (*Harpia harpyja*)			X	
Spanish Imperial Eagle (*Aquila adalberti*)			X	
Golden Eagle (*A. chrysaetos*)	X		X	
Lesser Kestrel (*Falco naumanni*)	X	X	X	
Mauritius Kestrel (*F. punctatus*)	X		X	
Seychelles Kestrel (*F. araeus*)				X
Aplomado Falcon (*F. femoralis*)			X	
Eurasian Hobby (*F. subbuteo*)	X		X	
Bat Falcon (*F. rufigularis*)			X	
Lanner Falcon (*F. biarmicus*)			X	
Peregrine Falcon (*F. peregrinus*)	X	X	X	

management practice aimed at supporting vulture populations that are threatened by declining food resources.

Supplemental feeding also has been used to provide a contaminant-free food resource in areas where poisoning is suspected (Terrasse 1985), and as a way to supplement essential nutrients lacking in depauperate natural food resources (Friedman and Mundy 1983).

Feeding sites should be located to minimize human disturbance and to assure high visibility of food and easy flight access for participants (Knight and Anderson 1990). Stations may be placed at a site that allows researchers an unobstructed view of feeding individuals, facilitating the monitoring of activities at the site (McCollough et al. 1994).

The amount and type of carrion, as well as its frequency of replenishment should vary according to the target species involved and its population characteristics. Friedman and Mundy (1983), for example, estimated that 500 kg of carrion per day are needed to maintain a population of 1,000 Cape Vultures (*Gyps coprotheres*). When determining the daily amount of carrion needed, consideration should be given to seasonal changes in daily energy requirements of the target species (i.e., during breeding [when adults are feeding nestlings] or during winter when metabolic needs increase due to declines in temperature). Carrion provided at feeding stations can come from carcasses of game species (Wilbur 1974, Knight and Anderson 1990, McCollough et al. 1994) or from surplus livestock (Friedman and Mundy 1983, McCollough et al. 1994). The frequency of food supplementation, the amount of food provided during each feeding, and the size of the carcasses may depend on the targeted scavenger species and the age-class that is to be supplemented. For example, Meretsky and Mannan (1999) found that small carcasses favored visits of adult Egyptian Vultures (*Neophron pernocterus*), which dominated younger birds during feeding bouts. These authors suggested use of small carcasses (e.g., chickens) to feed small vulture species when other non-target vulture species that specialize in larger carcasses are present in the area.

A special type of vulture restaurant for the Bearded Vulture — a scavenger that feeds mainly on large bones that are broken when carried aloft and dropped on rocks — consists of the bones of large domestic animals and wild ungulates. Twenty-six "official" Bearded Vulture restaurants are currently maintained in the Spanish Pyrenees, for a population of about 90 breeding pairs of this species (Carrete et al. 2006). Large and small feeding points differ in the number of birds that they attract.

Large supplementary feeding points (n = 5) are provided artificially with >5,000 kg of lamb legs each year, and as many as 80 birds may congregate there during early spring. On the other hand, small supplementary feeding points (n = 21) may see only 6–12 birds at once because the food supply is intermittent and less abundant (<3,000 kg of legs of lambs at year). Bone restaurants also have been established in the French Pyrenees, and on the Mediterranean islands of Corsica and Crete, each of which supports populations of fewer than 10 pairs of Bearded Vultures (Godoy et al. 2004). Nevertheless, there are suggestions (Carrete et al. 2006) that supplementary feeding in the Pyrenees should be reviewed given that its usefulness in reducing pre-adult mortality has yet to be proved, and its effect on productivity is negative due to density-dependent effects.

When choosing a location for a vulture restaurant, several things should be kept in mind. First, opportunistic mammalian scavengers, including foxes, wolves (*Canis lupus*), and dogs (*C. familiaris*) may visit the restaurant and be "supplemented" as well. A feeding site originally devised for supplementing a breeding population of Cinereous Vultures (*Aegypius monachus*) in southern Spain, for example, attracted a group of about 1,000 young Griffon Vultures (*G. fulvus*), which were not present in large numbers before.

In Spain, a large network of vulture feeding stations is in place, most of which are maintained by different regional governments or non-governmental organizations. Special permits are required to operate a vulture feeding station in Spain. In most regions, stations are fenced off to deter mammalian scavengers, are located far from water courses that may become contaminated with infectious agents, and are situated in open areas that enable landings and take-offs. Sites near human settlements are discouraged. Few attempts have been made to evaluate how much food is available through feeding locations versus other sources (but see Donázar and Fernández 1990). However, the combination of food provisioning, an increase of both large-game populations and free-ranging domestic animals such as cattle, horses, and sheep, along with poison-control measures are fostering an explosive growth of several populations of vultures in the country. In the last two decades, numbers of Griffon Vultures in Spain have increased from fewer than 12,000 breeding pairs to more than 30,000, establishing themselves as the densest *Gyps* population in the Western Palearctic.

Supplemental Feeding During Breeding

Attempts to increase reproductive success by supplemental feeding have been successful for several species of raptors. A 2-year supplemental feeding program in the Sespe Condor Sanctuary in California that used carcasses of California mule deer (*Odocoileus hemionus*) apparently increased the productivity of California Condors in the area (Wilbur et al. 1974). The creation of artificial feeding sites near the most productive areas of Egyptian Vultures in the Italian peninsula has been proposed as the most effective way to stop declining populations there (Liberatori and Penteriani 2001). In Catalonia, northeastern Spain, breeding pairs of Bonelli's Eagle (*Hieraaetus fasciatus*) have been supplemented with domestic chickens (J. Real, pers. comm.). Spanish Imperial Eagles with a history of low breeding success currently are being supplemented across their breeding range in Spain (González et al. 2006). In the latter instance, birds are provided with carcasses of domestic rabbits on elevated platforms or on high, visible branches of trees within the eagles' territory every two days for broods of three nestlings, and every four days for broods of two nestlings. Feeding begins a few days after hatching and stops when the young have fledged. Platforms are inaccessible to carnivores and carrion-eating mammals such as red foxes (*Vulpes vulpes*) and dogs.

NEST-SITE IMPROVEMENTS

Protecting Natural Nests

Raptors, particularly those using stick nests in trees, may suffer from losses of eggs or nestlings if the nest falls or collapses due to wind, storms or because the branches can no longer support the weight of the nest. Sometimes it is necessary to prevent damage of the nest or its contents by securing the nest using supports or reinforcing the branch or branches carrying the nest. If a tree or branch containing a raptor's nest falls with nestlings still in it, the parents of some species may still provide care to the nestlings placed in a hand-made nest in a nearby tree. This procedure has been used successfully with Black-winged Kites (*Elanus caeruleus*) (R. Sánchez-Carrión, pers. comm.) and Spanish Imperial Eagles (Ferrer and Hiraldo 1991) in southern Spain.

Occasionally, managers may need to translocate a nest because it is situated on a power line or located in a place not conducive to successful nesting. A nest of Peregrine Falcons was successfully translocated from one skyscraper to another skyscraper a few blocks away by removing the fertile eggs for artificial incubation and placing a set of fake eggs in the desired nest site. Once the female resumed incubation of the fake eggs, her real clutch was returned to her (D. Bird, unpubl. data). Osprey nests containing young have been successfully shifted from utility poles to artificial nesting poles (Ewins 1994).

Artificial Nest Sites

Wooden, plastic, and concrete boxes have been used for many cavity-nesting species, including kestrels and owls (Hamerstrom et al. 1973, Collins and Landry 1977). These species readily accept nest-boxes and numerous researchers have taken advantage of this to carry out long-term studies of birds using them (Korpimäki 1988, Dijkstra et al. 1990, Smith and Belthoff 2001, Bortolotti et al. 2002).

Breeding raptors often use artificial structures, including power pylons and utility poles, to support their nests. For such species, nesting platforms can be used to increase nest availability. A large number of falcons, hawks, and eagles, have bred on such platforms (see Bird et al. 1996). The construction of artificial platforms, along with the implementation of other management practices, has been responsible for the successful increase of Ospreys in U.S. from 1981 to 1994 (Houghton and Rymon 1997).

For species that nest in the abandoned nests of other species (e.g., Great Grey Owl [*Strix nebulosa*]) and Saker Falcon; Bull et al. 1988), this management technique is particularly useful, not only for increasing bird populations, but also for maintaining stable populations.

Numerous publications and web sites describe different models suited for every species. Readers are referred to details in Giron Pendleton et al. (1987), Ewins 1994, Dewer and Shawyer (2001), and Smith and Belthoff (2001).

PREDATOR PROTECTION AT NESTS

The Case of the Mauritius Kestrel

In the 1970s, with only two known pairs surviving in a patch of remnant native forest of approximately 4,000 ha on Mauritius Island, the Mauritius Kestrel (*F. punctatus*) was the most endangered bird of prey in the world (Cade and Jones 1993). Conservation and management actions taken since that time have included most of those presented in this chapter, including the use of artificial nest boxes, food supplementation at the nest, fostering and captive breeding (Jones et al. 1991). Together, these efforts resulted in one of the most impressive population recoveries of a critically endangered species anywhere. By 1993–1994, the population of Mauritius Kestrels reached 222–286 individuals with an estimated 56–58 pairs having established territories in the wild (Jones et al. 1994).

A common threat for island fauna is the introduction of exotic predators that become the dominant predators of indigenous species, and thus regulate the latter's populations. In the case of the Mauritius Kestrel, eggs, nestlings and recently fledged young were vulnerable to introduced black rats (*Rattus rattus*), mongooses (*Herpestes auropunctatus*), and feral cats (*Felis catus*) (Cade and Jones 1993). An important component of the Mauritius Kestrel conservation program has been intensive trapping of these predators in release areas and in breeding territories to safeguard kestrel nests (Jones et al. 1994). Predator control included both live trapping and the use of poisons. Although the effectiveness of this management practice has yet to be evaluated, it is believed to have reduced predation on kestrels in some areas (Jones et al. 1991).

Nest-guarding

The goal of nest-guarding is to protect the nests of target species from depredation by both wildlife and humans as well as from natural disturbances (e.g., flooding of nesting cavities) by actively monitoring individual nests. Nest-guarding has been employed successfully in conservation programs for the Saker Falcon in Hungary (Bagyura et al. 1994), where the number of young fledged in warden-protected nests was almost twice that of nests of routinely monitored nests (2.55 versus 1.66 young, respectively). And indeed, 12 breeding pairs that had consistently failed during preceding years, bred successfully during the 1986–87 breeding season. Based on their experience with Saker Falcons, Bagyura et al. (1994) remarked on the importance of 24-hour nest-guarding. Depending on the behavior and habits of potential nest predators, many nest predation episodes take place at night and cannot be prevented if monitoring is conducted sporadically or even intensively but only during the day.

Nest-guarding programs also have been established to protect cliff-nesting Egyptian Vultures from human disturbance in the Italian peninsula (Liberatore and Penteriani 2001). Human disturbance near nest sites during the incubation period has been responsible for about 8% of vulture breeding failure. Protecting nest sites and supplemental feeding likely has helped to stop the decline of vulture populations there since the early 1990s.

An alternative management strategy is to establish buffer zones around raptor nests aimed at protecting nests from the effects of recreational activities, human development, or habitat management activities. In this case, care and protection of the nest site, although passive, can be as effective as active guarding to stop nest losses for some species. The size of the buffer zone depends both on site-specific considerations and the species involved (Postovit and Postovit 1987, Richardson and Miller 1997).

RELEASE OF REHABILITATED INDIVIDUALS

Each year, thousands of birds of prey are recovered from the wild and placed in rehabilitation centers and wildlife refuges. Many of these individuals eventually are released into natural habitats and, although this practice may not lead to the recovery of wild populations, not releasing them can be detrimental to these populations. Between 15,000 and 26,000 birds of prey are received and treated in 65 recovery centers in Spain annually, and about half are returned to the wild (Fajardo et al. 2000). Individuals released from rehabilitation centers can be used to augment wild populations far from the areas from which the birds were originally recovered. This method was used in Spain when 64 Eurasian Eagle-Owls (*Bubo bubo*) from the central and southern part of the country were released to successfully augment local Eurasian Eagle-Owl populations in northern Spain (Zuberogoitia et al. 2003).

LITERATURE CITED

ANDERSON, D.W. AND J.J. HICKEY. 1972. Eggshell changes in certain North American birds. *Proc. Int. Ornithol. Cong.* 15:514–540.

ARCHIBALD, G.W. 1978. Supplemental feeding and manipulation of feeding ecology of endangered birds: a review. Pages 131–134 in S.A. Temple [ED.], Endangered birds: management techniques for preserving threatened species. University of Wisconsin Press, Madison WI U.S.A.

ARMBRUSTER, H.J. 1978. Current Peregrine Falcon populations in Canada and raptor management programs. Pages 47–54 in P.P.

Schaeffer and S.M. Ehlers [EDS.], Proceedings of the National Audubon symposium on the current status of Peregrine Falcon populations in North America. National Audubon Society, Tiburon CA U.S.A.

BAGYURA, J., L. HARASZTHY AND T. SZITTA. 1994. Methods and results of Saker Falcon *Falco cherrug* management and conservation in Hungary. Pages 391–395 in B.-U. Meyburg and R.D. Chancellor [EDS.], Raptor Conservation Today. World Working Group on Birds of Prey and Owls, Berlin, Germany.

BARCLAY, J.H. 1987. Augmenting wild populations. Pages 239–247 in B.A. Giron Pendleton, B.A. Millsap, K.W. Cline, and D.M. Bird [EDS.], Raptor management techniques manual. National Wildlife Federation, Washington, DC U.S.A.

BENNETT, E. 1974. Eagle transplant successful. *Def. Wildl. Int.* 49:429.

BIRD, D.M. AND P.C. LAGUË. 1982a. Influence of forced renesting and handrearing on growth of young captive kestrels. *Can. J. Zool.* 60:89–96.

——— AND P.C. LAGUË. 1982b. Fertility, egg weight loss, hatchability and fledging success in replacement clutches of captive kestrels. *Can. J. Zool.* 60:80–88.

——— AND P.C. LAGUË. 1982c. Influence of forced-renesting, seasonal date of laying, and female characteristics on clutch size and egg traits in captive American Kestrels. *Can. J. Zool.* 60:71–79.

———, W. BURNHAM AND R.W. FYFE. 1985. A review of cross-fostering in birds of prey. Pages 433–438 in I. Newton and R.D. Chancellor [EDS.], Conservation studies on raptors. International Council for Bird Preservation Tech. Publ. 5. Cambridge, United Kingdom.

———, D. VARLAND AND J.J. NEGRO. 1996. Raptors in Human Landscapes. Academic Press, San Diego, CA U.S.A.

BORTOLOTTI, G.R., R.D. DAWSON AND G.L. MURZA. 2002. Stress during feather development predicts fitness potential. *J. Anim. Ecol.* 71:333–342.

BRETAGNOLLE, V., P. INCHAUSTI, J.F. SEGUIN AND J.C. THIBAULT. 2004. Evaluation of the extinction risk and conservation alternatives for a very small insular population: the Bearded Vulture *Gypaetus barbatus* in Corsica. *Biol. Conserv.* 120:19–30.

BULL, L.E., M.G. HENJUM AND R.S. ROHWEDER. 1988. Nesting and foraging habitat of Great Gray Owls. *J. Raptor Res.* 22:107–115.

BURNHAM, W.A., J. CRAIG, J.H. ENDERSON AND W.R. HEINRICH. 1978. Artificial increase in reproduction of wild Peregrine Falcons. *J. Wildl. Manage.* 42:625–628.

CADE, T.J. 1980. The husbandry of falcons for return to the wild. *Int. Zoo Yearb.* 20:23–35.

———. 2000. Progress in translocation of diurnal raptors. Pages 343–372 in R.D. Chancellor and B.-U. Meyburg [EDS.], Raptor at risk. World Working Group on Birds of Prey and Owls, Berlin, Germany; and Hancock House, London, United Kingdom.

——— AND C.G. JONES. 1993. Progress in the restoration of the Mauritius Kestrel. *Conserv. Biol.* 7:169–175.

CARRETE, M., J.A. DONÁZAR AND A. MARGALIDA. 2006. Density-dependent productivity depression in Pyrenean Bearded Vultures: implications for conservation. *Ecol. Applic.* 16:1674–1682.

COLLINS, C.T. AND R.E. LANDRY. 1977. Artificial nest burrows for Burrowing Owls. *N. Am. Bird Bander* 2:151–154.

DEWER, S.M. AND C. SHAWYER. 2001. Boxes, baskets and platforms: artificial nest sites for owls and other birds of prey. New Edition, Hawk and Owl Trust, Newton Abbot, United Kingdom.

DIJKSTRA, C., A. BULT, S. BIJLSMA, S. DAAN, T. MEIJER AND M. ZIJLSTRA. 1990. Brood size manipulations in the Kestrel (*Falco tinnunculus*): effects on offspring and parent survival. *J. Anim. Ecol.* 59:269–285.

DONÁZAR, J.A. AND C. FERNÁNDEZ. 1990. Population trends of the Griffon Vulture *Gyps fulvus* in northern Spain between 1969 and 1989 in relation to conservation measures. *Biol. Conserv.* 53:83–91.

ENGEL, J.M. AND F.B. ISAACS. 1982. Bald Eagle translocation techniques. USDI Fish and Wildlife Service, Twin Cities, MN U.S.A.

EWINS, P.J. 1994. Artificial nest structures for Ospreys: a construction manual. Canadian Wildlife Service, Environment Canada, Toronto, Ontario, Canada.

FAJARDO, I., G. BABILONI AND Y. MIRANDA. 2000. Rehabilitated and wild Barn Owls (*Tyto alba*): dispersal, life expectancy and mortality in Spain. *Biol. Conserv.* 94:287–295.

FERRER, M. AND F. HIRALDO. 1991. Evaluation of management techniques for the Spanish Imperial Eagle. *Wildl. Soc. Bull.* 19:436–442.

FRIEDMAN, R. AND P. MUNDY. 1983. The use of "restaurants" for the survival of vultures in South Africa. Pages 345–355 in S.R. Wilbur and J.A. Jackson [EDS.], Vulture biology and management. University of California Press, Berkeley CA U.S.A.

FYFE, R.W. AND H.I. ARMBRUSTER. 1977. Raptor research and management in Canada. Pages 282–293 in R.D. Chancellor [ED.], World conference on birds of prey: report of proceedings. International Council for Bird Preservation, Vienna, Austria.

———, H.I. ARMBRUSTER, U. BANASCH AND L.J. BEAVER. 1978. Fostering and cross-fostering of birds of prey. Pages 183–193 in S.A. Temple [ED.], Endangered birds: management techniques for preserving threatened species. University of Wisconsin Press, Madison WI U.S.A.

GIRON PENDLETON, B.A., B.A. MILLSAP, K.W. CLINE AND D.M. BIRD [EDS.]. 1987. Raptor management techniques manual. National Wildlife Federation, Washington, D.C. U.S.A.

GODOY, J.A., J.J. NEGRO, F. HIRALDO AND J.A. DONÁZAR. 2004. Phylogeography, genetic structure and diversity in the endangered Bearded Vulture (*Gypaetus barbatus*, L.) as revealed by mitochondrial DNA. *Mol. Ecol.* 13:371–390.

GONZÁLEZ, L.M., A. MARGALIDA, R. SÁNCHEZ AND J. ORIA. 2006. Supplementary feeding as an effective tool for improving breeding success in the Spanish Imperial Eagle (*Aquila adalberti*). *Biol. Conserv.* 129:477–486.

HAMERSTROM, F., F.N. HAMERSTROM AND J. HART. 1973. Nest boxes: an effective management tool for kestrels. *J. Wildl. Manage.* 42:400–403.

HELANDER, B. 1978. Feeding White-tailed Sea Eagles in Sweden. Pages 149–159 in S.A. Temple [ED.], Endangered birds: management techniques for preserving threatened species. University of Wisconsin Press, Madison WI U.S.A.

HICKEY, J.J. AND D.W. ANDERSON. 1968. Chlorinated hydrocarbons and eggshell changes in raptorial and fish-eating birds. *Science* 162:271–273.

HIRALDO, F., J.J. NEGRO, J.A. DONÁZAR AND P. GAONA. 1996. A demographic model for a population of the endangered Lesser Kestrel in southern Spain. *J. Appl. Ecol.* 33:1085–1093.

HOUGHTON, L.M. AND L. RYMON. 1997. Nesting distribution and population of U.S. Ospreys 1994. *J. Raptor Res.* 31:44–53.

JEFFERIES, D.J. 1973. The effects of organochlorine insecticides and their metabolites on breeding birds. *J. Reprod. Fertil. Suppl.* 19:337–352.

JONES, C.G., W. HECK, R.E. LEWIS, Y. MUNGROO AND T. CADE. 1991. A summary of the conservation management of the Mauritius Kestrel *Falco punctatus* 1973–1991. *Dodo, J. Jersey Wildl. Preserv. Trust* 27:81–99.

———, W. HECK, R.E. LEWIS, Y. MUNGROO, G. SLADE AND T. CADE. 1994. The restoration of the Mauritius Kestrel *Falco punctatus* population. *Ibis* 137:S173–S180.

KENNEDY, R.S. 1977. A method of increasing Osprey productivity. Pages 35–42 in J.C. Ogden [ED.], Transactions of the North American Osprey Research Conference. *U. S. Nat. Park Serv. Trans. Proc. Ser.* 2, Washington, DC U.S.A.

KNIGHT, R.L. AND D.P. ANDERSON. 1990. Effects of supplemental feeding on an avian scavenging guild. *Wildl. Soc. Bull.* 18:388–394.

KORPIMÄKI, E. 1988. Cost of reproduction and success of manipulated broods under varying food conditions in Tengmalm's Owl. *J. Anim. Ecol.* 57:1027–1039.

LANDE, R. 1988. Demographic models of the Northern Spotted Owl (*Strix occidentalis caurina*). *Oecologia* 75:601–607.

LIBERATORI, F. AND V. PENTERIANI. 2001. A long-term analysis of the declining population of the Egyptian Vulture in the Italian peninsula: distribution, habitat preference, productivity and conservation implications. *Biol. Conserv.* 101:381–389.

MCCOLLOUGH, M.A., C.S. TODD AND R.B. OWEN, JR. 1994. Supplemental feeding program for wintering Bald Eagles in Maine. *Wildl. Soc. Bull.* 22:147–154.

MERETSKY, V.J. AND R.W. MANNAN. 1999. Supplemental feeding regimes for Egyptian Vultures in the Negev Desert, Israel. *J. Wildl. Manage.* 63:107–115.

MONNERET, R.J. 1974. Experimental double-clutching of wild peregrines. *Captive Breeding Diurnal Birds Prey* 1:13.

———. 1977. Project peregrine. Pages 56–61 in T.A. Greer [ED.], Bird of prey management techniques. British Falconers' Club, Oxford, United Kingdom.

MUNDY, P.J., D. BUTCHART, J.A. LEDGER AND S.E. PIPER. 1992. The vultures of Africa. Academic Press Inc., San Diego, CA U.S.A.

NEWTON, I. 1979. Population ecology of raptors. Buteo Books, Vermillion, SD U.S.A.

——— AND M. MARQUISS. 1981. Effect of additional food on laying dates and clutch-sizes of sparrowhawks. *Ornis. Scand.* 12:224–229.

OAKS J.L., M. GILBERT, M.Z. VIRANI, R.T. WATSON, C.U. METEYER, B.A. RIDEOUT, H.L. SHIVAPRASAD, S. AHMED, M.J.I. CHAUDHRY, M. ARSHAD, S. MAHMOOD, A. ALI AND A.A. KHAN. 2004. Diclofenac residues as the cause of vulture population decline in Pakistan. *Nature* 427:630–633.

PEAKALL, D.B. 1976. The Peregrine Falcon (*Falco peregrinus*) and pesticides. *Can. Field-Nat.* 90:301–307.

POSTOVIT, H.R. AND B.C. POSTOVIT. 1987. Impacts and mitigation techniques. Pages 183–213 in B.A. Giron Pendleton, B.A. Millsap, K.W. Cline, and D.M. Bird [EDS.], Raptor management techniques manual. National Wildlife Federation, Washington, DC U.S.A.

RATCLIFFE, D.A. 1970. Change attributable to pesticides in egg breakage frequency and eggshell thickness in some British birds. *J. Appl. Ecol.* 7:67–115.

RICHARDSON, C.T. AND C.K. MILLER. 1997. Recommendations for protecting raptors from human disturbance: a review. *Wildl. Soc. Bull.* 25:634–638.

SERRANO, D. AND J.L. TELLA. 2003. Dispersal within a spatially structured population of Lesser Kestrel: the role of spatial isolation and conspecific attraction. *J. Anim. Ecol.* 72:400–410.

———, M.G. FORERO, J.A. DONÁZAR AND J.L. TELLA. 2004. Dispersal and social attraction affect colony selection and dynamics of Lesser Kestrels. *Ecology* 85:3438–3447.

SHERROD, S.K., W.R. HEINRICH, W.A. BURNHAM, J.H. BARCLAY AND T.J. CADE. 1981. Hacking: a method for releasing Peregrine Falcons and other birds of prey. The Peregrine Fund, Inc., Ithaca, NY U.S.A.

SMITH, B.W. AND J.R. BELTHOFF. 2001. Effects of nest dimensions of use of artificial burrow systems by Burrowing Owls. *J. Wildl. Manage.* 65:318–326.

SPITZER, P.R. 1978. Osprey egg and nestling transfers: their value as ecological experiments and as management procedures. Pages 171–182 in S.A. Temple [ED.], Endangered birds: management techniques for preserving threatened species. University of Wisconsin Press, Madison WI U.S.A.

STATTERSFIELD, A.J. AND D.R. CAPPER. 2000. Threatened birds of the world. BirdLife International, Cambridge, United Kingdom.

TELLA, J.L. 2001. Action is needed now, or BSE crisis could wipe out endangered birds of prey. *Nature* 410:408.

TERRASSE, J.F. 1985. The effects of artificial feeding on Griffon, Bearded and Egyptian vultures in the Pyrenees. Pages 429–430 in I. Newton and R.D. Chancellor [EDS.], Conservation studies on raptors. International Council for Bird Preservation Tech. Publ. 5. Cambridge, United Kingdom.

WALTON, B.J. 1977. Development of techniques for raptor management, with emphasis on the Peregrine Falcon. *Calif. Dep. Fish Game Adm. Rep.* 77–4.

——— AND C.G. THELANDER. 1983. Peregrine Falcon nest management, hack site and cross-fostering efforts. Santa Cruz Predatory Bird Research Group and The Peregrine Fund, Inc., University of California, Santa Cruz, Santa Cruz CA U.S.A.

WELLICOME, T.I. 1997. Reproductive performance of Burrowing Owls (*Speotyto cunicularia*): effects of supplemental food. Pages 68–73 in J.L. Lincer and K. Steenhof [EDS.], The Burrowing Owl, its biology and management: including the proceedings of the First International Burrowing Owl Symposium. Raptor Research Report 9.

———. 2000. Effects of food on reproduction in Burrowing Owls (*Athene cunicularia*) during three stages of the breeding season. Ph.D. dissertation. University of Alberta, Alberta, Canada.

WIEMEYER, S.N. 1981. Captive propagation of Bald Eagles at Patuxent Wildlife Research Center and introduction into the wild, 1976–1980. *Raptor Res.* 15:68–82.

WILBUR, S.R. 1978. Supplemental feeding of California Condors. Pages 135–140 in S.A. Temple [ED.], Endangered birds: management techniques for preserving threatened species. University of Wisconsin Press, Madison WI U.S.A.

———, W.D. CARRIER AND J.C. BORNEMAN. 1974. Supplemental feeding program for California Condors. *J. Wildl. Manage.* 38:343–346.

ZUBEROGOITIA, I., J.J. TORRES AND J.A. MARTÍNEZ. 2003. Reforzamiento poblacional del búho real *Bubo bubo* en Bizkaia (España). *Ardeola* 50:237–244.

Rehabilitation

23

Patrick T. Redig
Lori Arent
Hugo Lopes
Luis Cruz
The Raptor Center, College of Veterinary Medicine,
University of Minnesota,
1920 Fitch Avenue, St. Paul, MN 55108 U.S.A.

INTRODUCTION

Reconstituting an injured raptor to a state of recovery sufficient to release it back to the wild consists of two major elements: (1) specialized veterinary care that ranges from first aid, emergency procedures, and internal medicine, to specialized diagnostics and orthopedic surgery; and (2) long-term recovery and reconditioning for release to the wild, or rehabilitation. Associated with these core activities are: (1) recovery, (2) convalescent husbandry and management, (3) disease- and injury-prevention, (4) preparation for release, and, finally, (5) the release itself. The aim of this chapter is to provide readers with an overview of the legal and organizational framework in which rehabilitation is conducted and information concerning the expectations for equipment, facilities, knowledge, and access to veterinary resources that people conducting rehabilitation should have at their disposal. We also provide information useful in dealing with members of the public who may be calling on such professionals for advice or assistance in resolving an encounter with an injured raptor. Owing to the sheer volume and detailed nature of the information, this chapter is not intended to be a "how-to" manual for rehabilitation. Nor is it a medical-procedures manual. That said, some material relating to medical procedures will be presented for the purpose of defining the contemporary state of the art for reference purposes.

LEGAL FRAMEWORK

In the United States, all birds of prey are protected by one or more pieces of Federal Law including the Migratory Bird Treaty Act, the Endangered Species Act, and the Eagle Protection Act (see Chapter 25). Depending on the species of raptor under consideration, one or more federal permits are required in order to conduct rehabilitation work. Elsewhere in the world, applicable pertinent legislation specifies what is needed for compliance. In the U.S., individual states have regulations pertaining to rehabilitation, and permits must be acquired. Some states (e.g., Minnesota, New Jersey, New York, Pennsylvania, and Wisconsin) have well-developed permitting procedures that require testing, sponsorship, education, define different levels of participation that specify certain allowable activities, may limit the species that are allowed to be held by a given individual, and require continuing education and annual reporting procedures in order to maintain the permit. These permitting systems have been developed with input from wildlife management interests as well as organizations such as the National Wildlife Rehabilitators Association (NWRA; www.nwrawildlife.org) and the International Wildlife Rehabilitation Council (www.iwrc-online.org). These organizations have promulgated codes of ethics that guide rehabilitators in their decision-making. An example of guidelines for the state of Minnesota can be found at www.dnr.state.mn.us/ecological_services/nongame/rehabilitation/permits.html.

ASPECTS OF RECOVERY OF INJURED RAPTORS — PROCEDURES AND RECOMMENDATIONS TO BE MADE TO THE PUBLIC

The Recovery Encounter

Often, the initial encounter a wildlife professional has with rehabilitation is a call from a citizen who has unexpectedly encountered an injured raptor during their everyday activities. In most cases, they are awed by, as well as apprehensive about, the circumstances, but feel strongly compelled to have something done in the interest of the bird's health and well-being. Thus, begins the process of recovery.

Equipment for Recovery

The essential requirement at the outset is to get the bird into protective custody as quickly as possible. This is necessary to prevent disappearance of the bird, injury by dogs or other predators, or exposure to the elements. Depending on the size of the bird and the ability and confidence of the person reporting the problem, advice on containment may be given. General recommendations are to corner the bird to prevent its escape by running, to throw a blanket, jacket, large towel or a landing net over it, and finally to place the bird in a confined area, be it a quiet room in a building, a box, or some kind of animal carrier. If persons are judged unwilling or unable to perform this task, they should be advised to keep the bird under direct observation until someone with necessary equipment and expertise can arrive on the scene.

Capture of an Injured Raptor

Methods of capture vary depending on the species of bird, the extent of its mobility, the location (e.g., alongside a road, on a window sill, in an open field, etc.). Resources that should be available in order to provide reasonable coverage of possible circumstances include a pair of leather gloves (welding gloves for most of the larger birds, hand gloves for smaller birds), blanket or bath towel, protective eye wear to reduce risk of a scratched cornea from a flapping wing, a landing net of the type used in sport fishing, an appropriate container, and one or two assistants.

The actual capture should be done with speed, agility, and due concern for not adding to the injury of the bird or injuring a person. Rescuers should avoid prolonged chases of the bird as this can lead to a condition known as "capture myopathy" which can cause serious damage and sometimes ruptured muscles. It occurs as a result of extreme physical exertion coupled with the physiological responses to fear of being captured. The rescuer should bear in mind that the talons are a raptor's primary defense. Most injured raptors, when approached and, especially, if cut off from escape routes by objects or other people, will assume a defensive posture with wings spread and facing what they perceive to be the greatest threat. Many raptors, particularly juveniles, will "surrender" by lying down, often rolling on their back and presenting their feet. Once pushed to this point, presentation of a towel or blanket often results in them grabbing it intensely, burying their talons into the fabric. In doing so, the bird renders itself relatively defenseless and it is now possible to wrap a free end of the fabric around their body. Alternatively, another towel can be draped over them. Once the head is covered, wings can be folded against the body. The towel can now be used to wrap the bird up like a "burrito" and restrain it for placement in a suitable container. A landing net or dip net also can be used to good effect, particularly if the bird is able to evade capture by running or flying short distances. Removal from the net can be difficult as they typically grasp at the mesh and entangle themselves. Ideally, the captured bird would be extracted from the net and wrapped in a towel for placement in a transport container. Once in container, the bird can be left wrapped in the towel or the towel gently removed, providing they have loosened their grasp with their talons. The latter is preferred in order to prevent overheating, although sometimes it is much less stressful on all parties to leave the bird in the box clinging to the towel. The bird is now ready for transport. Be certain that the container has adequate ventilation and, in the case of a cardboard box, that the flaps are secured.

Recommendations about Capture

Some dos and don'ts regarding recovery include the following:

1. Do not place or wrap the bird in a gunnysack.
2. Do not place the bird in a chicken-wire cage or Havahart® trap.
3. Do not place the bird in a container bedded with straw, hay, ground corncobs or other organic material that may contain spores of yeasts and fungi.
4. Do not wrap the body in any kind of elastic bandage.
5. Do not attempt to feed the bird unless it must be held for more than 24 hours (see below).
6. Do not "exhibit" the bird or otherwise cause additional stress by having it looked at unnecessarily by other people or domestic animals (e.g., dogs).
7. Do place the bird on soft towels or absorbent blankets.
8. Do use shredded paper as a bed (especially if the bird is unable to stand due to a broken leg or broken back).
9. Do offer the bird water to drink, either in a bowl or from an eyedropper or basting syringe if transport is going to take several hours. Remove the water bowl from carrier during transport to prevent head trauma or sternally recumbent birds from drowning.

Information Collection

An important part of rehabilitation that contributes to addressing the larger issues of morbidity and mortality factors that affect raptors is collecting as much information as possible about the circumstances that led to the injury or debilitation of the bird (Sleeman and Clark 2003). Important pieces of information to be collected at recovery include: (1) the time when the bird was first found, (2) the geographic location (at least to the county and preferably to the township level), and (3) the presence of objects that may have been involved in the event (e.g., overhead wires, highways, windows, barbed wire fences, chain-link fences, wind turbines, sources of intoxicants and contaminants such as pesticides or oil, etc.). It also is important to note the condition of the bird (e.g., whether it is alert and actively resisting capture, or lying prone and not resisting, sitting erect on its hocks, seizuring, gasping for air, etc.). A crime-scene investigation approach to collecting information will aid in caring for the individual bird, and contribute to the overall base of knowledge of factors that adversely affect raptors.

Transport

Transportation of injured raptors most often occurs by car and, occasionally, by aircraft. Rapid transport to a facility for medical care is one of the major factors contributing to the success or failure of rehabilitation. For transport by car or truck, the major considerations are protecting the bird from further stress as much as possible and providing a suitable environment (e.g., adequate ventilation, protection from heat, cold, wind, and minimizing exposure to extraneous noise such as radios or other audio devices). Although sometimes not possible, it is preferable to avoid transport in an open bed pick-up truck for any great distances. In addition, backs of pick-up trucks covered with a tarp or a topper sometimes trap carbon monoxide and should be used with caution.

For transport by aircraft, use of a private airplane typically presents no additional concerns beyond those given to transport by car. Transport by commercial aircraft requires use of a container approved by and outfitted in accordance with regulations for live animal transport (see Live Animal Regulations, 31st Ed., International Air Transport Association, 2004 www.iata.org/ps/publications). Briefly, a solid-walled container (not a cardboard box) is the basic requirement. A suitably sized fiberglass animal carrier is adequate. The container must be outfitted with foam padding on the inside of the top and the ventilation grates on the sides and door should be covered partially with opaque material (e.g., duct tape, burlap, mosquito screen, or muslin) to darken the inside. Carpeting affixed to the floor with duct tape or other adhesive will provide an absorbent surface as well as a good footing. No perches, water or food containers should be placed in the container. The door should be secured with a nylon tie-wrap. Current airline security operations may result in airport personnel opening the door for inspection. They must be advised of the contents beforehand in order to reduce the likelihood of escape. Birds shipped by air must arrive at the cargo facility 2 hours before scheduled flight time. Some carriers require a health certificate issued by a veterinarian. In most cases, it is possible to waive this requirement on the basis of the fact that the bird is being shipped for critical emergency care. In hot weather, airlines will refuse to accept live animals for shipping when the forecast temperature at the place and time of landing exceeds 85°F (29°C). All aspects of getting an injured bird into rehabilitation can be accomplished by working with a licensed rehabilitation facility, of which there are many throughout the U.S. and elsewhere. Contact information for rehabilitation centers in the U.S. as well as the phone numbers of contact people in each of the Fish and Wildlife Regions that are in charge of rehabilitation permits in their region is available online from NWRA under "Need Help?" on their menu. In the U.S., if all else fails, call The Raptor Center at the University of Minnesota (612-624-4750) to help make appropriate arrangements for the bird. To find rehabilitation clinics in regions or countries outside North America, the internet is the best recourse.

CONSIDERATIONS FOR MEDICAL CARE OF INJURED RAPTORS

Triage

Triage is an inevitable component of raptor rehabilitation. While decisions as to which birds should be forwarded for rehabilitation are best made on the basis of establishing a minimum database about the bird, there are some situations where it is patently clear that rehabilitation is not a viable option. Given the effort and expense involved in transport, it is useful to define those few situations where the most expeditious option would be humane euthanasia of the bird at the time of recovery with all due consideration to the sensitivities of any members of the public that are involved. Those conditions that are beyond medical treatment leading to release of the bird include missing all or a significant part of a wing or leg, severe beak damage or destruction, fractures of long bones of the wing where there are obvious large open wounds containing whole fragments or shards of broken, dry, devitalized bone, and one or both eyes severely damaged or destroyed. If in doubt, or methods of euthanasia suitable for the circumstances are not available, the interests of the bird and the public will be served better by having the bird transported to a facility where a more informed decision can be made on the basis of physical examination and radiographic evaluation.

Medical Treatment

State-of-the-art delivery of health care to injured raptors entails application of veterinary skill sets, therapeutic products, defined procedures, and technology (Redig 2003). These include, but are not limited to: (1) gas anesthesia to be used for restraint for conducting thorough physical examination as well as analgesia for surgical procedures, (2) radiology equipment, (3) materials

and equipment for diagnostic sample collection, including syringes and tubes for blood samples, swabs and culture media for bacteriology, microscope and ancillary equipment for conducting fecal parasite exams, (4) materials, equipment, and skill sets for proper application of bandages and splints as well as orthopedic surgery, (5) reagents and skill sets for conducting critical care of extremely debilitated patients, (6) housing for immediate, post-admission/post-surgical care and for long-term convalescent care and reconditioning for release to the wild, and (7) adequate food supplies and personnel to conduct the entire process. Other pieces of useful equipment include endoscopic units, cardiac monitoring equipment, and general surgical gear. Most rehabilitation facilities will have access to the majority of these items either intrinsically or by virtue of affiliation with an experienced veterinarian.

Gas anesthesia is indispensable for handling of injured raptors. Its use reduces stress on the patient and enables the clinician to conduct detailed physical examination and collection of diagnostic samples. The agent most in use is Isoflurane® (Minrad, Inc., Bethlehem, PA); Sevoflurane® (Abbot Laboratories, Abbot Park, IL) is used by some clinicians. The desirable characteristics of these agents include rapid induction, minimum depression of heart and respiratory rates, widespread tolerance, and rapid recovery. Severely injured and compromised birds may be anesthetized safely and often experience a reduction in the extreme state of their condition when rendered unconscious by the gas. Administration requires a precision vaporizer suitable for isoflurane or sevoflurane, an open-breathing system (e.g., Ayres T-piece, Banes circuit), and a mask that can be placed over the entire head (Fig. 1). Induction is accomplished by placing the head of the restrained bird in the mask with an elasticized-material dam (e.g., Vetrap® [3M Animal Care Product Division, St. Paul, MN]), sealing off the neck and setting the vaporizer to 5% and the oxygen flow meter to 1 liter/minute for birds in the 1–4 kg range. Smaller birds can be induced and maintained at lower oxygen flows. No pre-anesthetic agents, such as atropine, are used. After 1–2 minutes, the effects of the anesthetic may be seen as a general loss of consciousness and general-body relaxation. The gas concentration can be reduced to a maintenance level, typically between 2% and 3%. Respiratory rate and character should be monitored and gas concentrations adjusted in accordance with maintaining the required depth of anesthesia. Intubation is recommended for procedures lasting more than 30 minutes. Lubri-

Figure 1. A Bald Eagle's (*Haliaeetus leucocephalus*) head has been placed inside a cone for induction of gas anesthesia. Note the conforming elastic dam made from Vetrap® that provides a partial seal around the bird's neck.

in the entire body surface is palpated, long bones are felt for fractures, joints are checked for range of motion, and external orifices (mouth, glottis, ears and cloaca) are illuminated and examined. Owing to the high frequency of head injuries in raptors, a full ophthalmological examination, including a fundic exam using either direct or indirect ophthalmoscopy (Figs. 2–4), is essential. Mydriasis (dilation of pupil) cannot be accomplished with atropine in birds as it is in mammals, however either dimming of room lights, isoflurane anesthesia, or both, will provide sufficient opening of the pupil for examination. Often, birds with fully repairable fractures are ren-

cation should be applied to protect the eyes during anesthesia. Flattening of the globe may occur in the ventrally positioned eye when the bird is placed in lateral recumbency. The globe should re-inflate 15–30 minutes after recovery. For recovery, the gas is set to 0%, the delivery system purged with oxygen, and the bird is maintained on oxygen until it shows signs of regaining consciousness, whereupon the tube or mask is removed. The bird should be restrained vertically until completely recovered (5–20 minutes), with attention paid to preventing aspiration of regurgitated stomach fluids.

Some cautions about use of gas anesthetics include presence of food in the gastrointestinal tract, hyperexcitement, and extreme dehydration. If anesthesia is elective, the bird should be fasted for 6–12 hours. If that is not possible, care must be taken during recovery to prevent aspiration if stomach contents are regurgitated. Generally speaking, anesthesia is avoided completely if the bird has a large volume of food in its crop or stomach. Severe dehydration should be mitigated by intravenous or subcutaneous administration of fluids prior to induction.

Physical Examination and Establishment of Minimum Database

The minimum database required to diagnose the extent and degree of injury or debilitation in a raptor includes physical examination (a head-to-toe examination where-

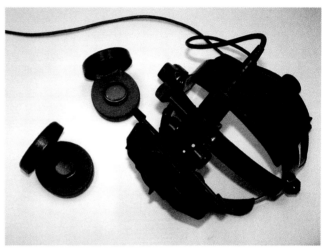

Figure 2. The headset and lenses used in "indirect" opthalmoscopy — see figure 4. This form of opthalmoscopy gives the operator a greater operating distance from the subject and allows for full viewing of the fundus.

Figure 3. An examination of the fundus with a focused light source called a transilluminator. The visual axis of the operator's eye is nearly parallel with the axis of the light source. Owing to the magnification provided by the patient's lens, very large and detailed views of the retina may be had.

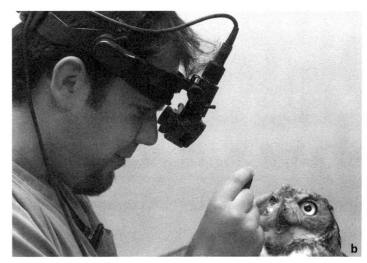

Figure 4. The use of the direct (a) and indirect (b) ophthalmoscopes in the examination of the fundus of a Great Horned Owl (*Bubo virginianus*). The direct scope has a series of lenses on a rotating disk, operated by the index finger of the examiner, which enables focusing on objects of varying location within the eye.

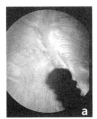

Figure 5. Normal (a) and traumatized (b) pectens.

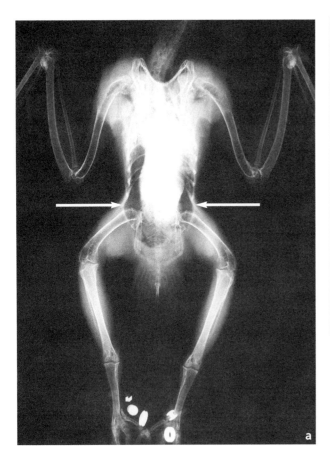

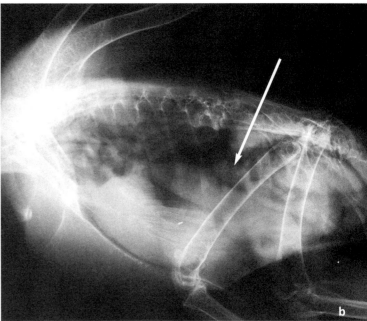

Figure 6. Ventro-dorsal (a) and lateral (b) radiographs. In a, the arrows point to hyperinflated abdominal airsacs, indicative of expiratory restriction in the upper respiratory system. In b, the arrow points to a swollen spleen, suggestive of an active viral infection.

dered unsuitable for release due to hemorrhage, detachment of deep structures, or both (Fig. 5). Vision impairment can be detected only by examination of the interior portions of the eye. In addition to these examination procedures, blood should be collected for (1) hematology (see chapter 16), minimally entailing determination of packed cell volume, total protein, along with total and differential white cell count, (2) collection of plasma for toxicological analysis (especially lead in Bald Eagles [*Haliaeetus leucocephalus*]) and (3) serology (detection of antibodies against specific diseases). Collection of microbiological samples from open wounds or scrapings from lesions (e.g., oral trichomoniasis lesions), and examination of freshly passed feces by direct smear and flotation methods for detection of eggs from internal parasites provides additional useful information. Lastly, full-body radiographs taken in both ventro-dorsal and lateral projections (Figs. 6a,b) are essential. With suitable facilities and the bird under anesthesia, the collection of this suite of samples can be accomplished in under 20 minutes. Radiological and basic hematology (PCV and TP), ophthalmological exam findings, fecal examination and physical examination are immediately available data that will allow decisions to be made about triage, treatment, or both.

Initial Critical Care

It should be assumed that any injured or ill raptor is in a state of dehydration. The minimum detection level of dehydration is around 5% of body weight. The upper end of the range compatible with life is in the range of 12–14%. Determination of dehydration is subjective, and is based on (1) skin elasticity, (2) appearance of the eye with dehydrated birds exhibiting a sunken globe and dullness to the cornea, and (3) moisture content of oral mucous membranes, usually assessed by palpating oral mucus membranes with the examiner's index finger. From a practical point of view, assuming a 10% level of dehydration is useful clinically. This means a 1-kg Red-tailed Hawk (*Buteo jamaicensis*) at 10% dehydration is missing roughly 100 cc of fluid from spaces within and without the vascular system. This volume needs to be replaced over several days, usually 4, while meeting contemporary daily fluid intake needs, generally 50 cc/kg.

Immediate replacement fluid needs are best met by intravenous or subcutaneous fluid administration. In extreme cases where subcutaneous absorption would be too slow and venous access is not possible owing to collapsed veins, fluids can be given intraosseously with uptake nearly equivalent to intravenous administration (Aguilar et al. 1993). This is accomplished by insertion of an 18- or 20-gauge needle through the distal end of the ulna and into the marrow cavity and infusing fluids via this route (Fig.7). Regardless of the route of administration, the goal is to replace 50% of the estimated deficit in 24 hours while at the same time meeting contemporary needs and assuming no further losses as in hemorrhaging. Thus, the 1-kg Red-tailed Hawk would need 100 cc of fluid in the first 24-hour period. A typical schedule would consist of four treatments each consisting of 12–15 cc given subcutaneously, and a similar volume given by oral infusion with a crop tube every six hours. The remaining deficit is made up on successive days, wherein one half of the remaining deficit is given daily along with meeting contemporary needs. All fluids should be heated to where they are warm to the

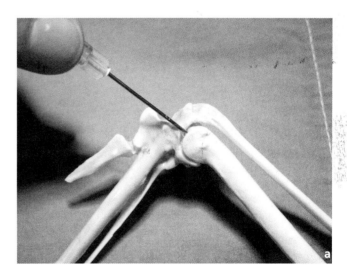

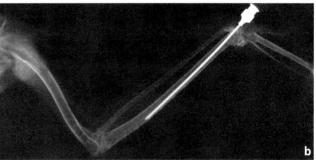

Figure 7. The installation of an intraosseous catheter. Above (a), the insertion point at the distal end of the ulna, lateral surface, is shown. Below (b) is a radiograph in which the full insertion of the needle into the marrow cavity is seen. For the short duration of time that such catheters are left in place (2–3 days), no permanent damage to the joint is typically seen.

touch before administration. In many cases, severely debilitated birds will show remarkable response to this simple regimen of fluid replacement within one to two hours of first administration.

In general, the major task of rehydration (i.e., re-establishing circulating blood volume), can be met by use of lactated Ringers solution for subcutaneous, intravenous, and intraosseous administration, and readily available rehydrating solutions such as Pedialyte® or Gatorade® for oral use. There are many refinements on this theme in terms of selection of fluids, determining state of dehydration and monitoring the response to treatment using Doppler blood pressure equipment, and tailoring the dosing schedule to exactly determined needs that will enhance the ability to deliver optimized treatment to the avian patient (Lichtenberger 2004).

Beyond fluid administration, calorie intake is a key ingredient in the treatment of all injured or debilitated raptors. If the bird is unable or unwilling to eat on its own, it is important to deliver food into the gastrointestinal tract immediately, and no later than 24 hours after admission. For debilitated birds, this generally requires feeding through a crop tube, a stainless steel feeding tube or a rubber catheter affixed to a syringe. Materials to be fed range from a puree of easily digested food stuffs (quail breast, liver) mixed with sufficient fluid to pass through the tube (this fluid becomes part of meeting the daily requirement for fluid). Commercially available products that are more suited toward the needs of the debilitated animal in terms of digestibility, absorbability, and known assay in terms of nutrient and calorie content include Oxbow Carnivore Care® (Oxbow, Murdock, NE), Lafeber's Critical Care for Raptors® (Lafeber Company, Cornell, IL), and Eukanuba Max-life® (The IAMS Company, Dayton, OH).

Debilitated birds require approximately 250 kcal/kg intake per day to meet the hypermetabolic needs associated with injury, stress, and illness. Assuming a caloric content of 2 kcal/ml (typical of prepared diets), a 1 kg bird would need 125 ml of such a diet. Typical crop volumes are in the order of 25–30 cc/kg, so, again, four treatments in a 24-hour period, each consisting of 30 cc of the material, would meet caloric requirements. Such treatment should be maintained until the bird has demonstrated a consistent daily weight gain and begins to show a keen interest in eating offered whole food items (e.g., mice, quail).

Fracture Treatment

State-of-the-art treatment for long-bone fractures entails surgical implantation of fixation hardware to attain the greatest overall success rate in recovery. A device known as the intramedullary-pin external skeletal fixator tie-in has a well-established track record of stabilizing fractures of the humerus, radius, ulna, femur, and tibiotarsal bones (Redig 2000). This device consists of a suitably sized and properly implanted intramedullary pin, 2 or 4 partially threaded positive profile acrylic interface half pins (IMEX, Imex Veterinary, Inc., Longview TX), and an acrylic connecting bar (Fig. 8). The key to this device's effectiveness lies in a link that is established between the intramedullary pin (IM) and the external skeletal fixator pins (ESF) by bending the end of the IM pin to 90 degrees and aligning it with ESF pins. A latex form is placed over the pins (Penrose drain), and while the fracture is held in appropriate reduction, the latex form is filled with liquid acrylic material (Technovit®, Jor-Vet, Loveland, CO). Properly applied, this fixator provides stabilization against rotational, bending, sheer, compression, and traction forces

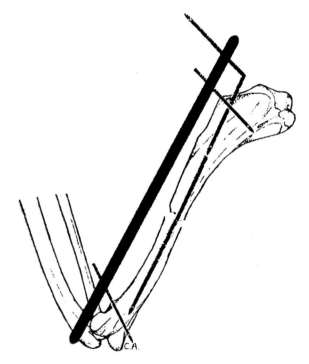

Figure 8. A line drawing (not to scale) of a tie-in fixator implanted on the humerus. The intramedullary pin should fill, but fit loosely into the marrow cavity. It is bent 90° to align it with the plane of the external skeletal fixator pins. The connecting bar is made of acrylic injected into a latex rubber mold (penrose drain) of a size roughly equivalent to the diameter of the bone. Note the placement of the ESF pins and the exit point of the intramedullary pin, none of which interferes with joint surfaces.

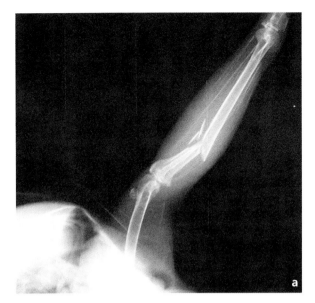

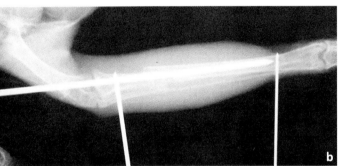

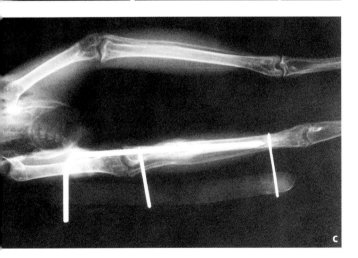

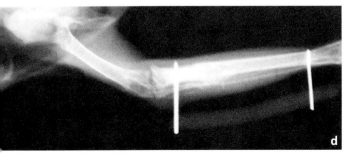

while allowing normal range of motion of the limb. Typical fracture healing times, to the point of removal of the hardware, range from 3 weeks for humeral fractures to 5 weeks for tibiotarsal fractures with an expected return to full function of 65–70% of all attempts (Redig 2000; Figs. 9a–d).

Fractures of the major metacarpus and the metatarsus require a different approach to treatment. Both are characterized by having scant soft tissue to provide blood supply to the bone and the metatarsus is rendered more difficult by virtue of being a weight-bearing bone. Fractures of both of these bones have a higher chance of healing if surgical repair is delayed several days to give the soft tissues time to recover from the insult of injury. Oral administration of peripheral vasodilating agents such as isoxuprine or pentoxyphylline is highly recommended. Light splinting of these limbs, a "figure-8" bandage for the metacarpal and a Robert-Jones or Schroeder-Thomas splint for the metatarsus will protect them prior to surgery. Stabilization of metacarpal fractures is best achieved with a Type I external skeletal fixator, while metatarsal fractures are managed with a Type II external skeletal fixator (Figs. 10a,b).

An essential component of fracture management is post-operative physical therapy. This is conducted under general gas anesthesia (to provide analgesia and prevent against uncontrolled movements that may cause the fixator to become loosened) and consists of passive-range-of-motion and stretch-and-hold procedures (Fig. 11). These are started within the first week post-operatively and continued on a twice-weekly basis, each session consisting of approximately 5 minutes of activity. These serve to improve blood flow to muscles, prevent ligaments and tendons from tightening, and to protect the integrity of joints. With the tie-in fixator or Type I external skeletal fixator, there is no impediment to movement of the joints or danger of unwanted motion of fracture fragments.

Figure 9. This series of radiographs depicts the repair of a proximal tibiotarsal fracture with a tie-in fixator. In this sequence, (a) pre-operative radiograph, (b) intra-operative radiograph to check alignment of pins, (c) 3 weeks post-operatively at which time abundant callus can be seen (arrow), and (d) 5 weeks post-operatively where the intramedullary pin has been removed, leaving the external skeletal fixator elements in place for another week to support the maturing callus.

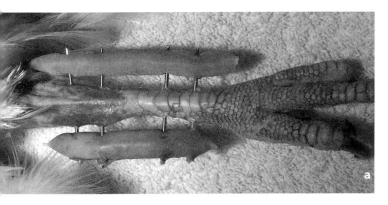

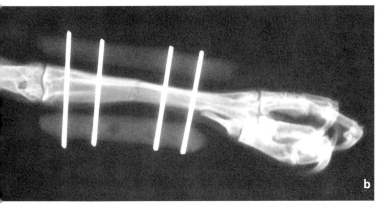

Figure 10. Gross (a) and radiographic images (b) of a Type II external-skeletal fixator (esf) as applied to the tarsometatarsus are shown. Note the acrylic bars on both sides of the leg and the placement of two pins proximal to the fracture and two pins distal to the fracture.

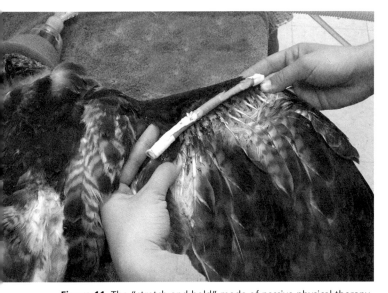

Figure 11. The "stretch-and-hold" mode of passive physical therapy. Note that the tie-in fixator, applied to the ulna in this case, does not interfere with the full extension of the elbow. These exercises are begun within the first post-operative week and are conducted two to three times per week for 5 minutes throughout the healing period. As shown, these exercises are conducted with the bird under general gas anesthesia.

Long-term Care

Another key component of convalescent management of injured or ill raptors is housing and feeding. Early convalescent housing is best provided by temperature- and light-controlled environments, in cages that are quiet (not stainless steel dog cages), padded, designed to prevent injury in the case of a hyperactive patient, and easily cleaned.

Later stages of convalescent recovery can be provided in large flight rooms or outdoor enclosures if weather conditions permit. Placed in these after fixation hardware has been removed, birds can begin to move around and regain some strength in the wings, legs, or both. Arent (2005) provides recommendations for which species of birds can be housed together in group settings. When individuals are capable of reaching perches placed at higher elevations on the walls, they are ready to enter the final stages of preparation for release, flight conditioning.

Flight Conditioning

Raptors depend on their athletic abilities to survive. A critical component to complete the rehabilitation process is to recondition the birds until they reach a level of fitness comparable to that prior to their injury. Two common methods used to provide reconditioning are termed "pen flying" and "creance flying". The former requires a long enclosure (100 ft [30.5 m] long recommended for birds the size of Red-tailed Hawks) with perches placed on both ends. Birds are then encouraged to fly back and forth repeatedly from one to several times per day until they are mechanically sound and have adequate endurance. Endurance needs vary by species, as some birds are fast-flyers (falcons), sprinters (accipiters), long-distance migrants (Ospreys [*Pandion haliaetus*], Broad-winged Hawks [*B. platypterus*]), or have a more static life-style (many species of owls).

Creance flying is a technique that involves removing birds from their housing enclosure, attaching leather straps called jesses to their lower legs and then adding a line (creance) to the straps (Arent 2001). The birds are then transported to a large open space with grass for soft gentle landings and encouraged to fly. The success of this technique has been measured using lactic acid production as an indicator of aerobic fitness (Chaplin et al. 1989, Chaplin 1990). Eight to ten 200-ft (60-m) flights without rest is the general goal for most species. Exercising a raptor 3–4 times per week for 4–6 weeks using this technique is an average reconditioning time following recovery from a wing fracture.

Long-term Convalescent Management Considerations

A variety of problems and maladies may befall the rehabilitation patient during the long period of convalescence. Included among them are aspergillosis, bumblefoot, abrasions of carpal joints, broken feathers and overgrown beaks and talons. It is incumbent upon those caring for birds to anticipate and proactively manage the cases so as to reduce the risk of occurrence of management-related problems. Simply put, cases are made or lost on the basis of the quality of the convalescent care. Comprehensive coverage of all of these, except aspergillosis, is presented by Arent (2005).

The fungal disease, Aspergillosis, is a serious complication among raptors held in captivity for rehabilitation purposes. While caused by a ubiquitous fungus in the environment, its occurrence is either management-related or due to lack of preventative measures. In particular, juvenile Bald Eagles, Northern Goshawks (*Accipiter gentilis*), Red-tailed Hawks, Roughlegs (*B. lagopus*), Golden Eagles (*Aquila chrysaetos*), Gyrfalcons (*Falco rusticolus*), Snowy Owls (*Bubo scandiaca*), and northern boreal forest owls with lead poisoning, are prone to developing a respiratory infection with this fungus. The disease does not manifest itself until it is markedly advanced and usually beyond treatment. The approach to prevention entails prophylactic treatment with antifungal drugs at the time of admission. The agent of choice is itraconazole (Sporonox®, Janssen Pharmaceutical Products, Titusville, NJ) administered orally at a dose of 7 mg/kg twice a day for 5 days, followed by once a day for the remaining time period. This schedule is continued for the first 3 weeks the animal is undergoing rehabilitation. Beyond prophylactic treatment, avoidance of steroids, especially dexamethasone, which is severely immunosuppressive, is a factor in reducing overall incidence of this disease.

Release Criteria

If the rehabilitation effort is to be deemed fully successful, then careful choices and decisions must be made as time for release is approached. The major criteria that should be met include (Arent 2001):

1. The bird's illness or injury must be resolved completely, and pose no sign of long-term physical threats (such as arthritis, a growing cataract, etc.).

2. The bird must have achieved an adequate level of fitness and proper flight mechanics.
3. The bird must have a full complement of flight and tail feathers.
4. The bird's feet must be in good condition and its talons sharp.
5. The bird's basic clinical values should fall within an acceptable range (packed cell volume, total protein, total and differential white cell counts, and fecal exam). Further care must be taken throughout rehabilitation to avoid exposure to novel pathogens from domestic sources that may be introduced into free-living birds (e.g., highly pathogenic strains of avian influenza).
6. A bird admitted at a young age (less than four months old) should demonstrate its ability to catch live prey.
7. Birds with unilateral visual deficit should be given careful consideration about their ability to catch prey and avoid objects before being released.
8. Bird's released in times of cold weather should be given 2–3 weeks of acclimatization in captivity prior to release.
9. If possible, consideration should be given to releasing the bird at its recovery location. This may not be possible owing to time of year, territoriality of resident breeders, replacement as a member of a breeding pair, and migration status. Regardless, the release site chosen must have appropriate habitat for the species and its normal prey, must not be in the territory of a known breeding pair during the breeding season, and if close to or during migration, the site should be near a normal migration route for that species.

SUMMARY

The successful recovery of injured raptors requires the application of complex medical procedures, long-term convalescent care, and active physical rehabilitation. The goal at all points is full restoration, rendering the patient virtually indistinguishable from a bird that has not sustained any injury. The release of a bird back to the wild that is not in full possession of all of its faculties or that is not athletically conditioned will likely end in a short survival period, with death resulting from starvation, accidental injury or predation. As rehabilitation of individual birds has little impact on the population of any raptor species, it is incumbent on those engaging in rehabilitation to (1) ensure that birds being released are in peak physical and mental condition, (2) extract information about causes of injury, parasites, and other medical conditions from each individual case for the sake of increasing our knowledge base, and (3) use examples drawn from actual rehabilitation case materials to educate the public about the types of problems raptors are facing in the ever more human-dominated environment.

Information gleaned from the rehabilitation of individual birds can have direct benefits for populations when that information is used to formulate public policy, pesticide regulations, disease prevention, and control plans. Many of the facilities conducting rehabilitation work are networked via the Internet, thereby making readily available sources of information accessible anywhere in the world. Much has been learned about medical problems and medical care of raptors in the last three decades. This knowledge and the ability to exchange it readily through informatics systems may be useful in addressing problems that affect raptors at the population level.

LITERATURE CITED

AGUILAR, R.F., G.R. JOHNSTON, C.J. CALIFOS, T. ROBINSON AND P.T. REDIG. 1993. Osseous-venous and central circulatory transit times of technietium-99m pertechnetate in anethetized raptors following intraosseous administration. *J. Zoo Wildl. Med.* 24:488–497.

ARENT, L. 2001. Reconditioning raptors: a training manual for the creance technique. The Raptor Center, College of Veterinary Medicine, University of Minnesota, Minneapolis MN U.S.A.

———. 2005. Care and management of captive raptors. The Raptor Center, College of Veterinary Medicine, Minneapolis, MN U.S.A.

CHAPLIN, S. 1990. Guidelines for exercise in rehabilitated raptors. *Wildl. J.* 12:17–20.

———, L. MUELLER AND L. DEGERNES. 1989. Physiological assessment of rehabilitated raptors prior to release. *Wildl. J.* 12:7–8, 17–18.

LICHTENBERGER, M. 2004. Shock and fluid therapy for the avian veterinarian. Pages 157–164 in Proceedings of the annual meeting of the Association of Avian Veterinarians. New Orleans, LA U.S.A.

REDIG, P.T. 2000. The use of an external skeletal fixator-intramedullary pin tie-in (ESF-IM Fixator) for treatment of longbone fractures in raptors. Pages 239–253 in J.T. Lumeij, J.D. Remple, P.T. Redig, M. Lierz, and J.E. Cooper [EDS.], Raptor Biomedicine III. Zoological Education Network, Inc., Lake Worth, FL U.S.A.

———. 2003. Falconiformes. Pages 150–161 in M.E. Fowler and R.E. Miller [EDS.], Zoo and wild animal medicine: current therapy 5, 5th Ed. W.B. Saunders/Elsevier, Philadelphia, PA U.S.A.

SLEEMAN, J.M. AND E.E. CLARK. 2003. Clinical wildlife medicine: a new paradigm for a new century. *J. Av. Med. Surg.* 17:33–37.

Public Education

24

JEMIMA PARRY-JONES
International Centre for Birds of Prey
Little Orchard Farm
Eardisland, Herefordshire, HR6 9AS United Kingdom

MIKE NICHOLLS
University of Greenwich at Medway
Chatham Maritime, Kent, ME4 4TB United Kingdom

GAIL C. FARMER
Acopian Center for Conservation Learning,
Hawk Mountain Sanctuary
410 Summer Valley Road, Orwigsburg, PA 17961 U.S.A.

THE ROLE OF PUBLIC EDUCATION

Conservation scientists have a responsibility to disseminate the results of the work they do to the public. This is so not only because science is "public knowledge," but also because doing so helps conservationists and scientists build approval and support for their work among the general public. Simply put, if raptor biologists and managers hope to continue being supported by the public and its institutions, they need to educate the public about birds of prey and the need to study and protect them.

Education is the product of a formal or informal learning experience. Public-education programs generally are informal learning experiences (Livingstone 2001), and it seems likely that most of the time most of the public will be exposed to raptors and gain information about them in an *ad hoc*, informal, and passive manner.

Learning has been defined as an active process in which learners construct meaning as they interact with and internalize the substance of the teaching they encounter (Driver and Easley 1978, Driver et al. 1996). The importance of learners' existing knowledge, skills, and attitudes are thus recognized as crucial, as they rely upon these when learning something new (Ausubel et al. 1978). Thus, learners bring with them the foundations upon which they construct further learning. These foundations may contain prejudices and misunderstandings, which may be difficult to reform in ways sympathetic to raptor conservation.

What sort of learning should be cultivated? Although undoubtedly there is a link between factual knowledge and conservation behavior (Ham and Kelsey 1998, Bradley et al. 1999), Ramsey and Rickson (1976) concluded that attitudes are one of the most important influences on "conservation behavior" (i.e., the way people behave with respect to conservation issues). Everitt et al. (2002) speculated that cultivation of positive public attitudes towards birds of prey is of paramount importance, more important than just distributing factual information, as positive attitudes towards wildlife appears to influence the public's willingness to act in favor of wildlife conservation (Aipanjiguly et al. 2002).

Here we discuss why and how public-education programs can be used to create positive attitudes among the public towards raptors. We introduce several conceptual and methodological approaches used in education programs to foster positive attitudes towards raptors and to encourage behavior that supports raptor conservation. We present case studies of raptor education programs that demonstrate how these conceptual and methodological approaches have been used. We also offer ways to design and evaluate education programs.

INFLUENCING PUBLIC ATTITUDES

In order to influence attitudes we first need to understand what an attitude is. For our purposes, it is useful to think of an attitude as having three major components: a cognitive component (reason), an affective component (emotion), and a change in behavior as a direct consequence of the cognitive and affective elements (Manzanal et al. 1999). What creates an attitude turns out to be quite complex because many extrinsic and intrinsic variables influence the cognitive and affective components (Kollmuss and Agyeman 2002).

It may seem intuitive that public attitudes towards wildlife will be negative when people and raptors are in conflict, and neutral or positive when no such conflict arises. Neutral attitudes and indifference towards birds of prey may result in indirect negative effects through land-use practices, environmental contamination, habitat loss, electrocutions, and road deaths. The public may be unaware or simply not care enough to change their behavior. Conflicts of interest, particularly from competition for prey, may generate negative attitudes (Valkama et al. 2005) leading to direct persecution by shooting, poisoning, and trapping. This relationship between negative attitudes and persecution with large predators and other vertebrate "pests" has been observed in many situations (e.g., Bandara and Tisdell 2003 for Asian elephants [*Elephas maximus*], Marker and Dickman 2004 for cheetahs [*Acinonyx jubatus*], Treves et al. 2004 for wolves [*Canis lupus*], Verdade and Campos 2004 for pumas [*Felis concolor*], Woodroffe and Ginsberg 1999 for African wild dogs [*Lycaon pictus*]).

The apparent inverse relationship between human attitude and the frequency of negative interaction with another species is not inevitable. There is evidence that many cultures have different attitudes towards similar wildlife (e.g., Kellert 1991, Bjerke et al. 1998, Seddon and Khoja 2003). Stock farmers in the Indian Himalayas, for example, have different attitudes towards wolves, which they persecute, than towards snow leopards (*Uncia uncia*), which they do not, even though both predators prey upon domestic livestock

(Mishra 1997). Similarly Greek fishermen persecute cormorants (*Phalacrocorax* spp.), but not pelicans (*Pelecanus* spp.), even though both groups of birds are competitors for commercial fisheries (Daoutopoulos and Pyrovetsis 1990). Similar instances occur between farmers and wildlife in Florida (Jacobson et al. 2002).

The challenge, therefore, is to create learning opportunities that will influence public attitudes in the desired way. Even in the face of prior misconceptions and prejudices, education can, with care, help cultivate positive attitudes in the general public. New attitudes, ultimately, can result in behavior that supports raptor conservation (Broun 1949, Fraser et al. 1996, Bildstein 2001).

CREATING AN EDUCATION PROGRAM

Identifying Education Aims, Objectives, and Outcomes

Identifying your education aims provides you with the "big picture" of what you are trying to accomplish. As such, it is your overall mission or goal. Education objectives are specific goals relating to a particular program. Objectives should outline specifically what it is you want to teach or what kind of experience you want to provide, and to whom, geographically and demographically, you want to "target" as your audience. Different audiences require different approaches, so it is important to identify the audience prior to designing your program. *Outcomes* are the changes in behavior, attitude, and understanding that you wish to achieve as a result of your program. This includes determining how you will know (i.e., measure) whether or not your objectives have been achieved. It is important to be as specific as possible in stating your aims, objectives, and outcomes, as these will guide all other steps of program development. Also, the more explicit and clear your objectives are, the easier it will be for you to evaluate whether or not the program actually meets its objectives.

For example, a raptor rehabilitator in India might want to reduce the number of cattle farmers using the veterinary drug, diclofenac, in a particular region because vultures in that region are dying from diclofenac poisoning. The rehabilitator's education objective would be "to provide cattle farmers with information and understanding on how diclofenac kills vultures and to give them a positive experience with a living vulture to discourage their use of diclofenac." In this case, the target audience is cattle farmers in the region, and the outcome would be to have farmers discontinue using diclofenac after experiencing the program.

In defining your aims, objectives, and outcomes, it is important to consider both the education and conservation needs in your region. Investing time, money, and other resources into an education program that is not needed can waste time and resources. Thus it is important to determine if there are other organizations or individuals providing similar types of expertise and experiences. For each organization, determine what they are teaching (i.e., their education objectives), whom they are teaching (i.e., their target audience), and how they are teaching (i.e., their education methods). Evaluating and understanding other programs at this stage allows you develop your own program in light of the existing education landscape. If similar programs already exist, you can tailor your program to complement them. This may mean choosing different objectives, targeting a different audience, using different methods, or all three. Sometimes it is useful to visit and experience the existing programs to get a better sense of what is being done and what works and what does not work, and so that you can meet the staff and, perhaps, collaborate with them.

Identifying Organizational or Personal Strengths and Weaknesses

What special interests, skills, or experiences can you or your organization offer the public? For example, the strengths of a research organization might be the application of science or field ecology as an information resource. A rehabilitator may have expertise in animal health and biology, and may have access to special resources, such as unreleasable raptors that can be used as ambassadors for their species during education programs. A public school teacher has pedagogical skills, specialized knowledge, and access to individual students on a regular and repeated basis.

Along with identifying your strengths, it is important to recognize your weaknesses with regards to your program and to seek out expert advice and involvement where necessary.

It is often the case that people involved with informal education programs have not had formal training in education. In these situations, we recommend developing a small network of education specialists who are willing to advise you on the development of your program and review your education and evaluation plans and materials.

Determining Conceptual and Methodological Approaches

I hear, and I forget
I see, and I remember
I do, and I understand

It seems almost superfluous to justify the wisdom of this proverb. Consensus from the education establishment supports this notion that understanding emerges as a result of participation in the process of learning. Merely receiving written or verbal information often fails to result in meaningful learning. Ham and Kelsey (1998) conclude that carefully targeted biodiversity education is far superior to mass-media campaigns designed to educate the general public. In reality, a public-education program often will be a compromise between what is desirable and what is practical given your abilities and the size of the intended audience.

Types of education programs can be grouped into two general categories: **passive participation**, in which learners merely receive information, and **active participation**, in which learners interact with the education experience. Both of these can be designed to operate *in-situ* (i.e., in sight or proximity of the raptors in their home range) and *ex-situ* (outside of such proximity). *Ex-situ* experiences include secondary (including virtual) learning opportunities, such as watching wildlife documentaries, reading books and journal articles and accessing web-based resources and simulation activities.

In *ex-situ*, passive program participants typically observe a bird of prey, perhaps while listening to an educator or reading interpretive information. Examples include live-raptor demonstrations, museum and zoo exhibits, and wildlife documentaries, especially on television. Active participatory programs, on the other hand, allow participants to actively engage in some aspect of the program, through games, projects, research, and other activities. Passive programs tend to be used with large audiences, and where it is not feasible to involve the audience actively in the program. Active participatory programs tend to work best with smaller groups. Both types of programs have a role in the "big picture" of raptor conservation. For example, a person might become interested in raptors for the first time by seeing a live-raptor demonstration at a zoo or bird of prey center. This type of one-time experience may not be enough to influence a person's attitudes or conservation behavior towards raptors, however, it may spark enough interest to inspire them to call a local wildlife organization to see if they offer any educational programs about birds of

Figure 1. Flight demonstrations using falconry techniques can include diurnal (a) and nocturnal (b) birds of prey. Good demonstrations involve experts (c) who can provide general overviews of what is being seen as well as answer questions specific to the species involved. Properly conducted, a good flying display can be both an exhilarating and educational experience *(Photos courtesy of J. Parry-Jones).*

prey. Smaller, more active programs can provide a more personal experience that may have a more lasting influence on a person's attitude.

In designing an education program targeting attitudes, the approach should provide a combination of cognitive and emotional experiences. Keep in mind, however, that what may seem like an emotional experience to the educator, may not have the same meaning to the audience. Understanding existing attitudes is an important aspect of deciding how to provide an emotional experience to the audience. Likewise, understanding the existing knowledge-base in the population is important for deciding what information to include in the program, as learning is a process that incorporates both existing and new knowledge (Ausubel et al. 1978, Driver and Easley 1978, Driver et al. 1996).

Here we provide a series of examples of the use of passive and active participatory programs, including case studies highlighting ways that they have been applied to public-education programs about raptors worldwide.

Watching bird-of-prey flying displays in zoos and bird-of-prey centers. Compared with many other countries, laws in the United Kingdom (U.K.) regarding captive raptors make it relatively easy to keep birds of prey and display them to the public. This and the ease with which raptors have been propagated in captivity have supported an enormous growth in specialist bird-of-prey centers (often and inappropriately called "falconry centers"). There are several hundred of these open to the public in the U.K., mostly in close proximity to large urban areas. Collectively, these facilities attract millions of visitors per year and, arguably, they are where the majority of the British public gain first-hand experiences with birds of prey (Fig. 1, Box 1 and 2).

Creating education exhibits involving predators is notoriously difficult, as many predators, including raptors, do not move much except when hungry. Not surprisingly, zoo exhibits that involve active, moving animals increase the amount of time zoo visitors spend observing the exhibit (Wolf and Tymitz 1981, Marcellini and Jenssen 1988). Zoo Atlanta (U.S.), for example, found that allowing the public to observe training sessions with Asian small clawed otters (*Aonyx cinera*) with an interpretive narration, resulted in longer "stay-times" and increased positive perceptions by the public compared with passive exhibition (Anderson et al. 2003).

Most bird-of-prey centers have their origins in falconry practice rather than traditional zoo animal management. Falconry techniques provide the means by which birds of prey can be trained for free flight, and bird-of-prey flying demonstrations for public view have occurred in the U.K. since the late 1960s (Fig. 1). A good flying display can be an exhilarating experience and will ideally focus a roving zoo audience. This, in turn, increases the opportunity for providing a meaningful educational experience. Furthermore, a flying display provides the ideal opportunity for verbal interactive communication, serving to interpret what is being seen. Such interpretation has been found to be of considerable importance in the educational value of zoo exhibits (Marshdoyle et al. 1982). The flying demonstration, with associated commentary, therefore, provides an unusual opportunity both to entertain and to convey positive attitudes.

Box 1. National Birds of Prey Centre.

Country: United Kingdom

Organization: National Birds of Prey Centre

Program: *Educational flying demonstration*

Methods: Use of free-flying trained birds to teach about different species and their flying and hunting techniques, habitat requirements, threats and species status worldwide

Target audience: Variety of audiences, casual, drop-in visitors, booked parties, specialist groups, schools and colleges

Summary: Four to five different species of raptors are flown at each demonstration, and demonstrations throughout the day differ (Fig.1). The birds are trained in natural behavior patterns. Falcons, for example, are trained to stoop to a swung lure representing a bird, kestrels are trained to hover, owls and buteos are asked to show typical flights from tree to tree, eagles are encouraged to soar if the environment is right for them, and vultures are trained to show their habits on the ground and in the air. Specialist birds such as Burrowing Owls (*Athene cunicularia*) disappear down artificial tunnels, and caracara's exhibit the digging and scratching behavioral pattern natural to them. Each demonstration is commented upon by the individual flying the bird. They describe the species, its particular flight and individual characteristics, and its needs and habitats. Two or three trainers take turns flying birds to bring a different pace to the commentary. Individual personalities of the birds are described to give the audience a personal interest in the bird, but this is handled carefully so as not to detract from the bird or the information given. The demonstration moves along with plenty of birds and action; too much standing with a bird and just talking can lose the audience quickly. There is no doubt that attitudes can be changed dramatically. Vultures, in particular, are misunderstood, and when seen up close and in flight, attitudes can turn 180 degrees in a few minutes.

Visitors are encouraged to ask questions at the end of each demonstration, and digital photography is now so popular that photographic opportunities are useful to create. There is no doubt that the flying demonstrations are the highlight of visits, and regular customers will become attached to certain birds and make special visits to see them fly.

That said, bird-of-prey flying demonstrations must be designed carefully and thoughtfully. Despite providing the viewing public with a spectacular, sometimes exhilarating display, the "human dominion over nature" interaction between handler and hawk may predominate in the public view and lead to misunderstandings of the role of captive collections in bird of prey conservation. Cromie and Nicholls (1995) and Horrocks (in Nicholls 1999) assessed the potential conservation educational value (Table 1) of both static and flying displays in zoos and bird-of-prey centers in the U.K. In general, they concluded that although the "potential" was used to good effect by many centers, others missed education opportunities, or sacrificed them in favor of sensationalist exploitation of birds of prey. Indeed, Foulds and Rubin (1999) showed that at one bird-of-prey center in the U.K., the experience of watching a flying display did nothing to persuade the viewing public to change their lack of intention to support bird-of-prey conservation.

It is quite easy to send unintended negative messages to an audience about the value of raptors because the attitude of the handler towards the captive raptor becomes a reflection of the value of raptors in general. For example, if a captive raptor appears uncared for (e.g., overgrown beak and talons, broken feathers, dirty

Box 2. The Hawk and Owl Trust.

Country: United Kingdom

Organization: The Hawk and Owl Trust

Methods: Experiential, inquiry

Target audience: School students and visitors of all ages

Summary: The Hawk and Owl Trust Conservation and Education Centre provides an exhibition for visitors on raptor ecology plus an outdoor trail showing nesting boxes for different raptor species located in the relevant habitat. Educators provide activities for visitors and school groups. Students visit the site for day visits and take part in activities that encourage them to experience and investigate habitats that are important for raptors. They might spend time in woodland or grassland experiencing the sounds, smells and sights of the habitat, investigate the creatures that live there through close observation, and construct food chains to see how raptors depend on other animals and plants. Other activities, including nest-building or pellet dissection and evening owl prowls, help to explain other aspects of raptor ecology. A closed-circuit camera provides a window for visitors into the lives of a Barn Owl family during the breeding season. Educators also travel to schools and youth groups and give activity workshops such as nest-box building or pellet dissection. The education service also supplies educational support materials for teachers and youth group leaders for use in their own settings.

Table 1. Scoring criteria used to evaluate the conservation-education value of bird-of-prey centers in the United Kingdom (adapted from Cromie and Nichols 1993).

Score	Broad education value of aviary exhibits	Interpretation of aviary exhibits	Education content of flying display commentary	Conservation content of flying display commentary
1 = poor	Aviary barren, giving no indication of the bird's natural habitat or ecology (e.g. Burrowing Owls on solid substrates with no opportunity to burrow), or presenting a potential health risk (e.g. Barn Owls [*Tyto alba*] in a simulated barn with straw floor covering, which is an *aspergillosis* risk).	Information is provided, but is wrong or misleading (usually a redundant label from a previous exhibit) or highly anthropomorphic (e.g., "Harry the Harris's Hawk [*Parabuteo unicinctus*] enjoys watching TV").	Commentary wholly misleading or anthropomorphic or sensationalist (e.g., "This Ferruginous Hawk [*Buteo regalis*] can 'take out' any other bird here."). Attitude towards birds may be poor, with jokes at the bird's expense (e.g., the bird is said to be "scared of heights" or "educationally sub-normal".)	No mention of conservation of birds of prey. Alternatively, boasts of a conservation role without merit, uses the word "conservation" without explanation or conservation is used inappropriately to justify the captivity of rare and endangered species.
2 = unsatisfactory	Enclosure contains some features or furnishing, but these are unsympathetic to or inappropriate for species displayed.	Label with species name, a simple distribution map or a few words of description.	Information is inaccurate or irrelevant. Difficult to separate fact from fiction. There is a set script irrespective of the audience.	Commentary may trivialize ease of keeping raptors as pets and imply that captivity is in the animal's best interest, (i.e., it is safer in captivity than in the wild).
3 = moderately good	An attempt to present the bird in a background that conveys aspects of the bird's natural history, but falls short of the ideal (e.g., an enclosure for a fishing owl [*Scotopelia* spp.] has a small pond, but the pond water has feces and debris in it).	Species name given, with distribution map and some information of natural history and conservation status, but style and format not readily assimilable by all ages and abilities.	Accurate information presented in an easily assimilable form. Popular topics were related to how birds hunt and how well they can fly, see or hear. However, may emphasize ease with which raptors may be kept as pets and is followed by an advertisement for the center.	Accurate information that informs public of threats to wild raptor populations and the role of zoos in conservation. However, information is not presented in an entertaining way or is not appropriate for all audiences.
4 = excellent	Attempts to recreate habitats from which species originates (e.g., appropriate trees for a forest species, approximation of a desert for a desert species). Clean enclosures.	Imaginative or interactive interpretation, perhaps using information technology. Information is accurate, relevant, and transferable for all ages and abilities.	Accurate and entertaining, commentary suited to audience and species of raptor flown without being patronizing. Makes audience feel privileged to share their world with raptors.	Accurate and honest information, commentary accompanies demonstration or display of suitable species whose captivity is wholly justifiable.

living conditions, handled roughly, etc.) then, irrespective of the content of the associated commentary, the message to the public may be, "this raptor is not worth my care, time or attention." If an educator gives a prepared 40-minute talk with a bird on the arm or in an exhibit, without notice or mention of any of the behavior the bird may exhibit during the lecture (muting, preening, rousing feathers, watching the audience), the unintended message may be, "this raptor is not really interesting."

Therefore, it is essential that the utmost care and thought be given to the management, health, and treatment of captive raptors in educational programs. Regardless of whether they are flown free, all captive raptors should have a period during each year where they are put into an enclosure sufficiently large enough for them to move round comfortably, rest from educa-

tion work and given a chance to molt, or even to breed. Birds should not be flown in displays year in and year out without a decent rest period, untethered in a large aviary. Nor should birds be kept away from natural light and air for long periods. Cromie and Nicholls (1995) showed that health and welfare practices in bird-of-prey centers in the U.K. were extremely varied and ranged from excellent to very poor. Hawks, falcons, and, sometimes, owls, for example, often were kept tethered in the pretext that they were flown free each day, whereas evidence suggested that they were not. Of particular concern was the correlation between poor management and welfare practice and poor education involvement. It seemed that those centers with poorest record of care and management were those that invested least in creating and conveying meaningful educational experiences. Arent and Martel (1996), Parry-Jones (1991, 1994,

Figure 2. View to the east from Hawk Mountain Sanctuary's North Lookout in the central Appalachian Mountains of eastern Pennsylvania, U.S.A. Hawk Mountain Sanctuary hosts 70,000 visitors annually, many of which are introduced to free-ranging raptors and their conservation needs for the first time (Photo courtesy of Hawk Mountain Sanctuary Archives).

1998, 1999, 2003) and Naisbitt and Holz (2004) provide excellent information on captive raptor care and management.

The use of live raptors in a program may not be suitable in all situations or in all countries. The laws of the country need to be known and understood, as it may not be legal to involve live animals for education. Live animals often provoke emotional responses, sometimes positive and sometimes negative. It is important to understand which type of response is likely in any given culture or audience so that the program can address and respond to the emotional reaction of the audience to the raptor. Discussions with falconers, rehabilitation groups, zoos, conservation organizations, and government departments may well help identify the values, dangers, or advantages of using live birds in a particular country.

Organized raptor watchsites. From a public-education point-of-view one advantage of zoos and bird-of-prey centers is that the public is guaranteed of seeing birds of prey. Furthermore, such collections of live birds of prey and other animals are usually sited close to large urban human populations, and traditionally are visited for entertainment rather than for education. This represents an opportunity to deliver a conservation education message to a true, cross-section of the general public. On the other hand, the spectacle of raptor migration as observed at raptor migration watch-sites provides an *"unparalleled opportunity to introduce the public to these secretive and normally widely dispersed birds of prey"* (Zalles and Bildstein 2000) (Fig. 2, Boxes 3–5).

Box 3. Hawk Mountain Sanctuary migration counts.

Country: United States

Organization: Hawk Mountain Sanctuary

Program: *Autumn and spring migration counts*

Methods: Flight interpretation and scheduled programs associated with the passage of thousands of migrating raptors

Target audience: General public, school groups

Summary: Hawk Mountain Sanctuary, the world's first refuge for birds of prey, is a 1,000-ha nature reserve in the central Appalachian Mountains of eastern Pennsylvania, United States, 170 km west of New York City. The "Mountain" is part of the Kittatinny Ridge, an internationally significant migration corridor for hawks, eagles, and falcons breeding in northeastern North America. Hawk Mountain was founded in 1934 by New York conservationist Rosalie Edge, who created the refuge to stop the slaughter of migrating raptors at the site. Each autumn, tens of thousands of raptors migrate past the watchsite. Occasionally, spectacular migrations of thousands of birds are recorded on single days. In the 62-year period from 1934 to 1995, an annual average of >17,000 diurnal raptors, representing 16 species, was recorded at the watchsite.

In the 1920s and early 1930s, hunters traditionally gathered on the ridge-tops of eastern Pennsylvania each autumn to shoot migrating hawks and eagles. Raptors were considered vermin at the time, and the state game commission had established bounties on several species. Each year, thousands of birds were killed as they traveled south along the central Appalachian Mountains. Hawk Mountain, in particular, became a favored shooting site. All of this changed in August 1934 when Maurice Broun, the sanctuary's first warden, posted the property and confronted local shooters. The next autumn, birdwatchers and naturalists began to flock to the new refuge in large numbers.

Today, Hawk Mountain exemplifies what grassroots conservation, environmental education, and ecological monitoring and research together can accomplish. The sanctuary maintains the longest and most complete record of raptor migration in the world. Its annual counts of migrating hawks and eagles have proved to be essential tools in assessing long-term trends in raptor populations throughout eastern North America. The extensive database played a key role in exposing first-generation organochlorine pesticides, including DDT, as causative agents for the precipitous declines in populations of several species of birds of prey that occurred earlier in the 20th century, as well as in measuring subsequent rebounds in raptor populations following decreases in the use of contaminants.

The sanctuary's extensive on- and off-site education programs, which have touched >1 million people, include weekend interpretive programs for the general public, weekday guided programs for primary and secondary school children, fully accredited college-level courses offered in cooperation with local colleges and universities, a science and mathematics based education curriculum focused on Turkey Vultures, and workshops for local educators (Fig. 2; Zalles and Bildstein 2000).

On spring and autumn weekends Hawk Mountain offers education programs about raptors and the phenomenon of migration for the public. Program subjects include hawk identification, the use of binoculars, the Sanctuary's culture and natural history, and raptor study techniques, all of which are conducted outdoors at various overlooks at the Sanctuary. At its two primary lookouts, interns and biologists spend the day interpreting the flight for the public, including spotting birds, highlighting interesting flight behavior, helping with identification, and answering questions. One of the most popular events is a live-raptor program in which the public has the chance to observe one or two local raptors up close and learn more about their biology, ecology, and conservation. The variety of approaches used and content levels available increases the number of visitors that can be reached during a visit to the Mountain.

Box 4. Peregrine-viewing point at Symonds Yat Rock.

Country: United Kingdom

Organization: Royal Society for the Protection of Birds (RSPB) and Forest Enterprise

Program: *Peregrine-viewing point at Symonds Yat Rock*

Methods: Guided observation with some interpretation

Target audience: General public

Summary: The Peregrine-viewing point at Symonds Yat Rock is a joint project run by the RSPB and Forest Enterprise, who own the site. Between April and August, telescopes are set up to give close views of the nesting peregrines. The rock is a spectacular site high above the River Wye. The location is in rural Gloucestershire, but is in close proximity to Bristol and Birmingham.

Box 5. Aren't Birds Brilliant.

Country: United Kingdom

Organization: Royal Society for the Protection of Birds (RSPB)

Program: *Aren't Birds Brilliant*

Methods: Guided observation with some interpretation

Target audience: General public

Summary: In the U.K. a scheme called *Aren't Birds Brilliant* (ABB) (www.rspb.org.uk/birds/brilliant/index.asp), which is managed by the Royal Society for the Protection of Birds (RSPB), invites the public-at-large to view birds during spring and summer at selected sites. These sites, often nesting areas, are distinct from RSPB-managed nature reserves and access to them depends on the cooperation of a range of landowners and other organizations. High-quality optical equipment is provided and the public is coached by a team of staff and volunteers in its use and both written and verbal information are provided on the species viewed. During 2005, 22 of 50 ABB sites featured birds of prey, including Ospreys (*Pandion haliaetus*), Red Kites (*Milvus milvus*), White-tailed Eagles (*Haliaeetus abicilla*), Northern Harriers (*Circus cyaneus*), and Peregrine Falcons (*Falco peregrinus*). Some sites, such as eagle-nesting sites on Isles of Skye and Mull off the west coast of Scotland are in relatively remote locations. However, other sites, such as Peregrine Falcon nests in central London and near to other major cities, are accessible to the public-at-large.

Aside from refining skills of observation, the public largely is passive in this kind of directed bird-watching. That said, an evaluation of the impact of the ABB scheme indicated that during 2006, the 22 ABB raptor sites were visited by nearly 384,000 people (P. Holden, pers. comm.). As a measure of the intention of these visitors to support raptor conservation, nearly 37,000 (10%) left contact details to receive additional information, and almost 1,500 (0.4%) joined the RSPB. The same pattern emerged at other ABB sites, which are devoted to sea birds, waders and passerines, and in total, the 50 sites attracted 480,000 visitors in 2006. Finally, over the past few years, just under 7% of AAB visitors joined the "Big Annual Bird watch" which is an annual census of garden birds carried out by volunteers.

That some members of the public are willing to visit more remote areas may mean such sites are visited by a self-selected sub-group of the public, a sub-group already with empathy for nature and living things. To persuade this group to support birds of prey may therefore be an easier job, but this does not detract from the opportunity to deliver accurate information and present a high-quality learning experience.

Inquiry-based learning activities. There is evidence that actively engaging participants during an educational experience increases learning outcomes and is more likely to influence attitudes than passive programs (Heimlich 1993, DeWhite and Jacobson 1994, Leeming et al. 1997, Manzanal et al. 1999). Here we briefly outline several approaches that can be used to create active educational experiences.

In many raptor-education programs, participants are given facts about raptors that are meant to capture their interest, inform them about an issue, or provide a knowledge base for other aspects of the program. One way to make this transfer of information more active is by using an inquiry-based learning approach. Inquiry learning is a process in which students address their own curiosity by seeking answers to their own questions (Pearce 1999, Minstrell and Van Ze 2000). This approach is perfect for exploration of the natural world. For example, rather than being told what is the most common prey for Barn Owls (*Tyto alba*), participants discover for themselves by dissecting Barn Owl pellets. Participants actively engage in the learning process, rather than simply being passive recipients of information. This is an important aspect to consider for conservation education, because there is a certain power to discovery, and a person may be more likely to have an emotional attachment to conclusions they came to through their own discovery process, than by simply being told.

Tafoya et al. (1980) define four types of inquiry-based activities: confirmation, structured, guided, and open. Confirmation activities require participants to verify concepts learned by participating in a given procedure. Structured inquiry activities provide participants with a guided question and procedure to follow. Guided inquiry activities are similar to structured activities in that they provide participants with a guiding question and suggested materials, but they allow participants to direct the investigation. Open-inquiry activities allow students to generate their own questions and design their own research project (Box 6).

Field projects and "citizen science." Positive experiences in nature predict positive attitudes towards

Box 6. Migrating Birds Know No Boundaries.

Country: Israel

Organizations: International Centre for the Study of Bird Migration, Museum for Technology, Science and Space, and the Davidson Institute for Scientific Education

Program: *Migrating Birds Know No Boundaries*

Method: Inquiry-learning

Target audience: 7th and 9th grades

Summary: Students participating in this program conduct research on migrating raptors (and other birds) using satellite telemetry and radar monitoring. Participants track raptors on their migrations from Israel to their winter nesting grounds in Africa, and back northward, via Israel, to their summer nesting grounds in Europe. During their projects, students are faced with solving real-world dilemmas, such as how to conserve nesting habitat or how to prevent dangerous collisions between migrating birds and man-made aircraft. The program uses technology and science to connect students to raptors and other students all over the Middle East and Africa.

Box 7. Starr Ranch Junior Biologists.

Country: United States

Organization: Audubon California, Starr Ranch Sanctuary

Program: *Starr Ranch Junior Biologists — Raptor Research*

Methods: Inquiry, field projects

Target audience: 8 to 14 year-olds

Summary: Starr Ranch Junior Biologists is a summer camp where children participate in field-ecology. During the raptor-research program, children spend 5 days learning about the biology and ecology of raptors by actually studying them. The Junior Biologists meet raptor biologists, and are introduced to why biologists study raptors and the different techniques they use. Participants then conduct their own study addressing an ecological question about raptors on the Sanctuary. During the summer of 2004, for example, the children wanted to know how many of the raptors that were nesting on the Sanctuary actually reused their nest from the previous year. The Junior Biologists visited nest sites from the previous year, determined whether they were active, used GPS and GIS to make a map of the nests on the Sanctuary, and compared their results with data on nesting raptors from the previous year. This not only increases the children's awareness of how raptors use the environment but, by actually studying raptors, students also gain a sense of intimacy with their subjects.

nature (Bogner 1998, Kals et al. 1999, Monroe 2003), although this may only be true in the absence of direct conflicts of interest. Positive experiences in nature, which involve repeated experiences that are personally rewarding, seem to have the most impact when they start during childhood and continue through adulthood (Kals et al. 1999). Nature-based programs can be made more active by engaging the participants in field projects. Manzanal et al. (1999) measured the effect of fieldwork on the ecological knowledge and environmental attitudes of 14–16 year-old students in Spain and found that fieldwork helped to clarify ecological concepts and directly improved attitudes in defense of the ecosystem wherein the students were working (Box 7).

Projects can be designed simply for education purposes (an inquiry-based field project), or they can be designed for long-term monitoring of local raptor populations, where there is a dual goal for conservation education and direct conservation output. Community members can be engaged as volunteers for raptor monitoring projects, such as local nest monitoring and local population monitoring (road surveys, etc.). This notion of using citizen scientists as a means to engage the public actively in conservation has been successfully employed by organizations such as the Audubon Society, The Peregrine Fund, Cornell Lab of Ornithology, HawkWatch International, Hawk Mountain Sanctuary (Bildstein 1998), the Hawk Migration Association of North America, the Royal Society for the Protection of Birds (RSPB), and the British Trust for Ornithology.

Coordinators of many raptor-migration watchsites have trained local community members to be volunteer counters (Bildstein 1998). During the first trans-continental raptor migration count in Panama in 2005, dozens of local community members and high school students participated as volunteer counters. Such experiences provide volunteers with an introduction to raptor biology and migration ecology, an opportunity to contribute to the understanding and conservation of raptors in their area, and the opportunity to continue participating year after year.

Action competence. Although it is important to inform and interest the public in raptor-conservation issues, it also is essential to empower them to act in favor of raptors. Recall that the third component of an attitude is "behavior as a direct consequence of the cognitive and affective elements." Action competence is an approach that aims to increase a person's belief in their participation and influence on solutions of environmental problems (Jensen and Schnack 1997, Bishop and

Box 8. Philippine Eagle Community-based Project.

Country: Philippines

Organization: Philippine Eagle Foundation

Program: *Philippine Eagle Community-based Project*

Methods: Action competence

Target audience: Adults and their families

Summary: The Pulangi Watershed Integrated Community-based Resource Management Project was established in 2001 to help build self-reliant and sustaining communities in the Upper Pulangi Watershed in Bukidnon and to protect pristine Philippine Eagle nesting habitat. The program adopts a participatory approach to local area planning, project development and management. Foundation staff facilitates training, planning, capacity building, and development of livelihood projects in consultation with partner communities. Since its inception, the community-based program has overseen the restoration of over 25 ha of denuded forest, all potential Philippine Eagle nesting habitat.

Scott 1998, Jensen 2002, Jensen and Nielsen 2003). Programs generally involve learning about a conservation problem, perhaps through an inquiry-based activity. On the basis of their conclusions, participants decide upon some type of social, political, or environmental action directed towards a solution of the environmental problem or a change in the conditions that caused it (Jensen and Schnack 1997, Bishop and Scott 1998, Jensen 2002, Jensen and Nielsen 2003). The leader of this type of program acts as the facilitator for enabling the group to act on their ideas (Box 8).

To engender action competence effectively, programs should provide a critical and thorough understanding of the problem in question, and of the nature of the action involved (Jensen and Schnack 1997). Furthermore, participants must be motivated enough to follow through on their solution. In some instances, participants will be self-selected and will likely come with the necessary motivation. In other instances a major goal of the program will be connecting the participants to the conservation problem and trying to inspire in them a drive to act. Once the action has been completed, the participants should be asked to evaluate the effectiveness of the action and critically examine reasons underpinning success or failure of the action (Bishop and Scott 1998).

Evaluating Program Effectiveness

To be effective, education programs should be developmentally dynamic. That means they should undergo constant evaluation and revision. Having an evaluation procedure enables an educator to determine how well their program is achieving their educational goals, and guides the revision process. If a program aims to encourage local landowners to adopt certain land-use practices that benefit birds of prey, then effectiveness may be relatively easy to measure in terms of the proportion of landowners who adopt these practices after participating in the program. Similarly, a program that involves school children building and monitoring owl nest-boxes could be assessed in terms of the proportion of schools in a region active in the program, average period of involvement in the program (1–3 years, 4–7 years, etc.), and so on. However, many public-education programs reach people in a passive and ephemeral fashion, and it is unlikely that individual responses to such programs can be tracked through time. Thus, the chance of following changes in attitudes is limited. In such cases measuring effectiveness in real terms may be impossible.

According to the Theory of Planned Behavior (Ajzen and Fishbein 1977) it may be practical and sufficient to question individuals to determine their "behavioral intention" (i.e., whether the educational experience has influenced the way they are likely to behave in the future). By asking participants about their perceived attitudes, subjective norms, and perceived behavioral control, behavioral intention may be predicted. Recently, this approach has been used to predict the efficacy of interpretive centers on modifying the food-storage behavior of visitors hoping to see black bears (*Ursus americanus*) (Lackey and Ham 2003), the watershed use of farmers (Beedel and Rehman 2000), and the behavior of boaters toward Florida manatees (*Trichechus manatus latirostris*) (Aipanjiguly et al. 2003). Questionnaires or other semi-quantitative surveys given at the time of the educational program can often be relied upon to give a good indication as to whether the experience has been effective.

Designing an evaluation procedure should be carefully researched and considered; the better the design, the better the feedback on a program. Many resources are available to assist with evaluation design. Nowak (1984), Patton (1990), Jacobson (1991) and Marciknowski (1993) provide good suggestions and strategies in this regard.

Table 2. Currently available raptor-curriculum guides.

Eye of the Falcon

Real-time satellite tracking projects are used as vehicles for people to learn about the ecology, behavior, evolution, and geography of raptors around the world. Inquiry based.

> Earthspan, Inc., U.S.A.
> michellefrankel@earthlink.net
> Available on the web at: www.earthspan.org/Education.htm
> (305) 604-8802

First Flight with Raptors

Raptor-based biology activities for kindergarten and grade one.

> Education Specialist, Hawk Mountain Sanctuary,
> 1700 Hawk Mountain Road, Kempton, PA 19529 U.S.A.
> Available on the web at: www.hawkmountain.org/education/
> resources_for_learning.htm
> (610) 756-6961

The Peregrine Project

Six-month unit on Peregrine Falcon biology and conservation for grades one to five.

> Maria Dubois, Conserve Wildlife Foundation of New Jersey,
> P.O. Box 400, 501 East State Street, Trenton, NJ 08625 U.S.A.
> (609) 984-0621

Hunters of the Sky

Science-based activity guide for birds of prey.

> Science Museum of Minnesota,
> 120 West Kellogg Boulevard, Saint Paul, MN 55102 U.S.A.
> schooloutreach@smm.org
> (651) 221-4748 or (800) 221-9444 ext. 4748

Raptor Ecology

Raptor ecology curriculum guide for grades four through eight.

> HawkWatch International, 1800 South West Temple,
> Suite 226, Salt Lake City, UT 84115 U.S.A.
> Available on the web at: www.hawkwatch.org
> (801) 484-6758

One Bird, Two Habitats

Migration ecology and research curriculum guide for grades five to eight. Not raptor-based, but can be applied to raptors.

> Illinois Department of Natural Resources,
> One Natural Resources Way, Springfield, IL 62702 U.S.A.
> Available on the web at: http://dnr.state.il.us/lands/education/
> CLASSRM/birds/1B2HFULL.PDF
> Teacher's Guide available on the web at: http://dnr.state.il.us/
> lands/education/CLASSRM/birds/onebird_activities.pdf

Available Resources

There are several good raptor-curriculum guides that can be modified and used in a variety of education contexts (schools, parks, watch-sites, etc.). Table 2 provides a list of curricula that we are most familiar with.

Finally, if you develop a program that has been successful, do not hesitate to share your success by publishing the program for others to use, particularly in a regions where raptor education resources are limited.

ACKNOWLEDGMENTS

We thank Chris Farmer for comments on earlier drafts of this chapter. Peter Holden (RSPB) provided information on the RSPB "Aren't birds brilliant" initiative. Rona Rubin contributed information on the theory of planned behavior.

LITERATURE CITED

AIPANJIGULY, S., S.K. JACOBSON AND R. FLAMM. 2003. Conserving manatees: knowledge, attitudes, and intentions of boaters in Tampa Bay, Florida. *Conserv. Biol.* 17:1098–1105.

AJZEN I. AND M. FISHBEIN. 1977. Attitude-behavior relations: a theoretical analysis and review of empirical research. *Psychol. Bull.* 84:888–918.

ANDERSON, A.S., R. PRESSLEY-KEOGH, M.A. BLOOMSMIT AND T.L. MAPLE. 2003. Enhancing the zoo visitor's experience by public animal training and oral interpretation at an otter exhibit. *Environ. Behav.* 35:826–841.

ARENT, L. AND M. MARTELL. 1996. Care and management of captive raptors. The Raptor Center at the University of Minnesota, St. Paul, MN U.S.A.

AUSUBEL, D.P., J.D. NOVAK AND H. HANESIAN. 1978. Education psychology: a cognitive view, 2nd Ed. Holt, Rinehart and Winston, New York, NY U.S.A.

BANDARA, R. AND C. TISDELL. 2003. Comparison of rural and urban attitudes to the conservation of Asian elephants in Sri Lanka: empirical evidence. *Biol. Conserv.* 110:327–342.

BEEDEL, J. AND T. REHMAN. 2000. Using social-psychology models to understand farmers' conservation behavior - the relationship of verbal and overt verbal responses to attitude objects. *J. Rural Stud.* 16:117–127.

BILDSTEIN, K. 1998. Long-term counts of migrating raptors: a role for volunteers in wildlife research. *J. Wildl. Manage.* 65:435–445.

———. 2001. Raptors as vermin: a history of human attitudes towards Pennsylvania's birds or prey. *Endangered Species Update* 18:124–128.

BISHOP, K. AND W. SCOTT. 1998. Deconstructing action competence: developing a case for a more scientifically attentive environmental education. *Public Understanding Science* 7:225–236.

BJERKE, T., T.S. ODEGARDSTUEN AND B.P. KALTENBORN. 1998. Attitudes toward animals among Norwegian adolescents. *Anthrozoos* 11:79–86.

BOGNER, F.X. 1998. The influence of short-term outdoor ecology education on long-term variables of environmental perspective. *J. Environ. Educ.* 29:17–29.

BRADLEY, J.C., T.M. WALICZEK AND J.M. ZAJICEK. 1999. Relationship between environmental knowledge and environmental attitude of high school students. *J. Environ. Educ.* 30:17–21.

BROUN, M. 1949. Hawks aloft: the story of Hawk Mountain. Cornwall Press, Cornwall, NY U.S.A.

CROMIE, R. AND M. NICHOLLS. 1995. The welfare and conservation aspects of keeping birds of prey in captivity. Report submitted to the Royal Society for the Prevention of Cruelty to Animals. Durell Institute of Conservation and Ecology, University of Kent, Canterbury, United Kingdom.

DAOUTOPOULOS, G.A. AND M. PYROVETSI. 1990. Comparison of conservation attitudes among fishermen in three protected lakes in Greece. *J. Environ. Manage.* 31: 83–92.

DEWHITE, T.G. AND S.K. JACOBSON. 1994. Evaluating conservation education programs at a South American zoo. *J. Environ. Educ.* 25:18–22.

DRIVER, R. AND J. A. EASLEY. 1978. Pupils and paradigms: a review of literature related to concept development in adolescent science. *Stud. Sci. Educ.* 5:61–84.

———, J. LEACH, R. MILLER AND P. SCOTT. 1996. Young people's images of science. Open University Press, Buckingham, United Kingdom.

EVERITT, P., R. RUBIN AND M.K. NICHOLLS. 2002. Changing public attitudes to birds of prey in the UK. Page 33 in R. Yosef, M.L. Miller, and D. Pepler [EDS.], Raptors in the new millennium. International Birding and Research Center, Eilat, Israel.

FOULDS, M. AND R. RUBIN. 1999. Flying displays, conservation, and the views of the general public. *The Falconers and Raptor Conservation Magazine,* Autumn 1999, p15.

FRASER, J.D., S.K. CHANDLER, D.A. BEUHLER AND J.K.D. SEEGAR. 1996. The decline, recovery, and future of the Bald Eagle population of the Chesapeake Bay, USA. Pages 181–188 in B.-U. Meyburg and R.D. Chancellor [EDS.], Eagle studies. World Working Group on Birds of Prey and Owls, Berlin, Germany.

HAM, L. AND E. KELSEY. 1998. Learning about biodiversity - a first look at the theory and practice of biodiversity education, awareness and training in Canada. Environment Canada, Quebec, Canada.

HEIMLICH, J. 1993. Nonformal environmental education: toward a working definition. The Environmental Outlook. ERIC/CSMEE Informational Bulletin, Columbus, OH U.S.A.

JACOBSON, S.K. 1991. Evaluation model for developing, implementing, and assessing conservation education programs: examples from Belize and Costa Rica. *Environ. Manage.* 15:143–150.

———, K.E. SIEVING, G.A. JONES AND A. VAN DOOR. 2002. Assessment of farmers' attitudes and behavioral intentions toward bird conservation on organic and conventional Florida farms. *Conserv. Biol.* 17:595–606.

JENSEN, B.B. 2002. Knowledge, action and pro-environmental behavior. *Environ. Educ. Res.* 8:325–334.

——— AND K. NIELSEN. 2003. Action-oriented environmental education: clarifying the concept of action. *J. Environ. Educ. Res.* 1:173–193.

——— AND K. SCHNACK. 1997. The action competence approach in environmental education. *Environ. Educ. Res.* 3:163–178.

KALS, E., D. SCHUMACHER AND L. MONTADA. 1999. Emotional affinity toward nature as a motivational basis to protect nature. *Environ. Behav.* 31:178–202.

KELLERT, S.R. 1991. Japanese perceptions of wildlife. *Conserv. Biol.* 5:297–301.

KOLLMUSS, A. AND J. AGYEMAN. 2002. Mind the gap: why do people act environmentally and what are the barriers to pro-environmental behavior? *Environ. Educ. Res.* 8:239–260.

LACKEY, B. AND S. HAM. 2003. Contextual analysis of interpretation focused on human-black bear conflicts in Yosemite National Park. *Appl. Environ. Educ. Commun.* 2:11–21.

LEEMING, F.C., B.E. PORTER, W.O. DWYER, M.K. COBERN AND D.P. OLIVER. 1997. Effects of participation in class activities on children's environmental attitudes and knowledge. *J. Environ. Educ.* 28:33–42.

LIVINGSTONE, D.W. 2001. Adults' informal learning: definitions, findings, gaps and future research. NALL Working Paper No. 21, Centre for the Study of Education and Work, University of Toronto, Toronto, Canada.

MANZANAL, R.F., L.M. RODRIGUEZ BARRIERO, M. CASAL JIMENEZ. 1999. Relationship between ecology fieldwork and student attitudes toward environmental protection. *J. Res. Sci. Teaching* 36:431–453.

MARCELLINI, D.L. AND T.A. JENSSEN. 1988. Visitor behavior in the National Zoo's reptile house. *Zoo Biol.* 7:329–338.

MARCIKNOWSKI, T. 1993. Assessment in environmental education. Pages 143–197 in Environmental education teacher handbook. Kraus International Publications, Millwood, NY U.S.A.

MARKER, L. AND A. DICKMAN. 2004. Human aspects of cheetah conservation: lessons learned from the Namibian farmlands. *Hum. Dimensions Wildl.* 9:297–305.

MARSHDOYLE, E., M.L. BOWMAN AND G. MULLINS. 1982. Evaluating programmatic use of a community resource: the zoo. *J. Environ. Educ.* 13:19–26.

MINSTRELL, J. AND E. VAN ZEE [EDS.]. 2000. Inquiring into inquiry learning and teaching in science. American Association for the Advancement of Science, Washington, DC U.S.A.

MISHRA, S.R. 1997. Livestock depredation by large carnivores in the Indian trans-Himalaya: conflict perceptions and conservation prospects. *Environ Conserv.* 24:338–343.

MONROE, M.C. 2003. Two avenues for encouraging conservation behaviors. *Hum. Ecol. Rev.* 10:113–125.

NAISBITT, R. AND P. HOLZ. 2004. Captive raptor management and rehabilitation. Hancock House, Blaine, WA U.S.A.

NICHOLLS, M.K. 1999. Education and conservation - the role of birds of prey centers. *The Falconers' and Raptor Conservation Magazine,* No. 41 Winter 1999/2000.

NOWAK, P.F. 1984. Direct evaluation: a management tool for program justification, evolution, and modification. *J. Environ. Educ.* 15:27–31.

PARRY-JONES, J. 1991. Jemima Parry-Jones' falconry: care, captive breeding and conservation. David & Charles Publishers, Devon, United Kingdom.

———. 1994. Training birds of prey. David & Charles Publishers, Devon, United Kingdom.

———. 1998. Understanding owls: biology, management, breeding, training. David & Charles Publishers, Devon, United Kingdom.

———. 1999. The really useful owl book. Kingdom Books, Havant, Hampshire, United Kingdom.

———. 2003. Jemima Parry-Jones' falconry: care, captive breeding and conservation, 2nd Ed. David & Charles Publishers, Devon, United Kingdom.

PATTON, M.Q. 1990. Qualitative evaluation and research methods. Sage Publications, Inc., Newbury Park, CA U.S.A.

PEARCE, C.S. 1999. Nurturing inquiry: real science for the elementary classroom. Heinemann, Portsmouth, NH U.S.A.

RAMSEY, C.E. AND R.E. RICKSON. 1976. Environmental knowledge and attitudes. *J. Environ. Educ.* 8:10–18.

SEDDON, P.J. AND A.R. KHOJA. 2003. Research note: youth attitudes to wildlife, protected areas and outdoor recreation in the kingdom of Saudi Arabia. *J. Ecotourism* 2:67–75.

TAFOYA, E., D. SUNAL AND P. KNECHT. 1980. Assessing inquiry potential: a tool for curriculum decision makers. *Sch. Sci. Math.* 80:43–48.

TREVES, A., L. NAUGHTON-TREVES, E.K. HARPER, D.J. MLADENOFF, R.A. ROSE, T.A. SICKLEY, A.P. WYDEVEN. 2004. Predicting human-carnivore conflict: a spatial model derived from 25 years of data on wolf predation on livestock. *Conserv. Biol.* 18:114–125.

VALKAMA, J., E. KORPIMAKI, B. ARROYO, P. BEJA, V. BRETAGNOLLE, E. BRO, R. KENWARD, S. MANOSA, S. REDPATH, S. THIRGOOD AND J. VINUELA. 2005 Birds of prey as limiting factors of gamebird populations in Europe. *Biol. Rev.* 80:171–203.

VERDADE, L.M. AND C.B. CAMPOS. 2004. How much is a puma worth? Economic conpensation as an alternative for the conflict between wildlife conservation and livestick production in Brazil. *Biota Neotropical* 4:1–4.

WOLF, R.L. AND B.L. TYMITZ. 1981. Studying visitor perceptions of zoo environments: a naturalistic view. *Int. Zoo Yearb.* 2:49–53.

WOODROFFE, R. AND J.R. GINSBERG. 1999. Conserving the African wild dog (*Lycaon pictus*). I. diagnosing and treating causes of decline. *Oryx* 33:132–142.

ZALLES, J.I. AND K.L. BILDSTEIN [EDS.]. 2000. Raptor watch: a global directory of raptor migration sites. Birdlife International, Cambridge, United Kingdom and Hawk Mountain Sanctuary, Kempton, PA, U.S.A.

Legal Considerations 25

BRIAN A. MILLSAP
U.S. Fish and Wildlife Service,
New Mexico Ecological Services Field Office,
2105 Osuna Road NE, Albuquerque, NM 87113 U.S.A.

MARGARET E. COOPER
School of Veterinary Medicine,
The University of the West Indies,
St. Augustine, Trinidad and Tobago

GEOFFREY HOLROYD
Canadian Wildlife Service, Environment Canada
Room 200, 4999-98 Avenue, Edmonton, AB, T6B 2X3 Canada

Here we provide an overview of the laws that regulate research and conservation of raptors, primarily the laws of the United States, Canada, and Europe (with an emphasis on Great Britain). Space limitations prevent us from detailing existing laws, which change frequently. We encourage researchers and managers to use the Internet and to consult with applicable government authorities to obtain detailed, current information on a case-by-case basis to ensure compliance with all applicable laws and regulations well in advance of initiating the work that may require permits or government authorizations. The list of issues described below provides a guide to researchers and managers in other countries, in their search for applicable laws and regulations.

INTRODUCTION

Laws and regulations have played a major positive role in raptor conservation globally. Raptors are high-profile wildlife, and their position in society both as animals to be revered and, in some cases, despised, has led to conservation and to persecution and over-exploitation at various times and places throughout history. To protect raptors, many nations, states, and local governments have created laws that regulate capture or killing of birds of prey, ensure their proper care in captivity, and protect wild raptors and habitats, especially for species at risk. Although these laws have been successful at furthering the conservation of raptors, they can be challenging if researchers and conservationists are not familiar with them, or worse, choose to ignore them.

INTERNATIONAL PERSPECTIVE

Many national conservation laws are based on international or regional obligations (e.g., international treaties). Because of this, the fundamental components of many wildlife laws are similar among countries. The prime example is the Convention on the International Trade in Endangered Species of Wild Fauna and Flora (CITES), signed in 1973 in Washington, D.C., U.S.A., and implemented by 169 countries as of July 2006. CITES provides a uniform system of control on the international movement of CITES-listed species, including raptors (see www.cites.org). Most countries have national laws that enact CITES-compliant movements of wildlife, including parts (e.g., tissues, feathers). Some species of raptors are listed in Appendix I as endangered species; other Falconiformes and Strigiformes (except Cathartidae) are included under Appendix II or III as look-a-like or potentially at-risk species. Thus, the international movement of raptors usually requires CITES compliance. Under CITES, two permits are required to move a raptor listed in Appendix I: an import permit from the destination country and an export permit from the country of origin. The import permit is required before the export permit will be issued. For Appendix II and III species, only an export permit is required, unless a national law states that an import permit is required in the country of destination.

Other international conventions, such as the Convention on Wetlands (Ramsar) and Convention on Biological Diversity, also have an impact on raptors, particularly in conservation and research. The Convention on the Conservation of Migratory Species of Wild Animals lists certain species of birds of prey in its Appendices as being in need of conservation (see www.biodiv.org/cooperation/joint.shtml.)

NATIONAL AND REGIONAL LAWS

Most countries have some regulations regarding raptors, though the extent and complexity of laws varies greatly. The variation in laws among countries is related to the stage of development, priorities, cultural attitudes, history and, in some cases, religion. Wildlife laws are designed primarily to protect free-living animals, but often they affect those in captivity and the process of taking or releasing them. Such laws frequently provide protection for specific species, for example, by prohibiting the killing, taking and injuring of an animal and extending protection to eggs, nests and young. In addition, many forms of exploitation are restricted. In many countries, hunting is regulated or prohibited, and where allowed, methods, seasons, and times of day when animals may be taken or controlled are specified.

This section of the chapter provides an overview of the most relevant areas of law in the U.S., Canada, and Great Britain. We provide Internet links to the most recent versions of the pertinent regulations, as well as to agency web sites with additional information (web site addresses were current as of 4 January 2007). Those working elsewhere can expect to find similar laws in many cases, and should consult with the wildlife management authority in the country of interest to ensure that necessary authorizations are obtained.

RAPTOR LAWS IN THE UNITED STATES

Raptor conservation in the U.S. has its foundation in law. Many important conservation advances have resulted from legislation; for example, the cessation of the slaughter of migrant hawks, elimination of bounties to encourage lethal control, suspension of general use of DDT, and provision of funding for research and management of threatened and endangered raptors. Prior to 1900, the federal government's only involvement with birds of prey was through predator control. Between 1900 and 1950, conservation organizations, backed by scientific information showing the beneficial nature of raptors, succeeded in obtaining protection for some birds of prey in 42 states (Millsap 1987). It was not until 1972, however, that most raptors received full protection at the federal level.

The objectives of this section are to (1) briefly review some of the U.S. laws that provide protection to raptors that raptor researchers and managers need to be aware of, and (2) describe permit requirements and procedures for raptor research and management activities. Implementing regulations discussed in this chapter are contained in the Code of Federal Regulations Title 50 (50 C.F.R.), Migratory Bird Treaty Act (Parts 10 and 21), Bald and Golden Eagle Protection Act (Part 22), and Endangered Species Act (Parts 17 and 23). Because implementing regulations and permitting procedures are subject to frequent changes, we provide links to World Wide Web pages that are maintained by agencies responsible for implementing the regulations and permits. We suggest that researchers check these sites for the most current information. Detailed information on migratory bird and eagle permits can be found on the Internet at www.fws.gov/permits/mbpermits/birdbasics.html, and for endangered species at www.fws.gov/endangered/permits/index.html.

Migratory Bird Treaty Act

Federal protection for migratory birds in the U.S. began when Congress enacted the Migratory Bird Act (MBA; 37 Stat. 878, ch. 45) in 1913. This act placed all migratory game and insectivorous birds under the protection of the U.S. government, and prohibited hunting of such species except pursuant to federal regulations (Bean 1983). The MBA was challenged successfully in federal court on the grounds that the property clause of the constitution granted states primary management authority over all wildlife (Bean 1983). In response, the State

Department concluded a treaty with Great Britain that protected birds migrating between the U.S. and Canada. That treaty was signed in March 1916, and implemented by the Migratory Bird Treaty Act (MBTA; 16 U.S.C. 703–711) in 1918. The Supreme Court upheld the constitutionality of the MBTA in 1920, and subsequently migratory bird treaties were enacted with Mexico, Japan and Russia. The original treaties provided no protection to birds of prey, but raptors were added in a 1972 amendment of the treaty with Mexico (Bond 1974). Currently, the MBTA makes it unlawful to take, possess, buy, sell, purchase, or barter any migratory bird listed in 50 C.F.R. Part 10, including feathers or other parts, nests, eggs, or products, except as allowed by implementing regulations (50 C.F.R. 21). The list of migratory birds covered by the MBTA includes all Falconiformes and Strigiformes that occur, other than accidentally, within the U.S.; the full list of species can be found at 50 C.F.R. 10.13 (http://migratorybirds.fws.gov/intrnltr/mbta/mbtintro.html). Implementing regulations provide for the issuance of permits that allow, among other things, banding and marking, scientific collecting, falconry, captive propagation, and control of depredating raptors (50 C.F.R. 21).

Bald and Golden Eagle Protection Act

In response to public concern over the plight of the Bald Eagle (*Haliaeetus leucocephalus*), Congress enacted protective legislation in 1940 to reduce human-caused mortality. As originally written, the Bald Eagle Protection Act (BEPA; 16 V.S.C. 668–688d) prohibited the taking or possession of Bald Eagles, their eggs, and their nests without a permit. The act included several prohibitions not found in the MBTA, the most important relating to molestation or disturbance. The BEPA has since been amended several times, most importantly in 1962 (P.L. 87–844), when the Act's protective provisions were extended to include the Golden Eagle (*Aquila chrysaetos*). Currently, the BGEPA makes it illegal to import, export, take, sell, purchase, or barter any Bald Eagle or Golden Eagle, including feathers or other parts, nests, eggs, or products, except as allowed by permit for scientific research, religious use, animal damage control, and falconry. Permits also may be issued for the taking of inactive Golden Eagle nests during the course of a resource recovery operation (50 C.F.R. 22).

Endangered Species Act

The Endangered Species Preservation Act (ESPA; P.L. 89–669) was passed by Congress in 1966. The ESPA directed the Secretary of the Interior to carry out a program to conserve, protect, restore, and propagate declining species of fish and wildlife. The scope of the ESPA was broadened in 1969 with passage of the Endangered Species Conservation Act (ESCA; P.L. 91–135), which expanded the land acquisition authority granted by the ESPA, directed the Secretary of the Interior to promulgate a list of wildlife species threatened with worldwide extinction, and prohibited importation of these species into the U.S. The ESCA also directed the Secretaries of State and Interior to convene an international ministerial meeting concerning the conservation of endangered species (Bean 1983). The international meeting was held on 3 March 1973, and led to the creation of CITES, as described previously.

The ESCA failed to provide the kinds of management tools necessary to conserve the majority of native endangered species. In particular, the ESCA contained no prohibitions on the taking of endangered species (this was left up to the states), and it did not adequately protect endangered wildlife from ongoing and proposed federal activities. To rectify this, Congress enacted the Endangered Species Act (ESA; 16 V.S.C. 1513–1543) in 1973. The ESA not only implements CITES, but it (1) defines species to include subspecies, as well as "distinct" populations; (2) formalizes the process for listing species as endangered or threatened (Section 4); (3) directs the Secretaries of the Interior and Agriculture to establish and implement a land conservation program for listed species (Section 5); (4) directs the Secretary of the Interior to cooperate with states by entering into management agreements and cooperative agreements with state agencies for the conservation of listed species, and authorizes the Secretary to provide financial assistance to states to carry out such agreements (Section 6); (5) directs all federal agencies to ensure that their actions and activities will not jeopardize the continued existence of any listed species, and formalizes a consultation process for making determinations of likely impacts (Section 7); (6) prohibits the import, export, taking, possession, transport, sale, and trade of any listed species (Section 9); (7) formalizes an exemption process, including provisions for permits that authorize activities prohibited under Section 9 (Section 10); and (8) prescribes civil and criminal penalties for violations of the Act (Section 11) (U.S. Congress 1983). The list of species protected under ESA can be found at 50 C.F.R. 17.11 and 17.12 (www.fws.gov/endangered/wildlife.html).

When Are U.S. Federal Permits Required?

Biologists and managers working with birds of prey protected under MBTA, BGEPA, or ESA (including CITES, if export and import are involved) must obtain federal permits if their activities violate provisions of the laws. "Hands-on" research (e.g., banding and marking, scientific collecting) clearly requires federal permits, but more subtle activities also may violate these federal laws (e.g., entering occupied nests of endangered species to retrieve prey remains). Conflicts between the activities of biologists and the law generally involve the prohibitions included under the term "take" in each of these laws. Because of the importance of understanding the scope of the take prohibition, its definition in each pertinent statute is given below:

- MBTA. — *"Take means to pursue, hunt, shoot, wound, kill, trap, capture, or collect"* (50 C.F.R. 10.12).
- BGEPA. — *"Take includes . . . pursue, trap, collect, molest or disturb"* (U.S.C. 668c).
- ESA. — *"The term 'take' means to harass, harm, pursue, hunt, shoot, wound, kill, trap, capture, or collect, or to attempt to engage in any such conduct"* (U.S. Congress 1983:4). *"Harass... means an intentional or negligent act or omission which creates the likelihood of injury to wildlife by annoying it to such an extent as to significantly disrupt normal behavioral patterns, including breeding, feeding, or sheltering"* (50 C.F.R. 17.3).

Many research and management techniques can result in violations of these prohibitions, especially as defined in the BGEPA and ESA where disturbance and harassment are prohibited acts. Biologists planning to work with species protected by these statutes should anticipate needing federal and state permits. When working with other species or when uncertain whether taking will occur, contact the state wildlife agency and the U.S. Fish and Wildlife Service's (FWS) Migratory Bird Management regional migratory bird permit offices (for MBTA and BGEPA species; contact information can be found at www.fws.gov/permits/mbpermits/addresses.html) or regional endangered species permit offices (for ESA protected species; contact infor-

mation is at www.fws.gov/endangered/permits/permitscontacts.html).

Types of Federal Permits and Application Procedures

Raptor-research and management activities typically involve five types of federal permits — banding or marking, scientific collecting, raptor propagation, endangered or threatened species, or import/export. These five main permit types are discussed below.

Banding or marking permit. A banding or marking permit is required before any person may capture any bird species protected by the MBTA for banding, marking, or radio- or satellite-tagging purposes. The U.S. Geological Survey Bird Banding Laboratory (BBL) issues banding permits. Contact information and permitting requirements can be found at www.pwrc.usgs.gov/BBL/default.htm. The BBL also maintains and manages all banding data and researchers wishing to access banding and recovery data for analysis should address their request to the BBL.

Scientific-collecting permit. A scientific-collecting permit is required to take or possess a protected bird, bird egg, bird part, or to possess a protected bird nest for scientific purposes. Permit-application procedures and requirements are given at www.fws.gov/forms/3-200-7.pdf for birds protected under MBTA, at www.fws.gov/forms/3-200-14b.pdf for species protected under the BGEPA, and www.fws.gov/endangered/permits/index.html for threatened and endangered raptor permits. One will soon be able to apply on-line for federal permits for scientific collecting. State permits also generally are required, and you should contact the state wildlife management agency in the state where work will occur to determine state permitting requirements and procedures.

Raptor-propagation permit. A raptor-propagation permit is required before any person may take, possess, transport, sell, purchase, barter, or transfer any raptor, raptor egg, or raptor semen for propagation purposes. The raptor-propagation permit was developed, in part, to encourage the captive production of raptors for conservation purposes. Raptor-propagation permits also can authorize the taking of non-threatened and non-endangered raptors and raptor eggs from the wild for propagation purposes, providing the state in which the activity is to occur also gives written authorization. Federally endangered and threatened species may be taken for propagation purposes under special circumstances, but such activities require both propagation and endangered species permits (discussed later). Additional details on this permit and application procedures can be found at www.fws.gov/forms/3-200-12.pdf.

Endangered and threatened species permits. An endangered and threatened species permit may be issued by the director of the FWS for scientific research or for enhancing the propagation or survival of an endangered or threatened species. FWS regional offices typically issue these permits. General permit application instructions and application forms are available at www.fws.gov/endangered/permits/index.html.

Import and export permits. The FWS's Division of Management Authority issues import and export permits under CITES, except that import/export permits involving Bald Eagles and Golden Eagles are processed by migratory-bird permit offices. Application instructions, and links to other important CITES permit information sites are at www.fws.gov/permits/; for eagles, go to www.fws.gov/forms/3-200-69.pdf.

There are several other types of permits available that authorize falconry, take of depredating migratory birds, and various forms of exhibition and education. Information and application instructions for these permits can be found at www.fws.gov/permits/mbpermits/birdbasics.html. In addition, many institutions now require researchers to develop animal-care protocols consistent with requirements of the Animal Welfare Act (www.aphis.usda.gov/ac/info.html). Typically, Animal Use and Care Committees at each institution oversee application of the requirements of this Act, under broad oversight of the U.S. Department of Agriculture.

The MBTA notes that states may enact and enforce laws or regulations that provide additional protection to migratory birds, including raptors. Additionally, many states list species as endangered or threatened that are not listed federally; often, these listings carry with them additional state permitting requirements. Because state laws and regulations vary, it is not possible to discuss all such laws and requirements here. However, researchers or managers planning to work with raptors should contact the pertinent state wildlife agency during the planning phase of their project to determine whether additional permits are required.

Timing of Permit Requests

The U.S. Fish and Wildlife Service and many state agencies require up to 3 months to process and issue permits, and especially complicated permits (or permits with incomplete applications) can take longer. Researchers and managers should apply for permits as early as possible to ensure that they are in hand before work needs to start. The U.S. Fish and Wildlife Service soon will have the capability to receive applications for scientific collecting on-line, an improvement that should reduce processing time.

RAPTOR LAWS IN CANADA

In Canada, raptors are not protected by any overarching federal legislation, as they are in the U.S. Rather, basic legal protection from disturbance and harassment is provided by provincial and territorial legislation. Raptors were not included in the Migratory Bird Convention with the U.S. in 1916, the enabling Canadian legislation in 1918, nor in any subsequent amendments to that Act. Consequently, each provincial and territorial government issues permits related to raptors. In 2003, the federal government did enact the Species at Risk Act (SARA), which protects all nationally listed raptors and requires permits for all research and conservation activities. In addition, international and inter-provincial movement of raptors is controlled under the Wild Animal and Plant Protection and Regulation of International and Interprovincial Trade Act (WAPRITTA). All projects that disturb or handle raptors are subject to approval by Canadian Council on Animal Care Committee. In addition, any project on crown land, federal, provincial or territorial must have the approval of the appropriate government authority. Consequently, any raptor researcher or manager must have several permits from different levels of government before any project can commence.

Migratory Bird Convention (MBC) Act

This act between Canada and the U.S., signed on 16 August 1916 and amended most recently on 14 December 1995, does not include raptors. Thus, the only part of the Canadian MBC Act (1917) that is relevant to raptors deals with banding permits. Raptor banders require a federal banding permit under this act to acquire and apply bands.

Species At Risk Act (SARA)

Regulations under this recent act are still evolving, but at the time of writing, its impact on raptor research and conservation is apparent. A few raptors are listed as endangered and threatened by the Act in Schedule 1 (www.dfo-mpo.gc.ca/species-especes/species/species_e.asp). The Act protects these listed raptors on federal lands and requires permits to be issued for any action that involves disturbance, including banding of the listed species, but only on federal lands. In National Parks the permits are issued by Parks Canada Agency. For all other federal lands, Environment Canada issues the permits under SARA. The Act does not apply to raptor research or conservation off federal lands, where provincial and territorial permits are required, nor does it apply to non-listed species of raptors on federal lands.

Canada Wildlife Act (CWA)

This act does not specifically mention raptors; however it does provide regulations for activities on National Wildlife Refuges and Migratory Bird Sanctuaries (http://laws.justice.gc.ca/en/w-9/265232.html). Thus, any raptor-related activities on these two types of protected areas require permits under the CWA.

The Wild Animal and Plant Protection and Regulation of International and Interprovincial Trade Act (WAPPRIITA) implements CITES in Canada and controls the inter-provincial and inter-territorial movement of raptors (http://laws.justice.gc.ca/en/w-8.5/265187.html). It came into force on 14 May 1996, when the Wild Animal and Plant Trade Regulations were announced. Any raptors or raptor parts that cross the international border require a CITES import/export permit (http://laws.justice.gc.ca/en/W-8.5/SOR-96-263/index.html). In addition, raptors and raptor parts that cross provincial borders require provincial or territorial permits usually for import and export. This second requirement needs special attention since most researchers would not realize that the inter-provincial transport of raptor parts require permits. Falconers are very aware of this somewhat onerous requirement to get import and export permits from all provinces if they want to move a falcon from their home province, even for a short visit.

Canadian Council on Animal Care (CCAC)

The Canadian Council on Animal Care, a national, peer-review organization founded in Ottawa in 1968, reviews all projects that use animals (www.ccac.ca/). Its mandate is straightforward and concise. To wit, *"to work for the improvement of animal care and use on a Canada-wide basis."* The mandate of the CCAC derives from

several federal and provincial laws. Basically any project approved by a local Animal Care Committee (ACC) has shown due diligence in respect to these laws. The Criminal Code of Canada, Section 446, Cruelty to Animals, forbids *"causing unnecessary suffering."* The century-old (1892) Code states that: *"Everyone commits an offence who willfully causes or, being the owner, willfully permits to be caused unnecessary pain, suffering or injury to an animal or bird . . ."* The Federal Health of Animals Act, C-66 (June, 1990, rev. March, 1992); 38–39 Elizabeth II, Chapter 21 is aimed at protecting Canadian livestock from contagious diseases, and keeping out foreign diseases. The Act states that *"the Governor in Council may make regulations for the purpose of protecting human and animal health . . . including regulations . . . governing the manner in which animals are transported within, into or out of Canada."* Some provincial acts also require ACC compliance. For example, in Saskatchewan, under the Veterinarians Act of 1987 (Chapter V-5.1) a person using an animal in research and employing procedures in studies approved by an Animal Care Committee (ACC) which includes a veterinarian, is exempt from the Act's provision that only a member of the Saskatchewan Veterinary Medical Association *"shall engage . . . in the practice of veterinary medicine."* The use of animals in a research facility in Ontario is governed by its Animals for Research Act (Revised Statutes of Ontario, 1980, Chapter 22 as amended by 1989, Chapter 72, s6 and Regulations 16,17,18,19. Revised Regulations of Ontario, 1980, March 1990), which is administered by the Ontario Ministry of Agriculture and Food, and requires annual registration of all research facilities in the province. It includes clauses requiring local ACCs, composed of a veterinarian and animal care authority, to assess and modify research projects in accordance with minimum standards for housing, procedures and care, and to inspect research premises. In addition, bird banding permits and provincial research permits require a project to be approved by an ACC.

Provincial and Territorial Legislation

Since raptors are not included in the Migratory Bird Convention Act, raptor management is vested in provincial and territorial governments. All provinces and territories have wildlife legislation that affects raptor researchers. A researcher should check with the provincial or territorial wildlife act where the study is planned for specific permits and application procedures. Activities that are regulated under provincial laws include research, collections, salvage of found-dead raptors, trapping, banding, telemetry, falconry, transportation of raptors within the province, import and export of raptors across provincial boundaries, and control of raptors that are damaging property or livestock.

Permit Requirements for Research and Management Activities in Canada

Federal and Provincial or Territorial banding permits. A federal bird banding permit is required to acquire and use bird bands that are issued by the Canadian Bird Banding office (www.cws-scf.ec.gc.ca/nwrc-cnrf/default.asp?lang=en&n=208B0F0B). However, to trap raptors a provincial permit also is required. Thus, a raptor bander must acquire a federal permit and a provincial or territorial permit for each jurisdiction where the research occurs. In addition, a requirement of these permits is that the trapping and banding activities must be reviewed and approved by a local Animal Care Committee. In some provinces the ACC proposal is built into the application form but can be left blank if the researcher attaches the approval of another ACC review (e.g., a university ACC).

Scientific-collection permits and research permits. Collecting and research permits for raptors are issued by provincial wildlife agencies. Each province has its own application process, and the researcher is encouraged to check with the provincial wildlife agency where the work will occur. Both trapping and banding are usually covered in these research permits.

Transportation of raptors. Some provinces require import and export permits for wildlife and wildlife parts moving across their provincial border, as well as a veterinarian inspection for live birds. Other provinces do not require permits. A researcher should determine the specific requirements of the provinces where the research occurs and the final destination of the specimens. In most cases, the permit requires a visit and inspection of specimens by a wildlife officer. Some provinces charge a fee for these permits. Some provinces also require that the collection permit be with the specimens while they are in transit within the province (each field staff should have a copy of the permit in their possession while they are collecting and moving specimens). If the specimens are transported across international borders, then CITES permits are required since most if not all Canadian raptors are listed in the appendices of CITES, whether at-risk or as look-alikes. Provincial permits may be required in addition to CITES permits. International transport with a CITES permit must be made at designated ports with inspection facilities.

Raptor propagation. Provincial permits are required for the possession and propagation of raptors, and to sell or barter raptors or raptor parts. If the transfer of raptors is international, then the facility needs to be registered by the CITES authority and restrictions apply to the movement of live raptors (e.g., they must be seamless-banded and be F2 or higher progeny).

Falconry. The sport of falconry is regulated by provincial permits, whereas hunting game birds with raptors requires the same federal and provincial hunting permits as does gun hunting. A provincial permit is required to acquire and possess a raptor and some provinces issue permits allowing limited wild harvest of certain species. Anyone interested in taking up falconry should contact his or her local Canadian Wildlife Service office to determine what is required. In Alberta, falconers must belong to the provincial falconry association as well. Recreational falconry is not allowed in all Canadian jurisdictions and the rules vary considerably from one province to the next.

RAPTOR LAWS IN EUROPE, WITH A FOCUS ON GREAT BRITAIN

Regional legislation has a major unifying influence on national legislation in the European Union (EU). The 25 Member States apply European Community (EC) directives and regulations (issued in 11 languages) on a wide range of matters that affect raptor management, such as conservation, animal health, health and safety at work, medicinal products and the veterinary profession. Directives (e.g., those on wild birds and on habitat protection) require implementation by national laws. Each Member State will, in its own way, provide legislation or administrative measures that will meet the requirements of the directive, such as which animals or plants are protected and the extent of protection provided. On the other hand, regulations take direct effect without further legislation on the part of the Member States, although the provision of enforcement (powers, offenses, and penalties) is a matter for national law. A prime example is the CITES regulations that provide uniform provisions in the EU for the importation and exportation of endangered species.

As a matter of terminology, many regulations and directives include "EC" or "EEC" in the title and are referred to as "EC legislation" because they are issued by the EC, which is the sector of the European Union that has legislative powers.

The fields of EC law that affect raptors are:

■ CITES and trade: The EC CITES Regulations and Directives can be found at http://ec.europa.eu/environment/cites/legislation_en.htm. A portal to the EU countries' CITES legislation is at www.eu-wildlifetrade.org/pdf/en/2_national_legislation_en.pdf.
■ Wildlife conservation: The Birds Directive and the Habitat Directives set out provisions for species and habitat protection (http://ec.europa.eu/environment/nature/biodiversity/current_biodiversity_policy/eu_biodiversity_legislation/habitats_birds_directives/index_en.htm).
■ Other directives deal with welfare in transport, scientific research, animal health, the veterinary profession, medicines and health and safety at work. EC legislation is available on EUR-lex: http://eur-lex.europa.eu/en/index.htm.

The Council of Europe (COE) is an entirely separate entity from the EU, having social and cultural aims and comprising 45 Member States in a much wider area of Europe than the EU. It also has produced conventions in fields relevant to raptors such as wildlife conservation, animal research, and welfare in transport of animals. States (including the EU) that ratify the Conventions incorporate the provisions in their national laws (see www.coe.int/DEFAULTEN.asp and http://conventions.coe.int/Treaty/Commun/ChercheSig.asp?NT=104&CM=8&DF=21/09/2005&CL=ENG).

As a rule, with the exception of CITES in the EU, the laws that affect a person working in raptor management in Europe will be the national law of the country where the work takes place. Although EU countries generally conform to the requirements of relevant directives, they will have separate legislation in national language(s). Most European countries should have legislation that implements the provisions of the EC legislation, the COE Conventions, or both.

Detailed regulation and attitudes vary from country to country. For example, Germany has extensive regulations on wildlife research and rehabilitation whereas British law allows any person to take even protected species of injured wildlife to tend and care for it until it is ready for release, although in some cases with raptors

it may be necessary to have the bird ringed and registered. Likewise, falconry is prohibited in some countries (e.g., Norway), but is hardly regulated at all in others (e.g., Britain, where it is only necessary to comply with more general rules that control the keeping of certain birds of prey, the taking of quarry species, recovery of lost or hacked birds, and general animal welfare and veterinary laws). Many countries have official government websites that may have information on legislation. A useful portal for EU Member States is http://europa.eu.int/abouteuropa/index_en.htm. Below we provide more detailed information on laws in Great Britain (i.e., England, Wales and Scotland). The United Kingdom comprises Great Britain and Northern Ireland.

Activities that involve the management of raptors in captivity include falconry, rehabilitation, raptor-keeping, captive-breeding, and research. In British legislation, only the last is subject to its own specific statute, whereas all are affected by a variety of laws (Cooper [ME] 2002, Cooper 2003a,b). The latter has separate but similar laws relating to raptors as does Britain, but they are not discussed here. In this section the terms "bird of prey" and "raptor" are used interchangeably to cover both falconiform and strigiform species. "Free-living" indicates birds that are not in captivity (i.e., living in the wild), but the term "wild bird" is used for species that are found in the wild, despite the fact that individual birds may be kept in captivity (after Cooper [JE] 2002). A veterinarian is referred to in British veterinary legislation (and in that of countries that follow this model) as a "veterinary surgeon."

It should be noted, in respect of British legislation, that since Devolution in 1999 and the transfer of some law-making powers to the Scottish Parliament and the Welsh Assembly, English, Welsh, and Scottish legislation are tending to diverge. Legislation since 1988 is available on the Internet at www.opsi.gov.uk/legislation/about_legislation.htmhttp://www.opsi.gov.uk/legislation/about_legislation.htm. Welsh legislation is provided on: www.wales-legislation.org.uk/scripts/home.php?lang=E. Most of the national law discussed is based on EC or COE legislation.

Wildlife Legislation

The Wildlife and Countryside Act 1981 (as substantially amended) (WCA) is the primary law relating to wildlife (www.jncc.gov.uk/page-1377, www.jncc.gov.uk/page-3614, and www.rspb.org/policy/wildbirdslaw/birdsandlaw/wca/index.asp). In Scotland the Nature

Conservation Act (Scotland) 2004 also applies (www.opsi.gov.uk/legislation/scotland/acts2004/20040006.htm). The government body primarily responsible for the WCA and other wildlife matters in England is the Department for Environment Farming and Rural Affairs (DEFRA).

The WCA provides legal protection for all birds (including raptors) that comprise a species that is resident in, or is a visitor to, the European territory of any EU member country in a wild state. It also affects the acquisition, disposition, and keeping of captive specimens of these species. The WCA makes it an offense to:

- Take, kill or injure any wild raptor,
- Take, damage or destroy a raptor's nest while it is being built or while it is in use,
- Disturb a Schedule 1 raptor when building its nest or when it is near a nest containing eggs or young,
- Disturb dependent young of a Schedule 1 raptor,
- Take or destroy a raptor egg,
- Possess a live or dead raptor or egg (including a part or derivative) unless it can be proved (by the possessor) to have been taken, killed or sold legally,
- Sell (other related activities such as advertise or transport for sale or barter) a live wild raptor,
- Use a wide range of methods to take or kill raptors,
- Keep any bird in a cage that does not allow it to spread its wings fully. This does not apply during transportation or when the bird is undergoing examination or treatment by a veterinary surgeon, or
- Release, intentionally, any non-indigenous (alien) raptor.

Additional protection and provisions in the WCA include:

- Offenses involving species listed on Schedule 1 receive higher penalties that those involving other species (around 12 of the rarer British raptors are listed under Schedule 1).
- Species listed on Schedule 4 originating from any source and kept in captivity for whatever purpose must be registered with DEFRA and ringed. These include a number of British raptors and some rare non-British species. However, the provision does not apply to the most common species (i.e., the Common Buzzard (*Buteo buteo*), Common Kestrel (*Falco tinnunculus*), Eurasian Sparrowhawk (*Accipiter nisus*), and owls. Although this requirement arises as soon as the raptor is taken into possession, there is an exception that allows a veterinary surgeon to keep a sick or injured Schedule 4 bird for treatment for up to 6 weeks. For details of the current species affected and the registration and ringing requirements see www.defra.gov.uk/wildlife-countryside/gwd/birdreg/index.htm.
- Schedule 4 and Article 10 (see CITES below) are monitored by Wildlife Inspectors, appointed by DEFRA.
- There is a range of exceptions from the basic provisions of the WCA outlined above whereby a license can be issued to authorize the taking of birds for scientific, educational, ringing (banding)/marking, re-introduction, falconry, or taxidermy purposes (www.defra.gov.uk/corporate/regulat/forms/cons_man/index.htm).
- Other exceptions relate to public health and safety, disease control, pest control, and the protection of agriculture. In most cases, a license is required to authorize such activities (www.defra.gov.uk/wildlife-countryside/vertebrates/default.htm).

Other legal factors affecting free-living raptors. If access is required to free-living raptors, it is likely that permission will be required to enter land, especially when it is a protected area (permit), a restricted (e.g., military) area (permit), or private land (owner's or occupier's permission).

Trade in Raptors

Under the WCA, within Britain the sale (and the allied activities of barter, advertising) of protected raptors is illegal. Exceptions are made for captive-bred raptors (provided that both parents can be shown to have been held legally in captivity when the egg was laid) and captive-breeding authorized by general or individual licenses.

The general provisions under EU CITES law can be found at: www.eu-wildlifetrade.org/html/en/wildlife_trade.asp, www.ukcites.gov.uk/intro/leg_frame.htm#The%20Commission, www.cites.org/, www.eu-wildlife

trade.org/pdf/en/6_marking_en.pdf, and www.ukcites. gov.uk/pdf_files/GN1%20General%20guidance%20 notes%20March06.pdf.

The EC Regulations automatically form part of the law of EU countries. They are listed at http://ec.europa. eu/environment/cites/legislation_en.htm and www.eu-wildlifetrade.org/pdf/en/1_international_legislation_ en.pdf. DEFRA is responsible for issuing permits, certificates and other authorization. It also is the CITES Management Authority in Great Britain. The main UK CITES website is at www.ukcites.gov.uk/intro/leg_frame.htm.

EU CITES provisions on external trade are as described for the U.S. and Canada. However, there also are additional requirements, and the EU status of many species has been upgraded from that of their Convention Appendices (http://ec.europa.eu/environment/cites/ pdf/diff_between_eu-cites.pdf). There are four Annexes on which the CITES species are listed. All raptor species are listed on Annex A. This gives all falconiforms and strigiforms a status equivalent to that under Appendix I of CITES within the EU territory. The movement of parts and derivatives of CITES species in or out of the EU also is controlled. Thus, a permit is required to import or export diagnostic and other biological specimens, including tissues and feathers.

There is free movement of legally acquired CITES species within the countries of the EU. Proof that the birds were obtained legally, either within the EU or from outside (e.g., evidence of legal importation, taken from the wild under license, taken in Britain as a sick or injured specimen, or from lawful captive breeding) must be available at all times. In circumstances where a permit is not required for acquisition, it is essential to keep good records and evidence, sufficient to prove legal acquisition.

There is a provision for the registration of raptor captive breeding facilities with the CITES Management Authority (www.eu-wildlifetrade.org/pdf/en/5_breeding_ en.pdf). Any commercial use of an Annex A species requires specific authorization. Such authorizations are known as Article 10 Certificates (or, for zoos, Article 60). The sale of captive-bred raptors and owls follows the CITES Convention in that F2 generation captive-bred offspring can be sold under an Article 10 certificate. No authorization is required if there is no commercial element (e.g., a pure gift), but the transaction and origin of the bird should be documented carefully, together with any evidence that is required to prove that the bird was legally obtained. Any captive-bred raptor to be used for commercial purposes must be ringed with

a closed ring. If this is not possible due to physical or behavioral attributes of the bird, a microchip should be used (www.ukcites.gov.uk/license/GN2%20Commercial %20Use%20Guidance_Nov%202005.doc). License information for bird of prey keepers can be found at www.ukcites.gov.uk/pdf_files/Sep05GN6%20Birds%2 0of%20Prey%20Keepers.pdf.

Requirements for commercial uses of wild disabled birds are described at www.ukcites.gov.uk/pdf_files/ Sep05GN13%20Commercial%20use%20of%20wild% 20disabled%20birds.pdf. An Article 10 or Article 60 certificate is required for any commercial exhibit of raptors, including display to the public and flying demonstrations. A summary of the permits available is provided at www.eu-wildlifetrade.org/html/en/wildlife_trade.asp. Other related legal aspects, such as animal and human welfare and health, are summarized at: www.eu-wildlife trade.org/pdf/en/4_welfare_en.pdf. Permit requirements can be found at www.eu-wildlifetrade.org/pdf/en/3_ permits_en.pdf.

Law Enforcement

Enforcement powers for CITES are contained in The Control of Trade in Endangered Species (Enforcement Regulations) 1997 (amended 2005) (COTES). Customs legislation also provides enforcement powers. The enforcement provisions are covered in detail in the Partnership Against Wildlife Crime's (PAW) "Wildlife Law Enforcer's Factfile" at www.defra.gov.uk/paw/publications/pdf/wildlifelaw-factfile-full.pdf. The CITES, WCA, and other laws are enforced by the Police, DEFRA, Inland Revenue, Customs Service, and local authorities, singly or cooperatively. Voluntary bodies undertake some prosecutions and also provide expert advice or evidence during crime investigation and prosecutions. Recently there has been a steady growth in the enforcement of wildlife laws. Legislatively authorized inspection and enforcement power, along with the severity of penalties, have been increased. A National Wildlife Crime Intelligence Unit was set up in 2002, and there is a Wildlife Liaison officer on all police forces.

PAW is a consortium of enforcement agencies, government, and voluntary organizations that works towards the improvement of wildlife protection through meetings and working groups. The PAW website also has a list of literature on British wildlife law at www.defra.gov.uk/paw/publications/default.htm.

Captive Management of Raptors

Falconry. Little legislation is directed specifically at raptor keepers aside from the species-specific laws on wildlife and the trade law mentioned above. Legislation on keeping animals, such as general welfare and treatment and the licensing of facilities in which they are kept, can be found at www.defra.gov.uk/wildlife-countryside/gwd/birdreg/02.htm#10 and www.defra.gov.uk/wildlife-countryside/gwd/birdreg/index.htm.

There is no specific regulation of the sport of falconry or of the falconers themselves. However, the WCA and CITES have important indirect implications for falconers (Irving 2006a, 2006b). For example, there may be a need for a permit to take prey species when hawking or when using a trap to recover a lost falconry raptor that has returned to the wild. Schedule 4 ringing and registration applies to falconry birds. In addition to government legislation, there is a measure of self-regulation among British falconers. The British Falconers' Club has a code of conduct for its members, backed by a Disciplinary Committee (www.british falconersclub.co.uk/code_conduct.htm). The Hawk Board and the Scottish Hawk Board represent individual raptor owners and bird of prey associations in dealings with the government (e.g., in matters of law, policy, and Schedule 4 of the WCA). It provides guidance for keepers and raptor displays (www.hawkboard-cff.org.uk/index.htm).

Rehabilitation. Wild raptors acquired in Britain for rehabilitation are taken under the WCA provision that allows anyone to take a sick or injured wild bird and tend it until it has recovered. No license or special qualifications are required on the part of the rehabilitator. A facility only requires a permit if it desires to acquire some other legal status, such as a zoo. These provisions may change under pending new animal welfare legislation. Schedule 4 listed species must be ringed and registered, although veterinary surgeons may keep Schedule 4 species for treatment for up to 6 weeks without applying for registration. The WCA provides that birds held for rehabilitation must be released when they have recovered fully. It may be necessary to have the readiness of the bird for release assessed by an appropriately experienced veterinarian or other raptor specialist. This evaluation can provide a justification for retaining a bird that is unfit for release in captivity. Record-keeping is of the utmost importance to provide evidence of compliance with the legislation.

Captive breeding. Occasionally, licenses are provided under the WCA to take raptors from the wild for captive breeding. EC-CITES provisions discussed above apply in these cases.

Raptor research. Scientific research that may cause harm requires authorization and veterinary supervision under the Animals (Scientific Procedures) Act 1986. This applies to *"any experimental or other scientific procedure . . . which may have the effect of causing that animal pain, suffering distress or lasting harm."* This includes causing *"death, disease, injury, physical or psychological stress, significant discomfort or any disturbance to normal health whether immediate or in the longer term."* Scientific studies on raptors that fall within this definition require licenses for the researcher, the project, and the premise(s) where the work is carried out. A cost–benefit analysis, justification for the animals used, and ethical review must be conducted. This applies to work in the field with wild raptors as well as research using captive raptors. Acquisition of raptors for research is subject to the wildlife and trade laws described previously. It may be possible to obtain raptors from the wild for scientific, conservation, or other purposes under a WCA license. Any take from the wild or, the use of a trapping method, other than for sick and injured animals, is subject to permit. The study of raptors in the wild requires a WCA permit if disturbance of a Schedule 1 species at the nest will occur, or if other prohibited offenses will result. The field study of raptors in the wild usually requires access to property. Entering or crossing land requires the landowner's or occupier's permission. Authorization is required if the land is a protected area or military zone.

Public display. If a raptor facility provides public access for viewing its birds on 7 or more days in a year, whether or not for payment, it falls within the definition of a zoo and must be licensed under the Zoo Licensing Act 1981 (as amended in 2002 to comply with EC legislation). Zoos must be licensed and are subject to regular inspection. They must conform to the Secretary of State's Standards of Modern Zoo Practice and demonstrate that the collection contributes to public education, conservation and science. The zoo must provide for the behavioral needs of its animals as well as veterinary care and record keeping (See: www.defra.gov.uk/wildlife-countryside/gwd/zoo.htm#direct, www.defra.gov.uk/wildlife-countryside/gwd/govt-circular022003.pdf, www.defra.gov.uk/wildlife-countryside/gwd/zoo.htm#stand). A CITES Article 60 certificate is required to authorize the display of Annex A species for commercial purposes. Flying demonstrations often are a feature of raptor centers and sometimes are given at special events such as fairs and agricultural shows. These require a CITES Article 10 certificate unless they are entirely non-commercial.

Animal health and welfare. Those keeping raptors are responsible for their welfare. Animal welfare is a strong issue in Britain and new legislation for England and Wales that passed through the Westminster Parliament during 2006 should now be in force (www.defra.gov.uk/animalh/welfare/bill/index.htm). The use of live prey to feed or train raptors (outside authorized hawking) is unlikely to be acceptable on ethical or animal-welfare grounds in Britain. The Welfare of Animals (Transport) Order 1997 (to be replaced in 2007 by EU Regulation 1 of 2005) provides that animals must be fit to travel and must not be caused unnecessary suffering or injury during transportation (www.defra.gov.uk/animalh/welfare/farmed/transport/summarywato.htm). This law also gives legal status to the CITES Guidelines on Transport (1980) and the International Air Transport Association Regulations which apply to raptors in transit.

The Veterinary Surgeon's Act 1966 requires that the diagnosis, medical and surgical treatment (whether for payment or not) of raptors (free-living or wild) must be carried out by a registered veterinary surgeon. There are some exceptions relevant to raptor management, including (1) research procedures licensed under ASPA are exempted from the Act, (2) first aid in an emergency may be carried out by anyone, (3) that the owner of a raptor may carry out minor medical treatment (making it therefore important that ownership is clearly determined in the case of raptors accepted for rehabilitation), and (4) that veterinary nurses and veterinary students may carry out limited procedures under supervision. There are extensive veterinary ethical and practice requirements and standards (www.rcvs.org.uk/).

The prescription, supply and administration of veterinary medicinal products are strictly governed by The Veterinary Medicines Regulations 2005 (www.rcvs.org.uk/shared_asp_files/uploadedfiles/8013AA6B-EEF3-4F54-A911-3CDBA703A56B_rcvsnews_nov05_pg6.pdf, www.rcvs.org.uk/Templates/ Internal.asp?NodeID=94060, and www.opsi.gov.uk/si/ si2005/uksi_20052745_en.pdf). Veterinary surgeons may only prescribe "POM-V" (veterinary prescription only) medicines for animals under their care and in accordance with the marketing authorization for a given drug. Since the range of medicines approved for use in birds is limited, the veterinary surgeon is likely to have to prescribe in accordance with the "cascade" which indicates the order of selection of drugs that are not specifically licensed for use in the given species or for the particu-

lar condition to be treated. The informed consent of the client should be obtained, preferably in writing for the use of this "off label" prescription. See (www.rcvs.org.uk/Templates/Internal.asp?NodeID=92574#choice and www.vmd.gov.uk/General/VMR/vmg_notes/VMNote15.pdf).

The import of raptors from outside the EU usually requires pre-departure quarantine, a license, health certification, and quarantine in approved premises on arrival. The import or export of diagnostic and biological samples may require authorization if they fall within the controls on pathogens (see www.defra.gov.uk/animalh/diseases/pathogens/index.htm). In-country legislation includes powers to control outbreaks of avian diseases such as psittacosis, Newcastle disease, and avian influenza (see www.defra.gov.uk/animalh/diseases/notifiable/disease/ai/wildbirds/index.htm#licence, www.defra.gov.uk/animalh/diseases/notifiable/disease/ai/keptbirds/index.htm, www.defra.gov.uk/animalh/diseases/notifiable/disease/ai/policy/index.htm#3, and www.defra.gov.uk/animalh/diseases/notifiable/disease/avianinfluenza.htm).

Raptor facilities that employ five or more staff are subject to health and safety at work (occupational health and safety) legislation. This legislation imposes a duty upon the employer to provide for the health welfare and safety of employees, volunteers, students, and visitors to premises (and the employer). Following EU legislation on the subject, the British law requires a risk assessment and codes of practice for the workplace. There are additional provisions for first aid, the reporting of accidents, and dealing with dangerous substances. The provision of information, training, and use of protective clothing are an integral part of health and safety provisions (see www.hse.gov.uk/pubns/hsc13.pdf and www.hse.gov.uk/pubns/leaflets.htm).

LITERATURE CITED

BEAN, M.J. 1983. The evolution of national wildlife law: revised and expanded edition. Praeger Publishers, New York, NY U.S.A.

BRITISH FALCONRY CLUB. 2005. The British Falconers' Club rule book. The British Falconers' Club, Tamworth, United Kingdom.

BOND, F.M. 1974. The law and North American raptors. Pages 1–3 in F.M. Hamerstrom, Jr., B.E. Harrell, and R.R. Olendorff [EDS.], Management of raptors. Raptor Research Reports 2.

BRITISH FIELD SPORTS SOCIETY FALCONRY COMMITTEE AND THE HAWK BOARD. Undated. Code of welfare and husbandry of birds of prey and owls. British Field Sports Society, London, United Kingdom.

CHITTY, J. 2006. The injured bird of prey: part 1: legal and logistical issues. *UK Vet* 11:88–94.

COOPER, J.E. 2002. Birds of prey: health and disease, 3rd Ed. Blackwell Publishing, Oxford, United Kingdom.

COOPER, M.E. 2002. Legislation and codes of practice relevant to working with raptors. Pages 284–293 in J.E. Cooper, Birds of prey: health and disease, 3rd Ed. Blackwell Publishing, Oxford, United Kingdom.

———. 2003a. Legislation for bird-keepers. Pages 96–105 in J.E. Cooper [ED.], Captive birds in health and disease. World Pheasant Association, Fordingbridge, United Kingdom and Hancock House Publishers, Surrey, British Columbia, Canada.

———. 2003b. The law affecting British wildlife casualties. Pages 42–48 in E. Mullineaux, D. Best, and J.E. Cooper [EDS.], BSAVA manual of wildlife casualties. British Small Animal Veterinary Association, Quedgeley, Gloucester, United Kingdom.

GAUNT, A.S. AND L.W. ORING [EDS.]. 1999. Guidelines to the use of wild birds in research. The Ornithological Council, Washington, DC U.S.A.

IRVING, G. 2006. Legislation and it's [*sic*] place in falconry. British Falconers' Club bi-annual newsletter, Issue 32, March 2006, 14–17. British Falconers' Club, Tamworth, United Kingdom.

———. 2006. Possession, movement, and registration of chicks and eggs of birds listed in schedule 4 to the wildlife and countryside act [*sic*]. British Falconers' Club Bi-annual Newsletter, Issue 32, March 2006, 18–19. British Falconers' Club, Tamworth, United Kingdom.

MILLSAP, B.A. 1987. Introduction to federal laws and raptor management. Pages 23–33 in B.A. Giron Pendleton, B.A. Millsap, K.W. Cline, and D.M. Bird [EDS.], Raptor management techniques manual. National Wildlife Federation, Washington, DC U.S.A.

UNITED STATES CONGRESS. 1983. The Endangered Species Act amended by Public Law 97-304 (the Endangered Species Act amendments of 1982). U.S. Government Printing Office, Washington, DC U.S.A.

Appendix

Common names and scientific names of raptors mentioned in the text

DIURNAL RAPTORS

African Fish Eagle
Haliaeetus vocifer

African Hawk-Eagle
Hieraaetus spilogaster

American Kestrel
Falco sparverius

Andean Condor
Vultur gryphus

Aplomado Falcon
Falco femoralis

Asian Imperial Eagle
Aquila heliaca

Bald Eagle
Haliaeetus leucocephalus

Barred Forest Falcon
Micrastur ruficollis

Bat Falcon
Falco rufigularis

Bateleur
Terathopius ecaudatus

Bearded Vulture
Gypaetus barbatus

Black Kite
Milvus migrans

Black Shaheen Falcon
Falco peregrinus peregrinator

Black Vulture
Coragyps atratus

Black-chested Buzzard-Eagle
Geranoaetus melanoleucus

Black-winged Kite
Elanus caeruleus

Bonelli's Eagle
Hieraaetus fasciatus

Broad-winged Hawk
Buteo platypterus

California Condor
Gymnogyps californianus

Cape Vulture
Gyps coprotheres

Cinereous Vulture
Aegypius monachus

Collared Forest Falcon
Micrastur semitorquatus

Collared Sparrowhawk
Accipiter cirrocephalus

Common Black Hawk
Buteogallus anthracinus

Common Buzzard
Buteo buteo

Common Kestrel
Falco tinnunculus

Cooper's Hawk
Accipiter cooperii

Crested Goshawk
Accipiter trivirgatus

Crested Serpent Eagle
Spilornis cheela

Egyptian Vulture
Neophron percnopterus

Eleonora's Falcon
Falco eleonorae

Eurasian Hobby
Falco subbuteo

Eurasian Sparrowhawk
Accipiter nisus

European Honey Buzzard
Pernis apivorus

Ferruginous Hawk
Buteo regalis

Galapagos Hawk
Buteo galapagoensis

Golden Eagle
Aquila chrysaetos

Great Black Hawk
Buteogallus urubitinga

Greater Spotted Eagle
Aquila clanga

Grey-faced Buzzard
Butastur indicus

Griffon Vulture
Gyps fulvus

Gyrfalcon
Falco rusticolus

Harpy Eagle
Harpia harpyja

Harris's Hawk
Parabuteo unicinctus

Javan Hawk-Eagle
Spizaetus bartelsi

King Vulture
Sarcoramphus papa

Lanner Falcon
Falco biarmicus

Lappet-faced Vulture
Aegypius tracheliotus

Lesser Kestrel
Falco naumanni

Lesser Spotted Eagle
Aquila pomarina

Levant Sparrowhawk
Accipiter brevipes

Madagascar Fish Eagle
Haliaeetus vociferoides

Mauritius Kestrel
Falco punctatus

Merlin
Falco columbarius

Mississippi Kite
Ictinia mississippiensis

Montagu's Harrier
Circus pygargus

Northern Crested Caracara
Caracara cheriway

Northern Goshawk
Accipiter gentilis

Northern Harrier
Circus cyaneus

Ornate Hawk-Eagle
Spizaetus ornatus

Osprey
Pandion haliaetus

Peregrine Falcon
Falco peregrinus

Philippine Eagle
Pithecophaga jefferyi

Pied Harrier
Circus melanoleucos

Plumbeous Kite
Ictinia plumbea

Prairie Falcon
Falco mexicanus

Red Kite
Milvus milvus

Red-headed Vulture
Sarcogyps calvus

Red-necked Falcon
Falco chicquera

Red-shouldered Hawk
Buteo lineatus

Red-tailed Hawk
Buteo jamaicensis

Roughleg
Buteo lagopus

Saker Falcon
Falco cherrug

Seychelles Kestrel
Falco araeus

Sharp-shinned Hawk
Accipiter striatus

Snail Kite
Rostrhamus sociabilis

Southern Crested Caracara
Caracara plancus

Spanish Imperial Eagle
Aquila adalberti

Steller's Sea Eagle
Haliaeetus pelagicus

Steppe Buzzard
Buteo b. vulpinus

Steppe Eagle
Aquila nipalensis

Swainson's Hawk
Buteo swainsoni

Swallow-tailed Kite
Elanoides forficatus

Swamp Harrier
Circus approximans

Turkey Vulture
Cathartes aura

Verreaux's Eagle
Aquila verreauxii

Wedge-tailed Eagle
Aquila audax

Western Marsh Harrier
Circus aeruginosus

White-bellied Sea Eagle
Haliaeetus leucogaster

White-rumped Vulture
Gyps bengalensis

White-tailed Eagle
Haliaeetus albicilla

White-tailed Hawk
Buteo albicaudatus

White-tailed Kite
Elanus leucurus

OWLS

Barn Owl
Tyto alba

Barred Owl
Strix varia

Blakiston's Fish Owl
Bubo blakistoni

Boreal Owl
Aegolius funereus

Burrowing Owl
Athene cunicularia

Eastern Screech Owl
Megascops asio

Eurasian Eagle-Owl
Bubo bubo

Eurasian Pygmy Owl
Glaucidium passerinum

Eurasian Scops Owl
Otus Scops

Ferruginous Pygmy Owl
Glaucidium brasilianum

Flammulated Owl
Megascops flammeolus

Great Grey Owl
Strix nebulosa

Great Horned Owl
Bubo virginianus

Little Owl
Athene noctua

Long-eared Owl
Asio otus

Morepork
Ninox novaeseelandiae

Northern Hawk-Owl
Surnia ulula

Northern Saw-whet Owl
Aegolius acadicus

Seychelles Scops Owl
Otus insularis

Short-eared Owl
Asio flammeus

Snowy Owl
Bubo scandiaca

Spotted Owl
Strix occidentalis

Tawny Owl
Strix aluco

Western Screech Owl
Megascops kennicottii

Index

The Editors

DAVID M. BIRD is regarded as one of the world's leading experts on birds of prey and he is often consulted by governments, universities, funding bodies, corporations, and the general public for his expertise. David has served as President (and Vice-President twice) of the Raptor Research Foundation Inc. (RRF), participated on numerous committees and organized several RRF symposia, three of which had published proceedings. He was also one of the editors on the original 1987 edition of this book.

After obtaining his M.Sc. in 1976 and being appointed as the curator of the Macdonald Raptor Research Centre, David quickly completed his Ph.D. in 1978. As Director of what is now called the Avian Science and Conservation Centre, David has published over 150 scientific papers on birds of prey, supervised 37 graduate students to completion, and is currently supervising nine. As a Full Professor of Wildlife Biology, he teaches several courses in ornithology, fish and wildlife management, scientific communication, and wildlife conservation.

David has served as Vice-President of the Society of Canadian Ornithologists twice and is currently the President-Elect. He is an elected Fellow of the American Ornithologists' Union and an elected member representing Canada on the prestigious International Ornithological Committee.

Over the last 30 years, David has given countless talks all over North America and made innumerable radio and television appearances both in Montreal and across Canada. He has written and co-edited seven books, including *City Critters: How to Live with Urban Wildlife*, *Bird's Eye-View: A Practical Compendium for Bird-Lovers*, and *The Bird Almanac: The Ultimate Guide to Facts and Figures on the World's Birds* He is also a regular columnist on birds for *The Gazette* of Montreal and *Bird Watcher's Digest* magazine.

Throughout his career, David's achievements have been recognized by various awards for wildlife conservation, the latest being the Quebec Education Award in 2007, the first ever given by Bird Protection Quebec.

KEITH L. BILDSTEIN is Sarkis Acopian Director of Conservation Science at Hawk Mountain Sanctuary in Kempton, Pennsylvania, where he oversees the Sanctuary's conservation science and education programs, and coordinates the activities of its graduate students, international interns, and visiting scientists

Bildstein received his B.S. in Biology at Muhlenberg College, in Allentown, Pennsylvania, in 1972, and his Masters and Ph. D. in Zoology from the Ohio State University, in Columbus, Ohio, in 1976 and 1978. He currently is Adjunct Professor of Wildlife Biology at the State University of New York-Syracuse. He was Visiting Assistant Professor of Biology at the College of William and Mary, in Williamsburg, Virginia, in 1978, and Distinguished Professor of Biology at Winthrop University in Rock Hill, South Carolina, from 1978 to 1992. He is a Fellow of the American Ornithologists' Union, and has been President of the Wilson Ornithological Society and the Waterbird Society, and Vice-president of the Raptor Research Foundation. Bildstein edited the *Wilson Bulletin*, a quarterly journal of ornithology, from 1984 through 1987, and was a member of the editorial board of The *Auk*, the AOU's journal, in 1997–2000. He has helped organize the scientific programs of seven national and seven international ornithological meetings.

Bildstein has authored or coauthored more than 100 papers in ecology and conservation, including 40 on raptors. His books include *White Ibis: wetland wanderer* (1993), *The raptor migration watch-site manual* (1995 [with Jorje Zalles]), *Raptor watch: a global directory of raptor migration sites* (2000 [with Jorje Zalles]), and *Migrating raptors of the world: their ecology and conservation* (2006). His co-edited works include *Conservation Biology of Flamingos* (2000), *Hawkwatching in the Americas* (2001), and *Neotropical Raptors* (2007).

Keith's current research involves the geography, ecology, and conservation of the world's migratory raptors; energy management in migrating raptors; the feeding and movement ecology of New and Old World vultures; and the wintering, breeding, and movement ecology of American Kestrels.

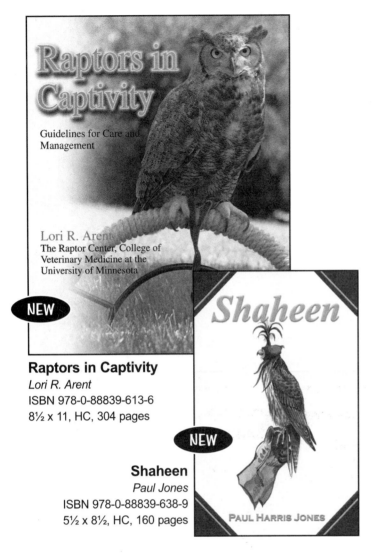
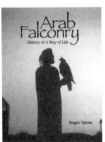
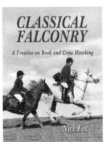

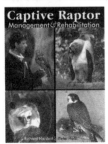